CANCER IMMUNOTHERAPY

CANCER IMMUNOTHERAPY: IMMUNE SUPPRESSION AND TUMOR GROWTH

Edited by

GEORGE C. PRENDERGAST
The Lankenau Institute for Medical Research
Lankenau Cancer Center
Wynnewood, PA

ELIZABETH M. JAFFEE
The Sidney Kimmel Comprehensive Cancer Center at Johns Hopkins Medical Institute
Bunting-Blaustein Cancer Research Building
Baltimore, MD

AMSTERDAM • BOSTON • HEIDELBERG • LONDON
NEW YORK • OXFORD • PARIS • SAN DIEGO
SAN FRANCISCO • SINGAPORE • SYDNEY • TOKYO

Academic Press is an imprint of Elsevier

Academic Press is an imprint of Elsevier
30 Corporate Drive, Suite 400, Burlington, MA 01803, USA
525 B Street, Suite 1900, San Diego, California 92101-4495, USA
84 Theobald's Road, London WC1X 8RR, UK

This book is printed on acid-free paper. ∞

Copyright © 2007, Elsevier Inc. All rights reserved.

Cover Image credit: Photo of a tumor cell being recognized by an T immune cell.
Courtesy of Marc Rubin, Center for Biologic Imaging, University of Pittsburgh

No part of this publication may be reproduced or transmitted in any form or by any means, electronic or mechanical, including photocopy, recording, or any information storage and retrieval system, without permission in writing from the publisher.

Permissions may be sought directly from Elsevier's Science & Technology Rights Department in Oxford, UK: phone: (+44) 1865 843830, fax: (+44) 1865 853333, E-mail: permissions@elsevier.com. You may also complete your request on-line via the Elsevier homepage (http://elsevier.com), by selecting "Support & Contact" then "Copyright and Permission" and then "Obtaining Permissions."

Library of Congress Cataloging-in-Publication Data
Application submitted.

British Library Cataloguing-in-Publication Data
A catalogue record for this book is available from the British Library.

ISBN: 978-0-12-372551-6

For information on all Academic Press publications
visit our Web site at www.books.elsevier.com

Printed in the United States of America
07 08 09 10   9 8 7 6 5 4 3 2 1

Working together to grow
libraries in developing countries

www.elsevier.com | www.bookaid.org | www.sabre.org

ELSEVIER   BOOK AID International   Sabre Foundation

Contents

Contributors xi

PART I: PRINCIPLES OF CANCER IMMUNOBIOLOGY

1. Introduction 3
George C. Prendergast and Elizabeth M. Jaffee

I. Overview 3
II. Historical Background 3
III. Looking Ahead: Marrying Chemotherapy and Immunotherapy 5
IV. Parts of the Book 6
References 8
Further Reading 8

2. Cancer Immunoediting: From Immune Surveillance to Immune Escape 9
Ryungsa Kim

I. Introduction 10
II. Cancer Immune Surveillance 10
III. Cancer Immunoediting 19
IV. Concluding Remarks 25
References 25

3. Immunosurveillance: Innate and Adaptive Antitumor Immunity 29
Masahisa Jinushi and Glenn Dranoff

I. Introduction 30
II. Innate Antitumor Responses 30
III. Innate Immune Cells 31
IV. Adaptive Antitumor Responses 33
V. The Interplay of Innate and Adaptive Antitumor Immunity 38
VI. Conclusion 39
References 39

4. Cytokine Regulation of Immune Tolerance to Tumors 43
Ming O. Li and Richard A. Flavell

I. Introduction 43
II. Cytokine Regulation of Immune Tolerance to Tumors 45
III. Summary and Future Perspectives 55
References 56

5. Immunological Sculpting: Natural Killer Cell Receptors and Ligands 63
David A. Sallman and Julie Y. Djeu

I. Introduction 64
II. Activating Human NK Receptors 65
III. Inhibitory NK Receptors 72
IV. The Ly49 Receptor Family 74
V. Immunotherapy Approaches 74
VI. Conclusion 77
References 78
Further Reading 80

6. Immune Escape: Immunosuppressive Networks 83
Shuang Wei, Alfred Chang, and Weiping Zou

I. Introduction 83
II. Imbalance Between Mature DCs and Immature DCs 84
III. Imbalance Between Stimulatory and Inhibitory B7 Family Molecules 87
IV. Imbalance Between Regulatory T Cells and Conventional T Cells 90
V. Concluding Remarks 92
References 92

PART II: CANCER THERAPEUTICS

7. Cytotoxic Chemotherapy in Clinical Treatment of Cancer 101
Rajesh Thirumaran, George C. Prendergast, and Paul B. Gilman

I. Introduction 101
II. DNA-Damaging Agents 103
III. Antimetabolites 109
IV. Antimitotics 112
V. Chemotherapy Regimens 113
References 115
Useful Web Sites 116

8. Targeted Therapeutics in Cancer Treatment 117
Colin D. Weekes and Manuel Hidalgo

I. Introduction 118
II. Cell Cycle 119
III. The MAPK Family 131
IV. Challenges in the Clinical Development of Signal Transduction Inhibitors 136
References 140

9. Concepts in Pharmacology and Toxicology 149
Richard A. Westhouse and Bruce D. Car

I. Introduction 150
II. Concepts in Pharmacokinetics 151
III. Concepts in Toxicology 159

IV. Clinical Concerns for Pharmacology and Safety 164
V. Conclusion 165
   References 165
   Further Reading 166

10. Cancer Immunotherapy: Challenges and Opportunities 167
Andrew J. Lepisto, John R. McKolanis, and Olivera J. Finn

I. Introduction 168
II. Prerequisites for Effective Cancer Immunotherapy: Identifying Tumor Antigens 168
III. Adoptive ("Passive") Immunotherapy 169
IV. Active-Specific Immunotherapy: Vaccines 171
V. Cancer-Induced Immunosuppression Impinges on Immunotherapy 172
VI. Cancer Immunotherapy in Mice Versus Humans 175
VII. Immunotherapy and Cancer Stem Cells 176
VIII. Autoimmunity Resulting from Cancer Immunotherapy 176
IX. Conclusion and Future Considerations 177
    References 178

11. Cancer Vaccines 183
Freda K. Stevenson, Gianfranco Di Genova, Christian Ottensmeier, and Natalia Savelyeva

I. Introduction 184
II. Tumor Antigens 185
III. Spontaneous Immunity to Cancer 187
IV. Toleragenic Pressure on Immunity to Cancer 187
V. Immune Responses to Conventional Vaccines 189
VI. Cancer Vaccine Strategies 194
VII. DNA Vaccines 195
VIII. Challenges of Translation to the Clinic 199
IX. Concluding Remarks 200
    References 200
    Further Reading 204

PART III: TARGETS AND TACTICS TO IMPROVE CANCER IMMUNOTHERAPY BY DEFEATING IMMUNE SUPPRESSION

12. Immunotherapy and Cancer Therapeutics: Why Partner? 207
Leisha A. Emens and Elizabeth M. Jaffee

I. Introduction: Why Immunotherapy for Cancer? 208
II. Immune Tolerance and Suppression: Multiple Layers of Negative Control 209
III. T Cell Activation: A Rheostat for Tuning Immune Responses 212
IV. Immune Modulation with Therapeutic Monoclonal Antibodies 219
V. Therapeutics that Mitigate the Influence of $CD4^+CD25^+$ Tregs 222
VI. Endocrine and Biologically Targeted Therapy 224
VII. Conclusion 225
     References 225

13. Immune Stimulatory Features of Classical Chemotherapy   235
Robbert G. van der Most, Anna K. Nowak, and Richard A. Lake

I. Introduction   236
II. Tumor Cell Death   236
III. Pathways to Immunogenicity   239
IV. Chemotherapy and the Immune System   243
V. A Practical Partnership: Chemotherapy and Immunotherapy   246
VI. Effects of Chemotherapy on Human Antitumor Immunity and Chemoimmunotherapy Clinical Trials   250
References   252

14. Dendritic Cells and Coregulatory Signals: Immune Checkpoint Blockade to Stimulate Immunotherapy   257
Drew Pardoll

I. Regulation of T Cell Responses to Antigen   258
II. Regulatory T Cells   261
III. Immune Checkpoints in the Tumor Microenvironment   262
IV. Monoclonal Antibodies that Interfere with Coinhibitory Receptors on T Cells   266
V. What Is the Most Effective Way to Use Checkpoint Inhibitors?   269
References   270

15. Regulatory T Cells in Tumor Immunity: Role of Toll-Like Receptors   277
Rong-Fu Wang

I. Introduction   278
II. Immune Cells in Immunosurveillance and Tumor Destruction   278
III. TLRs and Their Signaling Pathways   279
IV. TLRs in Innate Immunity, Inflammation, and Cancer Development   280
V. Tumor-Infiltrating Immune Cells in the Tumor Microenvironment   281
VI. Molecular Marker for $CD4^+$ Tregs   282
VII. Antigen Specificity of $CD4^+$ Tregs   282
VIII. Suppressive Mechanisms of Tregs   283
IX. Functional Regulation of Tregs and Effector Cells by TLR Signaling   283
X. Implications for Enhancing Antitumor Immunity   284
XI. Conclusion   285
References   285

16. Tumor-Associated Macrophages in Cancer Growth and Progression   289
Alberto Mantovani, Paola Allavena, and Antonio Sica

I. Introduction   289
II. Macrophage Polarization   290
III. Macrophage Recruitment at the Tumor Site   291
IV. Tam Expression of Selected M2 Protumoral Functions   294
V. Modulation of Adaptive Immunity by Tams   296
VI. Targeting Tams   297
VII. Concluding Remarks   300
References   302

17. Tumor-Associated Myeloid-Derived Suppressor Cells 309
Stephanie K. Bunt, Erica M. Hanson, Pratima Sinha, Minu K. Srivastava, Virginia K. Clements, and Suzanne Ostrand-Rosenberg

I. Introduction 310
II. Multiple Suppressive Mechanisms that Contribute to Immunosuppression in Individuals with Tumors 310
III. MDSCs as a Key Cell Population that Mediates Tumor-Induced Immunosuppression 311
IV. MDSCs' Use of Mechanisms to Mediate Effects on Multiple Target Cells 317
V. MDSC Induction by Tumor-Derived Cytokines and Growth Factors 321
VI. MDSC Linking of Inflammation and Tumor Progression 322
VII. Agents Responsible for Reducing MDSC Levels 323
VIII. Conclusions: Implications for Immunotherapy 326
References 327
Further Reading 331

18. Programmed Death Ligand-1 and Galectin-1: Pieces in the Puzzle of Tumor-Immune Escape 333
Gabriel A. Rabinovich and Thomas F. Gajewski

I. Programmed Death Ligand 1 and Programmed Death 1 Interactions 334
II. Galectin 1 338
References 344
Further Reading 346

19. Indoleamine 2,3-Dioxygenase in Immune Escape: Regulation and Therapeutic Inhibition 347
Alexander J. Muller and George C. Prendergast

I. Introduction 348
II. IDO Function in T Cell Regulation 351
III. Complex Control of IDO by Immune Regulatory Factors 351
IV. Immune Tolerance Via IDO in Dendritic Cells 353
V. IDO Dysregulation in Cancer Cells 357
VI. IDO as a Target for Therapeutic Intervention 359
VII. Discovery and Development of IDO Inhibitors 360
VIII. Conclusion 361
References 362
Further Reading 368

20. Arginase, Nitric Oxide Synthase, and Novel Inhibitors of L-Arginine Metabolism in Immune Modulation 369
Susanna Mandruzzato, Simone Mocellin, and Vincenzo Bronte

I. Introduction 370
II. NOS: Genes, Regulation, and Activity 371
III. ARG: Genes, Regulation, and Activity 372
IV. Immunoregulatory Activities of ARG and NOS 374
V. Possible Physiological Role for L-ARG Metabolism in Immunity Control 381
VI. NOS in Cancer 382

VII. ARG in Cancer   384
VIII. ARG and NOS Inhibitors: A Novel Class of Immune Adjuvants?   386
IX. Conclusion and Perspectives   388
    References   389
    Further Reading   399

Index   401

Contributors

Numbers in parentheses indicate the page(s) on which the authors' contributions begin.

Paola Allavena (289) Istituto Clinico Humanitas, Milan, Italy

Vincenzo Bronte (369) Istituto Oncologico Veneto, Padua, Italy

Stephanie K. Bunt (309) Department of Biological Sciences, University of Maryland Baltimore County, Baltimore, MD, USA

Bruce D. Car (149) Bristol-Myers Squibb Co., Princeton, NJ, USA

Alfred Chang (83) Department of Surgery, University of Michigan School of Medicine, Ann Arbor, MI, USA

Virginia K. Clements (309) Department of Biological Sciences, University of Maryland Baltimore County, Baltimore, MD, USA

Gianfranco Di Genova (183) Molecular Immunology Group, Cancer Sciences Division, Southampton University Hospitals Trust, Southampton, UK

Julie Y. Djeu (63) University of South Florida, Cancer Biology Ph.D. Program, H. Lee Moffitt Cancer Center, Tampa, FL, USA

Glenn Dranoff (29) Harvard Medical School, Dana Farber Cancer Institute, Boston, MA, USA

Leisha A. Emens (207) The Sidney Kimmel Comprehensive Cancer Center at Johns Hopkins Medical Institute, Bunting-Blaustein Cancer Research Building, Baltimore, MD, USA

Olivera J. Finn (167) Department of Immunology, University of Pittsburgh, Pittsburgh, PA, USA

Richard A. Flavell (43) Section of Immunobiology, Yale University School of Medicine, New Haven, CT, USA

Thomas F. Gajewski (333) University of Chicago, Chicago, IL, USA

Paul B. Gilman (101) The Lankenau Institute for Medical Research, Lankenau Cancer Center, Wynnewood, PA, USA

Erica M. Hanson (309) Department of Biological Sciences, University of Maryland Baltimore County, Baltimore, MD, USA

Manuel Hidalgo (117) The Sidney Kimmel Comprehensive Cancer Center at Johns Hopkins Medical Institute, Division of Medical Oncology, Baltimore, MD, USA

Elizabeth M. Jaffee (3, 207) The Sidney Kimmel Comprehensive Cancer Center at Johns Hopkins Medical Institute, Bunting-Blaustein Cancer Research Building, Baltimore, MD, USA

Masahisa Jinushi (29) Harvard Medical School, Dana Farber Cancer Institute, Boston, MA, USA

Ryungsa Kim (9) Department of Surgical Oncology, Hiroshima University, Research Institute for Radiation Biology and Medicine, Hiroshima University, Hiroshima, Japan

Richard A. Lake (235) Asbestos-Related Diseases Institute of Australia (ADRIA), University of Western Australia, Sir Charles Gairdner Hospital, Nedlands, Perth, Australia

Andrew J. Lepisto (167) Department of Immunology, University of Pittsburgh, Pittsburgh, PA, USA

Ming O. Li (43) Section of Immunobiology, Yale University School of Medicine, New Haven, CT, USA

Susanna Mandruzzato (369) Department of Oncology and Surgical Sciences, Oncology Section, Institute Oncologico Veneto, Padua, Italy

Alberto Mantovani (289) Istituto Clinico Humanitas, Milan, Italy

John R. McKolanis (167) Department of Immunology, University of Pittsburgh, Pittsburgh, PA, USA

Simone Mocellin (369) Istituto Oncologico Veneto, Padua, Italy

Alexander J. Muller (347) Lankenau Institute for Medical Research, Lankenau Cancer Center, Wynnewood, PA, USA

Anna K. Nowak (235) University Department of Medicine, Western Australian Institute for Medical Research, Queen Elizabeth II Medical Centre, Nedlands, Perth, Australia

Suzanne Ostrand-Rosenberg (309) Department of Biological Sciences, University of Maryland Baltimore County, Baltimore, MD, USA

Christian Ottensmeier (183) Molecular Immunology Group, Cancer Sciences Division, Southampton University Hospitals Trust, Southampton, UK

Drew Pardoll (257) The Sidney Kimmel Comprehensive Cancer Center at Johns Hopkins Medical Institute, Bunting-Blaustein Cancer Research Building, Baltimore, MD, USA

George C. Prendergast (3, 101, 347) The Lankenau Institute for Medical Research, Lankenau Cancer Center, Wynnewood, PA, USA

Gabriel A. Rabinovich (333) Instituto de Biología y Medicina Experimental, Consejo Nacional de Investigaciones, Científicas y Técnicas de Argentina, Buenos Aires, Argentina

David A. Sallman (63) University of South Florida, Cancer Biology Ph.D. Program, H. Lee Moffitt Cancer Center, Tampa, FL, USA

Natalia Savelyeva (183) Molecular Immunology Group, Cancer Sciences Division, Southampton University Hospitals Trust, Southampton, UK

Antonio Sica (289) Istituto Clinico Humanitas, Milan, Italy

Pratima Sinha (309) Department of Biological Sciences, University of Maryland Baltimore County, Baltimore, MD, USA

Minu K. Srivastava (309) Department of Biological Sciences, University of Maryland Baltimore County, Baltimore, MD, USA

Freda K. Stevenson (183) Molecular Immunology Group, Cancer Sciences Division, Southampton University Hospitals Trust, Southampton, UK

Rajesh Thirumaran (101) The Lankenau Institute for Medical Research, Lankenau Cancer Center, Wynnewood, PA, USA

Robbert G. van der Most (235) University Department of Medicine, Western Australian Institute for Medical Research, Queen Elizabeth II Medical Centre, Nedlands, Perth, Australia

Rong-Fu Wang (277) Department of Immunology, Baylor College of Medicine, Houston, TX, USA

Colin D. Weekes (117) The Sidney Kimmel Comprehensive Cancer Center at Johns Hopkins Medical Institute, Division of Medical Oncology, Baltimore, MD, USA

Shuang Wei (83) Department of Surgery, University of Michigan School of Medicine, Ann Arbor, MI, USA

Richard A. Westhouse (149) Bristol-Myers Squibb Co., Princeton, NJ, USA

Weiping Zou (83) Department of Surgery, University of Michigan School of Medicine, Ann Arbor, MI, USA

PART I

PRINCIPLES OF CANCER IMMUNOBIOLOGY

CHAPTER 1

Introduction

GEORGE C. PRENDERGAST AND ELIZABETH M. JAFFEE

I. Overview
II. Historical Background
III. Looking Ahead: Marrying Chemotherapy and Immunotherapy
IV. Parts of the Book

I. OVERVIEW

Limited scientific collaborations have been conducted historically between immunologists, geneticists, and cell biologists who study cancer. This situation is beginning to change with an emerging consensus among all cancer researchers that inflammation and immune escape play key causal roles in the development and progression of malignancies. This finding is particularly true for adult solid tumors, the causes of which are complex and multifactoral, posing a major clinical challenge. This book aims at cross-fertilizing ideas and concepts among investigators who are striving to develop combinatorial immunological or pharmacological agents as cancer therapeutics. The specific goals are (1) to highlight emerging principles of immune suppression in cancer patients and (2) to discuss how to combine immunotherapeutic and chemotherapeutic agents to defeat mechanisms of immune or inflammatory suppression and improve cancer treatment. Many immune-based therapies have focused on activating the immune system.

However, it has become increasingly clear that these therapies are thwarted by the ability of cancers to evade or suppress the immune system. In this book, contributors with variety of perspectives and diverse experience provide an overview of how tumors evolve to evade the immune system, describe the nature of standard and experimental approaches used to treat cancer in oncology clinics, and explain how these approaches might be enhanced by inhibiting important mechanisms of tumoral immune tolerance and suppression.

II. HISTORICAL BACKGROUND

"I can't understand why people are frightened of new ideas. I'm frightened of the old ones."
—John Cage (1912–1992)

Starting about 1980, research investigations in cancer genetics and cell biology began to assume the prominence in cancer research that they still hold in the twenty-first century. Hatched initially from studies of animal tumor viruses, the field of cancer genetics has contributed significantly to the understanding of the biologic pathways involved in tumor initiation and progression and has identified specific targets for therapeutic intervention. With the discovery of cellular oncogenes, the once radical idea that cancer was a disease of normal cellular genes "gone wrong" not only was established as the dominant idea in the field but also strongly influenced the development of new drugs to treat cancer, with the goal of attacking the products of those genes. At the same time, these developments outpaced other concepts of cancer as a systemic disease involving perturbations in the immune system. After decades of mutual skepticism, an historically important consensus among cancer researchers is emerging in the twenty-first century about the causality of chronic inflammation and altered immunity in driving malignant development and progression. Ironically, this synthesis is having the effect of making the "new" genetic ideas of the past two decades about cancer seem somewhat dated; in particular, it is becoming apparent that the tumor cell-centric focus championed by cancer geneticists is unlikely to give a full understanding of clinical disease because the systemic and localized tissue conditions that surround and control the growth and activity of the tumor cell remain unknown. Perhaps contributing to some consternation about the conceptual weight of the "new" ideas, few of the molecular therapeutics developed from them have had much major clinical impact (e.g., the Bcr-Abl kinase inhibitor Gleevec® represents perhaps the only impressive success case among molecular cancer therapeutics).

Among the earliest pathohistological descriptions of cancer is Rudolf Virchow's, written in 1863. Virchow first noted the surfeit of inflammatory cells in many tumors. From this root, tumor immunologists have for many years struggled to fully understand the precise relationships between inflammation, immunity, and cancer and to develop principles that can robustly impact the diagnosis, prognosis, and treatment of cancer. With the emergence of cancer genetics and tumor cell-centric concepts of disease as major conceptual drivers in the late twentieth century, roles identified for tumor stromal cells and immunity in cancer became marginalized or simply ignored by many investigators. Indeed, old skepticisms about whether immunity was important in cancer patients have persisted until quite recently, as can be illustrated by the omission in a prominent review in 2000 of immune escape as a critical trait of cancer (Hanahan and Weinberg, 2000). However, since the turn of the twenty-first century, perspectives in the field have once again undergone a radical shift, with many cancer researchers now focused intensely on how tumorigenesis and tumor dormancy versus progression are shaped by the stromal microenvironment, inflammation, and alterations in the immune system. These recent perspectives are based on emerging data that have come about since technological advances from the molecular revolution have been adopted to better understand immune system and host interactions.

During the past 25 years, as a result of historical and scientific divisions, limited communication, understanding, and collaboration have existed between tumor immunologists, molecular geneticists, and cell biologists working in the field. On one hand, this situation has been exaggerated by what now seems like an overly narrow focus of geneticists and cell biologists on tumor cell-centric concepts of cancer, which continues to persist to some degree. On the other hand, immunologists have struggled to establish a clear understanding of how inflammation and immune cells can contribute to promoting or controlling cancer. Biases rooted to some extent in old controversies, which have been transmitted to younger scientists entering each field, have further limited communication and interaction between the two camps. Many of these old issues have been happily put to rest in the twenty-first century as a result of experiments conducted with modern transgenic animal model systems and the emergence of carefully controlled clinical observations, such that the key pathophysiological foundations of inflammation and immune dysfunction in cancer are now firmly established (Dunn et al., 2004). Contributing to the new perspective is a wider appreciation of both the critical role of the tumor microenvironment in malignant development and the power of immune suppression mechanisms in licensing cancer cell proliferation, survival, and metastasis. In terms of immunotherapeutic responses, it seems increasingly clear that to "push on the gas" of immune activation, it will be necessary to "get off the brakes" of immune suppression—an idea that cancer geneticists may recognize as reminiscent of the concept in their field of oncogenes driving neoplastic cell proliferation only when the blockades imposed by tumor suppressor genes are relieved.

III. LOOKING AHEAD: MARRYING CHEMOTHERAPY AND IMMUNOTHERAPY

Clearly, the goal of cancer therapy is to kill a residual tumor that cannot be excised surgically. However, the inherent nature of the cancer cell limits the full effectiveness of therapies that have been developed or that arguably can be developed. Being of host origin, cancer cells share features of the host that make effective treatment difficult due to side effects that limit the therapeutic window. Moreover, the plastic nature of tumors makes them remarkably resilient in rebounding from clinical regimens of radiotherapy and chemotherapy that are traditionally used. For example, even when the vast majority of cancer cells are killed by a cytotoxic chemotherapeutic drug, a small number of residual cells that are resistant to the agent can be sufficient to seed the regrowth of a tumor. Making matters worse, this tumor may no longer respond to the previously successful therapy because of the capacity of tumor cells to evolve resistance under selective pressures applied by cytotoxic agents. Indeed, the concept of selection is integral to understanding this disease: development and progression in cancer are driven by the selection of cells that survive normally lethal conditions. Resistance to any normally lethal pressure can be selected by evolution in a cancer cell population because of the cells' genetic plasticity, an important characteristic of cancer cells. As demonstrated in the treatment of other diseases caused by a highly mutable entity, such as HIV, successful targeting of tumor cells may require the application of multiple agents that target different survival mechanisms. However, compared to HIV, the genetic space available for the evolution of a cancer cell is far larger due to the greater size of the cancer cell genome. Thus, effective eradication of tumors has proven—and may continue to prove to be—quite challenging because of the cancer genome's ability to evolve survival mechanisms in response to agents' multiple selection pressures, even when multiple agents in combination are used. The oncology field's best chance is to identify as many of these mechanisms as possible and to discover approaches that synergize to inhibit these many mechanisms. The war on cancer has really just begun.

Two general solutions to this apparently dismal situation may be to redirect the focus of attack from the tumor cell itself to the environment that sustains its growth and survival or to engage the immune system in a way that allows it to eradicate tumor cells like an infection. The former strategy is essentially passive in nature insofar as cancer cells are killed by an indirect route. For example, by depriving tumors of blood supply, antiangiogenic therapies can indirectly kill cancer cells. Resistance to such therapies should be difficult to evolve, as the argument goes, because stromal cells in the tumor environment are not genetically plastic. However, due to their passive nature, such therapies are still prone to circumvention through tumor cell evolution (e.g., vascular mimicry in the case of antiangiogenesis therapies [Folberg et al., 2000]). In contrast, an active strategy to engage or "awaken" active immunity in the cancer patient has many appeals, the chief of which is its capability to "dodge and weave" with tumor heterogeneity, the inherent outcome of the tumor cells' response to selection pressures. In this regard, the immune system may be particularly well suited to clear the small numbers of residual tumor cells (particularly dormant cells or cancer stem cells) that may be poorly eradicated by radiotherapy and chemotherapy, which could help lengthen remission periods. Indeed, even treatments that did not cure but rather converted cancer to a long-term subclinical condition, by analogy to HIV infections, would represent a resounding success. Therefore, the key question becomes how a tumor can outrun an activated immune system, given that precisely this event has occurred during tumorigenesis, so that the balance might be tipped back in favor of the immune system.

This book addresses the challenge of the former director of the U.S. National Cancer Institute to develop new strategies that can successfully manage cancer patients in oncology clinics by 2015. The authors first introduce the concept, definition, and significance of immunoediting and immune-suppression mechanisms in cancer progression and therapy. Two fundamental ideas here are that subclinical (occult) cancers occur commonly and that clinical cancer represents only those rare lesions that have escaped immune control. Second, given the likelihood that state-of-the-art treatment will involve combination treatment, we provide an overview of standard and experimental therapeutics that are used in the oncology clinic, including cytotoxic chemotherapy, molecular targeted therapeutics, immunotherapeutics, and cancer vaccines. Third, we introduce new ideas about treatment combinations that may prove effective, particularly combinations of chemotherapy and immunotherapy that have been little explored. Last, we summarize recent advances in the definition of key molecular mechanisms of immune suppression, the disruption of which may derepress antitumor immunity and thereby enhance the efficacy of other anticancer agents. In the culture of the early twenty-first century, cancer immunologists tend to be oriented to biological therapies and to have limited knowledge of cancer pharmacology or genetics. Conversely, cancer geneticists and pharmacologists tend to be more familiar with small molecule therapies and to have limited understanding of cancer immunology or immune-based therapies (other than perhaps passive therapies, e.g. antibodies). Overall, we hope to illustrate why interactions between these two comparatively disparate groups may be both intellectually and clinically beneficial.

IV. PARTS OF THE BOOK

Part I of the book introduces principles of cancer immunobiology. In Chapter 2, the central concept of immunoediting is introduced. This fundamental process has three parts, termed *immunosurveillance*, *immune equilibrium*, and *immune escape*, which lead to control, stasis, or outgrowth of a malignancy. Immunoediting starts with the immune recognition and destruction of cells that have acquired genetic and epigenetic alterations characteristic of tumor cells, but at the same time, the selective pressure produced by immunoediting drives tumor evolution and progression. In this process, the cell-intrinsic traits of cancer (immortalization, growth deregulation, apoptotic resistance, and tumor suppressor inactivation) lead to the development of subclinical or occult lesions that are not clinically important until cell-extrinsic traits (invasion, angiogenesis, metastasis, and immune escape) have been achieved. The complex roles for inflammatory cells and altered immunity in the development of cell-extrinsic traits represent an increasingly important area for investigation. Chapter 3 discusses key aspects of immunosurveillance, including the generation of innate and adaptive immune responses to tumor cells. Chapter 4 examines cytokines that promote immune tolerance to tumors. Chapter 5 explains immune "sculpting" processes that occur in the tumor from the evolution of the battle between immune cells and tumor cells; this discussion focuses key roles played in the inflammatory tumor microenvironment by natural killer cells (NK cells). Chapter 6 discusses the emergence of the diverse immunosuppressive networks ultimately evolved by tumors that successfully escape immune control.

Part II provides an overview of accepted and experimental classes of cancer therapeutics that clinics use. Chapter 7 introduces standard cytotoxic chemotherapeutics, which along with radiotherapy continue to represent the major part of the armentarium used by clinical oncologists. Chapter 8 introduces the new classes of small molecule drugs termed *molecular targeted therapeutics*, many of which are still in clinical trials, that take advantage of the latest understanding in modern cancer genetics and cell biology. Chapter 9 offers an overview of cancer pharmacology and safety assessment for both "classical" types of cytotoxic drugs as well as modern molecular-targeted therapeutics, where the central goal is learning whether "hitting the target" in the tumor cell can be achieved in a relatively safe and effective manner. In Chapters 10–11, various active immunotherapeutic approaches that are being tested clinically are summarized, and Chapter 11 also includes a special focus on various types of cancer vaccines.

Part III introduces a set of molecular targets and tactics that are hypothesized to improve cancer treatment by defeating immune suppression. In Chapters 12–13, a discussion of the rationale for interest in combining immunotherapy and cytotoxic chemotherapy is presented. Evidence that cytotoxic chemotherapy acts in part by stimulating immune activity has existed for some time, but this evidence is not widely known. Furthermore, while cancer pharmacology and cancer immunology are each very well-developed fields, there have been few efforts to investigate combinatorial regimens at either the preclinical or clinical level. These chapters highlight reasons why such investigations should be of interest to pursue. Chapters 14–17 provide an overview of several important cell-based mechanisms of immune suppression that appear to arise almost universally in cancer, including the production of inhibitory dendritic cells and coinhibitory molecules, T regulatory cells, tumor-associated macrophages, and suppressor myeloid cells. The importance of these mechanisms to tumor outgrowth is underscored by the fact that many tumors are actually composed of a large proportion if not a majority of these cell types, as compared to other tumor stromal cells or tumor cells themselves. Chapters 18–20 introduce three important molecular mechanisms of immune suppression in tumors, including those mediated by the cell surface molecules galectin-1 and programmed death ligand-1 (PDL-1), the catabolic enzymes indoleamine 2,3-dioxygenase (IDO) and arginase, and the vascular regulator inducible nitric oxide synthase (iNOS). Strategies for therapeutic inactivation of these mechanisms, which have been justified in preclinical models, are moving forward for evaluation in clinical trials.

Acknowledgments

Work in the authors' laboratories is supported by National Cancer Institute grants CA82222, CA100123, and CA109542 and the Lankenau Hospital Foundation (G.C.P.); by National Cancer Institute grants of the Specialized Program of Research Excellence in Gastrointestinal Cancer (CA62924) and Breast Cancer (CA88843); by the Avon Foundation for Breast Cancer Research and the National Cooperative Drug

Discovery Group (U19CA72108); and by generous donations from the Goodwin Family and The Sol Goldman Pancreatic Research Center (E.M.J.). Dr. Jaffee is the Dana and Albert "Cubby" Broccoli Professor of Oncology. G.C.P. would like to acknowledge contributions from long-term collaborator A.J. Muller toward development of the perspectives offered in this chapter. G.C.P. declares competing financial interests as a significant stockholder and scientific advisory member at New Link Genetics Corporation, a biotechnology company that has licensed intellectual property to develop inhibitors of the immune suppressive enzyme IDO for combinatorial cancer treatment, as described in patents WO 2004 093871 "Novel methods for the treatment of cancer" (pending) and WO 2004 094409 "Novel IDO inhibitors and methods of use" (pending).

References

Dunn, G. P., Old, L. J., and Schreiber, R. D. (2004). The three Es of cancer immunoediting. *Ann. Rev. Immunol.* **22**, 329–360.

Folberg, R., Hendrix, M. J., and Maniotis, A. J. (2000). Vasculogenic mimicry and tumor angiogenesis. *Am. J. Pathol.* **156**, 361–381.

Hanahan, D., and Weinberg, R. A. (2000). The hallmarks of cancer. *Cell* **100**, 57–70.

Further Reading

Emens, L. A., and Jaffee, E. M. (2005). Leveraging the activity of tumor vaccines with cytotoxic chemotherapy. *Cancer Res.* **65**, 8059–8064.

Muller, A. J., and Scherle, P. A. (2006). Targeting the mechanisms of tumoral immune tolerance with small-molecule inhibitors. *Nature Rev. Cancer* **6**, 613–625.

Prendergast, G. C. (ed.) (2004). *Molecular Cancer Therapeutics: Strategies for Drug Discovery and Development*. Wiley-Liss: New York.

CHAPTER 2

Cancer Immunoediting: From Immune Surveillance to Immune Escape

RYUNGSA KIM

I. Introduction
II. Cancer Immune Surveillance
   A. Historical Background
   B. Experimental Evidence for Immune Surveillance
   C. Clinical Evidence for Immune Surveillance
   D. Nonimmunological Surveillance
III. Cancer Immunoediting
   A. Elimination
   B. Equilibrium
   C. Escape
IV. Concluding Remarks

Immunoediting has emerged as a fundamental model in cancer immunology. As enunciated by Dunn *et al.* (2002, 2004, 2005), immunoediting proposes to explain the complex evolution of the "cat and mouse" relationship that exists between cancer cells and the immune system during the development of an overt malignancy. Immunoediting has three temporally distinct stages, termed *elimination*, *equilibrium*, and *escape*. Elimination is equivalent to immune surveillance, in which innate and adaptive components of the immune system act to eradicate cancers that arise in the organism. Due to the plastic nature of cancer, the progression of which is driven by a selection for variant cells that can survive normally lethal conditions, elimination can provide a selective pressure for immune "sculpting" that generates immune-resistant cells. As this selection proceeds, the second stage of equilibrium in immunoediting is reached when the immune system can still control tumors but no longer eradicate them. With continued selection, dormant tumors in equilibrium may evolve to the third stage of immune escape, in which cells can defeat, evade, or tolerize the immune system. This chapter provides an overview of how diverse mechanisms of immunoediting preserve positive features of the immune stromal environment that facilitate tumor progression as tumor immunity is progressively defeated.

I. INTRODUCTION

Since Ehrlich first proposed the idea in 1909 that nascent transformed cells arise continuously in human bodies and that the immune system scans for and eradicates these transformed cells before they are manifested clinically, immune surveillance has been a controversial topic in tumor immunology (Ehrlich, 1909). In the midtwentieth century, experimental evidence that tumors could be repressed by the immune system came from the use of tumor transplantation models. The findings from these models strongly suggested the existence of tumor-associated antigens (TAAs) that formed the basis of immune surveillance, as postulated by Burnet and Thomas (Burnet, 1957). The functional role of antigen-presenting cells (APCs) in crosspriming for T cell activation was subsequently demonstrated, and the cancer immune surveillance model was developed. However, the idea of cancer immune surveillance resisted widespread acceptance until the 1990s when experimental animal models using knockout mice validated the existence of cancer immune surveillance in chemically induced and spontaneous tumors. The central roles of immune effector cells, such as B cells, T cells, natural killer (NK) cells, and natural killer T (NKT) cells as well as type I and II interferons (IFNs) and perforin (pfp), have since been clarified in cancer immune surveillance (Dunn et al., 2002; Dunn et al., 2005).

As part of the twenty-first century concept of cancer immunoediting leading from immune surveillance to immune escape, three essential phases have been proposed (Dunn et al., 2002): (i) elimination; (ii) equilibrium; and (iii) escape. Nascent transformed cells can be eliminated initially by immune effector cells, such as NK cells, and by the secreted IFN-γ in innate immune response. Elimination of transformed cells results in immune selection and immune sculpting, which induce tumor variants that decrease immunogenicity and become resistant to immune effector cells in the equilibrium phase. Eventually, during tumor progression, when the increased tumor size can be detected by imaging diagnosis, tumor-derived soluble factors (TDSFs) can induce several escape mechanisms from immune attack in the tumor microenvironment (Kim et al., 2006). This chapter provides a general overview and basic principles of immunoediting from immune surveillance to escape as well as a discussion of the central role of immune effector cells in the process of immunoediting. A better understanding of the mechanisms of immunoediting during tumor progression may provide new insights for improving cancer immunotherapy.

II. CANCER IMMUNE SURVEILLANCE

A. Historical Background

In the early twentieth century, Ehrlich first proposed the existence of immune surveillance for eradicating nascent transformed cells before they are clinically detected (Ehrlich, 1909). Almost 50 years later, Burnet and Thomas postulated that the control of nascent transformed cells may represent the actions of an ancient immune system, which played a critical role in preventing malignant transformation (Burnet, 1957). The idea was supported by experimental results showing strong immune-mediated rejection of transplanted tumors into mice. Although there was excellent evidence in support of the belief that immune surveillance mechanisms prevent the outgrowth of tumor cells induced by horizontally transmitted, ubiquitous, potential oncogenic viruses, there was much less evidence for immune surveillance acting against chemically induced tumors in syngeneic mice (Klein, 1976). The use of genetically identical mice, however, generated tumor-specific protection from methylcholantrene (MCA) and virally induced tumors (Prehn and Main, 1957; Old and Boyse, 1964). These results from mouse models strongly suggested the existence of TAAs and immune surveillance for protection from transformed cells in the host, which was postulated by Burnet and Thomas (Burnet, 1957; Burnet, 1971; Thomas, 1982). Despite the fact that several lines of evidence from experimental mouse models showed the immune system played a critical role in dealing with transformed cells, it was argued that there was no increased incidence of spontaneous or chemically induced tumors in athymic nude mice compared to wild-type animals (Rygaard and Povlsen, 1974; Stutman, 1974). This evidence suggested that immune surveillance in mice targeted transforming viruses in tumors rather than the tumors themselves (Burnet, 1957). It is now known that athymic nude mice have NK cells and fewer T cells than wild-type mice, which can contribute to immune surveillance. Further, athymic mice have detectable populations of functional $\gamma\delta$-T cell receptor-bearing lymphocytes (Ikehara et al., 1984; Maleckar and Sherman, 1987). Years later, titration of the MCA dosage revealed that nude mice actually did form more tumors than the control mice population (Engel et al., 1996). Similarly, tumor formation induced by MCA was greater in severe combined immunodeficient (SCID) mice than in wild-type BALB/c mice (Engel et al., 1997). These observations put to rest one of the widely cited skepticisms about whether the immune system had any role in suppressing cancers that were not induced by transforming viruses.

B. Experimental Evidence for Immune Surveillance

The emergence of gene-targeted mice, such as knockout mice, has facilitated the gathering of evidence for immune-mediated surveillance of spontaneous epithelial tumors (Dunn et al., 2002). Strikingly, mice with a variety of immunodeficiencies produced by specific genetic ablations have been shown to be more susceptible to MCA-induced tumors and spontaneous lymphomas (Dunn et al., 2004).

1. NK, NKT, and $\gamma\delta$-T cells

During the mid-1970s to the 1990s, many investigators sought evidence to illustrate the immune surveillance concept. The discovery of NK cells provided a considerable stimulus for the possibility that they functioned as the effectors of immune surveillance (Herberman and Holden, 1978) even though a precise definition and understanding of these cells had not been confirmed. However, it was not until the 2000s that gene-targeted and lymphocyte subset-depleted mice were used to definitively establish the relative importance of NK and NK1.1$^+$ T cells in protecting against tumor initiation and metastasis. In these models, CD3$^+$ NK cells were responsible for tumor rejection and protection from metastasis in models where control of major histocompatibility complex (MHC) class I-deficient tumors was independent of interleukin 12 (IL-12) (Smyth et al., 2000a). C57BL/6 mice that were depleted of both NK and NKT cells by the anti-NK1.1 monoclonal antibody (mAb), which can eliminate both NK and NKT cells, were two to three times more susceptible to MCA-induced tumor formation than control mice (Smyth et al., 2001). Further, a similar result was observed in C57BL/6 mice treated with antiasialo-GM1, which selectively eliminates NK but not NKT cells, even though antiasialo-GM1 can also eliminate activated macrophages. A protective role for NKT cells was only observed when tumor rejection required endogenous IL-12 activity. In particular, T cell receptor (TCR) Jα281 gene-targeted mice confirmed a critical function for NKT cells in protecting against spontaneous tumors initiated by the chemical carcinogen, MCA. Jα281$^{-/-}$ mice, lacking Vβ14Jα281-expressing invariant NKT cells, formed MCA-induced sarcomas at a higher frequency than did wild-type mice (Smyth et al., 2000a). Another study showed that mice treated with the NKT cell-activating ligand α-galactosylceramide throughout MCA-induced tumorigenesis exhibited a reduced incidence of tumors and displayed a longer latency period to tumor formation than did control mice (Hayakawa et al., 2003).

Mice lacking γδ-T cells were also shown to be highly susceptible to multiple regimens of cutaneous carcinogenesis. After exposure to carcinogens, skin cells expressed Rae-1 and H60, MHC-related molecules structurally resembling human MHC class I chain-related A (MICA). Each of these is a ligand for NKG2D, a receptor expressed by cytolytic T cells and NK cells. In vitro, skin-associated NKG2D$^+$ γδ-T cells killed skin carcinoma cells by a mechanism that was sensitive to blocking NKG2D attachment (Girardi et al., 2001). The localization of γδ-T cells in epithelia may therefore contribute to helping prevent epithelial malignancies.

2. IFN-γ

Endogenously produced IFN-γ protected the host against transplanted tumors and the formation of chemically induced and spontaneous tumors. When the mice were treated with neutralizing monoclonal antibody to IFN-γ, the growth rate of immunogenic sarcomas transplanted into mice was greater than the growth rate in the control mice (Dighe et al., 1994). Further, overexpression of the truncated dominant negative form of the murine IFN-γ receptor γ subunit (IFNGR1) in Meth A fibrosarcoma completely abrogated tumor sensitivity to IFN-γ, and the tumors showed enhanced tumorigenicity and reduced immunogenicity when they were transplanted into syngeneic BALB/c mice (Dighe et al., 1994). These results showed that IFN-γ had direct effects on tumor cell immunogenicity and played an important role in promoting tumor cell recognition and elimination. In the experiment with MCA-induced tumor formation, compared with wild-type mice, mice lacking sensitivity to either IFN-γ (IFNGR-deficient mice) or all IFN family members (signal transducers and activators of transcription [Stat]1-deficient mice; Stat1 is the transcription factor that is important in mediating IFNGR signaling) developed tumors more rapidly and with greater frequency when challenged with different doses of the chemical carcinogen MCA. In addition, IFN-γ-insensitive mice developed tumors more rapidly than wild-type mice when bred onto a background deficient in the p53 tumor-suppressor gene (Kaplan et al., 1998). IFN-γ-insensitive p53$^{-/-}$ mice also developed a broader spectrum of tumors compared with mice lacking p53 alone. The importance of this experiment lay in the finding that certain types of human tumors become selectively unresponsive to IFN-γ. Thus, IFN-γ forms the basis for an extrinsic tumor-suppressor mechanism in immunocompetent hosts. Using experimental (B6, RM-1 prostate carcinoma) and spontaneous (BALB/c, DA3 mammary carcinoma) models of metastatic cancer, mice deficient in both pfp and IFN-γ were significantly less proficient than pfp- or IFN-γ-deficient mice in preventing metastasis of tumor cells to the lung. The pfp and IFN-γ-deficient mice were equally susceptible as mice depleted of NK cells in both tumor metastasis models; hence, IFN-γ appeared to play an early role in protection from metastasis (Street et al., 2001). Further analysis demonstrated that IFN-γ, but not pfp, controlled the growth rate of sarcomas arising in these mice; in addition, the host IFN-γ and direct cytotoxicity mediated by cytotoxic lymphocytes expressing pfp independently contributed antitumor effector functions that together controlled the initiation, growth, and spread of tumors in mice.

In another study, both IFN-γ and pfp were critical for suppression of lymphomagenesis, but the level of protection afforded by IFN-γ was strain specific. Lymphomas arising in IFN-γ-deficient mice were very nonimmunogenic compared with those derived from pfp-deficient mice, suggesting a comparatively weaker immune selection pressure by IFN-γ (Street et al., 2002). A significant incidence of late onset adenocarcinomas observed in both IFN-γ- and pfp-deficient mice indicated that some epithelial tissues were also subject to immune surveillance.

3. Perforin and Fas/FasL System

The cytotoxic factors perforin (pfp) and Fas/FasL are additional important factors involved in immune surveillance. In general, cell-mediated cytotoxicity attributed to cytotoxic T lymphocytes (CTLs) and NK cells is derived from either the granule exocytosis pathway or the Fas pathway. The granule exocytosis pathway utilizes pfp to direct the granzymes to appropriate locations in target cells, where they cleave critical substrates that initiate apoptosis. Granzymes A and B induce death via alternate, nonoverlapping pathways. The Fas/FasL system is responsible for activation-induced cell death but also plays an important role in lymphocyte-mediated killing under certain circumstances (Russell and Ley, 2002). The interplay between these two cytotoxic systems provides opportunities for therapeutic interventions to control malignant disease, but oversuppression of these pathways also leads to decreased tumor cell killing. In fact, C57/BL/6 mice lacking pfp$^{-/-}$ were more susceptible to MCA-induced tumor formation. In MCA-induced tumor formation, pfp$^{-/-}$ mice developed significantly more tumors compared with pfp-sufficient mice treated in the same manner (Street et al., 2001; Street et al., 2002). In addition, a previous study showed that pfp-dependent cytotoxicity is not only a crucial mechanism of both CTL- and NK-dependent resistance to injected tumor cell lines but that the cytoxicity also operates during viral and chemical carcinogenesis that were induced by MCA or 12-O-tetradecanoylphorbol-13-acetate (TPA) plus 7,12-dimethylbenzanthracene (DMBA) or induced by injection of oncogenic Moloney sarcoma virus in vivo (van den Broek et al., 1996). Experiments addressing the role of Fas-dependent cytotoxicity by studying resistance to tumor cell lines that were stably transfected with Fas did not provide evidence for a major role of Fas and did not exclude a minor contribution of Fas in tumor surveillance. Another study showed that pfp$^{-/-}$ mice have a high incidence of malignancy in distinct lymphoid cell lineages (T, B, NKT), indicating a specific requirement for pfp in protection against lymphomagenesis (Smyth et al., 2000a). The susceptibility to lymphoma was enhanced by the simultaneous lack of expression of the p53 gene. The pfp$^{-/-}$ mice were at least 1000-fold more susceptible to these lymphomas when transplanted, compared with immunocompetent mice in which tumor rejection was controlled by CD8$^+$ T lymphocytes (Smyth et al., 2000a). Taken together, these results indicate that components of the immune system were involved in controlling primary tumor development and showed the differential role of pfp and IFN-γ in protecting tumor formation between lymphoid and epithelial malignancies.

4. Lymphocytes

Although evidence suggested that the immune surveillance of cancer is dependent on both IFN-γ and lymphocytes, the critical demonstration for the involvement of lymphocytes came from the use of gene-targeted mice lacking the recombinase–activating gene 1 (RAG-1) or RAG-2. Homozygous mutants of RAG-2 are viable but fail to produce mature B or T lymphocytes (Shinkai et al., 1992). Loss of the RAG-2 function in vivo results in the total inability to initiate VDJ rearrangement, leading to a novel SCID phenotype. RAG-2 function and VDJ recombinase activity, per se, are not required for development of cells other than lymphocytes. Since nude mice do not completely lack functional T cells and the two components of the immune system, IFN-γ and pfp, to prevent tumor formation in mice, an elegant study using RAG-2$^{-/-}$ and Stat1$^{-/-}$ mice model showed for the first time that lymphocytes and IFN-γ collaborate to prevent the formation of carcinogen-induced sarcomas and spontaneous epithelial carcinomas (Shankaran et al., 2001). In detail, both the wild-type and RAG-2$^{-/-}$ mice had a pure 129/SvEv genetic background, and they were injected with MCA and monitored for tumor formation. RAG-2$^{-/-}$ mice formed tumors earlier than wild-type mice and with greater frequency. After 160 days, 9 out of 15 RAG-2$^{-/-}$ mice but only 2 out of 15 wild-type mice formed MCA-induced tumors. The increased tumor formation in RAG-2$^{-/-}$ was comparable to that in IFN-γ-insensitive mice that lacked either the IFNGR1 (12 out of 20) or Stat1 (17 out of 30) versus 11 out of 57 wild-type mice. In the collaboration between the lymphocytes- and IFN-γ/Stat1-dependent tumor suppressor mechanisms, mice lacking both genes (i.e., RAG-2$^{-/-}$ × Stat1$^{-/-}$ mice [RkSk mice]) showed increased susceptibility to MCA-induced tumors with 13 out of 18 mice compared to 11 out of 57 wild-type mice. However, RkSk mice did not show a significant increased incidence compared to mice that lacked either RAG-2$^{-/-}$ or Stat1$^{-/-}$. Thus, these findings indicated that T, NKT, and/or B cells are essential to suppress the formation of chemically induced tumors and to allow the presence of an extensive overlap between lymphocytes and Stat1-dependent IFN-γ-signaling.

As for the effect of tumor suppressor mechanisms on spontaneous tumors, 9 out of 11 wild-type mice were free of malignant disease, 2 had adenomas, but none had cancer. In contrast, 12 out of 12 RAG-2$^{-/-}$ mice showed malignant lesions in the intestinal tract and elsewhere. Half of these mice formed malignant diseases: 3 cecal adenocarcinoma, 1 ileocecal adenocarcinoma, 1 small intestinal adenocarcinoma, and 1 lung adenocarcinoma. In addition, 6 out of 11 RkSk mice developed mammary carcinomas, including 2 with adenocarcinomas and 1 with a distinct adenocarcinoma in the breast and cecum. The other 5 RkSk mice did not show palpable masses but the following were found at necropsy: 2 cecal adenocarcinomas, 1 cecal and lung adenocarcinoma, and 2 intestinal adenomas. Overall, 82% of the RkSk mice formed spontaneous cancers. Thus, these findings suggest the lack of lymphocytes, either alone or in combination with the IFN-γ-signaling defect, and indicate that the RAG-2$^{-/-}$ and RkSk mice are significantly more susceptible for spontaneous epithelial tumor formation than their wild-type counterparts. Moreover, RkSk mice form more spontaneous cancers than RAG-2$^{-/-}$ mice, suggesting that the overlap of the tumor suppressor mechanisms mediated by lymphocytes and IFN-γ/Stat1-signaling may only be partially effective (Shankaran et al., 2001).

In another report, the relative contributions of αβ- and γδ-T cells in blocking tumor formation by chemical carcinogens, such as MCA, DMBA, and TPA, were studied; this report also noted the effect of injecting the squamous cell carcinoma cell line phocine distemper virus (PDV) into TCRβ$^{-/-}$, TCRδ$^{-/-}$, and TCRβ$^{-/-}$δ$^{-/-}$ mice, which lack αβ-T cells, γδ-T cells, and all T cells, respectively (Girardi et al., 2001). Comparing tumor formation using PDV cells between wild-type and TCRβ$^{-/-}$ mice, 41 out of 110 sites developed tumors in TCRβ$^{-/-}$ mice, whereas 13 out of 134 sites developed tumors in wild-type mice. Although tumor latency accounted for a minor reduction, αβ-T cells reduced the number of tumors formed. In contrast, in TCRδ$^{-/-}$ and TCRβ$^{-/-}$δ$^{-/-}$ mice, nearly 100% of sites showed tumor formation, and the latency was substantially reduced. These findings indicate that αβ-T cells and γδ-T cells regulate the tumor growth of PDV cells in a distinct fashion and that the lack of αβ-T cells is not compensated by the presence of γδ-T cells and NK cells. In addition, the role of γδ-T cells in the development of MCA-induced sarcomas and spindle cell carcinomas was studied, and an increase in the number of tumors formed in TCRβ$^{-/-}$ and TCRδ$^{-/-}$ mice after MCA injection was observed compared to FVB mice (Girardi et al., 2001). Further, in naturally occurring human carcinomas induced by DMBA and TPA, 67% of TCRδ$^{-/-}$ mice showed tumor formation with increased tumor burden compared to 16% of wild-type mice. In contrast, TCRβ$^{-/-}$ and wild-type mice were equally susceptible to DMBA- and TPA-induced carcinogenesis. In addition, TCRδ$^{-/-}$ mice also showed a higher incidence of progression of papillomas into carcinomas. These findings indicate a distinct additional contribution in the regulation of tumor growth in αβ-T cells and γδ-T cells. In turn, it seems that αβ-T cells act to inhibit initial tumor formation that converts to malignant progression, whereas γδ-T cells directly inhibit tumor progression by using their cytotoxic mechanisms to kill tumor cells. Thus, the previous and recent data support the following basic concept of cancer immune surveillance originally proposed by Burnet and Thomas: the naturally existing immune system can recognize nascent transformed cells and can eliminate primary tumor formation by lymphocytes and secreted cytokines as important protective mechanisms in the host. The studies that used inbred mouse lines targeting disruptions in genes encoding critical components of the immune system are listed in Table 2.1, and these data support the control of tumor formation by the immune systems of both innate and adaptive immune compartments in cancer immune surveillance.

5. Type I IFNs

Much less is known about the involvement of type I interferons, IFN-α/β, which regulate immunological functions and

TABLE 2.1 Evidences of Experimental Studies on the Cancer Immune Surveillance Using Knockout Mice

Target Gene	Affected Target /Effector cell	Tumor Formation	Reference
TCR Jα281	NKT	MCA-induced sarcoma	Smyth et al. (2000a)
TCR-δ	γδ-T	MCA-induced sarcoma DMBA-induced skin tumor	Girardi et al. (2001)
TCRβ TCRβ/TCRδ	αβT T/γδT	MCA-induced sarcoma Reduced latency	Girardi et al. (2001)
IFN-γ	IFN-γ	MCA-induced sarcoma Spontaneous lymphoma Lung adenocarcinoma	Street et al. (2001) Street et al. (2002)
Stat1	IFN-γ-signaling	MCA-induced sarcoma	Kaplan et al. (1998) Shankaran et al. (2001)
Perforin	CTL/NK	MCA-induced sarcoma Spontaneous lymphoma TPA/DMBA-induced sarcoma	van den Broek et al. (1996) Smyth et al. (2000b) Street et al. (2001) Street et al. (2002)
RAG-2/Stat1	T/B/NKT/IFN-γ-signaling	MCA-induced sarcoma Spontaneous epithelial and mammary carcinoma	Shankaran et al. (2001)
IFNGR1 or Stat1/p53	IFN-γ-signaling/ tumor susceptibility	More rapid tumor formation/ broader spectrum of tumors	Kaplan et al. (1998)
Perforin/p53	CTL/NK/tumor susceptibility	Enhances susceptibility to lymphoma	Smyth et al. (2000b)

TCR, T cell receptor; IFN, interferon; Stat1, signal transducers and activators of transcription 1; RAG, recombinase-activating gene; NKT, natural killer T; IFNGR, interferon γ receptor; CTL, cytotoxic T lymphocyte; NK, natural killer; MCA, methylcholantrene; TPA, 12-O-tetradecanoylphorbol-13-acetate; DMBA, 7,12-dimethylbenzanthracene induce the same biologic effects as IFN-γ in the cancer immunoediting process. Some previous studies suggested a potential antitumor function for endogenously produced IFN-α/β. This was demonstrated by showing that neutralization of IFN-α/β using polyclonal antibodies in mice enhanced the growth of transplanted, syngeneic tumor cells in immunocompetent mice (Gresser et al., 1974), and the rejection was abrogated in the allografts or tumor xenografts (Gresser et al., 1988). In a study on the potential function of endogenously produced IFN-α/β in cancer immunoediting for tumor transplantation and primary tumor formation, endogenously produced IFN-α/β rejected highly immunogenic and syngeneic mouse sarcomas (Dunn et al., 2005). Further, although tumor cell immunogenicity was not influenced by the sensitivity to IFN-α/β, the requirement of IFN-α/β sensitivity in antitumor immune response for a host-protective effect depends on the level of hematopoietic cells. The host-protective effect of IFN-α/β was not completely overlapped for those of IFN-γ, indicating that IFN-α/β clearly played an important role and was a critical component in the process of cancer immunoediting. In this report, endogenously produced IFN-α/β rejected the tumor formation of highly immunogenic MCA-induced sarcomas and also inhibited the outgrowth of primary carcinogen-induced tumors in immunocompetent mice. Further, MCA-induced sarcomas derived from interferon-α receptor 1-deficient (IFNAR1$^{-/-}$) mice were rejected in a lymphocyte-dependent way in wild-type mice. This suggested that tumors formed in the absence of IFN-α/β responsiveness are more immunogenic than those formed in immunocompetent mice, results which differ from the poor immunogenicity found in tumors derived from IFNGR1$^{-/-}$ mice. Unlike the case of IFN-α/β, this poor immunogenicity can be rendered highly immunogenic and can be rejected when IFN-γ sensitivity is recovered by enforced expression of IFNGR1 (Kaplan et al., 1998). Thus, the finding that the functions of IFN-α/β and IFN-γ for cancer immunoediting do not completely overlap is supported by the differential effects of these cytokines on tumor cell immunogenicity.

Type I IFNs are considered as an important link between innate and adaptive immunity (Asselin-Paturel and Trinchieri, 2005), and this function acts primarily on several different bone marrow-derived cell subsets for tumor elimination. IFN-α/β have been shown to activate dendritic cells (DCs) and to increase the cytotoxic activity of NK cells through the induction of the TNF-related apoptosis-inducing ligand (TRAIL). In addition, IFN-α-derived tumor cells can promote antitumor immunity by preventing apoptotic cell death after stimulation of T lymphocytes. Furthermore, type I IFNs promote the development of memory-phenotype CD8$^+$ T (but not CD4$^+$) cells through the induction of IL-15. These findings indicate that the editing function of the immune system during tumor progression is served not only by lymphocytes and IFN-γ but also by IFN-α/β. Nevertheless, the involvement of the endogenously produced IFN-α/β in a host-protective function for naturally occurring antitumor immune responses to spontaneous tumor formation remains to be elucidated. Previous experimental and clinical studies in which exogenous IFN-α/β was administered may serve as evidence that IFN-α/β are an important immunostimulator to enhance antitumor immune responses that contribute to tumor reduction. Whether type I IFNs are actively and continuously induced during tumor progression by specific cells, such as tumors or nontumorous components, in host protection is still not known. Further molecular and cellular analyses to identify type I IFNs and to determine their responsiveness in cancer immunoediting will be needed.

C. Clinical Evidence for Immune Surveillance

A considerable number of experimental results, especially of gene-targeted studies, have made a convincing argument for the existence of cancer immunoediting. Two requisites for tumor immunity in clinical situations will now be discussed—both of these need to be satisfied so that immune effector cells and secreted cytokines can exert an antitumor immune response in the process of cancer immunoediting.

1. Tumor-Infiltrating Lymphocytes

It is readily accepted that tumor-infiltrating lymphocytes (TILs) in tumors can attack and eradicate tumor cells in the cancer patient. In fact, the presence of intratumor TILs is an important piece of evidence for immune response between tumor cells and immune effector cells. Several studies have shown that the high-grade density of $CD8^+$ T cells in cancer cell nests was correlated with prognosis, and the presence of TILs was able to predict a better survival as an independent prognostic factor in various types of cancers, including colon cancer (Naito et al., 1998), esophageal cancer (Yasunaga et al., 2000), oral squamous cell carcinoma (Reichert et al., 1998), breast cancer (Yoshimoto et al., 1993), ovarian cancer (Sato et al., 2005), and malignant melanoma (Haanen et al., 2006). Of importance, the T lymphocytes recruited around the tumor site (peritumor site) do not always contribute to antitumor immune response, but, rather, intratumor T lymphocytes are important for eradicating tumor cells (Naito et al., 1998). Other studies have shown a similar positive correlation between NK cell infiltration and the survival for gastric cancer (Ishigami et al., 2000), colorectal cancer (Kondo et al., 2003), and squamous cell lung cancer (Villegas et al., 2002). Thus, significant evidence has been presented for the link between the presence of TILs and the increased survival of cancer patients.

2. Organ Transplant-Related Cancer

The theory that cancer may arise under conditions of reduced immune capacity is supported by observations in humans with immune deficiencies, such as those that occur following organ transplants. Increased relative risk ratios for various types of cancers have been observed in immunosuppressed transplant recipients, and these cancers have no apparent viral origin. Information on 5692 Nordic recipients of renal transplants in 1964–1982 was linked with the national cancer registries in 1964–1986 and population registries (Birkeland et al., 1995). Significant overall excess risks of twofold to fivefold were seen in both sexes for cancers of the colon, larynx, lung, and bladder and in men for cancers of the prostate and testis. Notable high risks, ranging from 10- to 30-fold above expectations, were associated with cancers of the lip, skin (nonmelanoma), kidney, and endocrine glands; with non-Hodgkin's lymphoma; and with cancers of the cervix and vulva-vagina (Birkeland et al., 1995). Kidney transplantation increases the risk of cancer in the short- and in the long-term, consistent with the theory that an impaired immune system allows carcinogenic factors to act. Another study on the development of solid-organ tumors after cardiac transplantation reported that 38 solid tumors were identified in 36 (5.9%) of 608 cardiac transplant recipients who survived more than 30 days. The tumors included the following types: skin, lung, breast, bladder, larynx, liver, parotid, testicle, uterus, and nonmelanoma (Pham et al., 1995). A recent review reported a high frequency of skin cancers and lymphoproliferative diseases in renal transplant recipients (Morath et al., 2004). However, a survey of the literature showed that the relative frequency of malignancy after renal transplantation varied widely between different geographical regions. The type of malignancy is different in various countries and depends on genetic and environmental factors. The hypothesis that the action of immunosuppressive drugs is responsible for the increased incidence of cancers in transplant recipients is supported by the observation that patients also develop cancers if they receive immunosuppressive therapy for conditions other than transplantation (e.g., rheumatoid arthritis or systemic lupus erythematosus).

The other possibility for organ transplant-related cancer is the transmission of a tumor via micrometastasis of an undiagnosed malignancy in the donor after transplantation. According to data from the Organ Procurement and Transplantation Network/UNOS, 21 donor-related malignancies were reported out of 108,062 transplant recipients (Myron Kauffman et al., 2002). Except for 15 tumors that existed in donors at the time of transplantation, 6 tumors were *de novo* donor-derived tumors that developed in transplanted hematogenous or lymphoid cells of the donor. Similar donor-derived tumors have been reported in allografts obtained from donors with breast cancer and malignant melanoma (Zeier et al., 2002). These *de novo* tumors could be activated by the use of immunosuppressive drugs in the recipient even though the *de novo* tumors also could have been inactivated by immune surveillance in the donor before transplantation.

D. Nonimmunological Surveillance

The concept of cancer immune surveillance has been formulated based on the hypothesis that cancer cells are recognized as "nonself" and capable of inducing a rejection reaction. The immune system contributes to the surveillance of spontaneously developing tumors against virally induced tumors. However, in addition to the extrinsic actions of the immune system, there are intrinsic nonimmune surveillance systems that regulate the growth of tumor cells. There are two major forms of nonimmune surveillance. One is DNA repair, deficiencies that result in an increased incidence of tumors in xeroderma pigmentosum, which has several defects in mismatch repair enzymes. The other is intracellular surveillance, which is well documented in apoptotic cell death elicited by DNA damage or the activation of oncogenes. Since the definition of the term *nonimmune surveillance* is control of tumor cell growth, tumor progression and escape from nonimmune surveillance in the early stages and from subsequent immune surveillance in the late stages are associated with an increased resistance to apoptosis. The p53 pathway is a well-known example of genetic surveillance. Upon DNA damage, wild-type p53 is upregulated and binds DNA to induce growth arrest, allowing DNA repair. Since the p53 gene is inactivated in about 50% of human cancers that impair the DNA binding capacity of the protein, cell growth can continue despite DNA damage, resulting in tumor development. Inherited mutations in p53 seen in the Li-Fraumeni syndrome are associated with increased susceptibility to malignant diseases. The relative contribution of nonimmune surveillance compared with immune surveillance remains to be elucidated, but it is likely that they are complementary and not redundant. In fact, p53-deficient mice are susceptible to the formation of spontaneous tumors, demonstrated when the crossing of p53-deficient mice and pfp-deficient mice showed disseminated lymphomas, indicating a direct involvement of cytotoxic lymphocytes in cancer immune surveillance (Smyth et al., 2000b). Mice lacking p53 and the IFN-γ receptor or p53 and Stat1 showed a wider spectrum of tumors than those lacking only p53 (Kaplan et al., 1998).

As for other types of nonimmune surveillance, the existence of other mechanisms of intercellular surveillance have been documented. Such mechanisms are elicited by the interaction of cancer cells and surrounding normal cells in the tumor microenvironment that influence the probability of disseminated tumor cell growth. In addition, studies suggest that there is a genetically determined variation in the stringency of chromatin imprinting. More relaxed imprinting may lead to increased cancer risk and has been termed *epigenetic surveillance*. The immune evasion of tumors mediated by nonmutational epigenetic events involving chromatin and epigenetics collaborates with mutations in determining tumor progression.

III. CANCER IMMUNOEDITING

Although strong evidence has been presented supporting the existence of a functional cancer immune surveillance process against cancer in mice and humans, cancer continues to develop in intact immune systems and is refractory to many treatment approaches. These findings might be caused by the failure of early host tumor immunity to eradicate nascent transformed cells. Even in the presence of continued immune pressure, the failure to eradicate tumor cells results in tumor progression with reduced immunogenicity. Cancer immunoediting has been proposed in terms of the dual functions of host immunity not only for eliminating tumor cells but also for shaping malignant disease during the period of equilibrium between the tumor and host.

A. Elimination

Elimination is the hallmark of the original concept in cancer immune surveillance for the successful eradication of developing tumor cells, working in concert with the intrinsic tumor suppressor mechanisms of the nonimmunogenic surveillance processes. The process of elimination includes innate and adaptive immune responses to tumor cells. For the innate immune response, several effector cells, such as NK, NKT, and $\gamma\delta$-T cells, are activated by the inflammatory cytokines that are released by the growing tumor cells, macrophages, and stromal cells surrounding tumor cells. The secreted cytokines recruit more immune cells that produce other proinflammatory cytokines, such as IL-12 and IFN-γ. The pfp-, FasL-, and TRAIL-mediated killing of tumor cells by NK cells release tumor antigens (TAs), which lead to adaptive immune responses. In the cross talk between NK cells and DCs, NK cells promote the maturation of DCs and their migration to tumor-draining lymph nodes (TDLNs), resulting in the enhancement of antigen presentation to naive T cells for clonal expansion of CTLs. The TAs-specific T lymphocytes are recruited to the primary tumor site and directly attack and kill tumor cells with the production of cytotoxic INF-γ.

The following four phases have been proposed for the elimination process (Dunn et al., 2002). The first phase is the recognition of tumor cells by innate immune cells and their limited killing. When a solid tumor has grown to more than 2–3 mm, it requires a blood supply and stromal remodeling for tumor progression, which in turn induces proinflammatory signals leading to the recruitment of innate immune cells that produce IFN-γ (e.g., NK, NKT, $\gamma\delta$-T cells, macrophages, and DCs) into the tumor site. The transformed cells can be recognized by infiltrating lymphocytes, such as NK, NKT, and $\gamma\delta$-T cells, which produce IFN-γ. The second phase involves the maturation and migration of DCs and cross-priming for T cells. IFN-γ exerts a limited cytotoxicity via antiproliferative and antiangiogenic effects and induces apoptosis. Some of the chemokines derived from tumors and surrounding nontumorous tissues block the formation of new blood vessels even while continuing to induce tumor cell death. Necrotic tumor cells are ingested by immature DCs (iDCs), which have matured under proinflammatory conditions, and migrate to TDLNs. The third phase is characterized by the generation of TA-specific T cells. The recruited tumor-infiltrating NK and macrophages produce IL-12 and IFN-γ, which kill more tumor cells by activating cytotoxic mechanisms, such as pfp, TRAIL, and reactive oxygen. In the TDLNs, the migrated DCs present TAs to native CD4$^+$ T cells that differentiate to CD4$^+$ T cells, which develop TAs-specific CD8$^+$ T cells that lead to clonal expansion. The fourth phase involves the homing of TA-specific T cells to the tumor site and the elimination of tumor cells. Tumor antigen-specific CD4[+] and CD8[+] T cells home to the primary tumor site, where the CTLs eliminate the remaining TA-expressing tumor cells; this process is enhanced by the secreted IFN-γ and also selects for tumor cells with reduced immunogenicity.

The recognition of tumor cells and how the unmanipulated immune system can be activated in a developing tumor, even though tumor-specific antigens may be expressed as distinct recognition molecules on the surface of tumor cells, have been controversial topics in the oncology field. As a hypothesis of the so-called "danger theory," discussed in detail in Chapter 3, it was considered that cellular transformation did not provide sufficient proinflammatory signals to activate the immune system in response to a developing tumor. In the absence of such signals, there is often no immune response, and tolerance may develop. However, studies have indicated that danger signals, such as buildup of uric acid, presence of potential toll-like receptor ligands (e.g., heat shock proteins), the occurrence of a ligand transfer molecule in the signaling cascade induced by CpG DNA, and the presence of extracellular matrix (ECM) derivatives, may induce proinflammatory responses that activate innate immune responses to foreign pathogens. Danger signals are thought to act by stimulating the maturation of DCs so that they can present foreign antigens and stimulate T lymphocytes. Dying mammalian cells have also been found to release danger signals of unknown identity. Of note, although local limited inflammation may be involved in initiating immune responses, excessive inflammation may promote tumor progression in steady-state conditions. This may be in part due to the anti-inflammatory reactions in APCs, which release anti-inflammatory cytokines such as IL-10 and transforming growth factor β (TGF-β) that inhibits the activation of effector cells (Figure 2.1).

B. Equilibrium

The equilibrium phase, the next step in cancer immunoediting, is characterized by the continuous sculpting of tumor cells, which produces cells resistant to immune effector cells. This process leads to the immune selection of tumor cells with reduced immunogenicity. These cells are more capable of surviving in an immunocompetent host, which explains the apparent paradox of tumor formation in immunologically intact individuals. Although random gene mutation may occur in tumors that produce more unstable tumors, these tumor cell variants are less immunogenic, and the immune selection pressure also favors the growth of tumor cell clones with a nonimmmunogenic phenotype. Several experimental studies using mice with different deficiencies of effector molecules indicated various degrees of immune selection pressure.

Lymphomas formed in pfp-deficient mice were more immunogenic than those in IFN-γ-deficient mice, suggesting that pfp may be more strongly involved than IFN-γ in the immune selection of lymphoma cells (Street et al., 2002). In contrast, MCA-induced sarcomas in IFN-γ receptor-deficient mice are highly immunogenic (Shankaran et al., 2001). Further, chemically induced sarcomas in both nude and SCID mice were more immunogenic than similar tumors from immunocompetent mice (Engel et al., 1997; Smyth et al., 2000a; Shankaran et al., 2001). These findings suggest that the original tumor cells induced in normal mice and selected by a T cell-mediated selection process have been adapted to grow in a host with a functional T cell system, which has eliminated highly immunogenic tumor cells, leaving nonimmunogenic tumor cells to grow. There is, however, no connection between this loss of immunogenicity and loss of MHC class I

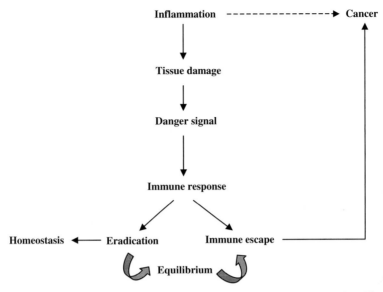

FIGURE 2.1 **Dual roles of inflammation in immune response and tumor progression.** The danger signals derived from tissue damage are able to activate immune effector cells to eradicate tumor cells in cellular homeostasis in the host. During tumor progression, immune effector cells are regulated by anti-inflammatory conditions due to an increase in tumor-derived soluble factors in the tumor microenvironment. Thus, the proinflammatory condition is dominantly shifted to tumor growth through the processes of equilibrium to escape. The balance of proinflammatory and anti-inflammatory conditions to immune effector cells plays a critical role in determining the antitumor immune response to eradicate tumor cells.

expression. Further, two important issues can be suggested: (i) the pfp-mediated cytotoxicity in T cells contributes more to the elimination of lymphoma cells than epithelial tumor cells, whereas IFN-γ-mediated cytotoxicity is directed more to the elimination of epithelial tumor cells; (ii) the higher immunogenicity of tumors derived from immunodeficient mice when compared to those from immunocompetent mice indicates less immune selection pressure in the tumors derived from immunodeficient mice than in those of immunocompetent mice. Thus, T cell-mediated elimination has adapted to highly immunogenic tumors, such as chemically and virally induced tumors. On the other hand, the immune selection pressure induces less immunogenic tumor variants that survive and grow in the tumor microenvironment. In cases of spontaneous tumors appearing for a long period of time, the immunogenic sculpting also produces fewer immunogenic tumors than chemically and virally induced tumors.

Since the equilibrium phase involves the continuous elimination of tumor cells and the production of resistant tumor variants by immune selection pressure, it is likely that equilibrium is the longest of the three processes in cancer immunoediting and may occur over a period of many years. In this process, lymphocytes and IFN-γ play a critical role in exerting immune selection pressure on tumor cells. During this period of Darwinian selection, many tumor variants from the original are killed, but new variants emerge carrying different mutations that increase resistance to immune attack. Since the equilibrium model persists for a long time in the interaction between cancer cells and the host, the transmission of cancer during organ transplantation can be considered. One report showed the appearance of metastatic melanoma 1–2 years after transplantation in two patients receiving renal transplants from the same donor. The donor had been previously treated for melanoma 16 years earlier and was considered tumor free (MacKie et al., 2003). Several similar observations have been reported in recipients of allografts from those considered as healthy donors (Cankovic et al., 2006). A possible explanation for the appearance of the transmitted cancer could be that the tumors were kept in equilibrium in the donor, but the continuous administration of immunosuppressive drugs activated and facilitated the growth of occult cancer in the recipient.

C. Escape

The final process in cancer immunoediting is the escape phase, which is shown by the outgrowth of tumor cell variants that escaped from immune recognition by effector cells. In this process, tumor variants that acquired insensitivity to immunologic detection and subsequent elimination through epigenetic and genetic alterations grew in an uncontrolled manner, resulting in clinically detectable malignant lesions. Surviving tumor variants are able to gain several escape mechanisms for immune surveillance during tumor progression, and the escape mechanisms are categorized by three principles (Malmberg, 2004): (i) lack of tumor antigen (TA) recognition, which is mediated by alterations on tumor cells or effector molecules that are crucial for recognition and activation by the immune system; (ii) lack of susceptibility for cell death, which is mediated by escape from the effector mechanism of cytotoxic lymphocytes; and (iii) induction of immune dysfunction, which is mediated by immunosuppressive factors derived from tumor cells and their inducing factors in the tumor microenvironment.

1. Alterations in Signal Transduction Molecules on Effector Cells

Given the lack of TA recognition, which is mediated by alterations of effector molecules that are important for immune system recognition and activation, the loss of signal transducer CD3-ζ chain (CD3-ζ) of TILs has been attributed to immune evasion in the cooperation of immunosuppressive cytokines and local impairment of TILs. The loss of CD3-ζ is reported to be correlated with increased levels of IL-10 and TGF-β, and downregulation of IFN-γ. The CD3-ζ chain is located as a large intracytoplasmic homodimer in the TCR that forms a part of the TCR–CD3 complex, which functions as a single transducer upon antigen binding. Since the TCR signal transduction through the formation of CD3 complex is one of three important signals for initiating a successful immune response as well as expressing tumor antigen and T helper 1 (Th1) polarization, any alterations in the CD3-ζ chain that are associated with the absence of p56lck tyrosine kinase (but not CD3-ζ) produce the changes in the signaling pathway for T cell activation. The alterations of TCR-ζ in several types of tumors, such as in pancreatic cancer (Schmielau et al., 2001), uveal malignant melanoma (Staibano et al., 2006), renal cell cancer (Riccobon et al., 2004), ovarian cancer (Lockhart et al., 2001), and oral cancer (Reichert et al., 1998), have been shown to be attributed to immune invasion that links to poor prognosis. Of importance, tumor-derived macrophages or tumor-derived factors led to a selective loss of TCR-ζ when compared with CD3-ζ (Lockhart et al., 2001). Given that the TCR/CD3-signaling led to lymphocyte proliferation, the poor proliferative responses of TILs could be explained by the defect in TCR-ζ expression. TIL underwent marked spontaneous apoptosis in vitro, which was associated with downregulation of the antiapoptotic Bcl-xL and Bcl-2 proteins. Further, since TCR-ζ is a substrate of caspase 3 leading to apoptosis, tumor cells can trigger in T lymphocytes caspase-dependent apoptotic cascades, which are not effectively protected by Bcl-2. In oral squamous cell carcinoma, a high proportion of T cells in the tumor undergo apoptosis, which correlates with FasL expression on tumor cells. FasL-positive microvesicles induced caspase-3 cleavage, cytochrome c release, loss of mitochondrial membrane potential, and reduced TCR-ζ chain expression in target lymphocytes.

2. Tumor-Derived Soluble Factors

Immunoediting provides a selective pressure in the tumor microenvironment that can lead to malignant progression. A variety of tumor-derived soluble factors, or TDSFs, contribute to the emergence of complex local and regional immunosuppressive networks, including vascular endothelial growth factor (VEGF), IL-10, TGF-β, prostaglandin E_2 (PGE$_2$), soluble phosphatidylserine (sPS), Fas (sFas), FasL (sFasL), and soluble MICA (sMICA) (Kim et al., 2006). Although deposited at the primary tumor site, these secreted factors can extend immunosuppressive effects into local lymph nodes and the spleen, thereby promoting invasion and metastasis.

VEGF plays a key role in recruiting immature myeloid cells from the bone marrow to enrich the microenvironment as tumor-associated immature DCs and macrophages (TiDCs and TAMs). Accumulation of TiDCs may cause roving dendritic cells and T cells to become suppressed through activation of indoleamine 2,3-dioxygenase and arginase I by tumor-derived growth factors, as discussed in Chapters 19 and 20. VEGF prevents DC differentiation and maturation by suppressing the NF-κB in hematopoietic stem cells. Blocking NF-κB activation in hematopoietic cells by tumor-derived factors is considered to be a mechanism by which tumor cells can directly downregulate the ability of the immune system to generate an antitumor response. In addition, since VEGF can be activated by Stat3, and DC differentiation requires decreasing activity of Stat3, neutralizing antibody specific for VEGF or dominant-negative Stat3 and its inhibitors can prevent Stat3 activation and promote DC differentiation and function. The increased serum levels of VEGF in cancers have been reported to be correlated with poor prognosis, which involves not only angiogenic properties but also the ability to induce immune evasion leading to tumor progression.

Soluble FasL and sMICA products also play important roles in immune evasion by inhibiting Fas- and NKG2D-mediated killing of immune cells. sPS, another TDSF, acts as an inducer of an anti-inflammatory response to TAMs, resulting in the release of anti-inflammatory mediators—such as IL-10, TGF-β, and PGE$_2$—that inhibit immune responses to DCs and T cells. The altered tumor surface antigen, such as FasL, also causes immune evasion by counterattacking immune cells, resulting in cell death. In addition, the soluble forms of FasL and MICA, sFasL and sMICA, are able to inhibit Fas and the NKG2D-mediated death of immune cells. Thus, it is likely that TDSFs play pivotal roles in constituting immunosuppressive networks that aid tumor progression and metastasis. Indeed, the immunosuppressive networks derived from these TDSFs can be critical factors in causing unsatisfactory clinical responses that are usually seen in immunotherapy of advanced cancer, and they remain an important obstacle to be overcome in the interaction between tumors and the immune system in the tumor microenvironment (Rosenberg et al., 2004; Zou, 2005).

3. Immunological Ignorance and Tolerance in Tumors

A tumor-specific immune response is regulated by tumor antigen levels and maturation stages of APCs, such as DCs. Many solid tumors, such as sarcomas and carcinomas, express tumor-specific antigens that can serve as targets for immune effector T cells. Nevertheless, the overall immune surveillance against such tumors seems relatively inefficient. Tumor cells are capable of inducing a protective cytotoxic T cell response if transferred as a single-cell suspension. However, if they are transplanted as small tumor pieces, tumors readily grow because the tumor antigen level can be modulated in the tumor microenvironment, which includes not only tumor cells but also bone marrow-derived cells, such as iDCs, and nonbone marrow-derived cells, such as fibroblasts, endothelium, and extracellular matrix (ECM). The ECM binds tumor antigen, and fibroblasts and endothelial cells compete with DCs for the antigen. Thus, many tumor antigens are downregulated, thereby facilitating tumor progression. Further, these stromal cells increase interstitial fluid pressure in the tumor, resulting in escape from immune attack by effector cells, as a result of greater difficulties in accessing the tumor. In these situations, insufficient levels of tumor antigen are largely ignored by T cells, adding to the suppressive effects of iDCs in the tumor microenvironment. In addition, iDCs stimulate CD4$^+$CD25$^+$ regulatory T cells, which inhibit T cell activation. It is known that sufficient levels of tumor antigen can produce an immune response, which is mediated by mature DCs presenting tumor antigens to T cells by cross-priming. However, even in the presence of sufficient levels of tumor antigen, iDCs inhibit maturation of DCs and T cell activation, resulting in immunological tolerance.

Thus, it is likely that tumor immune evasion is mediated not only by immunological ignorance due to decreased levels of tumor antigen but also by immunological tolerance due to inhibition of T cell activation by iDCs. Many important events and the central roles of effector cells in the process of immunoediting, from the phases of immune surveillance to escape, are summarized in Figure 2.2.

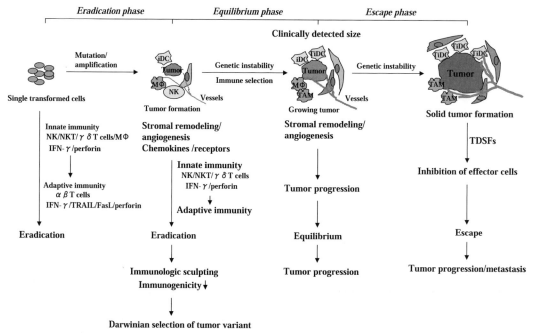

FIGURE 2.2 **Cancer immunoediting, from immune surveillance to escape.** When nascent transformed cells exist as single cells, they are readily eradicated by innate and adaptive immune responses. During tumor growth, tumor cells are required for angiogenesis and stromal remodeling, which produce tumor cell variants that have low immunogenicity and are resistant to immune attack; the equilibrium phase proceeds even though the elimination phase continues through immune selection pressure. Some tumor cells are eradicated in the equilibrium phase, however, and further tumor progression leads to the release of tumor-derived soluble factors that are involved in several mechanisms of immune evasion in the escape phase. Abbreviations: iDC, immature dendritic cell; IFN, interferon; Mϕ, macrophage; NK, natural killer; NKT, natural killer T; TAM, tumor-associated macrophage; TiDC, tumor-associated iDC; TDSFs, tumor-derived soluble factors; TRAIL, TNF-related apoptosis-inducing ligand.

IV. CONCLUDING REMARKS

Owing to the abundant experimental and clinical evidence, there should no longer be any doubt for the existence of cancer immunoediting from immune surveillance to escape. Cancer cells are gradually able to gain several mechanisms of immune evasion during tumor progression even though they are pursued by the initial and continuing phases of immune surveillance. Immunological sculpting contributes to immune selection pressure, which produces tumor cell variants that are resistant to immune effector cells due to their low immunogenicity. In advanced cancers, the marked shifting to immunosuppressive conditions due to the constitution of the immunosuppressive network in tumors makes it more difficult to provoke an immune activation to eliminate cancer cells. Given that adoptive immunotherapy using the peptide vaccine and DC transfer is not sufficient to reduce tumor volume and tumor elimination by direct priming for T cells in such conditions, indirect cross-priming for T cells, which can be induced by massive cell death in combination with anticancer drugs, will be required. Indeed, the modulation of anticancer drug-induced cell death as well as the activation of antitumor immune responses by using molecular targeting drugs, such as antibodies and small molecules, may provide remarkable enhancement of chemotherapeutic effects in cancer therapy. Further studies on cellular and molecular mechanisms to contribute to antitumor immune responses will be needed.

References

Asselin-Paturel, C., and Trinchieri, G. (2005). Production of type I interferons: Plasmacytoid dendritic cells and beyond. *J. Exp. Med.* **202**, 461–465.

Birkeland, S. A., Storm, H. H., Lamm, L. U., Barlow, L., Blohme, I., Forsberg, B., Eklund, B., Fjeldborg, O., Friedberg, M., Frodin, L., *et al.* (1995). Cancer risk after renal transplantation in the Nordic countries, 1964–1986. *Int. J. Cancer* **60**, 183–189.

Burnet, F. M. (1957). Cancer: A biological approach. *Br. Med. J.* **1**, 841–847.

Burnet, F. M. (1971). Immunological surveillance in neoplasia. *Transplant Rev.* **7**, 3–25.

Cankovic, M., Linden, M. D., and Zarbo, R. J. (2006). Use of microsatellite analysis in detection of tumor lineage as a cause of death in a liver transplant patient. *Arch. Pathol. Lab. Med.* **130**, 529–532.

Dighe, A. S., Richards, E., Old, L. J., and Schreiber, R. D. (1994). Enhanced *in vivo* growth and resistance to rejection of tumor cells expressing dominant negative IFN gamma receptors. *Immunity* **1**, 447–456.

Dunn, G. P., Bruce, A. T., Ikeda, H., Old, L. J., and Schreiber, R. D. (2002). Cancer immunoediting: From immunosurveillance to tumor escape. *Nat. Immunol.* **3**, 991–998.

Dunn, G. P., Old, L. J., and Schreiber, R. D. (2004). The immunobiology of cancer immunosurveillance and immunoediting. *Immunity* **21**, 137–148.

Dunn, G. P., Bruce, A. T., Sheehan, K. C., Shankaran, V., Uppaluri, R, Bui, J. D., Diamond, M. S., Koebel, C. M., Arthur, C., White, J. M., *et al.* (2005). A critical function for type I interferons in cancer immunoediting. *Nat. Immunol.* **6**, 722–729.

Ehrlich, P. (1909). Ueber den jetzigen stand der karzinomforschung. *Ned. Tijdschr. Geneeskd.* **5**, 273–290.

Engel, A. M., Svane, I. M., Mouritsen, S., Rygaard, J., Clausen, J., and Werdelin, O. (1996). Methylcholanthrene-induced sarcomas in nude mice have short induction times and relatively low levels of surface MHC class I expression. *APMIS* **104**, 629–639.

Engel, A. M., Svane, I. M., Rygaard, J., and Werdelin, O. (1997). MCA sarcomas induced in *scid* mice are more immunogenic than MCA sarcomas induced in congenic, immunocompetent mice. *Scand. J. Immunol.* **45**, 463–470.

Girardi, M., Oppenheim, D. E., Steele, C. R., Lewis, J. M., Glusac, E., Filler, R., Hobby, P., Sutton, B., Tigelaar, R. E., and Hayday, A. C. (2001). Regulation of cutaneous malignancy by gammadelta T cells. *Science* **294**, 605–609.

Gresser, I., Bandu, M. T., and Brouty-Boye, D. (1974). Interferon and cell division. IX. Interferon-resistant L1210 cells: Characteristics and origin. *J. Natl. Cancer Inst.* **52**, 553–559.

Gresser, I., Maury, C., Vignaux, F., Haller, O., Belardelli, F., and Tovey, M. G. (1988). Antibody to mouse interferon alpha/beta abrogates resistance to the multiplication of Friend erythroleukemia cells in the livers of allogeneic mice. *J. Exp. Med.* **168**, 1271–1291.

Haanen, J. B., Baars, A., Gomez, R., Weder, P., Smits, M., de Gruijl, T. D., von Blomberg, B. M., Bloemena, E., Scheper, R. J., van Ham, S. M., *et al.* (2006). Melanoma-specific tumor-infiltrating lymphocytes but not circulating melanoma-specific T cells may predict survival in resected advanced-stage melanoma patients. *Cancer Immunol. Immunother.* **55**, 451–458.

Hayakawa, Y., Rovero, S., Forni, G., and Smyth, M. J. (2003). Alpha-galactosylceramide (KRN7000) suppression of chemical- and oncogene-dependent carcinogenesis. *Proc. Natl. Acad. Sci. USA* **100**, 9464–9469.

Herberman, R. B., and Holden, H. T. (1978). Natural cell-mediated immunity. *Adv. Cancer Res.* **27**, 305–377.

Ikehara, S., Pahwa, R. N., Fernandes, G., Hansen, C. T., and Good, R. A. (1984). Functional T cells in athymic nude mice. *Proc. Natl. Acad. Sci. USA* **81**, 886–888.

Ishigami, S., Natsugoe, S., Tokuda, K., Nakajo, A., Che, X., Iwashige, H., Aridome, K., Hokita, S., and Aikou, T. (2000). Prognostic value of intratumoral natural killer cells in gastric carcinoma. *Cancer* **88**, 577–583.

Kaplan, D. H., Shankaran, V., Dighe, A. S., Stockert, E., Aguet, M., Old, L .J., and Schreiber, R. D. (1998). Demonstration of an interferon gamma-dependent tumor surveillance system in immunocompetent mice. *Proc. Natl. Acad. Sci. USA* **95**, 7556–7561.

Kim, R., Emi, M., Tanabe, K., and Arihiro, K. (2006). Tumor-driven evolution of immunosuppressive networks during malignant progression. *Cancer Res.* **66**, 5527–5536.

Klein, G. (1976). Immune surveillance—A powerful mechanism with a limited range. *Natl. Cancer Inst. Monogr.* **44**, 109–113.

Kondo, E., Koda, K., Takiguchi, N., Oda, K., Seike, K., Ishizuka, M., and Miyazaki, M. (2003). Preoperative natural killer cell activity as a prognostic factor for distant metastasis following surgery for colon cancer. *Dig. Surg.* **20**, 445–451.

Lockhart, D. C., Chan, A. K., Mak, S., Joo, H. G., Daust, H. A., Carritte, A., Douville, C. C., Goedegebuure, P. S., and Eberlein, T. J. (2001). Loss of T-cell receptor-CD3zeta and T-cell function in tumor-infiltrating lymphocytes but not in tumor-associated lymphocytes in ovarian carcinoma. *Surgery* **129**, 749–756.

MacKie, R. M., Reid, R., and Junor, B. (2003). Fatal melanoma transferred in a donated kidney 16 years after melanoma surgery. *N. Engl. J. Med.* **348**, 567–568.

Maleckar, J. R., and Sherman, L. A. (1987). The composition of the T cell receptor repertoire in nude mice. *J. Immunol.* **138**, 3873–3876.

Malmberg, K. J. (2004). Effective immunotherapy against cancer: A question of overcoming immune suppression and immune escape? *Cancer Immunol. Immunother.* **53**, 879–892.

Morath, C., Mueller, M., Goldschmidt, H., Schwenger, V., Opelz, G., and Zeier, M. (2004). Malignancy in renal transplantation. *J. Am. Soc. Nephrol.* **15**, 1582–1588.

Myron Kauffman, H., McBride, M. A., Cherikh, W. S., Spain, P. C., Marks, W. H., and Roza, A. M. (2002). Transplant tumor registry: Donor related malignancies. *Transplantation* **74**, 358–362.

Naito, Y., Saito, K., Shiiba, K., Ohuchi, A., Saigenji, K., Nagura, H., and Ohtani, H. (1998). CD8$^+$ T cells infiltrated within cancer cell nests as a prognostic factor in human colorectal cancer. *Cancer Res.* **58**, 3491–3494.

Old, L. J., and Boyse, E. A. (1964). Immunology of experimental tumors. *Annu. Rev. Med.* **15**, 167–186.

Pham, S. M., Kormos, R. L., Landreneau, R. J., Kawai, A., Gonzalez-Cancel, I., Hardesty, R. L., Hattler, B. G., and Griffith, B. P. (1995). Solid tumors after heart transplantation: Lethality of lung cancer. *Ann. Thorac. Surg.* **60**, 1623–1626.

Prehn, R. T., and Main, J. M. (1957). Immunity to methylcholanthrene-induced sarcomas. *J. Natl. Cancer Inst.* **18**, 769–778.

Reichert, T. E., Day, R., Wagner, E. M., and Whiteside, T. L. (1998). Absent or low expression of the zeta chain in T cells at the tumor site correlates with poor survival in patients with oral carcinoma. *Cancer Res.* **58**, 5344–5347.

Riccobon, A., Gunelli, R., Ridolfi, R., De Paola, F., Flamini, E., Fiori, M., Saltutti, C., Petrini, M., Fiammenghi, L., Stefanelli, M., *et al.* (2004). Immunosuppression in renal cancer: Differential expression of signal transduction molecules in tumor-infiltrating, near-tumor tissue, and peripheral blood lymphocytes. *Cancer Invest.* **22**, 871–877.

Rosenberg, S. A., Yang, J. C., and Restifo, N. P. (2004). Cancer immunotherapy: Moving beyond current vaccines. *Nat. Med.* **10**, 909–915.

Russell, J. H., and Ley, T. J. (2002). Lymphocyte-mediated cytotoxicity. *Annu. Rev. Immunol.* **20**, 323–370.

Rygaard, J., and Povlsen, C. O. (1974). The mouse mutant nude does not develop spontaneous tumours. An argument against immunological surveillance. *Acta. Pathol. Microbiol. Scand. Microbiol. Immunol.* **82**, 99–106.

Sato, E., Olson, S. H., Ahn, J., Bundy, B., Nishikawa, H., Qian, F., Jungbluth, A. A., Frosina, D., Gnjatic, S., Ambrosone, C., *et al.* (2005). Intraepithelial CD8$^+$ tumor-infiltrating lymphocytes and a high CD8$^+$/regulatory T cell ratio are associated with favorable prognosis in ovarian cancer. *Proc. Natl. Acad. Sci. USA* **102**, 18538–18543.

Schmielau, J., Nalesnik, M. A., and Finn, O. J. (2001). Suppressed T-cell receptor zeta chain expression and cytokine production in pancreatic cancer patients. *Clin. Cancer Res.* **7**, Suppl. 3, 933s–939s.

Shankaran, V., Ikeda, H., Bruce, A. T., White, J. M., Swanson, P. E., Old, L. J., and Schreiber, R. D. (2001). IFNgamma and lymphocytes prevent primary tumour development and shape tumour immunogenicity. *Nature* **410**, 1107–1111.

Shinkai, Y., Rathbun, G., Lam, K. P., Oltz, E. M., Stewart, V., Mendelsohn, M., Charron, J., Datta, M., Young, F., Stall, A. M., et al. (1992). RAG-2-deficient mice lack mature lymphocytes owing to inability to initiate VDJ rearrangement. *Cell* **68**, 855–867.

Smyth, M. J., Thia, K. Y., Street, S. E., Cretney, E., Trapani, J. A., Taniguchi, M., Kawano, T., Pelikan, S. B., Crowe, N. Y., and Godfrey, D. I. (2000a). Differential tumor surveillance by natural killer (NK) and NKT cells. *J. Exp. Med.* **191**, 661–668.

Smyth, M. J., Thia, K. Y., Street, S. E., MacGregor, D., Godfrey, D. I., and Trapani, J. A. (2000b). Perforin-mediated cytotoxicity is critical for surveillance of spontaneous lymphoma. *J. Exp. Med.* **192**, 755–760.

Smyth, M. J., Crowe, N. Y., and Godfrey, D. I. (2001). NK cells and NKT cells collaborate in host protection from methylcholanthrene-induced fibrosarcoma. *Int. Immunol.* **13**, 459–463.

Staibano, S., Mascolo, M., Tranfa, F., Salvatore, G., Mignogna, C., Bufo, P., Nugnes, L., Bonavolonta, G., and De Rosa, G. (2006). Tumor infiltrating lymphocytes in uveal melanoma: A link with clinical behavior? *Int. J. Immunopathol. Pharmacol.* **19**, 171–179.

Street, S. E., Cretney, E., and Smyth, M. J. (2001). Perforin and interferon-gamma activities independently control tumor initiation, growth, and metastasis. *Blood* **97**, 192–197.

Street, S. E., Trapani, J. A., MacGregor, D., and Smyth, M. J. (2002). Suppression of lymphoma and epithelial malignancies effected by interferon gamma. *J. Exp. Med.* **196**, 129–134.

Stutman, O. (1974). Tumor development after 3-methylcholanthrene in immunologically deficient athymic-nude mice. *Science* **183**, 534–536.

Thomas, L. (1982). On immunosurveillance in human cancer. *Yale J. Biol. Med.* **55**, 329–333.

Villegas, F. R., Coca, S., Villarrubia, V. G., Jimenez, R., Chillon, M. J., Jareno, J., Zuil, M., and Callol, L. (2002). Prognostic significance of tumor infiltrating natural killer cells subset CD57 in patients with squamous cell lung cancer. *Lung Cancer* **35**, 23–28.

Yasunaga, M., Tabira, Y., Nakano, K., Iida, S., Ichimaru, N., Nagamoto, N., and Sakaguchi, T. (2000). Accelerated growth signals and low tumor-infiltrating lymphocyte levels predict poor outcome in T4 esophageal squamous cell carcinoma. *Ann. Thorac. Surg.* **70**, 1634–1640.

Yoshimoto, M., Sakamoto, G., and Ohashi, Y. (1993). Time dependency of the influence of prognostic factors on relapse in breast cancer. *Cancer* **72**, 2993–3001.

Zeier, M., Hartschuh, W., Wiesel, M., Lehnert, T., and Ritz, E. (2002). Malignancy after renal transplantation. *Am. J. Kidney Dis.* **39**, E5.

Zou, W. (2005). Immunosuppressive networks in the tumour environment and their therapeutic relevance. *Nat. Rev. Cancer* **5**, 263–274.

van den Broek, M. E., Kagi, D., Ossendorp, F., Toes, R., Vamvakas, S., Lutz, W. K., Melief, C. J., Zinkernagel, R. M., and Hengartner, H. (1996). Decreased tumor surveillance in perforin-deficient mice. *J. Exp. Med.* **184**, 1781–1790.

CHAPTER

3

Immunosurveillance: Innate and Adaptive Antitumor Immunity

MASAHISA JINUSHI AND GLENN DRANOFF

I. Introduction
II. Innate Antitumor Responses
III. Innate Immune Cells
   A. NK Cells
   B. NKT Cells
   C. γδ-T Cells
   D. Macrophages
   E. Granulocytes
IV. Adaptive Antitumor Responses
   A. Adaptive Immunity in Immunosurveillance
   B. Targets of Antitumor T Cell Responses
   C. Antitumor Effector Mechanisms
   D. Cytotoxic Mechanisms
V. The Interplay of Innate and Adaptive Antitumor Immunity
VI. Conclusion

The interplay of innate and adaptive antitumor immunity dictates the intensity and outcome of the endogenous anticancer response. Stress-induced molecules on tumor cells trigger innate immune reactions, whereas the processing and presentation of tumor-associated antigens evoke adaptive immune recognition. Innate and adaptive antitumor responses may impact tumor development in different ways. In some cases, endogenous reactions suppress tumor formation while at the same time exerting a selective pressure that fosters the emergence of escape variants. Alternatively, some host responses directly promote tumor cell growth, invasion, and metastasis through the elaboration of inflammatory mediators and cytokines. Unique and overlapping roles for innate and adaptive antitumor immunity have been revealed in murine models through the application of gene-targeting techniques and the administration of neutralizing or depleting antibodies. These investigations have uncovered a complex network of interactions among tumor cells, immune elements, and stromal components in the tumor microenvironment, which together shape the direction, quality, and dynamics of the anticancer response. A more detailed understanding of the cross talk between tumor and host cells should advance the development of therapeutic strategies that stimulate tumor protection.

I. INTRODUCTION

The immune response to cancer can be broadly divided into innate and adaptive components. The cellular elements of innate immunity include granulocytes, macrophages, mast cells, dendritic cells (DCs), and natural killer (NK) cells. Innate immunity serves as a first line of defense against cancer, as germ line-encoded pattern recognition receptors, such as Toll-like receptors (TLRs), rapidly detect infected or stressed cells, thereby triggering potent effector mechanisms aimed at accomplishing tumor containment. In contrast, adaptive immunity, which includes humoral immunity composed of antibodies produced by B cells and cellular immunity composed of $CD4^+$ and $CD8^+$ T cells, evolves over the course of several days, reflecting the requirement for activation and expansion of rare lymphocytes that can specifically recognize tumor-associated antigens (TAAs). The great diversity in the antigen specificity of B and T cells is made possible by somatic rearrangements of the genes that encode immunoglobulins and T cell receptors, respectively. Also included in adaptive immunity are a third set of specialized T cells, the NKT cells and γδ-T cells, which function at the interface between innate and adaptive immunity.

The interplay of tumor cells and endogenous immunity is increasingly recognized to play a decisive role throughout the multiple stages of carcinogenesis (Dranoff, 2004). Clinicopathologic studies demonstrate that dense intratumoral lymphocyte infiltrates, especially those enriched in cytotoxic $CD8^+$ T cells, are associated with reduced frequencies of disease recurrence and prolonged survival in several types of cancer (Clark et al., 1989; Pages et al., 2005; Zhang et al., 2003). Nonetheless, in the setting of unresolved inflammation, tumor cells and stromal elements subvert host immunity to promote disease progression. These divergent outcomes illustrate the complexity of the host–tumor interaction. This chapter discusses the mechanisms whereby innate and adaptive antitumor immunity mediate these dual roles during tumor development and presents the ways by which the tumor microenvironment helps sculpt these responses.

II. INNATE ANTITUMOR RESPONSES

Normal cells are endowed with intricate machinery that affords protection against genotoxic stress induced by cell intrinsic and extrinsic insults, including DNA replication errors, oxidative damage, microbial infection, and inflammation. The failure of the DNA damage response to resolve single-stranded or double-stranded DNA breaks poses a significant risk for malignant transformation. In this context, the innate immune system functions as an extrinsic surveillance mechanism for genotoxic injury. NKG2D ligands, which include the major histocompatibility complex (MHC) class I-related molecules (MHC class I chain-related A [MICA] and MHC class I chain-related B [MICB]) and four UL16 binding proteins (ULBP1–4) in humans as well as the retinoic acid-early (RAE) inducible gene products and H60 in rodents, are induced by DNA damage through a pathway involving ataxia telangiectasia mutated (ATM), ataxia telangiectasia and Rad3 related (ATR), Chk-1, and Chk-2 (Gasser et al., 2005). The surface expression of these ligands on stressed cells triggers NKG2D-dependent activation of NK, NKT, and γδ-T cells, which inhibit tumor growth through cytotoxicity and production of interferon (IFN)-γ. In addition, the release of cytoplasmic stress-response molecules, such as heat-shock protein 70 (HSP70), HMGB1, and uric acid, activates macrophages and DCs in part through TLR engagement, resulting in interleukin 12 (IL-12) production and the transition to adaptive immunity (Medzhitov, 2001).

While these innate responses may contribute to tumor suppression, their aberrant activation may also prove deleterious when normal tissues are perturbed, as during autoimmunity or chronic inflammation. In these settings, the sustained expression of NKG2D ligands leads to downregulation of NKG2D surface expression through increased endocytosis and the consequent suppression of protective responses (Oppenheim et al., 2005). As a result, there may be an onset of a kind of low-level chronic inflammation process, termed *smoldering inflammation*, which preserves certain characteristics that actually promote tumor progression. Thus, depending on its context, expression of stress ligands may either trigger cytotoxic antitumor reactions or contribute to conditions that can facilitate immune escape.

III. INNATE IMMUNE CELLS

A. NK Cells

NK cells are key participants in innate antitumor responses and employ several effector mechanisms, including perforin (pfp), death receptor ligands, and IFN-γ production. NK cells were initially characterized by their ability to lyse MHC class I-deficient tumor cells without prior stimulation. In the tumor microenvironment, NK cell function is regulated through a combination of inhibitory and activating receptors, cytokines (e.g., IL-2 and IL-15), and costimulatory molecules, including CD80, CD86, CD40, CD70, and ICOS. The NK cell contribution to tumor defense is highlighted by the increased susceptibility of NK cell-deficient mice (rendered deficient through the administration of antibodies against the membrane proteins NK1.1 or asialo-GM1) to tumors induced by the chemical carcinogen methylcholanthrene (Smyth et al., 2001).

NK cells express several families of inhibitory receptors that deliver negative regulatory signals following engagement of target cell MHC class I molecules. These proteins include the killer cell immunoglobulin-like receptors (KIRs), which are expressed in primates; the Ly49 lectin-like homodimers, which are expressed in rodents; or the C-type lectin-like molecules (CD94 and NKG2A/E), which are expressed in both primates and rodents. Individual NK cells display varying patterns of inhibitory receptors, yielding an increased ability of the NK cell population as a whole to detect losses of individual MHC class I alleles.

NK cells are endowed with several families of activating receptors, including the natural cytotoxicity receptors (NKp46, NKp44, NKp30, and NKp80), additional Ly49 proteins, and NKG2D. Although the importance of NKp46 for protection against influenza virus was recently elucidated, a major role for this receptor in tumor immunity appears restricted to MHC class I-deficient tumor cells in the 129sv murine strain, which lacks the activating Ly49 receptor (Gazit et al., 2006). Functional redundancy with other natural cytotoxicity receptors might underlie this finding although further studies are required.

In contrast to the natural cytotoxicity receptors, a major role for the NKG2D pathway in tumor cell recognition by NK cells has been delineated (Raulet, 2003). NKG2D ligands are frequently expressed in transformed cells as part of the DNA damage response, raising the possibility that NKG2D-triggered responses are a critical link between target cell genotoxic stress and immune-mediated destruction. Indeed, the administration of blocking antibodies to NKG2D increases the susceptibility of mice to tumors caused by chemical carcinogens that act by damaging DNA, highlighting a potent NKG2D-dependent mechanism of tumor suppression (Smyth et al., 2005). These antitumor activities primarily involve pfp, whereas the production of IFN-γ and the existence of the tumor necrosis factor (TNF)-related apoptosis-inducing ligand (TRAIL) are more important for NK cell tumor suppression in other contexts. Consistent with these findings, chemically induced sarcomas arising in pfp-deficient mice express the NKG2D ligand Rae-1 and are rejected upon transplantation into wild-type mice. Moreover, chemically induced tumors in wild-type mice frequently fail to express NKG2D ligands, underscoring a selective pressure for escape from NKG2D surveillance during chemical carcinogenesis.

B. NKT Cells

NKT cells are a specialized type of T cell that express an invariant T cell receptor alpha chain (Vα14-Jα18 in mice and Vα24-Jα18 in humans) and particular NK cell markers, such as CD161 or NKR-P1 (Taniguchi et al., 2003). The invariant T cell receptors are specific for glycolipid antigens presented by CD1d, an MHC class I-related molecule expressed on antigen-presenting cells and some cancers. A role for NKT cells in tumor suppression was revealed by the increased susceptibility of Jα18-deficient mice, which lack invariant NKT cells, to chemically induced tumors and experimentally induced metastases (Smyth et al., 2000a). Correspondingly, the administration of α-galactosylceramide (α-GalCer), a natural lipid isolated from marine sponges that activates NKT cells through CD1d binding, augments antitumor immunity in multiple model systems. Since some glycolipid antigens (e.g., gangliosides and glycophosphatidylinositols) are expressed by tumor cells, it is conceivable that NKT cells might be directed to reject tumor cells through these antigens in some cases. NKT cell-mediated tumor destruction involves IFN-γ production, which contributes to the activation and cytotoxicity of NK and $CD8^+$ T cells. NKT cells are also required for the therapeutic effects of GM-CSF and IL-12-based cytokine strategies that have been explored widely for cancer treatment (Cui et al., 1997; Gillessen et al., 2003).

NKT cells produce both T helper 1 (Th1) and T helper 2 (Th2) cytokines, depending on their mode of activation, underscoring key regulatory roles for the cytokine milieu and glycolipid antigen repertoire present in the tumor microenvironment. Indeed, NKT cells can undermine tumor rejection in some tumor models through a mechanism that involves transforming growth factor β (TGF-β) production by $Gr-1^+$ myeloid suppressor cells (Terabe et al., 2003). Studies have indicated that the $CD4^-$ NKT cells effectuate tumor rejection in the MCA-induced fibrosarcoma and B16F10 melanoma models, whereas $CD4^+$ NKT cells contribute to the pathogenesis of inflammatory diseases (e.g., asthma) by the secretion of IL-4, IL-5, and IL-13 (Akbari et al., 2006; Crowe et al., 2005). A deeper understanding of the factors determining the induction of NKT cell subsets during tumor development is an important goal of further investigation.

C. γδ-T Cells

γδ-T cells are a small population of T lymphocytes that integrate features of innate and adaptive immunity. While these cells undergo VDJ recombination during thymic development, their TCR diversity is relatively limited compared to conventional αβ-T cells, and they function more in pattern recognition (Hayday, 2000). γδ-T cells constitute a significant proportion of intraepithelial lymphocytes (IELs) in the skin and gastrointestinal and genitourinary tract mucosa. Their importance for tumor surveillance has been revealed by the increased incidence of chemically induced fibrosarcomas and spindle cell carcinomas in γδ-T cell deficient mice (Girardi et al., 2003). For example, Vδ1-T cells are a type of γδ-T cell that is enriched in various tumors, and they may be activated through TCR signaling evoked by CD1-presented lipid antigens or NKG2D signaling triggered by MIC or ULBPs. γδ-T cells serve as a major early source of IFN-γ during disease development and may mediate direct antitumor cytotoxicity. In addition, these cells might function in antigen presentation, as activated Vγ2δ2-T cells have been shown to prime conventional αβ-T cells against soluble antigens following migration to regional lymph nodes (Brandes *et al.*, 2005). Collectively, these studies illustrate how γδ-T cells serve as a first-line of defense against epithelial tumors arising in the skin, gastrointestinal tract, and genitourinary tract.

D. Macrophages

Macrophages are a prominent component of the cellular response to tumors, where they mediate diverse functions. The release of stress-induced molecules, such as HSP70 or HMGB1, from necrotic tumor cells may trigger TLR-dependent macrophage activation, resulting in the production of cytotoxic reactive oxygen and nitrogen species and the secretion of inflammatory cytokines. Indeed, a spontaneous mutation that developed in SR/CR mice, which manifest a natural resistance to tumor growth, results in striking macrophage activation and cytotoxicity toward multiple tumor cell lines (Hicks *et al.*, 2006). Macrophages may further contribute to tumor protection by stimulating antitumor T cells while suppressing $CD4^+CD25^+$ regulatory T cells (Tregs) through IL-6 release.

In contrast to these beneficial activities, tumor-associated macrophages (TAMs) also play major roles in promoting tumor progression (see Chapter 16). In the context of unresolved inflammation, tumor cells exploit macrophage activities that are critical for wound healing, including the secretion of angiogenic molecules, growth factors, and matrix metalloproteinases (Condeelis and Pollard, 2006). Together, these products foster the breakdown of basement membranes and the establishment of a robust vascular network, thereby driving tumor cell invasion, expansion, and metastasis. Indeed, breast cancer progression was ameliorated in mice rendered macrophage deficient by virtue of a mutation in the important macrophage survival factor colony stimulating factor-1. The key factors that determine whether macrophages mediate tumor protection or promotion remain to be elucidated.

E. Granulocytes

Granulocytes might contribute to tumor destruction through the release of toxic moieties packaged in granules (e.g., cathepsin G and azurocidin), the generation of reactive oxygen species, and inflammatory cytokine secretion. Experimental tumors engineered to secrete granulocyte-colony stimulating factor were rejected through a pathway requiring neutrophils; moreover, this reaction stimulated the generation of adaptive T cell responses that eradicated subsequent tumor challenges (Colombo *et al.*, 1991). Neutrophils were similarly required for the antitumor effects of Her-2/neu-based DNA vaccinations in a transgenic breast cancer model (Curcio *et al.*, 2003). While eosinophils have been intensively studied for their roles in parasite infection and allergy, their local activation through T cell-derived IL-4, IL-5, and IL-13 in the tumor microenvironment may also contribute to tumor destruction through the release of granule components. Whether persistent granulocyte responses in chronic inflammation might promote tumor formation through tissue remodeling and angiogenesis stimulation remains to be explored further.

IV. ADAPTIVE ANTITUMOR RESPONSES

The adaptive antitumor response is typically initiated by DCs, which capture dying tumor cells, process the antigenic cargo for MHC class I and II presentation, migrate to draining lymph nodes, and stimulate antigen specific T and B lymphocytes. In the tumor microenvironment, DCs may be activated by "danger signals" released from stressed or necrotic tumor cells, thereby triggering a maturation program that includes expression of multiple costimulatory molecules and cytokines that result in effector T cell responses (Matzinger, 2002) (see Chapter 4). Alternatively, the production of immunosuppressive cytokines, such as TGF-β, IL-10, and VEGF, in the microenvironment may inhibit DC function, instead yielding abortive T cell effector responses and the augmented regulatory T cell function (Gabrilovich, 2004).

Productive antitumor CD4$^+$ T cell responses promote potent and long-lasting CD8$^+$ T cell responses and contribute to further DC maturation (Hung et al., 1998). These functions are accomplished in part through the secretion of a broad array of cytokines, including IFN-γ, IL-4, IL-5, IL-6, IL-10, and IL-13. CD4$^+$ T cell CD40 ligand expression may trigger CD40 signaling on DCs, resulting in enhanced IL-12 production and the differentiation of IFN-γ secreting Th1 cells and cytotoxic T lymphocytes.

CD4$^+$ T cells also stimulate B cell production of antitumor antibodies. These may include immunoglobulins that block growth or survival pathways dependent on cell surface receptors, such as Her-2/Neu. Antibodies may also target innate immune components to the tumor microenvironment, where specific cytolytic mechanisms may be unleashed. These include complement fixation, with resultant membrane disruption, and antibody-dependent cellular cytotoxicity mediated by NK cells, macrophages, and granulocytes. Antibodies might also opsonize tumor cells for DC Fc receptor-mediated cross-presentation of tumor antigens, leading to the induction of additional CD4$^+$ and CD8$^+$ T cell responses (Dhodapkar et al., 2002). Studies have illuminated differences in the affinities of IgG subclasses to bind the array of activating and inhibitory Fcγ receptors expressed on DCs, which influences the balance of effector versus tolerizing function (Nimmerjahn and Ravetch, 2005). In this regard, IgG2a and IgG2b display higher affinity than IgG1 and IgG3 for the activating Fcγ IV receptor and are more potent for antibody-dependent cellular cytotoxicity. Together, these properties resulted in superior suppression of B16 melanoma growth *in vivo*. These findings should advance the development of therapeutic monoclonal antibodies and provide a framework for clarifying the roles of endogenously produced antitumor antibodies.

A. Adaptive Immunity in Immunosurveillance

The importance of adaptive immunity for tumor immunosurveillance was first established through studies of mice harboring targeted mutations of the recombinase-activating gene 2 (RAG-2), which is required for immunoglobulin and T cell receptor gene rearrangement (Shankaran et al., 2001). Since the assembly of B and T lymphocyte antigen receptors requires RAG-mediated double-stranded DNA breaks to initiate V(D)J immunoglobulin gene recombination, RAG-2-deficient mice lack all B lymphocytes, αβ- and γδ-T cells, and NKT cells. These mice manifested an increased susceptibility to chemical carcinogen-induced tumors, and the fibrosarcomas arising in these animals were frequently rejected upon transplant to wild-type animals. Subsequent studies of mice deficient in αβ- or γδ-T cells alone revealed a similar enhanced susceptibility to chemical carcinogens, highlighting the key roles of T lymphocytes in tumor protection. Consistent with these findings, the development of intratumoral T cell infiltrates in multiple cancers is correlated with the absence of early metastasis and prolonged disease-free survival. The adaptive immune system, however, may also promote carcinogenesis within the background of chronic inflammation. In transgenic models of hepatitis B-induced liver cancer, smoldering CD4$^+$ and CD8$^+$ T cell responses are required for progression of hepatocellular carcinoma (Nakamoto et al., 1998). Similarly, CD4$^+$ T cells activated by normal cutaneous bacterial flora promote the evolution of squamous cell carcinoma in a human papilloma virus transgenic model (Daniel et al., 2003).

Tregs may play an important role in modulating the dual roles of adaptive immunity in tumor protection and promotion (Dranoff, 2005). Substantial evidence in multiple tumor models indicates that Tregs present a major impediment to cytotoxic T cell-mediated tumor rejection, particularly in the context of therapy-induced responses. Indeed, the presence of Tregs defined by the expression of the specific marker FoxP3 in ovarian cancer patients is tightly linked to inferior clinical outcomes (Curiel et al., 2004). On the other hand, Tregs function to maintain immune homeostasis, and their disruption leads to severe autoimmune disease and chronic inflammation. These functions may underlie the striking ability of Tregs to effectuate tumor destruction in murine models of inflammation-induced cancer (Erdman et al., 2005). Moreover, the presence of Tregs in the cellular infiltrates of Hodgkin's lymphomas has been linked to improved survival, perhaps reflecting a dependence of Reed-Sternberg cells (a unique characteristic of this disease) on particular components of the host response.

B lymphocytes similarly play dual roles in tumor immunosurveillance. While antibodies may promote tumor destruction through the mechanisms discussed earlier, in the HPV transgenic model of cutaneous squamous cell carcinoma, antibodies were required for disease development (de Visser et al., 2005). In this setting, antibodies promoted the recruitment of innate immune cells to the tumor microenvironment, where persistent inflammation promoted carcinogenesis. Additional investigations are required to delineate the key mechanisms that determine if adaptive immune responses will inhibit or accelerate tumor development.

B. Targets of Antitumor T Cell Responses

The principal mechanism involved in the priming of antitumor T cells appears to be cross-presentation of tumor-associated antigens by professional antigen presenting cells, particularly DCs. This pathway allows processing of exogenously acquired tumor antigens into the MHC class I pathway, whereas proteins captured through endocytosis or autophagy are typically processed for MHC class II-restricted presentation. Tumor-associated antigens can be broadly divided into several categories, including cancer-testes antigens, which show restricted expression in adult germ cells but frequent upregulation in cancers, mutated proteins, differentiation antigens, and pathogen-encoded sequences, such as Epstein-Barr virus in some B cell lymphomas (Boon and van der Bruggen, 1996). Studies have unveiled an unexpected complexity to tumor antigen processing for CD8$^+$ cytotoxic T cells. Tumor infiltrating lymphocytes were specific for epitopes generated from posttranslational splicing of peptides derived from fibroblast growth factor 5 in renal cell carcinoma or gp100 in melanoma (Hanada et al., 2004; Vigneron et al., 2004).

Notwithstanding these striking examples of tumor-specific antigens, most of the gene products that stimulate endogenous responses in tumor-bearing hosts are non-mutated and expressed in some normal tissues. In these cases, the tumor-reactive T cells are of lower affinity because thymic deletion often purges the repertoire of high-affinity T cells with potential autoreactivity. Overexpressed self-antigens may also stimulate CD4$^+$CD25$^+$ regulatory T cells, recently demonstrated for the heat shock protein J-like 2 and the cancer-testis antigen LAGE-1 (Nishikawa et al., 2005; Wang et al., 2004). Together, these mechanisms result in intrinsic and extrinsic modes of T cell tolerance, which limit the overall potency of the antitumor T cell response. One possible way to avoid tumor antigen-specific tolerance is to identify recurrent mutations or novel epitopes arising from protein splicing, which could be incorporated into therapeutic strategies to augment antitumor T cell responses. However, a high priority is the elucidation of these tolerance mechanisms and the development of methods to bypass them (as described in more detail in Chapters 14–20).

C. Antitumor Effector Mechanisms

1. Cytokines

IFNs IFN-γ plays a key role in tumor suppression (Dunn et al., 2004). NK, NKT, and γδ-T cells are major sources of IFN-γ early during tumor development, whereas $CD4^+$ and $CD8^+$ T cells may become additional sources as adaptive immunity evolves. IFN-γ contributes to tumor protection in multiple ways, including inhibition of angiogenesis, the induction of phagocyte cytotoxicity, and the stimulation of DC IL-12 production, which in turn promote Th1 and cytotoxic T cell responses. Mice with targeted mutations of IFN-γ or downstream signaling components established a major role for this cytokine in protection from chemically induced and spontaneous tumors. Moreover, IFN-γ functions as the master regulator of tumor cell immunogenicity. Whereas methylcholanthrene-induced tumors that arose in RAG-2-deficient mice were efficiently rejected upon transplantation into wild-type mice, tumors that developed in mice doubly deficient for RAG-2 and the IFN-γ receptor manifested robust growth after transplant into wild-type mice. The immunogenicity of these tumors was enhanced through the restoration of MHC class I presentation, identifying $CD8^+$ T cells as a major component of IFN-γ-stimulated tumor suppression.

A requirement for type I interferons (IFN-α/β) in tumor suppression has also recently been defined (Dunn et al., 2005). Mice with targeted mutations of the type I IFN receptors or wild-type animals administered neutralizing antibodies to type I IFNs both manifested enhanced susceptibility to chemical carcinogenesis or tumor transplantation. Protection in these systems involved host immunity and p53 tumor suppressor function in cancer cells (Takaoka et al., 2003). Thus, IFN-γ and IFN-α/β mediate critical but distinct functions in tumor immunosurveillance.

IL-12 and IL-18 Upon activation, phagocytes secrete IL-12 and IL-18, which in turn stimulate innate and adaptive cells to produce IFN-γ and thereby contribute to tumor suppression. Mice deficient in p40, a subunit shared by IL-12 and IL-23, manifest an increased susceptibility to chemical carcinogens. IL-12 enhances NK and NKT cell antitumor activities through NKG2D and pfp-dependent pathways (Smyth et al., 2005). In contrast, IL-18 augments NK cell cytotoxicity in a NKG2D-independent fashion that involves through Fas ligand-mediated killing. Thus, IL-12 and IL-18 elicit both overlapping and distinct pathways for tumor protection.

IL-2, IL-15, and IL-21 IL-2 potently activates the antitumor effector functions of both innate and adaptive cytotoxic cells. The systemic infusion of high doses of recombinant IL-2 or the adaptive transfer of NK cell and $CD8^+$ T lymphocytes stimulated *ex vivo* with IL-2 can evoke durable tumor regressions in a small minority of patients with advanced melanoma and renal cell carcinoma (Rosenberg, 2001). IL-2-induced tumor immunity involves both NKG2D-dependent pathways and pfp-mediated killing. In addition, IL-2 is critical for immune homeostasis, as mice deficient in the cytokine or signaling components succumb to chronic inflammatory disease due to defects in the maintenance of FoxP3-expressing regulatory T cells. Thus, IL-2 may also contribute to tumor protection through the control of inflammation-driven carcinogenesis.

The closely related cytokine IL-15 is critical for NK cell and memory CD8+ T cell homeostasis. The ability of IL-15 to amplify proximal TCR signaling in memory CD8+ T cells overcomes tolerance in a TCR transgenic model of tumor-induced anergy (Teague *et al.*, 2006). Constitutive IL-15 expression, however, may contribute to tumor promotion, as IL-15 transgenic mice succumbed to growth-factor-induced NKT cell leukemias (Fehniger *et al.*, 2001).

The IL-21 receptor shares the common γ-chain subunit that is also employed by the IL-2 and IL-15 receptors. IL-21 promotes NK cell differentiation from hematopoietic progenitors and induces their functional maturation, manifested by increased cell size, granularity, proliferation, and expression of activating receptors (Parrish-Novak *et al.*, 2000). The therapeutic administration of IL-21 enhances the rejection of B16 melanomas and MCA205 fibrosarcomas through NK cell activities that involve NKG2D and IFN-γ. Collectively, IL-2, IL-15, and IL-21 play complementary roles in tumor surveillance.

IL-23 and IL-17 IL-23 is a heterodimeric cytokine composed of a unique p19 subunit and a p40 subunit shared with IL-12. While activated macrophages and DCs produce both IL-12 and IL-23, these cytokines trigger distinct downstream effector pathways. IL-12 promotes the development of IFN-γ-secreting Th1 cells, whereas IL-23 supports the expansion and activation of Th17 cells, a recently discovered CD4+ T subset that is critical for tissue inflammation. Consistent with these differences, the production of IL-23 promotes tumor cell growth and invasion through upregulation of MMP9, COX-2, and angiogenesis, whereas IL-23 deficiency attenuates tumor formation through a reduction in inflammation (Langowski *et al.*, 2006). IL-23 also restrains protective immunity through inhibiting the intratumoral localization of Th1 cells and cytotoxic CD8+ lymphocytes. In some systems, though, IL-23 might also evoke protective antitumor responses, in part mediated by activated granulocytes, which as discussed earlier can trigger tumor cytotoxicity. Further investigations are required to understand the multiple roles of IL-23 in tumor surveillance.

2. GM-CSF

Granulocyte-macrophage colony-stimulating factor (GM-CSF) stimulates the production, proliferation, maturation, and activation of granulocytes, macrophages, and DCs. Vaccination with irradiated tumor cells engineered to secrete GM-CSF enhances tumor immunity in mice and patients through enhanced tumor antigen presentation; CD1d-restricted NKT cells, CD4 and CD8+ T cells, and antibodies are required for tumor protection (Hodi and Dranoff, 2006). Whereas GM-CSF-deficient mice manifest pulmonary alveolar proteinosis, autoimmune disease, and loss of some protective immune responses, they do not manifest increased spontaneous cancers. In contrast, mice doubly deficient in GM-CSF and IFN-γ develop at high frequency diverse hematological and solid tumors in the background of chronic inflammation and infection, illustrating the critical role of these cytokines as tumor suppressors (Enzler *et al.*, 2003). In this model, lifelong antibiotic therapy suppressed inflammation and prevented tumor formation, indicating that the interplay of microbial agents and immune homeostasis was critical to tumor susceptibility. More detailed analysis of these mice should help elucidate the mechanisms underlying inflammation-driven carcinogenesis.

D. Cytotoxic Mechanisms

1. *Perforin*

Cytotoxicity triggered by CD8+ T cells through the pfp-granzyme pathways is critical to tumor suppression (Smyth *et al.*, 2000b). The pfp-deficient mice display an enhanced susceptibility to chemical carcinogenesis and spontaneously develop lymphomas. These tumors are rejected upon transplantation to wild-type animals through a CD8$^+$ T cell-dependent mechanism, whereas mice doubly deficient in pfp and β2-microglobulin manifest increased tumor susceptibility compared to pfp-deficient mice. The pfp deficiency also compromises NK cell cytotoxicity against MHC class I-deficient tumor cells, which may be primarily dependent on NKG2D. Together, these findings underscore the importance of the pfp–granzyme pathway for innate and adaptive antitumor cytotoxicity.

2. TNF Family Members

Death receptors belonging to the TNF superfamily constitute the second major mechanism of cellular cytotoxicity. TNF-related apoptosis-inducing ligand (TRAIL) is a type II transmembrane protein that binds to five receptors in humans and one in mice. Among these, DR4 (TRAIL-R1) and DR5 (TRAIL-R2) transduce death signals through caspase-8, FADD, and Bax, while the remaining receptors (TRAIL-R3, TRAIL-R4, and osteoprotegerin) may attenuate killing as nonsignaling receptor decoys. Murine liver NK cells constitutively express TRAIL, and the ligand is upregulated in other innate and adaptive lymphocytes through interferons, IL-2, and IL-15. The IFN-γ-dependent induction of TRAIL on NK cells is critical for the antitumor effects of IL-12 and α-GalCer, whereas TRAIL deficiency augments the development of liver metastasis in pfp-deficient but not IFN-γ-deficient hosts. TRAIL-deficient mice manifest an increased susceptibility to methylcholanthrene induced tumors, and when these mice are crossed with p53-deficient mice, the offspring show accelerated tumor formation (Takeda et al., 2002).

The interaction of Fas and Fas ligand similarly triggers antitumor cytotoxicity. Mice deficient in this pathway manifest an increased incidence of spontaneous plasmacytoid lymphomas (Davidson et al., 1998). In transplantation models, Fas/Fas ligand interactions contribute to the control of metastases. Thus, tumor growth may select for escape from this pathway.

LIGHT is another TNF family member that participates in tumor suppression by functioning as a costimulatory molecule (Yu et al., 2004). The engineered expression of LIGHT triggers an antitumor T cell response that mediates destruction of established tumors, illustrating that appropriate manipulation of the tumor microenvironment may be therapeutic.

V. THE INTERPLAY OF INNATE AND ADAPTIVE ANTITUMOR IMMUNITY

As discussed earlier, innate immunity functions not only to direct antitumor effects but also to stimulate the generation of adaptive immune responses through the presentation of tumor antigens by DCs. Studies have shown that the type of stimuli present in the tumor microenvironment dictates the type of program for DC activation, thereby directing adaptive responses toward either protective immunity or tolerance (Gabrilovich, 2004). The detection of "danger signals" elicited by stress or cell death and interactions with other innate cells appears to be important for triggering DC activation. Thus, NK cells that recognize NKG2D ligand-positive, MHC class I-deficient tumors; NKT cells that detect CD1d-presented lipids; and γδ-T cells that recognize CD1-presented antigens can all interact productively with DCs. In this cross-talk, cytokines, B7 family members, and CD40 play critical roles.

Investigations have uncovered a new cell population termed *IFN-producing killer dendritic cells* (IKDCs) that underscore the close interplay of DCs and other innate elements (Chan et al., 2006; Taieb et al., 2006). IKDCs are a new subset of DCs that express some NK cell markers, produce type I IFNs, and are cytotoxic. The phenotype of these cells (CD11c$^+$B220$^+$CD49b$^+$NKG2D$^+$Gr-1$^-$) is distinct from NK cells and plasmacytoid DCs. IKDCs can be activated with NKG2D ligands and lyse tumor targets through TRAIL. However, following migration to draining lymph nodes, these cells display features of antigen-presenting cells, such as upregulation of MHC and costimulatory molecules, and they can stimulate T cell responses. Additional studies are required to clarify the role of IKDC in endogenous and therapeutic tumor immunity.

VI. CONCLUSION

Investigations of endogenous antitumor responses have unveiled a previously unappreciated complexity of interaction among tumor cells, stroma, and immune elements. Innate immune activation by stress-inducible signals on tumor cells triggers a response aimed at controlling disease development; effective responses may transition to adaptive immunity that manifests further specificity and memory. However, persistent immune activation may also result in an interplay of innate and adaptive elements that supports tumor cell proliferation, survival, invasion, angiogenesis, and metastasis. A central player that has emerged in coordinating these tumor-promoting inflammatory responses is the NF-κB transcription factor (Karin, 2006). A more detailed understanding of the mechanisms regulating the development of tumor-protective or tumor-promoting host responses is critical to the crafting of immunotherapeutic strategies.

References

Akbari, O., Faul, J. L., Hoyte, E. G., Berry, G. J., Wahlstrom, J., Kronenberg, M., DeKruyff, R. H., and Umetsu, D. T. (2006). CD4+ invariant T-cell-receptor+ natural killer T cells in bronchial asthma. *N. Engl. J. Med.* **354**, 1117–1129.

Boon, T., and van der Bruggen, P. (1996). Human tumor antigens recognized by T lymphocytes. *J. Exp. Med.* **183**, 725–729.

Brandes, M., Willimann, K., and Moser, B. (2005). Professional antigen-presentation function by human gammadelta T Cells. *Science* **309**, 264–268.

Chan, C. W., Crafton, E., Fan, H. N., Flook, J., Yoshimura, K., Skarica, M., Brockstedt, D., Dubensky, T. W., Stins, M. F., Lanier, L. L., et al. (2006). Interferon-producing killer dendritic cells provide a link between innate and adaptive immunity. *Nat. Med.* **12**, 207–213.

Clark, W., Elder, D., Guerry, D., Braitman, L., Trock, B., Schultz, D., Synnestvedt, M., and Halpern, A. (1989). Model predicting survival in stage I melanoma based on tumor progression. *J. Natl. Cancer Inst.* **81**, 1893–1904.

Colombo, M. P., Ferrari, G., Stoppacciaro, A., Parenza, M., Rodolfo, M., Mavillo, F., and Parmiani, G. (1991). Granulocyte-colony stimulating factor gene suppresses tumorigenicity of a murine adenocarcinoma *in vivo*. *J. Exp. Med.* **173**, 889–897.

Condeelis, J., and Pollard, J. W. (2006). Macrophages: Obligate partners for tumor cell migration, invasion, and metastasis. *Cell* **124**, 263–266.

Crowe, N. Y., Coquet, J. M., Berzins, S. P., Kyparissoudis, K., Keating, R., Pellicci, D. G., Hayakawa, Y., Godfrey, D. I., and Smyth, M. J. (2005). Differential antitumor immunity mediated by NKT cell subsets *in vivo*. *J. Exp. Med.* **202**, 1279–1288.

Cui, J., Shin, T., Kawano, T., Sato, H., Kondo, E., Toura, I., Kaneko, Y., Koseki, H., Kanno, M., and Taniguchi, M. (1997). Requirement for V_a14 NKT cells in IL-12-mediated rejection of tumors. *Science* **278**, 1623–1626.

Curcio, C., Di Carlo, E., Clynes, R., Smyth, M. J., Boggio, K., Quaglino, E., Spadaro, M., Colombo, M. P., Amici, A., Lollini, P. L., et al. (2003). Nonredundant roles of antibody, cytokines, and perforin in the eradication of established Her-2/neu carcinomas. *J. Clin. Invest.* **111**, 1161–1170.

Curiel, T. J., Coukos, G., Zou, L., Alvarez, X., Cheng, P., Mottram, P., Evdemon-Hogan, M., Conejo-Garcia, J. R., Zhang, L., Burow, M., et al. (2004). Specific recruitment of regulatory T cells in ovarian carcinoma fosters immune privilege and predicts reduced survival. *Nat. Med.* **10**, 942–949.

Daniel, D., Meyer-Morse, N., Bergsland, E. K., Dehne, K., Coussens, L. M., and Hanahan, D. (2003). Immune enhancement of skin carcinogenesis by CD4+ T cells. *J. Exp. Med.* **197**, 1017–1028.

Davidson, W., Giese, T., and Fredrickson, T. (1998). Spontaneous development of plasmacytoid tumors in mice with defective fas-fas ligand interactions. *J. Exp. Med.* **187**, 1825–1838.

de Visser, K. E., Korets, L. V., and Coussens, L. M. (2005). De novo carcinogenesis promoted by chronic inflammation is B lymphocyte dependent. *Cancer Cell* **7**, 411–423.

Dhodapkar, K. M., Krasovsky, J., Williamson, B., and Dhodapkar, M. V. (2002). Antitumor monoclonal antibodies enhance cross-presentation of cellular antigens and the generation of myeloma-specific killer T cells by dendritic cells. *J. Exp. Med.* **195**, 125–133.

Dranoff, G. (2004). Cytokines in cancer pathogenesis and cancer therapy. *Nat. Rev. Cancer* **4**, 11–22.

Dranoff, G. (2005). The therapeutic implications of intratumoral regulatory T cells. *Clin. Cancer Res.* **11**, 8226–8229.

Dunn, G. P., Bruce, A. T., Sheehan, K. C., Shankaran, V., Uppaluri, R., Bui, J. D., Diamond, M. S., Koebel, C. M., Arthur, C., White, J. M., and Schreiber, R. D. (2005). A critical function for type I interferons in cancer immunoediting. *Nat. Immunol.* **6**, 722–729.

Dunn, G. P., Old, L. J., and Schreiber, R. D. (2004). The immunobiology of cancer immunosurveillance and immunoediting. *Immunity* **21**, 137–148.

Enzler, T., Gillessen, S., Manis, J. P., Ferguson, D., Fleming, J., Alt, F. W., Mihm, M., and Dranoff, G. (2003). Deficiencies of GM-CSF and interferon-gamma link inflammation and cancer. *J. Exp. Med.* **197**, 1213–1219.

Erdman, S. E., Sohn, J. J., Rao, V. P., Nambiar, P. R., Ge, Z., Fox, J. G., and Schauer, D. B. (2005). CD4$^+$CD25$^+$ regulatory lymphocytes induce regression of intestinal tumors in ApcMin/+ mice. *Cancer Res.* **65**, 3998–4004.

Fehniger, T., Suzuki, K., Ponnappan, A., VanDeusen, J., Cooper, M., Florea, S., Freud, A., Robinson, M., Durbin, J., and Caligiuri, M. (2001). Fatal leukemia in interleukin 15 transgenic mice follows early expansions in natural killer and memory phenotype CD8$^+$ T cells. *J. Exp. Med.* **193**, 219–231.

Gabrilovich, D. (2004). Mechanisms and functional significance of tumour-induced dendritic-cell defects. *Nat. Rev. Immunol.* **4**, 941–952.

Gasser, S., Orsulic, S., Brown, E. J., and Raulet, D. H. (2005). The DNA damage pathway regulates innate immune system ligands of the NKG2D receptor. *Nature* **436**, 1186–1190.

Gazit, R., Gruda, R., Elboim, M., Arnon, T. I., Katz, G., Achdout, H., Hanna, J., Qimron, U., Landau, G., Greenbaum, E., *et al.* (2006). Lethal influenza infection in the absence of the natural killer cell receptor gene Ncr1. *Nat. Immunol.* **7**, 517–523.

Gillessen, S., Naumov, Y. N., Nieuwenhuis, E. E., Exley, M. A., Lee, F. S., Mach, N., Luster, A. D., Blumberg, R. S., Taniguchi, M., Balk, S. P., *et al.* (2003). CD1d-restricted T cells regulate dendritic cell function and antitumor immunity in a granulocyte-macrophage colony-stimulating factor-dependent fashion. *Proc. Natl. Acad. Sci. USA* **100**, 8874–8879.

Girardi, M., Glusac, E., Filler, R. B., Roberts, S. J., Propperova, I., Lewis, J., Tigelaar, R. E., and Hayday, A. C. (2003). The distinct contributions of murine T cell receptor (TCR)gammadelta+ and TCRalphabeta+ T cells to different stages of chemically induced skin cancer. *J. Exp. Med.* **198**, 747–755.

Hanada, K., Yewdell, J. W., and Yang, J. C. (2004). Immune recognition of a human renal cancer antigen through post-translational protein splicing. *Nature* **427**, 252–256.

Hayday, A. C. (2000). Gamma-delta cells: A right time and a right place for a conserved third way of protection. *Annu. Rev. Immunol.* **18**, 975–1026.

Hicks, A. M., Riedlinger, G., Willingham, M. C., Alexander-Miller, M. A., Von Kap-Herr, C., Pettenati, M. J., Sanders, A. M., Weir, H. M., Du, W., Kim, J., *et al.* (2006). Transferable anticancer innate immunity in spontaneous regression/complete resistance mice. *Proc. Natl. Acad. Sci. USA* **103**, 7753–7758.

Hodi, F. S., and Dranoff, G. (2006). Combinatorial cancer immunotherapy. *Adv. Immunol.* **90**, 337–360.

Hung, K., Hayashi, R., Lafond-Walker, A., Lowenstein, C., Pardoll, H., and Levitsky, H. (1998). The central role of CD4$^+$ T cells in the antitumor immune response. *J. Exp. Med.* **188**, 2357–2368.

Karin, M. (2006). NF-kappaB and cancer: Mechanisms and targets. *Mol. Carcinog.* **45**, 355–361.

Langowski, J. L., Zhang, X., Wu, L., Mattson, J. D., Chen, T., Smith, K., Basham, B., McClanahan, T., Kastelein, R. A., and Oft, M. (2006). IL-23 promotes tumour incidence and growth. *Nature* **442**, 461–465.

Matzinger, P. (2002). The danger model: A renewed sense of self. *Science* **296**, 301–305.

Medzhitov, R. (2001). Toll-like receptors and innate immunity. *Nat. Rev. Immunol.* **1**, 135–145.

Nakamoto, Y., Guidotti, L. G., Kuhlen, C. V., Fowler, P., and Chisari, F. V. (1998). Immune pathogenesis of hepatocellular carcinoma. *J. Exp. Med.* **188**, 341–350.

Nimmerjahn, F., and Ravetch, J. V. (2005). Divergent immunoglobulin g subclass activity through selective Fc receptor binding. *Science* **310**, 1510–1512.

Nishikawa, H., Kato, T., Tawara, I., Saito, K., Ikeda, H., Kuribayashi, K., Allen, P. M., Schreiber, R. D., Sakaguchi, S., Old, L. J., and Shiku, H. (2005). Definition of target antigens for naturally occurring CD4(+) CD25(+) regulatory T cells. *J. Exp. Med.* **201**, 681–686.

Oppenheim, D. E., Roberts, S. J., Clarke, S. L., Filler, R., Lewis, J. M., Tigelaar, R. E., Girardi, M., and Hayday, A. C. (2005). Sustained localized expression of ligand for the activating NKG2D receptor impairs natural cytotoxicity *in vivo* and reduces tumor immunosurveillance. *Nat. Immunol.* **6**, 928–937.

Pages, F., Berger, A., Camus, M., Sanchez-Cabo, F., Costes, A., Molidor, R., Mlecnik, B., Kirilovsky, A., Nilsson, M., Damotte, D., *et al.* (2005). Effector memory T cells, early metastasis, and survival in colorectal cancer. *N. Engl. J. Med.* **353**, 2654–2666.

Parrish-Novak, J., Dillon, S. R., Nelson, A., Hammond, A., Sprecher, C., Gross, J. A., Johnston, J., Madden, K., Xu, W., West, J., *et al.* (2000). Interleukin 21 and its receptor are involved in NK cell expansion and regulation of lymphocyte function. *Nature* **408**, 57–63.

Raulet, D. H. (2003). Roles of the NKG2D immunoreceptor and its ligands. *Nat. Rev. Immunol.* **3**, 781–790.

Rosenberg, S. A. (2001). Progress in human tumour immunology and immunotherapy. *Nature* **411**, 380–384.

Shankaran, V., Ikeda, H., Bruce, A. T., White, J. M., Swanson, P. E., Old, L. J., and Schreiber, R. D. (2001). IFNg and lymphocytes prevent primary tumour development and shape tumour immunogenicity. *Nature* **410**, 1107–1111.

Smyth, M., Thia, K., Street, S., Cretney, E., Trapani, J., Taniguchi, M., Kawano, T., Pelikan, S., Crowe, N., and Godfrey, D. (2000a). Differential tumor surveillance by natural killer (NK) and NKT cells. *J. Exp. Med.* **191**, 661–668.

Smyth, M., Thia, K., Street, S., MacGregor, D., Godfrey, D., and Trapani, J. (2000b). Perforin-mediated cytotoxicity is critical for surveillance of spontaneous lymphoma. *J. Exp. Med.* **192**, 755–760.

Smyth, M. J., Cretney, E., Takeda, K., Wiltrout, R. H., Sedger, L. M., Kayagaki, N., Yagita, H., and Okumura, K. (2001). Tumor necrosis factor-related apoptosis-inducing ligand (TRAIL) contributes to interferon gamma-dependent natural killer cell protection from tumor metastasis. *J. Exp. Med.* **193**, 661–670.

Smyth, M. J., Swann, J., Cretney, E., Zerafa, N., Yokoyama, W. M., and Hayakawa, Y. (2005). NKG2D function protects the host from tumor initiation. *J. Exp. Med.* **202**, 583–588.

Taieb, J., Chaput, N., Menard, C., Apetoh, L., Ullrich, E., Bonmort, M., Pequignot, M., Casares, N., Terme, M., Flament, C., et al. (2006). A novel dendritic cell subset involved in tumor immunosurveillance. *Nat. Med.* **12**, 214–219.

Takaoka, A., Hayakawa, S., Yanai, H., Stoiber, D., Negishi, H., Kikuchi, H., Sasaki, S., Imai, K., Shibue, T., Honda, K., and Taniguchi, T. (2003). Integration of interferon-alpha/beta signalling to p53 responses in tumour suppression and antiviral defence. *Nature* **424**, 516–523.

Takeda, K., Smyth, M. J., Cretney, E., Hayakawa, Y., Kayagaki, N., Yagita, H., and Okumura, K. (2002). Critical role for tumor necrosis factor-related apoptosis-inducing ligand in immune surveillance against tumor development. *J. Exp. Med.* **195**, 161–169.

Taniguchi, M., Harada, M., Kojo, S., Nakayama, T., and Wakao, H. (2003). The regulatory role of Valpha14 NKT cells in innate and acquired immune response. *Annu. Rev. Immunol.* **21**, 483–513.

Teague, R. M., Sather, B. D., Sacks, J. A., Huang, M. Z., Dossett, M. L., Morimoto, J., Tan, X., Sutton, S. E., Cooke, M. P., Ohlen, C., and Greenberg, P. D. (2006). Interleukin-15 rescues tolerant $CD8^+$ T cells for use in adoptive immunotherapy of established tumors. *Nat. Med.* **12**, 335–341.

Terabe, M., Matsui, S., Park, J. M., Mamura, M., Noben-Trauth, N., Donaldson, D. D., Chen, W., Wahl, S. M., Ledbetter, S., Pratt, B., et al. (2003). Transforming growth factor-beta production and myeloid cells are an effector mechanism through which CD1d-restricted T cells block cytotoxic T lymphocyte-mediated tumor immunosurveillance: Abrogation prevents tumor recurrence. *J. Exp. Med.* **198**, 1741–1752.

Vigneron, N., Stroobant, V., Chapiro, J., Ooms, A., Degiovanni, G., Morel, S., van der Bruggen, P., Boon, T., and Van den Eynde, B. J. (2004). An antigenic peptide produced by peptide splicing in the proteasome. *Science* **304**, 587–590.

Wang, H. Y., Lee, D. A., Peng, G., Guo, Z., Li, Y., Kiniwa, Y., Shevach, E. M., and Wang, R. F. (2004). Tumor-specific human CD4+ regulatory T cells and their ligands: Implications for immunotherapy. *Immunity* **20**, 107–118.

Yu, P., Lee, Y., Liu, W., Chin, R. K., Wang, J., Wang, Y., Schietinger, A., Philip, M., Schreiber, H., and Fu, Y. X. (2004). Priming of naive T cells inside tumors leads to eradication of established tumors. *Nat. Immunol.* **5**, 141–149.

Zhang, L., Conejo-Garcia, J., Katsaros, D., Gimotty, P., Massobrio, M., Regnani, G., Makrigiannakis, A., Gray, H., Schlienger, K., Liebman, M., et al. (2003). Intratumoral T cells, recurrence, and survival in epithelial ovarian cancer. *N. Engl. J. Med.* **348**, 203–213.

CHAPTER

4

Cytokine Regulation of Immune Tolerance to Tumors

MING O. LI AND RICHARD A. FLAVELL

I. Introduction
II. Cytokine Regulation of Immune Tolerance to Tumors
   A. TGF-β
   B. IL-10
   C. IL-23
   D. Vascular Endothelial Growth Factor
III. Summary and Future Perspectives
   References

Tumors arise from genetic and epigenetic changes of self tissues. This chapter proposes that the default immune response to tumors is tolerance, which utilizes similar tolerance pathways to normal self tissues under steady state or during wound healing. Cytokines appear to be important regulators of these processes, which inhibit antitumor immune responses by immune suppression or immune deviation. This chapter reviews the roles of several families of cytokines in immune tolerance to tumors, including transforming growth factor β (TGF-β), interleukin 10 (IL-10) and interleukin 23 (IL-23), and vascular endothelial growth factor (VEGF). Effective immunotherapy of cancers may be achieved by manipulating these cytokine-mediated tolerance pathways.

I. INTRODUCTION

Cancer develops as a result of intimate interactions between tumor cells and their environment. How the immune system responds to tumors and regulates their progression to cancer is of interest not only for the understanding of disease mechanisms but also for the immunotherapy of cancer. Immune regulation of cancer occurs at multiple levels and engages both the innate and adaptive arms of the immune system, affecting virtually all steps of tumor formation and progression. One fundamental question of such regulation is whether recognition of tumors by the immune system leads to tumor rejection or tolerance. The cancer immunosurveillance hypothesis proposed by F. M. Burnet has been at the center of debate for decades (Burnet, 1970). Most tumors possess tumor-specific or tumor-associated antigens recognizable by the immune system, often as a consequence of genetic and epigenetic changes associated with tumor development (Fearon and Vogelstein, 1990; Jones and Baylin, 2002). However, the sole presence of neoantigens does not result in immune activation. In fact, the adaptive immune response is tightly regulated by the status of the innate immune system and the presence of regulatory cell types and soluble factors, including cytokines (Li et al., 2006; Medzhitov and Janeway, 1997; Sakaguchi, 2004). Indeed, it appears that steady-state recognition of antigens leads to immune tolerance, while pathogen infection breaks the barriers of immune suppression to induce immunity. This controlled reactivity is likely selected during evolution to enable the immune system to fight off foreign pathogens and preserve tolerance to self tissues.

Does the immune system see tumors as self or foreign? This is an unanswered question that likely depends on tumor type. For many years, there has been a consensus on the immune system's active role in inhibiting virus-induced tumors, an unsurprising situation considering viruses are infectious entities. In humans, this assertion is supported by increased virus-associated tumors in AIDS patients or immunosuppressed recipients of organ transplants (Straathof et al., 2003). As for tumors of nonviral origins, the role of adaptive immunity is not well understood. Since tumors originate from normal tissues, various self-tolerance mechanisms operate to inhibit the development of tumor immunity, unless tumor-associated alterations break the suppressive mechanisms. In other words, the default pathway of tumor recognition is likely to be immune tolerance. A study with an SV40 T antigen (Tag)-induced spontaneous tumor model reveals that highly immunogenic tumors induce T cell tolerance without being rejected (Willimsky and Blankenstein, 2005). Although the underlying mechanisms need further investigation, it is conceivable that the pathways engaged to tolerate tumor antigens are similar to those of self antigens. Good support for this concept is provided by the observations that removal of self-tolerance checkpoints, including $CD4^+CD25^+$ regulatory T cells (Tregs) (Yamaguchi and Sakaguchi, 2006), the inhibitory costimulatory molecule CTLA4 (Chambers et al., 2001), and the immunoregulatory cytokine transforming growth factor β (TGF-β) (Li and Flavell, 2006a), results in enhanced tumor immunity.

Tumors are obviously a form of altered self. They possess characteristics resembling infectious agents, such as unbridled proliferation, induction of tissue damage, and life-threatening dissemination (metastasis). It has been postulated that the invasive activity of tumors induces tissue damage and inflammation that may trigger adaptive immunity against them (Pardoll, 2003). Although the full impact of inflammation associated with tumor invasion remains an increasingly important area for investigation, it is important to note that the features of tumor-induced inflammation are more reminiscent of wound healing than inflammation caused by acute infection that induces adaptive immunity (Dvorak, 1986). Consequences of tumor growth and invasion include tissue damage and cell lysis. Studies have shown that cell lysate from necrotic cells induces the expression of genes involved in tissue remodeling and angiogenesis, such as MMP9 and VEGF (Li et al., 2001). Interestingly, in contrast to bacterial lipopolysaccharide (LPS), cell lysate does not induce the expression of interleukin (IL)-12, the cytokine that activates natural killer (NK) cells, $CD8^+$ T cells, and $CD4^+$ T helper 1 (Th1) cells promoting cellular immune responses, although both agents activate the Toll-like receptor and NF-κB pathways (Li et al., 2001). It appears that tissue damage (cell lysis) primarily initiates a healing process without inducing the cellular arm of adaptive immunity. In fact, chronic inflammation fails to trigger tumor rejection but rather fosters tumor growth (Coussens and Werb, 2002). Notably, the normal wound healing process takes place at the resolution phase of infection or trauma, and it is often immunosuppressive. It is therefore plausible that during tumor invasion, the immune suppression associated with wound healing will induce tolerance to tumors.

Nevertheless, it is possible that the immune system may have evolved strategies to differentiate tumors from normal self tissues and that it is actively involved in the control of malignant development. With this in mind, it is interesting to note that the DNA damage pathway, which is often associated with oncogenic transformation, induces expression of major histocompatibility complex (MHC) class I chain-related A (MICA) family of proteins that promote immunity via the engagement of NKG2D receptors on NK cells, $\gamma\delta$-T cells, and $CD8^+$ T cells (Gasser et al., 2005). In some tumor models. including chemical carcinogen-induced tumor models, there is strong evidence in support of the immunosurveillance hypothesis (Smyth et al., 2006). For example, in immunodeficient hosts, such as recombinase-activating gene 1 $(RAG-1)^{-/-}$ mice, tumors develop at higher frequency and are more immunogenic compared to those developed in wild-type hosts (Shankaran et al., 2001). In a similar vein, administration of blocking antibody to the tumor necrosis factor-related apoptosis-inducing ligand (TRAIL), an effector molecule of immune cells, promotes carcinogen-induced tumors (Takeda et al., 2002). Moreover, a higher proportion of these tumors are TRAIL-sensitive compared to tumors developed in wild-type mice (Takeda et al., 2002). In the aforementioned SV40 Tag-induced tumor model, while T cells are tolerated, B cells produce antibodies against Tag (Willimsky and Blankenstein, 2005). Thus, tumors can differentially regulate T and B cell responses. The precise molecular mechanisms of immune regulation by tumors, and how they may override default immune suppressive pathways, remain to be fully explored.

In summary, the ultimate outcome of immune recognition of tumors depends on the balance of the tolerance and the immunity pathways. Successful establishment of a tumor in an immune-competent host results from either (i) possession of the dominant tolerogenic feature that is inherited by tumor cells as the default pathway, which is indistinct from immune tolerance to normal self tissues, or (ii) emergence resulting as a consequence of immune selection as proposed in the immunoediting model (Dunn et al., 2002). To uncover the molecular mechanisms of immune tolerance to tumors, it is therefore vital to achieve an understanding of tumor–host interactions. Hence, this chapter discusses the function of cytokines in the regulation of immune tolerance with an emphasis on T cell tolerance to tumors and explains the possibility of enhancing tumor immunotherapy via the manipulation of the regulatory cytokine network.

II. CYTOKINE REGULATION OF IMMUNE TOLERANCE TO TUMORS

Cytokines are regulatory peptides that control virtually all physiological processes through the modulation of cell proliferation, differentiation, migration, and survival. It is a rule rather than an exception that most cytokines have pleiotropic functions and sometimes opposite influences on disease pathology, depending on their dose, timing of action, and targets. This role cannot be overemphasized when considering cytokine regulation of tumor development. The following sections focus on cytokine regulation of immune tolerance to tumors and describe their function in other aspects of tumor development.

In most cases, an effective immune response against tumors is dependent on the cellular arm of the immune system, the major players of which include CD8+ cytotoxic T lymphocytes (CTLs), CD4+ Th1 cells, γδ-T cells, and NK cells. The effector functions of these cells are responsible for direct target cell killing as well as production of cytokines and other modulators that regulate the function of various cell types, including tumor cells. The cellular immune response is responsible for the host defense of intracellular pathogens; however, when the cellular immune response is dysregulated, it can also trigger autoimmune diseases (e.g., type I diabetes). As a self-tolerance mechanism, several cytokines are involved in the suppression of cellular immune responses either under steady state or during the resolution phase of infection or trauma, which impedes the development of tumor immunity. In addition, cytokines that skew immune responses to T helper 2 (Th2) cells or to the newly discovered Th17 pathway can also influence immune responses to tumors. Therefore, cytokine inhibition of tumor immunity can be caused by immune suppression or deviation.

A. TGF-β

1. Overview

TGF-βs (TGF-β1, 2, 3) are regulatory molecules that affect multiple biological processes, including carcinogenesis, wound healing, and immune homeostasis (Blobe et al., 2000). Active TGF-β mediates its biological functions via binding to type I (TGF-βRI) and II (TGF-βRII) serine/threonine kinase receptors (Huse et al., 2001). TGF-β binding of TGF-βRII activates TGF-βRI, which phosphorylates Smad2 and Smad3 and leads to their nuclear translocation in a complex with Smad4. Smad complex binds to target promoters in association with other transcription factors and regulates gene expression (ten Dijke and Hill, 2004). In addition to Smads, TGF-β also activates other signaling pathways, including various MAPKs and phosphatidylinositol-3 kinase (PI3K). The precise roles of these pathways in the immune systems remain to be established.

The function of TGF-β in cancer is complex. TGF-β can act as a tumor suppressor or a tumor promoter depending on the stages of tumor development and the target cells on which TGF-β acts (Wakefield and Roberts, 2002). TGF-β inhibits the proliferation of epithelial and hematopoietic cell lineages, a key mechanism for TGF-β-mediated tumor suppression. As tumors evolve, they often become resistant to TGF-β-induced growth inhibition. In contrast, tumors undergo epithelial-to-mesenchymal transition and in response to TGF-β acquire invasive properties. TGF-β also modulates the functions of cells present in the tumor environment. The potent regulatory activity of TGF-β on immune cell functions represents an important mechanism of immune tolerance to tumors.

2. TGF-β Regulation of T Cells

Active immune suppression by regulatory cell types, such as CD4+CD25+ Tregs and cytokines (including TGF-β1), is an essential mechanism of self tolerance (Li et al., 2006; Sakaguchi, 2004). Among the three isoforms of TGF-β, TGF-β1 is predominantly expressed in the immune system. Deficiency of TGF-β1 in mice results in an early lethal autoimmune phenotype (Kulkarni et al., 1993; Shull et al., 1992), which is dependent on T cells (Kobayashi et al., 1999; Letterio et al., 1996). Since TGF-β1 can modulate the activity of almost all lineages of leukocytes, it was not clear in these early studies whether T cells are direct targets of TGF-β1. To help determine if T cells are direct targets, several groups have used transgenic approaches to inhibit TGF-β signaling in T cells and have observed disturbance of T cell homeostasis. Expression of a dominant negative TGF-βRII from the CD4 promoter (CD4-DNRII) that lacks the CD8 silencer inhibits TGF-β signaling in both CD4+ and CD8+ T cells. This leads to the generation of autoantibodies and the development of an inflammatory disease (Gorelik and Flavell, 2000). In another report, expression of the dominant negative TGF-βRII from the CD2 promoter results in a CD8⁺ T cell lymphoproliferative disorder with little inflammation (Lucas et al., 2000). In both models, the phenotypes are less severe than that developed in TGF-β1-deficient mice. This can be explained by TGF-β effects on cells other than T cells. Alternatively, the weak phenotype could be due to incomplete blockade of TGF-β signaling in T cells in the transgenic mice.

To define a definitive function of TGF-β signaling in T cells, Li and Flavell (2006b) generated T cell-specific TGF-βRII-deficient mice by crossing a strain of floxed TGF-βRII mice with CD4-Cre transgenic mice. These mice developed a severe multifocal inflammatory disease and succumbed to death in a period of 5 weeks. The severity of the phenotype resembles that of TGFβ1-deficient mice, which highlights T cells as a central direct target of TGF-β1 *in vivo*. Both CD4⁺ and CD8⁺ T cells in these mice are expanded, exhibit an activated phenotype, and produce effector cytokines. In addition, TGF-β signaling in T cells is essential for the maintenance of peripheral CD4⁺CD25⁺ Tregs. This study has further shown that both the inhibition of T cell activation and the maintenance of Tregs represent cell autonomous activity of TGF-β signaling in T cells. These observations reveal a pivotal function for TGF-β in regulating T cell tolerance via the inhibition of T cell activation and the maintenance of Tregs.

In addition to inducing T cell tolerance, Li and Flavell (2006b) have uncovered an important role for TGF-β signaling in regulating T cell survival. When T cell receptor transgenic CD4⁺ OT-II T cells are rendered insensitive to TGF-β, they are partially activated but not differentiated to Th1 or Th2 cells. Strikingly, TGF-βRII knockout OT-II T cells are greatly depleted in peripheral lymphoid organs, a finding that is associated with a high rate of cell apoptosis. It is not yet understood why, in the absence of TGF-β signaling, some CD4⁺ T cells are activated and differentiated into effector T cells, as seen on the polyclonal background, while other T cells (e.g., OT-II T cells) undergo apoptosis. The implication of this dual T cell effect by TGF-β on tumor immunity is also an open question. Nevertheless, these studies reveal a pleiotropic function for TGF-β in T cell tolerance and T cell homeostasis. Li and Flavell (2006b) speculate that TGF-β promotion of T cell survival and inhibition of T cell activation may ensure a diverse and self-tolerant T cell repertoire *in vivo*.

3. TGF-β Regulation of CD8⁺ T Cell Responses to Tumors

Compared to T cell-specific TGF-βRII knockout mice, CD4-DNRII mice develop a less severe inflammatory phenotype and survive to adulthood, which has made it possible to study TGF-β regulation of immune responses to tumors. Significantly, these mice are resistant to challenge with EL-4 thymoma and B16-F10 melanoma (Gorelik and Flavell, 2001). Inhibition of TGF-β signaling in CD8⁺ T cells is essential for the eradiation of tumors in this model, which is associated with the expansion of tumor-specific CTLs (Gorelik and Flavell, 2001). Similarly, transfer of CD8⁺ T cells expressing DNRII inhibits the growth of endogenous prostate tumors and enhances survival of tumor-bearing mice (Zhang et al., 2005). These studies highlight CD8⁺ T cells as important targets of TGF-β in tumor tolerance.

The molecular mechanisms of TGF-β inhibition of CD8⁺ T cell responses have begun to be addressed. CD8⁺ T cells activated in the presence of TGF-β are unable to kill target cells (Ranges *et al.*, 1987), which is likely due to TGF-β inhibition of perforin expression (Smyth *et al.*, 1991). TGF-β also suppresses the expression of transcription factor c-Myc and Fas ligand expression in CD8⁺ T cells (Genestier *et al.*, 1999). In addition, TGF-β inhibits granzyme B expression through direct promoter association of Smad2 and Smad3, which synergize with CREB/ATF-2 to block granzyme B transcription (Thomas and Massague, 2005).

Smad2 and Smad3 complexes are also recruited to the IFN-γ promoter upon TGF-β treatment (Thomas and Massague, 2005). Therefore, TGF-β utilizes multiple pathways to suppress the expression of cytotoxic genes in CD8$^+$ T cells.

What are the cellular sources of TGF-β that inhibit CD8$^+$ T cell responses to tumors? As tumors arise from normal self tissues, it is conceivable that similar T cell tolerance mechanisms to self antigens are also involved in the induction of tolerance to tumors by TGF-β, which may involve TGF-β produced by T cells, antigen-presenting cells, or stromal cells. An important self-tolerance pathway is mediated by CD4$^+$CD25$^+$ Tregs, as mice or humans devoid of this lineage develop severe autoimmunity (Sakaguchi, 2004). Tregs also impede immune responses to tumors (Yamaguchi and Sakaguchi, 2006). One study has shown that transfer of Tregs could inhibit the effector function of CTLs to tumors (Chen et al., 2005). Interestingly, the inhibition of antitumor responses by Tregs requires TGF-β signaling in CD8$^+$ T cells (Chen et al., 2005). It remains to be determined whether TGF-β is produced by Tregs, by "Treg-modified" antigen-presenting cells, or by other cell types to suppress CD8$^+$ T cells. Nevertheless, this TGF-β-dependent regulatory circuit likely represents a normal self-tolerance pathway. In support of this hypothesis, TGF-β signaling in CD8$^+$ T cells is also essential for the suppression of autoimmune diabetes by Tregs (Green et al., 2003).

During tumor progression, most tumors will evolve to overexpress TGF-β. This common event has been postulated as a mechanism for tumor evasion of immune surveillance (Wojtowicz-Praga, 2003). Consistent with this, overexpression of TGF-β1 in a highly immunogenic murine tumor suppresses the antitumor immune response (Torre-Amione et al., 1990). The involvement of endogenous tumor cell-produced TGF-β in inhibiting antitumor immune responses is likely dependent on the tumor type. On one hand, RNAi inhibition of TGF-β1 expression in glioma cells led to restoration of NKG2D expression in CD8$^+$ T cells and decreased tumor growth in mice (Friese et al., 2004). On the other hand, RNAi inhibition of TGF-β1 in EL4 cells (the major TGF-β isoform produced by these cells) had minimal effect on the development of antitumor immunity, whereas expression of a soluble form of TGF-βRII in tumors (as a tactic to trap TGF-β produced by tumor cells and proximal stromal cells) enhances the CTL response against EL4 cells (Thomas and Massague, 2005). It is possible that TGF-β2 and TGF-β3 produced by EL4 cells are involved in suppressing CTL responses. Alternatively, TGF-β1 produced by host cells induces immune tolerance to these tumors. Future studies with TGF-β conditional knockout mouse models will be necessary to elucidate the function of TGF-β produced by tumor cells and host cells in the regulation of CTL responses to tumors.

4. TGF-β Regulation of CD4$^+$ T Cell Responses to Tumors

Compared to regulation of tumor immunity in CD8$^+$ T cells, regulation of tumor immunity by TGF-β signaling in CD4$^+$ T cells is more complicated and less well understood. TGF-β potently inhibits differentiation of Th1 and Th2 cells, apparently by inhibiting the expression of T-bet and Gata-3 (lineage specification transcription factors for Th1 and Th2 cells) (Gorelik and Flavell, 2002). Consistent with this mechanism, CD4$^+$ T cells from the CD4-DNRII mice expressing dominant negative TGF-βIIR produce high levels of Th1 and Th2 cytokines (Gorelik and Flavell, 2000). Since Th1 cells promote cellular immune responses, enhanced Th1 cell differentiation will likely augment tumor immunity. On the other hand, increasing the Th2 response that promotes humoral rather than cellular immunity will impede immune rejection of tumors. In fact, Th2 cytokines may even precipitate oncogenesis of epithelial cells. For example, inhibiting TGF-β signaling in T cells has led to enhanced IL-6 production and accelerated azoxymethane-induced colon cancer in mice (Becker et al., 2004). In addition, T cell-specific deletion of Smad4 has resulted in gastrointestinal epithelial cancers in mice, which is associated with enhanced production of Th2 cytokines (Kim et al., 2006). Therefore, the impact of TGF-β signaling on Th1 or Th2 cell differentiation can have opposite influences on tumor development.

In contrast to its inhibitory effects on Th1 and Th2 cells, TGF-β promotes the differentiation of the newly characterized Th17 cells (Bettelli et al., 2006; Mangan et al., 2006; Veldhoen et al., 2006) and maintains the homeostasis of Tregs (Marie et al., 2005). TGF-β induces expression of the critical Treg marker FoxP3 as well as generation of Tregs from activated CD4+ T cells in vitro (Chen et al., 2003). Conversely, the inflammatory cytokine IL-6 inhibits generation of Tregs but induces de novo differentiation of Th17 cells, which produce the signature cytokine IL-17 (Bettelli et al., 2006; Veldhoen et al., 2006). Consistent with in vitro studies, peripheral Tregs are reduced, and differentiation of Th17 cells is inhibited in TGF-β1-deficient mice (Mangan et al., 2006). Tregs and possibly Th17 cells (as discussed in further detail in the following paragraphs) inhibit the development of tumor immunity. Indeed, increased numbers of Tregs are associated with tumor progression and reduced survival of ovarian cancer patients (Curiel et al., 2004). Although a definitive role for TGF-β in the regulation of Th17 differentiation and Treg homeostasis in cancer-bearing hosts remains to be established, the positive effects of TGF-β on these two lineages of T cells likely represent important mechanisms of TGF-β-mediated immune tolerance to tumors.

5. TGF-β Regulation of NK Cell Responses to Tumors

Natural killer (NK) cells are innate immune cells that participate in early defense against microbial infections (Hamerman et al., 2005). NK cells express receptors that allow them to respond to microbial products and cell-stress signals, such as the MICA family of proteins. NK cells can directly kill target cells. In addition, NK cells produce high levels of IFN-γ that regulates the cellular arm of the adaptive immunity (Martin-Fontecha et al., 2004). To avoid potential injury to normal self tissues and to prevent inappropriate activation of immune responses, NK cells are subjected to negative regulation by a variety of inhibitory receptors (Raulet and Vance, 2006).

TGF-β is an important regulator of NK cell function, being a potent antagonist of IL-12-induced production of IFN-γ in NK cells (Bellone et al., 1995; Hunter et al., 1995). Consistent with this, inhibition of TGF-β signaling in NK cells leads to enhanced IFN-γ production and potentiated Th1 cell responses, in response to *Leishmania* infection in mice (Laouar et al., 2005). TGF-β inhibits the activity of NK cells by limiting expression of the activating receptors NKp30 and NKG2D (Castriconi et al., 2003). In fact, reduced expression of NKG2D is associated with elevated levels of TGF-β1 in human cancer patients (Lee et al., 2004). Although the function of TGF-β signaling in NK cell antitumor activity remains to be fully established, it is conceivable that TGF-β promotes NK tolerance to tumors.

6. TGF-β and Immunotherapy of Cancer

The potent regulatory activity of TGF-β on T cells and NK cells represents an important mechanism of immune tolerance to tumors. In mice, elevated serum levels of TGF-β were found in developing tumors that had induced T cell tolerance (Willimsky and Blankenstein, 2005). High levels of circulating TGF-β are also associated with poor prognosis in a number of human cancers, including gastric, lung, and prostate cancers (Hasegawa *et al.*, 2001; Saito *et al.*, 1999; Shariat *et al.*, 2001). Over the years, there has been growing interest in developing inhibitors of the TGF-β pathway to augment antitumor immune responses.

TGF-β inhibitors under development for clinical applications include antisense oligonucleotides to block TGF-β expression, anti-TGF-β antibodies, soluble TGF-β receptors, and small-molecule inhibitors of TGF-β receptor kinases (Yingling *et al.*, 2004). Several studies have targeted tumor cell-produced TGF-β to enhance antitumor immune responses. In a preclinical study, rats with established intracranial gliomas were immunized with modified gliosarcoma cells that expressed an antisense RNA against TGF-β2. All rats treated with the TGF-β2 antisense-modified tumor cells underwent remission of preinoculated tumors in a manner that was associated with enhanced activity of cytotoxic effector cells (Fakhrai *et al.*, 1996). This tumor cell vaccine approach has entered phase I/II clinical trials by NovaRx. Antisense Pharma has also developed a TGF-β2-specific antisense oligonucleotide AP-12009 to treat gliomas. In several phase I/II clinical trials, intratumoral administration of AP-12009 resulted in a significant increase of survival time (Bogdan, 2004). Preclinical studies show that AP-12009 prevented glioma cell proliferation and reversed tumor-induced T cell suppression (Platten *et al.*, 2001).

The elucidation of the TGF-β receptor complex and its signaling pathways has generated great interest in targeting TGF-β receptors for therapeutics. Several small-molecule inhibitors targeting the TGF-β receptor I kinase domain have been developed. One of these inhibitors, SD-208 (Scios), inhibits the growth of intracranial SMA-560 gliomas in syngeneic mice (Uhl *et al.*, 2004). Interestingly, the antitumor effects of SD-208 do not correlate with changes in tumor cell viability or proliferation, but they are associated with increased tumor infiltration by natural killer cells, CD8[+] T cells, and macrophages (Uhl *et al.*, 2004). Oral SD-208 also appears effective in inhibiting TGF-β-induced Smad phosphorylation in spleens (Uhl *et al.*, 2004). This study illustrates the potential of using small molecule inhibitors of TGF-β to relieve immune suppression and augment antitumor immune responses.

B. IL-10

1. Overview

IL-10 was first identified as a molecule produced by Th2 cells that inhibits production of Th1 cytokines (Fiorentino *et al.*, 1989). IL-10 exerts pleiotropic effects on hemopoietic and nonhemopoietic cells by binding to its receptor complexes IL-10R1 and IL-10R2 (Moore *et al.*, 2001). Activated IL-10 receptors initiate multiple signaling pathways, the best characterized of which is the Jak1-Tyk2/Stat3 pathway (Finbloom and Winestock, 1995). The function of IL-10 in cancer is complex: depending on the experimental model, IL-10 displays both immunosuppressive and immunostimulating activities (Mocellin *et al.*, 2005), which can be explained by differential effects of IL-10 on target cells.

2. IL-10 Regulation of T Cells and NK Cells

Genetic studies in mice have established an essential role for IL-10 in regulating T cell tolerance to self or innocuous environmental antigens. IL-10-deficient mice develop lethal colitis (Kuhn *et al.*, 1993), which is dependent on T cells. Consistent with earlier studies, maintenance of gut mucosal tolerance is associated with IL-10 inhibition of Th1 cell differentiation.

What are the cellular targets of IL-10 in its blockade of Th1 cell differentiation? This area is incompletely understood but will be resolved by the generation of cell type-specific IL-10 receptor knockout mice. Studies have suggested that the immunosuppressive activity of IL-10 on T cells is largely mediated by its effects on antigen-presenting cells (de Waal Malefyt et al., 1991; Fiorentino et al., 1991). IL-10 inhibits monocyte differentiation into dendritic cells (DCs) (Allavena et al., 1998; Buelens et al., 1997), the professional antigen-presenting cells that prime the activation of naive T cells (Banchereau and Steinman, 1998). IL-10 also blocks the production of IL-12 and expression of costimulatory molecules in DCs (Allavena et al., 1998), a role which is associated with IL-10 inhibition of Th1 cell differentiation. Although IL-10 supports the differentiation of macrophages (Allavena et al., 1998), it inhibits the production of inflammatory cytokines, such as TNF-α, IL-1, and IL-12; downregulates MHC class II expression; and blocks the expression of costimulatory molecules (de Waal Malefyt et al., 1991; Fiorentino et al., 1991; Willems et al., 1994). Collectively, these processes result in impaired antigen presentation and CD4$^+$ T cell activation by macrophages. The pivotal role of IL-10 signaling on macrophages in immune tolerance is corroborated by the study of macrophage/neutrophil-specific Stat3 knockout mice. Macrophages from these mice are refractory to the effects of IL-10 (Takeda et al., 1999), and these mice develop chronic colitis (Takeda et al., 1999).

In addition to regulating antigen-presenting cells, IL-10 also directly modulates T cell responses. IL-10 inhibits proliferation and cytokine synthesis of CD4$^+$ T cells (Del Prete et al., 1993; Groux et al., 1996). This effect is more profound on naive T cells than activated T cells, probably explained by IL-10R downregulation upon T cell activation (Liu et al., 1994). In contrast to the inhibition of Th1 cells, IL-10 does not block (and sometimes even stimulates) CD8$^+$ T cell proliferation and cytotoxic activity (Groux et al., 1998; Santin et al., 2000). Similar to CD8$^+$ T cells, IL-10 also promotes NK cell function. For example, IL-10 potentiates IL-2-induced NK cell proliferation and cytokine production (Carson et al., 1995). The molecular mechanisms by which IL-10 regulates the function of CD4$^+$ T cells, CD8$^+$ T cells, and NK cells remain to be determined. Nonetheless, these pleiotropic effects of IL-10 complicate an understanding of its role in regulating immune responses to tumors.

3. IL-10 Regulation of Antitumor Immune Responses

IL-10 can act on both tumor cells and immune cells to inhibit antitumor immune responses. Pretreatment of tumor cells with IL-10 downregulates MHC I expression and inhibits CD8$^+$ T cell-mediated tumor cell lysis (Kurte et al., 2004). Suppression of MHC I expression by IL-10 is likely due to the reduced expression of the transporter associated with antigen processing 1 (TAP-1) and TAP-2 genes (Kurte et al., 2004), resulting in diminished transport of peptides to the endoplasmic reticulum and consequent loading of MHC I–peptide complexes. The potent inhibitory effect of IL-10 on antigen-presenting cells is also involved in the induction of immune tolerance to tumors. The cytotoxic activity of CD8$^+$ T cells is blunted when they are stimulated with IL-10-treated DCs (Steinbrink et al., 1999). This result is likely relevant in human cancers because DCs expressing low levels of IL-12 but high levels of IL-10 have been found in human melanoma metastasis (Enk et al., 1997). In addition, IL-10-producing monocytes are present in ascites of patients with ovarian carcinoma (Loercher et al., 1999). In a transgenic mouse model, expression of IL-10 under the control of the IL-2 promoter results in a higher growth rate of the immunogenic lung carcinomas than in control animals (Hagenbaugh et al., 1997). In a mouse model of plasmacytoma, anti-IL-10 antibody treatment also potentiates antitumor responses (Jovasevic et al., 2004). Similar studies with anti-IL-10 receptor antibody or IL-10 antisense oligodeoxynucleotides have also revealed that inhibition of IL-10 expression or function can result in enhanced immune responses to tumors (Kim et al., 2000; Vicari et al., 2002). Although the precise cellular and molecular mechanisms still need to be established, these *in vivo* studies demonstrate that endogenous IL-10 can be involved in the induction of tolerance to tumors.

As a pleiotropic cytokine, IL-10 has also been shown to promote antitumor immune responses in animal models. Ectopic expression of IL-10 in multiple tumor cell lines, including mammary adenocarcinoma, colon carcinoma, and melanoma, resulted in tumor rejection in syngeneic mice (Adris *et al.*, 1999; Gerard *et al.*, 1996; Giovarelli *et al.*, 1995), apparently depending on T and/or NK cells. Administration of IL-10 to tumor-bearing mice also triggers T-cell- or NK-cell-mediated tumor regression (Berman *et al.*, 1996; Kundu *et al.*, 1996; Zheng *et al.*, 1996). It remains to be established how IL-10-regulated immune responses lead to tumor tolerance or tumor immunity in the animal models. It is possible that this dual effect originates from the differential roles of IL-10 on tumor cells, antigen-presenting cells, CD4+ T cells, CD8+ T cells, and NK cells. In addition, IL-10 may influence tumorigenesis independent on its regulation of immune responses. For example, IL-10 potently inhibits the growth of transplanted tumors in SCID or nude mice, which is associated with the inhibition of angiogenesis at tumor sites (Stearns *et al.*, 1999).

C. IL-23

1. Overview

IL-23 is a dimeric cytokine composed of the two subunits p19 and p40 (Oppmann *et al.*, 2000). IL-23 is closely related to IL-12, which is composed of the same p40 subunit plus a distinct p35 polypeptide. The dimeric receptors for IL-12 and IL-23 contain a common IL-12Rβ1 chain plus unique subunits IL-12Rβ2 or IL-23R. IL-12 and IL-23 are produced by innate immune cells but regulate the development of adaptive immunity (Hunter, 2005). IL-12 promotes cellular immune responses and is important for host defense against intracellular pathogens and for the induction of antitumor immune responses. On the other hand, IL-23 promotes the development of Th17 cells that secrete inflammatory cytokines IL-17, IL-17F, IL-6, and TNF-α, as well as proinflammatory chemokines (Hunter, 2005). Although the precise physiological functions of Th17 cells remain to be established, recent studies have unraveled a pivotal function for IL-23 and IL-17 in the pathogenesis of adjuvant-induced autoimmune diseases (Cua *et al.*, 2003; Langrish *et al.*, 2005; Murphy *et al.*, 2003) as well as in host defense against extracellular pathogens (Khader *et al.*, 2005; Lieberman *et al.*, 2004). The activity of IL-23 and IL-17 in the regulation of antitumor immune responses is just beginning to be elucidated (Langowski *et al.*, 2006).

2. IL-23 and Th17 Cells

The key function of IL-23 in regulating T cell immunity was elucidated by the seminal findings that deficiency of p19 (IL-23 specific) but not p35 (IL-12 specific) alleviates adjuvant-induced autoimmune diseases, including experimental allergic encephalomyelitis (EAE), and collagen-induced arthritis (CIA) (Cua *et al.*, 2003; Murphy *et al.*, 2003). *In vitro* stimulation of activated and/or memory CD4+ T cells in the presence of IL-23 leads to the production of IL-17 but not Th1 or Th2 cytokines. Significantly, in an adoptive transfer model, IL-23-stimulated encephalitogenic T cells that produce a unique subset of cytokines, IL-17, IL-17F, IL-6, and TNF-α and that are sufficient to induce disease (Langrish *et al.*, 2005). In addition, IL-17 blockade attenuates neurological disease (Langrish *et al.*, 2005). Additional studies have further established that IL-17-producing T cells represent a distinct lineage of helper T cells (Th17) (Harrington *et al.*, 2005; Park *et al.*, 2005) whose development is suppressed by Th1 or Th2 cytokines.

IL-23 is a poor inducer of Th17 cell differentiation from naive CD4+ T cells, likely explained by the lack of IL-23R expression in these T cells (Harrington *et al.*, 2005).

For this reason, IL-23 is probably involved in the survival and/or expansion of committed Th17 cells. Recent studies have revealed that in the presence of IL-6, Th17 cell differentiation from naive T cells can be driven by TGF-β (Bettelli *et al.*, 2006; Veldhoen *et al.*, 2006). TGF-β is expressed constitutively by multiple cell lineages (Li *et al.*, 2006). On the other hand, IL-6 and IL-23 expression is induced only after pathogen infection or other inflammatory stimuli. Although a central role for the IL-23/Th17 pathway in the pathogenesis of adjuvant-induced pathological autoimmunity has been established, the physiological functions of this pathway remain to be fully explored. IL-23 is induced by bacterial products, such as lipopolysaccharides (Oppmann *et al.*, 2000). IL-17 triggers the production of proinflammatory mediators, including IL-8, TNF-α, and granulocyte colony-stimulating factor that promote neutrophil recruitment (Kolls and Linden, 2004). It is conceivable that the IL-23/Th17 axis may have evolved to combat a subset of extracellular pathogens that are controlled by neutrophils during acute infection. This is supported by findings that IL-23/Th17 is involved in the protection of a number of bacterial infections, including *Klebsiella pneumoniae*, *Borrelia burgdorferi*, and *Citrobacter rodentium* (Fedele *et al.*, 2005; Happel *et al.*, 2005; Mangan *et al.*, 2006).

Intriguingly, IL-23 but not IL-12 is also preferentially induced by endogenous signals that are associated with tissue damage, such as prostaglandin E2 (PGE_2) and extracellular ATP (Schnurr *et al.*, 2005; Sheibanie *et al.*, 2004). This fits with fact that neutrophils are active participants in the removal of damaged tissues. IL-17 also promotes angiogenesis *in vivo* (Numasaki *et al.*, 2003). Together, these observations suggest that IL-23/Th17 may help control trauma and initiate repair by inducing neutrophil recruitment and angiogenesis. This likelihood would also be consistent with the well-known function of TGF-β in the regulation of wound healing (Li *et al.*, 2006).

3. IL-23 Regulation of Antitumor Immune Responses

Chronic inflammation has long been associated with increased incidence of tumors. A 2006 study highlighted a critical role for IL-23-regulated inflammation in tumor growth (Langowski *et al.*, 2006). The expression of IL-23 subunits p19 and p40, but not the IL-12 subunit p35, was selectively elevated in a number of human carcinomas compared to adjacent normal tissues (Langowski *et al.*, 2006). IL-23 p19 is localized in the tumor-infiltrating $CD11c^+$ DCs but not in the tumor stroma, suggesting that the tumor microenvironment supports IL-23 production. Since tumor development is associated with tissue stress and damage, endogenous signals such as PGE_2 and extracellular ATP may be involved in the induction of IL-23. In fact, PGE_2 is highly expressed in tumors as a result of enhanced cyclooxygenase 2 expression (Ristimaki *et al.*, 1997; Sano *et al.*, 1995; Wolff *et al.*, 1998). Consistent with the enrichment of IL-23 in tumors, IL-17 was also found upregulated in these samples (Langowski *et al.*, 2006). In addition, in human colorectal tumors, increased IL-23 p19 was also associated with increased infiltration of neutrophils (Langowski *et al.*, 2006).

What is the function of IL-23 in tumors? To address this question, a chemical carcinogen-induced papilloma model was used. Significantly, mice deficient in IL-23 p19 were highly resistant to the induction of papillomas (Langowski *et al.*, 2006). On the other hand, mice deficient in the IL-12 subunit p35 displayed accelerated induction of tumors (Langowski *et al.*, 2006). Therefore, it appears that IL-23 promotes carcinogenesis, while IL-12 suppresses it. Since IL-23 and IL-12 share the same p40 subunit, deficiency in the unique subunit of IL-23 might result in elevated production of IL-12. Interestingly, mice deficient in p40 are also resistant to tumor induction (Langowski *et al.*, 2006). These observations suggest that the tumor-promoting effect of

IL-23 is dominant over the tumor-suppressing effect of IL-12.

What are the mechanisms by which IL-23 supports tumor development? Consistent with a role for IL-23 in Th17 development, IL-17 expression in the skin is greatly diminished in IL-23 p19$^{-/-}$ or p40$^{-/-}$ mice compared to IL-17 expression in wild-type or p35$^{-/-}$ mice (Langowski et al., 2006). Infiltrating granulocytes are also decreased in IL-23 p19$^{-/-}$ or p40$^{-/-}$ mice (Langowski et al., 2006). IL-17 is known to promote tumor angiogenesis (Numasaki et al., 2003). Consistent with this finding, angiogenesis is significantly inhibited in the skin of IL-23 p19$^{-/-}$ or p40$^{-/-}$ mice and is associated with reduced expression of the proangiogenic factor MMP9 (Langowski et al., 2006). Interestingly, increased CD8$^+$ T cells that express cytotoxic T cell markers were present in the hyperplastic skin of IL-23 p19$^{-/-}$ or p40$^{-/-}$ mice (Langowski et al., 2006). Downregulation of MMP9 and increased infiltration of CD8$^+$ T cells were also detected when mice were treated with a neutralization antibody to IL-23 p19 (Langowski et al., 2006). Taken together, these observations suggest that IL-23 promotes tumor development by inducing Th17 differentiation and angiogenesis and by concomitantly inhibiting CD8$^+$ T cell responses. It remains an open question whether suppression of CD8$^+$ T cells by IL-23 is secondary to IL-23 induction of Th17 differentiation or whether it represents an independent pathway. IL-23 is unlikely to be involved in initiating Th17 cell differentiation, so roles for TGF-β and IL-6 in promoting tumor-associated Th17 cell differentiation might be entertained. In any case, these findings highlight a strong association between the IL-23/Th17 pathway, tumor-associated chronic inflammation, and immune suppression.

D. Vascular Endothelial Growth Factor

1. Overview

Vascular endothelial growth factor (VEGF) is a chief proangiogenic factor in development, wound healing, and pathogenic processes, such as carcinogenesis and rheumatoid arthritis (Hoeben et al., 2004). Seven VEGF protein isoforms have been identified, all of which share a common VEGF homology domain. VEGF functions by binding to VEGF tyrosine kinase receptors that initiate multiple signaling pathways affecting cell proliferation, survival, migration, and tissue permeability (Hoeben et al., 2004). VEGF is highly expressed by tumor cells and tumor-associated stromal cells (Toi et al., 1996), enabling tumor neoangiogenesis and promoting tumoral immune tolerance.

2. VEGF Regulation of Antitumor Immune Responses

Strong evidence exists of an important function for VEGF in suppressing antitumor immune responses through its regulation of antigen-presenting cells. VEGF blocks DC differentiation and maturation from hematopoietic stem cells *in vitro* (Gabrilovich et al., 1998; Oyama et al., 1998). This effect is associated with inhibition of transcription factor NF-κB (Oyama et al., 1998) which is essential for DC development, activation, and survival (Ouaaz et al., 2002). In addition, antibody-mediated neutralization of VEGF in tumor-bearing mice promotes DC differentiation and function (Gabrilovich et al., 1998, 1999). In human patients with gastric, breast, lung, heart, or neck cancers, there is a reverse association between VEGF expression and the number and function of mature DCs present (Almand et al., 2000). These observations reveal that enhanced VEGF production may lead to impaired DC activity in cancer patients.

Concomitant with the inhibition of DC maturation, VEGF promotes the generation of immature myeloid cells (iMCs), also known as myeloid suppressor cells. iMCs are present at high frequencies in tumor-bearing mice and human cancer patients (Young et al., 1987, 1997), and they potently suppress immune activity (Bronte et al.,

2001; Kusmartsev and Gabrilovich, 2002). Upon adoptive transfer to mice, antigen-loaded iMCs induce antigen-specific T cell tolerance (Kusmartsev *et al.*, 2004). iMCs also promote the development of Tregs in tumor-bearing mice (Huang *et al.*, 2006). The precise molecular mechanisms by which iMCs suppress T cell activity are not well understood but may involve the indoleamine 2,3-dioxygenase and arginase I immunosuppressive pathways (Serafini *et al.*, 2006).

III. SUMMARY AND FUTURE PERSPECTIVES

Since tumors arise from normal self tissues, the default pathway to tumor-associated antigens is likely to be tolerance, especially tolerance that prevents induction of destructive cellular immune responses. The barriers that prevent antitumor immune responses are related to, if not the same as, those that inhibit autoimmune diseases, which include mechanisms of immune suppression or immune deviation under steady state and during wound healing.

Cytokines are important regulators of immune tolerance to tumors (Table 4.1). TGF-β is the most potent and pleiotropic regulatory cytokine identified as of this writing. Under steady state conditions, TGF-β inhibits destructive cellular immune responses directed at self antigens by inhibiting activation and/or differentiation of Th1, CTL, and NK cells and by maintaining peripheral Tregs. This is a pivotal pathway for maintaining tumor tolerance. Under the condition of wound healing, TGF-β may also ensure immune suppression while promoting the normal healing process. This same process may have a critical role in inhibiting antitumor immune responses, as inflammation associated with tumor invasion resembles that of wound healing (Dvorak, 1986).

TGF-β also induces the differentiation of Th17 cells that inhibit cellular immune responses and enhance tumor progression. Taken together, the TGF-β pathway appears to be a major obstacle against the generation of an effective antitumor immune response. IL-10 is another key pleiotropic cytokine. Under some circumstances, regulation of Th1 cells, tumor cells, and antigen-presenting cells by IL-10 induces tumor tolerance; however, IL-10 can also trigger tumor immunity by stimulating $CD8^+$ T cells and NK cells. IL-23 promotes the generation of Th17 cells and is closely associated with malignant development. The IL-23/Th17 pathway induces angiogenesis and inhibits $CD8^+$ T cells. This pathway offers a good example of how antitumor immune responses are blunted through immune deviation. In addition, VEGF signals induce angiogenesis and inhibit antitumor immune responses by inducing the production of immature myeloid cells. As is the case for TGF-β, the VEGF-induced inhibition of antitumor immune responses likely has its origins in the immune suppression processes that characterize normal wound healing.

TABLE 4.1 Cytokine Regulation of Tumor Tolerance

	Immune Suppression	Immune Deviation
TGF-β	⊣ CTL ⊣ Th1 ⊣ NK → Treg	→ Th17
IL-10	⊣ Th1 ⊣ APC	
IL-23		→ Th17
VEGF	→ iMC	

Several families of cytokines induce tumor tolerance by immune suppression and/or immune deviation. TGF-β inhibits the differentiation and effector function of cells involved in cellular immunity, including CTLs, Th1, and NK cells. In addition, TGF-β induces immune suppression by maintaining Tregs and triggers immune deviation by promoting Th17 differentiation. IL-10 inhibits the differentiation of Th1 cells and the function of antigen-presenting cells (APCs). IL-23 impedes the development of tumor immunity by enhancing Th17 cells. VEGF induces iMCs, which suppresses antitumor immune responses.

Immune regulatory cytokines are part of a regulatory network that inhibits the development of antitumor immune responses. The concept that tumor tolerance relates to natural self-tolerance makes tumor immunity a special kind of autoimmunity. Unraveling the interactions between the immune system and tumors awaits the development of better animal models that more accurately recapitulate the features of spontaneous cancers that represent the main clinical challenge. While there is evidence for cancer immunosurveillance in some tumor models, the default suppressive pathways that may exist may continue challenge efforts to trigger tumor-associated immune stimulation for cancer therapy. Defining mechanisms of immune suppression and immune tolerance will continue to represent an important goal for cancer researchers.

Acknowledgments

We thank F. Manzo for her assistance with manuscript preparation. R. A. F. is an investigator of Howard Hughes Medical Institute. This work is supported by an NIH career development award K01 AR053595-01 (M. O. L.), American Diabetes Association (R. A. F.), and NIH grant R01 DK51665 (R. A. F.).

References

Adris, S., Klein, S., Jasnis, M., Chuluyan, E., Ledda, M., Bravo, A., Carbone, C., Chernajovsky, Y., and Podhajcer, O. (1999). IL-10 expression by CT26 colon carcinoma cells inhibits their malignant phenotype and induces a T cell-mediated tumor rejection in the context of a systemic Th2 response. *Gene Ther.* **6**, 1705–1712.

Allavena, P., Piemonti, L., Longoni, D., Bernasconi, S., Stoppacciaro, A., Ruco, L., and Mantovani, A. (1998). IL-10 prevents the differentiation of monocytes to dendritic cells but promotes their maturation to macrophages. *Eur. J. Immunol.* **28**, 359–369.

Almand, B., Resser, J. R., Lindman, B., Nadaf, S., Clark, J. I., Kwon, E. D., Carbone, D. P., and Gabrilovich, D. I. (2000). Clinical significance of defective dendritic cell differentiation in cancer. *Clin. Cancer Res.* **6**, 1755–1766.

Banchereau, J., and Steinman, R. M. (1998). Dendritic cells and the control of immunity. *Nature* **392**, 245–252.

Becker, C., Fantini, M. C., Schramm, C., Lehr, H. A., Wirtz, S., Nikolaev, A., Burg, J., Strand, S., Kiesslich, R., Huber, S., Ito, H., Nishimoto, N., Yoshizaki, K., Kishimoto, T., Galle, P. R., Blessing, M., Rose-John, S., and Neurath, M. F. (2004). TGF-beta suppresses tumor progression in colon cancer by inhibition of IL-6 trans-signaling. *Immunity* **21**, 491–501.

Bellone, G., Aste-Amezaga, M., Trinchieri, G., and Rodeck, U. (1995). Regulation of NK cell functions by TGF-beta 1. *J. Immunol.* **155**, 1066–1073.

Berman, R. M., Suzuki, T., Tahara, H., Robbins, P. D., Narula, S. K., and Lotze, M. T. (1996). Systemic administration of cellular IL-10 induces an effective, specific, and long-lived immune response against established tumors in mice. *J. Immunol.* **157**, 231–238.

Bettelli, E., Carrier, Y., Gao, W., Korn, T., Strom, T. B., Oukka, M., Weiner, H. L., and Kuchroo, V. K. (2006). Reciprocal developmental pathways for the generation of pathogenic effector TH17 and regulatory T cells. *Nature* **441**, 235–238.

Blobe, G. C., Schiemann, W. P., and Lodish, H. F. (2000). Role of transforming growth factor beta in human disease. *N. Engl. J. Med.* **342**, 1350–1358.

Bogdan, U. (2004). Specific therapy for high-grade glioma by convection-enhanced delivery of the TGF-b2 specific antisense oligonucleotide AP 12009. In *Am. Soc. Clin. Oncol. Ann. Meet. Abstract*, pp. 1514.

Bronte, V., Serafini, P., Apolloni, E., and Zanovello, P. (2001). Tumor-induced immune dysfunctions caused by myeloid suppressor cells. *J. Immunother.* **24**, 431–446.

Buelens, C., Verhasselt, V., De Groote, D., Thielemans, K., Goldman, M., and Willems, F. (1997). Interleukin-10 prevents the generation of dendritic cells from human peripheral blood mononuclear cells cultured with interleukin-4 and granulocyte/macrophage-colony-stimulating factor. *Eur. J. Immunol.* **27**, 756–762.

Burnet, F. M. (1970). The concept of immunological surveillance. *Prog. Exp. Tumor Res.* **13**, 1–27.

Carson, W. E., Lindemann, M. J., Baiocchi, R., Linett, M., Tan, J. C., Chou, C. C., Narula, S., and Caligiuri, M. A. (1995). The functional characterization of interleukin-10 receptor expression on human natural killer cells. *Blood* **85**, 3577–3585.

Castriconi, R., Cantoni, C., Della Chiesa, M., Vitale, M., Marcenaro, E., Conte, R., Biassoni, R., Bottino, C., Moretta, L., and Moretta, A. (2003). Transforming growth factor beta 1 inhibits expression of NKp30 and NKG2D receptors: Consequences for the NK-mediated killing of dendritic cells. *Proc. Natl. Acad. Sci. USA* **100**, 4120–4125.

Chambers, C. A., Kuhns, M. S., Egen, J. G., and Allison, J. P. (2001). CTLA-4-mediated inhibition in regulation of T cell responses: Mechanisms and manipulation in tumor immunotherapy. *Annu. Rev. Immunol.* **19**, 565–594.

Chen, M. L., Pittet, M. J., Gorelik, L., Flavell, R. A., Weissleder, R., von Boehmer, H., and Khazaie, K. (2005). Regulatory T cells suppress tumor-specific CD8 T cell cytotoxicity through TGF-beta signals *in vivo. Proc. Natl. Acad. Sci. USA* **102**, 419–424.

Chen, W., Jin, W., Hardegen, N., Lei, K. J., Li, L., Marinos, N., McGrady, G., and Wahl, S. M. (2003). Conversion of peripheral CD4$^+$CD25-naive T cells to CD4$^+$CD25$^+$ regulatory T cells by TGF-beta induction of transcription factor Foxp3. *J. Exp. Med.* **198**, 1875–1886.

Coussens, L. M., and Werb, Z. (2002). Inflammation and cancer. *Nature* **420**, 860–867.

Cua, D. J., Sherlock, J., Chen, Y., Murphy, C. A., Joyce, B., Seymour, B., Lucian, L., To, W., Kwan, S., Churakova, T., Zurawski, S., Wiekowski, M., Lira, S. A., Gorman, D., Kastelein, R. A., and Sedgwick, J. D. (2003). Interleukin-23 rather than interleukin-12 is the critical cytokine for autoimmune inflammation of the brain. *Nature* **421**, 744–748.

Curiel, T. J., Coukos, G., Zou, L., Alvarez, X., Cheng, P., Mottram, P., Evdemon-Hogan, M., Conejo-Garcia, J. R., Zhang, L., Burow, M., Zhu, Y., Wei, S., Kryczek, I., Daniel, B., Gordon, A., Myers, L., Lackner, A., Disis, M. L., Knutson, K. L., Chen, L., and Zou, W. (2004). Specific recruitment of regulatory T cells in ovarian carcinoma fosters immune privilege and predicts reduced survival. *Nat. Med.* **10**, 942–949.

de Waal Malefyt, R., Haanen, J., Spits, H., Roncarolo, M. G., te Velde, A., Figdor, C., Johnson, K., Kastelein, R., Yssel, H., and de Vries, J. E. (1991). Interleukin 10 (IL-10) and viral IL-10 strongly reduce antigen-specific human T cell proliferation by diminishing the antigen-presenting capacity of monocytes via downregulation of class II major histocompatibility complex expression. *J. Exp. Med.* **174**, 915–924.

Del Prete, G., De Carli, M., Almerigogna, F., Giudizi, M. G., Biagiotti, R., and Romagnani, S. (1993). Human IL-10 is produced by both type 1 helper (Th1) and type 2 helper (Th2) T cell clones and inhibits their antigen-specific proliferation and cytokine production. *J. Immunol.* **150**, 353–360.

Dunn, G. P., Bruce, A. T., Ikeda, H., Old, L. J., and Schreiber, R. D. (2002). Cancer immunoediting: From immunosurveillance to tumor escape. *Nat. Immunol.* **3**, 991–998.

Dvorak, H. F. (1986). Tumors: Wounds that do not heal. Similarities between tumor stroma generation and wound healing. *N. Engl. J. Med.* **315**, 1650–1659.

Enk, A. H., Jonuleit, H., Saloga, J., and Knop, J. (1997). Dendritic cells as mediators of tumor-induced tolerance in metastatic melanoma. *Int. J. Cancer* **73**, 309–316.

Fakhrai, H., Dorigo, O., Shawler, D. L., Lin, H., Mercola, D., Black, K. L., Royston, I., and Sobol, R. E. (1996). Eradication of established intracranial rat gliomas by transforming growth factor beta antisense gene therapy. *Proc. Natl. Acad. Sci. USA* **93**, 2909–2914.

Fearon, E. R., and Vogelstein, B. (1990). A genetic model for colorectal tumorigenesis. *Cell* **61**, 759–767.

Fedele, G., Stefanelli, P., Spensieri, F., Fazio, C., Mastrantonio, P., and Ausiello, C. M. (2005). *Bordetella pertussis*-infected human monocyte-derived dendritic cells undergo maturation and induce Th1 polarization and interleukin-23 expression. *Infect. Immun.* **73**, 1590–1597.

Finbloom, D. S., and Winestock, K. D. (1995). IL-10 induces the tyrosine phosphorylation of tyk2 and Jak1 and the differential assembly of STAT1 alpha and STAT3 complexes in human T cells and monocytes. *J. Immunol.* **155**, 1079–1090.

Fiorentino, D. F., Bond, M. W., and Mosmann, T. R. (1989). Two types of mouse T helper cell. IV. Th2 clones secrete a factor that inhibits cytokine production by Th1 clones. *J. Exp. Med.* **170**, 2081–2095.

Fiorentino, D. F., Zlotnik, A., Vieira, P., Mosmann, T. R., Howard, M., Moore, K. W., and O'Garra, A. (1991). IL-10 acts on the antigen-presenting cell to inhibit cytokine production by Th1 cells. *J. Immunol.* **146**, 3444–3451.

Friese, M. A., Wischhusen, J., Wick, W., Weiler, M., Eisele, G., Steinle, A., and Weller, M. (2004). RNA interference targeting transforming growth factor-beta enhances NKG2D-mediated antiglioma immune response, inhibits glioma cell migration and invasiveness, and abrogates tumorigenicity *in vivo. Cancer Res.* **64**, 7596–7603.

Gabrilovich, D., Ishida, T., Oyama, T., Ran, S., Kravtsov, V., Nadaf, S., and Carbone, D. P. (1998). Vascular endothelial growth factor inhibits the development of dendritic cells and dramatically affects the differentiation of multiple hematopoietic lineages *in vivo. Blood* **92**, 4150–4166.

Gabrilovich, D. I., Ishida, T., Nadaf, S., Ohm, J. E., and Carbone, D. P. (1999). Antibodies to vascular endothelial growth factor enhance the efficacy of cancer immunotherapy by improving endogenous dendritic cell function. *Clin. Cancer Res.* **5**, 2963–2970.

Gasser, S., Orsulic, S., Brown, E. J., and Raulet, D. H. (2005). The DNA damage pathway regulates innate immune system ligands of the NKG2D receptor. *Nature* **436**, 1186–1190.

Genestier, L., Kasibhatla, S., Brunner, T., and Green, D. R. (1999). Transforming growth factor beta1 inhibits Fas ligand expression and subsequent activation-induced cell death in T cells via downregulation of c-Myc. *J. Exp. Med.* **189**, 231–239.

Gerard, C. M., Bruyns, C., Delvaux, A., Baudson, N., Dargent, J. L., Goldman, M., and Velu, T. (1996). Loss of tumorigenicity and increased immunogenicity induced by interleukin-10 gene transfer in B16 melanoma cells. *Hum. Gene Ther.* **7**, 23–31.

Giovarelli, M., Musiani, P., Modesti, A., Dellabona, P., Casorati, G., Allione, A., Consalvo, M., Cavallo, F., di Pierro, F., De Giovanni, C., et al. (1995). Local release of IL-10 by transfected mouse mammary adenocarcinoma cells does not suppress but enhances antitumor reaction and elicits a strong cytotoxic lymphocyte and antibody-dependent immune memory. *J. Immunol.* **155**, 3112–3123.

Gorelik, L., and Flavell, R. A. (2000). Abrogation of TGFbeta signaling in T cells leads to spontaneous T cell differentiation and autoimmune disease. *Immunity* **12**, 171–181.

Gorelik, L., and Flavell, R. A. (2001). Immune-mediated eradication of tumors through the blockade of transforming growth factor-beta signaling in T cells. *Nat. Med.* **7**, 1118–1122.

Gorelik, L., and Flavell, R. A. (2002). Transforming growth factor-beta in T-cell biology. *Nat Rev. Immunol.* **2**, 46–53.

Green, E. A., Gorelik, L., McGregor, C. M., Tran, E. H., and Flavell, R. A. (2003). CD4$^+$CD25$^+$ T regulatory cells control anti-islet CD8$^+$ T cells through TGF-beta-TGF-beta receptor interactions in type 1 diabetes. *Proc. Natl. Acad. Sci. USA* **100**, 10878–10883.

Groux, H., Bigler, M., de Vries, J. E., and Roncarolo, M. G. (1996). Interleukin-10 induces a long-term antigen-specific anergic state in human CD4$^+$ T cells. *J. Exp. Med.* **184**, 19–29.

Groux, H., Bigler, M., de Vries, J. E., and Roncarolo, M. G. (1998). Inhibitory and stimulatory effects of IL-10 on human CD8$^+$ T cells. *J. Immunol.* **160**, 3188–3193.

Hagenbaugh, A., Sharma, S., Dubinett, S. M., Wei, S. H., Aranda, R., Cheroutre, H., Fowell, D. J., Binder, S., Tsao, B., Locksley, R. M., Moore, K. W., and Kronenberg, M. (1997). Altered immune responses in interleukin 10 transgenic mice. *J. Exp. Med.* **185**, 2101–2110.

Hamerman, J. A., Ogasawara, K., and Lanier, L. L. (2005). NK cells in innate immunity. *Curr. Opin. Immunol.* **17**, 29–35.

Happel, K. I., Dubin, P. J., Zheng, M., Ghilardi, N., Lockhart, C., Quinton, L. J., Odden, A. R., Shellito, J. E., Bagby, G. J., Nelson, S., and Kolls, J. K. (2005). Divergent roles of IL-23 and IL-12 in host defense against *Klebsiella pneumoniae*. *J. Exp. Med.* **202**, 761–769.

Harrington, L. E., Hatton, R. D., Mangan, P. R., Turner, H., Murphy, T. L., Murphy, K. M., and Weaver, C. T. (2005). Interleukin 17-producing CD4$^+$ effector T cells develop via a lineage distinct from the T helper type 1 and 2 lineages. *Nat. Immunol.* **6**, 1123–1132.

Hasegawa, Y., Takanashi, S., Kanehira, Y., Tsushima, T., Imai, T., and Okumura, K. (2001). Transforming growth factor-beta1 level correlates with angiogenesis, tumor progression, and prognosis in patients with nonsmall cell lung carcinoma. *Cancer* **91**, 964–971.

Hoeben, A., Landuyt, B., Highley, M. S., Wildiers, H., Van Oosterom, A. T., and De Bruijn, E. A. (2004). Vascular endothelial growth factor and angiogenesis. *Pharmacol. Rev.* **56**, 549–580.

Huang, B., Pan, P. Y., Li, Q., Sato, A. I., Levy, D. E., Bromberg, J., Divino, C. M., and Chen, S. H. (2006). Gr-1+CD115+ immature myeloid suppressor cells mediate the development of tumor-induced T regulatory cells and T-cell anergy in tumor-bearing host. *Cancer Res.* **66**, 1123–1131.

Hunter, C. A. (2005). New IL-12-family members: IL-23 and IL-27, cytokines with divergent functions. *Nat. Rev. Immunol.* **5**, 521–531.

Hunter, C. A., Bermudez, L., Beernink, H., Waegell, W., and Remington, J. S. (1995). Transforming growth factor-beta inhibits interleukin-12-induced production of interferon-gamma by natural killer cells: A role for transforming growth factor-beta in the regulation of T cell-independent resistance to *Toxoplasma gondii*. *Eur. J. Immunol.* **25**, 994–1000.

Huse, M., Muir, T. W., Xu, L., Chen, Y. G., Kuriyan, J., and Massague, J. (2001). The TGF beta receptor activation process: An inhibitor- to substrate-binding switch. *Mol. Cell* **8**, 671–682.

Jones, P. A., and Baylin, S. B. (2002). The fundamental role of epigenetic events in cancer. *Nat. Rev. Genet.* **3**, 415–428.

Jovasevic, V. M., Gorelik, L., Bluestone, J. A., and Mokyr, M. B. (2004). Importance of IL-10 for CTLA-4-mediated inhibition of tumor-eradicating immunity. *J. Immunol.* **172**, 1449–1454.

Khader, S. A., Pearl, J. E., Sakamoto, K., Gilmartin, L., Bell, G. K., Jelley-Gibbs, D. M., Ghilardi, N., deSauvage, F., and Cooper, A. M. (2005). IL-23 compensates for the absence of IL-12p70 and is essential for the IL-17 response during tuberculosis but is dispensable for protection and antigen-specific IFN-gamma responses if IL-12p70 is available. *J. Immunol.* **175**, 788–795.

Kim, B. G., Joo, H. G., Chung, I. S., Chung, H. Y., Woo, H. J., and Yun, Y. S. (2000). Inhibition of interleukin-10 (IL-10) production from MOPC 315 tumor cells by IL-10 antisense oligodeoxynucleotides enhances cell-mediated immune responses. *Cancer Immunol. Immunother.* **49**, 433–440.

Kim, B. G., Li, C., Qiao, W., Mamura, M., Kasperczak, B., Anver, M., Wolfraim, L., Hong, S., Mushinski, E., Potter, M., Kim, S. J., Fu, X. Y., Deng, C., and Letterio, J. J. (2006). Smad4 signalling in T cells is required for suppression of gastrointestinal cancer. *Nature* **441**, 1015–1019.

Kobayashi, S., Yoshida, K., Ward, J. M., Letterio, J. J., Longenecker, G., Yaswen, L., Mittleman, B., Mozes, E., Roberts, A. B., Karlsson, S., and Kulkarni, A. B. (1999). Beta 2-microglobulin-deficient background ameliorates lethal phenotype of the TGF-beta 1 null mouse. *J. Immunol.* **163**, 4013–4019.

Kolls, J. K., and Linden, A. (2004). Interleukin-17 family members and inflammation. *Immunity* **21**, 467–476.

Kuhn, R., Lohler, J., Rennick, D., Rajewsky, K., and Muller, W. (1993). Interleukin-10-deficient mice develop chronic enterocolitis. *Cell* **75**, 263–274.

Kulkarni, A. B., Huh, C. G., Becker, D., Geiser, A., Lyght, M., Flanders, K. C., Roberts, A. B., Sporn, M. B., Ward, J. M., and Karlsson, S. (1993). Transforming growth factor beta 1 null mutation in mice causes excessive inflammatory response and early death. *Proc. Natl. Acad. Sci. USA* **90**, 770–774.

Kundu, N., Beaty, T. L., Jackson, M. J., and Fulton, A. M. (1996). Antimetastatic and antitumor activities of interleukin 10 in a murine model of breast cancer. *J. Natl. Cancer Inst.* **88**, 536–541.

Kurte, M., Lopez, M., Aguirre, A., Escobar, A., Aguillon, J. C., Charo, J., Larsen, C. G., Kiessling, R., and Salazar-Onfray, F. (2004). A synthetic peptide homologous to functional domain of human IL-10 down-regulates expression of MHC class I and transporter associated with antigen processing 1/2 in human melanoma cells. *J. Immunol.* **173**, 1731–1737.

Kusmartsev, S., and Gabrilovich, D. I. (2002). Immature myeloid cells and cancer-associated immune suppression. *Cancer Immunol. Immunother.* **51**, 293–298.

Kusmartsev, S., Nefedova, Y., Yoder, D., and Gabrilovich, D. I. (2004). Antigen-specific inhibition of CD8$^+$ T cell response by immature myeloid cells in cancer is mediated by reactive oxygen species. *J. Immunol.* **172**, 989–999.

Langowski, J. L., Zhang, X., Wu, L., Mattson, J. D., Chen, T., Smith, K., Basham, B., McClanahan, T., Kastelein, R. A., and Oft, M. (2006) IL-23 promotes tumour incidence and growth. *Nature* **442**, 461–465.

Langrish, C. L., Chen, Y., Blumenschein, W. M., Mattson, J., Basham, B., Sedgwick, J. D., McClanahan, T., Kastelein, R. A., and Cua, D. J. (2005). IL-23 drives a pathogenic T cell population that induces autoimmune inflammation. *J. Exp. Med.* **201**, 233–240.

Laouar, Y., Sutterwala, F. S., Gorelik, L., and Flavell, R. A. (2005). Transforming growth factor-beta controls T helper type 1 cell development through regulation of natural killer cell interferon-gamma. *Nat. Immunol.* **6**, 600–607.

Lee, J. C., Lee, K. M., Kim, D. W., and Heo, D. S. (2004). Elevated TGF-beta1 secretion and down-modulation of NKG2D underlies impaired NK cytotoxicity in cancer patients. *J. Immunol.* **172**, 7335–7340.

Letterio, J. J., Geiser, A. G., Kulkarni, A. B., Dang, H., Kong, L., Nakabayashi, T., Mackall, C. L., Gress, R. E., and Roberts, A. B. (1996). Autoimmunity associated with TGF-beta1-deficiency in mice is dependent on MHC class II antigen expression. *J. Clin. Invest.* **98**, 2109–2119.

Li, M., Carpio, D. F., Zheng, Y., Bruzzo, P., Singh, V., Ouaaz, F., Medzhitov, R. M., and Beg, A. A. (2001). An essential role of the NF-kappa B/Toll-like receptor pathway in induction of inflammatory and tissue-repair gene expression by necrotic cells. *J. Immunol.* **166**, 7128–7135.

Li, M. O., and Flavell, R. A. (2006a). TGF-beta, T cell tolerance and immunotherapy of autoimmune diseases and cancer. *Expert Rev. Clin. Immunol.* **2**, 257–265.

Li, M. O., and Flavell, R. A. (2006b). Transforming growth factor-beta controls of development, homeostasis, and tolerance of T cells by regulatory T cell-dependent and -independent mechanisms. *Immunity* **25**, 455–471.

Li, M. O., Wan, Y. Y., Sanjabi, S., Robertson, A. K., and Flavell, R. A. (2006). Transforming growth factor-beta regulation of immune responses. *Annu. Rev. Immunol.* **24**, 99–146.

Lieberman, L. A., Cardillo, F., Owyang, A. M., Rennick, D. M., Cua, D. J., Kastelein, R. A., and Hunter, C. A. (2004). IL-23 provides a limited mechanism of resistance to acute toxoplasmosis in the absence of IL-12. *J. Immunol.* **173**, 1887–1893.

Liu, Y., Wei, S. H., Ho, A. S., de Waal Malefyt, R., and Moore, K. W. (1994). Expression cloning and characterization of a human IL-10 receptor. *J. Immunol.* **152**, 1821–1829.

Loercher, A. E., Nash, M. A., Kavanagh, J. J., Platsoucas, C. D., and Freedman, R. S. (1999). Identification of an IL-10-producing HLA-DR-negative monocyte subset in the malignant ascites of patients with ovarian carcinoma that inhibits cytokine protein expression and proliferation of autologous T cells. *J. Immunol.* **163**, 6251–6260.

Lucas, P. J., Kim, S. J., Melby, S. J., and Gress, R. E. (2000). Disruption of T cell homeostasis in mice expressing a T cell-specific dominant negative transforming growth factor beta II receptor. *J. Exp. Med.* **191**, 1187–1196.

Mangan, P. R., Harrington, L. E., O'Quinn, D. B., Helms, W. S., Bullard, D. C., Elson, C. O., Hatton, R. D., Wahl, S. M., Schoeb, T. R., and Weaver, C. T. (2006). Transforming growth factor-beta induces development of the T(H)17 lineage. *Nature* **441**, 231–234.

Marie, J. C., Letterio, J. J., Gavin, M., and Rudensky, A. Y. (2005). TGF-β1 maintains suppressor function and Foxp3 expression in CD4$^+$CD25$^+$ regulatory T cells. *J. Exp. Med.* **201**, 1061–1067.

Martin-Fontecha, A., Thomsen, L. L., Brett, S., Gerard, C., Lipp, M., Lanzavecchia, A., and Sallusto, F. (2004). Induced recruitment of NK cells to lymph nodes provides IFN-gamma for T(H)1 priming. *Nat. Immunol.* **5**, 1260–1265.

Medzhitov, R., and Janeway, C. A., Jr. (1997). Innate immunity: The virtues of a nonclonal system of recognition. *Cell* **91**, 295–298.

Mocellin, S., Marincola, F. M., and Young, H. A. (2005). Interleukin-10 and the immune response against cancer: A counterpoint. *J. Leukoc. Biol.* **78**, 1043–1051.

Moore, K. W., de Waal Malefyt, R., Coffman, R. L., and O'Garra, A. (2001). Interleukin-10 and the interleukin-10 receptor. *Annu. Rev. Immunol.* **19**, 683–765.

Murphy, C. A., Langrish, C. L., Chen, Y., Blumenschein, W., McClanahan, T., Kastelein, R. A., Sedgwick, J. D., and Cua, D. J. (2003). Divergent pro- and anti-inflammatory roles for IL-23 and IL-12 in joint autoimmune inflammation. *J. Exp. Med.* **198**, 1951–1957.

Numasaki, M., Fukushi, J., Ono, M., Narula, S. K., Zavodny, P. J., Kudo, T., Robbins, P. D., Tahara, H., and Lotze, M. T. (2003). Interleukin-17 promotes angiogenesis and tumor growth. *Blood* **101**, 2620–2627.

Oppmann, B., Lesley, R., Blom, B., Timans, J. C., Xu, Y., Hunte, B., Vega, F., Yu, N., Wang, J., Singh, K., Zonin, F., Vaisberg, E., Churakova, T., Liu, M., Gorman, D., Wagner, J., Zurawski, S., Liu, Y., Abrams, J. S., Moore, K. W., Rennick, D., de Waal-Malefyt, R., Hannum, C., Bazan, J. F., and Kastelein, R. A. (2000). Novel p19 protein engages IL-12p40 to form a cytokine, IL-23, with biological activities similar as well as distinct from IL-12. *Immunity* **13**, 715–725.

Ouaaz, F., Arron, J., Zheng, Y., Choi, Y., and Beg, A. A. (2002). Dendritic cell development and survival require distinct NF-kappaB subunits. *Immunity* **16**, 257–270.

Oyama, T., Ran, S., Ishida, T., Nadaf, S., Kerr, L., Carbone, D. P., and Gabrilovich, D. I. (1998). Vascular endothelial growth factor affects dendritic cell maturation through the inhibition of nuclear factor-kappa B activation in hemopoietic progenitor cells. *J. Immunol.* **160**, 1224–1232.

Pardoll, D. (2003). Does the immune system see tumors as foreign or self? *Annu. Rev. Immunol.* **21**, 807–839.

Park, H., Li, Z., Yang, X. O., Chang, S. H., Nurieva, R., Wang, Y. H., Wang, Y., Hood, L., Zhu, Z., Tian, Q., and Dong, C. (2005). A distinct lineage of CD4 T cells regulates tissue inflammation by producing interleukin 17. *Nat. Immunol.* **6**, 1133–1141.

Platten, M., Wild-Bode, C., Wick, W., Leitlein, J., Dichgans, J., and Weller, M. (2001). N-[3,4-dimethoxycinnamoyl]-anthranilic acid (tranilast) inhibits transforming growth factor-beta release and reduces migration and invasiveness of human malignant glioma cells. *Int. J. Cancer* **93**, 53–61.

Ranges, G. E., Figari, I. S., Espevik, T., and Palladino, M. A., Jr. (1987). Inhibition of cytotoxic T cell development by transforming growth factor beta and reversal by recombinant tumor necrosis factor alpha. *J. Exp. Med.* **166**, 991–998.

Raulet, D. H., and Vance, R. E. (2006). Self-tolerance of natural killer cells. *Nat. Rev. Immunol.* **6**, 520–531.

Ristimaki, A., Honkanen, N., Jankala, H., Sipponen, P., and Harkonen, M. (1997). Expression of cyclooxygenase-2 in human gastric carcinoma. *Cancer Res.* **57**, 1276–1280.

Saito, H., Tsujitani, S., Oka, S., Kondo, A., Ikeguchi, M., Maeta, M., and Kaibara, N. (1999). The expression of transforming growth factor-beta1 is significantly correlated with the expression of vascular endothelial growth factor and poor prognosis of patients with advanced gastric carcinoma. *Cancer* **86**, 1455–1462.

Sakaguchi, S. (2004). Naturally arising CD4[+] regulatory T cells for immunologic self-tolerance and negative control of immune responses. *Annu. Rev. Immunol.* **22**, 531–562.

Sano, H., Kawahito, Y., Wilder, R. L., Hashiramoto, A., Mukai, S., Asai, K., Kimura, S., Kato, H., Kondo, M., and Hla, T. (1995). Expression of cyclooxygenase-1 and -2 in human colorectal cancer. *Cancer Res.* **55**, 3785–3789.

Santin, A. D., Hermonat, P. L., Ravaggi, A., Bellone, S., Pecorelli, S., Roman, J. J., Parham, G. P., and Cannon, M. J. (2000). Interleukin-10 increases Th1 cytokine production and cytotoxic potential in human papillomavirus-specific CD8([+]) cytotoxic T lymphocytes. *J. Virol.* **74**, 4729–4737.

Schnurr, M., Toy, T., Shin, A., Wagner, M., Cebon, J., and Maraskovsky, E. (2005). Extracellular nucleotide signaling by P2 receptors inhibits IL-12 and enhances IL-23 expression in human dendritic cells: A novel role for the cAMP pathway. *Blood* **105**, 1582–1589.

Serafini, P., Borrello, I., and Bronte, V. (2006). Myeloid suppressor cells in cancer: Recruitment, phenotype, properties, and mechanisms of immune suppression. *Semin. Cancer Biol.* **16**, 53–65.

Shankaran, V., Ikeda, H., Bruce, A. T., White, J. M., Swanson, P. E., Old, L. J., and Schreiber, R. D. (2001). IFNgamma and lymphocytes prevent primary tumour development and shape tumour immunogenicity. *Nature* **410**, 1107–1111.

Shariat, S. F., Shalev, M., Menesses-Diaz, A., Kim, I. Y., Kattan, M. W., Wheeler, T. M., and Slawin, K. M. (2001). Preoperative plasma levels of transforming growth factor beta(1) (TGF-beta(1)) strongly predict progression in patients undergoing radical prostatectomy. *J. Clin. Oncol.* **19**, 2856–2864.

Sheibanie, A. F., Tadmori, I., Jing, H., Vassiliou, E., and Ganea, D. (2004). Prostaglandin E2 induces IL-23 production in bone marrow-derived dendritic cells. *Faseb J.* **18**, 1318–1320.

Shull, M. M., Ormsby, I., Kier, A. B., Pawlowski, S., Diebold, R. J., Yin, M., Allen, R., Sidman, C., Proetzel, G., Calvin, D., *et al.* (1992). Targeted disruption of the mouse transforming growth factor-beta 1 gene results in multifocal inflammatory disease. *Nature* **359**, 693–699.

Smyth, M. J., Dunn, G. P., and Schreiber, R. D. (2006). Cancer immunosurveillance and immunoediting: The roles of immunity in suppressing tumor development and shaping tumor immunogenicity. *Adv. Immunol.* **90**, 1–50.

Smyth, M. J., Strobl, S. L., Young, H. A., Ortaldo, J. R., and Ochoa, A. C. (1991). Regulation of lymphokine-activated killer activity and pore-forming protein gene expression in human peripheral blood CD8+ T lymphocytes. Inhibition by transforming growth factor-beta. *J. Immunol.* **146**, 3289–3297.

Stearns, M. E., Garcia, F. U., Fudge, K., Rhim, J., and Wang, M. (1999). Role of interleukin 10 and transforming growth factor beta1 in the angiogenesis and metastasis of human prostate primary tumor lines from orthotopic implants in severe combined immunodeficiency mice. *Clin. Cancer Res.* **5**, 711–720.

Steinbrink, K., Jonuleit, H., Muller, G., Schuler, G., Knop, J., and Enk, A. H. (1999). Interleukin-10-treated human dendritic cells induce a melanoma-antigen-specific anergy in CD8(+) T cells resulting in a failure to lyse tumor cells. *Blood* **93**, 1634–1642.

Straathof, K. C., Bollard, C. M., Rooney, C. M., and Heslop, H. E. (2003). Immunotherapy for Epstein-Barr virus-associated cancers in children. *Oncologist* **8**, 83–98.

Takeda, K., Clausen, B. E., Kaisho, T., Tsujimura, T., Terada, N., Forster, I., and Akira, S. (1999). Enhanced Th1 activity and development of chronic enterocolitis in mice devoid of Stat3 in macrophages and neutrophils. *Immunity* **10**, 39–49.

Takeda, K., Smyth, M. J., Cretney, E., Hayakawa, Y., Kayagaki, N., Yagita, H., and Okumura, K. (2002). Critical role for tumor necrosis factor-related apoptosis-inducing ligand in immune surveillance against tumor development. *J. Exp. Med.* **195**, 161–169.

ten Dijke, P., and Hill, C. S. (2004). New insights into TGF-beta-Smad signalling. *Trends Biochem. Sci.* **29**, 265–273.

Thomas, D. A., and Massague, J. (2005). TGF-beta directly targets cytotoxic T cell functions during tumor evasion of immune surveillance. *Cancer Cell* **8**, 369–380.

Toi, M., Taniguchi, T., Yamamoto, Y., Kurisaki, T., Suzuki, H., and Tominaga, T. (1996). Clinical significance of the determination of angiogenic factors. *Eur. J. Cancer* **32A**, 2513–2519.

Torre-Amione, G., Beauchamp, R. D., Koeppen, H., Park, B. H., Schreiber, H., Moses, H. L., and Rowley, D. A. (1990). A highly immunogenic tumor transfected with a murine transforming growth factor type beta 1 cDNA escapes immune surveillance. *Proc. Natl. Acad. Sci. USA* **87**, 1486–1490.

Uhl, M., Aulwurm, S., Wischhusen, J., Weiler, M., Ma, J. Y., Almirez, R., Mangadu, R., Liu, Y. W., Platten, M., Herrlinger, U., Murphy, A., Wong, D. H., Wick, W., Higgins, L. S., and Weller, M. (2004). SD-208, a novel transforming growth factor beta receptor I kinase inhibitor, inhibits growth and invasiveness and enhances immunogenicity of murine and human glioma cells *in vitro* and *in vivo*. *Cancer Res.* **64**, 7954–7961.

Veldhoen, M., Hocking, R. J., Atkins, C. J., Locksley, R. M., and Stockinger, B. (2006). TGFbeta in the context of an inflammatory cytokine milieu supports *de novo* differentiation of IL-17-producing T cells. *Immunity* **24**, 179–189.

Vicari, A. P., Chiodoni, C., Vaure, C., Ait-Yahia, S., Dercamp, C., Matsos, F., Reynard, O., Taverne, C., Merle, P., Colombo, M. P., O'Garra, A., Trinchieri, G., and Caux, C. (2002). Reversal of tumor-induced dendritic cell paralysis by CpG immunostimulatory oligonucleotide and anti-interleukin 10 receptor antibody. *J. Exp. Med.* **196**, 541–549.

Wakefield, L. M., and Roberts, A. B. (2002). TGF-beta signaling: Positive and negative effects on tumorigenesis. *Curr. Opin. Genet. Dev.* **12**, 22–29.

Willems, F., Marchant, A., Delville, J. P., Gerard, C., Delvaux, A., Velu, T., de Boer, M., and Goldman, M. (1994). Interleukin-10 inhibits B7 and intercellular adhesion molecule-1 expression on human monocytes. *Eur. J. Immunol.* **24**, 1007–1009.

Willimsky, G., and Blankenstein, T. (2005). Sporadic immunogenic tumours avoid destruction by inducing T-cell tolerance. *Nature* **437**, 141–146.

Wojtowicz-Praga, S. (2003). Reversal of tumor-induced immunosuppression by TGF-beta inhibitors. *Invest. New Drugs* **21**, 21–32.

Wolff, H., Saukkonen, K., Anttila, S., Karjalainen, A., Vainio, H., and Ristimaki, A. (1998). Expression of cyclooxygenase-2 in human lung carcinoma. *Cancer Res.* **58**, 4997–5001.

Yamaguchi, T., and Sakaguchi, S. (2006). Regulatory T cells in immune surveillance and treatment of cancer. *Semin. Cancer Biol.* **16**, 115–123.

Yingling, J. M., Blanchard, K. L., and Sawyer, J. S. (2004). Development of TGF-beta signalling inhibitors for cancer therapy. *Nat. Rev. Drug Discov.* **3**, 1011–1022.

Young, M. R., Newby, M., and Wepsic, H. T. (1987). Hematopoiesis and suppressor bone marrow cells in mice bearing large metastatic Lewis lung carcinoma tumors. *Cancer Res.* **47**, 100–105.

Young, M. R., Wright, M. A., and Pandit, R. (1997). Myeloid differentiation treatment to diminish the presence of immune-suppressive CD34+ cells within human head and neck squamous cell carcinomas. *J. Immunol.* **159**, 990–996.

Zhang, Q., Yang, X., Pins, M., Javonovic, B., Kuzel, T., Kim, S. J., Parijs, L. V., Greenberg, N. M., Liu, V., Guo, Y., and Lee, C. (2005). Adoptive transfer of tumor-reactive transforming growth factor-beta-insensitive CD8+ T cells: Eradication of autologous mouse prostate cancer. *Cancer Res.* **65**, 1761–1769.

Zheng, L. M., Ojcius, D. M., Garaud, F., Roth, C., Maxwell, E., Li, Z., Rong, H., Chen, J., Wang, X. Y., Catino, J. J., and King, I. (1996). Interleukin-10 inhibits tumor metastasis through an NK cell-dependent mechanism. *J. Exp. Med.* **184**, 579–584.

CHAPTER

5

Immunological Sculpting: Natural Killer Cell Receptors and Ligands

DAVID A. SALLMAN AND JULIE Y. DJEU

  I. Introduction
 II. Activating Human NK Receptors
     A. NKG2D
     B. NKG2C
     C. Natural Cytotoxicity Receptor Family
     D. Activating Killer Immunoglobulin-Like Receptors
     E. DNAM-1 (CD226)
     F. 2B4 Receptor (CD244)
III. Inhibitory NK Receptors
     A. Inhibitory KIRs
     B. NKG2A
     C. Other Inhibitory Receptors
 IV. The Ly49 Receptor Family
  V. Immunotherapy Approaches
     A. Retention of NKG2D Ligands on Tumor Cells
     B. Antibody-Dependent Cell Cytotoxicity
     C. Cytokines for Enhancement of NK Receptors
     D. Allograft Transplantation
 VI. Conclusion
     References
     Further Reading

In the twentieth and twenty-first centuries, there have been many advances in the oncology field's understanding of natural killer (NK) cell biology. Most of the research thus far has focused on the family of receptors expressed on NK cells as well as the ligands that these receptors recognize. In addition, the signaling pathways leading to activation or inhibition are now being elucidated. It is now well-known that NK cells are primary effector cells that attack transformed cells. Therefore, studying the processes of NK activation and inhibition will be important to creating innovative therapeutic treatments that utilize a patient's own immune system to counteract cancer. NK cells are able to bind a very diverse family of receptors. NK cells have an advantage over other immune cells in attacking transformed cells because NK receptors, specifically NKG2D, can recognize ligands that become expressed upon transformation. NK cells are prevented from attacking self through the engagement of human leukocyte antigens (HLA) by killer immunoglobulin-like receptors (KIRs). The balance of the expression of activating and inhibiting receptors thus controls NK lytic function. Tumors are able to evade immune detection through shedding of NK receptor ligands. Ligand shedding has been found to be dependent on metalloproteinases. This chapter explains the structure and function of NK receptors and how cooperation between multiple receptors determines NK activation or inhibition. Studies reported in this chapter emphasize that proper NK receptor ligand expression in tumors is essential for tumor rejection. The information presented will hopefully lead to further therapeutic innovations to treat cancer.

I. INTRODUCTION

Natural killer (NK) cells are bone-marrow-derived cells that constitute one of the major components of the innate immune system. They were first discovered by their intrinsic ability to lyse transformed or virally infected cells without prior deliberate stimulation. NK cells function similarly to cytotoxic T lymphocytes (CTL) in that they use the same effector molecules, perforin and granzymes, to induce lysis and apoptosis in target cells. Like CTLs, NK cells also react to target cells by secreting cytokines and chemokines to expand the inflammatory response to the offending target. However, NK cells provide a much more immediate immune response upon recognition of a target cell. This advantage occurs because NK cells possess preformed stores of lytic granules as well as cytokine transcripts. Another intriguing difference is their range of targets. While it was clearly defined how CTL are preprogrammed to find a specific target cell, it was not understood how NK cells recognize different unrelated targets. Because they do not possess clonally distributed antigen receptors, such as the T cell receptor (TCR), the molecular mechanisms governing NK cells remained a mystery for several decades. The discoveries of inhibitory receptors and then activation receptors have clarified the orchestration of receptors that fine-tune NK cell function.

Unlike T cells, which survey a specific target antigen in association with major histocompatibility complex (MHC) class I molecules, NK cells appeared to preferentially kill tumor cells lacking MHC class I. The "missing self" hypothesis was based on findings that cells expressing normal levels of MHC class I are more resistant to NK lysis than those lacking MHC class I and that NK cells could kill diseased "self" cells as long as MHC expression was downregulated or altered. Further research has refined the role of MHC class I molecules in NK cells, demonstrating that NK cells are prevented from activation by normal tissues expressing self MHC class I molecules and that they can recognize diseased cells with reduced MHC class I molecules due to transformation or viral infection. This selective property is conferred on NK cells by activating and inhibiting receptors that function cooperatively to determine the fate of the target cell.

Both activation and inhibitory pathways in NK cells depend on phosphorylation. Inhibitory NK receptors, which primarily recognize MHC class I molecules as ligands, contain within their cytoplasmic tails immunoreceptor tyrosine inhibition motifs (ITIMs) that become phosphorylated upon receptor/ligand interaction to recruit a phosphatase called Src homology 2 domain-containing phosphatase, also referred to as Src-homology-2 (SHP-1). Recruitment of SHP-1 results in shutdown of the lytic signal cascade. On the other hand, activating NK receptors bind a wide array of ligands other than classical MHC class I molecules, utilizing various adapter proteins containing either PI3K-binding sites or immunoreceptor tyrosine activatory motifs (ITAMs). When phosphorylated, these sites bind Syk/Zap70 tyrosine kinases that trigger a signal cascade, which ultimately leads to granule exocytosis and death of a target cell.

Clearly, NK function depends on its receptors and the ligands that are expressed on potential target cells. The purpose of this chapter is to describe the receptor families in order to shed light on the mechanisms of NK activation and inhibition. Many infections and tumors are able to progress by evading immune detection. This chapter will discuss evasion techniques that tumor cells have developed to escape NK cell detection and will provide information on new approaches that can exploit NK biology to treat malignancies.

II. ACTIVATING HUMAN NK RECEPTORS

In spite of being prevented from killing "normal self" cells that display MHC class I molecules, NK cells can readily respond to altered self induced by transformation or viral and bacterial infection (Blery et al., 2000; Yokoyama and Plougastel, 2003; Moretta et al., 2006). In the absence of MHC class I molecules on target cells, the ability of NK cells to recognize and lyse these targets indicate that there are activating signals to carry out the lytic function. The effectors of this function are distributed among a number of unique receptors displayed on NK cells (Figure 5.1). These receptors on NK cells function to survey the host to locate and destroy any pathologically affected or stressed cell expressing the appropriate ligand. This section describes the activating NK receptor/ligand interactions and the signaling pathways involved in lytic function. All activating NK receptors lack endogenous kinase function and must associate with adapter proteins to transduce the lytic signal cascade (Figure 5.2). Specifically, they must utilize adapters DAP10 and DAP12 as well as two other adapter proteins, CD3-ζ, which usually associates with the TCR, and Fcϵ-RIγ, which normally associates with the Fcγ-RIII (CD16) receptor.

A. NKG2D

NKG2D is a homodimeric C-type lectin receptor that is expressed on all NK cells. This receptor is expressed on certain T cells, particularly NKT cells and activated CD8$^+$ T cells (Bauer et al., 1999, Cosman et al., 2001). NKG2D has the unusual property of binding to multiple ligands derived from diverse genes in the target cell that are only at best distantly related. Two chief ligands are the MHC class I chain-related A and B (MICA and MICB) proteins, which are transmembrane proteins containing α1, α2, and α3 domains with some homology to MHC class I molecules but differing from these molecules by lack of binding to β-microglobulin expressed on the cell surface. NKG2D also recognizes the unique long-16 binding proteins (ULBPs), also called retinoic acid early transcript 1 (RAET-1), a family of glycosylphosphatidylinositol (GPI)-anchored proteins containing α1 and α2 domains with relatively little homology to each other (Cosman et al., 2001).

Two isoforms of NKG2D, termed NKG2D-long (NKG2DL) and NKG2D-short (NKG2DS), have been reported. Like other NK receptors, these isoforms each require adapter proteins to transduce the activating signal in NK cells (Diefenbach et al., 2002). NKG2DL has been found to associate only with the adapter protein DAP10 and not DAP12, in contrast to NKG2DS, which associates with both adapter proteins. However, as of this writing, humans only have been found to express NKG2DL

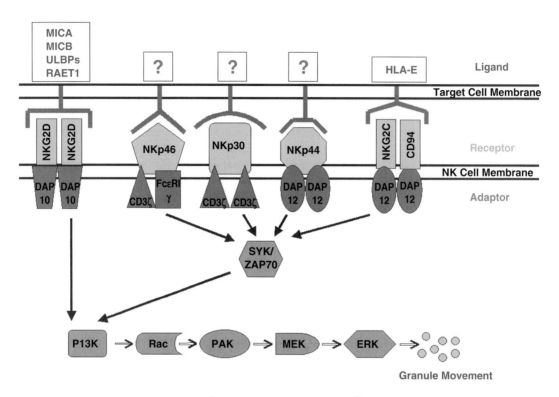

family are the key effectors in executing NK-mediated cytotoxicity. Upon receptor engagement of its ligand, adaptor proteins that associate with the receptor through charged interactions activate a signal cascade either via PI3K or Syk.Zap70, which ultimately results in the phosphorylation of ERK. ERK activation leads to granule exocytosis and lysis of the target cell.

to induce cytotoxicity through the DAP10 protein.

NKG2D is a highly conserved and important receptor on NK cells for lysis of altered self. Its ligands are present in embryonal tissues and are lost in mature organs. On the other hand, these ligands are upregulated in cellular distress situations, as occurs during viral infection and transformation (Cosman et al., 2001; Moretta et al., 2006). Heat shock proteins produced during transformation can promote MICA and MICB gene expression through the heat shock elements present on the promoter regions of these genes (Bauer et al., 1999). Indeed, a variety of tumors, including leukemia, hepatoma, melanoma, and prostate cancer, have been demonstrated to express MICA and MICB, thereby providing ligands for NK recognition (Groh et al., 2002; Salih et al., 2003; Holdenrieder et al., 2006). Other equally important ligands of NKG2D were discovered through studies of human cytomegalovirus (HCMV). HCMV expresses the protein UL16 that functions in immune evasion by blocking the expression of ULBPs, such as RAET1 that bind to UL16 (Cosman et al., 2001). Notably, these proteins were found to be expressed on tumor cells (Radosavljevic et al., 2002). All NKG2D ligands, although diverse in nature, can associate with NKG2D because they express the same amino acid residues in key binding regions as well as have a similar alpha domain structure (McFarland et al., 2003). Thus, NK cell promiscuity in killing unre-

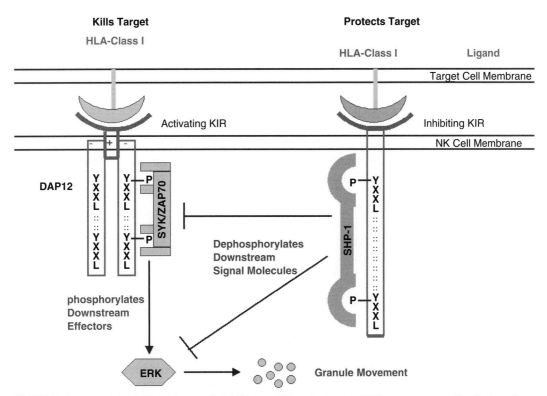

FIGURE 5.2 **Model of activating and inhibitory KIRs.** Activating KIRs are structurally distinct from inhibiting receptors in that activating receptors have a short cytoplasmic tail. As a result of the shorter cytoplasmic tail, activating KIRs do not contain an ITIM. The ITAM-containing adaptor protein DAP12 associates with the activating KIR. Upon ligand engagement, the tyrosines in the ITAM ($YxxL \times_{6-8} YxxL$) of DAP12 are phosphorylated and become a binding site for the two SH2 domains of SYK/ZAP70 kinase. Activation of SYK/ZAP70 ultimately leads to ERK activation and granule movement. In the case of inhibitory KIRs, binding to an HLA-class I ligand leads to phosphorylation of the tyrosines in the ITIM ($YxxL \times_{28-32} YxxL$) in the cytoplasmic tail. These phosphotyrosines specifically bind the SH2 domains of the tyrosine phosphatase SHP-1. Recruitment of SHP-1 leads to its activation and results in inhibition of NK-mediated cytotoxicity by dephosphorylating essential signal molecules, such as Syk/ZAP70 and ERK. For an inhibiting KIR to prevent NK stimulation, it must be in close proximity to the activating KIR. This occurs during a process known as *KIR clustering*, and the initial kinases that are triggered by an activating receptor are responsible for phosphorylation of the nearby ITIM in the inhibitory receptor. Thus, inhibitory receptors work to shut down the function of an activating receptor.

lated target cells can be explained by various target cells expressing different NKG2D ligands that trigger NK cells via a common receptor.

In addition to viral infection and tumor transformation, other pathological conditions can significantly increase the expression of NKG2D ligands. DNA damage in cells has been shown to have a key role in this expression. DNA damaging agents and DNA replication inhibitors induced the expression of NKG2D ligands in cells, while inhibition of ATM or ATR, which are involved in detecting double-stranded DNA breaks or stalled DNA replication, respectively, blocked induction of NKG2D ligands (Gasser *et al.*, 2005). This DNA damage response may also operate in precancerous lesions, resulting in cells that display NKG2D ligands that tag themselves for NK recognition. Genotoxic stress, microbial infection, and cell transformation all constitute types of cellular alterations in which induction of NKG2D ligands acts as an indicator of cell stress. Therefore, elevating the expression of NKG2D ligands is a common natural means to alert the immune system to the presence of disease. It is also important to note that NKG2D can trigger NK cytotoxicity despite MHC class I expression on the target cell, suggesting that this receptor, when activated, can override the protective signal of self MHC class I as long as the danger signal provided by the NKG2D ligand is available (Diefenbach et al., 2001).

In the mouse system, NKG2D is able to bind three subfamilies of RAET1 genes, Rae1, histocompatibility 60 (H60), and mouse UL-16 binding protein 1 (Mult1) (Dienfenbach et al., 2001). Although these genes are structurally related and localized on chromosome 10, they share little amino acid sequence homology. Rae1 is a lipid-linked membrane protein, and H60 is a transmembrane protein. No MIC analog has been found in mice. As in humans, tumor formation in mice is associated with upregulation of these NKG2D ligands. Interestingly, IFN-γ has been well-known to induce NK resistance in tumor cells. Indeed, it was found that IFN-γ downregulates the expression of the NKG2D ligand H60 on methylcholanthrene (MCA)-induced sarcomas concurrently with enhancing the expression of MHC class I molecules (Bui et al., 2006). These cooperative effects resulted in a dramatic decrease in the sensitivity of MCA-induced sarcomas to NK-mediated cytotoxicity. On the other hand, IFN-γ treatment has the opposite effect of activating CTLs, an event that requires the specific interaction of MHC class I with the target antigen. Therefore, it has been proposed that IFN-γ may serve to inhibit innate immunity while activating adaptive immunity, thereby facilitating progression toward a long-term T cell memory response (Bui et al., 2006). In contrast to this mouse model, IFN-γ has been found to have a different effect on human acute myeloid lymphoma (AML). AML cells are typically very resistant to lysis by NK cells. In vivo, treatment of these cells with IFN-γ and growth factors that promote myeloid differentiation resulted in a robust upregulation of ULBP1 and other NK receptor ligands, whereas treatment with the growth factors alone had only a minimal effect (Nowbakht et al., 2005). The combination treatment resulted in gain of sensitivity of AML cells to lysis by NK cells. However, increased sensitivity to NK-mediated cell lysis was accomplished only by blocking MHC class I expression. Therefore the response to IFN-γ may be complex depending on the type of tumor and the factors that are present in the environment.

Like all other NK activation receptors, NKG2D has a positively charged amino acid in its transmembrane region that is vitally important for its association with the adapter protein, DAP 10, which contains a negative charge in its transmembrane domain. NKG2D is the only NK receptor that associates with DAP10 (Wu et al., 1999). Interestingly, DAP10 does not contain an ITAM like the other adapter proteins but rather an *YxxM* motif (*x* donates any amino acid) that serves as a consensus PI3K binding site. Upon ligand interaction with NKG2D, the tyrosine in the *YxxM* motif of DAP10 becomes phosphorylated to bind the SH2 domain of PI3K. Binding to DAP triggers PI3K activation, leading to downstream signal events that result in lytic granule movement (Jiang et al., 2000). A sequential phosphorylation of Rac, p21-activated kinase (PAK), extracellular signal-regulated kinase (ERK), and mitogen-activated protein kinase/ERK (MEK) occurs after PI3K activation, and it is ERK that appears to mobilize the lytic granules, as shown in Figure 5.2 (Djeu et al., 2002).

Besides directly inducing target cell lysis, NKG2D engagement also results in the production of multiple cytokines and chemokines. Specifically, binding of NKG2D to MICA/B or ULBP binding results in the production of IFN-γ, TNF-α, lymphotoxin, and granulocyte-macrophage colony-stimulating factor (GM-CSF) (Bauer et al., 1999; Cosman et al., 2001). In addition, the chemokines CCL4 and CCL1 are secreted upon NKG2D–ULBP binding. These cytokines and chemokines augment the immune processes to eliminate microbial pathogens and transformed cells.

B. NKG2C

The C-type lectin receptor family, NKG2, consists of four other members in addition to NKG2D: NKG2A, NKG2C, NKG2E, and NKG2F. With the exception of NKG2D, all NKG2 receptors form heterodimers with CD94 via disulphide bridges and bind the nonclassical MHC class I molecule human leukocyte antigen E (HLA-E) (Yokoyama and Plougastel, 2003; Moretta et al., 2006). HLA-E is a poorly polymorphic MHC class Ib molecule whose surface expression is dependent on its association with peptides derived from the leader sequence of various MHC class I alleles. Thus, HLA-E expression is closely correlated with the overall expression of MHC class I on cells and can act as a sensor for MHC class I levels in cells. Among these HLA-E-specific receptors, NKG2C is the only activating receptor. The NKG2C/CD94 heterodimer associates with the adapter DAP12, which exists as a homodimer (Lanier et al., 1998a). A Lys residue in the transmembrane region of NKG2C is the essential amino acid in its association with DAP12. Upon ligand interaction with NKG2C/CD94, DAP12 becomes phosphorylated in its two tyrosine residues in the $YxxL \times_{6-8} YxxL$ motif and is embedded in the ITAM. These phosphorylated sites recruit Syk/Zap70 kinases through their dual SH2 domains, triggering a signal cascade similar to that with NKG2D (see Figure 5.2). Syk/Zap70 activation results in PI3K activation, and, once PI3K is activated, the same signal process via Rac, PAK, and MEK (described for NKG2D) occurs, ultimately leading to ERK activation, which drives lytic granules toward the target cell (Jiang et al., 2002).

Although the role of NKG2C in NK activation has not been clearly defined, it does seem to play an important role in response to certain viral infections. In support of this, a study has shown that NKG2C expression increases in response to HCMV infection (Guma et al., 2006). NKG2E has a Lys residue in its transmembrane domain that has not been shown to associate with any adapter protein. Although the structure of NKG2E appears to give it an activating role, further research needs to be performed to determine its precise function.

C. Natural Cytotoxicity Receptor Family

Besides the NKG2 family of receptors, three other activation receptors were discovered through biological assays using monoclonal antibodies (mAbs) that, upon binding to the NK cell, induce a strong cytotoxic reaction. NKp46, NKp30, and NKp44, which are all members of the natural cytotoxicity receptor (NCR) family, are mostly restricted to NK cells, and are able to independently activate NK cells (Moretta et al., 2006). With the exception of NKp44, NCRs are expressed on all subsets of NK cells, independent of their activation state. The density of NCRs on the cell surface of NK cells is directly correlated with the cytolytic ability of NK cells, and they require adapter proteins to activate the lytic pathway. NCR-mediated killing is effectively blocked by the expression of various MHC class I molecules. However, as MHC expression is decreased through either viral infection or tumor transformation, NCRs can subsequently activate NK-mediated cytolysis. Although all NCRs have a similar activation function, they differ in structure and the molecular mechanism for inducing activation. The ligands for NCRs are unknown, but their existence on tumor cells can be inferred by the ability of antibodies against the three NCRs to block NK lysis of numerous solid and leukemic tumor cells (Moretta *et al.*, 2006). The varying degree of involvement of NCRs among cells of different histiotypes also suggests that tumor cells may express different NCR ligands with different densities.

1. NKp46

A novel 46 kDa receptor, Nkp46, was cloned using an mAb that could elicit cytoxicity and cytokine release against interleukin (IL)-2-cultured human NK cells (Sivori *et al.*, 1997). NKp46 is a transmembrane receptor with two C2-type Ig-like domains in its extracellular region. The transmembrane region has a positively charged arginine residue that associates with a negative charge located in the transmembrane region of its adapter proteins, CD3-ζ and Fcε-RIγ (Moretta *et al.*, 2006). It is this association that enables NKp46 to transmit its killing signal inside the NK cell. The adapter proteins for NKp46 have multiple ITAMs located on its cytoplasmic tail. Upon receptor binding, the ITAM becomes phosphorylated by the Src family kinases and transmits its signal through the recruitment and phosphorylation of Syk tyrosine kinase.

2. NKp30

NKp30 is a 30-kDa activating NK receptor that was discovered using another NK-directed mAb (Pende *et al.*, 1999). NKp30 initiates NK activation through binding to its ITAM-bearing adapter protein CD3-ζ that becomes phosphorylated upon receptor engagement. Binding of NKp30 to its adapter protein occurs through the association between the positively charged arginine residue of the receptor and the negatively charged transmembrane region of the adapter protein. In contrast to NKp46, NKp30 has not been shown to associate only with Fcε-RIγ. NKp30 is structurally distinct in that it contains only a single, small V-type extracellular domain.

3. NKp44

The third NCR has a molecular weight of 44 kDa (Vitale *et al.*, 1998). Whereas NKp46 and NKp30 are expressed on all NK cells, NKp44 is only expressed on activated NK cells that are cultured in IL-2. Many studies have shown that IL-2 increases the cytolytic ability of NK cells. Therefore, NKp44 appears to provide part of the molecular explanation for the increased cytotoxicity of NK cells treated with IL-2. Like NKp30, NKp44 contains a single V-type extracellular domain that functions in protein–protein interactions (Moretta *et al.*, 2006). In contrast to the other NCRs, NKp44 has a lysine residue in its transmembrane region that associates with its adapter protein DAP12, which contains a single ITAM motif to transduce a lytic signal cascade (Blery *et al.*, 2000, Moretta *et al.*, 2006).

D. Activating Killer Immunoglobulin-Like Receptors

Killer immunoglobulin-like receptors (KIRs) were first discovered by their unique property of preventing destruction of self. This effect is mediated by KIR recognition of the classical MHC class I ligands (i.e., HLA-A, HLA-B, and HLA-C). KIRs are separated into two classes based on the number of extracellular Ig domains they possess (designated either 2D or 3D) and on whether they have a short (S) or long (L) cytoplasmic tail (Blery *et al.*; Yokoyama and Plougastel, 2003; Moretta *et al.*, 2006). Notably, the S forms have cytotoxic properties, whereas the L forms block NK lysis. Amazingly, the distinct activating or inhibiting effects of these different KIRs relate not to the extracellular domains, which are highly homologous, but rather to the small differences in the S and L cytoplasmic tails. Why would evolution sculpt two similar receptors with contrasting functions? It is well-known that the L form of KIRs is important in recognizing self MHC and preventing autoimmunity. In addition, these inhibitory KIRs are believed to have evolved first, serving as ancestors to the activating KIRs (Abi-Rached and Parham, 2005). It is postulated that at some point during evolution, a mutation with the short KIR form found a functional niche. Although KIR function is still a controversial topic, it is believed that activating KIRs could play an important role in recognizing nonself MHC molecules that are relevant to allotransplantation. Indeed, emerging data on the role of NK cells in bone marrow transplantation tend to support this hypothesis.

The KIR family is composed of a large number of receptors that have differing specificity for allelic polymorphisms in HLA molecules (human MHC class I). KIR expression is highly diverse throughout the human population and is distinct in each NK cell within an individual (Yawata et al., 2006). Interestingly, Yawata et al. (2006) noted that women often have either an overexpressed or underexpressed KIR phenotype that correlates with the health of the individual. In addition, KIR regulation seems to depend on the HLA expression of the host, and the receptor/ligand interactions that operate may dictate the KIR phenotype of each person.

As mentioned earlier, activating KIRs are distinct from inhibiting KIRs in that the cytoplasmic tail of activating KIRs is truncated and lacks an ITIM. For KIR to activate NK cell-mediated lysis, the positively charged lysine residue in the transmembrane domain of the KIR must associate with the negatively charged adapter protein DAP12 (Blery et al., 2000). As in the NKG2C/CD94 receptor, the associated DAP12 becomes phosphorylated in its ITAM to form a binding site for Syk/Zap70, which results in triggering of the NK lytic pathway (Lanier et al., 1998b). As might be imagined, the balance between activating and inhibiting KIRs partly determines the lytic capacity of a NK cell. Of the short forms of KIRs that have been characterized in detail, KIR3DS1 recognizes HLA-Bw7, KIR 2DS1 recognizes HLA-C^{Ly80}, and KIR2DS2 recognizes HLA-C^{Asn80} (Blery et al., 2000; Moretta et al., 2006). Out of the entire KIR family, the short form of KIR2DL4 may have the most unique characteristics. Although it does contain a functional ITIM domain, it is an activating receptor that is found only on the $CD56^{bright}$ or cytokine-producing population of peripheral blood NK cells (Kikuchi-Maki et al., 2005). KIR2DL4 is structurally distinct from other activating KIR in that it contains an Arg residue instead of Lys residue in its extracellular domain. Although controversial, some studies have shown that KIR2DL4 can bind HLA-G, which is a ligand that is only found on trophoblast cells during pregnancy (Blery et al., 2000). KIR2DL4 is also distinct from other activation KIRs in that it does not associate with DAP12 but associates with the adapter protein Fcγ-RIα (Kikuchi-Maki et al., 2005). Unlike the other KIRs, upon binding its ligand, KIR2DL4 stimulates robust production of IFN-γ but does not induce cytotoxicity (Rajagopalan et al., 2001).

E. DNAM-1 (CD226)

An adhesion molecule belonging to the immunoglobulin superfamily with 2 Ig-like domains, called the DNAX-accessory molecule 1 (DNAM-1), was first discovered to enhance cytotoxicity and cytokine release from NK and T cells. DNAM-1 has been found to bind two members of the nectin family, the poliovirus receptor (PVR/CD155) and nectin-2 (Bottino, et al., 2003). These nectin proteins are highly expressed in carcinoma, melanoma, and neuroblastoma tumor cell lines and also some freshly isolated neuroblastoma cells from late-stage cancers (Castriconi et al., 2004). NK lysis of these neuroblastoma cells required DNAM-1/PVR interaction, while PVR-negative neuroblastoma cells were resistant to NK cells. Because neuroblastoma cells often downregulate MHC class I expression and can escape T cell detection, PVR is likely to play a key role in immune destruction of these tumor cells by NK cells. It is interesting to note that nectins are widely expressed in normal neuronal, epithelial, endothelial, and fibroblastic cells, albeit at a lower level than in tumor cells. In these normal cells, the simultaneous expression of MHC class I with PVR or nectin-2 precludes NK cells from recognizing self.

F. 2B4 Receptor (CD244)

2B4 is a member of the CD2 family expressed on both NK and T cells with a robust affinity for its ligand CD48 (Brown et al., 1998; Eissmann et al., 2005). Upon receptor engagement, 2B4 significantly increases NK-mediated cytotoxicity and granule exocytosis and produces IFN-γ and IL-2. In the absence of high NCR expression, 2B4 is unable to activate NK-mediated cytotoxicity, indicating that it functions as a coreceptor. The human 2B4 receptor contains one V-type and one C2 type extracellular domain as well as four immunoreceptor tyrosine-based switch motifs (ITSMs) on its long cytoplasmic tail. However, unlike other NK receptors, 2B4 does not contain a charged amino acid in its transmembrane region. Phosphorylation of the ITSMs of 2B4 is essential for its function. Phosphorylated 2B4 is able to bind signaling lymphocytic activating molecule (SLAM)-associated protein (SAP) and the Src-related kinase Fyn, both of which are essential for 2B4-mediated NK activation (Bloch-Queyrat et al., 2005). Fyn is not only recruited to the SLAM receptor, but it is also able to phosphorylate SLAM (Eissmann et al., 2005). Once Fyn is activated, downstream targets, such as LAT, Vav, and PLC-γ, are turned on, leading to the lytic process.

2B4 can also exert an inhibitory function in human NK cells. This phenomenon was explained by the finding that each of the phosphorylated ITSMs of 2B4 has distinct and cooperative functions (Eissmann et al., 2005). The first, second, and fourth ITSMs of 2B4 were shown to activate NK-mediated cytotoxicity by binding SAP, while the third ITSM had an inhibitory function. Interestingly, the third ITSM was found to be able to bind numerous phosphatases (i.e., SHP-1, SHP-2, SHIP, and Csk), all of which can inhibit NK cytotoxicity. However, the third ITSM could only bind these negative regulatory molecules in the absence of SAP, as occurs in patients with X-linked lymphoproliferative (XLP) disease. Therefore, another important function of SAP is preventing inhibitory phosphatases from binding the 2B4 receptor and shutting down the NK lytic pathway. The availability of the SLAM-associated adapter protein SAP thus dictates whether an activating or inhibitory signal occurs with 2B4.

III. INHIBITORY NK RECEPTORS

NK cells have long been known to resist killing of tumor cells that express MHC class I. The molecular basis for this restricted behavior lies in the expression of surface receptors that specifically recognize MHC class I alleles on target cells to deliver a signal cascade that inhibits, rather than activates, NK lytic function. It is only in the absence of such inhibitory receptors that NK cells can muster the killing of target cells mediated by activating receptors. Some exceptions exist where NK cells, triggered by a combination of activating receptors including NKG2D, can lyse tumor cells that are positive for MHC class I. In humans, the inhibitory role specific for MHC class I molecules is undertaken by KIRs and NKG2A/CD94 (Blery et al., 2000; Moretta et al., 2006).

A. Inhibitory KIRs

Inhibitory KIRs belong to a highly diverse Ig-like gene family, with the capacity to recognize polymorphic determinants of the human MHC class I molecules HLA-A, HLA-B, and HLA-C. Although they are highly similar to activating KIRs in their extracellular domains, they differ at their long cytoplasmic tails that contain one or more ITIM motifs (Blery *et al.*, 2000; Moretta *et al.*, 2006). The ITIM is similar to ITAM, but the motif is $I/V/LxYxxL/V \times _{28-32} I/V/LxYxxL/V$, with the long intervening amino acid sequence providing an appropriate spacing to accommodate the two SH2 domains of SHP-1. Binding of the KIR inhibitory receptor to its MHC class I ligand will result in phosphorylation of the tyrosines in the ITIM motif within the cytoplasmic tail (see Figure 5.2). This phosphorylated ITIM then recruits the protein phosphatase SHP-1 via its SH2 domains, after which the activated SHP-1 dephosphorylates essential kinases of the lytic granule pathway (Gupta *et al.*, 1997). It is noteworthy that inhibitory KIRs function primarily to shut down the lytic signal cascade mediated by activating KIRs. For KIR to inhibit NK-mediated lysis, it must be in close proximity to the activating receptor. This requirement is dictated by the fact that SHP-1 only has high enzymatic activity when its own SH2-domain is bound, as occurs when it associates with phosphorylated ITIM; this bound SHP-1 can only dephosphorylate nearby substrates. Therefore, clustering of activating KIRs on the cell surface must first take place to allow the essential Src kinases to phosphorylate the ITIM of the neighboring inhibitory KIRs. After this event, inhibition of the activating signal can then occur through the phosphatase SHP-1 (Blery *et al.*, 2000; Moretta *et al.*, 2006). Whether activation or inhibition occurs may therefore depend not only on the presence of these reciprocal receptors but also on the affinity of the two types of KIRs for the corresponding MHC class I ligand. All of the KIR2DL receptors recognize HLA-C, while KIR3DL1 and KIR3DL2 have been shown to recognize HLA-B and HLA-A, respectively.

B. NKG2A

The NKG2 family of C-type lectin receptors mirrors the KIR family in possessing both activating and inhibitory isoforms. A key difference from KIRs is that the nonclassical HLA-E is the only ligand for this family of receptors. The NKG2A/CD94 heterodimer constitutes an inhibitory receptor and, like inhibitory KIRs, the NKG2A molecule expresses two ITIM motifs on its cytoplasmic tail that recruit the SHP-1 phosphatase. The downstream events mediated by SHP-1 are identical to those triggered by inhibitory KIRs.

It is noteworthy that inhibitory receptors are primarily targeted toward classical and nonclassical HLAs, suggesting an important role in preventing self-reactivity against normal cells, which are all consistently MHC class I positive. Normal NK cells, upon maturation, express a clonally-distributed set of various inhibitory KIRs and NKG2A/CD94. While each mature NK cell may differ in the repertoire of KIR expression from another NK cell in the same individual, they each should express at least one self-reactive inhibitory KIR. In this way, the host is protected from self attack by its own NK cells. The diversity of the KIR repertoire is regulated by DNA methylation of CpG island clusters in the promoter regions of these KIR genes (Santourlidis *et al.*, 2002).

C. Other Inhibitory Receptors

B7-H3 is a member of the B7 family of coreceptors and is located primarily on antigen-presenting cells. A study has revealed that B7-H3 binds NK cells and inhibits tumor lysis (Castriconi *et al.*, 2004). Interestingly, B7-H3 appears to be expressed on various tumor cells, such as neuroblastomas, and masking its effects by binding specific antibodies could potently upregulate the lytic effect by NK cells. The receptor for B7-H3 is unknown as of this writing, and its role in the overall target recognition by NK cells has also yet to be defined.

A similar phenomenon is also seen with the tetraspanin CD81, which specifically binds to hepatitis C virus envelope glycoprotein E2 (Crotta *et al.*, 2006). NK cells express CD81, and its cross-linking blocks CD16 and IL-12-induced NK cell activation against tumor cells. NKG2D-mediated lysis was also reduced by CD81 ligation. Thus, both B7-H3 and CD81 receptor types can inhibit the activation signal in NK cells.

IV. THE LY49 RECEPTOR FAMILY

Ly49 genes encode a large group of receptors that are located on the surface of mouse cells but that are not found in humans (Yokoyama and Plougastel, 2003). However, the Ly49 receptors are functionally similar to human KIRs despite their lack of sequence similarity. This receptor family is constituted of inhibiting receptors (Ly49 A, C, G, I) and activating receptors (Ly49D), which are restricted to recognition of murine MHC class I (i.e., H-2 molecules as they are termed in the mouse). One exception is Ly49H, which binds the murine CMV-derived m157 protein that is somewhat related to H-2 (Arase *et al.*, 2002). As in the KIRs, the inhibiting signal overrides any activating signal that is generated by an Ly49 receptor.

Ly49 genes are type II transmembrane homodimeric receptors with a single lectin-like extracellular domain that recognizes its MHC ligand H-2. Inhibitory Ly49 receptors possess ITIM motifs in their cytoplasmic regions, which function mechanistically like human ITIM motifs in inhibiting NK activation and granule movement. These receptors also recruit SHP-1 phosphatase to prevent phosphorylation of the essential kinases of NK activation. Ly49-activating receptors, as with human KIRs, lack an ITIM motif but instead associate with the ITAM-bearing DAP12 protein (Smith *et al.*, 1998). This association is also similarly dependent on the interaction of the positively charged arginine residue of the transmembrane domain of the Ly49 receptor and the negatively charged aspartic acid residue of DAP12. Interestingly, the specific Ly49 repertoire and the amount of each Ly49 receptor directly depend on the H-2 expression of the host (Yokoyama and Plougastel, 2003). IL-12 and IL-18 treatment enables NK cells to mediate cytotoxicity and produce IFN-γ despite expression of inhibitory receptors (Ortaldo and Young, 2003). These NK cells need two activating signals (activating Ly49 receptor and IL-12/IL-18) to overcome inhibition. These observations support the notion that temporary autoreactivity could occur during acute inflammation from IL12/IL18 production by dendritic cells (DCs) that can induce IFN-γ expression in NK cells, ultimately leading to expansion of the immune response. Another possible explanation for this phenomenon is that the main activating ligand for the Ly49 receptors is not H-2 but some other nonself protein. This assumption has been supported by multiple studies revealing that the murine CMV viral glycoprotein m157 is the specific ligand for Ly49H (Arase *et al.*, 2002; Yokoyama and Plougastel, 2003). Recognition of this protein by NK cells provides *in vivo* protection against MCMV. As a balancing act, murine cytomegalovirus (MCMV) susceptibility is controlled by an inhibitory receptor Ly49I, which also recognizes the m157 glycoprotein. Thus, like human KIRs, there is a delicate interplay of activating and inhibitory receptors among the Ly49 family, dictated either by MHC class I or viral proteins to mediate NK function in mice.

V. IMMUNOTHERAPY APPROACHES

Cancer immunotherapy is part of a growing research trend in attempting to harness the cytolytic function of immune cells against cancer. Difficulties in tumor immunotherapy arise due to the evasiveness of the cancer. Tumors are able to hide from NK cell immune surveillance by downregulating or shedding ligands of activating NK receptors. In addition, activating NK receptors succumb to the control of tumor cells and their released factors. Therefore, multiple studies have focused on the biological processes of tumor evasion to create therapies to counteract the tumor's progression.

A. Retention of NKG2D Ligands on Tumor Cells

MICA was the first NK receptor ligand that was found to be shed during tumor development. MICA is commonly expressed on multiple tumor types of epithelial origin, including gastrointestinal, lung, breast, kidney, ovary, prostate, colon, and hematopoietic malignancies. The mechanism of MICA shedding is dependent on metalloproteinases (Salih et al., 2002). Use of a matrix metalloproteinase inhibitor (MMPI) prevented MICA shedding and led to the accumulation of ligands on the cell surface, while a serine protease inhibitor had no effect.

One study has found that another NKG2D ligand, ULBP2, is shed in a similar fashion as MICA (Waldhauer and Steinle, 2006). Elevated soluble ULBP2 levels were found in the sera of patients with hematopoietic malignancies. ULBP2 is structurally distinct from MICA in that it is linked to the cell surface by a glycosylphosphatidylinositol molecule. Therefore, it was suggested that a phospholipase was essential for proteolytic cleaving of ULBP2. However, it was found that metalloproteinases were also the key enzymes of ULBP2 shedding, suggesting a common tumor evasive mechanism for shedding NK receptor ligands (Waldhauer and Steinle, 2006). More important, prevention of ULBP2 shedding, by treatment with MMPI, correlated with an increase in NK-mediated cytotoxicity.

Another mechanism for lack of target killing could be due to negative control of the NKG2D itself on NK cells. Using an MHC class I-deficient lung adenocarcinoma model, it was shown that the tumor was completely resistant to NK-mediated cytolysis, due not only to loss of NKG2D ligands but also due to the downregulation in NK receptor expression (Le Maux Chansac et al., 2005). In fact, binding of soluble MICA/B can downregulate NKG2D expression and function in NK cells (Groh et al., 2002). The release of cytokines, such as transforming growth factor β (TGF-β), can also downregulate NKG2D expression (Moretta et al., 2006). Thus, for therapeutic purposes, it would be valuable to identify some means to prevent the release of NK ligands from tumor cells.

In light of this need, studies in preclinical models are focusing on achieving greater ligand expression on tumors. In mice, it was shown that multiple types of tumor cells became susceptible to NK-mediated cytotoxicity upon transduction of the tumor cells with NKG2D ligands (Diefenach et al., 2001). Interestingly, if mice were exposed to any of the NKG2D ligand-transduced tumor cells, metastasis did not occur. In addition, these mice resisted rechallenge with the parental tumor cells, and the wild-type tumor cells lacking NKG2D ligands were rejected. This finding indicates that immunization with NKG2D ligand-transfected tumor cells can prevent *in vivo* growth of nontransfected tumor cells, and this strategy might be an attractive one to bring to the clinic.

Another strategy that might be exploited to prevent ligand shedding by tumor cells could be the use of MMPIs. It will be relevant for clinical studies to determine which MMPI is most effective in retaining NK ligands on various tumor types.

B. Antibody-Dependent Cell Cytotoxicity

One class of innovative therapeutic agents includes antibody compounds that specifically target receptors on the cell surface of tumors. This type of drug represents a new trend in individualizing treatment specifically for the phenotype of a patient's tumors. Herceptin and Rituxan are relatively new humanized monoclonal antibody agents that have been very clinically significant in the treatment of cancer. Herceptin is an effective therapy against breast cancer that targets the proto-oncogene p185HER-2/neu. Rituxan is indicated for non-Hodgkin's lymphoma and targets CD20 on B cells. Both antibodies have been found to inhibit growth and induce apoptosis in target cells. Besides these direct effects, antibody-dependent cell-mediated cytotoxicity (ADCC) plays a very important role in the *in vivo* efficacy of these two antibodies (Clynes *et al.*, 2000). Receptors for the Fc portion of these antibodies reside in immune cells, and both the activating receptor, Fcγ-RIII (CD16), and the inhibitory receptor, Fcγ-RIIB, can control the outcome of antibody therapy. It was observed that expression of Fcγ-RIIB in mice greatly reduced the efficacy of Herceptin or Rituxan treatment, while expression of the Fcγ-RIII receptor was important for antibody-mediated prevention of tumor growth. More important, mutant Fcγ-RIII$^{-/-}$ mice that were treated with Herceptin could not reject the transplanted tumors efficiently compared to wild-type mice, demonstrating a contributing role of ADCC in efficacy of these drugs. However, a slight prevention in tumor formation did occur even without the presence of the activating Fc receptor, indicating that apoptotic/growth inhibitory pathways are also components of the overall effect of these drugs. Binding of the Fcγ-RIII on NK cells was essential for ADCC to occur. Notably, NK cells do not constitutively express Fcγ-RIIB and are not subject to downregulation by antibodies. Mutation of key amino acids in the binding region of the antibody ablated all cytotoxic effects, indicating that NK cells are crucial for ADCC.

In summary, the goal in designing antibody therapeutic agents should be to maximize the binding between the antibody and the NK cell while minimizing its interaction with inhibitory receptors.

C. Cytokines for Enhancement of NK Receptors

For maximal optimization of NK cell function, the activating NK receptors should be enhanced or maintained on NK cells. IL-2 is known to induce NKp44, while IL-12 and IL-15 are potent inducers of NKG2D (Cosman *et al.*, 2001; Moretta *et al.*, 2006). Other cytokines, such as IL-18, can also synergize with IL-12 to increase NK receptor expression (Ortaldo, 2003). Ligation of Toll-like receptors 3 and 9 on NK cells can also enhance NK activation against tumor cells, and this could be due to certain activating receptors that are induced (Moretta *et al.*, 2006). Cytokines such as IL-2, IL-12, and IL-21 are also able to enhance ADCC in NK cells; it would be important to define if this action is through the upregulation of CD16 (Roda *et al.*, 2006). Thus, a careful evaluation of cytokines and the identification of the NK receptors they can induce, either by themselves or in concert with other cytokines, could help to bring a new approach of targeting NK receptor expression for therapy of cancer.

D. Allograft Transplantation

NK cells are inhibited from killing self because of the high expression of MHC class I molecules in all epithelial cells and other cells of the body. While this is true in an autologous setting under normal conditions, NK cells can be turned against allogeneic cells that represent nonself. Alloreactivity can occur if there is a mismatch between the KIRs expressed on NK cells and the MHC class I expressed on the allogeneic cells. This property has been exploited in the allogeneic bone marrow transplantation setting, where NK cells have been used successfully to treat certain hematological malignancies. Specifically, studies have shown that mismatched hematopoietic transplants have a powerful graft-versus-leukemia (GVL) effect in acute myeloid leukemia (AML) patients (Ruggeri et al., 2002). The GVL effect occurs during the transplant by the production of alloreactive NK cells from the haploidentical donors. Under these conditions, a mismatch is likely to occur, whereby a portion of the donor NK cells will not express the inhibitory KIR for the MHC class I on the tissues of the allograft recipient. Without the inhibitory signal, this fraction of NK cells is released from suppression and can target allogeneic diseased cells in the AML patients. It has been shown that hematological tumors express MICA/B and ULBPs, which could be the trigger for GVL. Another study discovered that alloreactive NK cells could be used in the *ex vivo* purging of tumor cells from patient bone marrow cells prior to autologous transplantation to minimize tumor cell contamination in the transferred stem cells (Koh et al., 2003). This strategy could greatly increase the efficacy of bone marrow transplants by preventing relapse. This study also showed that alloreactive NK cells were more potent than syngeneic NK cells in eliminating tumor cells. Alloreactive NK cells have a significant advantage over alloreactive T cells, which are known to cause graft-versus-host disease (GVHD), which results from classical T cell responses to alloantigens through the TCR. Exploiting KIR mismatch for GVL without GVH in cancer patients is an attractive strategy that is being tested in the clinic and that could prove to be a potent therapeutic regimen for transplantation.

VI. CONCLUSION

As presented in this chapter, NK cells are a major component of immune surveillance against tumor formation and viral infection. NK cells are a unique effector of the immune system in that they are always equipped to attack target cells without needing antigen presentation or cytokine stimulation. NK cells have a distinct advantage over $CD8^+$ T cells in tumor immunotherapy because NK cells do not require tumor-specific antigen recognition. To carry out the effector function, NK cells possess a wide array of receptors that recognize both self and nonself ligands. Out of NK cells' many receptors, NKp30, NKp44, NKp46, and NKG2D are the key players in triggering NK-mediated cytotoxicity. Upon engagement, these receptors induce granule movement and target lysis through a specific phosphorylation cascade that is initiated by the phosphorylation of an ITAM or PI3K. NK-mediated cytotoxicity is inhibited by the recruitment of SHP-1 phosphatase to the ITIM of an inhibitory receptor. Inhibitory receptors are essential in protecting self from autoimmune reactions. KIRs, which specifically recognize MHC class I molecules, such as HLA-A, HLA-B, and HLA-C as ligands, are the primary receptors involved in this protection process. The balance between activation and inhibition is carefully mediated through the overall signal based on stimulatory receptors, inhibitory receptors, coreceptors, and cytokines.

Many malignancies evolve strategies to evade NK detection through means such as ligand shedding, ligand downregulation, constitutive immunosuppressive cytokine production, and/or the upregulation of MHC. The ligands for the activating NKG2D receptor, MICA/MICB and ULBPs, would normally signal NK cells that a cellular anomaly has occurred. However, as a tumor progresses, tumors are able to shed these ligands and evade attack by NK cells. Therefore, MICA/MICB and ULBPs could prove to be valuable prognostic indicators of the stage of many malignancies. Retaining these NKG2D ligands to maintain immune detection of certain tumors is a worthwhile goal to pursue as a means of immunotherapy.

Although the NCR family of receptors has been found to be essential for NK effector function, its ligands have still not been elucidated. Identifying NCR ligands will shed further light on how NK cells are able to detect transformed and virally infected cells. Are NCR ligands upregulated under cellular stress conditions as occurs with NKG2D ligands? If so, what mechanisms control their expression? Understanding these processes will be vital in developing innovative treatment against cancer as well as in expanding the overall understanding of NK cell biology.

The discovery of 2B4 as a coreceptor has provided insight on how the balance between activating/inhibiting signals can be tilted. This coreceptor can provide cross talk between NK cells and other immune cells in addition to enhancing NK cell activation via SAP or inhibition via SHP-1. SAP mediates binding of NK cells to DCs and T cells via CD48, and this interaction can potentiate an immune response. 2B4 can also greatly enhance NK-mediated cytotoxicity by inducing cytokine production. The importance of 2B4 is best exemplified in XLP patients who do not express SAP. Without SAP, 2B4 is nonfunctional, and this defect corresponds to low NK cytolysis and disease progression in the XLP patients.

Our knowledge of NK biology has been greatly expanded in recent years. For decades, NK cells were thought to be nonspecific and promiscuous. Cloning of the key receptors has revealed a multitude of activating and inhibitory receptors, helping shed light on how NK cells induce cytolysis in tumors and virally infected cells. This chapter has provided some overview of the diverse structures of NK receptors as well as the molecular mechanisms behind NK activation and inhibition. Many new tumor immunotherapy studies are attempting to exploit this newfound understanding of NK activation to treat many types of malignancies. As many NK receptors and ligands are also found on DCs and T cells, it will be important to gain greater information on what roles they play elsewhere in the immune system. Clearly, further research is needed to understand how NK cells cooperate with other immune cells in the fight against cancer.

References

Abi-Rached, L., and Parham, P. (2005). Natural selection drives recurrent formation of activating killer cell immunoglobulin-like receptor and Ly49 from inhibitory homologues. *J. Exp. Med.* **201**, 1319–1332.

Arase, H., Mocarski, E. S., Campbell, A. E., Hill, A. B., and Lanier, L. L. (2002). Direct recognition of cytomegalovirus by activating and inhibitory NK cell receptors. *Science* **296**, 1323–1326.

Bauer, S., Groh, V., Wu, J., Steinle, A., Phillips, J. H., Lanier, L. L., and Spies, T. (1999). Activation of NK cells and T cells by NKG2D, a receptor for stress-inducible MICA. *Science* **285**, 727–729.

Blery, M., Olcese, L., and Vivier, E. (2000). Early signaling via inhibitory and activating NK receptors. *Human Immunol.* **61**, 51–64.

Bloch-Queyrat, C., Fondaneche, M. C., Chen, R., Yin, L., Relouzat, F., Veillette, A., Fischer, A., and Latour, S. (2005). Regulation of natural cytotoxicity by the adaptor SAP and the Src-related kinase Fyn. *J. Exp. Med.* **202**, 181–192.

Bottino, C., Castriconi, R., Pende, D., Rivera, P., Nanni, M., Carnemolla, B., Cantoni, C., Grassi, J., Marcenaro, S., Reymond, N., et al. (2003). Identification of PVR (CD155) and Nectin-2 (CD112) as cell surface ligands for the human DNAM-1 (CD226) activating molecule. *J. Exp. Med.* **198**, 557–567.

Brown, M. H., Boles, K., van der Merwe, P. A., Kumar, V., Mathew, P. A., and Barclay, A. N. (1998). 2B4, the natural killer and T cell immunoglobulin superfamily surface protein, is a ligand for CD48. *J. Exp. Med.* **188**, 2083–2090.

Bui, J. D., Carayannopoulos, L. N., Lanier, L. L., Yokoyama, W. M., and Schreiber, R. D. (2006). IFN-dependent down-regulation of the NKG2D ligand H60 on tumors. *J. Immunol.* **176**, 905–913.

Carretero, M., Cantoni, C., Bellon, T., Bottino, C., Biassoni, R., Rodriguez, A., Perez-Villar, J. J., Moretta, L., Moretta, A., and Lopez-Botet, M. (1997). The CD94 and NKG2-A C-type lectins covalently assemble to form a natural killer cell inhibitory receptor for HLA class I molecules. *Eur. J. Immunol.* **27**, 563–567.

Castriconi, R., Dondero, A., Corrias, M. V., Lanino, E., Pende, D., Moretta, L., Bottino, C., and Moretta, A. (2004). Natural killer cell-mediated killing of freshly isolated neuroblastoma cells: Critical role of DNAX accessory molecule-1-poliovirus receptor interaction. *Cancer Res.* **64**, 9180–9184.

Castriconi, R., Dondero, A., Augugliaro, R., Cantoni, C., Carnemolla, B., Sementa, A. R., Negri, F., Conte, R., Corrias, M. V., Moretta, L., et al. (2004a). Identification of 4Ig-B7-H3 as a neuroblastoma-associated molecule that exerts a protective role from an NK cell-mediated lysis. *Proc. Natl. Acad. Sci. USA* **101**, 12640–12645.

Clynes, R. A., Towers, T. L., Presta, L. G., and Ravetch, J. V. (2000). Inhibitory Fc receptors modulate *in vivo* cytotoxicity against tumor targets. *Nat. Med.* **6**, 443–446.

Cosman, D., Mullberg, J., Sutherland, C. L., Chin, W., Armitage, R., Fanslow, W., Kubin, M., and Chalupny, N. J. (2001). ULBPs, novel MHC class I-related molecules, bind to CMV glycoprotein UL16 and stimulate NK cytotoxicity through the NKG2D receptor. *Immunity* **14**, 123–133.

Crotta, S., Ronconi, V., Ulivieri, C., Baldari, C. T., Valiante, N. M., Abrignani, S., and Wack, A. (2006). Cytoskeleton rearrangement induced by tetraspanin engagement modulates the activation of T and NK cells. *Eur. J. Immunol.* **36**, 919–929.

Diefenbach, A., Jensen, E. R., Jamieson, A. M., and Raulet, D. H. (2001). Rae1 and H60 ligands of the NKG2D receptor stimulate tumour immunity. *Nature* **413**, 165–171.

Diefenbach, A., Tomasello, E., Lucas, M., Jamieson, A. M., Hsia, J. K., Vivier, E., and Raulet, D. H. (2002). Selective associations with signaling proteins determine stimulatory versus costimulatory activity of NKG2D. *Nat. Immunol.* **3**, 1142–1149.

Djeu, J. Y., Jiang, K., and Wei, S. (2002). A view to a kill: Signals triggering cytotoxicity. *Clin. Cancer Res.* **8**, 636–640.

Eissmann, P., Beauchamp, L., Wooters, J., Tilton, J. C., Long, E. O., and Watzl, C. (2005). Molecular basis for positive and negative signaling by the natural killer cell receptor 2B4 (CD244). *Blood* **105**, 4722–4729.

Gasser, S., Orsulic, S., Brown, E. J., and Raulet, D. H. (2005). The DNA damage pathway regulates innate immune system ligands of the NKG2D receptor. *Nature* **436**, 1186–1190.

Groh, V., Wu, J., Yee, C., and Spies, T. (2002). Tumour-derived soluble MIC ligands impair expression of NKG2D and T-cell activation. *Nature* **419**, 734–738.

Guma, M., Cabrera, C., Erkizia, I., Bofill, M., Clotet, B., Ruiz, L., and Lopez-Botet, M. (2006). Human cytomegalovirus infection is associated with increased proportions of NK cells that express the CD94/NKG2C receptor in aviremic HIV-1-positive patients. *J. Infect. Dis.* **194**, 38–41.

Gupta, N., Scharenberg, A. M., Burshtyn, D. N., Wagtmann, N., Lioubin, M. N., Rohrschneider, L. R., Kinet, J. P., and Long, E. O. (1997). Negative signaling pathways of the killer cell inhibitory receptor and Fc gamma RIIb1 require distinct phosphatases. *J. Exp. Med.* **186**, 473–478.

Holdenrieder S. A., Stieber, P., Peterfi, A., Nagel, D., Steinle, A., and Salih, H. R. (2006). Soluble MICB in malignant diseases: Analysis of diagnostic significance and correlation with soluble MICA. *Cancer Immunol. Immunother.* **55**, 1584–1589.

Jiang, K., Zhong, B., Gilvary, D. L., Corliss, B. C., Hong-Geller, E., Wei, S., and Djeu, J. Y. (2000). Pivotal role of phosphoinositide-3 kinase in regulation of cytotoxicity in natural killer cells. *Nat. Immunol.* **1**, 419–425.

Jiang, K., Zhong, B., Gilvary, D. L., Corliss, B. C., Vivier, E., Hong-Geller, E., Wei, S., and Djeu, J. Y. (2002). Syk regulation of phosphoinositide 3-kinase-dependent NK cell function. *J. Immunol.* **168**, 3155–3164.

Kikuchi-Maki, A., Catina, T. L., and Campbell, K. S. (2005). Cutting edge: KIR2DL4 transduces signals into human NK cells through association with the Fc receptor gamma protein. *J. Immunol.* **174**, 3859–3863.

Koh, C. Y., Ortaldo, J. R., Blazar, B. R., Bennett, M., and Murphy, W. J. (2003). NK-cell purging of leukemia: Superior antitumor effects of NK cells H2 allogeneic to the tumor and augmentation with inhibitory receptor blockade. *Blood* **102**, 4067–4075.

Lanier, L. L., Corliss, B., Wu, J., and Phillips, J. H. (1998a). Association of DAP12 with activating CD94/NKG2C NK cell receptors. *Immunity* **8**, 693–701.

Lanier, L. L., Corliss, B. C., Wu, J., Leong, C., and Phillips, J. H. (1998b). Immunoreceptor DAP12 bearing a tyrosine-based activation motif is involved in activating NK cells. *Nature* **391**, 703–707.

Le Maux Chansac, B., Moretta, A., Vergnon, I., Opolon, P., Lecluse, Y., Grunenwald, D., Kubin, M., Soria, J. C., Chouaib, S., and Mami-Chouaib, F. (2005). NK cells infiltrating a MHC class I-deficient lung adenocarcinoma display impaired cytotoxic activity toward autologous tumor cells associated with altered NK cell-triggering receptors. *J. Immunol.* **175**, 5790–5798.

McFarland, B. J., Kortemme, T., Yu, S. F., Baker, D., and Strong, R. K. (2003). Symmetry recognizing asymmetry: Analysis of the interactions between the C-type lectin-like immunoreceptor NKG2D and MHC class I-like ligands. *Structure* **11**, 411–422.

Moretta, L., Bottino, C., Pende, D., Castriconi, R, Mingari, M.C., and Moretta, A. (2006). Surface NK receptors and their ligands on tumor cells. *Sem. Immunol.* **18**, 151–158.

Nowbakht, P., Ionescu, M. C., Rohner, A., Kalberer, C. P., Rossy, E., Mori, L., Cosman, D., De Libero, G., and Wodnar-Filipowicz, A. (2005). Ligands for natural killer cell-activating receptors are expressed upon the maturation of normal myelomonocytic cells but at low levels in acute myeloid leukemias. *Blood* **105**, 3615–3622.

Ortaldo, J. R., and Young, H. A. (2003). Expression of IFN-gamma upon triggering of activating Ly49D NK receptors *in vitro* and *in vivo*: Costimulation with IL-12 or IL-18 overrides inhibitory receptors. *J. Immunol.* **170**, 1763–1769.

Pende, D., Cantoni, C., Rivera, P., Vitale, M., Castriconi, R., Marcenaro, S., Nanni, M., Biassoni, R., Bottino, C., Moretta, A., and Moretta, L. (2001). Role of NKG2D in tumor cell lysis mediated by human NK cells: Cooperation with natural cytotoxicity receptors and capability of recognizing tumors of nonepithelial origin. *Eur. J. Immunol.* **31**, 1076–1086.

Pende, D., Parolini, S., Pessino, A., Sivori, S., Augugliaro, R., Morelli, L., Marcenaro, E., Accame, L., Malaspina, A., Biassoni, R., *et al.* (1999). Identification and molecular characterization of NKp30, a novel triggering receptor involved in natural cytotoxicity mediated by human natural killer cells. *J. Exp. Med.* **190**, 1505–1516.

Radosavljevic, M., Cuillerier, B., Wilson, M. J., Clement, O., Wicker, S., Gilfillan, S., Beck, S., Trowsdale, J., and Bahram, S. (2002). A cluster of ten novel MHC class I related genes on human chromosome 6q24.2–q25.3. *Genomics* **79**, 114–123.

Rajagopalan, S., Fu, J., and Long, E. O. (2001). Cutting edge: Induction of IFN-gamma production but not cytotoxicity by the killer cell Ig-like receptor KIR2DL4 (CD158d) in resting NK cells. *J. Immunol.* **167**, 1877–1881.

Roda, J. M., Parihar, R., Lehman, A., Mani, A., Tridandapani, S., and Carson, W. E., 3rd (2006). Interleukin-21 enhances NK cell activation in response to antibody-coated targets. *J. Immunol.* **177**, 120–129.

Ruggeri, L., Capanni, M., Urbani, E., Perruccio, K., Shlomchik, W. D., Tosti, A., Posati, S., Rogaia, D., Frassoni, F., Aversa, F., *et al.* (2002). Effectiveness of donor natural killer cell alloreactivity in mismatched hematopoietic transplants. *Science* **295**, 2097–2100.

Salih, H. R., Antropius, H., Gieseke, F., Lutz, S. Z., Kanz, L., Rammensee, H. G., and Steinle, A. (2003). Functional expression and release of ligands for the activating immunoreceptor NKG2D in leukemia. *Blood* **102**, 1389–1396.

Salih, H. R., Rammensee, H. G., and Steinle, A. (2002). Cutting edge: Down-regulation of MICA on human tumors by proteolytic shedding. *J. Immunol.* **169**, 4098–4102.

Santourlidis, S., Trompeter, H. I., Weinhold, S., Eisermann, B., Meyer, K. L., Wernet, P., and Uhrberg, M. (2002). Crucial role of DNA methylation in determination of clonally distributed killer cell Ig-like receptor expression patterns in NK cells. *J. Immunol.* **169**, 4253–4261.

Sivori, S., Vitale, M., Morelli, L., Sanseverino, L., Augugliaro, R., Bottino, C., Moretta, L., and Moretta, A. (1997). p46, a novel natural killer cell-specific surface molecule that mediates cell activation. *J. Exp. Med.* **186**, 1129–1136.

Smith, K. M., Wu, J., Bakker, A. B., Phillips, J. H., and Lanier, L. L. (1998). Ly-49D and Ly-49H associate with mouse DAP12 and form activating receptors. *J. Immunol.* **161**, 7–10.

Vitale, M., Bottino, C., Sivori, S., Sanseverino, L., Castriconi, R., Marcenaro, E., Augugliaro, R., Moretta, L., and Moretta, A. (1998). NKp44, a novel triggering surface molecule specifically expressed by activated natural killer cells, is involved in non-major histocompatibility complex-restricted tumor cell lysis. *J. Exp. Med.* **187**, 2065–2072.

Waldhauer, I., and Steinle, A. (2006). Proteolytic release of soluble UL16-binding protein 2 from tumor cells. *Cancer Res.* **66**, 2520–2526.

Welte, S. A., Sinzger, C., Lutz, S. Z., Singh-Jasuja, H., Sampaio, K. L., Eknigk, U., Rammensee, H. G., and Steinle, A. (2003). Selective intracellular retention of virally induced NKG2D ligands by the human cytomegalovirus UL16 glycoprotein. *Eur. J. Immunol.* **33**, 194–203.

Wu, J., Song, Y., Bakker, A. B., Bauer, S., Spies, T., Lanier, L. L., and Phillips, J. H. (1999). An activating immunoreceptor complex formed by NKG2D and DAP10. *Science* **285**, 730–732.

Yawata, M., Yawata, N., Draghi, M., Little, A. M., Partheniou, F., and Parham, P. (2006). Roles for HLA and KIR polymorphisms in natural killer cell repertoire selection and modulation of effector function. *J. Exp. Med.* **203**, 633–645.

Yokoyama, W. M., and Plougastel, B. F. (2003). Immune functions encoded by the natural killer gene complex. *Nat. Rev. Immunol.* **3**, 304–316.

Further Reading

Farag, S. S., and Caligiuri M. (2006). Human natural killer cell development and biology. *Blood Rev.* **20**, 123–137. *This review focuses on cytokine regulation of the development and maturation of NK cells and* *includes discussions on their potential use in tumor immunotherapy.*

Lanier, L. L. (2005). NK cell recognition. *Annu. Rev. Immunol.* **23**, 225–274. *This study provides a thorough review of the structure and function of all NK receptors and ligands that are involved in NK effector functions.*

Raulet, D. H. (2004). Interplay of natural killer cells and their receptors with the adaptive immune response. *Nature Immunol.* **10**, 996–1002. *This article reviews the interaction of natural killer cells with dendritic cells that leads to either positive and negative regulation of the adaptive immune response.*

CHAPTER 6

Immune Escape: Immunosuppressive Networks

SHUANG WEI, ALFRED CHANG, AND WEIPING ZOU

I. Introduction
II. Imbalance Between Mature DCs and Immature DCs
   A. Mature Versus Immature DCs in the Tumor Microenvironment
III. Imbalance Between Stimulatory and Inhibitory B7 Family Molecules
   A. B7-H1 in the Tumor Microenvironment
   B. B7-H4 in the Tumor Environment
   C. Cytokine Networks and B7 Family Members in the Tumor Microenvironment
IV. Imbalance Between Regulatory T Cells and Conventional T Cells
   A. Imbalance Between Treg Cells and Conventional T Cells in the Bone Marrow
   B. Imbalance Between Treg Cells and Conventional T Cells in Human Tumors
V. Concluding Remarks
   References

The tumor microenvironment is a battlefield between the tumor and host immune system. The myriad of cellular interactions in the tumor microenvironment determines immune response versus tolerance to tumor cells. Recent advances in molecular and cellular tumor immunology have demonstrated that the tumor actively recruits and alters immune cell phenotypes and functions, promoting either immune suppression or tolerance of tumor-associated antigens. In this chapter, we provide a framework for understanding how the immune response is altered in favor of forming a suppressive network in the tumor microenvironment, integrating different immune cell types and molecules.

I. INTRODUCTION

The tumor microenvironment is a battleground during the neoplastic process, where strife between immune cells and other stromal cell types can either restrain or foster (depending on conditions) the proliferation, survival, and migration of tumor cells. As they get the upper hand in this battlefield, tumor cells cannot only survive and disseminate, but more importantly, they can mimic some of the signaling pathways of the immune system to propagate tolerable conditions and thus escape antitumor immunity.

In the twenty-first century, there have been significant advances in defining the molecular interactions between tumor and immune cells in the tumor microenvironment, in particular those interactions that define the suppressive network by which tumor cells defeat the immune system. Indeed, it has become increasingly evident that poor tumor-associated antigen (TAA)-specific immunity is not due to a passive process whereby adaptive immunity is shielded from detecting TAA; rather, there is an active process of tolerability taking place in the tumor microenvironment.

Studies have demonstrated multiple layers of suppressive network formed in cancer patients. The network includes at least three imbalances in tumor environment: (i) an imbalance between mature dendritic cells (DCs) and immature DCs, (ii) an imbalance between stimulatory and inhibitory B7 family molecules, and (iii) an imbalance between conventional T cells and regulatory T cells (Tregs). These cellular and molecular levels of dysfunction contribute importantly to tumor immunopathogenesis. This chapter takes human ovarian cancer as an illustrative example, discussing how these imbalances form the immunosuppressive network in patients with cancer, and explains how to apply the knowledge gained to develop and refine novel immune-boosting strategies.

II. IMBALANCE BETWEEN MATURE DCs AND IMMATURE DCs

Antigen-presenting cells (APCs) include DCs, monocytes or macrophages, and B lymphocytes. DCs are a heterogeneous group of APCs that display differences in anatomic localization, cell surface phenotype, and function (Banchereau and Steinman, 1998; Lanzavecchia and Sallusto, 2001). To induce tumor immunity, sufficient numbers of functional APCs must be present *in situ*; be able to capture, process, and present TAAs; migrate to secondary lymphoid organs; and then stimulate TAA-specific T cells. TAA-specific T cells must then home to the tumor, recognize TAA on tumor cells, and eradicate the tumor cells. Defects in any of these processes will cause suboptimal or no immune response against the tumor.

A. Mature Versus Immature DCs in the Tumor Microenvironment

DCs are the most potent among APCs. The biology of DCs and DC-based clinical trials have been extensively reviewed (Banchereau and Steinman, 1998; Cerundolo et al., 2004; Figdor et al., 2004; O'Neill et al., 2004; Schuler et al., 2003). DCs' maturation status may determine their immunogenic versus tolerogenic properties (Lutz and Schuler, 2002). Maturation status is basically defined by the phenotype and cytokine profile (Lutz and Schuler, 2002). Mature DCs highly express CD40, CD80, CD83, and CD86 on their surface and produce high levels of interleukin (IL)-12. It is thought that mature DCs are functional and capable of inducing potent TAA-specific T cell immunity.

Myeloid DCs arise from the same progenitor cells that also give rise to monocytes and macrophages and express, at least in humans, cell-surface markers characteristic of the myeloid lineage (Banchereau and Steinman, 1998). However, the presence of functional immunogenic (mature) myeloid DCs is rare in human ovarian tumors (Zou et al., 2001), breast tumors (Bell et al., 1999; Iwamoto et al., 2003), renal cell carcinomas, and prostate cancer (Troy et al., 1998a, 1998b). Multiple factors, such as a defective DC recruitment, differentiation, maturation, and survival, could be the cause. Abundant evidence documents that the differentiation and maturation of myeloid DCs are profoundly suppressed by factors present in the tumor environment. Cancer cells, including ovarian tumors, are major producers of vascular endothelial growth factor (VEGF) (Carmeliet and Jain, 2000; Kryczek et al., 2005), causing high VEGF levels in the tumor microenvironment. VEGF was the first tumor-derived molecule reported to suppress DC differentiation and maturation (Gabrilovich *et al.*, 1996). Subsequently, IL-6 and macrophage colony-stimulating factor (MCSF) derived from tumor cells and tumor environmental macrophages were shown to switch DC differentiation toward macrophage differentiation by upregulating the expression of MCSF receptor in monocytes (Menetrier-Caux *et al.*, 1998). High levels of IL-6 and C-reactive protein can be found in the peripheral blood and malignant ascites of patients with ovarian cancer (Carmeliet and Jain, 2000; Freedman *et al.*, 2004; Kryczek *et al.*, 2000, 2005). Notably, IL-6 can promote B cell differentiation and has been suggested to block the suppressive activity of regulatory T cells (Pasare and Medzhitov, 2003). Moreover, a variety of human tumors highly express cyclooxygenase 2 (COX-2) (Joki *et al.*, 2000; Shono *et al.*, 2001; Wolff *et al.*, 1998). COX-2 promotes prostaglandin E2 (PGE$_2$) production in the tumor environment. PGE$_2$ in turn suppresses DC differentiation and function (Kalinski *et al.*, 1998). In addition, tumor cells, tumor-associated macrophages, and regulatory T cells (see below) often produce IL-10 and transforming growth factor β (TGF-β), which also suppress DC maturation and function. Human neuroblastomas, melanoma, and many other tumors highly express gangliosides, which also suppress human DC differentiation (Peguet-Navarro *et al.*, 2003; Shurin *et al.*, 2001). Strikingly, DC differentiation cytokines, such as granulocyte-macrophage colony-stimulating factor (GM-CSF) and IL-4, as well as the Th1-type cytokines IL-12 and interferon (IFN)-γ, are rare in the human tumor environment, at least in human ovarian cancer. Abundance of VEGF, IL-6, MCSF, TGF-β, IL-10, COX-2, PGE$_2$, gangliosides, and other suppressive molecules versus negligible GM-CSF, IL-4, IL-12, and IFN-γ cause an aberrant cytokine pattern in the tumor environment (Freedman *et al.*, 2004; Kryczek *et al.*, 2000). This tumor environmental cytokine imbalance blocks DC differentiation and maturation. While functional mature myeloid DCs can induce potent TAA-specific immunity *in vivo* (Dhodapkar *et al.*, 1999; Labeur *et al.*, 1999), immature or partially differentiated myeloid DCs induce either suppressive T cells (Dhodapkar *et al.*, 2001; Jonuleit *et al.*, 2000) or T cell unresponsiveness (Hawiger *et al.*, 2001) in the tumor environment. The induced suppressive T cells would home to draining lymph nodes and, in turn, systemically disable TAA-specific immunity.

Human plasmacytoid DCs arise from circulating cells lacking lineage markers and CD11c but expressing HLA–DR, and these DCs are further identified phenotypically by expression of CD123, the IL-3Ra chain (Colonna *et al.*, 2004). PDC precursor cells are the principal circulating cell of the plasmacytoid DC lineage and were identified as high type I IFN-producing cells following virus infection. When activated and matured through CpG motifs, plasmacytoid DCs may induce T cell IFN-γ through type I IFN. Plasmacytoid DCs are found in the tumor environment of patients with ovarian cancer (Zou *et al.*, 2001), melanoma (Salio *et al.*, 2003), and head and neck squamous cell carcinoma (HNSCC) (Hartmann *et al.*, 2003). Tumor-associated plasmacytoid DCs exhibited a typical immature cell phenotype with CD83$^-$ and low expression of CD40, CD80, and CD86. Tumor cells produce the chemokine ligand CXCL12 (SDF-1), and plasmacytoid DCs express CXCR4, the receptor for CXCL12. Tumor-derived CXCL12 mediates plasmacytoid DC tumor trafficking (Zou *et al.*, 2001). CXCL12 protects tumor plasmacytoid DCs from apoptosis (Zou *et al.*, 2001). *In vitro*-activated peripheral blood plasmacytoid DCs stimulate tumor-specific T cell IFN-γ production in patients with melanoma (Salio *et al.*, 2003). Interestingly, tumor environmental plasmacytoid DCs exhibit a reduced Toll-like receptor 9 (TLR9) expression (Hartmann *et al.*, 2003), which is the most specific TLR pathway for inducing IFN-γ. IFN-γ not only triggers innate immunity, such as activating natural killer (NK) cells, but also promotes adaptive Th1 T cell responses. It suggests that plasmacytoid DCs can be phenotypically and functionally modulated in the tumor environment. In support of this, tumor environmental plasmacytoid DCs induce significant T cell IL-10 production by T cells, which suppresses myeloid DC-induced TAA-specific T cell effector functions (Zou et al., 2001). A study has published data revealing that tumor plasmacytoid DCs induce IL-10$^+$CCR7$^+$CD8$^+$ T cells to home to draining lymph nodes and suppress TAA-specific central priming (Wei et al., 2005). The fact that allogeneic plasmacytoid DCs are able to induce CD4$^+$ (Moseman et al., 2004) and CD8$^+$ (Gilliet and Liu, 2002) suppressive regulatory T cells supports these data. A large amount of plasmacytoid DCs, but not functional mature myeloid DCs, accumulate in the ovarian tumor environment (Zou et al., 2001).

In addition to immature myeloid DCs and plasmacytoid DCs, two other antigen-presenting cell subsets are also implicated significantly in tumor immune suppression: myeloid suppressor cells and indoleamine 2,3-deoxygenase (IDO$^+$) myeloid DCs. The antigen specificity, phenotype, mechanisms of action, and function of myeloid suppressor cells are discussed in detail elsewhere (Bronte et al., 2001, 2003, 2005; Bronte and Zanovello, 2005; Bunt et al., 2006; Sinha et al., 2005), including in Chapters 19 and 21 of this book. This chapter briefly discusses IDO$^+$ myeloid DCs (Muller et al., 2005a; Muller et al., 2005b; Muller and Prendergast, 2005). IDO$^+$ myeloid DCs are also found in the tumor environment and draining lymph nodes. IDO expression has been documented in human and murine myeloid DCs (Munn et al., 2002). IDO catalyzes the oxidative catabolism of tryptophan, an amino acid essential for T cell proliferation and differentiation. IDO$^+$ DCs reduce access to free tryptophan and thus block T cell cycle progression. This, in turn, prevents the clonal expansion of T cells and promotes T cell death by apoptosis, anergy, or immune deviation (Munn et al., 2002).

IDO$^+$ DCs can be generated in vitro from monocyte-derived DCs, and this has also been found in vivo, in breast tumor tissue, and tumor-draining lymph nodes in patients with melanoma, breast, colon, lung, and pancreatic cancers (Munn et al., 2002). Emerging evidence suggests that IDO$^+$ DCs contribute to tumor-mediated immunosuppression (Munn et al., 2002). Moreover, tumor cells frequently overexpress IDO, rendering them immunosuppressive in vitro and in vivo, and systemic treatment with a bioactive inhibitor of IDO, 1 methyltryptophan (1-MT), significantly delays tumor growth in mouse models (Uyttenhove et al., 2003). Notably, in murine DCs, IDO expression is upregulated by cytotoxic T lymphocyte-associated antigen 4 immunoglobulin (CTLA-4–Ig) (Grohmann et al., 2002). Furthermore, CTLA-4-expressing T regulatory cells (see following paragraphs) can induce IDO expression in tumor environmental DCs, effectively converting them into regulatory DCs (Curiel et al., 2004b). It is unclear whether myeloid DCs expressing the coregulatory molecules B7-H1 or B7-H4 are phenotypically identical to IDO-expressing myeloid DCs in the human tumor environment. Nonetheless, the presence of IDO$^+$ myeloid DCs in tumor-draining lymph nodes and in the tumor environment would further tip the balance of tumor environmental DC subsets in favor of tolerable conditions.

DCs were initially thought to be exclusively immunogenic, actively inducing or upregulating immune responses. However, studies have demonstrated that DCs possess dual functions, including both stimulatory and regulatory (suppressive) activities. In essence, regulatory DCs are able to actively downregulate an immune response or to induce immune tolerance by influencing the activity of other cell types. It appears that DCs present in the tumor microenvironment primary possess regulatory function. Although it is possible that professional "regulatory" DCs exist, the authors of this chapter favor the view that a regulatory function is not the intrinsic nature of a distinct DC subset (Curiel et al., 2003, 2004a; Gabrilovich et al., 1996; Munn and Mellor, 2004; Zou et al., 2001). In this view, DCs recruited to the tumor microenvironment undergo changes that endow them with regulatory functions that are favorable for the tumor. Thus, it can be hypothesized that through a different "instruction," this process could be reversed for therapeutic purposes.

In summary, immature or partially differentiated DCs can function as regulatory DCs and comprise an important component of the immunosuppressive network in the tumor microenvironment. The discussion in this chapter suggests that the imbalanced mature and immature DC subset distribution in this microenvironment further contributes to tumor immune evasion. An important focus of present research is exploration of the underlying mechanisms by which tumor DCs induce suppressive T cells.

III. IMBALANCE BETWEEN STIMULATORY AND INHIBITORY B7 FAMILY MOLECULES

A. B7-H1 in the Tumor Microenvironment

B7-H1 (PD-L1) is a B7 family member with approximately 25% homology between B7.1, B7.2, and B7-H1 (Choi et al., 2003; Dong et al., 1999; Tamura et al., 2001). IL-10 and VEGF stimulate B7-H1 expression in myeloid DCs present in ovarian tumors and their draining lymph nodes. A significant fraction of tumor-associated T cells are regulatory T cells (CD4$^+$CD25$^+$FOXP3$^+$) (Curiel et al., 2004b) (see following paragraphs), which express PD-1, the ligand for B7-H1 (Chen, 2004). Tumor-associated T cells can then, through reverse signaling via B7-H1, suppress IL-12 production by myeloid DCs, thus reducing their immunogenicity (Curiel et al., 2003). Blocking B7-H1 enhances myeloid DC-mediated T cell activation immunity (Brown et al., 2003; Curiel et al., 2003; Strome et al., 2003) and reduces the growth of a transplanted human ovarian carcinoma in non-obese diabetic/severe immune deficient (NOD/SCID) mice with adoptively transferred autologous human TAA-specific T cells (Curiel et al., 2003). Induction of B7-H1 on myeloid DCs by tumor microenvironmental factors is a novel mechanism for tumor immune evasion (Curiel et al., 2003). On the other hand, expression of B7-H1 on human tumors, such as in ovarian cancer, lung cancer, melanoma, glioblastoma, and squamous cell carcinoma (Dong et al., 2002), also contributes to immune evasion by inducing effector T cell apoptosis (Dong et al., 2002), thus facilitating tumor growth (Strome et al., 2003). PD-1 is one of the ligands for B7-H1. PD-1 blockade by genetic manipulation (PD-1$^{-/-}$) or antibody treatment efficiently inhibits mouse B16 melanoma and CT26 colon cancer dissemination and metastasis accompanied with enhanced effector T cell number and function (Iwai et al., 2002, 2005).

It is noteworthy that the complex interactions among B7-H1, its receptors, and other ligands, as well as between B7-H1 expressing cells and effector T cells, still remain largely unknown. Some experiments demonstrate that B7-H1 positively regulates T cell activation. Ligation of T cells by B7-H1 stimulates CD28-dependent T cell activation (Dong et al., 1999, 2002; Petroff et al., 2002; Tamura et al., 2001). Consistent with this observation, transgenic expression of B7-H1 on mouse pancreatic islets promotes islet allograft rejection in a minor mismatch setting (Subudhi et al., 2004). There are several possibilities to explain the plausible mechanisms of this dual nature of B7-H1 in T cell activation. The first possibility is that PD-1, so far the only identified receptor of B7-H1, contains an immunoreceptor tyrosine inhibition motif (ITIM) and an immunoreceptor tyrosine-based switch motif (ITSM) domain that can transduce dual signals. The second possibility is that B7-H1 interacts with an additional unidentified receptor different from PD-1. Several groups have provided evidence indicating the existence of another putative receptor for B7-H1. The third possibility is that the quantitative and kinetic expression of B7-H1, as well as the potentially different affinity of B7-H1 bound to its receptors, would account for different outcomes of T cell activation. Nonetheless, tumor and tumor environmental DCs apparently contribute to immune evasion through B7-H1.

In summary, epithelial tumor cells and associated myeloid DCs highly express B7-H1 and mediate T cell apoptosis or attenuate T cell activation. The significant influence of tumor environmental B7-H1 and PD-1 on the interaction between T cells, tumor cells, and APCs constitutes a novel target for tumor immunotherapy.

B. B7-H4 in the Tumor Environment

APCs are critical for initiating and maintaining TAA-specific T cell immunity. Tumor-associated macrophages (TAMs) function as one class of APC in the tumor microenvironment. TAMs markedly outnumber dendritic cells (DCs) and other APCs and they represent an abundant population of APCs in solid tumors (Mantovani *et al.*, 2002; Pollard, 2004; Vakkila and Lotze, 2004; Wyckoff *et al.*, 2004). Numerous studies have investigated the phenotypes and functions of DCs in tumor immunity (Cerundolo *et al.*, 2004; Curiel *et al.*, 2003; Finn, 2003; Gabrilovich, 2004; Gilboa, 2004; Munn and Mellor, 2004; O'Neill *et al.*, 2004; Pardoll, 2003; Zou, 2005; Zou *et al.*, 2001). TAMs are thought to suppress TAA-specific immunity in cancer patients (Mantovani *et al.*, 2002; Pollard, 2004; Vakkila and Lotze, 2004; Wyckoff *et al.*, 2004) and studies in mice show that TAMs directly act on tumor cells to promote growth and metastasis (Pollard, 2004; Vakkila and Lotze, 2004; Wyckoff *et al.*, 2004). However, immunohistochemical assessment of the number and the distribution of TAMs in human tumors has yielded scant, and often contradictory results regarding their potential role in tumor pathogenesis (Bingle *et al.*, 2002; Ohno *et al.*, 2003; Zavadova *et al.*, 1999). Thus, immune functional data are essential for understanding the roles and potential suppressive mechanisms of macrophages in human tumor microenvironment. In recent work, B7-H4 and TAM-mediated immune suppression have been linked in ovarian cancer.

B7-H4 (B7x, B7S1) is a recently-identified member of the B7 family (Chen, 2004; Prasad *et al.*, 2003; Sica *et al.*, 2003; Zang *et al.*, 2003). Although B7-H4 mRNA expression was found in multiple tissues and organs in normal donors, B7-H4 protein expression is rare. Strikingly, human ovarian tumor-associated macrophages (TAMs) and human cancers of the lung, breast, ovary, and renal cell have been shown to aberrantly express the B7-H4 protein (Choi *et al.*, 2003; Krambeck *et al.*, 2006; Kryczek *et al.*, 2006b; Sica *et al.*, 2003). However, it appears that TAMs express both intracellular and surface B7-H4 protein, whereas primary ovarian tumor cells and established ovarian tumor cell lines exclusively express intracellular B7-H4 (Kryczek *et al.*, 2006a, 2006b).

B7-H4 is a negative regulator of T cell responses *in vitro* by inhibiting T cell proliferation, cell cycle progression, and cytokine production (Chen, 2004; Prasad *et al.*, 2003; Sica *et al.*, 2003; Zang *et al.*, 2003). Antigen-specific T cell responses were impaired in mice treated with a B7-H4-Ig fusion protein (Sica *et al.*, 2003). Kryczek *et al.* document that B7-H4$^+$ TAMs significantly inhibit TAA-specific T cell proliferation, cytokine production, and cytotoxicity *in vitro* (2006b). These B7-H4$^+$ TAMs also inhibit TAA-specific immunity *in vivo* and foster tumor growth in chimeric NOD/SCID mice bearing autologous human tumors, despite the presence of potent TAA-specific effector T cells. The notion that B7-H4 TAM signals contribute to immunopathology is supported by several lines of evidence. First, B7-H4$^+$ TAMs are significantly more suppressive than B7-H4⁻ TAMs. Second, blocking B7-H4 on tumor-conditioned macrophages disables their suppressive capacity. Third, forced B7-H4 expression renders normal macrophages suppressive. Fourth, blocking B7-H1 and inhibiting iNOS and arginase have minor effects on B7-H4⁺ macrophage-mediated T cell suppression. It remains to be defined whether ovarian tumor-associated myeloid suppressor cells and ovarian tumor cells would release the immunosuppressive agents iNOS and arginase, causing T cell suppression. These findings establish B7-H4⁺ TAMs as a novel immune regulatory population in human ovarian cancer. The data here indicate that the suppressive potency of B7-H4⁺ TAM is similar to that of CD4⁺ Tregs (Curiel et al., 2004b). Interestingly, B7-H4⁺ TAMs and CD4⁺ Tregs, the two functionally suppressive immune cell populations, are not only physically localized in the ovarian tumor environment (Curiel et al., 2004b), but observations also suggest a mechanistic link between B7-H4⁺ APCs (including macrophages) and CD4⁺ Tregs (Kryczek et al., 2006a, 2006b). In fact, CD4⁺ Tregs stimulate APC B7-H4 expression and enable APC suppressive activity through inducing B7-H4 (Kryczek et al., 2006a). As B7-H4⁺ TAMs significantly outnumber CD4⁺ Tregs in ascites and the solid tumor mass, their contribution to tumor immune evasion is likely significant. Findings clearly establish the presence of a novel suppressor cell population in human cancer that forces a reexamination of the relative importance of regulatory T cells in the immunopathogenesis of cancer and a rethinking of strategies to improve TAA-specific immunity through abrogation of suppressor cell function (Cerundolo et al., 2004; Finn, 2003; Gilboa, 2004; Khong and Restifo, 2002; Munn and Mellor, 2004; Pardoll, 2003; Schreiber et al., 2002; Yu et al., 2004; Zou, 2006).

Kryczek et al. found that recombinant and tumor environmental IL-6 and IL-10 stimulate monocyte/macrophage B7-H4 expression and that GM-CSF and IL-4 reduce B7-H4 expression. Kryczek et al. also observed a similar control mechanism for B7-H4 regulation on myeloid DCs (Kryczek et al., 2006a, 2006b). Tumor cells, TAMs, and Tregs may be the source for IL-6 and IL-10 (Curiel et al., 2004b; Zou et al., 2001). These data provide mechanisms for how tumor environmental IL-6 and IL-10 induce immune dysfunction. GM-CSF has been used to boost TAA-specific immunity in mouse cancer models (Levitsky et al., 1994; van Elsas et al., 2001). A proposed mechanism for GM-CSF efficacy in these models is differentiation or attraction of DCs that boost TAA-specific immunity. In light of the work of Kryczek et al., it will be interesting and worthwhile to reexamine these GM-CSF studies to determine whether a GM-CSF-mediated reduction in B7-H4 APC expression accounts for efficacy. Further, cytokines IL-4, GM-CSF, IL-6, and IL-10 have no regulatory effects on tumor B7-H4 expression. The data suggest that tumor B7-H4 and B7-H4 APC may be functionally distinct and be differentially regulated. Although renal cell carcinoma B7-H4 expression was proposed to be associated with clinical progression and patient outcome (Krambeck et al., 2006), the role and functional mechanism of tumor B7-H4 remain to be defined. Furthermore, the potential complex interactions among B7-H4, its receptors, and other ligands, as well as between B7-H4 expressing cells and effector T cells, remain unknown (Chen, 2004).

In summary, B7-H4 is highly expressed in macrophages and epithelial tumor cells in the tumor environment. B7-H4-expressing APCs suppress T cell activation, and in cancers such as renal cell carcinoma, B7-H4 is associated with tumor progression and affects patient outcome. B7-H4 in the tumor microenvironment contributes to tumor immune evasion, and targeting B7-H4 or the signaling pathway it controls may offer novel modalities to reverse cancer immunosuppression.

C. Cytokine Networks and B7 Family Members in the Tumor Microenvironment

Pathological cytokine and chemokine networks that support tumor cell growth, survival, and movement clearly exist in the tumor microenvironment. The dysfunctional chemokine network has been reviewed and discussed in the literature (Balkwill, 2004; Balkwill and Mantovani, 2001; Scotton et al., 2001) as well as Chapters 2, 4, and 16 of this book. This chapter briefly discusses the cytokine networks that contribute to immune suppression in the microenvironment of human tumors (Dranoff, 2004). In this setting, there is a paucity of molecules promoting DC differentiation and function (e.g., GM-CSF and IL-4) and cytokines inducing Th1-type responses (e.g., IL-12, IL-18, and IFN-γ), but there is also an abundance of molecules suppressing DC differentiation and function (e.g., VEGF, IL-6, IL-10, TGF-β, MCSF, arginase, IDO, PGE$_2$, COX-2, and NOS). These suppressive molecules are largely produced by tumor cells, macrophages, and other tumor stromal cells. This aberrant pattern profoundly affects tumor-specific immunity through regulating the expression of B7 family members on tumor-associated APCs (Zou, 2005). As discussed previously, tumor cells and tumor-associated APCs highly express the inhibitory B7 family members B7-H1 and B7-H4. In contrast, the expression of the stimulatory B7 family members CD80 and CD86 is limited. Besides inhibiting DC differentiation and maturation, tumor environmental factors such as IL-10 selectively modulate the expression of B7 family members so as to tilt the balance toward immune suppression: inhibitory molecules, including B7-H1 and B7-H4, are upregulated, whereas stimulatory molecules, including CD80/B7-1 or CD86/B7-2, are downregulated. Considering that B7-H1 can exhibit immune stimulatory function (Subudhi et al., 2004; Tamura et al., 2001), it remains to be defined why tumor environmental B7-H1 exclusively mediates an immunosuppressive effect. Nonetheless, an imbalance between costimulatory molecules (B7.1, B7.2) and coinhibitory molecules (B7H1, B7H4) is created in the tumor environment, spreading into tumor-draining lymph nodes and "instructing" immunogenic APCs to become regulatory APCs. Coinhibitory molecules become the tumor's "mask and weapon," initially to avoid immune attack and then to reduce T cell priming and defeat the invasion of effector T cells.

IV. IMBALANCE BETWEEN REGULATORY T CELLS AND CONVENTIONAL T CELLS

Regulatory T cells or Tregs are functionally defined as T cells that inhibit an immune response by influencing the activity of another cell type (Shevach, 2004). The most well-defined Treg cells are CD4$^+$CD25$^+$ T cells (Sakaguchi et al., 2001; Shevach, 2002; Von Herrath and Harrison, 2003). The antigen specificity, phenotype, mechanisms of action, and function of Treg cells have been discussed in the literature (Chen et al., 2005; Green et al., 2003; Peng et al., 2004; Wang et al., 2004; Wang and Wang, 2005) and are discussed in Chapter 15 of this book. The immunosuppressive cell surface molecule CTLA-4 is conceptually related to Treg cell function. The role of CTLA-4 and the effects of CTLA-4 blockade in mouse and human cells have been discussed (Chapters 10 and 12). In this chapter, the focus will be the distribution and trafficking of Treg cells in human tumors and the imbalance between Treg cells and conventional T cells in the tumor environment.

A. Imbalance Between Treg Cells and Conventional T Cells in the Bone Marrow

Naturally occurring murine CD4$^+$CD25$^+$ Treg cells differentiate in the thymus

(Francois Bach, 2003; Shevach, 2002; Von Herrath and Harrison, 2003; Wood and Sakaguchi, 2003). In homeostatic conditions in the human and mouse, CD4$^+$CD25$^+$ Treg cells are found primarily in thymus, peripheral blood, lymph nodes, and spleen (Sakaguchi, 2005; Shevach, 2002). Under these conditions, it appears that Treg cells exhibit an equal distribution among the different lymphoid compartments. Notably, more than 25% of CD4$^+$ T cells are phenotypically and functionally Treg cells in normal bone marrow (Zou et al., 2004). The chapter authors' unpublished data further show that the prevalence of Treg cells reach up to 50% in the bone marrow in patients with prostate cancer. The data suggest that in a homeostatic situation, there exists an imbalance between Treg cells and conventional T cells in the bone marrow, and this imbalance is further enforced in cancer patients.

Interestingly, a number of reports have shown that functional memory T cells exist in bone marrow, where they can serve as a site for naive TAA-specific T cell priming (Becker et al., 2005; Feuerer et al., 2003; Mazo et al., 2005; Tripp et al., 1997). TAA-specific T cells isolated from bone marrow in tumor-bearing mice and cancer patients are functional *in vitro* and are able to prevent tumor growth upon transfer to another host. These observations suggest that these TAA-specific T cells are functionally suppressed in the bone marrow, possibly by Tregs (Feuerer et al., 2001a, 2001b, 2003; Mazo et al., 2005; Tripp et al., 1997). Therefore, it is evident that the bone marrow plays an active role in humoral and cellular lymphocyte immunity.

The notion has been further supported by the observation that large numbers of functional CD4$^+$ Treg cells accumulate in the bone marrow of healthy volunteers and mice (Zou et al., 2004). The observation is confirmed in a mouse model in which a red fluorescent protein reporter was knocked into the endogenous genomic locus for the specific Treg marker FOXP3 (Wan and Flavell, 2005). Further, bone marrow CD4$^+$ Treg cells express functional CXCR4, the receptor for CXCL12, and granulocyte-colony-stimulating factor (G-CSF) mobilizes CD4$^+$ Treg release from bone marrow through reducing marrow CXCL12 (Zou et al., 2004). Interestingly, activation upregulates CXCR4 expression and enables CD4$^+$ Treg cells to migrate to the bone marrow through CXCL12 (Zou et al., 2004). Thus, CXCR4/CXCL12 signals are crucial for activated CD4$^+$ Treg cell bone marrow trafficking. This finding suggests that bone marrow is a preferential site for migration and/or selective retainment of CD4$^+$ Treg cells, and bone marrow may function as an immunoregulatory organ (Zou et al., 2004).

In summary, Treg cells accumulate to high levels in the bone marrow, and their prevalence appears to be substantially elevated in cancer patients. Bone marrow is a common site for metastasis of many human epithelial tumors, such as those associated with prostate cancer. At present, findings suggest that bone marrow provides an immunosuppressive environment for tumor retention and growth. Imbalance between Treg cells and conventional T cells in the bone marrow may provide an immune shield to facilitate metastatic spread at this site.

B. Imbalance Between Treg Cells and Conventional T Cells in Human Tumors

In the twenty-first century, numerous studies have reported a much higher frequency of CD4$^+$CD25$^+$ T cells in peripheral blood and tumors in patients with a variety of major cancers, including cancers of the breast, colon, rectum, esophagus, stomach, liver, lung, ovary, and pancreas; this finding is also true for individuals with different types of melanomas, leukemias, and lymphomas (Liyanage et al., 2002). This list continues to grow with the rapidly emerging interest of how Tregs can contribute to human tumorigenesis. Together, these data offer compelling evidence of an imbalance between Treg cells and conventional T cells in the tumor environment.

How are elevated levels of Tregs generated in the tumor microenvironment? Evidence suggests several sources, including trafficking of Tregs into the tumor or generation of Tregs at the tumor site by differentiation, DC-induced expansion, or cell conversion. Tregs from the thymus, lymph nodes, bone marrow, and peripheral blood can traffic into the tumor. Regulatory T cells express CCR4, and abundant expression of CCL22 (the ligand for CCR4) in the tumor environment stimulates tumor infiltration by Treg cells (Curiel et al., 2004b). The tumor environment contains molecules capable of suppressing APC differentiation and function, including dysfunctional APCs, which in turn stimulate differentiation of Treg cells. DCs can also stimulate Treg cell expansion in the tumor environment and draining lymph nodes (Sakaguchi, 2001; Shevach, 2002; Zou, 2005). Last, conventional T cells can be converted into regulatory T cells by TGF-β, and high levels of TGF-β are commonly found in the tumor environment. In human ovarian tumor cells, CD4$^+$CD25$^+$ Treg cells block tumor-specific immunity, foster tumor growth, and predict poor patient survival (Curiel et al., 2004b).

In summary, Treg cells abnormally accumulate to high levels in many human tumors, often comprising a major fraction of immune cells present in the tumor environment. Their role is very important since they are likely to be critical in disabling TAA-specific effector T cell immunity in the cancer patient. Manipulation of regulatory T cells, including depletion, blocking trafficking into tumors, or reducing their differentiation and suppressive mechanisms, all represent potential innovative strategies for cancer treatment.

V. CONCLUDING REMARKS

The tumor microenvironment presents an overwhelming arsenal of active immune imbalances, including an imbalance of mature versus immature DCs (Zou et al., 2001), stimulatory versus inhibitory B7 family molecules (Chen, 2004; Curiel et al., 2003), and regulatory versus conventional T cells (Curiel et al., 2004b). The pathological interaction between tumor and host immune system results in formation of a "tolerance network" through establishing these irregular immune balances in the tumor environment. Exciting prospects exist for effective immune-based therapies against human cancer that are based on exploiting these mechanisms to provide new interventional strategies (Dunn et al., 2004; Finn, 2003; Laheru and Jaffee, 2005; Muller and Prendergast, 2005; Pardoll, 2003).

References

Balkwill, F. (2004). Cancer and the chemokine network. *Nat. Rev. Cancer* **4**, 540–550.

Balkwill, F., and Mantovani, A. (2001). Inflammation and cancer: Back to Virchow? *Lancet* **357**, 539–545.

Banchereau, J., and Steinman, R. M. (1998). Dendritic cells and the control of immunity. *Nature* **392**, 245–252.

Becker, T. C., Coley, S. M., Wherry, E. J., and Ahmed, R. (2005). Bone marrow is a preferred site for homeostatic proliferation of memory CD8 T cells. *J. Immunol.* **174**, 1269–1273.

Bell, D., Chomarat, P., Broyles, D., Netto, G., Harb, G. M., Lebecque, S., Valladeau, J., Davoust, J., Palucka, K. A., and Banchereau, J. (1999). In breast carcinoma tissue, immature dendritic cells reside within the tumor, whereas mature dendritic cells are located in peritumoral areas. *J. Exp. Med.* **190**, 1417–1426.

Bingle, L., Brown, N. J., and Lewis, C. E. (2002). The role of tumour-associated macrophages in tumour progression: Implications for new anticancer therapies. *J. Pathol.* **196**, 254–265.

Bronte, V., Kasic, T., Gri, G., Gallana, K., Borsellino, G., Marigo, I., Battistini, L., Iafrate, M., Prayer-Galetti, T., Pagano, F., and Viola, A. (2005). Boosting antitumor responses of T lymphocytes infiltrating human prostate cancers. *J. Exp. Med.* **201**, 1257–1268.

Bronte, V., Serafini, P., Apolloni, E., and Zanovello, P. (2001). Tumor-induced immune dysfunctions caused by myeloid suppressor cells. *J. Immunother.* **24**, 431–446.

Bronte, V., Serafini, P., De Santo, C., Marigo, I., Tosello, V., Mazzoni, A., Segal, D. M., Staib, C., Lowel, M., Sutter, G., Colombo, M. P., and Zanovello, P. (2003). IL-4-induced arginase 1 suppresses alloreactive T cells in tumor-bearing mice. *J. Immunol.* **170**, 270–278.

Bronte, V., and Zanovello, P. (2005). Regulation of immune responses by L-arginine metabolism. *Nat. Rev. Immunol.* **5**, 641–654.

Brown, J. A., Dorfman, D. M., Ma, F. R., Sullivan, E. L., Munoz, O., Wood, C. R., Greenfield, E. A., and Freeman, G. J. (2003). Blockade of programmed death-1 ligands on dendritic cells enhances T cell activation and cytokine production. *J. Immunol.* **170**, 1257–1266.

Bunt, S. K., Sinha, P., Clements, V. K., Leips, J., and Ostrand-Rosenberg, S. (2006). Inflammation induces myeloid-derived suppressor cells that facilitate tumor progression. *J. Immunol.* **176**, 284–290.

Burnet, F. M. (1957). Cancer: A biological approach. *Br. Med. J.* **1**, 841–847.

Carmeliet, P., and Jain, R. K. (2000). Angiogenesis in cancer and other diseases. *Nature* **407**, 249–257.

Cerundolo, V., Hermans, I. F., and Salio, M. (2004). Dendritic cells: A journey from laboratory to clinic. *Nat. Immunol.* **5**, 7–10.

Chen, L. (2004). Co-inhibitory molecules of the B7-CD28 family in the control of T-cell immunity. *Nat. Rev. Immunol.* **4**, 336–347.

Chen, M. L., Pittet, M. J., Gorelik, L., Flavell, R. A., Weissleder, R., von Boehmer, H., and Khazaie, K. (2005). Regulatory T cells suppress tumor-specific CD8 T cell cytotoxicity through TGF-beta signals *in vivo. Proc. Natl. Acad. Sci. USA* **102**, 419–424.

Choi, I. H., Zhu, G., Sica, G. L., Strome, S. E., Cheville, J. C., Lau, J. S., Zhu, Y., Flies, D. B., Tamada, K., and Chen, L. (2003). Genomic organization and expression analysis of B7-H4, an immune inhibitory molecule of the B7 family. *J. Immunol.* **171**, 4650–4654.

Colonna, M., Trinchieri, G., and Liu, Y. J. (2004). Plasmacytoid dendritic cells in immunity. *Nat. Immunol.* **5**, 1219–1226.

Curiel, T. J., Cheng, P., Mottram, P., Alvarez, X., Moons, L., Evdemon-Hogan, M., Wei, S., Zou, L., Kryczek, I., Hoyle, G., Lackner, A., Carmeliet, P., and Zou, W. (2004a). Dendritic cell subsets differentially regulate angiogenesis in human ovarian cancer. *Cancer Res.* **64**, 5535–5538.

Curiel, T. J., Coukos, G., Zou, L., Alvarez, X., Cheng, P., Mottram, P., Evdemon-Hogan, M., Conejo-Garcia, J. R., Zhang, L., Burow, M., Zhu, Y., Wei, S., Kryczek, I., Daniel, B., Gordon, A., Myers, L., Lackner, A., Disis, M. L., Knutson, K. L., Chen, L., and Zou, W. (2004b). Specific recruitment of regulatory T cells in ovarian carcinoma fosters immune privilege and predicts reduced survival. *Nat. Med.* **10**, 942–949.

Curiel, T. J., Wei, S., Dong, H., Alvarez, X., Cheng, P., Mottram, P., Krzysiek, R., Knutson, K. L., Daniel, B., Zimmermann, M. C., David, O., Burow, M., Gordon, A., Dhurandhar, N., Myers, L., Berggren, R., Hemminki, A., Alvarez, R. D., Emilie, D., Curiel, D. T., Chen, L., and Zou, W. (2003). Blockade of B7-H1 improves myeloid dendritic cell-mediated antitumor immunity. *Nat. Med.* **21**, 21.

Dhodapkar, M. V., Steinman, R. M., Krasovsky, J., Munz, C., and Bhardwaj, N. (2001). Antigen-specific inhibition of effector T cell function in humans after injection of immature dendritic cells. *J. Exp. Med.* **193**, 233–238.

Dhodapkar, M. V., Steinman, R. M., Sapp, M., Desai, H., Fossella, C., Krasovsky, J., Donahoe, S. M., Dunbar, P. R., Cerundolo, V., Nixon, D. F., and Bhardwaj, N. (1999). Rapid generation of broad T-cell immunity in humans after a single injection of mature dendritic cells. *J. Clin. Invest.* **104**, 173–180.

Dong, H., Strome, S. E., Salomao, D. R., Tamura, H., Hirano, F., Flies, D. B., Roche, P. C., Lu, J., Zhu, G., Tamada, K., Lennon, V. A., Celis, E., and Chen, L. (2002). Tumor-associated B7-H1 promotes T-cell apoptosis: A potential mechanism of immune evasion. *Nat. Med.* **8**, 793–800.

Dong, H., Zhu, G., Tamada, K., and Chen, L. (1999). B7-H1, a third member of the B7 family, co-stimulates T-cell proliferation and interleukin-10 secretion. *Nat. Med.* **5**, 1365–1369.

Dranoff, G. (2004). Cytokines in cancer pathogenesis and cancer therapy. *Nat. Rev. Cancer* **4**, 11–22.

Dunn, G. P., Old, L. J., and Schreiber, R. D. (2004). The immunobiology of cancer immunosurveillance and immunoediting. *Immunity* **21**, 137–148.

Feuerer, M., Beckhove, P., Bai, L., Solomayer, E. F., Bastert, G., Diel, I. J., Pedain, C., Oberniedermayr, M., Schirrmacher, V., and Umansky, V. (2001a). Therapy of human tumors in NOD/SCID mice with patient-derived reactivated memory T cells from bone marrow. *Nat. Med.* **7**, 452–458.

Feuerer, M., Beckhove, P., Garbi, N., Mahnke, Y., Limmer, A., Hommel, M., Hammerling, G. J., Kyewski, B., Hamann, A., Umansky, V., and Schirrmacher, V. (2003). Bone marrow as a priming site for T-cell responses to blood-borne antigen. *Nat. Med.* **9**, 1151–1157.

Feuerer, M., Rocha, M., Bai, L., Umansky, V., Solomayer, E. F., Bastert, G., Diel, I. J., and Schirrmacher, V. (2001b). Enrichment of memory T cells and other profound immunological changes in the bone marrow from untreated breast cancer patients. *Int. J. Cancer* **92**, 96–105.

Figdor, C. G., de Vries, I. J., Lesterhuis, W. J., and Melief, C. J. (2004). Dendritic cell immunotherapy: Mapping the way. *Nat. Med.* **10**, 475–480.

Finn, O. J. (2003). Cancer vaccines: Between the idea and the reality. *Nat. Rev. Immunol.* **3**, 630–641.

Francois Bach, J. (2003). Regulatory lymphocytes: Regulatory T cells under scrutiny. *Nat. Rev. Immunol.* **3**, 189–198.

Freedman, R. S., Deavers, M., Liu, J., and Wang, E. (2004). Peritoneal inflammation—A microenvironment for epithelial ovarian cancer (EOC). *J. Transl. Med.* **2**, 23.

Gabrilovich, D. (2004). Mechanisms and functional significance of tumour-induced dendritic-cell defects. *Nat. Rev. Immunol.* **4**, 941–952.

Gabrilovich, D. I., Chen, H. L., Girgis, K. R., Cunningham, H. T., Meny, G. M., Nadaf, S., Kavanaugh, D., and Carbone, D. P. (1996). Production of vascular endothelial growth factor by human tumors inhibits the functional maturation of dendritic cells [published erratum; appears in *Nat. Med.* 1996 2(11):1267]. *Nat Med* **2**, 1096–1103.

Gilboa, E. (2004). The promise of cancer vaccines. *Nat. Rev. Cancer* **4**, 401–411.

Gilliet, M., and Liu, Y. J. (2002). Generation of human CD8 T regulatory cells by CD40 ligand-activated plasmacytoid dendritic cells. *J. Exp. Med.* **195**, 695–704.

Green, E. A., Gorelik, L., McGregor, C. M., Tran, E. H., and Flavell, R. A. (2003). CD4$^+$CD25$^+$ T regulatory cells control anti-islet CD8$^+$ T cells through TGF-beta-TGF-beta receptor interactions in type 1 diabetes. *Proc. Natl. Acad. Sci. USA* **100**, 10878–10883.

Grohmann, U., Orabona, C., Fallarino, F., Vacca, C., Calcinaro, F., Falorni, A., Candeloro, P., Belladonna, M. L., Bianchi, R., Fioretti, M. C., and Puccetti, P. (2002). CTLA-4-Ig regulates tryptophan catabolism *in vivo*. *Nat. Immunol.* **3**, 1097–1101.

Hartmann, E., Wollenberg, B., Rothenfusser, S., Wagner, M., Wellisch, D., Mack, B., Giese, T., Gires, O., Endres, S., and Hartmann, G. (2003). Identification and functional analysis of tumor-infiltrating plasmacytoid dendritic cells in head and neck cancer. *Cancer Res.* **63**, 6478–6487.

Hawiger, D., Inaba, K., Dorsett, Y., Guo, M., Mahnke, K., Rivera, M., Ravetch, J. V., Steinman, R. M., and Nussenzweig, M. C. (2001). Dendritic cells induce peripheral T cell unresponsiveness under steady state conditions *in vivo*. *J. Exp. Med.* **194**, 769–779.

Iwai, Y., Ishida, M., Tanaka, Y., Okazaki, T., Honjo, T., and Minato, N. (2002). Involvement of PD-L1 on tumor cells in the escape from host immune system and tumor immunotherapy by PD-L1 blockade. *Proc. Natl. Acad. Sci. USA* **99**, 12293–12297.

Iwai, Y., Terawaki, S., and Honjo, T. (2005). PD-1 blockade inhibits hematogenous spread of poorly immunogenic tumor cells by enhanced recruitment of effector T cells. *Int. Immunol.* **17**, 133–144.

Iwamoto, M., Shinohara, H., Miyamoto, A., Okuzawa, M., Mabuchi, H., Nohara, T., Gon, G., Toyoda, M., and Tanigawa, N. (2003). Prognostic value of tumor-infiltrating dendritic cells expressing CD83 in human breast carcinomas. *Int. J. Cancer* **104**, 92–97.

Joki, T., Heese, O., Nikas, D. C., Bello, L., Zhang, J., Kraeft, S. K., Seyfried, N. T., Abe, T., Chen, L. B., Carroll, R. S., and Black, P. M. (2000). Expression of cyclooxygenase 2 (COX-2) in human glioma and *in vitro* inhibition by a specific COX-2 inhibitor, NS-398. *Cancer Res.* **60**, 4926–4931.

Jonuleit, H., Schmitt, E., Schuler, G., Knop, J., and Enk, A. H. (2000). Induction of interleukin 10-producing, nonproliferating CD4($^+$) T cells with regulatory properties by repetitive stimulation with allogeneic immature human dendritic cells. *J. Exp. Med.* **192**, 1213–1222.

Kalinski, P., Schuitemaker, J. H., Hilkens, C. M., and Kapsenberg, M. L. (1998). Prostaglandin E$_2$ induces the final maturation of IL-12-deficient CD1a+CD83+ dendritic cells: The levels of IL-12 are determined during the final dendritic cell maturation and are resistant to further modulation. *J. Immunol.* **161**, 2804–2809.

Khong, H. T., and Restifo, N. P. (2002). Natural selection of tumor variants in the generation of "tumor escape" phenotypes. *Nat. Immunol.* **3**, 999–1005.

Krambeck, A. E., Thompson, R. H., Dong, H., Lohse, C. M., Park, E. S., Kuntz, S. M., Leibovich, B. C., Blute, M. L., Cheville, J. C., and Kwon, E. D. (2006). B7-H4 expression in renal cell carcinoma and tumor vasculature: Associations with cancer progression and survival. *Proc. Natl. Acad. Sci. USA* **103**, 10391–10396.

Kryczek, I., Grybos, M., Karabon, L., Klimczak, A., and Lange, A. (2000). IL-6 production in ovarian carcinoma is associated with histiotype and biological characteristics of the tumour and influences local immunity. *Br. J. Cancer* **82**, 621–628.

Kryczek, I., Lange, A., Mottram, P., Alvarez, X., Cheng, P., Hogan, M., Moons, L., Wei, S., Zou, L., Machelon, V., Emilie, D., Terrassa, M., Lackner, A., Curiel, T. J., Carmeliet, P., and Zou, W. (2005). CXCL12 and vascular endothelial growth factor synergistically induce neoangiogenesis in human ovarian cancers. *Cancer Res.* **65**, 465–472.

Kryczek, I., Wei, S., Zou, L., Zhu, G., Mottram, P., Xu, H., Chen, L., and Zou, W. (2006a). Cutting Edge: Induction of B7-H4 on APCs through IL-10: Novel suppressive mode for regulatory T cells. *J. Immunol.* **177**, 40–44.

Kryczek, I., Zou, L., Rodriguez, P., Zhu, G., Wei, S., Mottram, P., Brumlik, M., Cheng, P., Curiel, T., Myers, L., Lackner, A., Alvarez, X., Ochoa, A., Chen, L., and Zou, W. (2006b). B7-H4 expression identifies a novel suppressive macrophage population in human ovarian carcinoma. *J. Exp. Med.* **203**, 871–881.

Labeur, M. S., Roters, B., Pers, B., Mehling, A., Luger, T. A., Schwarz, T., and Grabbe, S. (1999). Generation of tumor immunity by bone marrow-derived dendritic cells correlates with dendritic cell maturation stage. *J. Immunol.* **162**, 168–175.

Laheru, D., and Jaffee, E. M. (2005). Immunotherapy for pancreatic cancer—science driving clinical progress. *Nat. Rev. Cancer* **5**, 459–467.

Lanzavecchia, A., and Sallusto, F. (2001). The instructive role of dendritic cells on T cell responses: Lineages, plasticity and kinetics. *Curr. Opin. Immunol.* **13**, 291–298.

Levitsky, H. I., Lazenby, A., Hayashi, R. J., and Pardoll, D. M. (1994). In vivo priming of two distinct antitumor effector populations: The role of MHC class I expression. *J. Exp. Med.* **179**, 1215–1224.

Liyanage, U. K., Moore, T. T., Joo, H. G., Tanaka, Y., Herrmann, V., Doherty, G., Drebin, J. A., Strasberg, S. M., Eberlein, T. J., Goedegebuure, P. S., and Linehan, D. C. (2002). Prevalence of regulatory T cells is increased in peripheral blood and tumor microenvironment of patients with pancreas or breast adenocarcinoma. *J. Immunol.* **169**, 2756–2761.

Lutz, M. B., and Schuler, G. (2002). Immature, semimature and fully mature dendritic cells: Which signals induce tolerance or immunity? *Trends Immunol.* **23**, 445–449.

Mantovani, A., Sozzani, S., Locati, M., Allavena, P., and Sica, A. (2002). Macrophage polarization: Tumor-associated macrophages as a paradigm for polarized M2 mononuclear phagocytes. *Trends Immunol.* **23**, 549–555.

Mazo, I. B., Honczarenko, M., Leung, H., Cavanagh, L. L., Bonasio, R., Weninger, W., Engelke, K., Xia, L., McEver, R. P., Koni, P. A., Silberstein, L. E., and von Andrian, U. H. (2005). Bone marrow is a major reservoir and site of recruitment for central memory CD8$^+$ T cells. *Immunity* **22**, 259–270.

Menetrier-Caux, C., Montmain, G., Dieu, M. C., Bain, C., Favrot, M. C., Caux, C., and Blay, J. Y. (1998). Inhibition of the differentiation of dendritic cells from CD34($^+$) progenitors by tumor cells: Role of interleukin-6 and macrophage colony-stimulating factor. *Blood* **92**, 4778–4791.

Moseman, E. A., Liang, X., Dawson, A. J., Panoskaltsis-Mortari, A., Krieg, A. M., Liu, Y. J., Blazar, B. R., and Chen, W. (2004). Human plasmacytoid dendritic cells activated by CpG oligodeoxynucleotides induce the generation of CD4$^+$CD25$^+$ regulatory T cells. *J. Immunol.* **173**, 4433–4442.

Muller, A. J., DuHadaway, J. B., Donover, P. S., Sutanto-Ward, E., and Prendergast, G. C. (2005a). Inhibition of indoleamine 2,3-dioxygenase, an immunoregulatory target of the cancer suppression gene Bin1, potentiates cancer chemotherapy. *Nat. Med.* **11**, 312–319.

Muller, A. J., Malachowski, W. P., and Prendergast, G. C. (2005b). Indoleamine 2,3-dioxygenase in cancer: Targeting pathological immune tolerance with small-molecule inhibitors. *Expert Opin. Ther. Targets* **9**, 831–849.

Muller, A. J., and Prendergast, G. C. (2005). Marrying immunotherapy with chemotherapy: Why say IDO? *Cancer Res.* **65**, 8065–8068.

Munn, D. H., and Mellor, A. L. (2004). IDO and tolerance to tumors. *Trends Mol. Med.* **10**, 15–18.

Munn, D. H., Sharma, M. D., Lee, J. R., Jhaver, K. G., Johnson, T. S., Keskin, D. B., Marshall, B., Chandler, P., Antonia, S. J., Burgess, R., Slingluff, C. L., Jr., and Mellor, A. L. (2002). Potential regulatory function of human dendritic cells expressing indoleamine 2,3-dioxygenase. *Science* **297**, 1867–1870.

O'Neill, D. W., Adams, S., and Bhardwaj, N. (2004). Manipulating dendritic cell biology for the active immunotherapy of cancer. *Blood* **104**, 2235–2246.

Ohno, S., Suzuki, N., Ohno, Y., Inagawa, H., Soma, G., and Inoue, M. (2003). Tumor-associated macrophages: Foe or accomplice of tumors? *Anticancer Res.* **23**, 4395–4409.

Pardoll, D. (2003). Does the immune system see tumors as foreign or self? *Annu. Rev. Immunol.* **21**, 807–839.

Pasare, C., and Medzhitov, R. (2003). Toll pathway-dependent blockade of CD4$^+$CD25$^+$ T cell-mediated suppression by dendritic cells. *Science* **299**, 1033–1036.

Peguet-Navarro, J., Sportouch, M., Popa, I., Berthier, O., Schmitt, D., and Portoukalian, J. (2003). Gangliosides from human melanoma tumors impair dendritic cell differentiation from monocytes and induce their apoptosis. *J. Immunol.* **170**, 3488–3494.

Peng, Y., Laouar, Y., Li, M. O., Green, E. A., and Flavell, R. A. (2004). TGF-beta regulates in vivo expansion of Foxp3-expressing CD4$^+$CD25$^+$ regulatory T cells responsible for protection against diabetes. *Proc. Natl. Acad. Sci. USA* **101**, 4572–4577.

Petroff, M. G., Chen, L., Phillips, T. A., and Hunt, J. S. (2002). B7 family molecules: Novel immunomodulators at the maternal-fetal interface. *Placenta* **23 Suppl A**, S95–S101.

Pollard, J. W. (2004). Tumour-educated macrophages promote tumour progression and metastasis. *Nat. Rev. Cancer* **4**, 71–78.

Prasad, D. V., Richards, S., Mai, X. M., and Dong, C. (2003). B7S1, a novel B7 family member that negatively regulates T cell activation. *Immunity* **18**, 863–873.

Sakaguchi, S. (2005). Naturally arising Foxp3-expressing CD25$^+$CD4$^+$ regulatory T cells in immunological tolerance to self and non-self. *Nat. Immunol.* **6**, 345–352.

Sakaguchi, S., Sakaguchi, N., Shimizu, J., Yamazaki, S., Sakihama, T., Itoh, M., Kuniyasu, Y., Nomura, T., Toda, M., and Takahashi, T. (2001). Immunologic tolerance maintained by CD25$^+$CD4$^+$ regulatory T cells: Their common role in controlling autoimmunity, tumor immunity, and transplantation tolerance. *Immunol. Rev.* **182**, 18–32.

Salio, M., Cella, M., Vermi, W., Facchetti, F., Palmowski, M. J., Smith, C. L., Shepherd, D., Colonna, M., and Cerundolo, V. (2003). Plasmacytoid dendritic cells prime IFN-gamma-secreting melanoma-specific CD8 lymphocytes and are found in primary melanoma lesions. *Eur. J. Immunol.* **33**, 1052–1062.

Schreiber, H., Wu, T. H., Nachman, J., and Kast, W. M. (2002). Immunodominance and tumor escape. *Semin. Cancer Biol.* **12**, 25–31.

Schuler, G., Schuler-Thurner, B., and Steinman, R. M. (2003). The use of dendritic cells in cancer immunotherapy. *Curr. Opin. Immunol.* **15**, 138–147.

Scotton, C. J., Wilson, J. L., Milliken, D., Stamp, G., and Balkwill, F. R. (2001). Epithelial cancer cell migration: A role for chemokine receptors? *Cancer Res.* **61**, 4961–4965.

Shevach, E. M. (2002). CD4$^+$CD25$^+$ suppressor T cells: More questions than answers. *Nat. Rev. Immunol.* **2**, 389–400.

Shevach, E. M. (2004). Fatal attraction: Tumors beckon regulatory T cells. *Nat. Med.* **10**, 900–901.

Shono, T., Tofilon, P. J., Bruner, J. M., Owolabi, O., and Lang, F. F. (2001). Cyclooxygenase-2 expression in human gliomas: Prognostic significance and molecular correlations. *Cancer Res.* **61**, 4375–4381.

Shurin, G. V., Shurin, M. R., Bykovskaia, S., Shogan, J., Lotze, M. T., and Barksdale, E. M., Jr. (2001). Neuroblastoma-derived gangliosides inhibit dendritic cell generation and function. *Cancer Res.* **61**, 363–369.

Sica, G. L., Choi, I. H., Zhu, G., Tamada, K., Wang, S. D., Tamura, H., Chapoval, A. I., Flies, D. B., Bajorath, J., and Chen, L. (2003). B7-H4, a molecule of the B7 family, negatively regulates T cell immunity. *Immunity* **18**, 849–861.

Sinha, P., Clements, V. K., and Ostrand-Rosenberg, S. (2005). Reduction of myeloid-derived suppressor cells and induction of M1 macrophages facilitate the rejection of established metastatic disease. *J. Immunol.* **174**, 636–645.

Strome, S. E., Dong, H., Tamura, H., Voss, S. G., Flies, D. B., Tamada, K., Salomao, D., Cheville, J., Hirano, F., Lin, W., Kasperbauer, J. L., Ballman, K. V., and Chen, L. (2003). B7-H1 blockade augments adoptive T-cell immunotherapy for squamous cell carcinoma. *Cancer Res.* **63**, 6501–6505.

Subudhi, S. K., Zhou, P., Yerian, L. M., Chin, R. K., Lo, J. C., Anders, R. A., Sun, Y., Chen, L., Wang, Y., Alegre, M. L., and Fu, Y. X. (2004). Local expression of B7-H1 promotes organ-specific autoimmunity and transplant rejection. *J. Clin. Invest.* **113**, 694–700.

Tamura, H., Dong, H., Zhu, G., Sica, G. L., Flies, D. B., Tamada, K., and Chen, L. (2001). B7-H1 costimulation preferentially enhances CD28-independent T-helper cell function. *Blood* **97**, 1809–1816.

Tripp, R. A., Topham, D. J., Watson, S. R., and Doherty, P. C. (1997). Bone marrow can function as a lymphoid organ during a primary immune response under conditions of disrupted lymphocyte trafficking. *J. Immunol.* **158**, 3716–3720.

Troy, A., Davidson, P., Atkinson, C., and Hart, D. (1998a). Phenotypic characterisation of the dendritic cell infiltrate in prostate cancer. *J. Urol.* **160**, 214–219.

Troy, A. J., Summers, K. L., Davidson, P. J., Atkinson, C. H., and Hart, D. N. (1998b). Minimal recruitment and activation of dendritic cells within renal cell carcinoma. *Clin. Cancer Res.* **4**, 585–593.

Uyttenhove, C., Pilotte, L., Theate, I., Stroobant, V., Colau, D., Parmentier, N., Boon, T., and Van den Eynde, B. J. (2003). Evidence for a tumoral immune resistance mechanism based on tryptophan degradation by indoleamine 2,3-dioxygenase. *Nat. Med.* **9**, 1269–1274.

Vakkila, J., and Lotze, M. T. (2004). Inflammation and necrosis promote tumour growth. *Nat. Rev. Immunol.* **4**, 641–648.

van Elsas, A., Sutmuller, R. P., Hurwitz, A. A., Ziskin, J., Villasenor, J., Medema, J. P., Overwijk, W. W., Restifo, N. P., Melief, C. J., Offringa, R., and Allison, J. P. (2001). Elucidating the autoimmune and antitumor effector mechanisms of a treatment based on cytotoxic T lymphocyte antigen-4 blockade in combination with a B16 melanoma vaccine: Comparison of prophylaxis and therapy. *J. Exp. Med.* **194**, 481–489.

Von Herrath, M. G., and Harrison, L. C. (2003). Regulatory lymphocytes: Antigen-induced regulatory T cells in autoimmunity. *Nat. Rev. Immunol.* **3**, 223–232.

Wan, Y. Y., and Flavell, R. A. (2005). Identifying Foxp3-expressing suppressor T cells with a bicistronic reporter. *Proc. Natl. Acad. Sci. USA* **102**, 5126–5131.

Wang, H. Y., Lee, D. A., Peng, G., Guo, Z., Li, Y., Kiniwa, Y., Shevach, E. M., and Wang, R. F. (2004). Tumor-specific human CD4$^+$ regulatory T cells and their ligands: Implications for immunotherapy. *Immunity* **20**, 107–118.

Wang, H. Y., and Wang, R. F. (2005). Antigen-specific CD4$^+$ regulatory T cells in cancer: Implications for immunotherapy. *Microbes Infect.* **7**, 1056–1062.

Wei, S., Kryczek, I., Zou, L., Daniel, B., Cheng, P., Mottram, P., Curiel, T., Lange, A., and Zou, W. (2005). Plasmacytoid dendritic cells induce CD8$^+$ regulatory T cells in human ovarian carcinoma. *Cancer Res.* **65**, 5020–5026.

Wolff, H., Saukkonen, K., Anttila, S., Karjalainen, A., Vainio, H., and Ristimaki, A. (1998). Expression of cyclooxygenase-2 in human lung carcinoma. *Cancer Res.* **58**, 4997–5001.

Wood, K. J., and Sakaguchi, S. (2003). Regulatory lymphocytes: Regulatory T cells in transplantation tolerance. *Nat. Rev. Immunol.* **3**, 199–210.

Wyckoff, J., Wang, W., Lin, E. Y., Wang, Y., Pixley, F., Stanley, E. R., Graf, T., Pollard, J. W., Segall, J., and Condeelis, J. (2004). A paracrine loop between tumor cells and macrophages is required for tumor cell migration in mammary tumors. *Cancer Res.* **64**, 7022–7029.

Yu, P., Lee, Y., Liu, W., Chin, R. K., Wang, J., Wang, Y., Schietinger, A., Philip, M., Schreiber, H., and Fu,

Y. X. (2004). Priming of naive T cells inside tumors leads to eradication of established tumors. *Nat. Immunol.* **5**, 141–149.

Zang, X., Loke, P., Kim, J., Murphy, K., Waitz, R., and Allison, J. P. (2003). B7x: A widely expressed B7 family member that inhibits T cell activation. *Proc. Natl. Acad. Sci. USA* **100**, 10388–10392.

Zavadova, E., Loercher, A., Verstovsek, S., Verschraegen, C. F., Micksche, M., and Freedman, R. S. (1999). The role of macrophages in antitumor defense of patients with ovarian cancer. *Hematol. Oncol. Clin. North Am.* **13**, 135–144, ix.

Zou, L., Barnett, B., Safah, H., Larussa, V. F., Evdemon-Hogan, M., Mottram, P., Wei, S., David, O., Curiel, T. J., and Zou, W. (2004). Bone marrow is a reservoir for CD4$^+$CD25$^+$ regulatory T cells that traffic through CXCL12/CXCR4 signals. *Cancer Res.* **64**, 8451–8455.

Zou, W. (2005). Immunosuppressive networks in the tumour environment and their therapeutic relevance. *Nat. Rev. Cancer* **5**, 263–274.

Zou, W. (2006). Regulatory T cells, tumour immunity and immunotherapy. *Nat. Rev. Immunol.* **6**, 295–307.

Zou, W., Machelon, V., Coulomb-L'Hermin, A., Borvak, J., Nome, F., Isaeva, T., Wei, S., Krzysiek, R., Durand-Gasselin, I., Gordon, A., Pustilnik, T., Curiel, D. T., Galanaud, P., Capron, F., Emilie, D., and Curiel, T. J. (2001). Stromal-derived factor-1 in human tumors recruits and alters the function of plasmacytoid precursor dendritic cells. *Nat. Med.* **7**, 1339–1346.

PART II

CANCER THERAPEUTICS

CHAPTER 7

Cytotoxic Chemotherapy in Clinical Treatment of Cancer

RAJESH THIRUMARAN, GEORGE C. PRENDERGAST, AND PAUL B. GILMAN

I. Introduction
II. DNA-Damaging Agents
   A. Alkylating Agents
   B. DNA Topoisomerase Inhibitors
   C. Plant Alkaloids
   D. Antitumor Antibiotics
   E. Platinum Compounds
III. Antimetabolites
   A. Folate Antagonists
   B. Purine Antagonists
IV. Antimitotics
   A. Vinca Compounds
   B. Taxanes
V. Chemotherapy Regimens
References
Useful Web Sites

Cytotoxic chemotherapy offers the main clinical tool to control invasive malignancy. The general classes of modalities used most widely in the oncology clinic are DNA-damaging agents, antimetabolites, and antimitotics. This chapter summarizes the actions and clinical uses of the most common members of these three classes of cancer chemotherapeutic agents.

I. INTRODUCTION

As of this writing, cytotoxic chemotherapy by and large remains the standard of care for clinical treatment of invasive neoplasia. The first agents in this class were developed and characterized starting in the 1940s and 1950s at various private and public medical centers in the United States and in Europe. During the last half of the twentieth century, many such small molecule cytotoxic agents have been developed and extensively used in the oncology clinic. As a group, these agents fall into the three broad categories of DNA-damaging agents, antimetabolites, and antimitotics. While there is a large body of literature describing the diverse actions of these agents, as a group, the agents generally tend to take advantage of the deregulated cell division cycle that is a central trait of the cancer cell. Because they selectively target dividing cells, these agents produce side effects in the patient due to effects on normal growing cells; many of these growing cells are found in the immune system, gastrointestinal tract, hair follicles, and tissues that contain a significant fraction of dividing cells. However, despite these side effects, which are controlled much better than they were years ago, cytotoxic chemotherapy is highly effective in controlling cancer. Indeed, combination chemotherapy regimens developed over years of mainly trial-and-error testing in the clinic can achieve significant cure rates in many, although unfortunately by no means all, pediatric and adult cancer patients.

A major limitation in the more successful use of cytotoxic chemotherapy is the development of drug resistance in patients. This phenomenon is a particular challenging aspect of cancer treatment generally because it results from the inherent capability of tumors to evolve variant cells when confronted with a negative selection of any kind. Cancer cells in a patient at the time of clinical presentation are typically highly heterogeneous at the genetic level because of the accumulation of different chromosomal mutations in different cell populations present in the tumor. The genetic plasticity that is inherent to cancer means that evolution of variant cells is readily achieved by the negative selection that is imposed by the cytotoxic drug treatment. Thus, the same tactic that eliminates cancer cells can also impose a selection for drug resistance. The problem of drug resistance, which commonly leads to a patient's demise, remains one of most pressing challenges in clinical oncology.

Clinical trials of the new classes of molecular targeted agents developed for cancer treatment suggest that their most effective use is in combination with cytotoxic regimens. The testing of such combination chemotherapy regimens is a major focus of many clinical trials. In particular, many combination chemotherapy trials involve monoclonal antibodies to growth factor receptors and other cell surface proteins on cancer cells. This class of passive immunotherapeutic agents is showing significant promise. In contrast, very few if any clinical trials to date have explored the combination of cytotoxic agents with active immunotherapeutic agents, which the chapter authors believe will begin to occur shortly as a result of basic and clinical research advances described elsewhere in this book.

This chapter offers a general overview of how cytotoxic agents are used and provides readers with a "bird's eye view" of the oncology field's standard agents, mechanisms, and clinical regimens. The chief realm of action of cytotoxic chemotherapeutic agents is the cell division cycle, or cell cycle for short. The cell cycle is characterized by the five phases: G_0, G_1, S, G_2, and M. The gap phases G_0, G_1, and G_2 are time periods in which cells are preparing for active DNA replication or mitosis, which occur during the S and M phases, respectively. The so-called resting phase, when cells have exited the cell cycle and are replicatively quiescent, is termed the G_0 phase. When quiescent cells are stimulated by growth factors to enter the cell cycle, they transit from the G_0 into the G_1 phase, which is characterized by molecular changes in the cell that allow it to grow and to prepare for DNA synthesis, which begins with the start of the S phase. Many drugs in the DNA-damaging and antimetabolite classes of cancer drugs act during the G_1 and S phases to disrupt processes needed for DNA synthesis or for DNA synthesis itself, triggering cell death. The second gap phase, G_2, occurs after the S phase and is a second growth period when the cell prepares to enter mitosis or the M phase. The mitotic spindle apparatus produced during the M phase is the target of spindle poisons that are part of the antimitotic class of cancer drugs. The physical process of cell division or cytokinesis occurs at the end of M phase when the cell completes chromosomal segregation and divides into two daughter cells. After completion of this process, cells will either reenter the cell cycle at G_1 to undergo further maturation and replication or enter the resting G_0 phase and stop dividing.

As mentioned earlier, cytotoxic chemotherapeutics affect both normal and malignant cells by acting at one or more phases of the cell cycle. Although both types of cells can die as a result of damage caused, normal cells have a greater ability to repair minor damage, and, as a result, they can survive at higher rates compared to neoplastic cells. The relatively increased vulnerability of malignant cells is the central factor exploited by chemotherapy to achieve beneficial effects in the clinic. Chemotherapy agents can act in a phase-specific manner, or they may affect cycling cells more generally throughout the cell cycle. In some cases, resting cells may also be vulnerable to the cytotoxic effects of various agents, so not all the beneficial effects of chemotherapy may be cell cycle-specific.

In preclinical and clinical studies, certain cytotoxic chemotherapeutic agents have been described to have immune stimulatory activities, particularly cyclophosphamide, paclitaxel, and cisplatin. In particular, cyclophosphamide is effective at culling T regulatory immune cells, and its use in combination with immunotherapeutics has been tested and used clinically in a variety of settings for immunomodulation. These immune stimulatory features may result from relieving mechanisms of immune suppression mediated by T regulatory cells and other immune cells that limit immune rejection of tumors. This area of cancer research represents a particularly intriguing area of investigation, given the clear utility of cytotoxic agents for clinical management of cancer.

II. DNA-DAMAGING AGENTS

Cancer chemotherapeutic agents that act by damaging DNA fall into three general subclasses: alkylating agents, topoisomerase inhibitors, and platinum compounds. Other cancer drugs that act by blocking transcription often also have associated DNA-damaging effects (e.g., mitomycin C). One class of modern antitumor agents that has been explored in clinical trials is called the "targeted cytotoxics." These cytotoxics have often been composed of a DNA-damaging agent chemically linked to an antibody or peptide designed to target the drug to cancer cells more efficiently. Since their initial discovery and development in the 1950s, DNA-damaging agents have become the most widely used class of drugs for clinical treatment of cancer. Combination chemotherapy regimens for treatment of many tumors often include at least one modality that damages DNA.

A. Alkylating Agents

1. History

The first nonhormonal small molecule to demonstrate significant antitumor activity in the clinic was a nitrogen mustard alkylating agent. The evaluation of nitrogen mustards as antitumor agents actually evolved from observed effects of sulfur mustard gas used as a weapon in World War I. Mustard gas had been used because of its vesicant effect on the skin and mucous membranes, especially the eyes and the respiratory tract. In addition to this deadly effect, however, depression of the hematopoietic and lymphoid systems was observed in victims. These observations led to further studies using less volatile nitrogen mustards. The first studies published in 1946 reported tumor regressions, a finding virtually unknown at the time, leading to the introduction of nitrogen mustard into clinical practice. Subsequently, less toxic and more clinically effective nitrogen mustard derivatives and other types of alkylating agents were developed for cancer treatment.

2. Chemical Action

Alkylating agents react with electron-rich atoms to form covalent bonds. The common alkane transferred by classical alkylating agents is a single-carbon methyl group that also includes longer hydrocarbons. The most important reactions with regard to the agents' antitumor activities are reactions with DNA bases. Some alkylating agents are monofunctional and react with only one strand of DNA. Others are bifunctional and react with an atom on both strands of DNA, producing a cross-link that covalently links the two strands of the DNA double helix. Unless repaired, this lesion will prevent the cell from replicating effectively. The lethality of the monofunctional alkylating agents results from the recognition of the DNA lesion by the cell and by the response of the cell to that lesion.

3. Melphalan

Melphalan (4-[bis(2-chloroethyl)amino]-L-phenylalanine) is an L-phenylalanine derivative of classical nitrogen mustard first synthesized by Berget and Stock in 1953. The cytotoxicity of this compound relates to the extent of its interstrand cross-linking with DNA, probably mainly through binding to the N7 position of guanine. Melphalan is not specific for a specific phase of the cell cycle and is active against both resting and rapidly dividing tumor cells. Melphalan is used mainly for treatment of myeloma and for palliative treatment of ovarian cancer. The major adverse reactions associated with the use of melphalan are bone marrow suppression and hypersensitivity reactions (~2% of patients). Melphalan can also cause nausea, vomiting, diarrhea, and oral ulcers. This compound is teratogenic, and, as with other DNA-damaging compounds that cause DNA mutations, there is a risk of developing secondary malignancies. Resistance to melphalan can develop in cancer cells by several mechanisms, including decreased cellular uptake of the drug, reduction in activity due to increased expression of sulfhydryl proteins (e.g., glutathione and glutathione-associated enzymes), and enhanced DNA repair.

4. Busulfan

Busulfan (1,4-butanediol dimethane sulfonate) is a bifunctional alkylating agent. In aqueous solution, this compound undergoes hydrolysis to release the methane sulfonate groups, thereby producing reactive carbonium ions that alkylate DNA. Its cytotoxic effects are thought to be due primarily to the DNA damage it causes. Busulfan is used mainly in combination with cyclophosphamide (Cytoxan®) as a conditioning regimen prior to bone marrow transplants of allogenic hematopoietic progenitor cells, but it is also used in treatment of chronic myelogenous leukemia (CML). The chief adverse reactions are myelosuppression, nausea, vomiting, diarrhea, anorexia, and mucositis. Hyperpigmentation of the skin is also seen, especially in the creases of hands and nail beds. Busulfan may cause impotence, male sterility, amenorrhea, ovarian suppression, menopause, and infertility. Busulfan lung is a rare entity in which the patient experiences a cough, breathlessness, and fever due to interstitial pulmonary fibrosis. Like other agents that cause DNA mutations, there is risk for causing secondary malignancies. Drug resistance develops through the same mechanisms as mentioned earlier for melphalan.

5. Chlorambucil

Chlorambucil (4-[bis(2-chlorethyl)amino] benzene butanoic acid) is a bifunctional alkylating agent that also forms cross-links with DNA. This compound is indicated for use by patients with chronic lymphocytic leukemia, Hodgkin's and non-Hodgkin's lymphomas, and Waldenström's macroglobulinemia. Adverse reactions include myelosuppression (white cells and platelets equally affected), nausea, and vomiting. Hyperuricemia also has been observed. Chlorambucil can rarely cause pulmonary fibrosis, seizures, amenorrhea, oligospermia, and secondary malignancies. Drug resistance mechanisms are similar to other mustard agents.

6. Mechlorethamine

Mechlorethamine (2-chloro-N-(2-chloroethyl)-N-methyl ethanamine hydrochloride) is an analog of mustard gas. This alkylating agent forms both interstrand and intrastrand DNA cross-links and has activity in all phases of the cell cycle. Mechlorethamine is indicated for use in the treatment of Hodgkin's and non-Hodgkin's lymphomas, and cutaneous T cell lymphoma. Adverse reactions and resistance mechanisms are similar to those of other mustards. Special risks include alopecia (hair loss), amenorrhea, and azoospermia. Mechlorethamine is also a potent vesicant that can cause severe skin necrosis if there is extravasation under the skin during administration.

7. Temozolomide

Temozolomide (3,4-dihydro-3-methyl-4-oxoimidazo [5,1-d]-as-tetrazine-8-carboxamide) is an acid-stable compound that can be given as an oral medication. This agent is not directly active but undergoes rapid nonenzymatic conversion at physiologic pH to a reactive compound, monomethyl triazeno imidazole carboxamide (MTIC), that alkylates DNA mainly at the O6 and N7 positions of guanine. Temozolomide is indicated for use by patients with brain tumors and metastatic melanoma. Adverse effects include myelosuppression, nausea, vomiting, headaches, and fatigue. Temozolomide is teratogenic, mutagenic, and carcinogenic. Resistance can develop through increased activity of DNA repair enzymes, particularly the O6-alkylguanine-DNA alkyltransferase.

8. Cyclophosphamide (Cytoxan®)

Cyclophosphamide (N,N-bis(2-chloroethyl)-2-oxo-1-oxa-3-aza-2u{5}-phosphacyclohexan-2-amine) is a mustard compound that is activated in the liver. Upon activation by the cytochrome P450 microsomal system, the cytotoxic metabolites phosphoramide mustard and acrolein are formed. Both metabolites can cause DNA cross-links, resulting in an inhibition of DNA synthesis. Cyclophosphamide is a cell cycle-nonspecific agent that is active in all phases of the cell cycle. It is used widely in the treatment of breast cancer, non-Hodgkin's lymphoma, chronic lymphocytic leukemia, ovarian cancer, bone and soft-tissue sarcoma, rhabdomyosarcoma, and neuroblastoma. Major adverse effects include myelosuppression (mainly affecting white cells and platelets), bladder toxicity in the form of hemorrhagic cystitis, dysuria, and increased urinary frequency. Alopecia develops about 3 weeks after therapy is initiated. Other known side effects are nausea, vomiting, amenorrhea, ovarian failure, sterility, cardiotoxicity, and immunosuppression with increased risk of infections and increased risk of secondary malignancies. Resistance mechanisms are similar to those of other mustards but also include increased expression of aldehyde dehydrogenase, which causes enhanced enzymatic detoxification of the drug.

9. Ifosfamide

Ifosfamide (3-(2-chloroethyl)-2-[(2-chloroethyl)amino]tetrahydro-2H-1,3,2-oxazaphosphorine 2-oxide) is similar in character to cyclophosphamide, becoming activated by the liver cytochrome P450 microsomal system to produce the active cytotoxic metabolites ifosfamide mustard and acrolein. Ifosfamide is specially indicated for patients with recurrent germ cell tumors, lung cancer, bladder cancer, head and neck cancer, cervical cancer, and Ewing's sarcoma. Adverse reactions and resistance mechanisms are similar to cyclophosphamide, but also include for ifosfamide neurotoxicities that can produce seizures, cerebellar ataxia, confusion, and hallucinations. Like cyclophosphamide, ifosfamide produces alopecia as well as infertility, and the agent is teratogenic.

10. Carmustine

Carmustine (1,3-bis(2-chloroethyl)-1-nitroso-urea) is a nitrosourea analog that is cell cycle-nonspecific in action. Chloroethyl metabolites of this compound interfere with DNA synthesis and function. Carmustine is indicated for the treatment of patients with brain tumors, Hodgkin's and non-Hodgkin's lymphomas, and myeloma. Adverse reactions and resistance mechanisms are similar to other mustards, but additional side effects include facial flushing, hepatotoxicity, nephrotoxicity, and pulmonary toxicity.

11. Dacarbazine and Procarbazine

Dacarbazine (5-(3,3-Dimethyl-1-triazenyl) imidazole-4-carboxamide, N-(1-methylethyl)-4-[(N'-methylhydrazino)methyl]benzamide) and procarbazine are hydrazine analogs that broadly methylate DNA, RNA, and protein and that inhibit the synthesis of each. Dacarbazine is indicated in the treatment of patients with malignant melanoma, soft-tissue sarcomas, and Hodgkin's lymphoma. Procarbazine is indicated in the treatment of patients with Hodgkin's and non-Hodgkin's lymphomas, brain tumors, and cutaneous T cell lymphoma. Adverse effects are similar to those of other mustards but also include neuropathy, confusion, seizures, photosensitivity, hypersensitivity reactions, amenorrhea, azoospermia, and teratogenicity. An important resistance mechanism is the increased activity of DNA repair enzymes, including O6-alkylguanine-DNA transferase.

B. DNA Topoisomerase Inhibitors

DNA topoisomerases unravel twists in DNA that occur as a result of DNA transcription and replication. The DNA topoisomerases I and II present in cells act through scission of the DNA backbone on one or two strands, respectively, followed by relief of torsional stress and then relegation of the broken DNA backbone. These enzymes are present in large complexes in the cell nucleus and control and carry out transcription and replication; they are also essential to maintain chromatin organization and cell survival. Inhibition of DNA topoisomerases by small molecules is an effective method for causing DNA damage due to the formation of irreversible covalent cross-links between the topoisomerase and DNA, stalling its replication and thereby leading to cell death. Among the inhibitors of DNA topoisomerase, which are widely used for cancer treatment, are the plant alkaloids and antibiotic compounds listed in the following paragraphs.

C. Plant Alkaloids

Plant alkaloids can inhibit the enzymatic action of topoisomerases and have been used to treat a broad spectrum of human cancers. Camptothecin is one of the older natural products in this class from which more modern semisynthetic derivatives, such as irinotecan and topotecan, have been chemically generated.

1. Etoposide

Etoposide is an alkaloid from the mandrake plant *Podophyllum peltatum* that has cell cycle-specific activity in the late S phase and G_2 phase. Etoposide inhibits topoisomerase II by stabilizing the enzyme–DNA complex and preventing the unwinding of DNA. Etoposide is indicated for the treatment of patients with germ cell tumors, lung cancer, Hodgkin's and non-Hodgkin's lymphomas, gastric cancer, breast cancer, and testicular cancer. Adverse effects include myelosuppression, nausea/vomiting, anorexia, alopecia, mucositis, diarrhea, hypersensitivity reactions, fever, bronchospasm, dyspnea, hypotension, and radiation recall skin changes. Etoposide poses an increased risk of secondary malignancies, particularly myelodysplasia and acute myelogenous leukemia associated with chromosome 11:23 translocations within 5 to 8 years of treatment. Resistance mechanisms include decreased expression of DNA topoisomerase II and the multidrug-resistant (MDR) phenotype produced by increased expression of the p170 glycoprotein that increases drug efflux, decreasing intracellular accumulation.

also derived from the actinobacteria *Streptomyces peucetius var. caesius*. This compound is also a cell cycle-nonspecific agent that acts through topoisomerase II inhibition and DNA intercalation. Daunorubicin is indicated in the treatment of patients with acute lymphoblastic leukemia and acute myelogenous leukemia. Adverse effects and resistance mechanisms are similar to doxorubicin. The vesicant properties of daunorubicin are stronger, however, and they can cause severe skin necrosis if there is extravasation under the skin during intravenous administration of the drug.

3. Idarubicin

Idarubicin is a semisynthetic analog of daunorubicin that is mechanistically similar in action but that exhibits some specificity for the late S and G_2 phases of the cell cycle. Indications, adverse reactions, and resistance mechanisms are similar to those of doxorubicin.

4. Mitoxantrone

Mitoxantrone is an anthracenedione derivative originally synthesized in 1970. This compound is also a DNA-intercalating chemical, acting to inhibit topoisomerase II by forming a cleavable enzyme–DNA complex. Mitoxantrone is indicated in the treatment of patients with hormone-refractory prostate cancer, breast cancer, acute myelogenous leukemia, and non-Hodgkin's lymphoma. Resistance mechanisms are similar to the those of other antitumor antibiotics. Adverse effects include myelosuppression, nausea/vomiting, mucositis, diarrhea, cardiomyopathy, alopecia, and elevation of the liver enzymes. Risk of secondary leukemia exists for patients undergoing treatment.

5. Dactinomycin

Dactinomycin was the first actinomycin antibiotic to be isolated from a *Streptomyces* species in the 1940s. This agent is a DNA-intercalating compound that inhibits both DNA and RNA synthesis. Its major mechanism of action is thought to be through the generation of DNA strand breaks via interaction with topoisomerase II. Dactinomycin is indicated in the treatment of patients with Wilm's tumor, rhabdomyosarcoma, germ cell tumors, gestational trophoblastic disease, and Ewing's sarcoma. Adverse effects and resistance mechanisms are similar to the other antitumor antibiotics listed previously.

E. Platinum Compounds

Platinum compounds that are used widely in cancer treatment damage DNA by creating intrastrand and interstrand cross-links. These compounds are cell cycle-nonspecific in action.

1. Cisplatin

Cisplatin (cis-diamminedichloroplatinum) is a small molecule that reacts with two different sites on DNA to produce strand cross-links, leading to an inhibition of DNA synthesis and transcription. Cisplatin is indicated in the treatment of ovarian, testicular, bladder, head and neck, esophageal, and lung cancers. It is also used to treat non-Hodgkin's lymphoma and gestational trophoblastic disease. Adverse effects include nephrotoxicity, myelosuppression, nausea/vomiting, alopecia, neurotoxicity, ototoxicity, increased liver enzymes, sterility, and vascular events (myocardial infarction or cerebrovascular thrombosis). Resistance mechanisms include decreased drug accumulation secondary to alterations in cellular transport, increased inactivation by glutathione and glutathione-related enzymes, and increased DNA repair activity.

2. Carboplatin

Carboplatin is a cisplatin-related compound in which the chloride-leaving groups have been substituted, leading to diminished nephrotoxicity but otherwise similar mechanism of action, antitumor activity, and adverse effects. The indications of carboplatin are similar to cisplatin.

2. Irinotecan and Topotecan

Irinotecan and topotecan are structurally related semisynthetic derivatives of the natural product camptothecin, an alkaloid derived from extracts of the tree *Camptotheca acuminata*. This compound is inactive in the parent form and is converted by intracellular carboxylesterase activity to its active metabolite termed SN-38. This metabolite binds to DNA topoisomerase I, stabilizing the enzyme–DNA complex and acting to prevent DNA ligation after cleavage and torsional relief by DNA topoisomerase I. The collision between this stable complex and the advancing replication fork results in double-stranded DNA breaks and cellular death. Both irinotecan and topotecan are cell cycle-nonspecific agents with activity in all phases of the cell cycle. Irinotecan is indicated for use in patients with colorectal cancers, for whom DNA topoisomerase I is expressed at relatively high levels, and in patients with lung cancers. Topotecan is used to treat ovarian cancer, small cell lung cancer, and acute myelogenous leukemia. Adverse reactions for both agents include myelosuppression, diarrhea, vomiting, liver enzyme elevations, asthenia, and fever; for topotecan, side effects also include microscopic hematuria and alopecia. It has been reported that irinotecan toxicity is modified by genetic variations in UDP-Glucuronosyltransferase (UGT1A1), the enzyme that glucuronidates SN-38 and converts it to an inactive product in bile. Since irinotecan toxicity is inversely related to the efficiency of converting SN-38, individuals with relatively lower UGT1A1 expression may experience more severe toxicity. Resistance mechanisms are similar to those for etoposide but for irinotecan include decreased activity of carboxylesterase activity that is needed for formation of the cytotoxic active SN-38 metabolite.

D. Antitumor Antibiotics

Antitumor antibiotics are the natural products of microbial metabolism. Most were initially isolated from fermentation broths of various species of *Streptomyces* based observed cytotoxic activities present in cru microbial extracts. Efforts to improve t antiproliferative actions and to reduce to icities of the original compounds were mad by genetic manipulations of parent organisms to produce analogs and by synthesis of rationally engineered analogs. A large number of variants have been explored, and antitumor antibiotics remain a mainstay of clinical anticancer therapy. While there is structural diversity among the compounds used in the clinical environment, there are remarkable similarities in mechanisms of action, effects on cellular intermediary metabolism, and toxicity profiles.

1. Doxorubicin

Doxorubicin is an anthracycline antibiotic derived from the actinobacteria *Streptomyces peucetius var. caesius*. This compound is a cell cycle-nonspecific agent that acts by blocking topoisomerase II activity and by intercalating into the flat space between the bases of the DNA double helix, where it can act further to disrupt DNA replication and transcription. Doxorubicin is indicated in the treatment of many human cancers, including breast, ovarian, lung, bladder, thyroid, liver, and gastric cancers; Hodgkin's and non-Hodgkin's lymphomas; Wilm's tumor; soft-tissue sarcoma; neuroblastoma; and acute lymphoblastic leukemia. Adverse effects related to the use of doxorubicin are myelosuppression, nausea/vomiting, mucositis, diarrhea, alopecia, and urine discoloration. Doxorubicin is also a strong-tissue vesicant. Cardiomyopathy is a special concern with the use of doxorubicin, and a lifetime limit for patient exposure is dictated. Resistance mechanisms include gp170-dependent MDR, decreased expression or mutations of topoisomerase II, and increased expression of sulfhydryl enzymes (e.g., glutathione and glutathione-dependent enzymes).

2. Daunorubicin

Daunorubicin is an anthracycline antibiotic related to doxorubicin that is

3. Oxaliplatin

Oxaliplatin acts similarly to cisplatin and carboplatin, but DNA mismatch repair enzymes are unable to recognize the oxaliplatin–DNA adducts due to their larger size. Oxaliplatin is an important agent in the treatment of patients with metastatic colorectal, pancreatic, and gastric cancers. Adverse effects include sensory neuropathy, laryngopharyngeal spasm, cold dysesthesias, myelosuppression, nausea/vomiting, diarrhea, allergic reactions, and, in rare cases, nephrotoxicity. Resistance mechanisms are similar to those of other platinum compounds.

III. ANTIMETABOLITES

Antimetabolites are similar in structure to naturally occurring compounds that are required for the viability and division of a cell. This structural similarity allows this class of antitumor agents to serve as substrates for important cellular enzymes. The efficacy of the most important antimetabolites against a range of tumor cells is based on the inhibition of purine or pyrimidine nucleoside synthesis pathways that are required for DNA synthesis. Thus, the antimetabolites as a class are most effective against tumors that have a high growth fraction (i.e., a high proportion of cells in the tumor that are undergoing DNA synthesis). Since cells in the S phase are particularly sensitive to antimetabolites, toxicity is common in any tissue that experiences a high rate of DNA synthesis, such as tissues of the gastrointestinal tract and bone marrow.

Most antimetabolites are inactive in the native form and require conversion through a series of enzymatic steps to an active compound. Antimetabolites inhibit the replication or the repair of DNA either by the direct inhibition of enzymes needed for DNA replication or by the incorporation of the antimetabolite (or a compound derived from the antimetabolite) directly into the DNA. Antimetabolites can be further divided into several subclasses, including:

- Folate antagonists: methotrexate
- Purine antagonists: 6-mercaptopurine, fludarabine
- Pyrimidine antagonists: 5-fluorouracil (5-FU), capecitabine, cytarabine, gemcitabine

A. Folate Antagonists

1. Methotrexate

Methotrexate (N-[4[[2,4-diamino-6-pteridinyl)methyl]methyl-amino]benzoil]-L-glutamic acid) is the most extensively used and characterized agent, exhibiting cell cycle-specific activity during the S phase. This compound inhibits the enzyme dihydrofolate reductase (DHFR), thereby limiting the production of reduced folates that are needed for *de novo* synthesis of purines used in DNA replication. Methotrexate is widely used for the treatment of patients with breast, colon, bladder, and head and neck cancers and for patients with osteosarcoma, acute lymphocytic leukemia, non-Hodgkin's lymphoma, and gestational trophoblastic disease. Adverse reactions include myelosuppression, mucositis, renal failure, pneumonitis, cerebral dysfunction, skin rash, abnormal liver function, fetal death, and oligospermia. Resistance mechanisms include amplification or increased expression of the DHFR gene, mutations that alter drug-binding affinity, and decreased expression of DNA mismatch repair enzymes.

B. Purine Antagonists

1. 6-Mercaptopurine

The 6-mercaptopurine compound (1,7-dihydro-6H-purine-6-thione monohydrate) is a cell cycle-specific purine analog with activity in the S phase. This compound is inactive until phosphorylated intracellularly by the enzyme hypoxanthine-guanine phosphoribosyltransferase (HGPRT), which converts the compound to a cytotoxic monophosphate derivative. Upon conversion, the derivative acts to inhibit *de novo* purine synthesis by inhibiting the enzyme 5-phosphoribosyl-1-pyrophosphate amidotransferase. The analog 6-mercaptopurine is indicated for use in acute lymphocytic leukemia. Adverse reactions include myelosuppression, mucositis, diarrhea, abnormal liver function, nausea/vomiting, urticaria, and teratogenicity. Resistance mechanisms include decreased expression of HGPRT or DNA mismatch repair enzymes or increased expression of alkaline phosphatase, which catabolizes the compound.

2. Fludarabine

Fludarabine (9H-purin-6-amine,2-fluro-9-(5-O-phosphono-β-D-arabinofuranosyl)2-fluro-ara-AMP) is a 5-monophosphate derivative of the adenosine analog arabinofuranosyl-adenosine (AraA). This compound has a high degree of specificity for lymphoid cells. It is a "prodrug" that is rapidly dephosphorylated upon administration, first to its monophosphate form and eventually to an active 5-trisphosphate metabolite. The triphosphate metabolite is incorporated into DNA during replication, inhibiting further synthesis of the chain. Fludarabine is indicated for the treatment of patients with chronic lymphocytic leukemia, non-Hodgkin's lymphoma, cutaneous T cell lymphoma, and hairy cell leukemia. Adverse reactions include myelosuppression, autoimmune hemolytic anemia, aplastic anemia, immunosuppression, fever, tumor lysis syndrome, abnormal liver function, neurotoxicity, and nausea/vomiting. An important resistance mechanism is decreased expression of the prodrug-activating enzyme deoxycytidine kinase.

3. Cladribine

Cladribine (2-chloro-2'-deoxy-β-D-adenosine) is a related adenosine analog that also has a high degree of specificity for lymphoid cells. It has antitumor activity against both dividing and resting cells. The compound is a prodrug that in its triphosphate form is incorporated into DNA, resulting in a blockade to DNA synthesis and transcription. Cladribine also depletes the central cellular cofactor nicotine adenine dinucleotide (NAD), causing a depletion of ATP. Cladribine is an effective treatment for patients with hairy cell leukemia, chronic lymphocytic leukemia, and non-Hodgkin's lymphoma. Adverse reactions and resistance mechanisms are similar to fludarabine with the addition of resistance through increased expression of 5'-nucleotidases that can dephosphorylate cladribine nucleotide metabolites Cld-AMP and Cld-ATP.

4. 5-Fluorouracil

The 5-fluorouracil analog (5-fluro-2,4(1H,3H)-pyrimidinedione) is a uracil nucleotide that has been one of the most widely used cancer chemotherapeutics. The 5-FU, as it is commonly referred to, becomes metabolized *in vivo* to fluoro-dUMP, which inhibits the enzyme thymidylate synthase. Incorporation of metabolites into RNA and DNA alters RNA processing and DNA synthesis and transcription. The 5-FU is indicated for use in patients with colorectal, breast, anal, esophageal, gastric, pancreatic, head and neck, ovarian, and liver cancers. Adverse reactions include myelosuppression, mucositis, diarrhea, hand-foot syndrome, cerebellar ataxia, chest pain, and blepharitis. Reactions differ significantly with bolus versus infusional administration. For example, myelosuppression is more common when 5-FU is administered as a bolus, whereas diarrhea and hand-foot syndrome are more common with infusional treatment. Resistance mechanisms include increased expression or alterations of thymidylate synthase and increased activity of DNA repair enzymes uracil glycolyase and dUTPase.

5. Capecitabine

Capecitabine (5'-deoxy-5-fluro-N [(pentyloxy) carbonyl]-cytidine) is an oral fluoropyrimidine carbamate prodrug form of 5-FU. Capecitabine is inactive until metabolized through several steps carried out in the liver to generate 5-FU. Capecitabine is useful in the treatment of patients with breast and colorectal cancers. Adverse effects are similar to 5-FU, but myelosuppression is reduced. Resistance mechanisms are similar to those of 5-FU.

6. Cytabarine

Cytabarine (4-amino-1-β-D-arabinofuranosyl-2(1H)-pyrimidone) is a deoxycytidine analog isolated originally from the sponge *Cryptothethya crypta*. This compound requires intracellular activation to the nucleotide metabolite ara-dCTP. Incorporation of the ara-dCTP into DNA results in chain termination and inhibition of DNA replication and transcription. Cytarabine is indicated for patients with acute myelogenous leukemia, acute lymphocytic leukemia, chronic myeloid leukemia, non-Hodgkin's lymphoma, and leptomeningeal carcinomatosis. Adverse reactions include myelosuppression, nausea/vomiting, cerebellar ataxia, confusion, abnormal liver function, acute pancreatitis, noncardiogenic pulmonary edema, conjunctivitis, keratitis, and seizures. Resistance mechanisms include decreased expression of the anabolic enzyme deoxycytidine kinase and increased expression of the catabolic enzymes cytodine deaminase or dCMP deaminase (both enzymes catabolize the drug) or increased expression of CTP synthetase (which increases levels of competing dCTP levels).

7. Gemcitabine

Gemcitabine (2'-deoxy-2'2'-diflurocytidine monohydrochloride) is a fluorine-substituted dCTP analog. The compound is a prodrug that requires intracellular activation to the cytotoxic triphosphate nucleotide metabolite dFdCTP. Incorporation of this metabolite into DNA terminates DNA synthesis and transcription, whereas incorporation of the metabolite into RNA blocks processing and translation. Gemcitabine is used in the treatment of patients with pancreatic cancer; non-small cell lung cancer; and bladder, breast, ovarian cancer, and soft-tissue sarcomas. Adverse reactions include myelosuppression, nausea/vomiting, diarrhea, flu-like syndrome, increased liver function, drug-induced pneumonitis, hematuria, rash, and an infusion reaction. Resistance mechanisms include decreased expression of the prodrug-activating enzyme dCTP kinase, decreased intracellular drug transport, and increased drug catabolism.

8. Hydroxyurea

Hydroxyurea inhibits the enzyme ribonucleotide reductase and thereby blocks the formation of nucleotides needed for DNA synthesis and repair. Hydroxyurea is indicated in the treatment of patients with chronic myelogenous leukemia, essential thrombocytosis, polycythemia vera, acute myelogenous leukemia (blast crisis), and head and neck cancer. Adverse effects include myelosuppression, nausea/vomiting, rash, headache, confusion, radiation recall skin changes, increased liver enzymes, and teratogenicity. The major resistance mechanism is amplification or increased expression of the ribonucleotide reductase gene.

9. Pemetrexed

Pemetrexed is a pyrollopyrimidine antifolate analog that inhibits the folate-dependent enzyme thymidylate synthase, thereby inhibiting synthesis of thymidine triphosphate (dTTP) and DNA replication. This compound is indicated in the treatment of patients with mesothelioma, NSCLC, colorectal cancer, breast cancer, and pancreatic cancer. Adverse effects include myelosuppression, skin rash, nausea, vomiting, mucositis, diarrhea, transient elevation of liver enzymes, and fatigue. Important resistance mechanisms include increased expression or alterations in the target enzyme thymidylate synthase.

IV. ANTIMITOTICS

In the 1950s, the identification of antimicrotubule agents that possess novel mechanisms of cytotoxic action and unique spectra of antitumor activity represented a major advance in cancer treatment. Based on their mechanism of action, antimitotics' primary effect is to disrupt the organization and dynamics of the mitotic spindle, preventing the M phase transit and cell division. The evident benefits of these drugs continue to promote significant interest in therapeutic agents directed against microtubule function. Most agents in use are structurally complex plant alkaloids possessing pharmacological activity. The first widely used class of microtubule directed agents were the vinca alkaloids, which have been a mainstay of chemotherapy regimens since they came into use in 1959. The development of taxanes as another subclass of antimicrotubule compounds has been important because these agents have unique mechanisms of action and unique spectra of activity. Semisynthetic strategies to produce taxanes have been important in wider use and study of these agents.

A. Vinca Compounds

1. Vinblastine, Vincristine, and Vinorelbine

Vinblastine and vincristine are alkaloids derived from the periwinkle plant *Catharanthus roseus*. These compounds have cell cycle-specific activity in the M phase, which is consistent with their ability to inhibit tubulin polymerization and prevent formation of the mitotic spindle. In this manner, they induce a terminal mitotic arrest that ultimately leads to cell death. Vinblastine is indicated in the treatment of patients with Hodgkin's and non-Hodgkin's lymphomas, breast cancer, Kaposi's sarcoma, renal cell cancer, and testicular cancer. Vincristine is more widely indicated, including for the treatment of patients with myeloma, acute lymphocytic leukemia, Hodgkin's and non-Hodgkin's lymphomas, rhabdomyosarcoma, neuroblastoma, Ewing's sarcoma, Wilm's tumor, chronic leukemia, thyroid cancer, brain tumors, and trophoblastic neoplasia. Adverse reactions are broad, including typical side effects of cytotoxic chemotherapeutics, such as myelosuppression, mucositis, fever, anemia, and alopecia. Vincristine also causes additional side effects, such as hypertension, neuropathy, depression, Raynaud's phenomenon, myocardial infarction, and pulmonary edema. Resistance mechanisms include gp170-mediated MDR and mutations in tubulin subunit proteins that decrease drug binding.

Vinorelbine is a semisynthetic derivative of vinblastine that also inhibits tubulin polymerization and disrupts spindle assembly in the M phase. This compound has a higher specificity for mitotic microtubules and a lower affinity for axonal microtubules, reducing neuropathy. Vinorelbine is indicated in the treatment of lung cancer, breast cancer, and ovarian cancer. Adverse reactions are similar to those produced by vinblastine include myelosuppression, nausea, vomiting, and constipation; altered liver function, requiring more frequent liver function tests; alopecia; neurotoxicity; hypersensitivity; and syndrome of inappropriate antidiuretic hormone secretion (SIADH).

B. Taxanes

Taxanes are complex molecules with a unique mechanism of action that is coupled to broad antitumor activity. Like vinca compounds, the primary action of taxanes is based on their ability to target microtubules; however, the action of taxanes is associated with microtubule stabilization rather than microtubule disruption. Since their discovery in the 1960s, taxanes have become one of the most important classes of anticancer agents available to the clinical oncologist.

1. Paclitaxel

Paclitaxel (Taxol®) is derived from extracts of the Pacific yew tree *Taxus brevifolia*. This compound is cell cycle-specific acting in the M phase of the cell cycle. Paclitaxel binds to microtubules and enhances tubulin polymerization, leading to microtubule stabilization. During mitosis, paclitaxel acts to disrupt the dynamics of the mitotic spindle, preventing its ability to function normally and blocking cell division and survival as a result. Paclitaxel has a significant role in the treatment of ovarian, breast, lung, head and neck, esophageal, prostate, and bladder cancers. It is also used in treating Kaposi's sarcoma. The utility of this compound is likely to continue to expand. Adverse effects of paclitaxel include the common side effects of other cytotoxic drugs, but it can also cause transient bradycardia, nail destruction, and mild elevation of liver enzymes. An important and notable side effect is neurotoxicity in the form of sensory neuropathy. In addition, a hypersensitivity reaction is observed in 20–40% of patients (attributable in part to the oil-based drug vehicle Cremophor). This reaction is commonly prevented by premedicating patients with steroids and antihistamine drugs. Resistance mechanisms include gp170-mediated MDR and tubulin mutations that reduce drug binding.

2. *Docetaxel*

Docetaxel is a semisynthetic taxane derived from extracts of the European yew tree *Taxus baccata*. Docetaxel is similar to paclitaxel in its mechanism of action, adverse reactions, and resistance mechanisms. However, docetaxel is more water soluble than paclitaxel, and it has a slightly higher affinity than paclitaxel for that site. In addition, hypersensitivity reactions occur in less than 5% of patients treated with docetaxel, and neuropathy is also observed less frequently. Docetaxel is effective in the treatment of breast, lung, head and neck, gastric, ovarian, and bladder cancers.

V. CHEMOTHERAPY REGIMENS

As of this writing, the standard of care for cancer patients involves combinatorial regimens that often include agents belonging to separate classes of cytotoxic chemotherapy described in this chapter. Doses and scheduling have been determined over many years of experience and through use of evidence-based medicine; through trial and error (experiments based on the inherently empirical way), classical cytotoxic agents have been developed (i.e., without direct reference to molecular mechanisms). Many cytotoxic chemotherapeutic agents are given via intravenous infusions. With regard to optimizing efficacy and reducing side effects, scheduling of administration can have significant effects. This aspect of clinical cancer trials is less widely appreciated than the effects of dose. The empirical effects of scheduling make developing combination regimens particularly challenging because complexities can mount exponentially with the addition of novel agents to an established regimen. To optimize ratios of efficacy to adverse effects (therapeutic window), historical experience has shown it is useful to administer regimens of cytotoxic drugs in multiple cycles with rest intervals for the patient of 2–6 weeks to obtain the maximal kill of cancer cells while allowing the patient to recover from drug side effects. The development of biological agents that boost blood counts (e.g., GM-CSF or Neupogen®), antiemetics that reduce nausea and vomiting, and other supportive treatments have helped limit the most common undesirable side effects of cytotoxic chemotherapy.

Information on the use of cytotoxic drugs in some typical clinical chemotherapeutic regimens is provided in Table 7.1 to illustrate several points. First, drugs from different categories of cancer chemotherapeutic agents are commonly found in many regimens that are believed to produce the best patient outcomes. Second, clinical cancer varies quite widely, and there may not be a consensus among different oncology practices on what constitutes a standard treatment. Thus, the clinical oncologist deals with the inherent challenge of how to tailor therapy for patients who, for reasons

TABLE 7.1 Examples of Chemotherapy Regimens for Different Tumors

Non-Hodgkin's lymphoma

CHOP regimen
Cyclophosphamide: 750 mg/m^2 via IV on day 1
Doxorubicin: 50 mg/m^2 via IV on day 1
Vincristine: 1.4 mg/m^2 via IV on day 1 (maximum dose of 2 mg)
Prednisone: 100 mg per day orally on days 1–5
Repeat cycle every 21 days (McKelvey et al., 1976)

FND regimen
Fludarabine: 25 mg/m^2 via IV on days 1, 2, and 3
Mitoxantrone: 10 mg/m^2 via IV on day 1
Dexamethasone: 20 mg orally per day for days 1–5
Repeat cycle every 21 days (Mclaughlin et al., 1996)

Breast cancer

Doxorubicin combined with cyclophosphamide followed by paclitaxel
Doxorubicin: 60 mg/m^2 via IV on day 1
Cyclophosphamide: 600 mg/m^2 via IV on day 1
Repeat cycle every 21 days
Paclitaxel: 175 mg/m^2 via IV on day 1
Repeat cycle every 21 days (Hudis et al., 1999)

CMF regimen
Cyclophosphamide: 600 mg/m^2 via IV on day 1
Methotrexate: 40 mg/m^2 via IV on day 1
5-Fluorouracil: 600 mg/m^2 via IV on day 1
Repeat cycle every 21 days (Weiss et al., 1987)

FEC regimen
5-Fluorouracil: 500 mg/m^2 via IV on day 1
Epirubicin: 50 mg/m^2 via IV on day 1
Cyclophosphamide: 500 mg/m^2 via IV on day 1
Repeat cycle every 21 days (Coombes et al., 1996)

Prostate cancer

Combination of docetaxel and prednisone
Docetaxel: 75 mg/m^2 via IV on day 1
Prednisone: 5 mg orally daily
Repeat cycle every 21 days (Eisenberger et al., 2004)

Combination of mitoxantrone and prednisone
Mitoxantrone: 12 mg/m^2 via IV on day 1
Prednisone: 5 mg orally twice daily
Repeat cycle every 21 days (Tannock et al., 1996)

Colorectal cancer

Folfox 4 regimen
Oxaliplatin: 85 mg/m^2 via IV on day 1
5-Fluorouracil: 400 mg/m^2 via IV bolus then 600 mg/m^2 continuous IV infusion for 22 hours on days 1 and 2
Leucovorin: 200 mg/m^2 via IV on days 1 and 2
Repeat cycle every 2 weeks (deGramont et al., 2003)

Folfox 6 regimen
Oxaliplatin: 100 mg/m^2 via IV on day 1
5-Fluorouracil: 400 mg/m^2 via IV bolus injection followed by 2400 mg/m^2 continuous IV infusion for 46 hours
Leucovorin: 400 mg/m^2 via IV on days 1 and 2
Repeat cycle every 2 weeks (Tournigand et al., 2001)

IFL Saltz regimen
Irinotecan: 125 mg/m^2 via IV weekly for 4 weeks
5-Fluorouracil: 500 mg/m^2 via IV weekly for 4 weeks
Leucovorin: 20 mg/m^2 via IV weekly for 4 weeks
Repeat cycle every 6 weeks (Saltz et al., 2004)

IFL Douillard regimen
Irinotecan: 180 mg/m^2 via IV on day 1
5-Fluorouracil: 400 mg/m^2 via IV bolus injection followed by a 600 mg/m^2 continuous IV infusion for 22 hours on days 1 and 2
Leucovorin: 200 mg/m^2 via IV on days 1 and 2
Repeat cycle every 2 weeks (Douillard et al., 2000)

IFL Folfiri regimen
Irinotecan: 180 mg/m^2 via IV on day 1
5-Fluorouracil: 400 mg/m^2 IV bolus injection followed by 2400 mg/m^2 continuous IV infusion for 46 hours
Leucovorin: 200 mg/m^2 via IV on days 1 and 2
Repeat cycle every 2 weeks (Andre et al., 1999)

Mayo Clinic regimen
5-Fluorouracil: 425 mg/m^2 via IV on days 1–5
Leucovorin: 20 mg/m^2 via IV on days 1–5
Repeat cycle every 4 weeks (Wasserman et al., 2001)

Roswell Park regimen
5-Fluorouracil: 500 mg/m^2 via IV weekly for 6 weeks
Leucovorin: 500 mg/m^2 via IV weekly for 6 weeks
Repeat cycle every 8 weeks (Poon, 1989)

Anal cancer

Wayne State regimen
5-Fluorouracil: 1000 mg/m^2 per day, continuously via IV infusion on days 1–4 and 29–32
Mitomycin C: 15 mg/m^2 via IV on day 1
Radiation therapy: 200 cGy per day on days 1–5, 8–12, and 5–19 (total dose 3000 cGy)
Administer chemotherapy concurrently with radiation therapy (Nigro et al., 1983)

MD Anderson regimen
5-Fluorouracil: 250 mg/m^2 per day via continuous IV infusion on days 1–5 of each week of radiotherapy
Cisplatin: 4 mg/m^2 per day via continuous IV infusion on days 1–5 of each week of radiotherapy
Radiation therapy: total dose of 5500 cGy over a period of 6 weeks
Administer chemotherapy concurrently with radiation therapy (Hung et al., 2003)

that are unclear, may have different reactions and levels of sensitivity to the same treatment regimens. For the clinical researcher, the absence of a consensus concerning a "best treatment" can confound the ability to conduct a comparative phase III trial of a novel agent, whether it is a novel cytotoxic, biological, or molecular targeted therapeutic agent; while the benefits of the new agent are typically accrued by combining it with an existing regimen, the suitable standard for the regimen may not be obvious or fully accepted in practice. Last, it may be illustrative to note the organization of typical therapeutic regimens, in which radiotherapy is sometimes additionally combined, underscoring the empirical nature of treatments used as benchmarks in the oncology clinic. Therefore, how to combine immunotherapeutic reagents with existing combinatorial treatment regimens poses a challenge to clinical researchers who wish to develop more effective approaches to managing cancer.

References

Andre T., Louvet, C., Maindrault-Goebel, F., Couteau, C., Mabro, M., Lotz, J. P., Gilles-Amar, V., Krulik, M., Carola, E., Izrael, V., de Gramont, A. (1999). CPT-11 (irinotecan) addition to bimonthly, high dose leucovorin and bolus and continuous infusion 5-fluorouracil (FOLFIRI) for pretreated metastatic colorectal cancer. *Eur. J. Cancer* **35**, 1343–1347.

Coombes, R. C., Bliss, J. M., Wils, J., Morvan, F., Espie, M., Amadori, D., Gambrosier, P., Richards, M., Aapro, M., Villar-Grimalt, A., McArdle, C., Perez-Lopez, F. R., Vassilopoulos, P., Ferreira, E. P. (1996). Adjuvant cyclophosphamide, methotrexate, and fluorouracil versus fluorouracil, epirubicin, and cyclophosphamide chemotherapy in premenopausal women with axillary node positive operable breast cancer: Results of a randomized trial. *J. Clin. Oncol.* **14**, 35–45.

de Gramont, A., Banzi, M., Navarro, M., Tabernero, J., Hickish, T., Bridgewater, J., Rivera, F., Fountzilas, G., and Andre, T. (2003). Oxaliplatin/5-FU/LV in adjuvant colon cancer: Results of the international randomized mosaic trial. *Proc. Am. Soc. Clin. Oncol.* **22**, 253.

Douillard, J. Y., Cunningham, D., Roth, A. D., Navarro, M., James, R. D., Karasek, P., Jandik, P., Iveson, T., Carmichael, J., Alakl, M., Gruia, G., Awad, L., and Rougier, P. (2000). Irinotecan combined with fluorouracil compared with fluorouracil alone as first line treatment for metastatic colorectal cancer: A multicenter randomized trial. *Lancet* **355**, 1041–1047.

Eisenberger, M. A., De Wit, R., Berry, W., Chi, K., Oudard, S., Christine, T., James, N., and Tannock, (2004). A multicenter phase III comparison of docetaxel plus prednisone versus mitoxantrone plus prednisone in patients with hormone refractory prostate cancer. *Proc. Am. Soc. Oncol.* **22**, 4.

Hudis, C., Seidman, A., Baselga, J., Raptis, G., Lebwohl, D., Gilewski, T., Moynahan, M., Sklarin, N., Fennelly, D., Crown, J. P., Surbone, A., Uhlenhopp, M., Riedel, E., Yao, T. J., and Norton, L. (1999). Sequential dose-dense doxorubicin, paclitaxel, and cyclophosphamide for resectable high-risk breast cancer:feasibility and efficacy. *J. Clin. Oncol.* **17**, 93–100.

Hung, A., Crane, C., Delclos, M., Ballo, M., Ajani, J., Lin, E., Feig, B., Skibber, J., and Janjan, N. (2003). Cisplatin-based combined modality therapy for anal carcinoma: A wider therapeutic index. *Cancer* **97**, 1195–1202.

McKelvey, E. M., Gottlieb, J. A., Wilson, H. E., Haut, A., Talley, R. W., Stephens, R., Lane, M., Gamble, J. F., Jones, S. E., Grozea, P. N., Gutterman, J., Coltman, C., and Moon, T. E. (1976). Hydroxydaunomycin combination chemotherapy in malignant lymphoma. *Cancer* **38**, 1484–1493.

McLaughlin, P., Hagemeister, F. B., Romaguera, J. E., Sarris, A. H., Pate, O., Younes, A., Swan, F., Keating, M., and Cabanillas, F. (1996). Fludarabine, mitoxantrone, and dexamethasone: An effective new regimen for indolent lymphoma. *J. Clin. Oncol.* **14**, 1262–1268.

Nigro, N. D., Seydel, H. G., Considine, B., Vaitkevicius, V. K., Leichman, L., and Kinzie, J. J. (1983). Combined preoperative radiation and chemotherapy for squamous cell carcinoma of the anal canal. *Cancer* **51**, 1826–1829.

Poon, M. A., O'Connell, M. J., Moertel, C. G., Wieand, H, S., Cullinan, S. A., Everson, L. K., Krook, J. E., Mailliard, J. A., Laurie, J. A., and Tschetter, L. K. (1989). Biochemical modulation of fluorouracil: Evidence of significant improvement of survival and quality of life in patients with advanced colorectal carcinoma. *J. Clin. Oncol.* **7**, 1407–1418.

Saltz, L. B., Cox, J. V., Blanke, C., Rosen, L. S., Fehrenbacher, L., Moore, M. J., Maroun, J. A., Ackland, S. P., Locker, P. K., Pirotta, N., Elfring, G. L., and Miller, L. L. (2004). Irinotecan plus fluorouracil and leucovoin for metastatic colorectal cancer. *N. Engl. J. Med.* **350**, 2335–2342.

Tannock, I. F., Osoba, D., Stockler, M. R., Ernst, D. S., Neville, A. J., Moore, M. J., Armitage, G. R., Wilson, J. J., Venner, P. M., Coppin, C. M., and Murphy, K. C. (1996). Chemotherapy with mitoxantrone plus prednisone or prednisone alone for symptomatic hormone-resistant prostate cancer: A Canadian randomized trial with palliative end points. *J. Clin. Oncol.* **14**, 1756–1764.

Tournigand, C. Louvet, C., Quinaux, E., Andre, T., Lledo, G., Flesch, M., Ganem, G., Landi, B., Colin, P., Denet, C., Mery-Mignard, D., Risse, M.-L., Buyse, M., and de Gramont, A. (2001). FOLFIRI followed by FOLFOX versus FOLFOX followed by FOLFIRI in metastatic colorectal cancer: Final results of a phase III study. *Proc. Am. Soc. Clin. Oncol.* **20**, 124a.

Wasserman, E., Sutherland, W., Cvitkovic, E. (2001). Irinotecan plus oxaliplatin: A promising combination for advanced colorectal cancer. *Clin. Colorectal Cancer* **1**, 149–153.

Weiss, R. B., Valagussa, P., Moliterni, A., Zambetti, M., Buzzoni, R., Bonadonna, G. (1987). Adjuvant chemotherapy after conservative surgery plus irradiation versus modified radical mastectomy. Analysis of drug dosing and toxicity. *Am. J. Med.* **83**, 455–463.

Useful Web Sites

1. FDA Oncology Tools: http://www.fda.gov/cder/cancer/index.htm
   This Web site is maintained by the Food and Drug Administration and includes information on approved oncology drugs, disease summaries, regulatory issues related to cancer, oncology reference tools, and other resources. This site also provides a useful drug dose converter (http://www.fda.gov/cder/cancer/animalframe.htm) for performing chemotherapy drug concentration conversions (e.g., mg/m^2 <–> mg/kg).

2. National Comprehensive Cancer Network: www.nccn.org

3. Oncolink: http://oncolink.upenn.edu
   This Web site is maintained by the Abramson Cancer Center of the University of Pennsylvania and is an excellent resource for information on all aspects of clinical cancer research.

CHAPTER

8

Targeted Therapeutics in Cancer Treatment

COLIN D. WEEKES AND MANUEL HIDALGO

I. Introduction
II. Cell Cycle
   A. Basic Concepts
   B. Agents Targeting the Cell Cycle
   C. Protein Kinases
   D. EGFR
   E. Anti-EGFR Strategies
   F. Mammalian Target of Rapamycin
III. The MAPK Family
   A. Compounds in Development
   B. Future Directions
   C. Src Kinase Inhibitors
IV. Challenges in the Clinical Development of Signal Transduction Inhibitors
   A. Clinical Trials: Design Issues
   B. Patient Selection
   C. Study Endpoints
   D. Combination Therapy
   References

The accumulated advances in the understanding of basic tumor biology have heralded the early twenty-first century state of modern chemotherapy. Modern chemotherapy has evolved into targeting discrete molecular signaling pathways to achieve an anticancer effect. Both monoclonal antibodies targeting the ligand as well as the extracellular domains of transmembrane receptors and small molecules designed to inhibit the kinase function of intracellular signaling proteins have been developed to achieve clinical effects. These molecules have the potential to improve efficacy while limiting toxicity. The efficacy of small molecule inhibitors depends on the relevance of the targeted signal to the oncogenesis of the tumor. This is best exemplified by the overwhelming cytologic response observed in patients with chronic myelogenous leukemia (CML) who are treated with imatinib, which targets the bcr/abl fusion protein that is essential for CML oncogenesis. Unfortunately, few solid tumors are dependent on a single pathway for oncogenesis. Therefore, it may be more efficacious to inhibit multiple targets with a single small molecule inhibitor. The second generation compounds are being designed to inhibit multiple signaling pathways simultaneously. Additional efforts are in development to rationally combine small molecule inhibitors with conventional cytotoxic chemotherapy. This strategy will likely require the use of cytotoxic chemotherapy in a manner that renders the tumor cell dependent on the small molecule targeted pathway for survival.

Publications have demonstrated that response to small molecule inhibitors is linked to the presence or acquisition of either activating or inactivating mutations in the target protein. However, these data were obtained only after treatment. Greater effort is required to develop novel strategies to determine which patients will respond to a given small molecule inhibitor. This will be imperative to improve patient selection as it pertains to the clinical development of small molecule inhibitors. In addition, the development of these agents will required novel approaches to clinical study designs. Many of the small molecules will not result in classic tumor shrinkage; therefore, functional imaging techniques may be a better measurement of response to therapy than classic Response Evaluation Criteria in Solid Tumors (RECIST). Furthermore, molecular imaging techniques may need to be developed to assess real-time pathway function as taking biopsies of tumors is of little practical use for most patients with visceral tumors. Last, clinical trial design will need to be modified, and the classic phases of compound development may also need to be modified to ensure that active compounds are not discarded inappropriately based on poor study design. In closing, the development of small molecule inhibitors as a therapeutic strategy represents a significant advancement in cancer therapy but has created new challenges for the therapy's advancement.

I. INTRODUCTION

Chemotherapy used for the treatment of malignancy historically has been restricted to cytotoxic agents. Perturbation of DNA synthesis and the events regulating cell division are the primary targets of traditional cytotoxic drugs, as outlined in Chapter 7. Unfortunately, the events regulating cell division are not specific to cancer cells; therefore, these medications result in a broad range of toxic side effects as a result of damage to normal cells. The narrow therapeutic window of traditional cytotoxic drugs is quite troublesome given that palliation is the primary goal of most types of oncology therapy. Oncology therapy has migrated to the use of small molecules targeting intracellular events specific to tumor cells. The era of "targeted therapy" was heralded by the approval of the antibody trastuzumab for the treatment of HER2/neu positive metastatic breast cancer (Slamon et al., 2001). The use of small molecule inhibitors of intracellular pathways as a therapeutic principle was validated by the efficacy of imatinib mesylate for the treatment of bcr/abl$^+$ CML and gastrointestinal stromal tumors (GIST) due to its inhibition of mutated c-kit signaling (Druker et al., 2001; Demetri et al., 2002).

The development of small molecule inhibitors for cancer therapy has paralleled the scientific understanding of cellular processes resulting in the initiation and growth of cancerous cells. Cellular differentiation and proliferation as well as apoptosis represent the critical cellular events mediating cancer cell initiation and survival. Cell proliferation and differentiation are regulated by a number of hormones, growth factors, and cytokines. These molecules interact with cellular receptors and communicate with the cell nucleus through a network of intracellular signaling pathways. In cancer cells, key components of these pathways may be altered through overexpression or mutation, acting as oncogenes that lead to dysregulated cell signaling, inhibition of apoptosis, metastasis, and cell proliferation. The silencing of tumor suppressor genes either by mutation or epigenetic modification represses the negative regulation of cellular processes, thereby contributing to the oncogenic process. The components of abnormal signaling pathways specific to neoplastic cells represent potential selective targets for new anticancer therapies. These potential targets include ligands (typically growth factors), cellular receptors, intracellular second messengers, and nuclear transcription factors. Additional targets for small molecule inhibitors include proteins involved in metabolic pathways, protein transport, and protein degradation.

Conceptually, there are multiple potential key points of intervention that allow attack of signaling pathways for cancer therapy. The first obvious point of intervention in a signaling cascade is the neutralization of ligands before they can associate with their receptors. This approach has been successfully validated with bevacizumab, a humanized monoclonal antibody targeting circulating vascular endothelial growth factor (VEGF) (Susman et al., 2005). The second approach to abrogating signaling pathways is the direct inhibition of receptors. This can be achieved by (i) preventing the binding of growth factors to their receptors, exemplified by the success of cetuximab, a chimeric antibody directed against the epidermal growth factor receptor (EGFR) (Meyerhardt et al., 2005), or (ii) by inhibiting the kinase activity of receptors with small molecule inhibitors of receptor phosphorylation, such as with erlotinib (Woodburn, 1999). The final approach relates to the inhibition of signaling by cytoplasmic secondary messengers, most of which are protein kinases. An example of this last approach is the use of imatinib, an inhibitor of the kinase activity of bcr-abl, c-kit, and platelet-derived growth factor receptor (PDGFR) (Druker et al., 2001; Demetri et al., 2002). Bortezomib inhibition of IκB proteosomal degradation provides an excellent example of a protein degradation modulator as a therapeutic target (Mitsiades et al., 2005). This chapter's discussion is not meant to be all inclusive; however, the chapter will focus on demonstrating how various cellular processes regulating oncogenesis can be targeted by small molecule inhibitors. Table 8.1 summarizes the key novel targets and provides a limited listing of some agents in development against these targets. In addition, this chapter provides insight into the challenges of incorporating small molecule inhibitors into the cache of general oncology practice either as single agents or in combination with other small molecule inhibitors or cytotoxic chemotherapy.

II. CELL CYCLE

A. Basic Concepts

The cell cycle forms the platform where all cellular processes are governed. Figure 8.1 illustrates the basic understanding of the cell cycle. Perturbations of the cell cycle are commonly described in carcinogenesis. The improved knowledge of cell cycle regulation combined with the understanding of its critical role in carcinogenesis have resulted in the targeting of cell cycle regulators as a potential therapeutic strategy. The cell cycle regulates the process of cellular proliferation and growth, as well as cell division after DNA damage. The cell cycle can be divided into four phases, the transition of which is regulated by a series of checkpoints. These checkpoints ensure the fidelity of the genetic code. The quiescent phase (G_0) represents the period of cellular

TABLE 8.1 Intracellular Targets and Compounds in Development

Intracellular Target	Compound	Generic Name	Trade Name
EGFR[1]	OSI774	Erlotinib	Tarceva
	ZD1839	Gefitinib	Iressa
	GW572016	Lapatinib	
	HKI-272		
	EKB-569		
	CI-1033		
MEK[2]	CI-1040		
	PDO325901		
	ARRY-142886 (AZD6244)		
Src	BMS354825	Dasatinib	Sprycel
	AZM475271		
	SKI-606		
Apoptosis	Apo2/TRAIL		
	YM155		
Aurora kinase	AZD1152		
	MK-0457		
Polo-like kinase	ON01910		
	BI 2536		
N-cadherin	Exherin		
HSP-90[3]	17-DMAG		
	IPI504		
KSP[4]	SB-743921		
	MK-0731		
BCR-ABL, cKit, PDGFR[5]	STI571	Imatinib mesylate	Gleevec
	AMN107	Nilotinib	
	BMS-354825	Dasatinib	Sprycel
Multikinase	BAY 43-9006	Sorafenib	Nexavar
	SU11248	Sunitinib	Sutent
	ZD6474	Zactima	
VEGFR[6]	PTK787/ZK222584		
mTOR	AY 22989	Rapamycin	Sirolimus
	CCI-779	Temsirolimus	
	RAD001	Everolimus	Certican
	AP23573		
FTI[7]	R115777	Tipifarnib	Zarnestra
CDK[8] inhibitors	L86-8275	Flavopiridol	
	UCN-01		
	Bryostatin-1		
	CYC202	Roscovitine	
	BMS387032		
	E7070	Indisulam	
	PD0332991		
Proteosome	PS-341	Bortezomib	Velcade

[1]Epidermal growth factor receptor
[2]Methyl ethyl ketone
[3]Heat shock protein 90
[4]Kinesis spindle protein
[5]Platelet-derived growth factor receptor
[6]Vascular endothelial cell growth factor receptor
[7]Farnesyl transferase inhibitor
[8]Cyclin-dependent kinase

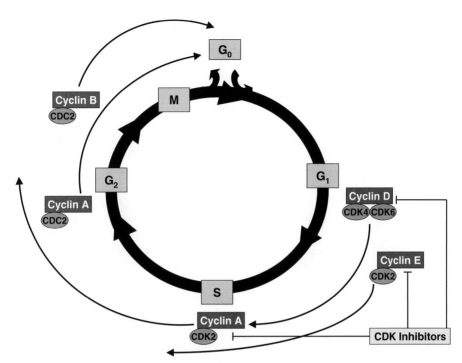

FIGURE 8.1 Cell Cycle.

dormancy during which the cell is not actively in the cell cycle. Preparation for DNA synthesis occurs during the initial gap 1 (G_1) phase of the cell cycle. The S phase is characterized by DNA synthesis, which is followed by a second preparatory phase known as gap 2 (G_2). Mitosis (M) culminates the cell cycle during which the newly synthesized DNA and nuclear contents migrate to the resultant daughter cells and complete the cycle. Progression through the cycle is facilitated by complexes formed between cyclin-dependent kinases (CDKs) and the regulatory protein cyclin. Inhibition of cell cycle progression is coordinated by a series of negative regulatory proteins known as CDK inhibitors (CDKIs). A detailed review of the cell cycle is beyond the context of this chapter, and the reader is referred to a more detailed review of the cell cycle by Scherr et al. (1999).

Targeting of the cell cycle and, in particular, CDKs stems from the prevalent perturbation of CDKs during carcinogenesis as well as the observation that CDK inhibition can induce apoptosis. Cancer cells are characterized by the loss of checkpoint control, which can occur by either cyclin overexpression or the inactivation of CDKIs (Malumbres and Barbacid, 2001). Cyclin D is commonly overexpressed in breast cancer, and the translocation event between chromosomes 11 and 14 (t11:14) prominent in mantle cell lymphoma results in the overexpression of cyclin D by its juxtaposition to the immunoglobulin promoter region (Buckley et al., 1993; Yatabe et al., 2000). The p16 gene is an INK4 gene that frequently undergoes hypermethylation, resulting in the epigenetic silencing of this CDK. This event is commonly observed in melanoma, colorectal, breast, and lung cancers (Shahjehan et al., 2001). Therefore, the goal of targeting the CDKs would be to reinstitute appropriate checkpoint regulation and resultant apoptosis.

CDKIs have been developed as cell-cycle specific agents due to the direct binding of the catalytic domain as well as their indirect impact on regulatory pathways of

CDKs (Senderowicz et al., 2002). The importance of transcription regulation by CDK-dependent phosphorylation of the C-terminal domain (CTD) of RNA polymerase II has been appreciated. CDK-dependent CTD phosphorylation results in RNA pol II inactivation, which suppresses mRNA production and transcription during mitosis (Senderowicz et al., 2002; Oelgeschlager, 2002). CDKI-mediated inhibition of CTD phosphorylation may contribute to its pro-apoptotic effect (Kobor and Greenblatt, 2002; Koumenis and Giaccia, 1997).

B. Agents Targeting the Cell Cycle

A number of compounds are in various stages of development; these include flavopiridol, UCN-01, bryostatin-1, roscovitine, YC202, BMS 387032, E7070, and PD0332991 (Table 8.2).

Flavopiridol is the most extensively studied molecule of this group of CDKIs, and the results of its development will be used to illustrate the utility of this class of agents in cancer therapy. Flavopiridol is a pan-CDKI of CDK2, CDK4, and CDK6 at nanomolar concentrations. It effectively induces cell cycle arrest at both the G_1/S and G_2/M checkpoints. In addition to its effects as a cell cycle modulator, its administration has been shown to induce apoptosis by a variety of mechanisms (Te Poele, 1999; Losiewicz, 1994). First, caspase activation via the activation of the mitogen-activated protein kinase (MAPK) family of proteins is associated with flavopiridol administration (Carlson et al., 1996). Furthermore, it has been demonstrated to abrogate CDK9-mediated transcriptional regulation, resulting in the depletion of cyclin D and XIAP mRNAs (Cartee et al., 2001; Carlson et al., 1999; Motwani et al., 2001). Collectively, the effects on transcription are thought to promote apoptosis and inhibit cell proliferation. The antitumor properties of flavopiridol have been validated in vivo with xenograft models of prostate and squamous head and neck cancers (Drees et al., 1997; Patel et al., 1998).

Clinical evaluation of flavopiridol indicates that its apoptotic effects are highly dose and schedule dependent. Flavopiridol demonstrates greater than 90% of protein binding to human plasma proteins; thus, its activity depends on achieving free drug levels of 250–300 nmol/L. These levels are achievable by administering the drug as a 30-minute bolus followed by a 4-hour infusion and are maintained for 4 hours. The efficacy of this regimen has been demonstrated by achieving nearly a 50% response rate in patients with fludarabine refractory chronic lymphocytic leukemia, resulting in prolonged survival (Byrd et al., 2007).

Flavopiridol has been combined with paclitaxel, irinotecan, and gemcitabine. Combination with paclitaxel 175 mg/m^2

TABLE 8.2 Cyclin-Dependent Kinase Inhibitors

Drug	Cell Cycle Phase	Target	IC$_{50}$ (nmol/L)	Phase of Development
Flavopiridol	G$_1$/S arrest, G$_2$	CDK1, CDK2, CDK4/CDK6, CDK9-Cyclin T-RNA Pol II	100	Phase II
UCN-01	S, G$_2$	CDK2	<260	Phase II
Bryostatin-1	G$_2$ arrest	CDK2		Phase II
Roscovitine	S, G$_2$	CDK2, CDK9-Cyclin T-RNA Pol II	100	Phase II
E7070	S, G$_2$	CDK2, CDK1	11	Phase II
BMS387032	S, G$_2$	CDK2, CDK1	48	Phase II
PD0332991	G$_1$ arrest	CDK4/6	11	Phase II

with escalating doses of flavopiridol given every 3 weeks in the phase I setting identified flavopiridol dose-limiting toxicities (DLTs) of neutropenia and pulmonary toxicity at 94 mg/m^2. Peak mean levels of flavopiridol ranged from 400 to 600 nmol/L. Major responses were observed in patients with esophageal and prostate carcinoma (Schwartz et al., 2001). The combination of weekly docetaxel 35 mg/m^2 infused for 4 hours followed by escalating doses of flavopiridol as a 1-hour infusion was able to achieve a flavopiridol dose of 80 mg/m^2 without DLT, while attaining peak flavopiridol levels in the micromolar range. Furthermore, partial responses (PR) were observed in patients with pancreatic, breast, and ovarian carcinomas (Rathkopf et al., 2004). In contrast, a phase I study in patients with breast cancer, combining weekly docetaxel followed in 24 hours with the administration of flavopiridol given either as a continuous infusion for 72 hours or as a 1-hour bolus infusion on 3 consecutive days, demonstrated DLTs of neutropenia and hypotension (Tar et al., 2004). Although the combination of flavopiridol with taxanes does demonstrate clinical activity in early phase studies, the toxicities of this combination are highly schedule dependent.

Flavopiridol and bryostatin-1 have been combined with cytotoxic agents to overcome cell cycle-mediated drug resistance. Cell cycle-mediated drug resistance is characterized by the insensitivity to a chemotherapeutic due to the induction of checkpoint activation, thus resulting in cell cycle arrest. This finding was exemplified by the failure of irinotecan to demonstrate antitumor activity in a xenograft model after exposure to flavopiridol. This is particularly due to the induction of cell cycle arrest at G_2 by flavopiridol. As would be expected, the sequencing of paclitaxel prior to flavopiridol administration demonstrated appropriate antitumor properties. Similar effects have been observed with preclinical testing of SN-38 (the active metabolite of irinotecan) and flavopiridol (Motwani, 2001). This strategy has been utilized in the clinical development of cytotoxic agent combinations with flavopiridol and other CDKIs. The sequential administration of irinotecan followed by flavopiridol divided by a 7-hour interval administered 4 of 6 weeks has been evaluated in a phase I study in 51 patients with solid tumors. No DLTs were observed, and peak flavopiridol concentrations were greater than 2 µmol/L at dose of 50 mg/m^2. Clinical activity was documented with 36% of patients attaining a PR in a variety of tumors and an addition to those with prolonged stable disease (Shah et al., 2002). A phase I study of the combination of flavopiridol with irinotecan and cisplatin is ongoing as of this writing (Shah et al., 2004). Similarly, a phase I trial is evaluating the combination of flavopiridol with gemcitabine. The evaluation of flavopiridol demonstrates the ability of preclinical models to direct strategy for rational combinations of small molecules and traditional cytotoxic agents.

Targeting mitosis with small molecule inhibitors has become a practical option with the understanding of the function of Aurora kinase and polo-like kinase 1 (PLK-1), as shown in Table 8.3. The proper segregation of DNA and activation of the anaphase-promoting complex (APC) requires the binding of the kinetochores to tubulin fibers. Aurora kinase B and survivin are integral components of the chromosome passenger complex (CPC) found in the kinetochore. Aurora kinase B activation via phosphorylation allows the appropriate binding of tubulin to the kinetochore (Lampson et al., 2004; Lampson and Kapoor, 2005). Aurora kinase B inhibition is associated with impaired chromosomal alignment, leading to abnormal chromosomal segregation, polyploidy, and eventually apoptosis. Several inhibitors of Aurora kinases are undergoing clinical investigation (Nair et al., 2004; Harrington et al., 2004). Thus far, these agents appear to be well tolerated, with bone marrow toxicity being the principal side effect.

TABLE 8.3 Mitosis-Specific Inhibitors

Drug	Intracellular Target	Patient Number	Toxicity	Phase of Development
ON01910	PLK-1[1]	N/A	N/A	Phase I
SB-743921	KSP[2]	44	Neutropenia, elevated liver function test	Phase I
MK-0731	KSP	17	Neutropenia, elevated liver function test	Phase I
AZD1152	Aurora kinase B	19	Neutropenia	Phase I
MK-0457	Aurora kinases B and C	22	Neutropenia	Phase I

[1] Polo-like kinase 1
[2] Kinesis spindle protein

The PLK protein family consists of at least four members in mammalian cells (Duncan et al., 2001). These proteins are characterized by a conserved c-terminal POLO box domain in addition to a serine–threonine kinase domain. PLK-1 has been demonstrated to be pivotal for mitotic progression by facilitating centrosome maturation, mitotic spindle assembly, and APC activation (Kotani et al., 1998). ON01910 is a small molecule inhibitor of PLK-1 that has been demonstrated to induce mitotic arrest and apoptosis in a variety of tumor cells (Gumireddy et al., 2005). It also augments the antitumor effects of a variety of cytotoxic agents. ON01910 is in the early phase clinical development. BI-2536 is a second compound in clinical development in the phase I setting. To date, two phase I studies evaluating different administration schedules have been performed, with neutropenia being the greatest observed toxic effect. One partial response has been observed in a patient with metastatic squamous cell carcinoma of the head and neck (Hofheinz et al., 2006; Munzert et al., 2006).

C. Protein Kinases

Protein tyrosine kinases (PTKs) comprise a large fraction of the approximately 40 tumor-suppressor genes and over 100 dominant oncogenes described to date (Futreal et al., 2001). PTKs are also the largest group of dominant oncogenes with structural homology. PTKs evolved to mediate aspects of multicellular communication and development. Somatic mutations in this very small group of genes cause a significant fraction of human cancers, again emphasizing the inverse relationship between normal developmental regulation and oncogenesis (Blume-Jensen and Hunter, 2001). There are more than 90 known PTK genes in the human genome; 58 encode transmembrane receptor PTKs distributed into 20 subfamilies, and 32 encode cytoplasmic, nonreceptor PTKs in 10 subfamilies (Plowman et al., 1999; Robinson et al., 2000).

Receptor tyrosine kinases (RTKs) have been organized into families based on sequence homology, structural characteristics, and distinct motifs in the extracellular domain. There are 20 known families in vertebrates. The various subfamilies include the receptors for epidermal growth factor (EGF), platelet-derived growth factor (PDGF), VEGF, fibroblast growth factor (FGF), and hepatocyte growth factor (HGF). RTKs share several structural features. They are glycoproteins possessing an extracellular ligand-binding domain, which conveys ligand specificity, and a single hydrophobic transmembrane domain, anchoring the receptor to the membrane. Intracellular sequences typically contain regulatory regions in addition to the catalytic domain. Ligand binding induces activation of the intracellular tyrosine kinase domain, leading to the initiation of signaling events specific for the receptor. RTK phosphorylation induces receptor dimerization with conformational changes that result in intermolecular phosphorylation at tyrosine residues at multiple sites. Receptor heterodimerization can also occur, as reported with transforming growth factor alpha interaction with receptor heterodimers comprising human epidermal growth factor receptor 2 (HER2) and EGFR (Kolibaba and Druker, 1997). In malignant tumors, a number of these receptors are overexpressed or mutated, leading to abnormal cell proliferation.

Nonreceptor tyrosine kinases are cytoplasmic proteins that transduce extracellular signals to downstream intermediates in pathways that regulate cell growth, activation, and differentiation. Many nonreceptor tyrosine kinases are linked to transmembrane receptors, including those for peptide hormones and cytokines. Unlike receptor tyrosine kinases, they lack transmembrane domains and ligand binding. They are activated by ligand binding to their associated receptors or events, such as cell adhesion, calcium influx, or cell cycle progression. More than 30 members are classified in 10 families, including Src, Abl, JAK, MKK2, and FES (Yamaguchi et al., 1995). The serine–threonine kinases are almost all intracellular and include key mediators of carcinogenesis such as raf, Akt/protein kinase B, and MEK. Effector molecules recruited after ligand-induced phosphorylation of RTKs include phospholipase C (PLC), phosphatidylinositol 3-kinase (PI3K), and Ras.

Table 8.1 summarizes the most relevant PK targeted for cancer treatment as well as the key agents in development. The detailed description of these compounds is beyond the scope of this general chapter. In the following section, the development of small molecule inhibitors of the EGFR MAPK/extracellular signal-regulated kinase (ERK) and mammalian target of rapamycin inhibitors will be discussed to illustrate strategies for targeting intracellular signal transduction pathways.

D. EGFR

The HER family consists of four receptors: HER1, (also known as EGFR), HER2 (Erb2 or HER2/neu), HER3 (Erb3), and also HER4. This receptor family shares the same extracellular molecular structure, a cysteine-rich ligand-binding domain, a single alpha-helix transmembrane domain, and an intracellular domain with tyrosine kinase activity contained in the carboxy-terminal tail of all family members except HER3 (Wells, 1999). The tyrosine kinase domain of EGFR shares 80% homology with that of HER2 and HER4 (Arteaga, 2001). There are three ligands with exclusive EGFR binding properties; these ligands include EGF, transforming growth factor α (TGFα), and amphiregulin. Betacellulin and epiregulin can bind to both EGFR and HER4, resulting in their activation. Ligand binding results in both homodimerization and heterodimerization with other HER family members (Pinkas-Kramaski et al., 1996; Yarden et al., 2001). HER2 is the preferred partner for heterodimerization with EGFR although it lacks a known ligand (Graus-Porta et al., 1997). Homodimers of EGFR are unstable, whereas heterodimerization with HER2 results in EGFR stabilization and enhanced cell surface expression (Worthylake et al., 1999).

EGFR ligand binding results in tyrosine kinase activation, which in turn mediates the autophosphorylation of one or more of five tyrosine residues in the carboxy terminus of the receptor at positions Y992, Y1068, Y1086, Y1448, and Y1173 (Schlessinger et al., 2000). The phosphotyrosines serve as docking sites for adaptor proteins (shown in Figure 8.2). EGFR autophosphorylation results in the propagation of intracellular signals via two pathways: the phosphatidylinositol 3-kinase (PI3K)/Akt and the Ras-Raf-MAPK (also known as extracellular-regulated kinase ERK1 and ERK2) (Blenis et al., 1993; Burgering and Coffer, 1995). The stress-activated protein kinase

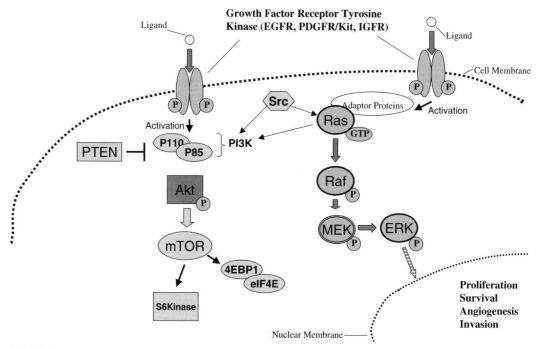

FIGURE 8.2 Receptor Tyrosine Kinase Intracellular Signal Cascade.

pathway involving protein kinase C is an alternative pathway of EGFR intracellular signaling (Tebar et al., 2002). These three pathways exert EGFR-mediated control of cellular proliferation, transformation, metastasis, angiogenesis, and decreased apoptosis (Lewis et al., 1998; Cantley et al., 2002). Tumor cells have developed alternative mechanisms for EGFR activation in addition to ligand binding that regulates EGFR activation under physiologic conditions. The first mechanism is receptor overexpression, resulting in ligand-independent dimerization of the receptor (Salomon et al., 1995). Second, tumor cells possess the ability to produce ligands (such as TGFα), resulting in autocrine activation of EGFR (Grandis et al., 1998). Last, truncated isoforms of EGFR, such as EGFRvIII, that lack the extracellular domain are constitutively activated and have been described in glioblastoma (Nishikawa et al., 1994). EGFR tyrosine kinase-activating mutations resulting from in-frame deletions within exon 19 and point mutations in codon 858 (exon 21) have been associated with response to tyrosine kinase inhibition mediated by gefitinib or erlotinib. Furthermore, these mutations preferentially segregate to adenocarcinomas in non-small-cell lung cancer (NSCLC) patients with minimal smoking exposure more than in patients who are smokers. EGFR overexpression has been associated with advanced disease states, chemotherapy resistance, and poor prognosis (Woodburn, 1999; Nicholson et al., 2001). The importance of EGFR as a regulator of signals propagated in tumor cells by the tumor microenvironment combined with its association with positive clinical outcomes has made EGFR an important target for therapy (Arteaga, 2002).

E. Anti-EGFR Strategies

The primary strategies employed for the targeting of EGFR have been to prevent the binding of the ligand to the receptor by monoclonal antibodies (mAb) directed against the extracellular domain of the

TABLE 8.4 EGFR Small Molecule Tyrosine Kinase Inhibitors

Drug	Type	EGFR IC$_{50}$ (μM)	HER2 IC$_{50}$ (μM)	Tumor	Phase of Development
Gefitinib	Selective, reversible	0.02	3.7	NSCLC	Approved
Erlotinib	Selective, reversible	0.02	3.5	NSCLC, Pancreas	Approved
EKB-569	Selective, irreversible	0.04	1.2	Colon	Phases II–III
Lapatinib	Bifunctional, reversible	0.01	0.009	Breast	Approved
CI-1033	Trifunctional, irreversible	0.0008	0.02	—	Phase II
HKI-272	Bifunctional, irreversible	0.092	0.059	NSCLC	Phase I

Rabindran, S. K. (2005). Antitumor activity of HER-2 inhibitors. *Cancer Lett.* **227**, 9–23

receptor or the use of small molecule inhibitors, known as tyrosine kinase inhibitors (TKIs), which are competitive antagonists to the binding of adenosine triphosphate (ATP) to the receptor's kinase pocket. This section describes in detail this second strategy. The use of monoclonal antibodies against the EGFR is described in Chapter 10.

The TKIs under clinical investigation include gefitinib, erlotinib, lapatinib (GW572016), and HKI-272 (Table 8.4). Both gefitinib and erlotinib selectively and reversibly target EGFR, preventing EGFR autophosphorylation and thereby locking the signal cascade. Preclinical *in vivo* testing demonstrated the antitumor effects of both agents in a variety of cancer models (Sirotnak *et al.*, 2000; Pollack *et al.*, 1999; Moyer *et al.*, 1997). Phase I clinical investigation of gefitinib showed favorable toxicities, primarily consisting of skin rash and diarrhea. DLTs were at doses far beyond doses required for antitumor effects (Baselaga *et al.*, 2002; Herbst *et al.*, 2002; Ranson *et al.*, 2002; Hidalgo *et al.*, 2001). Phase I studies combining gefitinib with cytotoxic agents were completed with a gefitinib dose of 250 or 500 mg combined with full-dose cytotoxic agents without incremental increase in toxicity.

Two phase II studies evaluated gefitinib either at 250 or 500 mg in NSCLC patients failing either one or two prior regimens. In this setting, response rates of 18.7% and 10.6% were documented and were associated with improved palliation of symptoms. The higher dose was not associated with improved response but caused more toxicity (Fukuoka *et al.*, 2003; Kris *et al.*, 2003). This led to the regulatory approval of gefitinib at a dose of 250 mg for NSCLC patients refractory to platinum-based or docetaxel chemotherapy. However, two subsequent placebo-controlled phase III studies in chemotherapy-naive NSCLC patients that evaluated the addition of gefitinib to doublet chemotherapy failed to demonstrate a benefit of gefitinib (Giaccone *et al.*, 2004; Herbst *et al.*, 2004). Furthermore, a placebo-controlled phase III study failed to demonstrate that use of single agent gefitinib is superior to use of a placebo in this group of patients. Subsequent to completion of these studies, activation mutations in the kinase domain of EGFR were found to predict response in patients receiving gefitinib and have been confirmed to mediate response to erlotinib as well (Lynch *et al.*, 2004; Paez *et al.*, 2004; Pao *et al.*, 2004). Although this information was not known at the time of the early phase evaluation of gefitinib, the experience to date with gefitinib development emphasizes the importance of appropriate patient selection. Gefitinib continues to undergo clinical evaluation in a variety of cancers despite the negative phase III results in NSCLC patients.

Phase I testing of erlotinib defined the daily dose of 150 mg to progress into phase II–III trials. DLTs of diarrhea and acneiform rash occurred at higher doses (Hidalgo et al., 2001). In contrast to gefitinib, erlotinib was developed at the maximum tolerated dose. Use of erlotinib is the first EGFR-targeted therapy to show significant survival benefit compared to results from use of placebo control, as demonstrated by a 6.7-month median survival compared to 4.7 months in patients with NSCLC in clinical progression or with chemotherapy refractory disease (Shepherd et al., 2004). In addition, erlotinib has received approval for use in metastatic pancreatic cancer therapy when combined with gemcitabine due to a modest improvement in overall survival (Moore et al., 2005).

Other agents include the irreversible inhibitors EKB-569 and CI-1033. These agents, in addition to lapatinib, are nonspecific and target several other members of the family. The development of EKB-569 and CI-1033 has been abandoned (Erlichman et al., 2006). Lapatinib targets the EGFR and reversibly inhibits the phosphorylation of the ErbB2 (Burris et al., 2003; Erlichman et al., 2001). Lapatinib has a typical TKI toxicity profile with diarrhea and rash (Burris et al., 2003). Early phase studies have demonstrated activity in herceptin refractory breast cancer (Blackwell et al., 2004). The dual EGFR–ErbB2 receptor inhibitor HKI-272 has shown to also inhibit the activation of the EGFR with the T790M mutations, which are known to induce resistance to erlotinib. Other strategies in clinical evaluation include dual targeting of EGFR with the simultaneous administration of monoclonal antibodies and TKI. This strategy both blocks receptor activation in addition to the propagation of intracellular signals. *In vivo* testing of this strategy demonstrated a synergistic antitumor effect of dual EGFR inhibition over either single agent (Jimeno et al., 2005).

F. Mammalian Target of Rapamycin

Mammalian target of rapamycin (mTOR) which is also named FRAP or RAFT1, is a serine–threonine kinase that belongs to the family of phosphatidylinositol kinase-like kinases involved in the regulation of a wide range of growth-related cellular functions, including transcription, translation, membrane trafficking, protein degradation, and reorganization of the actin cytoskeleton, as shown in Figure 8.2 (Brown et al., 1994; Sabatini et al., 1994; Schmelzle and Hall, 2000). mTOR functions as a nutrient and growth factor sensor, controlling cellular growth and proliferation. Abnormal activation of signaling pathways both proximal and distal to this kinase occur frequently in human cancer. The best known function of mTOR, in the context of cell proliferation, is the regulation of translation initiation, presumably mediated by the activation of the 40S ribosomal protein S6 kinase (p70^{s6k}) and the inactivation of 4E-binding protein 1 (4E-BP1). The increase in the translation of a subset of mRNAs brings about protein products that are required to traverse the G_1/S checkpoint of the cell cycle.

mTOR activity appears to be regulated by the PI3K/protein kinase B (PI3K/Akt), as shown in Figure 8.2 (Downward et al., 1998; Scott et al., 1998; Nave et al., 1999). PI3K/Akt is activated as a result of the ligand-dependent activation of RTKs, G-protein-coupled receptors, or integrins. Receptor-independent activation can also occur, for example, in cells expressing constitutively active Ras proteins (Rodriquez-Viciana et al., 1997; Kauffmann-Zeh et al., 1997). Activated PI3K catalyzes the conversion of phosphatidylinositol 4,5-biphosphate (PIP_2) to phosphatidylinositol-3, 4, 5-trisphosphate (PIP_3), whereas phosphatase and tensin homologue (PTEN) dephosphorylates PIP_3 acting as a negative regulator of PI3K signaling. The best characterized phosphorylation target of PI3K is the pleckstrin homology domain of Akt. This phosphorylation stimulates the catalytic activity of Akt, resulting in the phosphorylation of a host of other proteins that affect cell growth, cell cycle entry, and cell survival. The precise mechanism of mTOR regulation by PI3K or Akt, however, is still not well understood (Abraham, 2002). Studies have identified the tuberous sclerosis (TSC) complex as a modulator between PI3K/Akt and mTOR (Inoki K et al., 2002; Gao et al., 2002; Tee et al., 2002). Over the past few years, two new regulators of the response of mTOR to nutrients have been discovered. Ras homolog enriched in brain (Rheb) is a small G protein that appears to be involved in nutrient-mediated signaling. The prevailing hypothesis at this juncture is that Rheb promotes mTOR signaling when it is in an active GTP-bound form, whereas the tuberin-hamartin heterodimer inhibits Rheb by converting it to an inactive GDP bound state. mTOR forms a scaffold complex with other proteins, such as the regulatory associated protein of mTOR (raptor) and mLST8, resulting in the formation of mTOR complex 1 (mTORC1) (Manning et al., 2002). Raptor, a 150-kDa evolutionarily conserved protein, may act as a scaffold protein linking mTOR to $p70^{s6k}$ and 4E-BP1, which is also involved in the regulation of mTOR response to nutrients (Hara et al., 2002). mTORC1 is specifically inhibited by the binding of rapamycin to the 12-kDa FK506-binding protein (FKBP12). The mTOR complex 2 (mTORC2) comprises mTOR, LST8, rictor (rapamycin-independent companion of mTOR), and mSin1 (also known as MAPK-associated protein 1). Formation of the rapamycin–FKBP12 complex does not directly inhibit mTORC2 function but has been demonstrated to perturb mTORC2 complex formation in approximately 20% of cancer cell lines, resulting in inhibition of Akt signaling. The rapamycin–FKBP12 complex formation inhibits the mTORC2-specific binding of rictor and mSin1 (Sabatini, 2006). In summary, rapamycin and its analogs universally inhibit mTORC1 activation of $p70^{s6k}$ and 4EBP1, whereas rapamycin's effects on mTORC2 are cell specific.

The principal downstream effect of mTOR is the control of cellular translation machinery through two separate downstream pathways: the eukaryotic initiation factor 4E binding protein 1 (4E-BP1), also known as phosphorylated heat- and acid-stable protein (PHAS-1), and the $p70^{s6k}$ (Figure 8.2) (Tee et al., 2003; Brunn et al., 1997; Gingras et al., 2001; Sonenberg and Gingras, 1998). Whether this phosphorylation is directly or indirectly regulated by mTOR is in question because recent data suggest that mTOR may regulate $p70^{s6k}$ activation by inhibiting phosphatases rather than by directly phosphorylating $p70^{s6k}$ (de Groot et al., 1994; Wang et al., 2001). It is also important to note that $p70^{s6k}$ may also be activated by TOR-insensitive signaling pathways involving phosphoinositide-dependent kinase 1 (PDK1), MAPK, and SAPK.

1. mTOR Targeting Agents

Rapamycin, the prototypical mTOR inhibitor, functions by binding intracellularly to FKBP12, a member of the FKBP immunophilin family (Volarevic and Thomas, 2001; Peterson et al., 1999; Heitman et al., 1991). The resultant FKBP12–RAP complex interacts with and inhibits the mTOR kinase activity, which, in turn, blocks the activation of $p70^{s6k}$ (Koltin et al., 1991; Fruman et al., 1995). Rapamycin and its analogs also block phosphorylation of 4E-BP1, leading to slowing or arrest of cells in G_1 phase of the cell cycle (Figure 8.2). In addition to its well-characterized inhibitory actions on $p70^{s6k}$ and 4E-BP1, rapamycin interferes with other intracellular processes involved in cell cycle progression that contribute to its antiproliferative actions. Rapamycin increases the turnover of cyclin D1 at the mRNA and protein level, upregulates p27, and inhibits cyclin A-dependent kinase activity (Seufferlein and Rozengurt, 1996; Grewe et al., 1999). In addition to the induction of growth arrest, rapamycin can induce apoptosis in the absence of p53 or p21 (Hashemolhesseini et al., 1998). Similar effects were observed with the treatment of pancreatic cells with the rapamycin analog temsirolimus (Kawamata et al., 1998). Rapamycin also possesses important antiangiogenic properties. The antiangiogenic effects

TABLE 8.5 mTOR Inhibitors

Drug	Type	Toxicity	Tumor	Response	Phase of Development
Temsirolimus (CCI-779)	Prodrug	Hypercholesterolemia, hyperglycemia, mucositis, elevated liver function test levels	Squamous cell carcinoma of head and neck, mantle cell lymphoma	9.2–38% response rate	Phase II
			Renal cell carcinoma	3.6-month improved overall survival versus interferon	Phase III
Everolimus (RAD001)	Rapamycin derivative	Stomatitis, neutropenia	Lung, colon, breast, and gastric	11–36% PR	Phase II
Amplimexon (AP23573)	Rapamycin derivative	Elevated liver function test levels	Bone sarcoma, leiomyosarcoma	29% CBR	Phase II

are due to VEGF antagonism of VEGF production as well as blockage of VEGF-mediated stimulation of endothelial cells and tube formation (Huang et al., 2001). In addition, VEGF induction of the PI3K/Akt signaling pathway is important for endothelial cell survival (Asano et al., 2005; Guba et al., 2002). Rapamycin treatment can induce apoptosis of VEGF-stimulated endothelial cells, potentially leading to the tumor vessel thrombosis (Yu and Sato, 1999).

mTOR inhibitors in clinical development include rapamycin and the structurally related compounds temsirolimus (CCI-779), everolimus (RAD001), and Amplimexon (AP23573), as shown in Table 8.5. The clinical development of temsirolimus will be outlined to demonstrate the issues regarding targeting the mTOR pathway for cancer therapy. In the NCI human tumor cell line screen, temsirolimus and rapamycin demonstrated similar antitumor profiles and potencies (Pearson correlation coefficient, 0.86), with IC_{50} values frequently less than 10^{-8} M (Suhara et al., 2001; Bruns , 2004). In vitro, human prostate, breast, and SCLC; glioblastoma; melanoma; and T-cell leukemia human tumor cell lines were among the most sensitive to temsirolimus, with IC_{50} values in the nanomolar range (Asano et al., 2005). The treatment of a variety of human tumor xenografts with temsirolimus caused significant tumor growth inhibition rather than tumor regression, suggesting that subsequent disease-directed trials should be designed to assess this potential outcome (Huang et al., 2001; Asano et al., 2005; Guba et al., 2002). In addition, several intermittent temsirolimus dosing regimens were effective in these human tumor xenograft studies, which may be relevant since the immunosuppressive effects of rapamycin analogs have been demonstrated to resolve approximately 24 hours following treatment (Asano et al., 2005).

Temsirolimus was well tolerated in phase I clinical trials using different schedules of administration, had a predictable pharmacological behavior, and demonstrated preliminary antitumor activity (Dudkin et al., 2001; Gibbons et al., 1999; Raymond et al., 2004; Hidalgo et al., 2000; Forouzesh et al., 2002). Unexpectedly, major response by tumors (partial response of greater than 50% reduction in bidimensional product of measurable lesions) was seen in early phase trials, which was inconsistent with the delayed tumor growth and not regression observed in preclinical studies. The consistent observation of tumor response by temsirolimus at nontoxic doses suggests that the optimal therapeutic dose may be lower than the maximal tolerated dose (MTD). Disease-directed studies are ongoing in a variety of tumors. Temsirolimus has been studied in glioblastoma multiforme because the overexpression of Akt is common in these tumors (Chang et al., 2004; Witzig et al., 2005). Preclinical data have demonstrated that PTEN-defective tumors or those expressing high levels of Akt are exquisitely sensitive to mTOR-targeted therapy (Neshat et al., 2001). Unfortunately, this observation was not upheld in the clinical setting; however, high baseline $p70^{s6k}$ expression did correlate with radiographic response. The activity of temsirolimus was first demonstrated in renal cell carcinoma in a phase II study of 111 patients with advanced, previously treated disease (Atkins et al., 2004). Patients were randomized to receive intravenous doses of 25, 75, or 250 mg/m^2 per week of temsirolimus. An objective response rate of 7% and a minor response rate of 26% were observed, which did not differ between doses. The time to progression (TTP) of 5.8 months and median survival (15 months) were also positive findings for this heavily pretreated group of patients. Furthermore, assessment of $p70^{s6k}$ inhibition in peripheral blood mononuclear cells demonstrated a dose-independent correlation with TTP. These findings led to the initiation of a phase III trial comparing temsirolimus as a single agent and in combination with interferon (IFN) to IFN for the treatment of patients with metastatic renal cell carcinoma. The results of the study, presented at the 2006 American Society of Clinical Oncology meeting, demonstrated a statistically significant benefit in median survival (10.9 versus 7.3 months) and median progression-free survival (PFS) (3.7 versus 1.9 months) for patients treated with temsirolimus in comparison to those treated with interferon (Hudes et al., 2006). There was no advantage to combining temsirolimus with IFN. The interim results of a phase II study demonstrated the activity of temsirolimus in recurrent and metastatic endometrial cancer with 63% of patients attaining a partial response or disease stabilization (Oza et al., 2006). Another interesting clinical observation was obtained in a phase II study of temsirolimus in breast cancer patients refractory to taxane or anthracyclines (Chan et al., 2005). Clinical benefit was observed in 37% of patients; however, none of the HER2-negative tumors demonstrated a significant response to temsirolimus. Subsequently, a phase III clinical trial combining oral temsirolimus with letrazole versus letrazole alone in patients with estrogen-dependent breast cancer was initiated to exploit the observed association of mTOR pathway activation and hormone resistance. This study was discontinued due to the low likelihood of achieving the target efficacy criteria at interim analysis. Clinical studies combining mTOR inhibitors with traditional cytotoxic agents are ongoing (Punt et al., 2003). In addition, the role of pharmacodynamic assessment in predicting disease response to mTOR-targeted therapy is an area requiring further development.

III. THE MAPK FAMILY

Numerous critical growth factors and cytokines transduce their signals from the cell membrane to the nucleus via protein kinase networks called signal transduction pathways, which have become major targets for anticancer drugs. An important example is the mitogen-activated protein kinase (MAPK) pathway, also referred to as the extracellular signal-regulated protein (ERK) pathway, as described earlier in this chapter (Sebolt-Leopold, 2004). When extracellular growth factors, such as the EGF bind to receptors (e.g., EGFR), conformational changes are induced in the receptor and lead to autophosphorylation, receptor dimerization, and recruitment of proteins, such as Ras, at the inner surface of the cell membrane, as shown in Figure 8.2 (McCormick *et al.*, 1993). Ras stimulates Raf activation, which in turn phosphorylates MEK and then activates ERK. At each step in the pathway, phosphorylation of the next signaling member is required for activation and downstream phosphorylation of the next protein kinase. ERK coordinates responses to extracellular signals by regulating gene expression, cytoskeletal rearrangements, and metabolism as well as cell proliferation, differentiation, and apoptosis. This pathway has been shown to be constitutively activated in a number of human cancers. Activation of the pathway causes gene expression changes and changes in cell proliferation, survival, and differentiation (Seger *et al.*, 1993).

Multiple lines of evidence indicate that the MAPK pathway is important in human cancer. Inappropriate Ras activation is associated with nearly one-third of all human cancers (Downward, 2003). One of the Raf paralogs, B-Raf, is mutated in many cancers, including malignant melanoma (27–70%), papillary thyroid cancer (36–53%), ovarian cancer (30%), and colorectal cancer (5–22%); the mutations are frequently gain-of-function substitutions that result in constitutive activity (Garnett and Marias, 2004). ERK is elevated in nearly 50% of breast cancers and is associated with a poor prognosis (Maemura *et al.*, 1999; Mueller *et al.*, 2000; Sivaraman *et al.*, 1997). More than one-third of tumor cell lines have constitutive activation of the ERK pathway (Hoshino *et al.*, 1999).

MEK (MAP/ERK kinase or MAPK kinase) occupies a central role in the MAPK pathway. Expression of constitutively active forms of MEK leads to transformation of cell lines (Cowley *et al.*, 1994; Mansour *et al.*, 1994). MEK kinases have dual kinase activity, with phosphorylation of both serine–threonine and tyrosine. There are two MEK homologues, MEK1 and MEK2, which are 80% amino acid identical, have very similar three-dimensional structures, and are similarly targeted by known inhibitors (Brott *et al.*, 1993; Ohren *et al.*, 2004). Both kinases are highly specific, with no known substrates aside from ERK (Seger *et al.*, 2003). Despite these similarities, there are a few differences between the MEK homologs. MEK2 is approximately seven times more catalytically active than MEK1 (Zheng and Guan, 1993), yet MEK2 knockout mice are fully viable while MEK1 knockout is embryonically lethal (Giroux *et al.*, 1999; Belanger *et al.*, 2003). The general interpretation of these findings is that MEK1 is able to compensate for the absence of MEK2 (Sebolt-Leopold, 2004).

A. Compounds in Development

There are several known MEK inhibitors (MEKIs), as shown in Table 8.6. The vast majority of MEKIs in development are highly selective and demonstrate little cross-reactivity with other protein kinases. Most of the known MEKIs do not bind to the ATP binding site of the kinase and thus are noncompetitive inhibitors (Wallace *et al.*, 2005). Instead, structural analysis demonstrates that these inhibitors bind to a unique site adjacent to the ATP binding pocket (Ohren *et al.*, 2004). This property is believed to be responsible for the high degree of selectivity of the MEKIs despite the ATP binding site being highly conserved among different human protein kinases. PD98059 and U0126 represent the initials generation of specific MEKIs (Alessi *et al.*, 1995; Dudley *et al.*, 1995; Davies *et al.*, 2000; Favata, 1998). These molecules demonstrated potent *in vitro* effects but lack *in vivo* activity (Allen *et al.*, 2003). As such, the clinical evaluation of these compounds has been abandoned. The compounds under clinical evaluation include CI-1040 (PD184352), PD00325901, and ARRY-142886 (AZD6244), included in Table 8.6. CI-1040 is the most advanced in its development and will be discussed in detail to illustrate issues regarding the development of compounds targeting the MEK signal transduction pathway.

TABLE 8.6 MEK Inhibitors

Drug	Type	IC$_{50}$	Toxicity	Clinical Response	Phase of Development
PD184352 (CI-1040)	Noncompetitive inhibitor	2 μM	Fatigue, diarrhea, rash, N/V	1 PR (pancreas), 27 SD (28%)	Phases I–II
PD0325901	Noncompetitive inhibitor	<1 nM	Rash, syncope diarrhea, fatigue, vision changes	2 PR (melanoma), 8 SD	Phase I
ARRY-142886 (AZD6244)	Noncompetitive inhibitor	10 nM	Hypoxia, rash, diarrhea, fatigue, blurred vision	4 SD$^¥$	Phase I

1. CI-1040

CI-1040 is an orally bioavailable compound demonstrating adequate *in vitro* activity. Studies conducted with the breast cancer cell line MDA-MB-231 demonstrated that 1 μmol/L decreased p-ERK1 and p-ERK2 levels by 99 and 92%, respectively. In addition, the observed decrement in p-ERK1/2 was associated with growth inhibition as measured by soft agar colony assays. Furthermore, xenograft experiments involving pancreatic, colon, and breast cancer cell lines confirmed CI-1040 antitumor effects. Antitumor activity was associated with the reversible inhibition of p-ERK by 6 hour, returning to baseline levels by 24 hour. Antitumor effects correlated with a high level of baseline p-ERK expression as well as the decrement of p-ERK levels in treated tumor tissue. These positive preclinical studies led to CI-1040 clinical evaluation in the phase I setting (Lorusso *et al.*, 2005). A total of 27 patients were treated at doses ranging from 100 mg to 1600 mg. The compound was well tolerated with no grade 4 toxicities and only limited grade 3 toxicities being observed. Approximately 98% of patients experienced grade 1–2 toxicities, primarily consisting of diarrhea, fatigue, rash, and nausea/vomiting. One patient with pancreatic cancer attained a PR lasting 12 months and 19 (25%) patients achieved disease stabilization. High fat foods were found to increase the area under the curve (AUC) and C$_{max}$ by threefold to fivefold, respectively. Therefore, a dose of 800 mg BID with food was established as the dose for phase II testing.

Pharmacodynamics were also assessed in addition to traditional pharmacokinetic testing in the phase I study. Tumor biopsy evaluation demonstrated a decrease in intratumor p-ERK levels ranging from 46 to 100%. In addition, phorbol myristate acetate ERK activation in peripheral blood mononuclear cells was analyzed. Based on these results, a parallel arm phase II study of CI-1040 was performed in patients with advanced breast, colon, NSCLC, and pancreatic cancers (Rinehart *et al.*, 2004). The study was designed as essentially four simultaneous Simon two-staged phase II studies. Early stoppage criteria were set at 13 patients if one CR or PR was not observed or if four clinical benefit responses (CBRs = CR + PR + SD) were not observed. If early stoppage criteria were not met, a subsequent cohort of 30 patients would be enrolled in each arm of the various tumor types. These criteria were calculated on the basis of a "null" response rate of 5% and a "positive" response rate of 20% or CBR rate of 40%. Unfortunately, the study was terminated early due to lack of response.

A total of 67 patients were enrolled in the study at the time of study closure; 14 patients had breast cancer, 20 had colon cancer, 18 had NSCLC, and 15 had pancreatic cancer. The patients predominantly had received

0–1 prior chemotherapy regimens (75%), and 90% had ECOG performance status (PS) of 0–1. CI-1040 was well tolerated, with only 19% of patients experiencing grade 3 toxicity and no patients experiencing grade 4 toxicity. The observed toxicities ranged from nausea, abdominal pain, diarrhea, rash, fatigue, anorexia, edema, and facial edema. Eight patients experienced disease stabilization at 3 months; however, no patients experienced a CR or PR. Pharmacokinetic results were very similar to those in the phase I trial. Pharmacodynamic evaluation consisted of immunohistochemical staining with quantitative analysis of archival tumor specimens. Phosphorylated ERK was elevated in most tumor types, and a nearly significant relationship was observed between p-ERK and the possibility of stable disease ($p < .055$) as analyzed by logistic regression. Based on these results, it appears that CI-1040 will not be further developed in these tumor types.

2. PD0325901

PD0325901 is a second-generation MEKI demonstrating a 50% increase in potency compared to CI-1040, along with improved bioavailability and longer MEK suppression (24 hours compared to 6–8 hours for CI-1040) (Sebolt-Leopold and Herrera, 2004). An initial phase I study involving 41 patients with advanced cancer has been completed (Lorusso et al., 2005; Menon et al., 2005). After a single cohort at 1 mg orally daily, the dosing was changed to twice daily ranging from 1 to 30 mg, with continuous dosing at the 15 and 20 mg twice-daily levels. Acneiform rash, elevated liver function test levels, and syncope were identified as the dose-limiting toxicities. Pharmacokinetic studies indicated a dose-dependent increase in exposure, with doses of 15 mg or higher exhibiting a prolonged plasma concentration above 50 ng/mL. The IC_{50} at which most susceptible cell lines exhibit growth inhibition ranges from 5 to 53 ng/mL.

Pharmacokinetic studies indicated a dose-dependent increase in exposure, with doses of 15 mg or higher exhibiting a prolonged plasma concentration above 50 ng/mL. The IC_{50} at which most susceptible cell lines exhibit growth inhibition ranges from 5 to 53 ng/mL. PD0325901 exhibited modest antitumor activity against this fairly heavily pretreated patient population. Two melanoma patients had PRs, both at the 20 mg BID dose level. Eight subjects at various dose levels experienced stable disease lasting 3–7 months: five patients had melanoma, two had NSCLC, and one had colon cancer. This compound is currently undergoing phase II testing in these tumor types. Pharmacodynamic analyses of p-ERK in tumor specimens after 15 days of treatment compared to pretreatment demonstrated a decrement in pERK expression at low doses; however, this finding was not observed with higher doses.

B. Future Directions

Integrated genetic and pharmacological analysis has shown that BRAF (but not NRAS) mutations predict sensitivity to MEKIs (Solit et al., 2006). In these experiments, cell lines harboring the most common BRAF mutations (V600E) were exquisitely sensitive to CI-1040, with IC_{50} values of 0.024–0.11 µM. Cell lines with wild-type BRAF, or with either wild-type or mutant NRAS, were not sensitive despite showing effective inhibition of p-ERK. These results were recapitulated in xenograft experiments. The authors called for clinical trials with MEKIs in which patients are stratified based on BRAF mutational status. This is akin to stratifying subjects with EGFR mutations to treatment with erlotinib and holds the promise of individually tailored therapy based on genetic analysis of tumors. However, since MEK is downstream of multiple growth factor pathways, patient selection could be difficult. Combination strategies with MEKIs have shown preclinical promise. One of the original MEKIs, PD98059, was found to have synergistic effects on breast cancer cell lines when combined with the EGFR inhibitor gefitinib as well as showed additive effects with tamoxifen (Normanno et al., 2006; Djahansouzi et al., 2005). In addition, MEKIs augment the antitumor activity of both PI3K inhibitors and mTOR inhibitors (Smalley et al., 2006; Legrier et al., 2006). Successful combination strategies with conventional cytotoxic chemotherapy agents, such as docetaxel, have also been demonstrated in xenograft models, revealing both antiproliferative and antiangiogenic effects (McDaid et al., 2005). These preclinical observations provide initial evidence for ways in which both genetic and pharmacodynamic parameters may be clinically utilized to enrich the patient population who may be expected to benefit for therapy with MEKIs.

C. Src Kinase Inhibitors

Src kinase is a multifunctional intracellular tyrosine kinase that has been implicated in the regulation of a variety of physiological and oncogenic processes, such as proliferation, differentiation, survival, motility, and angiogenesis (Ishizawar et al., 2004). Src has significant structural homology to a family of about 10 proteins collectively known as the Src kinase family, including Lck, Fyn, Yes, Yrk, Blk, Fgr, Hck, Lyn, and Frk (Sawyer et al., 2001). The discovery of Src and its function has significant historical importance. Src was found to be the protein responsible for the transforming properties of the Rous sarcoma virus, the etiological agent of the chicken sarcomas. Initially what was thought to be a threonine kinase was later found by Hunter (1980). to be a tyrosine kinase; Src was one of the first tyrosine kinases to be discovered and became a lead example of the importance of tyrosine kinases in cancer. Furthermore, c-src was the first oncogene discovered, setting off an explosion of research in oncogenes (Martin, 2004).

The c-src gene is weakly oncogenic as compared with its viral homolog, v-src, which readily induces sarcomatous tumors in avian organisms (Biscardi et al., 2000). The differences in the two proteins lie in their structure and regulation, with c-src existing mainly in an inactive configuration by multiple intramolecular interactions. Mutations that disrupt this interaction, such as those in the C-terminal negative regulatory domain in v-src, result in constitutive activation of the kinase. While initial reports documented the existence of mutated src in colon cancer, more detailed studies failed to prove this finding (Irby et al., 1999). It appears that the fundamental mechanism by which src is altered in human tumors is by overexpression of its wild-type form, a finding that occurs in cancers of the colon, lung, skin, breast, ovarian, endometrial, and head and neck (Biscardi et al., 1999). Indeed, the current understanding of the role of src in cancer is that this protein is important by acting as a facilitator of other signaling molecules rather than by being oncogenic on its own (Biscardi et al., 1999).

Src kinase interacts with multiple cellular factors, such as membrane receptors (e.g., EGF family, colony-stimulating factor 1 [CSF-1], PDGF, and integrins), steroid hormone receptors, G-protein regulated pathways, STATs, focal adhesion kinase, and adaptor proteins, such as p130Cas and Shc (Biscardi et al., 2000; Duxbury, 2004; Zhang et al., 2004). One of the better characterized functions of Src kinase is its ability to activate the EGFR by phosphorylation of its residue Tyr 845, which is mediated by physical interaction between the two proteins (Biscardi et al., 1999). Tyr 845 phosphorylation is important for the activation of signal transducer and activator of transcription 5 (Stat5). Src also influences receptor endocytosis and degradation by phosphorylating clathrin, dynamin, and Cbl, which are proteins involved in that process (Zhang et al., 2004). Another important function is the regulation of focal adhesion kinase (FAK), a 120-kDa tyrosine kinase involved in cancer metastasis by modulating focal adhesions. These adhesion kinases are dynamic intracellular structures that link the extracellular matrix to the cell cytoskeleton through integrins. Src and FAK form transient dimers that result in phosphorylation of FAK at various residues and activation of the adaptor protein paxillin and subsequent activation of signaling pathways, such as ERK, Jun kinase, and Rho (Playford and Schaller, 2004).

Given its significant implications in cancer development and progression, Src kinase has become a target for drug development (Sawyer et al., 2001). Different strategies are being used, including inhibition of protein–protein interactions, protein stability, and kinase activity. AP22408 inhibits the binding of proteins to the SH2 domain of Src (Shakespeare et al., 2000). The agent exhibits osteoclast-selective antibone reabsorptive activity, suggesting that these agents may be useful for patients with bone metastasis. Protein stability has been targeted as part of the more general approach of inhibiting the chaperone Hsp90. Drugs such as the ansamycin class of drugs inhibit Hsp90, disrupting the association of this protein with other proteins such as Src (Workman, 2003).

The class of drugs more widely used to inhibit tyrosine kinases are small molecule inhibitors of the kinase ATP binding site. At least three of these compounds are in clinical development. SKI-606 is a 4-anilino-3-quinolinecarbonitrile that inhibits both Src and Abl kinase (Golas et al., 2003). The agent inhibits Src with an IC_{50} of 1.2 nM in enzyme assays, inhibits anchorage-independent growth of Src-transformed fibroblasts with an IC_{50} of 100 nM, and inhibits Src-dependent protein tyrosine phosphorylation at similar concentrations. By virtue of its inhibitory activity against Abl kinase, the drug exerts potent antitumor effects against CML models (Golas et al., 2003). AZM475271 is an anilinoquinazoline that targets Src kinase activity (Ple et al., 2004). In preclinical studies, the agent exerted potent Src kinase inhibitory activity and blocked the growth of transformed 3T3 and the NSCLC cell line A549 (Ple et al., 2004). A study in pancreatic cancer models shows that inhibition of Src kinase alone and particularly in combination with gemcitabine inhibited the growth and metastatic potential of pancreatic cancer cells (Yezhelyev et al., 2004). Biological studies demonstrated that treated tumors exhibited a decrease in cell proliferation and microvessel density as well as an increase in apoptosis. In addition, studies conducted with PP2, a commercially available Src kinase inhibitor in pancreatic cancer models, demonstrates that this agent has activity in gemcitabine-resistant cell lines, further supporting the notion that Src inhibition may be an interesting approach in pancreatic cancer (Duxbury, 2004). Finally, BMS-354825 is an orally bioavailable inhibitor of Src and Abl and has demonstrated antitumor activity in Gleevec-resistant CML cell lines and is currently in development in that tumor type (Ahmad, 2005; Burgess et al., 2005).

IV. CHALLENGES IN THE CLINICAL DEVELOPMENT OF SIGNAL TRANSDUCTION INHIBITORS

The development of small molecule signal transduction inhibitors (STIs) of biologic processes is quite complex. The first generation of these molecules in development has demonstrated that this avenue of therapy is a feasible approach to anticancer therapy. However, the development of these compounds bears an enormous financial burden, thus making it imperative to define more efficacious ways of administering these medications. The further development of small molecule inhibitors will require a novel phase I dose-finding study design, improved patient selection criteria, novel study endpoints, as well as the development of rational combination therapy.

A. Clinical Trials: Design Issues

The goal of early phase clinical studies is to determine the optimum dose and patient population best suited for definitive clinical investigation to determine efficacy. The classic endpoint of phase I trials is determining the maximum tolerated dose. A paradigm switch to attaining the optimal biologic dose will be needed for future clinical evaluation. For example, rapamycin and its analogues have been demonstrated to have an antitumor effect in a dose-independent manner, with these effects occurring well below the maximal tolerated dose (Dudkin et al., 2001; Gibbons et al., 1999; Raymond et al., 2004; Hidalgo et al., 2000; Forouzesh et al., 2002). This finding is likely the case for many other targeted agents. Pharmacodynamic assessment of efficacy needs to be incorporated into early phase clinical evaluation of new compounds. This will allow the determination of an optimal biologic dose as part of the drug approval process. It will be imperative to provide the optimal biologic dose of agents and avoid cumulative toxicity that may incur as a result of inhibiting multiple cellular pathways simultaneously. Pharmacodynamic response can be measured in tumor biopsies obtained prior to and after the initiation of treatment or in surrogate markers. In addition, surrogate tissues, such as peripheral blood mononuclear cells (PBMC) or skin biopsies, have been used for this purpose with varying degrees of success (Pelabra et al., 2003). Current criticisms of a pharmacodynamic approach to the development of biologic agents would include the poor reliability of the results obtained with these studies. In addition, results obtained in normal surrogate tissue may not reflect drug effects observed in malignant tissue. Last, target inhibition does not necessarily correlate with drug efficacy because there may be a myriad of intracellular mechanisms resulting in drug resistance.

Modification of phase II and phase III study design in addition to establishing new pharmacodynamic approaches will be necessary to efficiently evaluate biologic agents in the future. Novel ways of optimizing trial design include the multinomial method and the randomized discontinuation design. The clinical development of sorafenib for patients with metastatic renal cell carcinoma demonstrated the importance of novel study design in the development of biologic agents (Ratain et al., 2006). This phase II study utilized a placebo-controlled randomized discontinuation design in which 202 patients were initially enrolled to be treated with sorafenib for a 12-week run-in period. At the end of that period, patients achieving greater than 25% tumor shrinkage were continued on the sorafenib until evidence of disease progression. Those patients attaining disease stabilization were randomized to either sorafenib or placebo for an additional 24 weeks, while patients with evidence of disease progression or intolerable toxicity were taken off the study at the end of the 12 week run-in period. This strategy allowed the investigators to separate slow tumor growth from resistance as demonstrated by a 32% increase in progression-free survival (50 versus 18%) for those patients treated with sorafenib among those patients initially achieving disease stabilization. A criticism of the randomized discontinuation design is that the information lost on the nonrandomized patients may be of a significant magnitude to underpower these studies compared to studies with the classic design of randomizing all patients.

B. Patient Selection

The development of biologic therapy has forced the introduction of novel clinical trial design. Classic clinical evaluation of a therapeutic compound consisted of (i) treating an unselected group of patients with a given cancer with the compound and (ii) assessing objective response by traditional imaging techniques. Appropriate patient selection will be imperative for the development of future biologic agents. The importance of patient selection is exemplified by the clinical development of trastuzumab, gefitinib, and cetuximab. The registration trial for trastuzumab was performed in patients with metastatic breast cancer characterized by documented overexpression of HER2 (Slamon et al., 2001). The addition of trastuzumab to traditional chemotherapy was associated with an 18% (50 versus 32%) increase in objective response, which translated to a 5-month improvement in median survival (25.1 versus 20.3). HER2 overexpression occurs in 25–30% of metastatic breast cancer patients. Performing this study in a group of unselected metastatic breast cancer patients would be expected to produce objective responses in less than 5% of patients and would likely be associated with an increased rate of cardiotoxicity over the 27% observed in the published study. It is likely that trastuzumab would have been deemed too toxic for a minimal survival benefit to recommend FDA approval for the treatment of metastatic breast cancer patients. Patient selection issues regarding the clinical development of cetuximab and gefitinib relate to the receptor activation status. Cetuximab received FDA approval for the treatment of irinotecan refractory metastatic colorectal cancer independent of EGFR expression status (Cunningham et al., 2004). Similarly, gefitinib was approved by the FDA for chemotherapy refractory advanced stage NSCLC in the second line setting. Subsequently, it was determined that patients' tumor sensitivity to gefitinib and later erlotinib is due to the presence of activating mutations in the kinase domain of EGFR (Lynch et al., 2004; Paez et al., 2004; Pao et al., 2004). Furthermore, these mutations segregate to patients of Asian descent and nonsmokers. Clinical studies are ongoing to test the hypothesis that the presence of activating mutations in certain patient populations may provide a survival benefit. There is a loose association the development of rash and efficacy with cetuximab and other EGFR-targeted therapy (Saltz et al., 2003). This finding may be a reflection of EGFR activation status. Therefore, what may be important for EGFR-targeted therapy may not simply be the overexpression of the receptor but the receptor activation (phosphorylation) status. Similarly, $p70^{s6k}$ activity has been associated with radiologic response in patients with glioblastoma multiforme treated with temsirolimus (Witzig et al., 2005). Future clinical development of biologic agents targeting receptor kinases or intracellular kinases will need to focus on selecting patients based on evidence of target activation and not based on histology. Histology may become important in the development of therapeutic combinations with cytotoxic agents as well as in combinatorial therapy, including hormonal therapy, because there is clear evidence that perturbation of hormonal signals can impact RTK-mediated signal transduction.

C. Study Endpoints

Novel endpoints need to be explored in addition to novel study design and patient selection for early phase clinical studies. As stated previously, compounds targeting signaling events are not likely to result in tumor response in the classic sense. Therefore, alternative endpoints to evaluate efficacy must be validated and accepted to enhance the understanding of how to use these drugs and to avoid mislabeling compounds as ineffective. Possible alternative endpoints that have been proposed include time to progression, the proportion of patients without progressive disease as their best response, progression rate, symptomatic benefit, measures of target inhibition, positron emission tomography (PET) scanning, and reduction in tumor markers. The combination of novel study design and novel therapeutic endpoints will help to efficiently investigate the use of compounds targeting intracellular signal events.

The traditional model of tumor size reduction as assessed by classic radiologic test may not be the proper endpoint for the development of biologic agents. Many of these molecules are cytostatic and not cytocidal when evaluated *in vitro* and may be expected to result in disease stabilization rather than tumor response in human subjects. In addition, the results of the extensive early phase trials of various biologic agents have demonstrated evidence of disease stabilization associated with a significant palliative effect, suggesting the presence of a biologic effect. Furthermore, there are rare incidences where a given tumor is dependent on a single biologic pathway for oncogenesis. Further, the new biologic agents may preferentially impact tumor cell metabolism. These factors would suggest that functional imaging may better represent tumor response to biologic therapy rather than traditional radiologic methods of assessing response. This finding was demonstrated by the treatment of von Hippel-Lindau (VHL)-null renal cell carcinoma with temsirolimus in a xenograft model in which the stability of HIF-1a was correlated with temsirolimus sensitivity and tumor response as measured by [^{18}F] fluoro-2-deoxy-D-glucose (FDG) PET (Thomas *et al.*, 2006). The development of biologic agents targeting molecules regulating angiogenesis further exemplifies the need for the validation of nontraditional measurements of tumor response. Dynamic contrast enhanced (DCE) magnetic resonance imaging (MRI) and diffusion-weighted MRI are two examples of novel imaging techniques that allow the quantification of various aspects of angiogenesis, including perfusion, vascular permeability, and necrosis, to be assessed (Hylton, 2006). These are particularly important as it is postulated that a component of the antitumor effect of bevacizumab is due to the modulation of vascular permeability (Willet *et al.*, 2004).

D. Combination Therapy

The evolution of biologic therapy will require the use of rational combination therapy. This is in part due to the complex mechanisms of resistance to biologic therapies. First, cross-talk between membrane receptors is a known mechanism. This is evidenced by the emerging role of HER2 in EGFR signaling and vice versa. Overexpression of HER2 can result in enhanced EGFR signaling (Pinkas-Kramarski *et al.*, 1996; Graus-Porta *et al.*, 1997; Worthylake *et al.*, 1999). Therefore, one could envisage a strategy targeting both HER2 and EGFR either as a dual kinase inhibitor or by combining antibody therapy with TKI-based therapy. Nullification of antioncogenic downstream effectors may also result in resistance to biologic agents. EGFR inhibitors augment the protein levels of the CDKI p27, resulting in cell cycle arrest. Proteosomal degradation of p27 in response to EGFR inhibition may result in a resistant phenotype (Wu *et al.*, 1996). Conversely, the upregulation of oncogenic downstream effectors, either independent or in response to biologic agents, is a common mechanism to induce resistance. In the case of EGFR inhibition, loss of PTEN, overexpression of Akt, as well as the functional activation of the PI3K and ERK pathways may occur *de novo* or in response to EGFR inhibition, resulting in abrogation of the therapeutic effect of EGFR inhibition. The importance of this mode of resistance is demonstrated by the association between the reduction of Akt activity in response to EMD7200 therapy and therapeutic benefit (Salazar *et al.*, 2004).

Last, the augmentation of complementary pathways may also provide a critical mechanism to either bypass the growth inhibition effects or provide a secondary signal to propagate oncogenesis. This would be particularly important for epithelial tumors, which are less likely to be dependent on a single oncogenic pathway to mediate oncogenesis. A prime example of this phenomenon is the acquired overexpression of VEGFR in response to antibody therapy targeting EGFR. The use of next-generation agent ZD6474 targeting both EGFR and VEGFR is an example of a strategy that may allow compensation for such events (Ciardiello *et al.*, 2004). In addition, as the oncology community gains a better understanding of the predicted biologic responses to a given biologic agent and the timing of such events, sequential rotation of small molecule inhibitors combined with conventional cytotoxic agents may become a practical strategy to overcome the inevitable resistance to therapy that will eventually occur. Understanding the complex mechanisms of resistance to biologic-based therapy will be imperative to designing effective combinations of biologic, cytotoxic, and immunologic therapy in the future.

References

Abraham, R. T. (2002). Identification of TOR signaling complexes: More TORC for the cell growth engine. *Cell* **111**, 9–12.

Ahmad, K. (2005). New agent overcomes resistance to imatinib. *Lancet Oncol.* **6**, 137.

Alessi, D. R., Cuenda, A., Cohen, P., Dudley, D. T., and Saltiel, A. R. (1995). PD 098059 is a specific inhibitor of the activation of mitogen-activated protein kinase kinase *in vitro* and *in vivo*. *J. Biol. Chem.* **270**, 27489–27494.

Allen, L. F., Sebolt-Leopold, J., and Meyer M. B. (2003). CI-1040 (PD184352), a targeted signal transduction inhibitor of MEK (MAPKK). *Semin. Oncol.* **Suppl 16**, 105–116.

Arteaga, C. L. (2001). The epidermal growth factor receptor: From mutant oncogene in nonhuman cancers to therapeutic target in human neoplasia. *J. Clin. Oncol.* **Suppl. 18**, 32S–40S.

Arteaga, C. L. (2002). Epidermal growth factor receptor dependence in human tumors: More than just expression. *Oncologist* **Suppl. 4**, 31–39.

Asano, T., Yao, Y., Zhu, J., Li, D., Abbruzzese, J. L., and Reddy, S. A. (2005). The rapamycin analog CCI-779 is a potent inhibitor of pancreatic cancer cell proliferation. *Biochem. Biophys. Res. Commun.* **331**, 295–302.

Atkins, M. B., Hidalgo, M., Stadler, W. M., Logan, T. F., Dutcher, J. P., Hudes, G. R., Park, Y., Liou, S. H., Marshall, B., Boni, J. P. *et al.* (2004). Randomized phase II study of multiple dose levels of CCI-779, a novel mammalian target of rapamycin kinase inhibitor, in patients with advanced refractory renal cell carcinoma. *J. Clin. Oncol.* **22**, 909–918.

Baselga, J., Rischin, D., Ranson, M., Calvert, H., Raymond, E., Keiback, D. G., Kaye, S. B., Gianni, L., Harris, A., Bjork, T., *et al.* (2002). Phase I safety, pharmacokinetic, and pharmacodynamic trial of ZD1839, a selective oral epidermal growth factor receptor tyrosine kinase inhibitor, in patients with five selected solid tumor types. *J. Clin. Oncol.* **20**, 4292–4302.

Belanger, L. F., Roy, S., Tremblay, M., Brott, B., Steff, A. M., Mourad, W., Hugo, P., Erikson, R., and Charron, J. (2003). Mek2 is dispensible for mouse growth and development. *Mol. Cell. Biol.* **23**, 4778–4787.

Biscardi, J. S., Tice, D. A., and Parsons, S. J. (1999). c-Src, receptor tyrosine kinases, and human cancer. *Adv. Cancer Res.* **76**, 61–119.

Biscardi, J. S., Maa, M. C., Tice, D. A., Cox, M. E., Leu, T. H., and Parsons S. J. (1999). c-Src-mediated phosphorylation of the epidermal growth factor receptor on Tyr845 and Tyr1101 is associated with modulation of receptor function. *J. Biol. Chem.* **274**, 8335–8343.

Biscardi, J. S., Ishizawar, R. C., Silva, C. M., and Parsons, S. J. (2000). Tyrosine kinase signalling in breast cancer: Epidermal growth factor receptor and c-Src interactions in breast cancer. *Breast Cancer Res.* **2**, 203–210.

Blackwell, K. L., Kaplan, E. H., Franco, S. X., Marcom, P. K., Maleski, J. E., Sorensen, M. J., and Berger, M. S. (2004). A phase II, open-labeled, multicenter study of GW572016 in patients with trastuzumab-refractory metastatic breast cancer. *J. Clin. Oncol.* **Suppl. 14**, 3006.

Blenis, J. (1993). Signal transduction via the MAP kinases: Proceed at your RSK. *Proc. Natl. Acad. Sci. USA* **90**, 5889–5892.

Blume-Jensen, P., and Hunter, T. (2001). Oncogenic kinase signaling. *Nature* **411**, 355–365.

Brown, E. J., Albers, M. W., Shin, T. B., Ichikawa, K., Keith, C. T., Lane, W. S., and Schreiber, S. L. (1994). A mammalian protein targeted by G1-arresting rapamycin-receptor complex. *Nature* **369**, 756–758.

Brott, B. K., Alessandrini, A., Largaespada, D. A., Copeland, N. G., Jenkins, N. A., Crews, C. M., and Erikson, R. L. (1993). MEK2 is a kinase related to MEK1 and is differentially expressed in murine tissues. *Cell. Growth. Differ.* **4**, 921–929.

Brunn, G. J., Hudson, C. C., Sekulic, A., Williams, J. H., Hosoi, H., Houghton, P. J., Lawrence, J. C., and Abraham, R. T. (1997). Phosphorylation of the translational repressor PHAS-I by the mammalian target of rapamycin. *Science* **277**, 99–101.

Bruns, C. J. (2004). Rapamycin-induced endothelial cell death and tumor vessel thrombosis potentate cytotoxic therapy against pancreatic cancer. *Clin. Cancer Res.* **10**, 2109–2119.

Buckley, M., Sweeney, K. J., Hamilton, J. A., Sini, R. L., Manning, D. L., Nicholson, R. I., DeFazio, A., Watts, C. K., Musgrove, E. A., and Sutherland, R. L. (1993). Expression and amplification of cyclin genes in human breast cancer. *Oncogene* **8**, 2127–2133.

Burgering, B. M., and Coffer, P. J. (1995). Protein kinase B (c-Akt) in phosphatidylinositol-3-OH kinase signal transduction. *Nature* **376**, 599–602.

Burgess, M. R., Skaggs, B. J., Shah, N. P., Lee, F. Y., and Sawyers, C. L. (2005). Comparative analysis of two clinically active BCR-ABL kinase inhibitors reveals the role of conformation-specific binding in resistance. *Proc. Natl. Acad. Sci. USA* **102**, 3395–3400.

Burris, H., Taylor, C., Jones, S., Pandite, L., Smith, D. A., Versola, M., Stead, A., Whitehead B., Spector, N., Wilding, G., *et al.* (2003). A phase I study of GW572016 in patients with solid tumors. *J. Clin. Oncol.* **Suppl. 22**, 248.

Byrd, J. C., Lin, T., Dalton, J. T., Wu, B., Phelps, M. A., Fischer, B., Moran, M., Blum, K. A., Rovin, B., Brooker-McEldowney, M. *et al.* (2007). Flavopiridol administered using a pharmacologically derived schedule is associated with marked clinical efficacy in refractory, genetically high-risk chronic lymphocytic leukemia. *Blood* **109**, 399–404.

Cantley, L. C. (2002). The phosphoinositide 3-kinase pathway. *Science* **296**, 1655–1657.

Carlson, B. A., Dubay, M. M., Sausville, E. A., Brizuela, L., and Worland, P. J. (1996). Flavopiridol induces G1 arrest with inhibition of cyclin-dependent kinase (CDK)2 and CDK4 in human breast carcinoma cells. *Cancer Res.* **56**, 2973–2978.

Carlson, B., Lahusen, T., Singh, S., Loaiza-Perez, A., Worland, P. J., Pestell, R., Albanese, C., Sausville, E. A., and Senderowicz, A. M. (1999). Downregulation of cyclin D1 by transcriptional repression of MCF-7 human breast carcinoma cells induced by flavopiridol. *Cancer Res.* **59**, 4634–4641.

Cartee, L., Wang, Z., Decker, R. H., Chellappan, S. P., Fusaro, G., Hirsch, K. G., Sankala, H. M., Dent, P., and Grant, S. (2001). The cyclin-dependent kinase inhibitor (CDKI) flavopiridol disrupts phorbol 12-myristate 12-acetate-induced differentiation and CDKI expression while enhancing apoptosis in human myeloid leukemia cells. *Cancer Res.* **61**, 2583–2591.

Chan, S., Scheulen, M. E., Johnston, I., Delbado, C., Mross, K., Cardoso, F., Dittich, C., Eiermann, W., Hess, D., Morant, R., *et al.* (2005). Phase II study of temsirolimus (CCI-779), a novel inhibitor of mTOR, in heavily pretreated patients with locally advanced or metastatic breast cancer. *J. Clin. Oncol.* **23**, 5314–5322.

Chang, S. M., Kuhn, J., Wen, P., Greenberg, H., Schiff, D., Conrad, C., Fink, K., Robins, H. I, Cloughesy, T., De Angelis, L., *et al.* (2004). Phase I/Pharmacokinetic study of CCI-779 in patients with recurrent malignant glioma on enzyme-induced antiepileptic drugs. *Invest. New Drugs* **22**, 427–435.

Ciardiello, F., Bianco, R., Caputo, R., Caputo, R., Damiano, V., Troiani, T., Melisi, D., De Vita, F., De Placido, S., Bianco, A. R. *et al.* (2004). Antitumor activity of ZD6474, a vascular endothelial growth factor receptor tyrosine kinase inhibitor, in human cancer cells with acquired resistance to antiepidermal growth factor receptor therapy. *Clin. Cancer Res.* **10**, 784–793.

Cowley, S., Paterson, H., Kemp, P., and Marshall, C. J. (1994). Activation of MAP kinase kinase is necessary and sufficient for PC12 differentiation and for transformation of NIH 3T3 cells. *Cell* **77**, 841–852.

Cunningham, D., Humblet, Y., Siena, S., Khayat, D., Bleiberg, H., Santoro, A., Bets, D., Mueser, M., Harstrick, A., Verslype, C. *et al.* (2004). Cetuximab monotherapy and cetuximab plus Irinotecan in Irinotecan-refractory metastatic colorectal cancer. *N. Engl. J. Med.* **351**, 317–319.

Davies, S. P., Reddy, H., Caivano, M., and Cohen, P. (2000). Specificity and mechanism of action of some commonly used protein kinase inhibitors. *Biochem. J.* **351**, 95–105.

de Groot, R. P., Ballou, L. M., and Sassone-Corsi, P. (1994). Positive regulation of the cAMP-responsive activator CREM by the p70 S6 kinase: An alternative route to mitogen-induced gene expression. *Cell* **79**, 81–91.

Demetri, G. D., von Mehren, M., Blanke, C. D., Van den Abbeele, A. D., Eisenberg, B., Roberts, P. J., Heinrich, M. C., Tuvenson, D. A., Singer, S., Janicek, M., *et al.* (2002). Efficacy and safety of imatinib mesylate in advanced gastrointestinal stromal tumors. *N. Engl. J. Med.* **347**, 472–480.

Djahansouzi, S., Heimerzheim, T., Reinhardt, M., Hansein, B., Dall, P., Bender, H., and Niederacher, D. (2005). Therapeutic doses of Tamoxifen only partially inhibit the non-genomic effects of estrogen: Evidence for additive anti-proliferative effect of Tamoxifen with MEK inhibitor. *J. Clin. Oncol.* **Suppl. 23**, 16S.

Downward, J. (1998). Mechanisms and consequences of activation of protein kinase B/Akt. *Curr. Opin. Cell. Biol.* **10**, 262–267.

Downward, J. (2003). Targeting RAS signaling pathways in cancer therapy. *Nat. Rev. Cancer* **3**, 11–22.

Drees, M., Dengler, W. A., Roth, T., Labonte, H., Mayo, J., Malspeis, L., Grever, M., Sausville, E. A., and Fiebig, H. H. (1997). Flavopiridol (L86-8275): Selective antitumor activity *in vitro* and activity *in vivo* for prostate carcinoma cells. *Clin. Cancer Res.* **3**, 271–279.

Druker, B. J., Talpaz, M., Resta, D. J., Peng, B., Buchdunger, E., Ford, J. M., Lydon, N. B., Kantarjian, H., Capdeville, R., Ohno-Jones, S., and Sawyers, C. L. (2001). Efficacy and safety of a specific inhibitor of the BCR-ABL tyrosine kinase in chronic myeloid leukemia. *N. Engl. J. Med.* **344**, 1031–1037.

Dudkin, L., Dilling, M. B., Cheshire, P. J., Harwood, F. C., Hollingshead, M., Arbuck, S. G., Travis, R., Sausville, E. A., and Houghton, P. J. (2002). Biochemical correlates of mTOR inhibition by the rapamycin ester CCI-779 and tumor growth inhibition. *Clin. Cancer Res.* **7**, 1758–1764.

Dudley, D. T., Pang, L., Decker, S. J., Bridges, A. J., and Saltiel, A. R. (1995). A synthetic inhibitor of the mitogen-activated protein kinase cascade. *Proc. Natl. Acad. Sci. USA* **92**, 7686–7689.

Duncan, P., Pollet, N., Niehrs, C., and Nigg, E. A. (2001). Cloning and characterization of Plx2 and Plx3, two additional Polo-like kinases from *Xenopus laevis*. *Exp. Cell Res.* **270**, 78–87.

Duxbury, M. S. (2004). Inhibition of SRC tyrosine kinase impairs inherent and acquired gemcitabine resistance in human pancreatic adenocarcinoma cells. *Clin. Cancer Res.* **10**, 2307–2318.

Erlichman, C., Boerner, S. A., Hallgren, C. G., Spieker, R., Wang, X. Y., James, C. D., Scheffer, G. L., Maliepaard, M., Ross, D. D., Bible, K. C. *et al*. (2001). The HER tyrosine kinase inhibitor CI1033 enhances cytotoxicity of 7-ethyl-10-hydroxycamptothecin and topotecan by inhibiting breast cancer resistance protein-mediated drug efflux. *Cancer Res.* **61**, 739–748.

Erlichman, C., Hidalgo, M., Boni, J. P., Martins, P., Quinn, S. E., Zacharchuk, C., Amorusi, P., Adjei, A. A., and Rowinsky, E. K. (2006). Phase I study of EKB-569, an irreversible inhibitor of the epidermal growth factor receptor, in patients with advanced solid tumors. *J. Clin. Oncol.* **24**, 2252–2260.

Favata, M. F., Horiuchi, K. Y., Manos, E. J., Daulerio, A. J., Stradley, D. A., Feeser, W. S., Van Dyk, D. E., Pitts, W. J., Earl, R. A., Hobbs, F., *et al*. (1998). Identification of a novel inhibitor of mitogen-activated protein kinase kinase. *J. Biol. Chem.* **273**, 18623–18632.

Forouzesh, B., Buckner, J., Adjei, A., Marks, R., *et al*. (2002). Phase I, bioavailability and pharmacokinetic study of oral dosage of CCI-779 administered to patients with advanced solid malignancies. *Eur. J. Cancer.* **Suppl. 33**, 7.

Fruman, D. A., Wood, M. A., Gjertson, C. K., Katz, H. R., Burakoff, S. J., and Bierer, B. E. (1995). FK506 binding protein 12 mediates sensitivity to both FK506 and rapamycin in murine mast cells. *Eur. J. Immunol.* **25**, 563–571.

Fukuoka, M., Yano, S., Giaccone, G., Tamura, T., Nakagawa, K., Douillard, J. Y., Nishiwaki, Y., Vansteenkiste, J., Kudoh, S., Rischin, D., Eek, R., *et al*. (2003). Multi-institutional randomized phase II trial of gefitinib for previously treated patients with advanced non-small-cell lung cancer. *J. Clin. Oncol.* **21**, 2237–2246.

Futreal, P., Kasprzyk, A., Birney, E., Mullikin, J. C., Wooster, R., and Stratton, M. R. (2001). Cancer and genomics. *Nature* **409**, 850–852.

Gao, X., Zhang, Y., Arrazola, P., Hino, O., Kobayashi, T., Yeung, R. S., Ru, B., and Pan, D. (2002). Tsc tumor suppressor proteins antagonize amino-acid-TOR signaling. *Nat. Cell. Biol.* **4**, 699–704.

Garnett, M. J., and Marias, R. (2004). Guilty as charged: B-RAF in a human oncogene. *Cancer Cell* **6**, 313–319.

Giaccone, G., Herbst, R. S., Manegold, C., Scagliotti, G., Rosell, R., Miller, V., Natale, R. B., Schiller, J. H., Von Pawel, J., Pluzanska, A., *et al*. (2004). Gefitinib in combination with gemcitabine and cisplatin in advanced non-small-cell lung cancer: A phase III trial-INTACT 1. *J. Clin. Oncol.* **22**, 777–784.

Gibbons, J. J., Discafani, C., Peterson, R., Hernandez, R., Skotnicki, J., and Frost, J. (1999). The effect of CCI-779, a novel macrolide anti-tumor agent, on the growth of human tumor cells *in vitro* and in nude mouse xenograft *in vivo*. *Proc. Am. Assoc. Cancer Res.* **40**, 301.

Gingras, A. C., Raught, B., and Sonenberg, N. (2001). Regulation of translation initiation by FRAP/mTOR. *Genes Dev.* **15**, 807–826.

Giroux, S., Tremblay, M., Bernard, D., Cardin-Girard, J. F., Aubry, S., Larouche, L., Rousseau, S., Huot, J., Landry, J., Jeannotte, L., and Charron, J. (2002). Embryonic death of Mek1-deficient mice reveals a role for the kinase in angiogenesis in the labyrinthine region of the placenta. *Curr. Biol.* **9**, 369–372.

Golas, J. M., Arndt, K., Etienne, C., Lucas, J., Nardin, D., Gibbons, J., Frost, P., Ye, F., Boschelli, D. H., and Boschelli, F. (2002). SKI-606, a 4-anilino-3-quinolinecarbonitrile dual inhibitor of Src and Abl kinases, is a potent antiproliferative agent against chronic myelogenous leukemia cells in culture and causes regression of K562 xenografts in nude mice. *Cancer Res.* **63**, 375–381.

Grandis, J. R., Melhem, M. F., Gooding, W. E., Day, R., Holst, V. A., Wagener, M. M., Drenning, S. D., and Tweardy, D. J. (1998). Levels of TGF-alpha and EGFR protein in head and neck squamous cell carcinoma and patient survival. *J. Natl. Cancer Inst.* **90**, 824–832.

Graus-Porta, D., Beerli, R. R., Daly, J. M., and Hynes, N. E. (1997). ErbB-2, the preferred heterodimerization partner of all ErbB receptors, is a mediator of lateral signaling. *EMBO J.* **16**, 1647–1655.

Grewe, M., Gansauge, F., Schmid, R. M., Adler, G., and Seufferlein, T. (1999). Regulation of cell growth and cyclin D1 expression by the constitutively active FRAP-p70s6K pathway in human pancreatic cancer cells. *Cancer Res.* **59**, 3581–3587.

Guba, M., von Breitenbuch, P., Steinbauer, M., Koehl, G., Flegel, S., Hornung, M., Burns, C. J., Zuelke, C., Farkas, S., Anthuber, M., *et al*. (2002). Rapamycin inhibits primary and metastatic tumor growth by antiangiogenesis: Involvement of vascular endothelial growth factor. *Nat. Med.* **8**, 128–135.

Gumireddy, K., Reddy, M., Cosenza, S. C., Boominathan, R., Baker, S. J., Papathi, N., Jiang J., Holland, J., Reddy, E. P. (2005). ON01910, a non-ATP-competitive small molecule inhibitor of Plk1, is a potent anticancer agent. *Cancer Cell* **7**, 275–286.

Hara, K., Maruki, Y., Long, X ., Yoshino, K., Oshiro, N., Hidayat, S., Tokunaga, C., Avruch, J., and Yonezawa, K. (2002). Raptor, a binding partner of target of rapamycin (TOR), mediates TOR action. *Cell* **110**, 177–189.

Harrington, E. A., Bebbington, D., Moore, J., Rasmussen, R. K., Ajose-Adeogun, A. O., Nakayama, T., Graham, J. A., Demur, C., Hercend, T., Diu-Hercend, A., *et al.* (2004). VX-680, a potent and selective small molecule inhibitor of the Aurora kinases, suppresses tumor growth *in vivo*. *Nat. Med.* **10**, 262–267.

Hashemolhosseini, S., Nagamine, Y., Morley, S. J., Desrivieres, S., Mercep, L., and Ferrari S. (1998). Rapamycin inhibition of the G_1 to S transition is mediated by effects on cyclin D1 mRNA and protein stability. *J. Biol. Chem.* **273**, 14424–14429.

Heitman, J., Movva, N. R., and Hall, M. N. (1991). Targets for cell cycle arrest by the immunosuppressant rapamycin in yeast. *Science* **253**, 905–909.

Herbst, R. S., Maddox, A. M., Rothenberg, M. L., Small, E. J., Rubin, E. H., Baselga, J., Rojo, F., Hong, W. K., Swaisland, H., Averbuch, S. D., *et al.* (2002). Selective oral epidermal growth factor receptor tyrosine kinase inhibitor ZD1839 is generally well-tolerated and has activity in non-small-cell lung cancer and other solid tumors: Results of a phase I trial. *J. Clin. Oncol.* **20**, 3815–3825.

Herbst, R., Prager, D., Hermann, R., Fehrenbacher, L., Johnson, B. E., Sandler, A., Kris, M. G., Tran, H. T., Klein, P., Li, X., *et al.* (2005). TRIBUTE—a phase III trial of erlotinib HCl (OSI-774) combined with carboplatin and paclitaxel (CP) chemotherapy in advanced non-small cell lung cancer (NSCLC). *J. Clin. Oncol.* **23**, 5892–5899.

Hidalgo, M., Buckner, J. C., Erlichman, C., Pollack, M. S., Boni, J. P., Dukart, G., Marshall, B., Speicher, L., Moore, L., and Rowinsky, E. K. (2006). A phase I and pharmacokinetic study of temsirolimus (CCI-779) administered intravenously daily for 5 days every 2 weeks to patients with advanced cancer. *Clin. Cancer Res.* **12**, 5755–5763.

Hidalgo, M., Siu, L. L., Nemunaitis, J., Rizzo, J., Hammond, L. A., Takimoto, C., Eckhardt, S. G., Tolcher, A., Britten, C. D., Denis, L., *et al.* (2001). Phase I and pharmacologic study of OSI-774, an epidermal growth factor receptor tyrosine kinase inhibitor, in patients with solid malignancies. *J. Clin. Oncol.* **19**, 3267–3279.

Hofheinz, R., Hochhaus, A., Al-Batran, S., Nanci, A., Reichardt, V., Trommeshauser, D., Hoffmann, M., Steegmaier, M., Munzert, G., Jäger, E., *et al.* (2006). A phase I repeated dose escalation study of Polo-like kinase 1 inhibitor BI 2536 in patients with advanced solid tumors. *J. Clin. Oncol.* **Suppl. 24**, 18S.

Hoshino, R., Chatani, Y., Yamori, T., Tsuruo, T., Oka, H., Yoshida, O., Shimada, Y., Ari-I, S., Wada, H., Fujimoto, J., *et al.* (1999). Constitutive activation of the 41-/43-kDa mitogen-activated protein kinase signaling pathway in human tumors. *Oncogene* **18**, 813–822.

Huang, S., Liu, L. N., Hosoi, H., Dilling, M. B., Shikata, T., and Houghton, P. J. (2001). p53/p21^{CIP1} cooperate in enforcing rapamycin-induced G1 arrest and determine the cellular response to rapamycin. *Cancer Res.* **61**, 3373–3381.

Hudes, G., Carducci, M., Tomczak, P., Dutcher, J., Figlin, R., Kapoor, A., Staroslawska, E., O'Toole, T., Park, Y., and Moore, L. (2006). A phase 3, randomized, 3-arm study of temsirolimus (TEMSR) or interferon-alpha (IFN) or the combination of TEMSR + IFN in the treatment of first-line, poor-risk patients with advanced renal cell carcinoma (adv RCC). *J. Clin. Oncol.* **Suppl. 24**, 18S.

Hunter, T. (1980). Protein phosphorylated by the RSV transforming function. *Cell* **22**, 647–648.

Hylton, N. (2006). Dynamic contrast enhanced—magnetic resonance imaging as an imaging biomarker. *J. Clin. Oncol.* **24**, 3293–3298.

Inoki, K., Li, Y., Zhu, T., Wu, J., and Guan, K. L. (2002). TSC2 is phosphorylated and inhibited by Akt and suppresses mTOR signaling. *Nat. Cell Biol.* **4**, 648–657.

Irby, R. B., Mao, W., Coppola, D., Kang, L., Loubeau, J. M., Trudeau, W., Karl, R., Fujita D. J., Jove, R., and Yeatman, T. J. (1999). Activating SRC mutation in a subset of advanced human colon cancers. *Nat. Genet.* **21**, 187–190.

Ishizawar, R., and Parsons, S. J. (2004). c-Src and cooperating partners in human cancer. *Cancer Cell* **6**, 209–214.

Jimeno, A., Rubio-Viquiera, B., Amador, M. L., Oppenheimer, D., Bouraroud, N., Kuleza, P., Sebastiani, V., Maitra, A., and Hidalgo, M. (2005). Epidermal growth factor receptor dynamics influences response to epidermal growth factor targeted agents. *Cancer Res.* **65**, 3003–3010.

Kauffmann-Zeh, A., Rodriguez-Viciana, P., Ulrich, E., Gilbert, C., Coffer, P., Downward, J., and Evan, G. (1997). Suppression of c-Myc-induced apoptosis by Ras signaling through PI(3)K and PKB. *Nature* **385**, 544–548.

Kawamata, S., Sakaida, H., Hori, T., Maeda, M., and Uchiyama, T. (1998). The upregulation of p27Kip1 by rapamycin results in G1 arrest in exponentially growing T-cell lines. *Blood* **91**, 561–569.

Kobor, M., and Greenblatt J. (2002). Regulation of transcription elongation by phosphorylation. *Biochim. Biophys. Acta.* **13**, 261–275.

Kolibaba, K. S., and Druker, B. J. (1997). Protein tyrosine kinases and cancer. *Biochim. Biophys. Acta.* **1333**, F217–F248.

Koltin, Y., Faucette, L., Bergsma, D. J., Levy, M. A., Cafferkey, R., Koser, P. L., Johnson, R. K., and Livi, G. P. (1991). Rapamycin sensitivity in *Saccharomyces cerevisiae* is mediated by a peptidyl-prolyl cis-trans isomerase related to human FK506-binding protein. *Mol. Cell. Biol.* **11**, 1718–1723.

Koumenis, C., and Giaccia A. (1997). Transformed cells require continuous activity of RNA polymerase II to resist oncogene-induced apoptosis. *Mol. Cell. Biol.* **17**, 7306–7316.

Kotani, S., Tugendreich, S., Fujii, M., Jorgensen, P. M., Watanabe, N., Hoog, C., Hieter, P., and Todokoro, K. et al. (1998). PKA and MPF-activated polo-like kinase regulate anaphase-promoting complex activity and mitosis progression. *Mol. Cell* **1**, 371–380.

Kris, M. G., Natale, R. B., Herbst, R. S., Lynch, T. J., Jr., Prager, D., Belani, C. P., Schiller, J. H., Kelly, K., Spiridonidis, H., Sandler, A., et al. (2004). Efficacy of gefitinib, an inhibitor of the epidermal growth factor receptor tyrosine kinase, in symptomatic patients with non-small cell lung cancer: A randomized trial. *JAMA* **290**, 2149–2158.

Lampson, M. A., Renduchitala, K., Khodjakov, A., and Kapoor, T. M. (2004). Correcting improper chromosome-spindle attachments during cell division. *Nat. Cell. Biol.* **6**, 232–237.

Lampson, M. A., and Kapoor, T. M. (2005). The human mitotic checkpoint protein BubR1 regulates chromosome-spindle attachments. *Nat. Cell. Biol.* **7**, 93–98.

Legrier, M. E., D'Silva, S., Yan, H. G., Horwitz, S. B., and McDaid, H. M. (2006). Chemosensitization by dual pharmacological inhibition of MEK and MTRO signaling pathways in human non-small cell lung cancer (NSCLC). *Proc. Amer. Assoc. Can. Res. Abstrt,* 4867.

Lewis, T. S., Shapiro, P. S., and Ahn, N. G. (1998). Signal transduction through MAP kinase cascades. *Adv. Cancer Res.* **74**, 49–139.

Lorusso, P. M., Adjei, A. A., Varterasian, M., Gadgeel, S., Reid, J., Mitchell, D. Y., Hanson, L., DeLuca, P., Bruzek, L., Piens, J., et al. (2005). Phase I and pharmacodynamic study of the oral MEK inhibitor CI-1040 in patients with advanced malignancies. *J. Clin. Oncol.* **23**, 5281–5293.

Lorusso, P. M., Krishnamurthi, S., Rinehart, J. R., Nabell, L., Croghan, G., Varterasian, M., Sadis, S. S., Menon, S. S., Leopold, J., Meyer, M. B., et al. (2005). A phase 1–2 clinical study of a second generation oral MEK inhibitor, PD 0325901 in patients with advanced cancer. *J. Clin. Oncol.* **Suppl. 23**, 16S.

Losiewicz, M. D., Carlson, B. A., Kaur G., Sausville E.A. and Worland P.J. (1994). Potent inhibition of cdc2 kinase activity by the flavonoid L86-8275. *Biochem. Biophys. Res. Commun.* **201**, 589–595.

Lynch T. J., Bell, D. W., Sordella, R., Gurubhagavatula, S., Okimoto, R. A., Brannigan, B. W., Harris, P. L., Haserlat, S. M., Supko, J. G., Haluska, F. G., et al. (2004). Activating mutations in the epidermal growth factor receptor underlying responsiveness of non-small cell lung cancer to gefitinib. *N. Engl. J. Med.* **350**, 2129–2139.

Maemura, M., Iino, Y., Koibuchi, Y., Yokoe, T., and Morishita Y. (1999). Mitogen-activated protein kinase cascade in breast cancer. *Oncology* **Suppl 57**, 37–44.

Malumbres, M., and Barbacid, M. (2001). To cycle or not to cycle: A critical decision in cancer. *Nat. Rev. Cancer.* **1**, 222–231.

Manning, B. D., Tee, A. R., Logsdon, M. N., Blenis, J., and Cantley, L. C. (2002). Identification of the tuberous sclerosis complex-2 tumor suppressor gene product tuberin as a target of the phosphoinositide 3-kinase/akt pathway. *Mol. Cell.* **10**, 151–162.

Mansour, S. J., Matten, W. T., Hermann, A. S., Candia, J. M., Rong, S., Fukasawa, K., Vande Woude, G. F., and Ahn, N. G. (1994). Transformation of mammalian cells by constitutively active MAP kinase kinase. *Science* **265**, 966–970.

Martin, G. S. (2004). The road to Src. *Oncogene* **23**, 7910–7917.

McCormicK, F. (1993). Signal transduction. How receptors turn Ras on. *Nature.* **363**, 15–16.

Menon, S. S., Whitfield, L. R., Sadis, S. S., Meyer, M. B., Leopold J., Lorusso, P. M., Krishnamurthi, S., Rinehart, J. R., Nabell, L., Croghan, C., et al. (2005). Pharmacokinetics (PK) and pharmacodynamics (PD) of PD 0325901, a second generation MEK inhibitor after multiple oral doses of PD 0325901 to advanced cancer patients. *J. Clin. Oncol.* **Suppl. 23**, 16S.

Meyerhardt, J. A., and Mayer, R. J. (2005). Systemic therapy for colorectal cancer. *N. Engl. J. Med.* **352**, 476–487.

Mitsiades, C. S., Mitsiades, N., Hideshima, T., Richardson, P. G., and Anderson, K. C. (2005). Proteasome inhibitors as therapeutics. *Essays Biochem.* **41**, 205–218.

Moore, M. J., Goldstein, D., Hamm, J., Kotecha, J., Gallinger, S., Au, H. J., Nomikos, D., Ding, K., Ptaszynski, M., and Parulekar, W. (2005). Erlotinib improves survival when added to gemcitabine in patients with advanced pancreatic cancer. A phase III trial of the National Cancer Institute of Canada Clinical Trials Group [NCIC-CTG]. *Proc. GI ASCO Abstr.* 77.

Motwani, M., Jung, C., Sirotnak, F. M., She, Y., Shah, M. A., Gonen, M., and Schwartz, G. K. (2001). Augmentation of apoptosis and tumor regression by flavopiridol in the presence of CPT-11 in HCT116 colon cancer monolayers and xenografts. *Clin. Cancer Res.* **7**, 4209–4219.

Moyer, J. D., Barbacci, E. G., Iwata, K. K., Arnold, L., Boman, B., Cunningham, A., DiOrio, C., Doty, J., Morin, A. J., Moyer, M. P., et al. (1997). Induction of apoptosis and cell cycle arrest by CP-358,774, an inhibitor of epidermal growth factor receptor tyrosine kinase. *Cancer Res.* **57**, 4838–4848.

Mueller, H., Flury, N., Eppenberger-Castori, S., Kueng, W., David, F., and Eppenberger U. (2000). Potential prognostic value of mitogen-activated protein kinase activity for disease-free survival of primary breast cancer patients. *Int. J. Cancer* **89**, 384–388.

Munzert, G., Steinbild, A., Frost, A., Hedborn, S., Renschler, J., Kaiser, R., Trommeshauser, D., Hoffmann, M., Steegmaier, M., and Mross, K. (2006). A phase I study of two administration schedules of the Polo-like kinase 1 inhibitor BI 2536 in patients with advanced solid tumors. *J. Clin. Oncol.* **Suppl. 24**, 18S.

Nair, J. S., Tse, A., Keen, N., and Schwartz, G. K. (2004). A novel aurora B kinase inhibitor with potent anticancer activity either as a single agent or in combination with chemotherapy. *J. Clin. Oncol.* **Suppl. 22**, 14S.

Nave, B. T., Ouwens, M., Withers, D. J., Alessi, D. R., and Shepherd, P. R. (1999). Mammalian target of rapamycin is a direct target for protein kinase B: Identification of a convergence point for opposing effects of insulin and amino-acid deficiency on protein translation. *Biochem. J.* **344**, 427–431.

Neshat, M. S., Mellinghoff, I. K., Tran, C., Stiles, B., Thomas, G., Petersen, R., Frost, P., Gibbons, J. J., Wu, H., and Sawyers, C. L. (2001). Enhanced sensitivity of PTEN-deficient tumors to inhibition of FRAP/mTOR. *Proc. Natl. Acad. Sci. USA* **98**, 10314–10319.

Nicholson, R. I., Gee, J. M., and Harper, M. E. (2001). EGFR and cancer prognosis. *Eur. J. Cancer.* **Suppl. 37**, S9–S15.

Nishikawa, R., Ji, X. D., Harmon, R. C., Lazar, C. S., Gill, G. N., Cavenee, W. K., and Huang, H. J. (1994). A mutant epidermal growth factor receptor common in human glioma confers enhanced tumorigenecity. *Proc. Natl. Acad. Sci. USA* **91**, 7727–7731.

Normanno, N., De Luca, A., Maiello, M. R., Campiglio, M., Napolitano, M., Mancino, M., Cartenuto, A., Viglietto, G., and Menard, S. (2006). The MEK/MAPK pathway is involved in the resistance of breast cancer cells to the EGFR tyrosine kinase inhibitor gefitinib. *J. Cell. Physiol.* **207**, 420–427.

Oelgeschlager, T. (2002). Regulation of RNA polymerase II activity by CTD phosphorylation and cell cycle control. *J. Cell. Physiol.* **190**, 160–169.

Ohren, J. F., Chen, H., Pavlovsky, A., Whitehead, C., Zhang, E., Kuffa, P., Yan, C., McConnell, P., Spessard, C., Banotai, C., *et al.* (2004). Structures of human MAP kinase kinase 1 (MEK1) and MEK2 describe novel noncompetitive kinanse inhibition. *Nat. Struct. Mol. Biol.* **11**, 1192–1197.

Oza, A. M., Elit, L., Biagi, J., Chapman, W., Tsao, M., Hedley, D., Hansen, C., Dancey, J., and Eisenhauer, E. (2006). Molecular correlates associated with a phase II study of temsirolimus (CCI-779) in patients with metastatic or recurrent endometrial cancer— NCIC IND 160. *J. Clin. Oncol.* **24**, 121s.

Paez, J. G., Janne, P. A., Lee, J. C., Tracy, S., Greulich, H., Gabriel, S., Herman, P., Kaye, F. J., Lindeman, N., Boggon, T. J., *et al.* (2004). EGFR mutations in lung cancer: Correlation with clinical response to gefitinib therapy. *Science* **304**, 1497–1500.

Pao, W., Miller, V., Zakowski, M., Doherty, J., Politi, K., Sarkaria, I., Singh, B., Heelan, R., Rusch, V., Fulton, L., *et al.* (2004). EGF receptor gene mutations are common in lung cancers from never smokers and are associated with sensitivity of tumors to gefitinib and erlotinib. *Proc. Natl. Acad. Sci. USA* **101**, 13306–13311.

Patel, V., Senderowicz, A. M., Pinto, D., Igishi, T., Raffeld, M., Quintanilla-Martinez, L., Ensley, J. F., Sausville, E. A., and Gutkind, J. S. (1998). Flavopiridol, a novel cyclin-dependent kinase inhibitor, suppresses the growth of head and neck squamous cell carcinomas by inducing apoptosis. *J. Clin. Invest.* **102**, 1674–1681.

Peralba, J. M., DeGraffenried, L., Friedrichs, W., Fulcher, L., Grunwald, V., Weiss, G., and Hidalgo, M. (2003). Pharmacodynamic evaluation of CCI-779, an inhibitor of mTOR, in cancer patients. *Clin. Cancer Res.* **9**, 2887–2892.

Peterson, R. T., Desai, B. N., Hardwick, J. S., and Schreiber, S. L. (1999). Protein phosphatase 2A interacts with the 70-kDa S6 kinase and is activated by inhibition of FKBP12-rapamycinassociated protein. *Proc. Natl. Acad. Sci. USA* **96**, 4438–4442.

Pinkas-Kramarski, R., Soussan, L., Waterman, H, Levkowitz, G., Alroy, I., Klapper, L., Lavi, S., Seger, R., Ratzkin, B. J., Sela, M., and Yarden, Y. (1996). Diversification of Neu differentiation factor and epidermal growth factor signaling by combinatorial receptor interactions. *EMBO J.* **15**, 2452–2467.

Playford, M. P., and Schaller, M. D. (2004). The interplay between Src and integrins in normal and tumor biology. *Oncogene* **23**, 7928–7946.

Ple, P. A., Green, T. P., Hennequin, L. F., Curwen, J., Fennell, M., Allen, J., Lambert-Van Der Brempt, C., and Costello, G. (2004). Discovery of a new class of anilinoquinazoline inhibitors with high affinity and specificity for the tyrosine kinase domain of c-Src. *J. Med. Chem.* **47**, 871–887.

Plowman, G. D., Sudarsanam, S., Bingham, J., Whyte, D., and Hunter, T. (1999). The protein kinases of Caenorhabditis elegans: A model for signal transduction in multicellular organisms. *Proc. Natl. Acad. Sci. USA* **96**, 13603–13610.

Pollack, V. A., Savage, D. M., Baker, D. A., Tsaparikos, K. E., Sloan, D. E., Moyer, J. D., Barbacci, E. G., Pustilnik, L. R., Smolarek, T. A., Davis, J. A., *et al.* (1999). Inhibition of epidermal growth factor receptor-associated tyrosine phosphorylation in human carcinomas with CP-358,774: Dynamics of receptor inhibition *in situ* and antitumor effects in athymic mice. *J. Pharmacol. Exp. Ther.* **291**, 739–748.

Punt, C. J., Boni, J., Bruntsch, U., Peters, M., and Thielert, C. (2003). Phase I and pharmacokinetic study of CCI-779, a novel cytostatic cell-cycle inhibitor, in combination with 5-fluorouracil and leucovorin in patients with advanced solid tumors. *Ann. Oncol.* **14**, 931–937.

Ranson, M., Hammond, L. A., Ferry, D., Kris, M., Tullo, A., Murray, P. I., Miller, V., Averbuch, S., Ochs, J., Morris, C., *et al.* (2002). ZD1839, a selective oral epidermal growth factor receptor-tyrosine kinase inhibitor, is well tolerated and active in patients with solid, malignant tumors: Results of a phase I trial. *J. Clin. Oncol.* **20**, 2240–2250.

Ratain, M. J., Eisen, T., Stadler, W. M., Flaherty, K. T., Kaye, S. B., Rosner, G. L., Gore, M., Desai, A. A., Patnaik, A., Xiong, H. Q., et al. (2006). Phase II placebo-controlled randomized discontinuation trial of sorafenib in patients with metastatic renal cell carcinoma. *J. Clin. Oncol.* **24**, 2505–2512.

Rathkopf, D., Fornier, M., Shah, M. A., Kortmansky, J., O'Reilly, E., King, A., Winkelmann, J., Kelsen, D. P., Olsen, S., and Schwartz, G. K. (2004). A phase I dose finding study of weekly, sequential docetaxel (Doc) followed by flavopiridol (F) in patients with advanced solid tumors. *J. Clin. Oncol.* **Suppl. 22**, 213s.

Raymond, E., Alexander, J., Faivre, S., Vera, K., Materman, E., Boni, J., Leister, C., Korth-Bradley, J., Hanauske, A., and Jean-Pierre, A. (2004). Safety and pharmacokinetics of escalated doses of weekly intravenous infusion of CCI-779, a novel mTOR inhibitor, in patients with cancer. *J. Clin. Oncol.* **22**, 2336–2347.

Rinehart, J., Adjei, A. A., Lorusso, P. M., Waterhouse, D., Hecht, J. R., Natale, R. B., Hamid, O., Varterasian, M., Asbury, P., Kaldjian, P., et al. (2004). Multicenter phase II study of the oral MEK inhibitor, CI-1040, in patients with advanced non-small-cell lung, breast, colon, and pancreatic cancer. *J. Clin. Oncol.* **22**, 4456–4462.

Robinson, D. R., Wu, Y. M., and Lin, S. F. (2000). The protein tyrosine kinase family of the human genome. *Oncogene* **19**, 5548–5557.

Rodriguez-Viciana, P., Warne, P. H., Khwaja, A., Marte, B. M., Pappin, D., Das, P., Waterfield, M. D., Ridley, A., and Downward, J. (1997). Role of phosphoinositide 3-OH kinase in cell transformation and control of the actin cytoskeleton by Ras. *Cell* **89**, 457–467.

Sabatini, D. M., Erdjument-Bromage, H., Lui, M., Tempst, P., and Snyder, S. H. (1994). RAFT1: A mammalian protein that binds to FKBP12 in a rapamycin-dependent fashion and is homologous to yeast TORs. *Cell* **78**, 35–43.

Sabatini, D. M. (2006). mTOR and cancer: Insights into a complex relationship. *Nat. Rev. Cancer* **6**, 729–734.

Salazar, R., Tabernero, J., Rojo, F., Jimenez, E., Montaner, I., Casado, E., Sala, G., Tillner, J., Malik, R., Baselaga, J., et al. (2004). Dose-dependent inhibition of the EGFR and signaling pathways with the anti-EGFR monoclonal antibody (MAb) EMD 7200 administered every three weeks (q3w). A phase I pharmacokinetic/pharmacodynamic (PK/PD) study to define the optimal biological dose (OBD). *J. Clin. Oncol.* **22**, 14S.

Salomon, D. S., Brandt, R., Ciardiello, F., and Normanno, N. (1995). Epidermal growth factor-related peptides and their receptors in human malignancies. *Crit. Rev. Oncol. Hematol.* **19**, 183–232.

Saltz, L. B., Kies, M., Abbruzzese, J., Azarnia, N., and Needle, M. (2003). The presence and intensity of the cetuximab-induced acne-like rash predicts increased survival in studies across multiple malignancies. *J. Clin. Oncol.* **22**, 204.

Sawyer, T., Boyce, B., Dalgarno, D., and Iuliucci, J. (2001). Src inhibitors: Genomics to therapeutics. *Expert. Opin. Investig. Drugs.* **10**, 1327–1344.

Schlessinger, J. (2000). Cell signaling by receptor tyrosine kinases. *Cell* **103**, 211–225.

Schmelzle, T., and Hall, M. N. (2000). TOR, a central controller of cell growth. *Cell* **103**, 253–262.

Schwartz, G. K., Ilson, D., Saltz, L., O'Reilly, E., Tong, W., Maslak, P., Werner, J., Perkins, P., Stoltz, M., and Kelsen, D. (2001). Phase II: Study of the cyclin-dependent kinase inhibitor flavopiridol administered to patients with advanced gastric carcinoma. *J. Clin. Oncol.* **19**, 1985–1992.

Scott, P. H., Brunn, G. J., Kohn, A. D., Roth, R. A., and Lawrence, J. C., Jr. (1998). Evidence of insulin-stimulated phosphorylation and activation of the mammalian target of rapamycin mediated by a protein kinase B signaling pathway. *Proc. Natl. Acad. Sci. USA* **95**, 7772–7777.

Sebolt-Leopold, J. S., and Herrera, R. (2004). Targeting the mitogen-activated protein kinase cascade to treat cancer. *Nat. Rev. Cancer* **4**, 937–947.

Sebolt-Leopold, J. S. (2004). MEK inhibitiors: A therapeutic approach to targeting the Ras-MAP kinase pathway in tumors. *Curr. Pharm. Des.* **10**, 1907–1914.

Seger, R., and Krebs, E. G. (1995). The MAPK signaling cascade. *Faseb J.* **9**, 726–735.

Seger, R., Ahn, N. G., Posada, J., Munar, E. S., Jensen, A. M., Cooper, J. A., Cobb, M. H., and Krebs, E. G. (2003). Purification and characterization of mitogen-activated protein kinase activator(s) from epidermal growth factor-stimulated A431 cells. *J. Biol. Chem.* **267**, 14373–14381.

Senderowicz, A. M. (2002). Cyclin-dependent kinases as targets for cancer therapy. In Giaccone, G., Schilsky, R., and Sondel P. (Eds.), *Cancer Chemotherapy and Biological Response Modifiers*. New York: Elsevier Science, 169–188.

Seufferlein, T., and Rozengurt, E. (1996). Rapamycin inhibits constitutive p70s6k phosphorylation, cell proliferation, and colony formation in small cell lung cancer cells. *Cancer Res.* **56**, 3895–3897.

Shah, M. A., Kortmansky, J., Motwani, M., Drobnjak, M., Gonen, M., Yi, S., Weyerbacher, A., Cordon-Cardo, C., Lefkowitz, R., Brenner, B., et al. (2005). A phase I/pharmacologic study of weekly sequential Irinotecan (CPT) and flavopiridol. *Clin. Cancer Res.* **11**, 3836–3845.

Shah, M. A., Kortmansky, J., Gonen, M., Tse, A., Lefkowitz, R., Kelsen, D., Colevas, D., Winkelman, J., Yi, S., and Schwartz, G. (2004). A phase I study of weekly sequential Irinotecan (CPT), cisplatin (CIS) and flavopiridol (F). *J. Clin. Oncol.* **Suppl. 22**, 14S.

Shahjehan, W. A., Laird, P., and DeMeester, T. (2001). DNA methylation: An alternative pathway to cancer. *Ann. Surg.* **234**, 10–20.

Shakespeare, W., Yang, M., Bohacek, R., Cerasoli, F., Stebbins, K., Sundaramoorthi, R., Azimioara, M., Vu, C., Pradeepan, S., Metcalf, C., et al. (2000). Structure-based design of an osteoclast-selective, nonpeptide src homology 2 inhibitor with *in vivo* antiresorptive activity. *Proc. Natl. Acad. Sci. USA* **97**, 9373–9378.

Shepherd, F., Pereira, J., Ciuleanu, T., Tan, E. H., Hirsh, V., Thongprasert, S., Bezjak, A., Tu, D., Santabarbara, P., and Seymour, L. (2004). A randomized placebo-controlled trial of erlotinib in patients with advanced non-small cell lung cancer (NSCLC) following failure of 1st line or 2nd line chemotherapy. A National Cancer Institute of Canada Clinical Trials Group (NCIC CTG) trial. *J. Clin. Oncol.* **Suppl. 22**, 14S.

Sherr, C. J., and Roberts, J. M. (1999). CDK inhibitors: Positive and negative regulators of G1-phase progression. *Genes. Dev.* **13**, 1501–1512.

Sirotnak F. M., Zakowski M. F., Miller V. A., Scher H. I., and Kris M. G. (2000). Efficacy of cytotoxic agents against human tumor xenografts is markedly enhanced by coadministration of ZD1839 (Iressa), an inhibitor of EGFR tyrosine kinase. *Clin. Cancer Res.* **6**, 4885–4892.

Sivaraman V. S., Wang H., Nuovo G. J., and Malbon C. C. (1997). Hyperexpression of mitogen-activated protein kinase in breast cancer. *J. Clin. Invest.* **99**, 1478–1483.

Slamon D. J., Leyland-Jones B., Shak S., Fuchs H., Paton V., Bajamonde A., Fleming T., Eiermann W., Wolter J., Pegram M., Baselga J., and Norton L. (2001). Use of chemotherapy plus a monoclonal antibody against HER2 for metastatic breast cancer that overexpresses HER2. *N. Engl. J. Med.* **344**, 783–792.

Smalley K. S., Haass N. K., Brafford P. A., Lioni M., Flaherty K. T., and Herlyn M. (2006). Multiple signaling pathways must be targeted to overcome drug resistance in cell lines derived from melanoma metastases. *Mol. Cancer Ther.* **5**, 1136–1140.

Solit DB, Garraway L. A., Pratilas C. A., Sawai A., Getz G., Basso A., Ye Q., Lobo J.M., She Y., Osman I., et al. (2006). BRAF mutation predicts sensitivity to MEK inhibition. *Nature.* **439**, 358–362.

Sonenberg, N., and Gingras, A. C. (1998). The mRNA 5' cap-binding protein eIF4E and control of cell growth. *Curr. Opin. Cell. Biol.* **10**, 268–275.

Suhara, T., Mano, T., Oliveira, B. E., and Walsh, K. (2001). Phosphatidylinositol 3-kinase/Akt signaling controls endothelial cell sensitivity to Fas-mediated apoptosis via regulation of FLICE-inhibitory protein (FLIP). *Circ. Res.* **89**, 13–19.

Susman, E. (2005). Bevacizumab adds survival benefit in colorectal cancer. *Lancet Oncol.* **6**, 136.

Tan, A. R., Yang, X., Berman, A., Zhai, S., Sparreboom, A., Parr, A. L., Chow, C., Brahim, J. S., Steinberg, S. M., Figg, W. D., et al. (2004). Phase I trial of the cyclin-dependent kinase inhibitor flavopiridol in combination with docetaxel in patients with metastatic breast cancer. *Clin. Cancer Res.* **10**, 5038–5047.

Tebar, F., Llado, A., and Enrich, C. (2002). Role of calmodulin in the modulation of the MAPK signaling pathway and the transactivation of epidermal growth factor receptor mediated by PKC. *FEBS Lett.* **517**, 206–210.

Tee, A. R., Fingar, D. C., Manning, B. D., Kwiatkowski, D. J. Cantley, L. C., and Blenis, J. (2002). Tuberous sclerosis complex-1 and -2 gene products function together to inhibit mammalian target of rapamycin (mTOR)-mediated downstream signaling. *Proc. Natl. Acad. Sci. USA* **99**, 13571–13576.

Tee, A. R., Anjum, R., and Blenis, J. (2003). Inactivation of the tuberous sclerosis complex-1 and -2 gene products occurs by phosphoinositide 3-kinase/Akt-dependent and -independent phosphorylation of tuberin. *J. Biol. Chem.* **278**, 37288–37296.

Te Poele, R., Okorokov, A., and Joel, S. (1999). RNA synthesis block by 5,6-dichloro-1-beta-D-ribofuranosylbenzimidazole (DRB) triggers p53-dependent apoptosis in human colon carcinoma cells. *Oncogene* **18**, 5785–5772.

Thomas, G. V., Tran, C., Mellinghoff, I. K., Welsbie, D. S., Chan, E., Fueger, B., Czernin, J., and Sawyers, C. L. (2006). Hypoxia-inducible factor determines sensitivity to inhibitors of mTOR in kidney cancer. *Nat. Med.* **12**, 122–127.

Volarevic, S., and Thomas, G. (2001). Role of S6 phosphorylation and S6 kinase in cell growth. *Prog. Nucleic Acid. Res. Mol. Biol.* **65**, 101–127.

Wallace, E. M., Lyssikatos, J. P., Yeh, T., Winkler, J. D., and Koch, K. (2005). Progress towards therapeutic small molecule MEK inhibitors for use in cancer therapy. *Curr. Top. Med. Chem.* **5**, 215–229.

Wang, X., Li, W., Williams, M., Terada, N., Alessi, D. R., and Proud, C. G. (2001). Regulation of elongation factor 2 kinase by p90(RSK1) and p70 S6 kinase. *EMBO J.* **20**, 4370–4379.

Wells, A. (1999). EGF receptor. *Int. J. Biochem. Cell. Biol.* **31**, 637–643.

Willett, C. G., Boucher, Y., di Tomaso, E., Duda, D. G., Munn, L. L., Tong, R. T., Chung, D. C., Sahani, D. V., Kalva, S. P., Kozin, S. V., et al. (2004). Direct evidence that the VEGF- specific antibody bevacizumab has antivascular effects in human rectal cancer. *Nat. Med.* **10**, 145–147.

Witzig, T. E., Geyer, S. M., Ghobrial, I., Inwards, D. J., Fonseca, R., Kurtin, P., Ansell, S. M., Luyun, R., Flynn, P. J., Morton, R. F., et al. (2005). Phase II trial of single-agent temsirolimus (CCI-779) for relapsed mantle cell lymphoma. *J. Clin. Oncol.* **23**, 5347–5356.

Woodburn, J. R. (1999). The epidermal growth factor receptor and its inhibition in cancer therapy. *Pharmacol. Ther.* **82**, 241–250.

Workman, P. (2003). Overview: Translating Hsp90 biology into Hsp90 drugs. *Curr. Cancer Drug Targets.* **3**, 297–300.

Worthylake, R., Opresko, L. K., Wiley, H. S. (1999). ErbB-2 amplification inhibits down-regulation and induces constitutive activation of both ErbB-2 and epidermal growth factor receptors. *J. Biol. Chem.* **247**, 8865–8874.

Wu, X., Rubin, M., Fan, Z., DeBlasio, T., Soos, T., Koff, A., and Mendelsohn, J. (1996). Involvement of p27KIP1 in GI arrest mediated by an anti-epidermal growth factor receptor monoclonal antibody. *Oncogene* **12**, 1397–1403.

Yamaguchi, K., Shirakabe, T., Shibuya, H., Irie, K., Oishi, I., Ueno, N., Taniquchi, T., Nishida, E., and Matsumoto, K. (1995). Identification of a member of the MAPKKK family as a potential mediator of AS TGF-BETA signal transduction. *Science* **270**, 2008–2011.

Yarden, Y., and Sliwkowski, M. S. (2001). Untangling the ErbB signaling network. *Nat. Rev. Mol. Cell. Biol.* **2**, 127–137.

Yatabe, Y., Suzuki, R., Tobinai, K., Matsuno, Y., Ichinohasama, R., Okamoto, M., Yamaguchi, M., Tamaru, J., Uike, N., Hashimoto, Y., *et al.* (2000). Significance of cyclin D1 overexpression for the diagnosis of mantle cell lymphoma: A clinicopatholoic comparison of cyclin D1-positive ML and cyclin D1-negative MCL-like-B-cell lymphoma. *Blood* **95**, 2253–2261.

Yezhelyev, M. V., Koehl, G., Guba, M., Brabletz, T., Jauch, K. W., Ryan, A., Barge, A., Green, T., Fennell, M., and Bruns, C. J., *et al.* (2004). Inhibition of SRC tyrosine kinase as treatment for human pancreatic cancer growing orthotopically in nude mice. *Clin. Cancer Res.* **10**, 8028–8036.

Yu, Y., and Sato, J. D. (1999). MAP kinases, phosphatidylinositol 3-kinase, and p70 S6 kinase mediate the mitogenic response of human endothelial cells to vascular endothelial growth factor. *J. Cell. Physiol.* **178**, 235–246.

Zhang, Q., Thomas, S. M., Xi, S., Smithgall, T. E., Siegfried, J. M., Kamens, J., Gooding, W. E., and Grandis, J. R. (2004). SRC family kinases mediate epidermal growth factor receptor ligand cleavage, proliferation, and invasion of head and neck cancer cells. *Cancer Res.* **64**, 6166–6173.

Zheng, C. F., and Guan, K. L. (1993). Cloning and characterization of two distinct human extracellular signal-regulated kinase activator kinases, MEK1 and MEK2. *J. Biol. Chem.* **268**, 11435–11439.

CHAPTER

9

Concepts in Pharmacology and Toxicology

RICHARD A. WESTHOUSE AND BRUCE D. CAR

I. Introduction
II. Concepts in Pharmacokinetics
   A. Absorption
   B. Distribution
   C. Metabolism
   D. Excretion
   E. Clearance of Biopharmaceutical Agents
   F. Determination of Compound Pharmacokinetics: *In Vivo* Pharmacokinetic Studies
III. Concepts in Toxicology
   A. Toxicology in Preclinical Drug Development (Clinical Trial Enabling)
   B. Target Validation
   C. Off-Target Effects
   D. Cardiovascular Safety Pharmacology
   E. Genotoxicity
   F. Biopharmaceutical Agents
   G. Vaccines
   H. Risk Assessment
   I. Therapeutic Index
IV. Clinical Concerns for Pharmacology and Safety
V. Conclusion
   References
   Further Reading

An in-depth knowledge of key concepts in pharmacology and toxicology is necessary to optimize and develop any new therapeutic agent. The development of traditional cancer therapeutics (nonselective cytotoxic agents) has evolved little over many years. The traditional test systems as defined by regulatory requirements are increasingly recognized as less scientifically appropriate for the new class of targeted agents that are emerging. These test systems are also of markedly less value in assessing biopharmaceutical oncologic agents (e.g., antibodies, vaccines, proteins, gene therapy drugs, etc.). The routine application of traditional risk-assessment practices is also typically inadequate for these new agents. Thus, additional studies are frequently conducted to provide a more focused science-based risk assessment of the potential toxicology issues. Nevertheless, an understanding of traditional concepts in pharmacology and toxicology remains important because the optimization of targeted drugs must still achieve relevant pharmacology and safety goals, even in the growing number of cases where innovative technologies and nonstandard approaches are employed. This chapter focuses on general concepts and techniques for lead optimization and development of targeted small molecule agents and biopharmaceutical agents, and it is intended to be a primer and resource to direct further study. Specific applications and arising issues for these types of agents are also discussed for illustration.

I. INTRODUCTION

In the pharmaceutical industry, the drug discovery process involves identification of an interventional target and means to manipulate that target, selection of a drug candidate series, optimization within that series, and selection of a lead drug candidate for progression to exploratory development. Drug discovery involves multiple potential drug candidate molecules that, through optimization, will ultimately focus on one or two candidates that are selected as the clinical lead compounds to progress to exploratory development. Exploratory development is comprised of early clinical trials to determine pharmacokinetics and proof of concept/efficacy. The studies are supported and enabled by rigorous nonclinical testing for the purpose of identification of potential toxicities and liabilities of these one or two candidates. Throughout the continuum of discovery and development, the financial investment increases dramatically. From this perspective, if a drug candidate is "destined" to fail, for whatever reason, it is most advantageous that the actual time of the failure occurs at the earliest point possible: late-stage failures are not only costly from the loss of invested resources but also from the loss of future revenue resulting from delayed entry into the market. Pharmaceutical discovery and development traditionally have been organized by segregating tasks, such that safety and pharmacology issues are often noted later in development. This approach leads to a high attrition rate for compounds, particularly in the early synchronous phases of preclinical and exploratory clinical development. The pharmaceutical industry is increasingly recognizing the need for broader, more predictive, and higher throughput profiling of compounds for potential pharmacology and safety issues so that potential liabilities can be identified and resolved earlier in the discovery process. While an increase in such activities may very well increase the duration of drug development, this change can be reasonably viewed as an investment in the future success of compounds (specifically, by decreasing the attrition rate at the development phase).

While knowledge in pharmacology and toxicology is important at every stage of drug discovery and development, the biggest *positive* impact that can be made in these areas is in the drug discovery phase. In this chapter, concepts in pharmacokinetics, pharmacodynamics, and safety/toxicity will be presented from the perspective of early identification of liabilities and lead optimization in drug discovery to minimize identified liabilities. The effective scientific management of lead optimization requires an understanding of general concepts of pharmacology and toxicology, of typical issues in these areas, and of application of specific technologies. The extremely thorough characterization of safety and pharmacology issues in definitive studies that are included in a regulatory package to enable clinical testing is beyond the scope of this chapter; specific guidelines and necessary details can be obtained from each governmental regulatory agency. The concepts and rationale behind these regulatory studies are necessary to understand, however, because they also hold true in the drug discovery phase. Knowledge of the concerns of regulatory agencies will guide the screening and profiling in drug discovery. The studies typically conducted in the development phase are large, cumbersome, and more conducive to validated traditional assays. In contrast, studies in the discovery phase are characterized as higher throughput, more flexible in design, and more conducive to innovative technologies; they may be more effectively used to optimize desirable molecular or biologic properties. Although some issues may be specific to the earlier and later stages of compound progression, in general, similar scientific issues guide the conduct of pharmacology or toxicology studies in either arena.

The discovery and development of traditional (nonselective) cytotoxic small molecules for oncology have adhered to classically regulated paradigms, which allow little room for innovation. The regulatory guidelines for these agents have not significantly evolved since their inception. Because the biology and pharmacology underlying cancer immunotherapeutic agents differ fundamentally from that of cytotoxic oncologic agents, a markedly different profile of issues is observed in discovery and development. In addition to satisfying proscribed regulatory requirements, a sound scientific rationale and innovative thinking must be applied to provide appropriate risk assessment of agents entering clinical trials.

Types of agents used for immunotherapy for cancer could include either small drug-like molecules or biopharmaceuticals. In general, small drug-like molecules or new chemical entities (NCE) have a low molecular weight, are well-defined by physicochemical properties, and are chemically synthesized as a single compound that can be purified and manipulated by a medicinal chemist to alter specific activities. In contrast, biopharmaceutical agents generally have a large molecular weight and complex physicochemical properties, are synthesized by host cell lines (or the entity itself *is* the host cell, as for somatic cell vaccines), may not be highly purified, and cannot be readily manipulated by a medicinal chemist to alter activities. The main classes of biopharmaceutical agents are antibodies (antagonist and activating actions), recombinant proteins (such as cytokines, hormones, growth factors, and enzymes), synthetic oligonucleotides, gene transfer products, and vaccines (protein and cellular such as somatic cell therapy). The approach of this chapter will be to present concepts in pharmacology and toxicology for targeted small molecule therapeutics for cancer and to contrast them with the significant differences existing for specific biopharmaceutical agents. For small molecules, medicinal chemists play a significant role in altering substructures in an effort to alter activities (potency or toxicity), while for biopharmaceutical agents, this approach to improving compound qualities is not possible. Resolving issues with biopharmaceutical agents generally requires an intimate understanding of the perturbed biologic process.

II. CONCEPTS IN PHARMACOKINETICS

Pharmacokinetics (PK) is the study of the disposition of a drug after its delivery to an organism—in short, a study of what "the body does to a drug." Classical pharmacology developed from the investigation of the disposition of small molecule chemical entities, either therapeutic or toxic (for general texts, see Rowland and Tozer [1995] and Ho and Gilbaldi [2003]). The traditional concepts of pharmacokinetics are presented here from the perspective of small molecules. In addition, there is a discussion of the process whereby pharmacokinetic properties are optimized by manipulating the physicochemical properties of a compound. The same general concepts of pharmacology apply to biopharmaceutical agents; however, there are a number of differences in the details, concerns, potential issues, and means of optimization. In this case, the discussion will be limited to recombinant proteins and antibodies, which have been the most commonly developed biopharmaceutical agents requiring pharmacokinetic profiling. Plasma quantification is generally not performed for inactive, nonreplicating vaccines, such as protein/peptide-based vaccines or inactive whole cell vaccines. In general, options for altering the physicochemical characteristics and favorably changing the pharmacokinetics of antibodies and recombinant proteins are more limited than options for altering small molecule chemical entities, and furthermore, the pharmacokinetics of antibodies and recombinant proteins are more dependent on normal biological behavior.

Pharmacokinetic analysis must begin with validated analytical methods to quantify the drug in various *in vivo* and *in vitro* samples. For small molecule agents, this is most commonly, if not exclusively, done by a gas or liquid chromatography (GC or LC) in conjunction with single or dual mass spectrometry (MS). Specific bioanalytical methods should be validated for the relevant sample matrix from each species and should also utilize standard curves for the specific analyte and use relevant internal standards. When stability of the analyte in the matrix is unknown, samples should be analyzed as soon after collection as possible, but it is best to perform stability testing very early in the program. These tests generally involve analyzing samples with concentrations that bracket the relevant concentrations, then reanalyzing the samples after processing and storing. Sample handling should always be consistent across studies so as not to add extraneous variables. Due to the complex nature and inherent heterogeneity of many biopharmaceutical agents, they cannot be quantified typically by GC/MS/MS. Monoclonal antibodies and recombinant proteins are frequently quantified by enzyme-linked immunosorbent assay (ELISA), protein capture ELISA, bioactivity assays, and other such procedures. These assays, which often require more extensive development, should undergo the same rigorous validation as GC/MS/MS methods. Multiple assays may be required to gain a complete assessment of exposure; for example, capture assays for antibodies can bind the active site or the Fc portion, depending on the need to quantify specific circulating complexes bound at these and other sites. Various reviews and opinions on bioanalytical method validation have been published (DeSilva et al., 2003), including guidance by the U.S. Food and Drug Administration Center for Drug Evaluation and Research (FDA-CDER, 2001).

The foundational concept on which pharmacokinetics is built is exposure. Drug exposure of an agent varies over time and is dependent on the amount of drug that is administered, absorbed, and excreted. The amount of a drug that passes to systemic absorption following oral administration is termed the *oral bioavailability*. The variation of drug concentration over time is due to a number of factors grouped into absorption, distribution, metabolism, and excretion (ADME). Exposure is generally most clearly expressed by plasma pharmacokinetic profiles of drug concentration versus time over a 24-hour period. Exposure data is used in various ways depending on the stage of discovery or development. Basic pharmacokinetic investigation is necessary to identify features of ADME that are less than desirable and take steps to remedy them. These data are then used for predictive modeling in deriving a target dose for efficacy in clinical trials. Optimally, there should be a relationship between exposure and efficacy (or surrogate pharmacodynamic effect) and in vitro potency, or a rationale understanding for lack thereof.

A. Absorption

Drug absorption is the process of entry of the drug into the vasculature from the delivery site. Of first concern is the route of delivery. Cytotoxic chemotherapeutics and biopharmaceuticals are almost always administered by the intravenous (IV) route (e.g., cetuximab, rituximab, tositumomab, alemtuzumab, etc.) although administration via other parenteral routes, such as intralesional/intratumor, subcutaneous, and intramuscular, is also possible. The need for close regulation of the clinically administered dose of a cytotoxic chemotherapeutic necessitates the IV delivery route. For biopharmaceutical agents, however, this route is necessary because of a different set of issues. Oral administration is usually not a reasonable option for biopharmaceuticals because their tertiary structure is critical for activity; in addition, these macromolecules are inherently unstable in the acid environment of the stomach and are also vulnerable to proteolytic enzymes in the gastrointestinal (GI) tract (which cause cleavage or breakdown). For such reasons, these macromolecules are rarely, if ever, absorbed well after oral administration.

The IV route provides 100% absorption and bioavailability and is the most pharmacokinetically reproducible. Delivery of biopharmaceuticals by the IV route is rarely an issue except for the inconvenience of controlling the delivery rate by infusion. Occasionally, biopharmaceutical agents or vehicles also carry acute tolerability issues. For small molecule chemical entities that are more frequently administered by the oral route, this delivery is usually more problematic in that it involves passage of drugs through barriers. In oral administration, these barriers include gastroenteric mucus, enteric mucosal epithelial cells, basement membranes, fibrous matrices, and capillary endothelium. Drugs move across cell membranes by simple passive diffusion, filtration, and carrier-mediated transport (either passive or active). Passive carrier-mediated transport is driven by concentration gradients, similar to simple diffusion, but is facilitated by a transport mechanism. Active transport requires energy to move a molecule, and it will work against a concentration gradient. Generally, most drugs penetrate cells by passive transport or diffusion, and physicochemical properties are critical for this process.

Some of the physicochemical characteristics that affect absorption through the GI tract include solubility, the acid dissociation of ionization constant (pKa), and partition coefficient (lipophilicity). Physiological factors, such as gastric pH, gastric emptying, and intestinal motility, should also be considered when optimizing absorption. Optimization of absorption in drug discovery should consider these factors in concert and recognize the interplay among them. Minor substructural alterations to a molecule could have obvious significant impact on absorption. Other means of increasing absorption should include pharmaceutical strategies, such as changes in vehicle and pH, which are usually used to increase the solubility. All too frequently, poor solubility and solubility-limited absorption are the banes of drug discovery. In general, exposure will increase only minimally with increasing suspension doses, and there is little chance that the increased exposure will be linear. If core structures are very lipophilic, substructural changes may only minimally impact solubility. Some strategies for preventing poor solubility include using the so-called prodrug approach, using novel vehicles, or milling the compound to nano-sized particles (thereby increasing the surface area and facilitating absorption).

As part of the descriptive profiling of a lead candidate, cellular permeability can be identified in intestinal mucosal epithelial cells, such as in the widely used Caco-2 cell line. Bidirectional permeability is monitored with varying apical pH to reflect potential *in vivo* luminal differences. Interaction with P-glycoprotein (P-gp), either as an inhibitor or substrate, can also affect absorption. P-gp is an active transport pump in enterocytes that moves xenobiotics back out into the lumen (digoxin is a classic substrate, and verapamil is a classic inhibitor of P-gp). Investigation of potential

P-gp pump effects can be identified in Caco-2 studies by coadministration of digoxin and verapamil into the test system.

Bioavailability (F) is a calculated measurement of the extent to which the active drug is absorbed and available at the action site. Generally, plasma concentration is considered equivalent to availability at the action site. Absolute bioavailability is expressed relative to maximum attainable exposure (i.e., by IV administration, which gives 100% bioavailability):

$$F = \frac{(AUC/dose)\ oral}{(AUC/dose)_{IV}}$$

where AUC is the area under the drug concentration time curve for a specific period of time.

Dose levels for the above derivation should be selected that produce relatively similar AUCs so that similar apparent volumes of distribution and elimination rates are maintained. Differences of these two parameters between the oral and IV doses invalidate the derivation. For example, an IV dose that saturates metabolism, even only initially after dosing, will not give a reliable reflection of bioavailability.

B. Distribution

After a drug is absorbed or delivered directly to the blood, it is distributed throughout the body by the systemic circulation. Drug distribution throughout the circulation is rapid, based on the efficiency and effectiveness of the cardiovascular system. Local high concentrations of blood will only occur in cases of bolus IV injections. Movement of drug from the blood to tissues is relative to concentration in the blood, blood flow, protein binding, penetration, and tissue affinity. This movement or translocation of drug molecules from the blood to tissues throughout the body is the more relevant reflection of distribution. The *volume of distribution* at steady state (Vss) reflects how well the drug molecules leave the blood plasma (vascular fluid) and enter the interstitial fluid (extravascular extracellular fluid) and intracellular fluid. At some time after dosing, the concentration of drug in the bloodstream reaches equilibrium with the extravascular concentration in any given tissue. This is not to imply that the concentrations in these compartments are equal: while the volume of distribution is relative to circulating concentration, penetration and tissue affinity play significant roles in keeping the drug in the vasculature or concentrating it in specific tissues.

A volume of distribution equivalent to blood volume indicates no extravascular distribution to tissues. This may be acceptable if the target is circulating in the blood, such as may be the case with some immunomodulatory or angiogenic therapies. For small molecules, Vss is generally most dependent on physicochemical properties (e.g., charge, lipophilicity, etc.). The distribution of biopharmaceuticals, on the other hand, is usually limited to vascular space and interstitial tissues due to the large size. Some tissues are more difficult to penetrate than others. The brain, for example, has a special barrier called the blood-brain barrier that maintains a restricted environment. In certain diseases, such as cancer, certain barriers are somewhat compromised, and blood vessels may be leaky.

Distribution has implications for the following factors:

- Efficacy: the ability to reach the intended target
- Metabolism: the ability to reach cells that metabolize the drug
- Toxicity: the ability to reach off-target cells

Investigations into the distribution of a drug may be needed when there is a disconnect between expected efficacy, based on *in vitro* or biochemical potency and coverage of that concentration by *in vivo* circulating free fraction, and the actual observed *in vivo* efficacy.

Key factors affecting distribution and potential disconnects as above include hemodynamics, serum protein binding, tissue penetration, and tissue affinity. Hemodynamics is rarely a cause for lack of drug distribution, but it is important to recognize that the blood is the main "distributor" of drug throughout the body. Binding to serum proteins is a common confounding factor that unexpectedly alters the drug's pharmacokinetic profile and biologic activity.

Protein binding and the consequences or implications of such can be dramatically different for small molecules and biopharmaceutical agents. Small molecule therapeutics that are bound to proteins are generally sequestered from pharmacological activity and also clearance. Binding may be irreversible or reversible. Irreversible binding is generally a result of bioactivation of the molecule to a reactive intermediate with covalent binding. Binding of this nature, either to plasma proteins or more frequently to cellular proteins, is usually not of concern regarding sequestering a drug from activity but instead is of concern from the toxicological impact. Metabolic bioactivation to a reactive intermediate, such as in the liver, may result in indiscriminate covalent binding. The damage to intracellular proteins and DNA can result in cell dysfunction, cell death, or mutagenicity. Binding to blood proteins is more frequently reversible and of concern from the perspective of distribution. Binding of this nature by hydrogen bonds or van der Waals forces in the blood effectively holds the drug in the vascular space due to the impermeability of these large complexes to biomembranes. In addition, even if the target is intravascular, steric hindrance from these types of binding effectively sequesters the drug from activity.

High protein binding will not only affect clearance but also the minimum efficacious exposure. Plasma concentrations of small molecule drugs quantified by GS/MS/MS are reported by convention as total drug (i.e., bound and unbound). In order to correlate pharmacokinetics with pharmacodynamics or toxicity, the free unbound fraction must be derived. For example, if a drug is 98% protein bound, only 2% of the drug is free for activity (efficacy, toxicity, and clearance).

Biopharmaceutical agents have a different set of distribution concerns. These agents do not extensively bind to plasma proteins but instead frequently interact with specific endogenous binding proteins that may be involved with transport and regulation. A good starting point in understanding the pharmacokinetics and pharmacodynamics of a biopharmaceutical agent is to understand how the endogenous molecule is distributed and "processed." Recombinant cytokine, hormone, and growth factor therapy can be expected to interact with specific binding proteins and/or receptors similar to their endogenous counterpart. These interactions will alter the pharmacokinetic profile at least to some extent. The specific interactions that occur should be investigated (or known from the biology of the biopharmaceutical agent) in light of its effect on activity, pharmacokinetics, toxicity, bioanalytical method, and ultimate fate of the complex. Lastly, immunogenicity is a significant factor that may affect the pharmacokinetic profile and potentially result in significant secondary undesirable effects.

C. Metabolism

Drug metabolism of small-molecules chemical entities is comprised of a variety of enzymes that facilitate excretion (together referred to as clearance). The reactions of drug metabolism have a common goal of generally making products that have greater aqueous solubility than their precursors. Phase I reactions generally provide functional polar groups to molecules that either facilitate excretion or further metabolism. Phase II reactions are conjugation reactions that add large polar moieties via high energy cofactors or a chemically reactive substrate.

Small-molecule chemical entities can be profiled in assays utilizing recombinant cytochromes P450 (CYPs) of various isoforms and also in hepatocytes in an effort to predict in vivo metabolism via Phase I and Phase I/II reactions, respectively. Additional drug discovery profiling should include assays that alert for inhibition or induction of drug metabolizing enzymes.

D. Excretion

The main routes of excretion for small molecule chemical entities are through the liver and kidney. The biliary system is generally responsible for excretion of larger, less polar molecules. While the hepatic acinus or lobule is the physiologic unit to liver metabolism, the biliary system is the physiologic unit for excretion. The system extends from the canaliculi to the common bile duct emptying into the duodenum and also the intestine and usually works in close association with hepatic metabolism as a preparatory step for excretion.

In the urinary system, the nephron is the physiologic unit responsible for excretion. The glomerulus of the nephron receives blood with a high hydrostatic pressure, which results in ultrafiltration due to 80-Å interendothelial pores. Molecules greater than 69 kDa, such as those bound to plasma proteins, are too large to pass through the filter. Aside from the size and charge of the molecule, the biggest factor affecting the characteristic of the filtrate is glomerular filtration rate or blood flow. Downstream in the nephron, the tubular epithelium is responsible for passive and active reabsorption and the excretion of molecules, water, and ions and is also active in xenobiotic metabolism. Active secretion of xenobiotics (especially organic acid and bases) may take place in the proximal convoluted tubule through specialized transporters. These transporters act on relevant molecules regardless of protein binding. In addition, the proximal tubular epithelial cells contain many of the drug-metabolizing enzymes found in the liver, which results in the facilitation of secretion into the nephron lumen as well as the excretion in the urine or general metabolism and release back into the circulation for further metabolism. Generally, small molecules and polar molecules are preferentially excreted through the nephron.

E. Clearance of Biopharmaceutical Agents

For biopharmaceutical agents, the metabolic processes just described generally do not occur, and the profiling paradigm in discovery is correspondingly impacted. The ultimate disposition of these agents varies widely according to type of drug and the interventional pathway. As already pointed out, the first consideration is that the molecular makeup of the drug will dictate the disposition to some extent. For instance, proteins and peptides will undergo catabolic processing similar to endogenous or dietary proteins. Peptidases will eventually cleave the molecule to amino acids that will be recycled. Alternatively, these drugs may interact with endogenous molecules as part of their mechanism of activity, making them unavailable for proteolysis or recycling. Antibodies undergo this process, not as they relate to the target, but as they relate to normal turnover of endogenous antibodies. The vast majority of antibodies are eliminated by catabolism at sites that are in rapid equilibrium with plasma. Antibody clearance is generally thought to be associated with receptor- and nonreceptor-mediated endocytosis targeted toward the Fc portion of the immunoglobulin. The pharmacokinetics of monoclonal antibodies generated against cellular targets can often be impacted by another set of unique factors, specifically antigen and antibody concentrations, potentially resulting in nonlinear pharmacokinetics (Lobo et al., 2004). High antigen concentration relative to antibody could result in short half-life due to rapid clearance through antigen–antibody processing, while low antigen concentration relative to antibody could result in prolonged half-life. Immunogenicity will also impact the pharmacokinetics, as already pointed out.

Although the available options for manipulating a biopharmaceutical agent for the purpose of optimization of pharmacokinetics are not as numerous as those for the small molecule chemical entity, some strategies have proven useful, such as alterations focused around the Fc portion of an antibody by pegylation (Molineux, 2003; Kozlowski et al., 2001) and by glycosylation (Drickamer and Taylor, 1998) or change in the immunoglobulin subtype. Pegylation and glycosylation have also been used for recombinant proteins. Pegylation—the attachment of polyethylene glycol to a protein—can increase half-life, decrease clearance, lower toxicity, increase stability and solubility, and decrease immunogenicity. Glycosylation modification of proteins, while useful in enhancing protein stability, is limited in its usefulness because glycosylation is an endogenous process that is frequently very species specific, and alteration of it from the norm expressed in human cells will most likely alter the activity and immunogenicity.

F. Determination of Compound Pharmacokinetics: *In Vivo* Pharmacokinetic Studies

Compound pharmacokinetics is determined by collecting blood samples at various times after compound administration and determining plasma concentrations as a function of time. Multiple blood samples should be collected during the absorption phase of exposure and at the expected time of maximum plasma concentration (Tmax) to allow for robust pharmacokinetic derivations. For a small molecule chemical entity, this is typically from 0.5 to 2 hours; for a biologic agent, the Tmax can be days to weeks. Good estimations of Tmax will result in more accurate derivations of maximum circulating levels (Cmax). Initially, plasma is analyzed for quantification of the active pharmaceutical agent. For small molecules, GC/MS/MS is used; for biologic agents, immunoassays may be used (DeSilva et al., 2003). The method or assay should be sensitive enough to detect the molecule at therapeutic levels and at least 10–100 times below this level. The method or assay should also be validated to be useful for the specific matrix (e.g., plasma, serum, etc.,) and species from which the sample is obtained. Sample handling is important when stability of the compound is in question. If little is known about *ex vivo* stability (e.g., on the shelf), then samples should be analyzed as soon as possible after collection. At some point, stability needs to be evaluated and should include processed samples with aliquots stored for analysis for variable durations. In some cases, stabilizing agents may need to be added to samples. This is especially important for prodrugs, which can be metabolized to the active (parent) drug *ex vivo* or during sample handling and storage by plasma enzymes. All prodrugs should be analyzed for the prodrug and the active (parent) drug with sufficiently sensitive methods.

Study design for small molecule chemical entities generally includes an IV route and a route that is clinically relevant. For this discussion, the oral route will be used as the clinically relevant administration. For nonrodents, the same animals are used for both portions of the study (crossover design) for control of interanimal variability. The general range of starting doses should be 10 mg/kg per os (PO) or by way of mouth and 5 mg/kg IV for rats and mice; the doses should be 5 mg/kg PO and 1 mg/kg IV for dogs and monkeys. Food effects on absorption should be controlled by fasting the animals overnight.

The following is a list of data that are generated from pharmacokinetic studies. Note that concentration measurements are total concentration and include both free and bound fractions.

AUC$_{(0-Xh)}$ = area under the curve for specific time 0 to Xh expressed as "micromolar × hour" or "nanogram × hour/milliliter"
Cmax = maximum concentration
Tmax = time of maximum concentration
$t_{1/2}$ = half-life
Vss = volume of distribution at steady state
CL$_{tot}$ = total clearance in milliliters/minute/kilogram
F = bioavailability of the oral route with respect to IV dose as a percent

The pharmacokinetic profile should also be evaluated for linearity with respect to dose. Exposure levels (the AUC) should increase in a relatively linear manner with increasing dose, especially over the dose range of efficacy. If exposure reaches a maximum and cannot increase with dose, there is the chance that dose escalation in clinical trials may not produce increased exposure, and efficacy may not be achieved. Nonlinear pharmacokinetics can result from a variety of causes. All too frequently, maximal exposure is attained due to solubility constraints. Administration of suspensions may sometimes result in marginal increases in exposure, at best. If there is no promise for identifying chemotypes with better solubility, suspensions of nano-sized particles may offer better exposure; however, these techniques are heroic and frequently cost-prohibitive to carry forward to the clinical trials, especially when there is an "at risk" notion associated with them. Other causes of nonlinear kinetics include saturation of metabolic pathways or plasma protein binding resulting in higher than expected exposure with increasing dose. These situations offer the converse concern, specifically that escalation of dosing above a certain point will result in significant toxicity at a steeper curve. In other words, above a certain dose, exposure increases at a faster rate, resulting in more difficult titration of dose with respect to control of toxicity. Nonlinear pharmacokinetics with biopharmaceutical agents is much more common, frequently due in part to interanimal variability that can often be difficult to understand completely (Lobo et al., 2004; Mahmood et al., 2005; Tang et al., 2005).

The data generated from in vivo pharmacokinetic studies are an important factor used in predicting human pharmacokinetics and efficacious human dose. Other data that factor into this calculation include efficacious exposure in the animal model of disease and species differences in metabolism and protein binding that have been identified by *in vitro* methods. Human dose projection assumes that the exposure (generally as AUC$_{0-24h}$) that results in efficacy in a rodent xenograft study will equal the therapeutic exposure in patients. The exposure associated with the minimal efficacious dose in rodent xenograft studies is used to rank different compounds and is a frequent starting point for a variety of human dosing projections. In general, the plasma drug available for interacting with the pharmacologic target is the unbound fraction. Highly protein-bound drugs (e.g., binding at 99%) frequently require much greater total bound and unbound exposures compared to drugs for which plasma protein binding is less (e.g., 90%). It is generally most accurate in projecting activities across species, such as from rat to man, to refer to the unbound fraction of the drug because protein binding may differ substantially between species.

Although widely used, the minimum efficacious AUC is a simplistic and potentially flawed tool for comparing drugs. The more robust parameter used in drawing correlations between efficacy and drug exposure is the *minimal time for target inhibition* (receptor or otherwise) during the dosing period (e.g., 16 hours of 100% inhibition within a 24-hour period) that is sufficient to produce tumor shrinkage or regression. Given the practical difficulty in obtaining such information, the simple AUC is the most relevant because these data consider concentration and time. In some circumstances, Cmax may be more relevant, such as in cases of very tight and/

or irreversable binding to the target (where time is not as critical a factor), or in some cases of toxicity, such as cardiac ion channel liabilities.

III. CONCEPTS IN TOXICOLOGY

A. Toxicology in Preclinical Drug Development (Clinical Trial Enabling)

At the preclinical level, the concepts and relevant profiling techniques of toxicology are critical to the process of lead selection optimization in drug discovery. The ultimate purpose of preclinical safety testing in the pharmaceutical industry is to predict the safety of a compound in humans. The nonclinical studies that enable clinical trials (i.e., those that are submitted to governmental regulatory agencies to allow progression in clinical dosing) must assure physicians running these studies that the potential adverse events are kept to a minimum or can be identified with some confidence before they become life threatening (or otherwise significant based on the benefit that the patient will be receiving). These studies can be broad in the sense that there is a complete multisystem evaluation of animal physiology and pathology, or they can be very focused toward the identification of a specific potential liability. One example of this is the cardiovascular safety pharmacology studies (discussed in Section III. E). These, and other specialized studies, have evolved to be required in regulatory submissions due to the increasing occurrence of drug-related adverse events and have been associated with multiple drug withdrawals.

In the process of enabling clinical studies, these nonclinical studies should (i) define a *no-effect dose*, (ii) identify and characterize *safety issues*, and, if there are any, (iii) develop a robust *risk assessment plan* for human trials so that the risk-benefit can be carefully evaluated. Naturally, possible safety issues identified by nonclinical studies will drive the design and conduct of clinical studies. In addition, the no-effect dose level identified in nonclinical studies will be a significant factor in defining the starting dose in the initial clinical study. For cytotoxic chemotherapeutic agents, there are mathematical equations for a safe starting dose based on the severely toxic dose to 10% (STD10) of the animals or maximum tolerated dose (MTD); however, for the targeted or noncytotoxic cancer therapeutics, human starting doses are calculated based on the no-effect or no-adverse effect levels (similar to the derivation and strategy used in noncancer therapies). The rationale for this is that these cancer drugs are often chronically administered and must, therefore, be safer than cytotoxics.

Further, the use of novel biomarkers designed around a specific toxicity identified in the nonclinical studies is valuable for early detection in clinics and would potentially give the clinician more confidence to accelerate dose escalation. Therefore, pharmaceutical companies are increasingly focusing on the identification biomarkers for potential liabilities. Moreover, nonclinical data might suggest that certain populations of patients should be excluded from studies because they pose a higher risk of developing the adverse event, such as by having concurrent diseases or suboptimal organ function that might predispose or exacerbate a toxicity. The specific details of the content of the preclinical portion of a regulatory submission and its conduct are beyond the scope of this chapter; for further information in this area, the reader is encouraged to consult information published by various regulatory agencies and specialized review articles available (Snodin *et al.*, 2006).

When a safety concern is identified and characterized, nonclincal studies should also make at least an attempt to develop mechanism-based risk assessment (see Section III.I) and also identify potential biomarkers. Increasingly, the specific alerts for potential safety concerns to clinicians are not enough. Perspective from robust risk assessment is becoming necessary.

B. Target Validation

Target validation can be proactive and performed early in the discovery process as part of the process to understand potential undesirable effects, or it can be performed as a reactionary measure over concerns of a safety issue that arises later in the discovery/development process. The former is obviously more desirable. Supraphysiologic effects are expected effects that occur via the target but at extreme levels. Molecular targets for cancer therapy that have been identified by genetic analysis may be, if not frequently, expressed or active in other cells types or tissues. If these mechanisms are critical to normal functioning of these tissues, intervention in these pathways will produce an undesired target-related effect. Defects in cell-mediated immunity have been reported in patients with imatinib-treated chronic myelogenous leukemia, which may be a result of blockade c-ABL signaling in T lymphocytes (Zipfel et al., 2004). Similarly, cutaneous side effects have been observed and noted to correlate with efficacy in patients treated with epidermal growth factor receptor (EGFR) inhibitors, such as erlotinib and cetuximab. However, cutaneous effects have also been observed in patients treated with agents targeting other receptor tyrosine kinases, such as imatinib, sorafenib, and sunitinib (Robert et al., 2005). Target validation begins with an understanding of the current science of the target and signal transduction pathway. Phenotype characterization of the genetically engineered mice, such as knockout and transgenic mice, could prove useful, given caveats that target expression levels could be different in humans. Inducible knockout mice are thought to be more relevant to extrapolating the therapeutic scenario, but such strains of mice may be more difficult to obtain or develop.

If a target-related liability is identified in this manner, identification across species will be necessary for risk assessment. Target-related undesirable effects cannot be resolved by modification of drug molecular structure or exposure (unless one can gain selectivity through target subtypes). Targeted delivery approaches, such as conjugation or prodrug approaches, may sometimes help to concentrate the active drug in the desired cells (Cavallara et al., 2006).

If the effect is not significantly adverse, it can actually be exploited as a surrogate biomarker for efficacy in clinic and/or controlled by other means in clinic. One example of this situation is the pharmacological hypertension associated with inhibition of VEGF activity by bevacizumab (Gordo et al., 2005). Evaluation of tumor hemodynamics and perfusion (as by dynamic contrast enhanced magnetic resonance imaging) has been used as a surrogate biomarker for therapeutic response to VEGF inhibitors. With these data, the pharmacodynamic effect can be assessed immediately in the clinical study in lieu of a tumor response (which takes months to collect). This strategy offers the benefit of potentially accelerating clinical trials. Sometimes the management of adverse effects can be achieved by dosing schedules with "dose holidays" that allow some recovery of the normal tissues while activity is still maintained at the intended site.

In order to adequately identify potential target-related effects, the presence of the target in the test species is obviously critical. The activity of a compound designed against a human target may be substantially less in a species whose homolog differs significantly at the site of drug interaction. The activity of a biopharmaceutical agent can frequently be species specific although agents active in humans are often also active in primates. Regardless of the safety model used especially for biotherapeutics, some component of a pharmacodynamic effect should be confirmed.

C. Off-Target Effects

Another objective of drug safety testing is the obvious desire to identify and characterize off-target effects. Once an *in vivo* effect that is potentially not target related is identified, it is critical to investigate the effect from a mechanism-based approach to define the relative risk of that effect occurring in clinical trials (risk assessment). Off-target effects (for targeted noncytotoxic small molecule and biopharmaceutical agents) generally occur due to lack of specificity and selectivity or metabolic activation. For targeted noncytotoxic small molecules, such as kinase inhibitors, the mechanism of these effects can sometimes be identified, and the pharmaceutical chemist can modify substructural components of the molecule to achieve better selectivity. Performing this *in vivo* can be time-consuming and expensive, so identification of the mechanism and development of an *in vitro* assay or at least an acute *in vivo* platform will facilitate the process. Through trial and error, the pharmaceutical chemist and biologist can narrow down substructural requirements of the structure–activity relationships. General profiling of the selectivity of the lead compound in a standard receptor and/or kinase panel (where relevant) may help in suggesting a mechanism or even alert one to potential impending liabilities. A common tool to help discriminate whether an *in vivo* effect is target related or off-target (including through metabolic activation) is to administer the stereoisomer (if the small molecule is chiral).

D. Cardiovascular Safety Pharmacology

Drug-related effects on cardiovascular physiology (electrical conduction and hemodynamics) are increasingly responsible for drug withdrawals from the market (e.g., terfenadine and cisapride). Small molecule chemical entities are required to have rather extensive evaluation of cardiovascular physiology (known as safety pharmacology) as part of most regulatory submissions. There is a heightened level of regulatory concern for the potential of drugs, such as terfenadine, to induce potentially fatal arrhythmias, namely Torsades de Pointes. This arrhythmia has been associated with a specific electrocardiographic finding (termed the *prolonged QT interval*) that is secondary to inhibiting the potassium ion channel hERG (human ether-a-go-go, also known as IKr). Briefly, the hERG channel is responsible for repolarization of the cardiac ventricle. This repolarization is manifested on the electrocardiogram (ECG) as the T wave. When repolarization is delayed, the T wave is lengthened from the Q wave (ventricular depolarization) and leads to a prolonged QT interval. This ECG finding and mutations in the hERG channel have been associated with fatal arrhythmias in humans, a condition that manifests as the clinical syndrome of Torsades de Pointes. The definitive cardiovascular safety pharmacology study is the conscious telemetrized dog or nonhuman primate model. In discovery, *in vitro* hERG channel and Purkinje fiber electrophysiology assays have shown very good concordance with the risk of prolonged QT interval in humans. Regulatory agencies carry a heightened concern for the exceedingly rare risk of Torsades de Pointes in patients with drug-induced moderate QT prolongation. In addition, despite the focus on hERG channel interactions, their relevance to risk assessment compared to the risk assessment associated with the interactions of drugs with the cardiac myofiber Na^+ channel is minimal. While hERG binding predisposes an individual to a rare arrhythmia, Na^+ channel inhibition directly compromises cardiac output in an already seriously compromised patient group. Through the authors' experience at Bristol-Myers Squibb Company, hERG and Na^+ channel interactions of compounds tested occur with a comparable frequency (Westhouse and Car, 2007).

Cardiovascular safety pharmacology is generally not considered a significant concern for biopharmaceuticals and cancer; however, there may be increasing concern for potential nonspecific binding via the Fc portion of antibodies to cardiac tissue, given emerging evidence of significant adverse cardiovascular effects of the breast cancer drug Herceptin®.

E. Genotoxicity

Genotoxicity has historically not been a concern for the development of cytotoxic chemotherapeutic agents, which by nature are thought to eradicate tumor cells by causing DNA damage. Thus, these agents are frequently positive in *clastogenicity* and *mutagenicity* assays largely due to the DNA-alkylating and nucleotide-substituting mechanisms of action. Genotoxicity tests for such agents are still performed, not to protect the patient population but rather to evaluate the potential hazards for those who may be environmentally exposed. Targeted chemotherapeutics that are noncytotoxic are also frequently positive in clastogenicity assays, as agents that disrupt the cell cycle consistently yield false positives. These agents should not, however, be positive in mutagenicity assays. Given the risk–benefit of oncologic indications, genotoxic liabilities in cytostatic agents are likely to be tolerated; however, in the context of a chronic administration regimen, such features may be deemed undesirable in the future, especially as nongenotoxic alternatives become available. Two examples of well-tolerated small molecule drugs that are chronically administered are imatinib and dasatinib, used for the treatment of chronic myelogenous leukemia.

To evaluate genotoxicity, small molecule therapeutics should be screened in assays such as the reverse mutation assay in *Salmonella typhimurium* and *E. coli* (Ames test), the rat or mouse micronucleus assay, and the Chinese hamster ovary (CHO) cytogenetic assay (or other tests for cytogenetic aberrations). Other shorter, higher throughput assays can be used earlier in the discovery process to investigate structure–activity relationships for compound optimization to detour around unwanted genotoxic liabilities. In addition, computer models are available to evaluate chemical structures for potential mutagenicity and carcinogenicity liabilities based on substructural components and literature.

Biopharmaceutical agents are not expected to carry any genotoxicity liability. Consequently, these agents are not generally tested for genotoxic liabilities unless they contain "unnatural" amino acids or other substitutions that may have unknown properties.

F. Biopharmaceutical Agents

Biotechnology-derived and biopharmaceutical agents have different sets of toxicological concerns compared to small molecules. The vast array of products is so broad that it is beyond the scope of this chapter to discuss the details of typical safety concerns of each. A few general concepts, however, will be applicable to most of these agents and will provide an impetus to the reader to seek out any further details of interest. It is important to stress the need for all studies to have a scientifically sound and rational design that includes consideration of the biology of the agent, the model, and the endpoint. In particular, studies must be conducted in species that express the target and that show relevant pharmacologic effects. Government regulatory agencies have published guidelines for safety assessment for some of these specific types of agents and also text on more general thoughts covered under "Points to Consider." Other points to consider have been developed to cover this class of biopharmaceutical agents more generally. These guidelines focus on the details of what definitive studies should be performed in support of a regulatory submission for clinical studies for each agency (Snodin *et al.*, 2006). In drug discovery, although specific details and in-depth guidelines may not be necessary, the concepts and concerns of regulatory agencies are absolutely applicable to driving the discovery process of optimization.

A major safety concern for any biopharmaceutical agent (e.g., antibody, recombinant protein) in clinical studies is *immunogenicity*. The generation of human antibodies to the therapeutic agent has significant consequences regarding safety as well as pharmacokinetics. The immunogenic potential of a biopharmaceutical can be evaluated most readily as an ancillary part of an *in vivo* study. Immunogenicity is generally related to intrinsic properties of the protein (analogous or heterologous protein), addition of conjugates to the protein (pegylation, etc.), or impurities either in the protein or its formulation (aggregates, fragments, etc.). Evaluating immunogenicity in a preclinical species has significant implications in the design of later studies to support clinical testing. Formation of neutralizing antibodies would remove that species from repeat-dosing studies because this effect eliminates the activity of the protein. In addition, neutralizing antibodies may also alter the pharmacokinetics, cross-react with other endogenous proteins, form immune complexes that deposit in tissues, or cause anaphylaxis or injection site reactions. These are major concerns because experience unfortunately has shown that the extrapolation of immunogenicity from animal studies to humans is poor for all species.

Other safety concerns that are of special note to human clinical studies are cytokine release reactions and anaphylaxis. Although anaphylactic reactions can occur with both small molecule therapeutics and biopharmaceuticals, the latter tend to be of greater concern, especially involving agents that are immunomodulatory in nature. Making this issue more complicated, anaphylaxis may be difficult to distinguish from cytokine release reactions (Greenberger, 2006).

G. Vaccines

In general, the safety issues for vaccines in cancer—whether inactive protein subunits or somatic cell therapies—usually relate to the antibody- or T cell-mediated immune response or to the adjuvants/immunomodulators included rather than to the vaccine antigen itself (Brennan and Dougan, 2005). Traditional adjuvants (where there is a strong knowledge concerning safety) usually do not have to be independently evaluated for safety. Novel synthetic adjuvants, however, will likely have to be tested, along with excipients or preservatives in traditional safety evaluations used for new chemical entities (NCEs). Protein-based or biotechnology-driven adjuvants, such as cytokines, should be evaluated for safety based on their mechanism of action similar to these types of products that are used for any indication. One main focus of these tests is to identify the implications of supraphysiologic responses. As with any biopharmaceutical agent, the use of a selected animal model for autoimmunity assessment requires use of a species-selection based rationale where the animal homolog of the human target is present (e.g., a tumor-associated antigen that is also present on normal cells). Using transgenic animals or a product related to the preclinical species homolog may be useful; however, data obtained in such settings should be interpreted cautiously. Along with traditional endpoints in safety testing, special attention should be focused on characterizing the nature and duration of the immune response to the vaccine. Regulatory agencies are also increasingly requiring studies that demonstrate protection against a challenge, not just that the vaccine administration results in immunogenicity (CPMP, 1997).

H. Risk Assessment

Risk assessment of all therapeutic agents involves factors that are weighed against toxicity. These include the disease itself (concerning life expectancy, quality of life, etc.) and its current standard of treatment, the mechanism of activity, recoverability from the treatment course, and other such factors. Overall, the benefit of the drug for the patient in mitigating or curing the disease should outweigh the risk associated with the toxicity. In light of this, although classical cytotoxic chemotherapeutic agents carry significant adverse effects, the potential benefit gained from their use far exceeds the risk of side effects. This acceptance has been absolutely necessary because the adverse effects are mechanism based. Cytotoxic chemotherapeutic agents have widespread effects on actively dividing cells (neoplastic cells, enterocytes, bone marrow cells, hair follicle cells, etc.). Although the affected cell type is different, the mechanism is the same, and the severity of effects track across cell types. In other words, increased toxicity will correlate with increased efficacy. Such adverse effects are acceptable because of unmet medical need, and the effects are reversible after the treatment course is completed. Additional mechanism-based adverse effects of these drugs include genotoxicity and teratogenicity, which are acceptable based on life expectancy and ability to control the risk, respectively. Cancer immunotherapies differ in that they do not seek to indiscriminately kill cells but rather to elicit specific responses exclusively or predominantly in the tumor cell population or in a cell population that is non-neoplastic but that supports the tumors. Although these therapies may be perceived to be safer (or have less risks or side effects), all untoward effects must be considered in light of the clinical benefit.

I. Therapeutic Index

The therapeutic index is a useful means for judging the potential of a pharmaceutical agent to show adverse changes in a treated patient population at effective doses. The therapeutic index is a ratio of the exposure associated with toxicity to that associated with efficacy.

The therapeutic index range for cytotoxic chemotherapeutic agents is always less than or equal to one, given their mechanism of action. This was acceptable to the patient population because of the lack of alternative therapies, recoverability of the adverse effects, and the duration of treatment. For targeted noncytotoxic small molecule therapies, such as dasatinib, erlotinib, imatinib, and Sunitinib, where the patients are being treated chronically, serious adverse events could not be tolerated; however, these compounds also frequently exhibit narrow therapeutic indices. Thus, one needs to identify a suitable safety range considering various aspects of the clinical disease and the potential adverse effects.

The overall safety of the drug will dictate other aspects of the clinical development program, the most notable of which is the human population selected for phase I trials (first dosing of the drug in humans). Because cytotoxic agents lack therapeutic margins, these drugs were not dosed in healthy volunteers but were tested directly in patients. Noncytotoxic small molecules and biopharmaceutical agents with improved tolerability have employed healthy volunteers in phase I human studies, such as for dasatinib. The biggest advantage to including a study with normal individuals is the generation of reliable exposure and tolerability data without confounding factors, such as advanced disease or concurrent drug administration that can affect endpoints. Obviously, however, the drug must be demonstrated to be very safe so that normal subjects are not subject to unnecessary risk.

IV. CLINICAL CONCERNS FOR PHARMACOLOGY AND SAFETY

The composite nonclinical pharmacology and toxicology data sets generated in discovery and development support the clinical administration of oncologic agents. The efficient conduct of a clinical program is impacted by definition of the potential for efficacy, precision of the human dose projection, and a clear knowledge of the potential safety liabilities in patients. The preclinical models used to predict clinical liabilities differ widely in their accuracy. Therefore, concordance with clinical liability should be understood when possible, and an appropriate level of confidence should be determined based on the model(s) used. Preclinical assays will not always completely resolve an issue or put it into relevant perspective. However, the value of such assays may at least serve as an alert to the clinical development teams of potential issues in human testing. The potential for drug–drug interactions exemplifies such early potential liability assessment. Cancer therapeutic regimens almost always include multiple drugs. Interactions between drugs can have profound consequences by either decreasing the efficacy or increasing the toxicity. This finding can happen simply by two drugs causing similar target organ toxicity, even by different mechanisms; for example, trastuzumab and anthracyclines could both cause cardiotoxicity (Slamon et al., 2001). In this case, the toxicity is known to be additive. True drug–drug interactions arise when the one drug actually affects the exposure of the other. This finding occurs when one drug affects the metabolism of the other; for example, if both drugs are metabolized by the same cytochrome P450 isozymes, then administration of both may result in saturation of the metabolizing enzymes and increased exposure of one or both drugs. In addition, if a drug is an inducer or inhibitor of a cytochrome P450 enzyme, there will be relative consequences in the metabolic rate and resulting exposure. These interactions must be borne in mind in cases such as those involving cancer treatment, for which combination therapy represents the standard of care.

V. CONCLUSION

The key elements to initiating and developing clinical plans to test therapeutic oncologics depend on carefully conducted nonclinical pharmacokinetic, pharmacological, and toxicologic tests and on an integrated understanding of potential liabilities. Within the broader discipline of toxicology, knowledge from traditional cytotoxics and current accumulating practical concepts from targeted cytostatic agents increasingly demonstrates that the approach to safety assessment of novel agents should be based more on scientific soundness than traditional approaches, which have traditionally tended to be inflexible in cancer trials. Identifying potential target-based adverse reactions early in discovery, while compounds can be optimized to reduce development cost and time, is important to reduce attrition of compounds at much later stages when they advance through development.

References

Brennan, F. R., and Dougan, G. (2005). Non-clinical safety evaluation of novel vaccines and adjuvants: New products, new strategies. *Vaccine* **23**, 3210–3222.

Cavallara, G., Mariano, L., Salmoso, S., Caliceti, P., and Gaetano, G. (2006). Folate-mediated targeting of polymeric conjugates of gemcitabine. *Int. J. Pharm.* **307**, 258–269.

Center for Drug Evaluation and Research (CDER) and Food and Drug Administration (FDA). (2001). *Bioanalytical Method Validation: Guidance for the Industry*. Rockville, Maryland: CDER and FDA.

Committee for Proprietary Medicinal Products (CPMP). (1997). *Notes for Guidance on Preclinical Pharmacological and Toxicological Testing of Vaccines*. CPMP/SWP/465/95.

DeSilva, B., Smith, W., Weiner, R., Kelley, M., Smolec, J., Lee, B., Khan, M., Tacey, R., Hill, H., and Celniker, A. (2003). Recommendations for the bioanalytical method validation of ligand-binding assays to support pharmacokinetic assessments of macromolecules. *Pharm. Res.* **22**, 1425–1431.

Drickamer, K., and Taylor, M. E. (1998). Evolving views of protein glycosylation. *Trends Biochem. Sci.* **23**, 321–324.

Greenberger, P. A. (2006). Drug allergy. *J. Allergy Clin. Immunol.* **117 Suppl 2**, S464–S470.

Ho, R. J. Y., and Gibaldi, M. (2003). *Biotechnology and Biopharmaceuticals: Transferring Proteins and Genes into Drugs*. Hoboken, New Jersey: John Wiley & Sons.

Kozlowski, A., Charles, S. A., and Harris, J. M. (2001). Development of pegylated interferons for the treatment of chronic hepatitis C. *BioDrugs* **15**, 419–429.

Lobo, E. D., Hansen, R. J., and Balthasar, J. P. (2004). Antibody pharmacokinetics and pharmacodynamics. *J. Pharmaceut. Sci.* **93**, 2645–2668.

Mahmood, I., Green, M. D., and Fisher, J. E. (2003). Selection of the first-time dose in humans: Comparison of different approaches based on interspecies scaling of clearance. *J. Clin. Pharmacol.* **43**, 692–697.

Molineux, G. (2004). Pegylation: Engineering improved biopharmaceutical for oncology. *Pharmacother.* **8 Pt 2**, 3S–8S.

Robert, C., Soria, J. C., Spatz, A., Le Cesne, A., Malka, D., Pautier, P., Wechsler, J., Lhomme, C., Escudier, B., Boige, V., *et al.* (2005). Cutaneous side-effects of kinase inhibitors and blocking antibodies. *Lancet Oncol.* **6**, 491–500.

Rowland, M., and Tozer, T. (1995). *Clinical Pharmacokinetics: Concepts and Applications*. Manchester, United Kingdom: Lippincott, Williams, & Wilkins.

Slamon, D. J., Leyland-Jones, B., Shak, S., Fuchs, H., Paton, V., Bajamonde, A., Fleming, T., Eiermann, W., Wolter, J., Pegram, M., *et al.* (2001). Use of chemotherapy plus a monoclonal antibody against HER2 for metastatic breast cancer that overexpresses HER2. *New Engl. J. Med.* **344**, 783–792.

Tang, L., Persky, A. M., Hochhaus, G., and Meibohm, B. (2004). Pharmacokinetic aspects of biotechnology products. *J. Pharmaceut. Sci.* **93**, 2184–2204.

Westhouse, R. A., and Car, B. D. (2007). Personal observation.

Zipfel, P. A., Zhang, W., Quiroz, M., and Pendergast, A. M. (2004). Requirement for Abl kinases in T cell receptor signaling. *Curr. Biol.* **14**, 1222–1231.

Further Reading

European Medicines Agency http://www.emea.eu.int/

Flessner, M. F., Lofthouse, J., Zakaria, E. R. (1997). *In vivo* diffusion of immunoglobulin G in muscle: Effects of binding, solute exclusion, and lymphatic removal. *Am. J. Physiol.* **273**, H2783–H2793.

Gorgon, M. S., and Cunningham, D. (2005). Managing patients treated with bevacizumab combination therapy. *Oncology* **69 Suppl 3**, 25–33.

Mahmood, I. (2003). Interspecies scaling of protein drugs: Prediction of clearance from animals to humans. *J. Pharmaceut. Sci.* **93**, 177–185.

Muller, R. H., and Keck, C. M. (2004). Drug delivery to the brain—realization by novel drug carriers. *J. Nanosci. Nanotechnol.* **4**, 471–483.

Snoden, D. J., and Ryle, P. R. (2006). Understanding and applying regulatory guidance on the preclinical development of biotechnology-derived pharmaceuticals. *Biodrugs* **20**, 25–52.

United States Food and Drug Administration, Center for Drug Evaluation and Research. http://www.fda.gov/cder

CHAPTER

10

Cancer Immunotherapy: Challenges and Opportunities

ANDREW J. LEPISTO, JOHN R. MCKOLANIS, AND OLIVERA J. FINN

I. Introduction
II. Prerequisites for Effective Cancer Immunotherapy: Identifying Tumor Antigens
III. Adoptive ("Passive") Immunotherapy
   A. Monoclonal Antibody Therapy
   B. T Cell Therapy
IV. Active-Specific Immunotherapy: Vaccines
V. Cancer-Induced Immunosuppression Impinges on Immunotherapy
   A. T Cell Suppression in Cancer and Immunotherapy
   B. APC Dysfunction in Cancer and Immunotherapy
   C. Regulatory T Cells in Cancer Immunotherapy
VI. Cancer Immunotherapy in Mice Versus Humans
VII. Immunotherapy and Cancer Stem Cells
VIII. Autoimmunity Resulting from Cancer Immunotherapy
IX. Conclusion and Future Considerations
   References

This chapter covers approaches to cancer immunotherapy and how they are affected by the immunosuppression imposed on the cancer patient by the growing tumor, by standard therapies such as radiation and chemotherapy, and by the advanced age of the patient. Successes and failures of both passive and active immunotherapy are highlighted with special emphasis on monoclonal antibody therapy and cancer vaccines. The discussion addresses different types of immunosuppression that occur during tumor development, including T cell and antigen-presenting cell (APC) dysfunction, induction of regulatory T cells, and production of immunosuppressive cytokines, and explains the negative influence they exert on the therapeutic potential of immunotherapy. The chapter also reviews modifications to the immunotherapy protocols that have been proposed and, in some instances, have already been applied to lessen the negative effects of cancer and increase the therapeutic value of immunologic approaches to cancer. Modifications include applying immunotherapy early in disease or in the setting of minimal residual disease, targeting cancer stem cells as well as mature cancer cells, depleting regulatory T cells, and neutralizing immunosuppressive cytokines. Finally, this chapter discusses the need for commonly accepted criteria and surrogate markers that will allow for interlaboratory comparisons of new immunotherapy approaches and that will serve to identify effective antitumor immunotherapies for further development.

I. INTRODUCTION

Treatment of cancer remains a daunting task. Standard treatments, such as chemotherapy and radiation therapy, are either not effective or result in only a temporary tumor regression while causing excessive toxicity to normal tissues. Immunotherapy is a promising nontoxic cancer therapy that is being seriously considered either as a sole form of therapy or in conjunction with other therapies. A quandary with immunotherapy is that the tumors are clearly capable of suppressing and subverting host immune responses, making attempts to bolster host immunity very challenging. It is therefore imperative to develop new immunotherapy approaches in such a way that they contain specific components that can counteract the immunosuppressive nature of the tumor environment. Neutralization of suppressive cytokines (e.g., transforming growth factor β [TGF-β]), made by tumor cells, elimination of regulatory T cells, and restoration of defective T cell and antigen-presenting cell (APC) functionality are essential for successful immunotherapy.

Numerous studies in cancer patients have clearly shown the presence of cellular and humoral immune responses against their tumors, but these responses are, for the most part, weak and apparently ineffective. They nevertheless provide evidence for the existence of tumor antigens and provide cells and antibodies as reagents that can lead to their identification. The expectation is that tumor antigens can be targeted by various immunotherapy approaches or used to augment antitumor immunity by various immune manipulations, including vaccines. Many different immunotherapy approaches have been attempted over the years, and some, such as monoclonal antibody therapy and to some extent adoptive T cell therapy (both referred to as "passive immunotherapy"), have been fairly successful. Other approaches directed toward activating the patient's own immune system, such as therapeutic cancer vaccines ("active immunotherapy"), have not had as much success. The major barriers to a more successful passive and active cancer immunotherapy is the state of immunosuppression found in the cancer patient that is induced by the cancer itself, by the standard chemotherapy and radiation treatments, or simply through immunosenescence due to advanced age of most cancer patients. Current efforts in cancer immunotherapy research are directed toward improving the immunotherapeutic reagents but also understanding the various mechanisms of immunosuppression and discovering ways to improve the immune state of the patient.

II. PREREQUISITES FOR EFFECTIVE CANCER IMMUNOTHERAPY: IDENTIFYING TUMOR ANTIGENS

Early studies in rodent models showed that tumors can be immunogenic and can elicit a protective immune response in autologous hosts (Klein *et al.*, 1960). Many research studies were subsequently conducted to define the molecular nature of antigenic differences between cancer cells and normal cells, especially after the advent of monoclonal antibodies that allowed their isolation and purification. It took three decades, however, to characterize the first mouse (De Plaen *et al.*, 1988) and, shortly thereafter, human tumor antigens recognized by T cells (Barnd *et al.*, 1989; van der Bruggen *et al.*, 1991).

Close to 100 well-characterized tumor antigens are currently part of the immunotherapy armamentarium (Graziano and Finn, 2005). They have been broadly classified into several categories:

1. **Cancer testis antigens**, such as melanoma antigen (MAGE) family or NY-ESO1, found on the majority of histologically different tumors and in normal spermatocytes and placental cells
2. **Tissue differentiation antigens**, such as tyrosinase and prostate-specific antigen (PSA), shared between tumor cells and the normal cells from which the tumor arose
3. **Overexpressed antigens**, such as human epidermal growth factor receptor 2 (HER2)/neu and mucin 1 (MUC1), found at low levels in normal tissues and high levels in tumors
4. **Unique tumor antigens**, such as p53, ras, and bcr-abl, arising from mutations, deletions, and translocations in normal genes
5. **Viral antigens**, derived from oncogenic viruses, such as human papilloma virus (HPV), hepatitis B virus (HBV) and Epstein-Barr virus (EBV), known to cause human cancers

Immunotherapy based on many of these molecules has been tested in animal models, and many have also been targeted in clinical trials as well. Some of the more successful efforts will be described in the following paragraphs. While having good tumor antigens is a great advantage when considering the potential efficacy and safety of immunotherapies based on these antigens, it must be pointed out that they have all been defined by their expression on mature tumor cells. The realization of the existence of tumor stem cells (reviewed later in the chapter) that give rise to recurrent tumors, mandates confirmation that these tumor antigens are also expressed on tumor stem cells. This is especially critical in view of the resistance of quiescent, nondividing cancer stem cells to standard therapies that are directed against dividing mature tumor cells. Immunotherapy, on the other hand, does not target dividing cells but rather any and all cells that express a particular antigen. Immunotherapy may turn out to be the only therapy that is effective against cancer stem cells.

III. ADOPTIVE ("PASSIVE") IMMUNOTHERAPY

A. Monoclonal Antibody Therapy

With the development of hybridoma technology that could produce monoclonal antibodies (mAbs) to target a single epitope (Kohler and Milstein, 1975), there was much hope that such antibodies directed to tumor antigens could be used as therapeutic reagents to specifically target tumor cells for destruction. Initial clinical trials were performed with mouse antibodies, but induction in the patient of human anti-mouse antibodies lessened their effectiveness. The mouse antibody could be given only two to three times before host immunity prevented further treatments (Levene et al., 2005). Advances in recombinant DNA technology revived interest in monoclonal antibody therapy. New "chimeric" antibodies were generated that contained the variable, antigen-specific region of the murine antibody and the constant regions of human antibodies. These chimeric antibodies retained the tumor antigen specificity while being much less immunogenic to the patient's immune system. A technological development that will further facilitate and promote monoclonal antibody therapy is the cloning of the human immunoglobulin gene locus into a mouse. These genetically engineered transgenic mice produce completely human antibodies when immunized with a human tumor antigen (Green, 1999).

Host immunity was only one of many challenges that met the development of antibody therapeutics. Others included inability of antibodies to penetrate sufficiently into the tumor mass, short half-life in the circulation requiring frequent administrations, and more practical issues of the cost of production and thus the cost of therapy. Yet, even with many of these problems still unsolved, the clear effectiveness of this therapy has made it a popular choice of oncologists worldwide. A good example of a therapeutic antibody is the chimeric antibody rituximab, specific for the CD20 antigen, commonly found on B lymphoid malignancies. Rituximab has been used to treat non-Hodgkin's lymphoma, chronic lymphocytic leukemia, and small-cell lymphocytic leukemia. It has been used alone or coupled to radioisotopes with maximal clinical responses occurring when the antibody therapy was combined with conventional chemotherapy (Cheson, 2006).

Other targets of monoclonal antibodies for leukemia treatment include CD33 and CD52, markers commonly expressed on leukemic cells. Anti-CD33 mAb (gemtuzumab) has been successfully used to treat acute myeloid leukemia, whereas an anti-CD52 mAb (alemtuzumab) is licensed for use against B cell chronic lymphocytic leukemia (Carter, 2006).

Another successful approach for the treatment of B cell malignancies has been the development of anti-idiotype antibodies. B cell malignancies are a clonal expansion of cells with a uniquely rearranged cell surface immunoglobulin (idiotype), which can serve as a tumor-specific antigen. Recombinant DNA technology makes it possible to produce large amounts of this Ig idiotype that can be used to generate anti-idiotypic antibody for therapy or used as a vaccine to elicit such antibodies in the patient (Armstrong *et al.*, 2005).

Several possible mechanisms are thought to be involved in the antitumor effect of antibodies, including complement-dependent lysis of tumor cells, antibody-dependent cell-mediated cytotoxicity, and possible inhibition of the function of the target antigen (Levene *et al.*, 2005).

Growth factors and their receptors, adhesion molecules, and angiogenic molecules are all targets of choice for antibody therapy. Bevacizumab, an mAb against vascular endothelial growth factor (VEGF), has shown clinical efficacy in several cancers, including lung and colorectal, especially when combined with conventional chemotherapy (Jain *et al.*, 2006). Similarly, cetuximab targets epidermal growth factor receptor (EGFR) and has been shown to competitively inhibit ligand (EGF) binding to the receptor, resulting in retardation of tumor growth and promising clinical results when used in conjunction with standard chemotherapy in colorectal cancer (Italiano, 2006). Edrecolomab (also known as mAb 17-1A) binds with low affinity to an epithelial cell-adhesion molecule (Ep-CAM) that is involved in calcium transport across the cell membrane and is overexpressed on carcinomas of the pancreas, stomach, rectum, and colon. Several clinical trials using edrecolomab in colorectal cancer have resulted in increased disease-free and overall survival, especially in Dukes' stage C colorectal cancer (Riethmuller *et al.*, 1998).

B. T Cell Therapy

Early demonstrations of successful adoptive T cell immunotherapy came from animal models. Mice bearing disseminated leukemia were successfully treated by a combination of cyclophosphamide and adoptive transfer of syngeneic immune lymphocytes (Cheever *et al.*, 1980). Others have showed that an injection of sensitized spleen cells cultured in interleukin (IL)-2 resulted in tumor elimination (Eberlein *et al.*, 1982). Clinical studies have shown that adoptive transfer of highly selected tumor-reactive T cells could mediate tumor regression (Morgan *et al.*, 2006).

Not all adoptive T cell therapy trials experience equal successes. In a phase III trial of 178 patients with renal cell carcinoma, there was no improvement in the response rate or survival of patients following therapy with low-dose rIL-2 plus CD8$^+$ tumor-infiltrating lymphocytes, compared with low-dose rIL-2 therapy alone (Figlin et al., 1999). This revealed the need for additional understanding of the conditions required for adoptive T cell immunotherapy to work.

One proposed manipulation of recipients prior to adoptive transfer of T cells was depletion of endogenous lymphocytes. In a reported clinical trial, nonmyeloablative but lymphodepleting systemic chemotherapy prior to tumor-specific T cell adoptive transfer resulted in tumor regression in more than 50% of the treated patients (Gattinoni et al., 2006). The postulated mechanism is that lymphodepletion removes cytokine-responsive endogenous cells that would otherwise compete with adoptively transferred T cells for factors important for their growth and activation. Evidence has revealed that cytokines of the common cytokine-receptor gamma chain family (most notably IL-7 and IL-15) play an important role in the proliferation and survival of adoptively transferred lymphocytes (Gattinoni et al., 2006). Lymphodepletion also reduces regulatory T cell numbers and may increase the availability and function of APCs (Gattinoni et al., 2006). One important concern raised in regard to adoptive T cell therapy is the (in)ability of transferred cells to traffic to the tumor site and into the tumor. Some solutions that have been proposed to enhance migration of T cells toward tumors include transducing transferred T cells with CXCR2 (Kershaw et al., 2002), a receptor for CXCL2, a chemokine commonly produced by tumor cells.

IV. ACTIVE-SPECIFIC IMMUNOTHERAPY: VACCINES

The identification of tumor antigens provided immunologists with candidate immunogens for vaccine development. Most vaccines, as of this writing, have been tested in the presence of cancer in the hope of boosting an ongoing, albeit weak, antitumor immune response. The goal of therapeutic vaccines is to control tumor growth and prevent tumor recurrence. Most cancer vaccine trials have been carried out in phase I and phase II settings, where evaluation of toxicity has been the primary endpoint and the immune response a secondary endpoint and a surrogate marker of efficacy. Little toxicity has been seen, but the immune response elicited has been only marginally higher than prevaccination immunity. There have also been several well-publicized phase III trials, primarily in patients with melanoma but also in lung, breast, and prostate cancer patients. Depending on who is analyzing the data and what their biases are, the overall success based on objective clinical responses has ranged between 2.6 and 10% (Mocellin et al., 2004; Rosenberg et al., 2004). While this finding can be seen as a failure of cancer vaccines, the facts that no other vaccines for other diseases have worked in therapy and that the tumor patient is by and large an immunosuppressed individual, increases tremendously the significance of these small numbers.

A future goal for cancer vaccines is to move the time of administration to earlier stages of disease and ultimately to develop prophylactic vaccines to treat individuals at high risk for developing cancer. A significant movement in this direction has been the U. S. Food and Drug Administration's (FDA's) approval of the vaccine against HPV, the known cause of cervical cancer, the second most common cancer in women worldwide. In large clinical trials, Merck's Gardasil vaccine for HPV types 6, 11, 16, and 18 has been shown to prevent infection and also cervical preinvasive and noninvasive cancers associated with HPV. HBV vaccines, which have been licensed for many years for prevention of infection, can also prevent liver cancer that develops as a result of chronic HBV infection. This finding was clearly shown in the Taiwan's national program of HBV vaccination established in

1984, which in 10 years resulted in decreased incidence of hepatocellular carcinoma in children aged 6–9 years from 0.52 to 0.13 per 100,000 (Chang et al., 1997; Zuckerman, 1997).

Both HPV and HBV vaccines are directed against viral antigens; therefore, there is less concern that they might induce autoimmunity. However, very few cancers are known to have viral etiology, and thus the majority of cancer vaccines are based on tumor antigens that are expressed on normal cells as well but are mutated, overexpressed, or differentially posttranslationally modified by the tumor. In many cases, the immune response to these molecules is weaker due to various levels of self-tolerance.

Efforts to elicit strong immune responses against these antigens have the potential to elicit autoimmunity as well. While theoretically this potential is there, it has been observed very rarely in animal models, and it has been dependent on specific mechanisms of tolerance that needed to be overcome. One successful approach to enhance immune responses to peptides derived from such tumor antigens has been to change the amino acid sequence in such a way that higher affinity T cells are induced but they still recognize the native peptide sequence. These modified peptides are known as *altered peptide ligands* (Nishimura et al., 2004).

Vaccines using DNA encoding a tumor antigen have been shown to generate strong cellular and humoral immune responses in animal models but have had lower efficacy in clinical trials (Yu and Finn, 2006). Some successes were seen with a DNA vaccine encoding Ig idiotypes, which has elicited tumor-specific immunity in B cell lymphoma patients (Syrengelas et al., 1996), and a DNA vaccine encoding PSA, which has generated specific immunity in prostate cancer patients (Pavlenko et al., 2004).

Dendritic cells (DCs) are specialized APCs that play a central role in the development of adaptive T cell immunity. They can also aid in the induction of innate immune responses by natural killer (NK) cells and NKT cells. Tumors can inhibit DC differentiation and maturation, induce DCs to mature into other nonhematopoietic cell types, and release chemokines that influence DC migration. Inasmuch as effectiveness of a vaccine depends on properly functioning antigen-presenting cells, the problem of DC malfunction in cancer patients has been sidestepped by *in vitro* maturation, activation, and antigen loading of DCs and using them as a vaccine. A phase I/II clinical trial in stage IV melanoma patients showed that DCs derived from CD34 progenitor cells pulsed with purified tumor antigens were able to induce tumor-specific T cells (Palucka et al., 2003). Provenge, a vaccine based on DCs loaded with the prostate cancer antigen, prostatic acid phosphatase (PAP), demonstrated a definite clinical response in androgen-independent prostate cancer patients in a phase III trial (Burch et al., 2004; Lin et al., 2006).

Ex vivo expanded autologous DCs have also been loaded with tumor-derived RNA and used to induce antigen-specific immune responses (Kyte et al., 2006; Su et al., 2003). DC vaccine-containing renal cell carcinoma total RNA induced polyclonal T cell responses against renal tumor-associated antigens, including telomerase, reverse transcriptase, G250, and oncofetal antigen, but not against self-antigens (Su et al., 2003). Another study reported that intradermal injection of DCs loaded with melanoma tumor RNA could produce vaccine-specific immune response detected by *in vitro* assays for interferon gamma and *in vivo*-delayed hypersensitivity responses (Kyte et al., 2006).

V. CANCER-INDUCED IMMUNOSUPPRESSION IMPINGES ON IMMUNOTHERAPY

Tumors employ numerous immunosuppressive mechanisms to subvert a protective immune response, including causing

T cell malfunction, APC deficiencies, and induction of regulatory T cells. Immunosuppression by tumors is often mediated by soluble factors such as cytokines, chemokines, and, in some cases, by enzymes. The type of immunosuppression that exists in a patient or is known to be characteristic of a particular tumor type dictates the choice of the most appropriate immunotherapy. To be successful, the chosen immunotherapy must contain as one of its components an agent designed to interfere with one or more mechanisms of tumor-induced immunosuppression. This may range from the use of costimulatory antibodies to rescue tumor-specific T cells from tolerance (Abken et al., 2002) to the use of depleting antibodies to eliminate regulatory T cells (Zou, 2006). It is clear that in later stage cancers, immunosuppression is more severe compared to early stage cancer (Lollini et al., 2006). Immunotherapies administered in very early disease, or even better, as immunoprevention in high-risk individuals, could result in very effective protection from cancer. Such early stage interventions are already standard in chemoprevention (Vogel et al., 2006) and can be very instructive for the design of the first cancer immunoprevention clinical trials.

Another important consideration in choosing appropriate cancer immunotherapy is the age of the patient. It is well established that the thymus gradually involutes with age, which results in a drastic reduction in the numbers of naive T cells generated (Aspinall and Andrew, 2000). Therapies that attempt to stimulate naive T cells would then not be ideal for older patients who may instead benefit more from immunotherapies that aim to reactivate existing memory T cells or may benefit from adoptive T cell therapy.

A. T Cell Suppression in Cancer and Immunotherapy

T cell deficiencies present in cancer patients are seen both in the circulating T cells (Mizoguchi et al., 1992; Schmielau and Finn, 2001) and also in those within the tumor microenvironment (Finke et al., 1993; Guilloux et al., 1994; Kono et al., 1996; Mortarini et al., 2003; Mukherjee et al., 2001; Nakagomi et al., 1993; Zippelius et al., 2004). Examination of peripheral blood T cells from cancer patients has shown that CD8 T cells exhibit reduced or absent CD3-ζ chain expression and reduced interferon (IFN)-γ production compared to normal controls. Further work identified numerous mechanisms that are responsible for the loss of T cell receptor (TCR)-ζ, including H_2O_2 produced by activated granulocytes (Schmielau and Finn, 2001), arginase (Rodriguez et al., 2002; Taheri et al., 2001), and caspase cleavage. Similarly, T cells present in the tumor or the metastatic site contain less CD3-ζ (Finke et al., 1993; Nakagomi et al., 1993), are less capable of IFN production (Guilloux et al., 1994), contain less perforin and granzyme B (Zippelius et al., 2004), exhibit reduced cytotoxicity (Zippelius et al., 2004), and appear to be of a less differentiated phenotype (Mortarini et al., 2003). All of these deficiencies likely contribute to the inability of the host immune system to control growing tumors in vivo as well as to the failure of most immunotherapies to improve antitumor immunity. Reversing, if possible, these T cell defects prior to immunotherapy is one way to increase its effectiveness (Baniyash, 2004; Whiteside, 2004).

Several reports have described the imbalance between T helper 1 (Th1) and T helper 2 (Th2) cytokines, toward generation of primarily Th2 T cells as another form of immunosuppression during cancer (Olver et al., 2006). This was reported to be due to ganglioside production by tumors (Crespo et al., 2006). Gangliosides are sialylated glycosphingolipids that are expressed at high levels in multiple tumors, including gliomas and melanomas, and have been a target for immunotherapy (Fredman et al., 2003).

Adoptive T cell transfers represent a possible solution to endogenous T cell deficiencies in the cancer patient (Gattinoni et al.,

2006). This type of immunotherapy could replace the "suppressed" T cells with *in vitro* activated and expanded T cells that would in theory home to the tumor site and eliminate the growing tumor. This immunotherapy approach has been employed in melanoma patients with some success; however, durable clinical responses are still rarely seen. One must consider, among other reasons for the low success rate of this therapy, the fact that these *in vitro* functional T cells may also succumb to the immunosuppressive environment in the cancer patient to the same extent as the endogenous T cells. There is an opportunity to improve the results of this therapy by further research into various *in vitro* modifications of transferred T cells that would make them resistant to immunosuppression.

B. APC Dysfunction in Cancer and Immunotherapy

Functional impairment of DCs has been reported in pancreatic cancer (Yanagimoto *et al.*, 2005), colorectal cancer (Huang *et al.*, 2003), chronic lymphocytic leukemia (Orsini *et al.*, 2003), and hepatocellular carcinoma (Lee *et al.*, 2004). These studies conclude that tumor-derived factors, including TGF-β and IL-10, negatively affect DC differentiation, maturation, and function, at least in part through reduced costimulatory molecule expression and reduced inflammatory cytokine (IL-12) production as well as increased apoptosis (Ito *et al.*, 2006).

In addition to the tumor-derived soluble factors, some tumor antigens, such as MUC1 (Carlos *et al.*, 2005; Hiltbold *et al.*, 2000; Monti *et al.*, 2004), PSA (Aalamian *et al.*, 2003), and HER2/neu (Hiltbold *et al.*, 1999), can negatively affect DC maturation and function. MUC1 is readily taken up by DCs; however, the protein is kept in a "holding pattern" in the cells and is not effectively processed and displayed on the cell surface for presentation to T cells (Hiltbold *et al.*, 2000). This also occurs with HER2/neu, resulting in poor activation of the immune response against these antigens. DCs that take up MUC1 are unable to stimulate a Th1-type CD4 response (Carlos *et al.*, 2005) thought to be essential for generation of effective antitumor cytotoxic T lymphocyte (CTL) and antibody responses. Therefore, this tumor antigen is able to sabotage DCs' ability to stimulate effective responses to cancer.

Tumor-induced suppression of DC numbers and functionality will clearly lessen the efficacy of immunotherapy, especially if that immunotherapy involves immunization with antigens that must be processed and presented by endogenous DCs. This can be alleviated by adoptive transfer of tumor antigen-loaded DCs that have been extracted from the patient are expanded in a suppression-free *in vitro* system. There still exists the possibility, however, that adoptively transferred DCs will be affected by the same immunosuppressive mechanisms as the endogenous DCs, and some potential approaches to protect them have been suggested, such as coadministration of neutralizing antibody to TGF-β (Kobie *et al.*, 2003).

C. Regulatory T Cells in Cancer Immunotherapy

The normal function of regulatory T cells (Tregs) *in vivo* is to dampen or prevent autoimmunity and play a role in the resolution of an active immune response. Several varieties of Tregs exist *in vivo*, including both $CD8^+$ and $CD4^+$ T cells that suppress immune responses via the production of suppressive cytokines, through direct cell-to-cell contact, or indirectly by affecting APC function. Increase in the numbers of Tregs in the peripheral blood of cancer patients has been documented in several types of cancer, including breast cancer, colorectal cancer, lung cancer, pancreatic cancer, and melanoma (Zou, 2006). The implication is that Tregs contribute to the inability of the immune system to fully eliminate the growing tumor.

The mechanisms of immunosuppression by Tregs include secretion of suppressive cytokines (IL-10, TGF-β), competition for cytokines (such as IL-2), induction of tolerogenic DC, and even in some cases direct lysis of T cells or APCs. Several studies have explored how these Tregs are generated during cancer development; more specifically, the question asked has been can tumors actually induce Tregs and then use them to subvert the immune response? There is experimental evidence that certain cytokines produced by tumors, such as TGF-β, IL-10, and VEGF, may be responsible either directly or indirectly for "converting" T cells into Tregs. One study has suggested that TGF-β causes abnormal DC differentiation in the tumor, resulting in a population of immature DCs that stimulate the proliferation of Tregs (Zou, 2006). The chemokine CCL22 appears to play a pivotal role in the recruitment of Tregs to the tumor site, at least in ovarian cancer.

All of these factors combine to make a profoundly immunosuppressive environment that effector T cells must overcome to eradicate the growing tumor. This provides a very challenging environment for immunotherapy that aims to generate effective antitumor immune responses. It is thus apparent that effective immunotherapy should include approaches that target Tregs *in vivo*. Several strategies have been employed, including depletion with anti-CD25 antibodies (in mice and humans), treatment with antiglucocorticoid-induced tumor necrosis factor receptor (anti-GITR) family-related gene antibody in mice or anticytotoxic T lymphocyte-associated antigen 4 (anti-CTLA-4) mouse and human antibodies, and treatment with cyclophosphamide. Depletion of Tregs with anti-CD25 has resulted in a decrease in tumor burden in several mouse tumor models and also showed some efficacy in human cancer (Zou, 2006). Depletion of CD25 expressing Tregs in humans has been accomplished by treating patients with a fusion protein consisting of an anti-CD25 antibody coupled to diphtheria toxin, commercially known as Ontak®. This drug showed specific elimination of Tregs while sparing other cell types expressing intermediate levels of CD25, which includes effector T cells.

GITR is expressed on the surface of Tregs *in vivo*. Binding of the antibody to GITR eliminates the suppressive activity of Tregs *in vitro*, and it results in a robust increase in tumor-infiltrating CD4 and CD8 T cells without any evidence of autoimmunity when given *in vivo* in mice. Anti-GITR treatment therefore represents an additional method for inhibiting Tregs function during cancer, which may allow for more effective immunotherapies.

CTLA-4 is a negative regulator of effector T cell activation and blocking its function has been an attractive target of immunotherapy of cancer. It now appears that blockade of signaling through CTLA-4 by anti-CTLA-4 antibody treatment both reduces negative signaling in effector T cells and also signaling to Tregs that is necessary for their suppressive function (Zou, 2006). Treatment with an anti-CTLA-4 antibody resulted in tumor necrosis in some human studies; however, it also caused expected severe, but reversible, autoimmunity (Beck *et al.*, 2006).

Treatment with low-level cyclophosphamide was also shown to reduce the numbers and inhibit the function of Tregs in mice. Subsequent work in humans has shown that other chemotherapeutic agents, such as fludarabine, selectively deplete Tregs *in vivo* (Beyer *et al.*, 2005). This finding suggests changes in how chemotherapy may be used in the future to augment rather than suppress immune responses and improve immunotherapy.

VI. CANCER IMMUNOTHERAPY IN MICE VERSUS HUMANS

A better understanding of the numerous immunosuppressive mechanisms in the tumor-bearing host that prevent the immune system from recognizing and destroying the tumor will eventually lead to the design of better immunotherapies. Unfortunately, data on tumor-induced immunosuppression are relatively recent and have been collected by studying patient samples mostly from "failed" immunotherapy trials. These trials were and still are based to a great extent on successful immunotherapy of transplantable tumors in mice (Ostrand-Rosenberg, 2004). These fast-growing tumors do not interact with the immune system in the same way and duration to have the same detrimental influence on the immune system of the animal; therefore, many immunotherapy approaches in mice are unimpeded and able to elicit very strong antitumor immunity. Furthermore, many tumor antigens were foreign antigens to the mice, removing from the picture self-tolerance as one of the major challenges to successful immunotherapy. Last, most experiments were performed in relatively young animals with young immune systems that did not represent a cancer patient for whom the immunotherapy was intended. All of these realizations, coupled with the advent of new technologies, have encouraged development of better animal models in which the new generation of immunotherapies can be properly tested (Gendler and Mukherjee, 2001; Ostrand-Rosenberg, 2004). Many of these new mouse models carry human tumor antigens as transgenes as well as develop spontaneous tumors in various tissue sites These spontaneous cancer models provide researchers with an opportunity to test the efficacy of immunotherapy to prevent cancer or treat it at various times during its development.

VII. IMMUNOTHERAPY AND CANCER STEM CELLS

Studies conducted in 2006 have identified the existence of cancer stem cells that are responsible for tumor recurrence (Houghton *et al.*, 2006; Li and Neaves, 2006). The cancer stem cells have been postulated to be a very small population of quiescent or very slowly dividing cells within a growing tumor mass and are inherently resistant to treatments such as chemotherapy that target proliferating cells. Since the state of proliferation is not a prerequisite for recognition and destruction by the immune mechanisms, immunotherapy may be the most effective way to eliminate these cells (Figure 10.1). Identification of cancer stem cell-specific antigens is a current goal for the cancer vaccine field. Some of the tumor antigens previously identified on mature tumor cells are also candidates for targeting cancer stem cells. The relative role of cancer stem cells in metastatic disease is an area of intense research—several groups have postulated that cancer stem cells are the source of metastatic lesions (Brabletz *et al.*, 2005; Mehlen and Puisieux, 2006). Immunotherapy is likely to be applied in the setting of minimal residual disease with the goal of clearing the invisible but present metastases.

VIII. AUTOIMMUNITY RESULTING FROM CANCER IMMUNOTHERAPY

The constant drive to develop more potent immunotherapies must be tempered by the real possibility that at least some of them have the potential to induce undesired autoimmunity. On the other hand, it has been repeatedly observed in some trials that patients who had the best clinical responses were the same ones that exhibited signs of autoimmunity. This is an important observation and begs the question of whether effective tumor immunity must also include some degree of autoimmunity since the target antigen for tumor immunotherapy is almost always a modified version of a self protein (Kaufman and Wolchok, 2006). This uncoupling of the tumor-specific immune responses and the concomitant autoimmunity is yet another important consideration in the development of new immunotherapies.

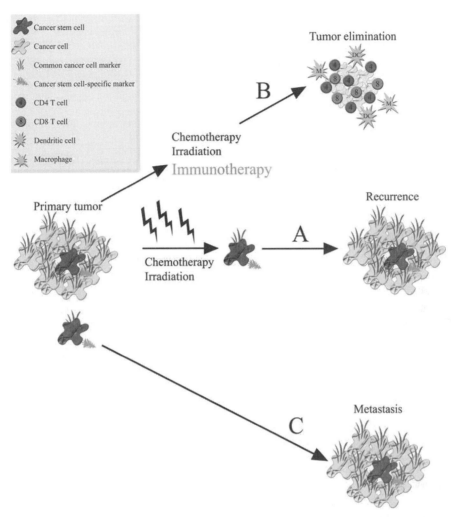

FIGURE 10.1 **Life of a tumor and opportunities for immunotherapy.** A growing tumor is a heterogeneous mix of mostly differentiated cancer cells and a very small population of cancer stem cells. The differentiated cancer cells exhibit the characteristic rapid proliferation, while the cancer stem cells are much slower at dividing, making them resistant to conventional therapy. (A) Following conventional treatment, such as chemotherapy, the majority of the tumor is eliminated; however, the cancer stem cells survive, differentiate, and cause tumor recurrence. (B) Immunotherapy that targets antigen(s) expressed by all cancer cells, including stem cells, could result in complete eradication of the tumor through immune effector mechanisms and prevention of tumor recurrence. (C) Elimination of cancer stem cells early in disease may have the added benefit of preventing metastasis because migration of these cancer stem cells may be the mechanism by which tumors spread from the primary site and establish themselves at distant sites. (See color plate.)

IX. CONCLUSION AND FUTURE CONSIDERATIONS

Immunotherapy represents one of the most promising approaches for future cancer treatment and prevention. Several types of common cancers are clearly resistant to conventional therapies; therefore, educating the body's immune system to eradicate cancerous cells represents a tremendous opportunity for a more effective treatment. Intense research into the reasons why immunotherapy trials have not been more successful has revealed a myriad of immunosuppressive strategies used by the growing tumor to circumvent and, in some cases, eliminate the natural host immune response; thus, it is clear that future therapies must take into consideration how to overcome these hurdles. Elimination of Tregs, adoptive transfer of effector CD8 T cells or tumor antigen-loaded DCs, and addition of agonist costimulatory antibodies all represent ways to fight back at the suppressive nature of the tumor; each method has shown very promising results in early phase clinical trials.

A common readout of an effective cancer immunotherapy is a postvaccine increase in tumor-specific T cells and antibodies that can be detected in the peripheral blood of immunized patients. The correlation between the numbers of tumor specific T cells and clinical prognosis has been put in doubt (Rosenberg et al., 2005), raising the question of how scientists and clinicians can predict the effectiveness of a particular immunotherapy approach. This is especially important in the setting of prevention or treatment of minimal residual disease for which time to occurrence or recurrence can be very long. So among other challenges to immunotherapy, defining markers of success remains a major research challenge. If and when the consensus is reached about what short-term measurements predict long-term successes, the field of immunotherapy and immunoprevention is likely to move with a higher speed and greater acceptance.

References

Aalamian, M., Tourkova, I. L., Chatta, G. S., Lilja, H., Huland, E., Huland, H., Shurin, G. V., and Shurin, M. R. (2003). Inhibition of dendropoiesis by tumor derived and purified prostate specific antigen. *J. Urol.* **170**, 2026–2030.

Abken, H., Hombach, A., Heuser, C., Kronfeld, K., and Seliger, B. (2002). Tuning tumor-specific T-cell activation: A matter of costimulation? *Trends Immunol.* **23**, 240–245.

Armstrong, A. C., Cheadle, E. J., and Hawkins, R. E. (2005). Toward personalized immunotherapy for non-Hodgkin lymphoma: Targeting the idiotypic immunoglobulin. *BioDrugs* **19**, 289–297.

Aspinall, R., and Andrew, D. (2000). Thymic involution in aging. *J. Clin. Immunol.* **20**, 250–256.

Baniyash, M. (2004). TCR zeta-chain downregulation: Curtailing an excessive inflammatory immune response. *Nat. Rev. Immunol.* **4**, 675–687.

Barnd, D. L., Lan, M. S., Metzgar, R. S., and Finn, O. J. (1989). Specific, major histocompatibility complex-unrestricted recognition of tumor-associated mucins by human cytotoxic T cells. *Proc. Natl. Acad. Sci. USA* **86**, 7159–7163.

Beck, K. E., Blansfield, J. A., Tran, K. Q., Feldman, A. L., Hughes, M. S., Royal, R. E., Kammula, U. S., Topalian, S. L., Sherry, R. M., Kleiner, D., Quezado, M., Lowy, I., Yellin, M., Rosenberg, S. A., and Yang, J. C. (2006). Enterocolitis in patients with cancer after antibody blockade of cytotoxic T-lymphocyte-associated antigen 4. *J. Clin. Oncol.* **24**, 2283–2289.

Beyer, M., Kochanek, M., Darabi, K., Popov, A., Jensen, M., Endl, E., Knolle, P. A., Thomas, R. K., von Bergwelt-Baildon, M., Debey, S., Hallek, M., and Schultze, J. L. (2005). Reduced frequencies and suppressive function of CD4+CD25hi regulatory T cells in patients with chronic lymphocytic leukemia after therapy with fludarabine. *Blood* **106**, 2018–2025.

Brabletz, T., Jung, A., Spaderna, S., Hlubek, F., and Kirchner, T. (2005). Opinion: Migrating cancer stem cells—an integrated concept of malignant tumour progression. *Nat. Rev. Cancer* **5**, 744–749.

Burch, P. A., Croghan, G. A., Gastineau, D. A., Jones, L. A., Kaur, J. S., Kylstra, J. W., Richardson, R. L., Valone, F. H., and Vuk-Pavlovic, S. (2004). Immunotherapy (APC8015, Provenge) targeting prostatic acid phosphatase can induce durable remission of metastatic androgen-independent prostate cancer: A Phase 2 trial. *Prostate* **60**, 197–204.

Carlos, C. A., Dong, H. F., Howard, O. M., Oppenheim, J. J., Hanisch, F. G., and Finn, O. J. (2005). Human tumor antigen MUC1 is chemotactic for immature dendritic cells and elicits maturation but does not promote Th1 type immunity. *J. Immunol.* **175**, 1628–1635.

Carter, P. J. (2006). Potent antibody therapeutics by design. *Nat. Rev. Immunol.* **6**, 343–357.

Chang, M. H., Chen, C. J., Lai, M. S., Hsu, H. M., Wu, T. C., Kong, M. S., Liang, D. C., Shau, W. Y., and Chen, D. S. (1997). Universal hepatitis B vaccination in Taiwan and the incidence of hepatocellular carcinoma in children. Taiwan Childhood Hepatoma Study Group. *N. Engl. J. Med* **336**, 1855–1859.

Cheever, M. A., Greenberg, P. D., and Fefer, A. (1980). Specificity of adoptive chemoimmunotherapy of established syngeneic tumors. *J. Immunol.* **125**, 711.

Cheson, B. D. (2006). Monoclonal antibody therapy of chronic lymphocytic leukemia. *Cancer Immunol. Immunother.* **55**, 188–196.

Crespo, F. A., Sun, X., Cripps, J. G., and Fernandez-Botran, R. (2006). The immunoregulatory effects of gangliosides involve immune deviation favoring type-2 T cell responses. *J. Leukoc. Biol.* **79**, 586–595.

De Plaen, E., Lurquin, C., Van Pel, A., Mariame, B., Szikora, J. P., Wolfel, T., Sibille, C., Chomez, P., and Boon, T. (1988). Immunogenic (tum-) variants of mouse tumor P815: Cloning of the gene of tum-antigen P91A and identification of the tum- mutation. *Proc. Natl. Acad. Sci. USA* **85**, 2274–2278.

Eberlein, T. J., Rosenstein, M., and Rosenberg, S. A. (1982). Regression of a disseminated syngeneic solid tumor by systemic transfer of lymphoid cells expanded in interleukin 2. *J. Exp. Med.* **156**, 385–397.

Figlin, R. A., Thompson, J. A., Bukowski, R. M., Vogelzang, N. J., Novick, A. C., Lange, P., Steinberg, G. D., and Belldegrun, A. S. (1999). Multicenter, randomized, phase III trial of CD8($^+$) tumor-infiltrating lymphocytes in combination with recombinant interleukin-2 in metastatic renal cell carcinoma. *J. Clin. Oncol.* **17**, 2521–2529.

Finke, J. H., Zea, A. H., Stanley, J., Longo, D. L., Mizoguchi, H., Tubbs, R. R., Wiltrout, R. H., O'Shea, J. J., Kudoh, S., Klein, E., et al. (1993). Loss of T-cell receptor zeta chain and p56lck in T-cells infiltrating human renal cell carcinoma. *Cancer Res.* **53**, 5613–5616.

Fredman, P., Hedberg, K., and Brezicka, T. (2003). Gangliosides as therapeutic targets for cancer. *BioDrugs* **17**, 155–167.

Gattinoni, L., Powell, D. J., Jr., Rosenberg, S. A., and Restifo, N. P. (2006). Adoptive immunotherapy for cancer: Building on success. *Nat. Rev. Immunol.* **6**, 383–393.

Gendler, S. J., and Mukherjee, P. (2001). Spontaneous adenocarcinoma mouse models for immunotherapy. *Trends Mol. Med.* **7**, 471–475.

Graziano, D. F., and Finn, O. J. (2005). Tumor antigens and tumor antigen discovery. *Cancer Treat. Res.* **123**, 89–111.

Green, L. L. (1999). Antibody engineering via genetic engineering of the mouse: XenoMouse strains are a vehicle for the facile generation of therapeutic human monoclonal antibodies. *J. Immunol. Methods* **231**, 11–23.

Guilloux, Y., Viret, C., Gervois, N., Le Drean, E., Pandolfino, M. C., Diez, E., and Jotereau, F. (1994). Defective lymphokine production by most CD8$^+$ and CD4$^+$ tumor-specific T cell clones derived from human melanoma-infiltrating lymphocytes in response to autologous tumor cells *in vitro*. *Eur. J. Immunol.* **24**, 1966–1973.

Hiltbold, E. M., Alter, M. D., Ciborowski, P., and Finn, O. J. (1999). Presentation of MUC1 tumor antigen by class I MHC and CTL function correlate with the glycosylation state of the protein taken up by dendritic cells. *Cell Immunol.* **194**, 143–149.

Hiltbold, E. M., Vlad, A. M., Ciborowski, P., Watkins, S. C., and Finn, O. J. (2000). The mechanism of unresponsiveness to circulating tumor antigen MUC1 is a block in intracellular sorting and processing by dendritic cells. *J. Immunol.* **165**, 3730–3741.

Houghton, J., Morozov, A., Smirnova, I., and Wang, T. C. (2007). Stem cells and cancer. *Semin. Cancer Biol.*

Huang, A., Gilmour, J. W., Imami, N., Amjadi, P., Henderson, D. C., and Allen-Mersh, T. G. (2003). Increased serum transforming growth factor-beta1 in human colorectal cancer correlates with reduced circulating dendritic cells and increased colonic Langerhans cell infiltration. *Clin. Exp. Immunol.* **134**, 270–278.

Italiano, A. (2006). Targeting the epidermal growth factor receptor in colorectal cancer: Advances and controversies. *Oncology* **70**, 161–167.

Ito, M., Minamiya, Y., Kawai, H., Saito, S., Saito, H., Nakagawa, T., Imai, K., Hirokawa, M., and Ogawa, J. (2006). Tumor-derived TGFbeta-1 induces dendritic cell apoptosis in the sentinel lymph node. *J. Immunol.* **176**, 5637–5643.

Jain, R. K., Duda, D. G., Clark, J. W., and Loeffler, J. S. (2006). Lessons from phase III clinical trials on anti-VEGF therapy for cancer. *Nat. Clin. Pract. Oncol.* **3**, 24–40.

Kaufman, H. L., and Wolchok, J. D. (2006). Is tumor immunity the same thing as autoimmunity? Implications for cancer immunotherapy. *J. Clin. Oncol.* **24**, 2230–2232.

Kershaw, M. H., Wang, G., Westwood, J. A., Pachynski, R. K., Tiffany, H. L., Marincola, F. M., Wang, E., Young, H. A., Murphy, P. M., and Hwu, P. (2002). Redirecting migration of T cells to chemokine secreted from tumors by genetic modification with CXCR2. *Hum. Gene Ther.* **13**, 1971–1980.

Klein, G., Sjogren, H. O., Klein, E., and Hellstrom, K. E. (1960). Demonstration of resistance against methylcholanthrene-induced sarcomas in the primary autochthonous host. *Cancer Res.* **20**, 1561–1572.

Kobie, J. J., Wu, R. S., Kurt, R. A., Lou, S., Adelman, M. K., Whitesell, L. J., Ramanathapuram, L. V., Arteaga, C. L., and Akporiaye, E. T. (2003). Transforming growth factor beta inhibits the antigen-presenting functions and antitumor activity of dendritic cell vaccines. *Cancer Res.* **63**, 1860–1864.

Kohler, G., and Milstein, C. (1975). Continuous cultures of fused cells secreting antibody of predefined specificity. *Nature* **256**, 495–497.

Kono, K., Salazar-Onfray, F., Petersson, M., Hansson, J., Masucci, G., Wasserman, K., Nakazawa, T., Anderson, P., and Kiessling, R. (1996). Hydrogen peroxide secreted by tumor-derived macrophages down-modulates signal-transducing zeta molecules and inhibits tumor-specific T cell-and natural killer cell-mediated cytotoxicity. *Eur. J. Immunol.* **26**, 1308–1313.

Kyte, J. A., Mu, L., Aamdal, S., Kvalheim, G., Dueland, S., Hauser, M., Gullestad, H. P., Ryder, T., Lislerud, K., Hammerstad, H., and Gaudernack, G. (2006). Phase I/II trial of melanoma therapy with dendritic cells transfected with autologous tumor-mRNA. *Cancer Gene Ther.* **13**, 905–918.

Lee, W. C., Chiang, Y. J., Wang, H. C., Wang, M. R., Lia, S. R., and Chen, M. F. (2004). Functional impairment of dendritic cells caused by murine hepatocellular carcinoma. *J. Clin. Immunol.* **24**, 145–154.

Levene, A. P., Singh, G., and Palmieri, C. (2005). Therapeutic monoclonal antibodies in oncology. *J. R. Soc. Med.* **98**, 146–152.

Li, L., and Neaves, W. B. (2006). Normal stem cells and cancer stem cells: The niche matters. *Cancer Res.* **66**, 4553–4557.

Lin, A. M., Hershberg, R. M., and Small, E. J. (2006). Immunotherapy for prostate cancer using prostatic acid phosphatase loaded antigen presenting cells. *Urol. Oncol.* **24**, 434–441.

Lollini, P. L., Cavallo, F., Nanni, P., and Forni, G. (2006). Vaccines for tumour prevention. *Nat. Rev. Cancer* **6**, 204–216.

Mehlen, P., and Puisieux, A. (2006). Metastasis: A question of life or death. *Nat. Rev. Cancer* **6**, 449–458.

Mizoguchi, H., O'Shea, J. J., Longo, D. L., Loeffler, C. M., McVicar, D. W., and Ochoa, A. C. (1992). Alterations in signal transduction molecules in T lymphocytes from tumor-bearing mice. *Science* **258**, 1795–1798.

Mocellin, S., Mandruzzato, S., Bronte, V., and Marincola, F. M. (2004). Cancer vaccines: Pessimism in check. *Nat. Med.* **10**, 1278–1280.

Monti, P., Leone, B. E., Zerbi, A., Balzano, G., Cainarca, S., Sordi, V., Pontillo, M., Mercalli, A., Di Carlo, V., Allavena, P., and Piemonti, L. (2004). Tumor-derived MUC1 mucins interact with differentiating monocytes and induce IL-10highIL-12low regulatory dendritic cell. *J. Immunol.* **172**, 7341–7349.

Morgan, R. A., Dudley, M. E., Wunderlich, J. R., Hughes, M. S., Yang, J. C., Sherry, R. M., Royal, R. E., Topalian, S. L., Kammula, U. S., Restifo, N. P., Zheng, Z., Nahvi, A., de Vries, C. R., Rogers-Freezer, L. J., Mavroukakis, S. A., and Rosenberg, S. A. (2006). Cancer regression in patients after transfer of genetically engineered lymphocytes. *Science.* **314**, 126–129.

Mortarini, R., Piris, A., Maurichi, A., Molla, A., Bersani, I., Bono, A., Bartoli, C., Santinami, M., Lombardo, C., Ravagnani, F., Cascinelli, N., Parmiani, G., and Anichini, A. (2003). Lack of terminally differentiated tumor-specific CD8+ T cells at tumor site in spite of antitumor immunity to self-antigens in human metastatic melanoma. *Cancer Res.* **63**, 2535–2545.

Mukherjee, P., Ginardi, A. R., Madsen, C. S., Tinder, T. L., Jacobs, F., Parker, J., Agrawal, B., Longenecker, B. M., and Gendler, S. J. (2001). MUC1-specific CTLs are non-functional within a pancreatic tumor microenvironment. *Glycoconj. J.* **18**, 931–942.

Nakagomi, H., Petersson, M., Magnusson, I., Juhlin, C., Matsuda, M., Mellstedt, H., Taupin, J. L., Vivier, E., Anderson, P., and Kiessling, R. (1993). Decreased expression of the signal-transducing zeta chains in tumor-infiltrating T-cells and NK cells of patients with colorectal carcinoma. *Cancer Res.* **53**, 5610–5612.

Nishimura, Y., Chen, Y. Z., Uemura, Y., Tanaka, Y., Tsukamoto, H., Kanai, T., Yokomizo, H., Yun, C., Matsuoka, T., Irie, A., and Matsushita, S. (2004). Degenerate recognition and response of human CD4+ Th cell clones: Implications for basic and applied immunology. *Mol. Immunol.* **40**, 1089–1094.

Olver, S., Groves, P., Buttigieg, K., Morris, E. S., Janas, M. L., Kelso, A., and Kienzle, N. (2006). Tumor-derived interleukin-4 reduces tumor clearance and deviates the cytokine and granzyme profile of tumor-induced CD8+ T cells. *Cancer Res.* **66**, 571–580.

Orsini, E., Guarini, A., Chiaretti, S., Mauro, F. R., and Foa, R. (2003). The circulating dendritic cell compartment in patients with chronic lymphocytic leukemia is severely defective and unable to stimulate an effective T-cell response. *Cancer Res.* **63**, 4497–4506.

Ostrand-Rosenberg, S. (2004). Animal models of tumor immunity, immunotherapy, and cancer vaccines. *Curr. Opin. Immunol.* **16**, 143–150.

Palucka, A. K., Dhodapkar, M. V., Paczesny, S., Burkeholder, S., Wittkowski, K. M., Steinman, R. M., Fay, J., and Banchereau, J. (2003). Single injection of CD34+ progenitor-derived dendritic cell vaccine can lead to induction of T-cell immunity in patients with stage IV melanoma. *J. Immunother.* **26**, 432–439.

Pavlenko, M., Roos, A. K., Lundqvist, A., Palmborg, A., Miller, A. M., Ozenci, V., Bergman, B., Egevad, L., Hellstrom, M., Kiessling, R., Masucci, G., Wersall, P., Nilsson, S., and Pisa, P. (2004). A phase I trial of DNA vaccination with a plasmid expressing prostate-specific antigen in patients with hormone-refractory prostate cancer. *Br. J. Cancer* **91**, 688–694.

Riethmuller, G., Holz, E., Schlimok, G., Schmiegel, W., Raab, R., Hoffken, K., Gruber, R., Funke, I., Pichlmaier, H., Hirche, H., Buggisch, P., Witte, J., and Pichlmayr, R. (1998). Monoclonal antibody therapy for resected Dukes' C colorectal cancer: Seven-year outcome of a multicenter randomized trial. *J. Clin. Oncol.* **16**, 1788–1794.

Rodriguez, P. C., Zea, A. H., Culotta, K. S., Zabaleta, J., Ochoa, J. B., and Ochoa, A. C. (2002). Regulation of T cell receptor CD3zeta chain expression by L-arginine. *J. Biol. Chem.* **277**, 21123–21129.

Rosenberg, S. A., Sherry, R. M., Morton, K. E., Scharfman, W. J., Yang, J. C., Topalian, S. L., Royal, R. E., Kammula, U., Restifo, N. P., Hughes, M. S., Schwartzentruber, D., Berman, D. M., Schwarz, S. L., Ngo, L. T., Mavroukakis, S. A., White, D. E., and Steinberg, S. M. (2005). Tumor progression can occur despite the induction of very high levels of self/tumor antigen-specific CD8+ T cells in patients with melanoma. *J. Immunol.* **175**, 6169–6176.

Rosenberg, S. A., Yang, J. C., and Restifo, N. P. (2004). Cancer immunotherapy: Moving beyond current vaccines. *Nat. Med.* **10**, 909–915.

Schmielau, J., and Finn, O. J. (2001). Activated granulocytes and granulocyte-derived hydrogen peroxide are the underlying mechanism of suppression of T-cell function in advanced cancer patients. *Cancer Res.* **61**, 4756–4760.

Su, Z., Dannull, J., Heiser, A., Yancey, D., Pruitt, S., Madden, J., Coleman, D., Niedzwiecki, D., Gilboa, E., and Vieweg, J. (2003). Immunological and clinical responses in metastatic renal cancer patients vaccinated with tumor RNA-transfected dendritic cells. *Cancer Res.* **63**, 2127–2133.

Syrengelas, A. D., Chen, T. T., and Levy, R. (1996). DNA immunization induces protective immunity against B-cell lymphoma. *Nat. Med.* **2**, 1038–1041.

Taheri, F., Ochoa, J. B., Faghiri, Z., Culotta, K., Park, H. J., Lan, M. S., Zea, A. H., and Ochoa, A. C. (2001). L-Arginine regulates the expression of the T-cell receptor zeta chain (CD3zeta) in Jurkat cells. *Clin. Cancer Res.* **7**, 958s–965s.

van der Bruggen, P., Traversari, C., Chomez, P., Lurquin, C., De Plaen, E., Van den Eynde, B., Knuth, A., and Boon, T. (1991). A gene encoding an antigen recognized by cytolytic T lymphocytes on a human melanoma. *Science* **254**, 1643–1647.

Vogel, V. G., Costantino, J. P., Wickerham, D. L., Cronin, W. M., Cecchini, R. S., Atkins, J. N., Bevers, T. B., Fehrenbacher, L., Pajon, E. R., Jr., Wade, J. L., 3rd, Robidoux, A., Margolese, R. G., James, J., Lippman, S. M., Runowicz, C. D., Ganz, P. A., Reis, S. E., McCaskill-Stevens, W., Ford, L. G., Jordan, V. C., and Wolmark, N. (2006). Effects of tamoxifen vs raloxifene on the risk of developing invasive breast cancer and other disease outcomes: The NSABP Study of Tamoxifen and Raloxifene (STAR) P-2 trial. *JAMA* **295**, 2727–2741.

Whiteside, T. L. (2004). Down-regulation of zeta-chain expression in T cells: A biomarker of prognosis in cancer? *Cancer Immunol. Immunother.* **53**, 865–878.

Yanagimoto, H., Takai, S., Satoi, S., Toyokawa, H., Takahashi, K., Terakawa, N., Kwon, A. H., and Kamiyama, Y. (2005). Impaired function of circulating dendritic cells in patients with pancreatic cancer. *Clin. Immunol.* **114**, 52–60.

Yu, M., and Finn, O. J. (2006). DNA vaccines for cancer too. *Cancer Immunol. Immunother.* **55**, 119–130.

Zippelius, A., Batard, P., Rubio-Godoy, V., Bioley, G., Lienard, D., Lejeune, F., Rimoldi, D., Guillaume, P., Meidenbauer, N., Mackensen, A., Rufer, N., Lubenow, N., Speiser, D., Cerottini, J. C., Romero, P., and Pittet, M. J. (2004). Effector function of human tumor-specific CD8 T cells in melanoma lesions: A state of local functional tolerance. *Cancer Res.* **64**, 2865–2873.

Zou, W. (2006). Regulatory T cells, tumour immunity and immunotherapy. *Nat. Rev. Immunol.* **6**, 295–307.

Zuckerman, A. J. (1997). Prevention of primary liver cancer by immunization. *N. Engl. J. Med.* **336**, 1906–1907.

CHAPTER

11

Cancer Vaccines

FREDA K. STEVENSON, GIANFRANCO DI GENOVA,
CHRISTIAN OTTENSMEIER, AND NATALIA SAVELYEVA

I. Introduction
II. Tumor Antigens
III. Spontaneous Immunity to Cancer
IV. Toleragenic Pressure on Immunity to Cancer
    A. T Cell Tolerance
    B. B Cell Tolerance
V. Immune Responses to Conventional Vaccines
    A. Memory T Cell Responses Induced by Conventional Vaccination
    B. Memory T Cell Subsets in Persistent Infection
    C. T Cell Vaccines
VI. Cancer Vaccine Strategies
    A. Peptide Vaccines
    B. DC-Based Vaccines
VII. DNA Vaccines
    A. Activation of T Cell Help for Antitumor Immunity
    B. Obstacles for Human Vaccines
    C. Physical Strategies to Improve Performance
    D. Prime/Boost Strategies
VIII. Challenges of Translation to the Clinic
IX. Concluding Remarks
    References
    Further Reading

This review of cancer vaccines focuses on problems faced in attempting to induce effective therapeutic immunity and describes promising strategies that might be employed to overcome these difficulties. The theme of this chapter is rational design of vaccines based on immunological knowledge. Mouse models provide a firm foundation for the research presented here, but information on immune responses in human subjects is also included. For the later, the chapter authors have partly turned to conventional vaccines that have been so successful in preventing infection but for which, until the twenty-first century, only limited analyses of immune pathways have been available. Dialogue between the immune system and coevolved persistent viruses in human subjects has also revealed data relevant for cancer treatments. In both settings, a memory response is required for continuous suppression. The authors' perspective on vaccine design is to use genetics not only to detect potential antigenic targets but also to engineer delivery vehicles that have the ability to penetrate to the immune processing machinery. This discussion argues that the key issue is flexible design tested in preclinical models and followed by rapid pilot clinical trials to ensure translational effectiveness. Objective immune readouts can then act as a surrogate for efficacy, prior to the longer road of testing clinical efficacy.

I. INTRODUCTION

It is almost impossible, and perhaps undesirable, to discuss the myriad of undefined cancer vaccines that have been given to patients in a generally vain attempt to suppress cancer. Instead, the focus here will be on vaccines aimed to induce immunity against specific target antigens. It is timely to review the strategies that apply known and emergent immunological principles to optimize vaccine performance. The review will illustrate the chapter authors' assessment of the direction of progress toward realistic vaccines for the future. Included in this review is the importance of implementing new objective criteria for monitoring immune responses.

Vaccines against cancer-associated pathogens, such as human papilloma virus (HPV) (Lowy and Schiller, 2006) or hepatitis B virus (HBV) (Zuckerman, 2006), are clearly successful in preventing the associated cancers. However, only 10–20% of cancers are known to be in this category, and, for some of these, preventative vaccination may be logistically difficult. The current twenty-first century need for cancer treatment is for therapeutic vaccines, ideally for patients in the early stage of disease who can be brought into remission by treatments that do not irreparably damage immune capacity. Therapeutic vaccination against infection is also required for persistent infections, such as hepatitis B and C, partly to ameliorate the later complications of chronic infection. This type of vaccination may also be required to deal with potential bioterrorism. Postexposure vaccination was in fact the setting for Louis Pasteur's successful first trial of vaccination against rabies in the late nineteenth century, so the concept has been tested. Vaccination following infection with smallpox, if given sufficiently early, is also effective in suppressing and modifying disease (Massoudi et al., 2003). For both rabies and smallpox, a window of time for induction of immunity is provided by the relatively slow kinetics of expansion of infected cells, a feature shared by most cancers.

The success of passive immunotherapy with monoclonal antibodies (Buske et al., 2006) or via allogeneic transplantation (Riddell et al., 2003) has demonstrated that the immune system is capable of attacking cancer cells. Active vaccination should provide continuous protection against emergent tumors, but induction of immunity faces challenges. For both cancer and persistent infections, immunity has to be generated in the face of increasing antigen production and before the immune system is compromised. This common goal has reunited the fields of microbiology and immunology, which had become separated. The desired immune effector mechanism will depend on circumstance, with most preventative vaccination aimed to induce antibody, whereas therapeutic vaccination will often aim to induce T cell attack. A further goal of therapeutic vaccination against cancer may be to contain rather than eliminate tumor cells. In this regard, it is similar to the strategies for some infections, possibly including the human immunodeficiency virus (HIV). While sterilizing immunity sounds attractive, the need for antigen to maintain T cell activity may be a factor. An understanding of the balances in the immune system and of evasion tactics is clearly the key to effective vaccination and can inform new vaccine designs. However, cancer vaccines have to solve the additional problem of poor immunogenicity of the majority of cancer antigens. In the authors' opinion, this can be overcome by judicious borrowing of immunogenic molecules from pathogens.

Vaccination against cancer has followed two distinct but overlapping strategies: (i) to modify cancer cells so that they act as their own source of undefined antigens and (ii) to present defined antigens in new vaccine formulations. The first has the advantage of not requiring knowledge of the antigen but has the disadvantage of difficulty in monitoring immune responses. Reliance must be on clinical effects, which, although ultimately critical, need large studies and can take years to evaluate. Unless the effect is dramatic, it will be difficult to glean the information from patients for intelligent modification of the vaccines. There is also a significant and often unacknowledged problem with mouse models, which are necessary to establish principles but for which limitations must also be recognized. Mouse tumors harbor retroviruses that, when released in an appropriate environment, produce highly immunogenic proteins (van Hall *et al.*, 2000). Protective immunity induced by nonspecific stimulation of tumor cells and/or antigen-presenting cells (APCs), therefore, often derives from retroviral antigens. Since these are considerably less common in human tumors, mouse models might generate overoptimistic conclusions.

The second more reductionist approach of delivering defined antigens in formulations aimed to activate specific pathways can build on emerging knowledge of immune pathways. The approach can harness modern genomics in a systematic manner, placing new antigens together with immunoenhancing molecules and in novel delivery systems. A major problem is that there are insufficient clear antigenic targets on cancer cells, and the knowledge of processing and presentation of component peptides for cytotoxic T cell attack is still developing. However, advances in gene expression profiling and proteomics as well as methods to elute processed peptides from tumor cells (Rammensee *et al.*, 2002) should fill this gap. This chapter focuses on this specific approach to cancer vaccines and particularly on the use of DNA delivery as a vaccine vehicle.

II. TUMOR ANTIGENS

Several reviews of the nature of tumor antigens have been published (Stevenson *et al.*, 2004b), making reiteration here unnecessary. Briefly, tumor antigens include proteins/glycoproteins, carbohydrates, and glycolipids. Since the latter two groups have been reviewed recently (Bitton *et al.*, 2002), the focus here will be on proteins/glycoproteins and their component immunogenic peptides. The group is large and diverse, comprising viral proteins, clonotypic proteins, lineage-specific proteins, differentiation antigens, aberrantly glycosylated mucins, mutated protooncogenes, products of chromosomal translocations, cancer-testis antigens, and overexpressed autoantigens. New categories of antigens may arise from splice variants, alternative reading frames, intron/exon sequences, and peptides arising in cells with impaired antigen processing function (van Hall *et al.*, 2006) as well as from additional products of genomic instability.

Each category has its advantages and disadvantages and is under investigation, in some cases in the clinic (Pardoll, 2002). Some general points of principle can be made. The first is that the molecular form of the target antigen should dictate the desired immune effector function. If the target is a glycoprotein expressed at the cell surface, antibodies may be effective. There is a strong interest in passive monoclonal antibody treatment for cancer. Examples include anti-CD20 antibodies for the treatment of B cell tumors (Buske *et al.*, 2006) and anti-HER2/neu (ErbB-2) (trastuzumab or

Herceptin®) for certain subtypes of breast cancer (McKeage and Perry, 2002). Active vaccination would ensure continual attack, but, unless the antigen is tumor specific, there is an accompanying danger of persistent cross-reactive attack on normal tissue. In addition, unless the target molecule is essential for tumor survival, it may be down modulated by the cancer cell.

Vaccines designed to induce antibodies should generally present whole proteins in a natural conformational form. There has been only limited success using B cell epitopes, which are often discontinuous. Expression and purification of tumor-derived proteins are not simple procedures, especially since many are glycosylated by a subtle process that is cell dependent. An example of an ideal cell surface antigenic target is the idiotypic (Id) Ig expressed by B cell tumors. Not only is it tumor specific, but it appears to be a requirement for survival of many B cell malignancies (McCarthy et al., 2003). Anti-Id antibody, delivered either passively or induced by active vaccination with Id protein conjugated to keyhole limpet hemocyanin (KLH), can suppress lymphoma (Press et al., 2001).

The problem for Id antigen is that preparation of individual Id Igs is difficult and expensive, commonly involving heterohybridization of lymphoma cells with myeloma-derived cell lines. Recombinant technology is now used, but multichain proteins present a challenge. Single-chain proteins, such as single-chain Fv (scFv), can substitute to some extent, but these are also difficult to express and purify (Hawkins et al., 1994). DNA vaccination should help to overcome this difficulty and is discussed later in this chapter. Individualized approaches to treatment of cancer have been unattractive to most companies, but this appears to be changing, with two multicenter trials of Id–KLH protein vaccination for follicular lymphoma proceeding (Hurvitz and Timmerman, 2005).

In addition to inducing the desired antibody, Id protein vaccines can also activate T cell responses. The potential for this depends on whether the individual Ig variable region sequences contain recombinatorial or mutated sequences able to bind to the host's major histocompatibility complex (MHC) class I or II molecules and activate T cells in the repertoire. Human Id-derived peptides can be presented via the MHC class I molecules (Neelapu et al., 2004), but the variability of variable region sequences between patients, together with the wide range of haplotypes, makes identification of such epitopes very difficult. However, $CD4^+$ T cells are clearly activated and provide help for antibody and cellular responses. T helper 1 (Th1) cells can also attack MHC class II-expressing lymphomas directly and, apparently in some cases, indirectly. Indirect attack was observed in a myeloma model when transgenic interferon (IFN)-γ-producing anti-Id $CD4^+$ T cells suppressed tumor via activation of macrophages (Corthay et al., 2005). This is of interest since myeloma cells only secrete Id protein and do not express it at the cell surface. If Id vaccination succeeds in human myeloma, induction of Th1 responses could be explored for other proteins secreted by tumor cells.

The majority of potential tumor antigens, however, are intracellular and expressed only as peptides in the groove of the MHC class I molecules. The goal is to induce a high level of both effector and memory cytotoxic T lymphocytes (CTLs) from a setting where tolerance may be operative. Passive immunotherapy again predicts success since injection of allogeneic T cells has shown that CTLs are capable of eliminating tumor cells (Riddell et al., 2003). In fact, this finding has led to attempts to engineer T cells in vitro to express the desired antitumor specificity (Stanislawski et al., 2001). While this adoptive transfer approach has merit, there is still much to learn about how injected T cells will survive and migrate to tumor sites.

Vaccination should produce a continuing response, but delivery of antigen is critical. Protein vaccines delivered via an exogenous route are generally less effective in inducing CTL responses. An alternative is to bypass processing by using a known peptide epitope, thereby inducing a focused CTL response. It is important here to know that the candidate peptide is expressed by the tumor cells, and this can be assessed by elution of peptides from the tumor cell MHC class I molecules followed by sequence analysis (Rammensee et al., 2002). Peptide can be delivered either directly, preloaded onto dendritic cells (DCs), or via genetic strategies. There are opportunities to add immunoenhancing agents to the various delivery systems and thereby take advantage of new knowledge of activation of immunity. Genetic vaccines delivered alone or via viral/bacterial vectors will facilitate access to the intracellular processing pathways and allow the presenting cell to select the target peptides. In some respects, these vaccines mimic infectious agents and are therefore able to engage the immune system efficiently. Details of approaches used as of this writing will be discussed in subsequent sections.

III. SPONTANEOUS IMMUNITY TO CANCER

There has long been a question as to whether the immune system controls cancer by recognizing and eliminating aberrant cells. This is not easy to answer, but evidence in favor is accumulating and is discussed in Part I of this book. Mouse models have implicated both the innate and the adaptive immune systems in controlling cancer development (Smyth et al., 2006). An interesting hypothesis, with some experimental support, is that the immune system is not only active against developing tumor cells but that it "sculpts" the cellular phenotype, forcing downregulation of immunogenicity. This hypothesis, which is included in the proposed process of immunoediting presented in Chapter 2, has important implications for cancer vaccines because it predicts that vaccination is most likely to succeed early in disease. In addition, it also argues for targeting molecules that are essential for cell survival.

In human subjects, chromosomal translocations characteristic of B cell malignancies can be detected in B cells from healthy individuals, indicating that such cells must be being eliminated although an immune pathway has not yet been demonstrated. There is considerable evidence, however, for spontaneous immunity against cancer antigens in patients, with both antibody (Sahin et al., 1997) and T cell responses (Nagorsen et al., 2003) recorded. While such responses have clearly proved inadequate in patients with cancer, they do provide insight into immune activation. A striking example of effective spontaneous immune control of epithelial cancers has been indicated in the disease paraneoplastic neuronal degeneration. Here, antitumor responses have been revealed by a cross-reactivity with neuronal antigens (Darnell and Posner, 2006). Not only does this finding implicate immunity in controlling cancer, but it provides an example of the tightrope that needs to be walked between this desired objective and the undesirable side effect of autoimmunity. For vaccination, the choice of target antigens is clearly critical.

IV. TOLERAGENIC PRESSURE ON IMMUNITY TO CANCER

There are many reasons why tumor antigens might fail to induce and maintain an effective immune response. Induction requires access of tumor antigens to the secondary lymphoid organs. This access may be via DCs that, following activation, migrate to the draining lymph node. The presence of inflammatory stimuli is critical for this process, but these stimuli may not be present in the environment of a slowly developing spontaneous tumor. In this setting, tolerance can result and could be common in tumor sites. It is becoming clear that tumor cells can produce immunosuppressive molecules that can act at all levels of immunity. These will be discussed in Part III of this book. Once in the lymph node, activated DCs can present antigen to naive $CD4^+$ T cells but only if a repertoire of appropriate T cell receptors has been generated by thymic selection. If not, there will be no induction of $CD4^+$ T cells, which are required to provide help for both B cells and $CD8^+$ T cells.

A. T Cell Tolerance

Multiple influences operate on peripheral T cells that have escaped deletion and may be activated. These are part of the normal mechanism of suppression of responses in the periphery and could operate against antitumor responses. There is also the possibility of simple exhaustion of $CD8^+$ T cells, which can apparently be blocked by inhibiting the interaction between programmed death 1 (PD-1) and its ligands PD-L1/L2 (see Part III and Chapter 18). Details of this mechanism of immune control and the potential roles of regulatory T cells (Tregs), T cell loss, tumor-associated macrophages, suppressor myeloid cells, and environmental enzymes in immune control are presented in other chapters of Part III in the context of their relevance to cancer vaccines. It is important to note that engaging certain Toll-like receptors (TLRs), such as TLR8, can overcome the suppressive activity of Tregs (Peng et al., 2005). The challenge for vaccination is to reverse peripheral tolerance by providing candidate antigens in a form to induce immunity. Encouragingly, it has been observed that interleukin (IL)-15 can change tolerant high-affinity $CD8^+$ T cells into cytotoxic T cells able to kill leukemic cells (Teague et al., 2006). The cancer antigens should ideally be "dressed up" to look like pathogens, and it is important to define the factors required.

B. B Cell Tolerance

For B cells, it is clear that self-reactive surface Ig^+ B cells can be detected in the periphery. However, two factors control activation of B cells: engagement of the B cell receptor (BCR) by antigen and provision of T cell help via CD40L–CD40 interaction and cytokines. In addition, for differentiation and proliferation of naive B cells, a third synergizing signal provided by microbial products acting via TLRs is required (Ruprecht and Lanzavecchia, 2006). Persistent exposure of B cells to self antigens, however, leads to tolerance (Nossal and Karvelas, 1990). Several toleragenic mechanisms have been implicated, including deletion, anergy, and Ig receptor editing (Ferry et al., 2006). Transgenic models have indicated that the form of antigen is important, with tolerance more likely with soluble rather than cell-bound antigen (Akkaraju et al., 1997).

For vaccines aimed to induce antibody against autologous tumor antigens, it is clearly necessary to include with the antigen a TLR signal and a means of inducing T cell help. This is commonly achieved by adjuvant and coupling antigens to KLH, as described earlier for Id Ig antigen in lymphoma. Conjugation with KLH has merit since T cell help from a large repertoire will provide help for induction of antibody. The same argument applies to DNA delivery of Id antigen, during which part of the tetanus toxin sequence has been used to generate T cell help (Stevenson et al., 2004a). The selected part is the nontoxic portion of tetanus toxin, the so-called fragment C (FrC), and this is highly effective in amplifying anti-Id antibody responses (Spellerberg et al., 1997). The fused FrC acts by engaging anti-FrC $CD4^+$ T cells to help the B cells producing anti-Id antibody via the classical hapten-carrier principle. Once induced, antibody is likely to be produced over time by long-lived plasma cells.

This long-term production may be important since the induced anti-Id antibody response may not be maintained via continuing differentiation of memory B cells. The reason for this is that memory B cells are vulnerable to ablation in the absence of adequate T cell help (Savelyeva et al., 2005). This was shown by injecting Id Ig alone, which caused complete loss of the

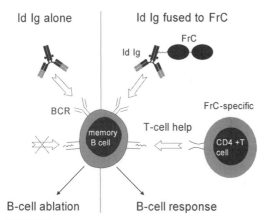

FIGURE 11.1 **Fate of memory B cells is determined by the availability of T-cell help.** Memory B cells specific for idiotypic Ig (Id Ig) were induced by vaccination with a DNA vaccine encoding idiotypic sequence fused to a gene encoding Fragment C (FrC) of tetanus toxin (DNA Id-FrC). Following priming, Id Ig protein either alone, or fused to FrC was injected. Id Ig alone caused rapid loss of anti-Id memory B cells, but Id-FrC protein stimulated the anti-Id B-cell response via provision of T-cell help from FrC-specific CD4+ T cells.

anti-Id response (Figure 11.1). In contrast, injection of Id Ig fused to FrC protein, which can reengage T cell help, boosted the anti-Id antibody response. Most autologous tumor antigens per se induce only limited T cell help. In addition, anti-Id T cells may be subjected to regulation, further reducing the availability of CD4+ T cells. This leaves vaccine-induced memory B cells sensitive to antigen-induced ablation. Similar observations have been made for carbohydrate antigens lacking T cell help (Mohiuddin et al., 2003). One way to overcome this result is to continue to vaccinate with antigen fused to either FrC or KLH, thereby protecting the B cell response by provision of T cell help.

Interestingly, there is some evidence that B cells can also have regulatory function (Mizoguchi and Bhan, 2006). Although less studied than Tregs, B cell studies have shown that the cells can produce IL-10, which directly inhibits inflammatory cascades, and can produce transforming growth factor β1 (TGF-β1), which induces apoptosis of effector T cells. Additional dampening mechanisms may involve antigen presentation and recruitment of Treg subsets and NKT cells. The default situation for both arms of the immune response is possibly regulation, and vaccination needs to reverse this.

V. IMMUNE RESPONSES TO CONVENTIONAL VACCINES

Vaccination against infectious organisms to prevent disease is one of the major achievements of medical science. Beginning with smallpox vaccines, with the disease now eradicated, successful vaccines are now in use against many common bacterial and viral infections, with a dramatic impact on public health worldwide. It might be thought therefore that cancer vaccines would be able to build on knowledge accrued from this success. However, in the past, vaccine development against infection had the single goal of protecting against infection, and the major pathway involved in this is antibody. For viruses associated with cancer development, induction of antibody-mediated immunity is a shared goal. HBV vaccination is clearly successful in suppressing subsequent hepatoma (Zuckerman, 2006). Trials of virus-like particles derived from HPV envelope protein are showing high levels of protection and an accompanying reduction in intraepithelial neoplasia and cervical cancer (Lowy and Schiller, 2006). However, at present, preventative vaccination can deal with only a small fraction of cancer.

For therapeutic vaccination, induction of antibody is certainly a desired goal for a minority of tumor cell surface antigens like Id Ig (see earlier) or HER2/neu (Dela Cruz et al., 2003). To plan effective vaccination protocols for patients, knowledge of the level and kinetics of antibody responses against pathogens could therefore be useful. However, these parameters appear to vary with the nature of the vaccine, and it is difficult to draw overall conclusions. The goals of investigators of responses to vaccination are mainly to establish the level and longevity of protective immunity. They have revealed that the peak of the primary response often takes several weeks to attain and that boosting is essential. In fact, multiple injections are generally required to reach levels of neutralizing antibodies sufficient for protection (Bocher et al., 1999). More recent studies are now emerging from the successful vaccinations against HPV where the viral-like particles can prime antibody responses or boost natural responses to infection (Villa et al., 2006). One factor that is evident from analyses of immune responses in human subjects is the influence of age.

Since most cancers develop in the older age group, knowledge of the apparent decline in immune capacity with age is required. In one study of antibody levels induced by influenza vaccination, a decreased response was clearly evident in the elderly (Goodwin et al., 2006). Maintenance of CD4$^+$ T cell responses to the influenza vaccine was also decreased in the older age group (Kang et al., 2004), possibly due to "hijacking" the response by reactivating cytomegalovirus (CMV) (see following paragraphs). In CMV-seropositive individuals, aging appears to be accompanied by a gradual increase in CMV-specific T cells together with a decrease in functionality (Khan et al., 2004). Interestingly, this can reduce concomitant immunity to Epstein-Barr virus (EBV) and could have a global detrimental effect on immune responsiveness. Clearly, immunosenescence is shaped by immunological experience (Vasto et al., 2006), and this experience is likely to impact on natural and induced immunity to cancer.

Although T cell responses are clearly important in controlling some infections, there is much less information available on T cell responses against conventional vaccines. For HBV vaccination, which uses a recombinant protein, proliferative responses to intramuscular injection appeared to develop within 2 weeks but then declined over the following months in spite of further booster injections. Intradermal injection appears to induce a delayed, but more sustained, cellular response (Rahman et al., 2000). Secondary responses against HBV and tetanus toxoid generated cytokine-producing T cells within 1 week after the booster injection and tended to reach peak levels 1–2 weeks later (De Rosa et al., 2004; Tassignon et al., 2005). Primary proliferative responses to mycobacterium bovis Bacille Calmette-Guerin (BCG) showed a rather slower response developing over several weeks and still rising after 8 weeks. The secondary response was again rapid, peaking at week 1 and being maintained for at least 8 weeks (Ravn et al., 1997). In summary, there are major differences in the kinetics of responses against each vaccine although the principle of a faster response after boosting holds. For priming, the nature and level of the antigen, the formulation, and site of injection are likely variables. Age is also an influence on both priming and maintenance of memory (Kang et al., 2004). These findings are lessons for cancer vaccines.

A. Memory T Cell Responses Induced by Conventional Vaccination

A key goal of all vaccinations is to induce immunological memory. The nature of memory is being unraveled, mainly in models, and this important topic is the subject of an issue of *Immunological Reviews* (June 2006, Vol. 211, Issue 1). While memory will not be discussed here, some observations in human subjects, who, in contrast to laboratory mice, are continuously buffeted by environmental pathogens, may have relevance. A study of immune activation induced by a boosting of normal human subjects with tetanus toxin (TT) (Di Genova et al., 2006) is an example of the value of information collected in this simple setting using modern assay techniques. It is built on previous observations that vaccination with one antigen can expand responses to other antigens by a "bystander" effect.

In the earlier study, injection of TT generated the expected increase in frequency of

TT-specific T cells. However, this finding was accompanied by a modest but significant increase in the frequency of T cells specific for herpes simplex virus (Donnenberg et al., 1984). Similarly, in another study, three healthy subjects responding to a booster injection of TT also showed an increased proliferative response to purified protein derivative (PPD) of tuberculin (Fernandez et al., 1994). In both studies, the bystander stimulation was of preexisting memory T (T_{mem}) cells.

The chapter authors took this experiment further by analyzing the kinetics of T cell responses against a booster injection of TT and comparing the results with responses against PPD and another antigen from *Candida albicans* (*C alb*). The study revealed that memory CD4[+] T cells against the antigens not in the TT vaccine expanded with kinetics similar to the TT-specific T cells (Di Genova et al., 2006). As illustrated for one representative patient from the study (Figure 11.2A), the proliferative responses against TT and PPD expanded together, with an accompanying slight increase in the response to *C alb*. Cytokine-producing T cells also rose and fell together. It is

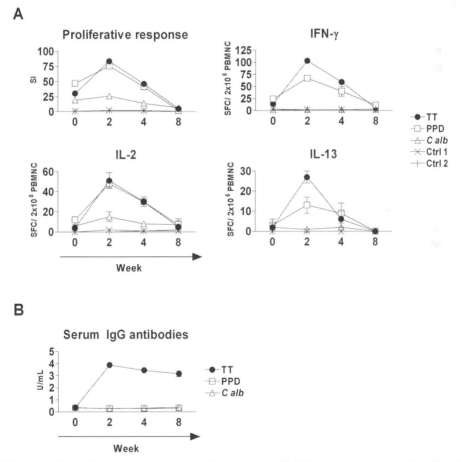

FIGURE 11.2 **Effects of booster vaccination with tetanus toxoid (TT) on human memory T cells and antibody responses against TT, purified protein derivative of tuberculin (PPD) and *Candida albicans* (*C alb*) antigen. A,** Vaccination with conventional TT led to expansion of specific CD4[+] T cells able to proliferate *in vitro* and to produce IFN-γ, IL-2 and IL-13. Bystander expansion of memory T cells specific for PPD occurred with parallel kinetics. A small expansion of the low level of memory T cells specific for *C alb* was also evident, but no response to control Ig antigens (Ctrl 1 and 2) was detected. **B,** Antibody levels increased only against the TT vaccine antigen. The data are from one representative patient from our study (Di Genova et al., 2006). *SI,* Stimulation index; *SFC,* spot forming cells.

important to note that naive T cells (controls 1 and 2) were unaffected, restricting the bystander activation to T_{mem} cells (Figure 11.2A). Interestingly, B cell responses were constrained at the level of antibody production that was confined to the immunizing TT antigen with no change in levels of antibodies against PPD or *C alb* (Figure 11.2B). However, bystander memory B cell stimulation can occur, leading to a transient expansion of plasma cells of unrelated specificities in the blood (Bernasconi et al., 2002). It appears, therefore, that T_{mem} and B cells are stimulated in a nonspecific manner during stimulation of specific T cells although no significant increase in antibody levels results. Bystander stimulation, however, may be a strategy for maintenance of memory, which for T cells in mice appears to involve IL-7 (Stockinger et al., 2006). This information is a cautionary tale particularly for those involved in monitoring T cell responses to vaccines because if T_{mem} cells are present, they might be stimulated by other environmental antigens rather than by the specific vaccine.

B. Memory T Cell Subsets in Persistent Infection

Phenotypic identification of functional subsets of T_{mem} cells is now developing. The most important markers so far are the CD45 isoforms, CD45RA and CD45RO, expressed by naive and T_{mem} cells, respectively. However, this is not absolute since CD45A can be detected on antigen-experienced CD4$^+$ and CD8$^+$ T cells (Akbar and Fletcher, 2005). The costimulatory molecules CD27 and CD28 and the chemokine receptor CCR7 also provide distinguishing markers. Since naive but not memory T cells require costimulation, CD27 and CD28 are expressed and then downregulated on maturation to effector cells. CCR7 expression is modulated according to migratory requirements, and CD62L is required for T cells to cross high endothelial venules. Expression of CD45RA and CCR7 has been used to define two main subsets of T_{mem} cells. These are (i) central memory (T_{CM}) cells (CD45RA$^-$CCR7$^+$), which home to secondary lymph nodes and proliferate strongly in response to antigen but lack immediate effector function, and (ii) effector memory T cells (T_{EM}) (CD45RA$^-$CCR7$^-$), located in extralymphoid organs, with immediate effector function (Sallusto et al., 2004).

The relationships between these subsets in the steady state and the perturbation by acute resolving or persistent infections are under investigation. However, even for persistent viruses, the pattern of expression appears to vary, with, for example, hepatitis C-specific CD8$^+$ T cells showing an early differentiation phenotype (CD27$^+$CD28$^+$), whereas CMV-specific CD8$^+$ T cells have a late differentiation phenotype (CD27$^-$CD28$^-$). Early intermediate and intermediate phenotypes characterize hepatitis B-specific or HIV-specific CD8$^+$ T cells respectively (Appay and Rowland-Jones, 2004). In EBV infection, the pattern varies according to whether T cells recognize lytic or latent antigens (Callan, 2004). There are also likely to be differences in T cell phenotype according to location (Klenerman and Hill, 2005), with almost all data in human subjects necessarily derived from blood cells. One influence on the level and phenotype of virus-specific CD8$^+$ T cells is likely to be the degree of antigen reencounter (Klenerman and Hill, 2005). Where frequent viral reactivation occurs, expansion and differentiation of T cells toward a late phenotype might occur. In immunocompetent hosts, these episodes are dealt with successfully. However, in patients with immune capacity compromised by disease or treatment or even in patients who are experiencing the natural aging process (Vasto et al., 2006), abnormal expansion, dysfunction, or exhaustion of CD8$^+$ T cells can occur. Clearly, cancer antigens could behave like persistent viral antigens in having the potential to perturb either a spontaneous or an induced immune response by similar pathways.

For CD4+ T cells, phenotypic and functional changes along the postthymic route of development also occur. Studies in persistent viral infections again show variability between different viruses (Appay and Rowland-Jones, 2004). CD4+ T cells certainly play a major role in long-term control of persistent infections, either by supporting CD8+ T cells or by targeting infected cells directly (Heller et al., 2006). Again, antigen load is a determinant, and continuous high levels can lead to loss of IL-2 secretion, or in some cases of hepatitis C, there could be the appearance of Tregs (Klenerman and Hill, 2005).

Persistent CMV infection perturbs memory T cell pools, with dominating expansions, as is evident in the elderly (see earlier discussion). Generally, the balance between subsets of both CD4+ and CD8+ T cells is important in mediating control of infection without immunopathology. An interesting emerging concept is that only part of the T cell response, comprising the so-called "driver T cells," is involved in protection (Klenerman and Hill, 2005). Relevance to cancer is clear in two respects: first, in terms of pathogenesis, since one of the consequences of failure to control infections such as EBV and hepatitis C is cancer; second, in understanding which immune pathways should be activated for control of cancer cell growth and progression. Information on the latter will assist monitoring strategies for cancer vaccines and help to strengthen the value of surrogate markers for evaluating efficacy.

C. T Cell Vaccines

Induction of T cell responses is clearly important for intracellular antigens. Interestingly, the prophylactic vaccines against HBV or HPV have poor effects on persistent infection (Autran et al., 2004), demonstrating the need to activate CD8+ T cell attack in this setting. The same applies to most chronic infections and to the majority of cancer antigens. Development of the so-called "T cell vaccines" has been expanded by the, so far unfulfilled, ambition to vaccinate against HIV (Robinson and Amara, 2005). While HIV, in its immense genetic variability and destructive effects on CD4+ T cells, may be an especially difficult case, it has helped to move T cell vaccines forward. One reason for this is that neutralizing antibodies have not proved effective so far, and CTLs are important in controlling viral spread (Robinson and Amara, 2005). Useful data have emerged from testing various vaccine designs against simian-human immunodeficiency virus (SHIV) in primates. The consensus appears to be that protein vaccines have not been effective and that gene-based vaccines are superior. The preferred strategy involves priming with a DNA vaccine and boosting with the same sequences delivered via a viral vector, usually modified vaccinia Ankara (MVA). This delivery combination is also able to generate some protection against malaria in human subjects (McConkey et al., 2003).

Gene-based vaccines allow manipulation of antigen and incorporation of additional sequences to modulate and amplify outcome (see following paragraphs). Evaluation of performance then requires objective measurement of responding T cells, including an analysis of phenotype, cytokine profile, and lytic ability. New technology (as described later in this chapter) is allowing a clear view of the kinetics and level of CD8+ T cells in large animals including human subjects. For SHIV, correlation with protection from infection is not perfect for two reasons: first, protection depends on the ability of memory T cells to expand on challenge, and second, measurement of T cells in blood does not detect effector T cells that have migrated to the site of infection (Robinson and Amara, 2005). Even so, it is already clear that the way antigen is delivered has a significant influence on induction and maintenance of immunity, and this information is vital for cancer vaccines.

VI. CANCER VACCINE STRATEGIES

Therapeutic vaccines against cancer face many of the difficulties associated with persistent infection, together with the damaging effects of cancer and/or treatment on immune capacity. The optimal setting for vaccination should be when disease load has been minimized by surgery or chemo/radiotherapy but where a functional immune system has been restored. Stem cell transplantation from either an autologous or allogeneic source can repopulate the lymphoid system, and, in fact, chemotherapy-induced lymphopenia encourages expansion of passively injected T cells to fill available niches (Dudley et al., 2005). For active vaccination, information on recovery of immune capacity in human subjects is limited. Studies are hampered by the immense variability in therapeutic schedules and in the subsequent course of disease. However, in the chapter authors' study of patients with multiple myeloma who had undergone intensive chemotherapy followed by an autologous transplant, the patients experienced substantial recovery of responsiveness to TT vaccination within 6 months to 1 year (McNicoll, 2005).

Melanoma has long been a chosen target for attempts to activate an antitumor immune response. One reason for this is the suggestive evidence for spontaneous immunity in some patients (Boon et al., 2006). However, immune regulation is likely to operate in this setting. Careful analysis of epitope-specific T cells has shown that this is the case, with T cells rendered ineffective due to local immunosuppression occurring at tumor sites (Boon et al., 2006). The goal of vaccination in immunogenic tumors exposed to the immune system should therefore be to reverse this situation, and this appears to be feasible in some cases. One encouraging aspect is that successful reversal can then induce additional immunity against other antigens that are not present in the vaccine (Boon et al., 2006).

A. Peptide Vaccines

For T cell vaccines, the objective is to induce high levels of $CD8^+$ T cells able to recognize specific epitopes bound to MHC class I molecules on the tumor cell surface. Candidate epitopes able to bind to specific class I haplotypes can now be predicted from antigen sequence analysis by using increasingly accurate algorithms (Rammensee et al., 2002). For induction and maintenance of $CD8^+$ T cells, support from $CD4^+$ T cells is required (Janssen et al., 2003), and this might account for the relative failure of short peptide vaccines. The other problem associated with that approach is that short peptides will not bind selectively to DCs, and binding to other lymphocytes could generate tolerance. This problem can be circumvented by the use of longer peptides that both target DCs and induce T cell help (Vambutas et al., 2005), and these are in clinical trial as of this writing. Peptide vaccines require adjuvant to activate DCs and may also benefit from slow release from an oil/water mixture of Freund's adjuvant. The range of adjuvants has increased in the twenty-first century, especially with the advent of synthetic oligonucleotides linked by phosphorothioate bonds, which include the CpG motif of cytosine linked to guanine. In human subjects, appear to stimulate plasmacytoid DCs and B cells via TLR9 (Krieg, 2006). No doubt a range of TLR ligands will emerge able to bind to selected TLRs, and exploration of the potential for specific adjuvants to direct and amplify immunity is expanding rapidly.

B. DC-Based Vaccines

Delivery of peptide or protein antigens can be improved by loading DCs *in vitro*. Regardless of the source of the DCs, the maturational status and mode of reinjection have to be considered. There are complexities in determining maturational status using phenotypic analysis, and, since DCs can also be tolerogenic, caution is needed (Reis e Sousa, 2006). In spite of this, several clinical trials of this approach have been carried out. The encouraging but variable results that have been reviewed in the literature (Pardoll, 2002) are discussed further in Chapter 14. A more convenient approach is to target DCs *in vivo* by linking the antigen to a molecule able to bind to a receptor expressed on a DC's surface. Fusion proteins with granulocyte-macrophage colony-stimulating factor (GM-CSF) have been used as well as monoclonal antibodies (MoAbs) directed against candidate receptors (Pardoll, 2002). One danger is that targeting can lead to tolerance, unless DC activation is ensured. This combined approach has been tested with a model antigen in mice, where a MoAb against DEC-205, an endocytic receptor on a DC, was able to induce high levels of $CD8^+$ T cells only if an activating anti-CD40 MoAb was coinjected (Bonifaz et al., 2004).

The strategy of loading DC with mRNA has also reached the clinic (Gilboa and Vieweg, 2004). Initially, a total tumor-derived RNA was used, but this was then refined to the use of specific mRNA-encoded antigens. Comparison with protein loading indicated that mRNA transfection is more effective, and a clinical trial using prostate-specific antigen (PSA) gave the striking result that most patients responded (Gilboa and Vieweg, 2004). There was also suggestive evidence of clinical effects. While these findings are very promising, the technical questions surrounding optimization and standardization of DC vaccines remain, and further clinical testing is awaited.

VII. DNA VACCINES

DNA vaccines have the advantage of simplicity and flexibility. They already contain immunostimulatory CpG sequences in the backbone of bacterial DNA, able to activate innate immunity via TLR9 and possibly by other pathways (Spies et al., 2003). It is important to note that DNA vaccines are cassettes of coding information for delivery to the transfected cell. Injection sites include muscle, skin, and mucosa, with several devices available to aid uptake (Wolff and Budker, 2005). The opportunity to place manipulated target gene sequences into plasmid vaccines to generate antibody, $CD4^+$ T cells, and/or $CD8^+$ T cells is unprecedented. The technology provides a direct genetic link from identification of tumor-derived sequence variants to induction of immunity. Added to this is the ability to codeliver selected molecules to activate specific immune pathways (Stevenson et al., 2004a). The generic design of a multicomponent DNA vaccine is shown in Figure 11.3. In addition to immunostimulatory CpG sequences, this design includes the antigen sequence that expresses a selected molecular form, which can be either a full-length protein or a peptide sequence (Stevenson et al., 2004a). A leader sequence can be added to direct protein to the endoplasmic reticulum. To activate T cell help, the chapter authors have chosen to fuse a sequence from tetanus toxin to the tumor antigen (see following paragraphs). Finally, there is the opportunity to add further sequences encoding immunomodulatory proteins within the backbone (Figure 11.3).

A. Activation of T Cell Help for Antitumor Immunity

The focus on vaccine designs has been on the importance of $CD4^+$ T cells in inducing and maintaining both antibody and $CD8^+$ T cell responses. For cancer patients, it appears that antitumor $CD4^+$ T cell responses may be inadequate and/or subject to regulation. For vaccination to succeed, the large antimicrobial $CD4^+$ T cell repertoire should be used. By fusing tumor gene sequences to microbial genes, it would then be possible to amplify antitumor immunity by capturing antimicrobial T cell help, even in the absence of antitumor $CD4^+$ T cells. The initial tumor antigen was the safe clonotypic Id antigen from B cell lymphoma (King et al., 1998). For these patient-specific

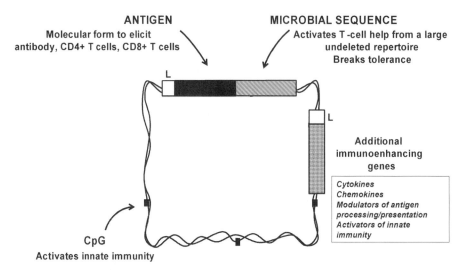

FIGURE 11.3 Multi-component DNA vaccine designs to induce and direct immune responses. DNA plasmid vaccines have CpG sequences in the backbone to activate innate immunity. Antigen sequences can be incorporated to express a selected molecular form, together with sequences to induce T-cell help. Additional genes can be included to further enhance and direct immune outcome. L, Leader sequence.

vaccines, DNA technology simplified vaccine manufacturing. The Id antigen was assembled as scFv, and the microbial sequence was the FrC of tetanus toxin. The fused scFv–FrC sequences raised high levels of anti-Id antibodies in mice and provided Id-specific protection against lymphoma, which is largely antibody mediated (King et al., 1998). Similar amplification of antitumor responses could be obtained by fusing other tumor antigen sequences with FrC, illustrating the generality of the principle for inducing antibody and CD4+ T cell responses (Padua et al., 2003; Thirdborough et al., 2002).

According to the chapter authors' experiences, induction of optimal CD8+ T cell responses requires a modified approach. Activation of the CD8+ T cell pathway is a known strength of DNA vaccines because encoded protein appears to gain access to the intracellular antigen processing machinery. However, T cell help is also required for optimal induction and maintenance of the response (Janssen et al., 2003). Provision of additional protein sequences, such as microbial genes, for this task has to face the problem of immunodominance (Yewdell and Bennink, 1999). This phenomenon operates naturally in human subjects: in viral infections, such as CMV, CTL activity in individuals appears to be highly focused onto only a few viral epitopes. These epitopes are likely to be the ones favored by the kinetics of antigenic expression and processing as well as by the level of T cells in the available repertoire. Superficially, this would appear to limit the range of immunity, but it is likely to reflect a need to control dangerous and potentially cross-reactive CTL activity. If the target epitope is lost by mutation, the T cell response can then refocus onto the next tier of epitopes, so protection should continue. The mechanism of immunodominance is not fully understood, but it is likely to be most serious at the time of boosting when CTLs induced at priming are able to eliminate APCs before any subdominant epitopes can be presented (Yewdell and Bennink, 1999). There is evidence of its operation in both mice and human subjects (Smith et al., 2005). To circumvent this, the Th-inducing sequence has been minimized from FrC to a single domain (Rice et al., 2002). Fortunately, all the MHC class I binding epitopes for mouse haplotypes and human HLA-A2 appear to be located in the second domain of FrC, which can be genetically excised (Rice et al., 2002). Target tumor epitopes have also been repositioned to the C terminus of the single domain (DOM), creating a p.DOM-epitope design. In several models, high levels of epitope-specific $CD8^+$ T cells could be induced (Rice et al., 2002; Stevenson et al., 2004a). These expanded on challenge and could kill target tumor cells efficiently. The chapter authors are testing the design in patients with prostate cancer using an epitope derived from prostate-specific membrane antigen. Electroporation is also being tested in this trial (see following paragraphs). Targeting a single epitope in cancer cells might appear limited, but there is the opportunity to inject two epitope-specific vaccines at different sites to generate dual attack in the absence of competition. It may be a good strategy to engage nondeleted CTLs against subdominant or aberrantly expressed epitopes. Novel design modifications are also being developed, with a focus on accessing antigen presentation pathways via either cross-presentation or direct presentation (Stevenson et al., 2004a). Preliminary data indicate that alternating designs aimed at each pathway can boost responses dramatically. Knowledge of the presentation route will inform new designs and help to plan delivery of additional factors, such as chemokines and cytokines (Ruffini et al., 2002). A range of additional molecules is being tested, including some targeted at the tumor environment (Zhou et al., 2005). There is now a stream of DNA vaccine designs aimed at a range of targets in infectious agents or cancer (Stevenson, 2004).

B. Obstacles for Human Vaccines

Clinical testing of individual DNA scFv–FrC fusion gene vaccines for patients with lymphoma is almost complete with evidence in a proportion of patients for induction of immune responses against both the FrC and the Id components of the fusion gene (Stevenson et al., 2004a). However, responses were weak and variable. A similar DNA vaccine encoding a chimeric Ig molecule with tumor-derived variable region sequences fused to the IgG2a and kappa mouse Ig constant regions induced apparently even lower levels of response (Timmerman et al., 2002). Results of these tentative trials have therefore strengthened the conclusion that it was not a vaccine design that was the problem, but that there was a deeper reason for difficulty in translating to human subjects.

A necessary pause for further investigation has temporarily slowed commercial development of DNA vaccines strategies, and there are in fact only two licensed for use, initially for fish and horses. In the clinic, while there have been indications of success in boosting antibody responses against hepatitis B using a skin delivery system (Tacket et al., 1999) and intriguing responses in individuals previously unresponsive to the licensed vaccine (Rottinghaus et al., 2003), the general levels of response using naked DNA vaccines, especially using an intramuscular route, have been low.

However, for intramuscular injection, one major problem has now become clear. From mouse models, it is known that the level of transfection is highly dependent on the volume injected (for a mouse, 30–50 µl). It appears that the hydrostatic pressure facilitates DNA delivery into the cells by increasing DNA distribution and uptake by myocytes (Otten et al., 2004). There may also be an added inflammatory stimulus. Reducing the volume in mice led to loss of response, and scaling up volume for human subjects would not be acceptable; therefore, other ways to circumvent this problem were required.

C. Physical Strategies to Improve Performance

The mechanism of uptake of naked DNA into tissue, and various ways to facilitate it, is of interest for both gene therapy and DNA vaccination (Wolff and Budker, 2005). For DNA vaccination, strategies to improve uptake and to induce local inflammation are attractive. Physical methods include encapsulation of DNA in polymeric microparticles or adsorption of DNA onto charged microparticles (Singh et al., 2006). For skin injection, delivery via needleless strategies or of DNA-coated gold particles by gene gun is being explored and could be especially useful for inducing immunity in that site (Dean, 2005).

Another promising strategy, especially for intramuscular injection, is electroporation. This is already a therapeutic modality for improving drug treatment of patients with cancer. Electroporation increases the number of muscle fibers that take up plasmid DNA. It has the advantage of transferring genes to both normal and regenerating muscle, leading to high levels of gene expression (Aihara and Miyazaki, 1998; Mathiesen, 1999). While electroporation raises the likelihood of integration into the genome (Wang et al., 2004), this is unlikely to be a major problem for cancer patients. Immune responses induced by adding electroporation to DNA injection in large animals (Mathiesen, 1999), including non-human primates, are increased to levels obtained previously and only attained by using live viral vectors (Otten et al., 2004). It seems, therefore, that there is at least one way to overcome the problem of DNA vaccine performance in human subjects, and trials are already under way (see the following paragraphs). Devices acceptable to patients have been developed (Tjelle et al., 2006), and although some degree of discomfort occurs, it is likely that this can be reduced. For skin injection, a parallel strategy is to use a tattooing device (Bins et al., 2005).

D. Prime/Boost Strategies

One strategy that has shown promise for enhancing the potency of DNA vaccines is the prime/boost vaccine approach. The requirements for priming and boosting of immune responses are not necessarily identical. For $CD8^+$ T cell responses, it appears that the low level of antigen expression following naked DNA vaccination is sufficient to prime responses, but boosting is relatively ineffective. Efficacy at boosting could be dramatically increased if the same antigen sequence was incorporated into a recombinant MVA (McConkey et al., 2003). If the DNA/MVA order was reversed, responses were abrogated. The detailed mechanism is unclear, but it is likely that boosting requires higher levels of antigen expression, possibly to overcome the loss of APCs by CTL attack. A higher level of inflammation induced by the MVA could also contribute. Other viral vectors have been found to be effective in boosting, and several trials of various combinations are underway against infectious disease, especially malaria (Hill, 2006) and HIV (Robinson and Amara, 2005). Bacterial vectors have also been explored and may have a particular place in inducing mucosal immunity.

One problem for using viral or bacterial vector delivery is the presence of preexisting immunity against vector-derived proteins. This is the case for adenoviral vectors and for MVA, and, although attempts are being made to denude viruses to evade this blocking immunity, it may be difficult. The use of lentiviruses could be an alternative but needs further investigation. Even if there is no preexisting immunity, it is likely to be generated after the first injection, thereby potentially blocking the effects of further boosters. For therapeutic vaccination against cancer, there is a likely requirement for continuing to offer booster vaccines; therefore, viral vectors may not be the right choice. Physical strategies are attractive to amplify antigen expression, and one possibility is to use electroporation. DNA priming followed by booster vaccines with the same vector plus electroporation can lead to very high levels of both antibody and CTL responses (Buchan et al., 2005). In fact, antibody responses achieved are comparable to those induced by the "gold standard" of protein plus complete Freund's adjuvant. Other physical strategies deserve investigation for similar effects.

VIII. CHALLENGES OF TRANSLATION TO THE CLINIC

Clinical trials of cancer vaccines are widespread, with only limited evidence for efficacy, mainly in hematological tumors (Rosenberg et al., 2004; Timmerman and Levy, 2004). One explanation may be the bias toward patients with advanced melanoma. As pointed out previously and as is the opinion of Mocellin et al. (2004), minimal residual disease is a more realistic setting for vaccine treatment, and it is too early to make judgments, especially with so many novel options for enhancing immunity becoming available. A very wide range of delivery strategies has been used, including peptides, DC-based vaccines, genetically modified whole tumor cell vaccines, viral vaccines, and various combinations. These have either been focused on defined antigens, such as CEA, MUC1, prostate, or melanoma antigens, or have used a less well-defined source of antigen (whole tumor cells or lysates).

According to a publicly available database (Wiley Interscience, *http://www.wiley.co.uk/genetherapy/clinical/*), 451 gene-based cancer vaccine studies are ongoing as of July 2006. For DNA vaccines, there are 75 clinical trials of DNA vaccination against cancer in progress internationally.

Several small trials are testing protocols of DNA priming and boosting with an MVA or other vectors. A trial of patients with melanoma using a string of peptide sequences induced apparent disease stabilization (Hawkins et al., 2006), but a similar protocol using the L523S antigen from lung cancer yielded disappointing results (Nemunaitis et al., 2006). A naked DNA vaccine trial against PSA, delivered together with cytokines, has also reported encouraging stabilization of PSA levels (Pavlenko et al., 2004) and some evidence of immune responses (Miller et al., 2005). These early investigations are difficult to summarize due to heterogeneity in vaccine design, disease, antigen, patient cohort, and clinical setting. The chapter authors' ongoing trial of an epitope-specific DNA fusion gene vaccine derived from prostate-specific membrane antigen and delivered ± electroporation will hopefully answer some questions.

Immune responses are likely to act as important surrogates for efficacy of cancer vaccines. It is vital, therefore, to have robust monitoring methods to make comparisons between different laboratories. Academic collaboratives, such as the Cancer Immunotherapy (CIMT) established in Germany, and academic/commercial collaboratives, such as the Cancer Vaccine Consortium in the United States, are addressing issues of standardization of assays. Antibody assays are relatively straightforward, but $CD8^+$ T cell assays require knowledge of the target peptides and usually have focused on HLA-A2 as a relatively common restriction element.

The range of available technology is rapidly expanding and now includes *ex vivo* or cultured enzyme-linked immunospot (ELISPOT) assay, flow cytometry-based cytokine secretion detection, single or multicytokine detection systems in enzyme-linked immunosorbent assay (ELISA)-type assays, and quantitative polymerase chain reaction (qPCR)-based detection systems of mRNA for cytokines (Hernandez-Fuentes et al., 2003; Hart and Heije, 2005). In some cases, tetramer binding can be used to measure levels of specific T cells, usually the $CD8^+$ subset. Correlation of immune phenotype with functional capacity of T cells and definition of markers that link cytotoxic function to phenotype will offer potent tools for assessing immune responses in clinical trials (Chattopadhyay et al., 2005; Rubio et al., 2003).

An emerging problem for all vaccine trials arises from the dramatic changes in the regulatory framework. Requirements for formally validated endpoints prior to beginning a trial are in danger of reducing the flexibility needed for developing technologies. The regulatory pressure on pilot clinical trials in academic centers is now reaching a point that virtually prevents the testing of any new approach. A dialogue between regulatory authorities and academic laboratories needs to take place so that promising strategies can be identified before larger trials are initiated, possibly with commercial involvement.

IX. CONCLUDING REMARKS

The desire to vaccinate cancer patients to prevent tumor growth and subsequent relapse is becoming realistic. However, many unknowns remain. Fast and flexible vaccine design and formulation are required, and preclinical models will still be required to test the importance of new immunological principles for breaking tolerance and activating immunity. It is likely that multiple designs will be needed to induce attack on antigenic targets of different molecular forms. Ideally, preclinical success should be quickly coupled to pilot clinical trials to test translational potential. Gene-based technologies provide obvious links from target molecule identification to vaccine vehicles. The concern that DNA vaccines, so successful in preclinical models, cannot translate to the clinic is being overcome by prime/boost strategies, some involving physical delivery techniques able to increase transfection *in vivo*. Clinical settings should be planned in partnership with current treatment modalities so that a recovering immune system can be mobilized when disease load is low. Efficacy in cancer patients should also be relevant for persistent infection, and the temporarily neglected collaboration between microbiology and immunology is now restored.

Acknowledgments

The authors gratefully acknowledge support from the Leukaemia Research Fund, Tenovus, and Cancer Research UK.

References

Aihara, H., and Miyazaki, J. (1998). Gene transfer into muscle by electroporation *in vivo*. *Nat. Biotechnol.* **16**, 867–870.

Akbar, A. N., and Fletcher, J. M. (2005). Memory T cell homeostasis and senescence during aging. *Curr. Opin. Immunol.* **17**, 480–485.

Akkaraju, S., Canaan, K., and Goodnow, C. C. (1997). Self-reactive B cells are not eliminated or inactivated by autoantigen expressed on thyroid epithelial cells. *J. Exp. Med.* **186**, 2005–2012.

Appay, V., and Rowland-Jones, S. L. (2004). Lessons from the study of T-cell differentiation in persistent human virus infection. *Semin. Immunol.* **16**, 205–212.

Autran, B., Carcelain, G., Combadiere, B., and Debre, P. (2004). Therapeutic vaccines for chronic infections. *Science* **305**, 205–208.

Bernasconi, N. L., Traggiai, E., and Lanzavecchia, A. (2002). Maintenance of serological memory by polyclonal activation of human memory B cells. *Science* **298**, 2199–2202.

Bins, A. D., Jorritsma, A., Wolkers, M. C., Hung, C. F., Wu, T. C., Schumacher, T. N., and Haanen, J. B. (2005). A rapid and potent DNA vaccination strategy defined by *in vivo* monitoring of antigen expression. *Nat. Med.* **11**, 899–904.

Bitton, R. J., Guthmann, M. D., Gabri, M. R., Carnero, A. J., Alonso, D. F., Fainboim, L., and Gomez, D. E. (2002). Cancer vaccines: An update with special focus on ganglioside antigens. *Oncol. Rep.* **9**, 267–276.

Bocher, W. O., Herzog-Hauff, S., Schlaak, J., Meyer zum Buschenfeld, K. H., and Lohr, H. F. (1999). Kinetics of hepatitis B surface antigen-specific immune responses in acute and chronic hepatitis B or after HBs vaccination: Stimulation of the *in vitro* antibody response by interferon gamma. *Hepatology* **29**, 238–244.

Bonifaz, L. C., Bonnyay, D. P., Charalambous, A., Darguste, D. I., Fujii, S., Soares, H., Brimnes, M. K., Moltedo, B., Moran, T. M., and Steinman, R. M. (2004). *In vivo* targeting of antigens to maturing dendritic cells via the DEC-205 receptor improves T cell vaccination. *J. Exp. Med.* **199**, 815–824.

Boon, T., Coulie, P. G., Van den Eynde, B. J., and van der Bruggen, P. (2006). Human T cell responses against melanoma. *Annu. Rev. Immunol.* **24**, 175–208.

Buchan, S., Gronevik, E., Mathiesen, I., King, C. A., Stevenson, F. K., and Rice, J. (2005). Electroporation as a "prime/boost" strategy for naked DNA vaccination against a tumor antigen. *J. Immunol.* **174**, 6292–6298.

Buske, C., Weigert, O., Dreyling, M., Unterhalt, M., and Hiddemann, W. (2006). Current status and perspective of antibody therapy in follicular lymphoma. *Haematologica* **91**, 104–112.

Callan, M. F. (2004). The immune response to Epstein-Barr virus. *Microbes Infect.* **6**, 937–945.

Chattopadhyay, P. K., Yu, J., and Roederer, M. (2005). A live-cell assay to detect antigen-specific CD4+ T cells with diverse cytokine profiles. *Nat. Med.* **11**, 1113–1117.

Corthay, A., Skovseth, D. K., Lundin, K. U., Rosjo, E., Omholt, H., Hofgaard, P. O., Haraldsen, G., and Bogen, B. (2005). Primary antitumor immune response mediated by CD4+ T cells. *Immunity* **22**, 371–383.

Darnell, R. B., and Posner, J. B. (2006). Paraneoplastic syndromes affecting the nervous system. *Semin. Oncol.* **33**, 270–298.

De Rosa, S. C., Lu, F. X., Yu, J., Perfetto, S. P., Falloon, J., Moser, S., Evans, T. G., Koup, R., Miller, C. J., and Roederer, M. (2004). Vaccination in humans generates broad T cell cytokine responses. *J. Immunol.* **173**, 5372–5380.

Dean, H. J. (2005). Epidermal delivery of protein and DNA vaccines. *Expert Opin. Drug Deliv.* **2**, 227–236.

Dela Cruz, J. S., Lau, S. Y., Ramirez, E. M., De Giovanni, C., Forni, G., Morrison, S. L., and Penichet, M. L. (2003). Protein vaccination with the HER2/neu extracellular domain plus anti-HER2/neu antibody-cytokine fusion proteins induces a protective anti-HER2/neu immune response in mice. *Vaccine* **21**, 1317–1326.

Di Genova, G., Roddick, J., McNicholl, F., and Stevenson, F. K. (2006). Vaccination of human subjects expands both specific and bystander memory T cells but antibody production remains vaccine specific. *Blood* **107**, 2806–2813.

Donnenberg, A. D., Elfenbein, G. J., and Santos, G. W. (1984). Secondary immunization with a protein antigen (tetanus toxoid) in man. Characterization of humoral and cell-mediated regulatory events. *Scand. J. Immunol.* **20**, 279–289.

Dudley, M. E., Wunderlich, J. R., Yang, J. C., Sherry, R. M., Topalian, S. L., Restifo, N. P., Royal, R. E., Kammula, U., White, D. E., Mavroukakis, S. A., *et al.* (2005). Adoptive cell transfer therapy following non-myeloablative but lymphodepleting chemotherapy for the treatment of patients with refractory metastatic melanoma. *J. Clin. Oncol.* **23**, 2346–2357.

Fernandez, V., Andersson, J., Andersson, U., and Troye-Blomberg, M. (1994). Cytokine synthesis analyzed at the single-cell level before and after revaccination with tetanus toxoid. *Eur. J. Immunol.* **24**, 1808–1815.

Ferry, H., Leung, J. C., Lewis, G., Nijnik, A., Silver, K., Lambe, T., and Cornall, R. J. (2006). B-cell tolerance. *Transplantation* **81**, 308–315.

Gilboa, E., and Vieweg, J. (2004). Cancer immunotherapy with mRNA-transfected dendritic cells. *Immunol. Rev.* **199**, 251–263.

Goodwin, K., Viboud, C., and Simonsen, L. (2006). Antibody response to influenza vaccination in the elderly: A quantitative review. *Vaccine* **24**, 1159–1169.

Hawkins, R. E., Dangoor, A., Keilholz, U., Schadendorf, D., Harris, A., Ottensmeier, C., Smyth, J., Hoffmnan, K., Anderson, R., and Pearce, G. (2006). Phase I/II trial of a PrimeBoost therapeutic vaccine in stage III/IV metastatic melanoma. Abstract No. 8030 in *ASCO Ann. Proc: Part I. J. Clin. Oncol.* **24(18S).**

Hawkins, R. E., Zhu, D., Ovecka, M., Winter, G., Hamblin, T. J., Long, A., and Stevenson, F. K. (1994). Idiotypic vaccination against human B-cell lymphoma. Rescue of variable region gene sequences from biopsy material for assembly as single-chain Fv personal vaccines. *Blood* **83**, 3279–3288.

Heller, K. N., Gurer, C., and Munz, C. (2006). Virus-specific CD4+ T cells: Ready for direct attack. *J. Exp. Med.* **203**, 805–808.

Hernandez-Fuentes, M. P., Warrens, A. N., and Lechler, R. I. (2003). Immunologic monitoring. *Immunol. Rev.* **196**, 247–264.

Hill, A. V. (2006). Pre-erythrocytic malaria vaccines: Towards greater efficacy. *Nat. Rev. Immunol.* **6**, 21–32.

Hurvitz, S. A., and Timmerman, J. M. (2005). Current status of therapeutic vaccines for non-Hodgkin's lymphoma. *Curr. Opin. Oncol.* **17**, 432–440.

Janssen, E. M., Lemmens, E. E., Wolfe, T., Christen, U., von Herrath, M. G., and Schoenberger, S. P. (2003). CD4+ T cells are required for secondary expansion and memory in CD8+ T lymphocytes. *Nature* **421**, 852–856.

Kang, I., Hong, M. S., Nolasco, H., Park, S. H., Dan, J. M., Choi, J. Y., and Craft, J. (2004). Age-associated change in the frequency of memory CD4+ T cells impairs long term CD4+ T cell responses to influenza vaccine. *J. Immunol.* **173**, 673–681.

Khan, N., Hislop, A., Gudgeon, N., Cobbold, M., Khanna, R., Nayak, L., Rickinson, A. B., and Moss, P. A. (2004). Herpesvirus-specific CD8 T cell immunity in old age: Cytomegalovirus impairs the response to a coresident EBV infection. *J. Immunol.* **173**, 7481–7489.

King, C. A., Spellerberg, M. B., Zhu, D., Rice, J., Sahota, S. S., Thompsett, A. R., Hamblin, T. J., Radl, J., and Stevenson, F. K. (1998). DNA vaccines with single-chain Fv fused to fragment C of tetanus toxin induce protective immunity against lymphoma and myeloma. *Nat. Med.* **4**, 1281–1286.

Klenerman, P., and Hill, A. (2005). T cells and viral persistence: Lessons from diverse infections. *Nat. Immunol.* **6**, 873–879.

Krieg, A. M. (2006). Therapeutic potential of Toll-like receptor 9 activation. *Nat. Rev. Drug Discov.* **5**, 471–484.

Lowy, D. R., and Schiller, J. T. (2006). Prophylactic human papillomavirus vaccines. *J. Clin. Invest.* **116**, 1167–1173.

Massoudi, M. S., Barker, L., and Schwartz, B. (2003). Effectiveness of postexposure vaccination for the prevention of smallpox: Results of a delphi analysis. *J. Infect. Dis.* **188**, 973–976.

Mathiesen, I. (1999). Electropermeabilization of skeletal muscle enhances gene transfer *in vivo*. *Gene Ther.* **6**, 508–514.

McCarthy, H., Ottensmeier, C. H., Hamblin, T. J., and Stevenson, F. K. (2003). Anti-idiotype vaccines. *Br. J. Haematol.* **123**, 770–781.

McConkey, S. J., Reece, W. H., Moorthy, V. S., Webster, D., Dunachie, S., Butcher, G., Vuola, J. M., Blanchard, T. J., Gothard, P., Watkins, K., *et al.* (2003). Enhanced T-cell immunogenicity of plasmid DNA vaccines boosted by recombinant modified vaccinia virus Ankara in humans. *Nat. Med.* **9**, 729–735.

McKeage, K., and Perry, C. M. (2002). Trastuzumab: A review of its use in the treatment of metastatic breast cancer overexpressing HER2. *Drugs* **62**, 209–243.

McNicoll, F. P. (2005). PhD thesis, University of Southampton, U.K.

Miller, A. M., Ozenci, V., Kiessling, R., and Pisa, P. (2005). Immune monitoring in a phase 1 trial of a PSA DNA vaccine in patients with hormone-refractory prostate cancer. *J. Immunother.* **28**, 389–395.

Mizoguchi, A., and Bhan, A. K. (2006). A case for regulatory B cells. *J. Immunol.* **176**, 705–710.

Mocellin, S., Mandruzzato, S., Bronte, V., and Marincola, F. M. (2004). Cancer vaccines: Pessimism in check. *Nat. Med.* **10**, 1278–1279; author reply, 1279–1280.

Mohiuddin, M. M., Ogawa, H., Yin, D. P., and Galili, U. (2003). Tolerance induction to a mammalian blood group-like carbohydrate antigen by syngeneic lymphocytes expressing the antigen, II: Tolerance induction on memory B cells. *Blood* **102**, 229–236.

Nagorsen, D., Scheibenbogen, C., Marincola, F. M., Letsch, A., and Keilholz, U. (2003). Natural T cell immunity against cancer. *Clin. Cancer Res.* **9**, 4296–4303.

Neelapu, S. S., Baskar, S., Gause, B. L., Kobrin, C. B., Watson, T. M., Frye, A. R., Pennington, R., Harvey, L., Jaffe, E. S., Robb, R. J., *et al.* (2004). Human autologous tumor-specific T-cell responses induced by liposomal delivery of a lymphoma antigen. *Clin. Cancer Res.* **10**, 8309–8317.

Nemunaitis, J., Meyers, T., Senzer, N., Cunningham, C., West, H., Vallieres, E., Anthony, S., Vukelja, S., Berman, B., Tully, H., *et al.* (2006). Phase I Trial of sequential administration of recombinant DNA and adenovirus expressing L523S protein in early stage non-small-cell lung cancer. *Mol. Ther.* **13**, 1185–1191.

Nossal, G. J., and Karvelas, M. (1990). Soluble antigen abrogates the appearance of anti-protein IgG1-forming cell precursors during primary immunization. *Proc. Natl. Acad. Sci. USA* **87**, 1615–1619.

Otten, G., Schaefer, M., Doe, B., Liu, H., Srivastava, I., zur Megede, J., O'Hagan, D., Donnelly, J., Widera, G., Rabussay, D., *et al.* (2004). Enhancement of DNA vaccine potency in rhesus macaques by electroporation. *Vaccine* **22**, 2489–2493.

Padua, R. A., Larghero, J., Robin, M., le Pogam, C., Schlageter, M. H., Muszlak, S., Fric, J., West, R., Rousselot, P., Phan, T. H., *et al.* (2003). PML-RARA-targeted DNA vaccine induces protective immunity in a mouse model of leukemia. *Nat. Med.* **9**, 1413–1417.

Pardoll, D. M. (2002). Spinning molecular immunology into successful immunotherapy. *Nat. Rev. Immunol.* **2**, 227–238.

Pavlenko, M., Roos, A. K., Lundqvist, A., Palmborg, A., Miller, A. M., Ozenci, V., Bergman, B., Egevad, L., Hellstrom, M., Kiessling, R., *et al.* (2004). A phase I trial of DNA vaccination with a plasmid expressing prostate-specific antigen in patients with hormone-refractory prostate cancer. *Br. J. Cancer* **91**, 688–694.

Peng, G., Guo, Z., Kiniwa, Y., Voo, K. S., Peng, W., Fu, T., Wang, D. Y., Li, Y., Wang, H. Y., and Wang, R. F. (2005). Toll-like receptor 8-mediated reversal of $CD4^+$ regulatory T cell function. *Science* **309**, 1380–1384.

Press, O. W., Leonard, J. P., Coiffier, B., Levy, R., and Timmerman, J. (2001). Immunotherapy of non-Hodgkin's lymphomas. *Hematology Am. Soc. Hematol. Educ. Prog.*, 221–240.

Rahman, F., Dahmen, A., Herzog-Hauff, S., Bocher, W. O., Galle, P. R., and Lohr, H. F. (2000). Cellular and humoral immune responses induced by intradermal or intramuscular vaccination with the major hepatitis B surface antigen. *Hepatology* **31**, 521–527.

Rammensee, H. G., Weinschenk, T., Gouttefangeas, C., and Stevanovic, S. (2002). Towards patient-specific tumor antigen selection for vaccination. *Immunol. Rev.* **188**, 164–176.

Ravn, P., Boesen, H., Pedersen, B. K., and Andersen, P. (1997). Human T cell responses induced by vaccination with *Mycobacterium bovis* bacillus Calmette-Guerin. *J. Immunol.* **158**, 1949–1955.

Reis e Sousa, C. (2006). Dendritic cells in a mature age. *Nat. Rev. Immunol.* **6**, 476–483.

Rice, J., Buchan, S., and Stevenson, F. K. (2002). Critical components of a DNA fusion vaccine able to induce protective cytotoxic T cells against a single epitope of a tumor antigen. *J. Immunol.* **169**, 3908–3913.

Riddell, S. R., Berger, C., Murata, M., Randolph, S., and Warren, E. H. (2003). The graft versus leukemia response after allogeneic hematopoietic stem cell transplantation. *Blood Rev.* **17**, 153–162.

Robinson, H. L., and Amara, R. R. (2005). T cell vaccines for microbial infections. *Nat. Med.* **11**, S25–32.

Rosenberg, S. A., Yang, J. C., and Restifo, N. P. (2004). Cancer immunotherapy: Moving beyond current vaccines. *Nat. Med.* **10**, 909–915.

Rottinghaus, S. T., Poland, G. A., Jacobson, R. M., Barr, L. J., and Roy, M. J. (2003). Hepatitis B DNA vaccine induces protective antibody responses in human non-responders to conventional vaccination. *Vaccine* **21**, 4604–4608.

Rubio, V., Stuge, T. B., Singh, N., Betts, M. R., Weber, J. S., Roederer, M., and Lee, P. P. (2003). Ex vivo identification, isolation and analysis of tumor-cytolytic T cells. *Nat. Med.* **9**, 1377–1382.

Ruffini, P. A., Neelapu, S. S., Kwak, L. W., and Biragyn, A. (2002). Idiotypic vaccination for B-cell malignancies as a model for therapeutic cancer vaccines: From prototype protein to second generation vaccines. *Haematologica* **87**, 989–1001.

Ruprecht, C. R., and Lanzavecchia, A. (2006). Toll-like receptor stimulation as a third signal required for activation of human naive B cells. *Eur. J. Immunol.* **36**, 810–816.

Sahin, U., Tureci, O., and Pfreundschuh, M. (1997). Serological identification of human tumor antigens. *Curr. Opin. Immunol.* **9**, 709–716.

Sallusto, F., Geginat, J., and Lanzavecchia, A. (2004). Central memory and effector memory T cell subsets: Function, generation, and maintenance. *Annu. Rev. Immunol.* **22**, 745–763.

Savelyeva, N., King, C. A., Vitetta, E. S., and Stevenson, F. K. (2005). Inhibition of a vaccine-induced anti-tumor B cell response by soluble protein antigen in the absence of continuing T cell help. *Proc. Natl. Acad. Sci. USA* **102**, 10987–10992.

Singh, M., Kazzaz, J., Ugozzoli, M., Malyala, P., Chesko, J., and O'Hagan, D. T. (2006). Polylactide-co-glycolide microparticles with surface adsorbed antigens as vaccine delivery systems. *Curr. Drug Deliv.* **3**, 115–120.

Smith, C. L., Mirza, F., Pasquetto, V., Tscharke, D. C., Palmowski, M. J., Dunbar, P. R., Sette, A., Harris, A. L., and Cerundolo, V. (2005). Immunodominance of poxviral-specific CTL in a human trial of recombinant-modified vaccinia Ankara. *J. Immunol.* **175**, 8431–8437.

Smyth, M. J., Dunn, G. P., and Schreiber, R. D. (2006). Cancer immunosurveillance and immunoediting: The roles of immunity in suppressing tumor development and shaping tumor immunogenicity. *Adv. Immunol.* **90**, 1–50.

Spellerberg, M. B., Zhu, D., Thompsett, A., King, C. A., Hamblin, T. J., and Stevenson, F. K. (1997). DNA vaccines against lymphoma: Promotion of anti-idiotypic antibody responses induced by single chain Fv genes by fusion to tetanus toxin fragment C. *J. Immunol.* **159**, 1885–1892.

Spies, B., Hochrein, H., Vabulas, M., Huster, K., Busch, D. H., Schmitz, F., Heit, A., and Wagner, H. (2003). Vaccination with plasmid DNA activates dendritic cells via Toll-like receptor 9 (TLR9) but functions in TLR9-deficient mice. *J. Immunol.* **171**, 5908–5912.

Stanislawski, T., Voss, R. H., Lotz, C., Sadovnikova, E., Willemsen, R. A., Kuball, J., Ruppert, T., Bolhuis, R. L., Melief, C. J., Huber, C., et al. (2001). Circumventing tolerance to a human MDM2-derived tumor antigen by TCR gene transfer. *Nat. Immunol.* **2**, 962–970.

Stevenson, F. K., Ottensmeier, C. H., Johnson, P., Zhu, D., Buchan, S. L., McCann, K. J., Roddick, J. S., King, A. T., McNicholl, F., Savelyeva, N., and Rice, J. (2004a). DNA vaccines to attack cancer. *Proc. Natl. Acad. Sci. USA* **101 Suppl 2**, 14646–14652.

Stevenson, F. K., Rice, J., and Zhu, D. (2004b). Tumor vaccines. *Adv. Immunol.* **82**, 49–103.

Stevenson, F. K. (2004c). DNA vaccines and adjuvants. *Immunol. Rev.* **199**, 5–8.

Stockinger, B., Bourgeois, C., and Kassiotis, G. (2006). $CD4^+$ memory T cells: Functional differentiation and homeostasis. *Immunol. Rev.* **211**, 39–48.

't Hart, B. A., and Heije, K. (2005). Broad spectrum immune monitoring in immune-mediated inflammatory disorders. *Drug Discov. Today* **10**, 1348–1351.

Tacket, C. O., Roy, M. J., Widera, G., Swain, W. F., Broome, S., and Edelman, R. (1999). Phase 1 safety and immune response studies of a DNA vaccine encoding hepatitis B surface antigen delivered by a gene delivery device. *Vaccine* **17**, 2826–2829.

Tassignon, J., Burny, W., Dahmani, S., Zhou, L., Stordeur, P., Byl, B., and De Groote, D. (2005). Monitoring of cellular responses after vaccination against tetanus toxoid: Comparison of the measurement of IFN-gamma production by ELISA, ELISPOT, flow cytometry and real-time PCR. *J. Immunol. Methods* **305**, 188–198.

Teague, R. M., Sather, B. D., Sacks, J. A., Huang, M. Z., Dossett, M. L., Morimoto, J., Tan, X., Sutton, S. E., Cooke, M. P., Ohlen, C., and Greenberg, P. D. (2006). Interleukin-15 rescues tolerant $CD8^+$ T cells for use in adoptive immunotherapy of established tumors. *Nat. Med.* **12**, 335–341.

Thirdborough, S. M., Radcliffe, J. N., Friedmann, P. S., and Stevenson, F. K. (2002). Vaccination with DNA encoding a single-chain TCR fusion protein induces anticlonotypic immunity and protects against T-cell lymphoma. *Cancer Res.* **62**, 1757–1760.

Timmerman, J. M., and Levy, R. (2004). Cancer vaccines: Pessimism in check. *Nat. Med.* **10**, 1279; author reply, 1279–1280.

Timmerman, J. M., Singh, G., Hermanson, G., Hobart, P., Czerwinski, D. K., Taidi, B., Rajapaksa, R., Caspar, C. B., Van Beckhoven, A., and Levy, R. (2002). Immunogenicity of a plasmid DNA vaccine encoding chimeric idiotype in patients with B-cell lymphoma. *Cancer Res.* **62**, 5845–5852.

Tjelle, T. E., Salte, R., Mathiesen, I., and Kjeken, R. (2006). A novel electroporation device for gene delivery in large animals and humans. *Vaccine* **24**, 4667–4670.

Vambutas, A., DeVoti, J., Nouri, M., Drijfhout, J. W., Lipford, G. B., Bonagura, V. R., van der Burg, S. H., and Melief, C. J. (2005). Therapeutic vaccination with papillomavirus E6 and E7 long peptides results in the control of both established virus-induced lesions and latently infected sites in a preclinical cottontail rabbit papillomavirus model. *Vaccine* **23**, 5271–5280.

van Hall, T., van Bergen, J., van Veelen, P. A., Kraakman, M., Heukamp, L. C., Koning, F., Melief, C. J., Ossendorp, F., and Offringa, R. (2000). Identification of a novel tumor-specific CTL epitope presented by RMA, EL-4, and MBL-2 lymphomas reveals their common origin. *J. Immunol.* **165**, 869–877.

van Hall, T., Wolpert, E. Z., van Veelen, P., Laban, S., van der Veer, M., Roseboom, M., Bres, S., Grufman, P., de Ru, A., Meiring, H., et al. (2006). Selective cytotoxic T-lymphocyte targeting of tumor immune escape variants. *Nat. Med.* **12**, 417–424.

Vasto, S., Malavolta, M., and Pawelec, G. (2006). Age and immunity. *Immun. Ageing* **3**, 2.

Villa, L. L., Ault, K. A., Giuliano, A. R., Costa, R. L., Petta, C. A., Andrade, R. P., Brown, D. R., Ferenczy, A., Harper, D. M., Koutsky, L. A., et al. (2006). Immunologic responses following administration of a vaccine targeting human papillomavirus types 6, 11, 16, and 18. *Vaccine* **24**, 5571–5583.

Wang, Z., Troilo, P. J., Wang, X., Griffiths, T. G., Pacchione, S. J., Barnum, A. B., Harper, L. B., Pauley, C. J., Niu, Z., Denisova, L., et al. (2004). Detection of integration of plasmid DNA into host genomic DNA following intramuscular injection and electroporation. *Gene Ther.* **11**, 711–721.

Wiley Interscience. Gene therapy clinical trials worldwide. *J. Gene Med.* Retrieved July 2006: *http://www.wiley.co.uk/genetherapy/clinical*

Wolff, J. A., and Budker, V. (2005). The mechanism of naked DNA uptake and expression. *Adv. Genet.* **54**, 3–20.

Yewdell, J. W., and Bennink, J. R. (1999). Immunodominance in major histocompatibility complex class I-restricted T lymphocyte responses. *Annu. Rev. Immunol.* **17**, 51–88.

Zhou, H., Luo, Y., Mizutani, M., Mizutani, N., Reisfeld, R. A., and Xiang, R. (2005). T cell-mediated suppression of angiogenesis results in tumor protective immunity. *Blood* **106**, 2026–2032.

Zuckerman, J. N. (2006). Protective efficacy, immunotherapeutic potential, and safety of hepatitis B vaccines. *J. Med. Virol.* **78**, 169–177.

Further Reading

Cancer Immunotherapy (CIMT). *http://www.c-imt.org/*
Cancer Vaccine Consortium. *http://www.sabin.org/CV_Consortium.htm*

PART III

TARGETS AND TACTICS TO IMPROVE CANCER IMMUNOTHERAPY BY DEFEATING IMMUNE SUPPRESSION

CHAPTER 12

Immunotherapy and Cancer Therapeutics: Why Partner?

LEISHA A. EMENS AND ELIZABETH M. JAFFEE

I. Introduction: Why Immunotherapy for Cancer?
II. Immune Tolerance and Suppression: Multiple Layers of Negative Control
   A. Central Tolerance
   B. Peripheral Tolerance
   C. Tumor Microenvironment
III. T Cell Activation: A Rheostat for Tuning Immune Responses
   A. Chemotherapy and Tumor Immunity
   B. Standard Chemotherapy and Tumor Vaccines: Altering the Tumor Microenvironment
   C. Immune Modulating Chemotherapy and Tumor Vaccines: Altering the Host Milieu
IV. Immune Modulation with Therapeutic Monoclonal Antibodies
   A. The mAbs Specific for Tumor Cell Biology
   B. The mAbs Specific for Immunologic Checkpoints
   C. The mAbs Specific for B7-DC
   D. The mAbs Specific for B7-H1, B7-H2, B7-H3, and B7-H4
   E. The mAbs Specific for CD40
   F. The mAbs Specific for OX40
   G. The mAbs Specific for 4-1BB
V. Therapeutics that Mitigate the Influence of $CD4^+CD25^+$ Tregs
   A. Drugs Specific for CD25
   B. Drugs Specific for GITR
   C. Drugs Specific for CTLA-4
VI. Endocrine and Biologically Targeted Therapy
VII. Conclusion
   References

Capitalizing on the power of the immune system to fight malignancy has enormous potential for cancer therapy. Immunotherapy has a mode of action that is completely distinct from both standard treatment modalities (surgery, chemotherapy, radiation, and endocrine therapy) and the newer biologically targeted small molecules. Accordingly, both the efficacy and the toxicity of immune-based therapy should have little overlap with any other approach to cancer treatment. Moreover, the unique, inherent capability of the immune system to respond to the development of a new primary tumor or disease relapse without additional treatment further distinguishes it from other management strategies. Despite these advantages, immunotherapy is potentially limited by systemic and local mechanisms of immune tolerance that function to turn the antitumor immune response off. Integrating immune-based therapy strategically with established and novel cancer therapies should generate a robust antitumor effect that takes advantage of the strengths of their individual modes of action and also leverages potential synergy between these very distinct modalities. The judicious use of standard and novel therapies in combination with immunotherapy can manipulate existing immunoregulatory pathways for therapeutic benefit, abrogating immune tolerance and amplifying pathways that push the immune response forward. The thoughtful combination of multiple treatment modalities should allow the full power of immunotherapy to be unleashed, resulting in increasing survival benefits and ultimately the eradication and prevention of malignant disease.

I. INTRODUCTION: WHY IMMUNOTHERAPY FOR CANCER?

Standard approaches to cancer treatment include surgery, radiation, chemotherapy, and endocrine therapy. Surgery and radiation are applied locally to the tumor site and are relatively precise in their effects. In contrast, chemotherapy and endocrine therapy exert systemic effects through oral or parenteral administration. They modulate aspects of tumor biology that may also be active in normal tissues and are used to cytoreduce established tumors or to eradicate micrometastatic disease. The proper integration of these local and systemic therapies can cure a significant number of hematologic malignancies and a smaller number of various solid tumors. However, many cancer patients continue to fail therapy. Multidisciplinary cancer therapy is primarily limited by two major factors.

First, systemic therapies frequently have collateral toxic effects on normal tissues that restrict dose and schedule and impair the patient's quality of life. Second, even optimized treatment regimens fail due to the outgrowth of tumor cells that are inherently resistant to therapy. Advances in the knowledge of the molecular basis and genetic basis of cancer have revolutionized cancer therapy, enabling the development of targeted therapeutics with exquisite (but still imperfect) specificity and limited toxicities. These molecularly targeted drugs are most often specific for tumor-associated molecules that are both requisite and sufficient for tumor growth, progression, and metastasis. Specifically targeted therapies, such as imatinib mesylate (Gleevec®) for chronic myelogenous leukemia (Giles *et al.*, 2005) and trastuzumab (Herceptin®) for breast cancer (Emens, 2005), have dramatically improved response rates, survival, and the patient's quality of life during and after therapy. However, even the impact of these novel agents will ultimately be limited by the emergence of drug-resistant tumor cells, which is an increasing clinical problem for both imatinib mesylate and trastuzumab. This fundamental limitation to all systemic cancer treatment strategies developed to date is a compelling argument for taking a radically different approach to therapy that is not plagued by drug resistance.

Manipulating the host–tumor interaction has emerged as a powerful way to complement traditional cancer therapies in an additive or synergistic fashion. One example of this is antiangiogenic therapy, which targets the tumor-associated vasculature rather than the transformed tumor cell itself. Immune-based therapy is another form of therapy that is particularly attractive because it recruits and activates the patient's own immune system to seek out and destroy cancers. Immunotherapy is designed to reconstruct host–tumor immunobiology, tipping the balance of immunologic homeostasis in favor of the cancer patient. A number of therapeutic monoclonal antibodies (mAbs), which have both direct and indirect antitumor effects, are already in widespread clinical use. These mAbs can modulate cellular interactions in the tumor microenvironment, in part by recruiting host innate immune effectors. Progress in cell-based immunotherapy has been slower, but viable cancer therapies based on the adoptive transfer of antigen-specific lymphocytes or the induction of an effective antitumor immune response by active immunization are under intense investigation. Like molecularly targeted therapies, immunotherapy is highly specific and well tolerated. It can target tumor antigens essential for transformation. The side effects of immune-based therapy are typically minimal, being limited to transient fever, flu-like symptoms, and, in the case of vaccines, local injection site reactions. Most important, immunotherapy is uniquely able to exert a durable therapeutic effect due to the induction of immunologic memory. This effect makes prolonged, repetitive cycles of therapy unnecessary and identifies immunotherapy as a promising modality for the secondary prevention of disease relapse and ultimately the prevention of primary tumor development.

II. IMMUNE TOLERANCE AND SUPPRESSION: MULTIPLE LAYERS OF NEGATIVE CONTROL

A. Central Tolerance

Immune-based therapy can be broadly divided into approaches that employ the passive administration of immunologic effectors like mAbs, cytokines, and lymphocytes and strategies that actively induce these same types of end immune effectors *in vivo* (i.e., vaccines). The design of active immunotherapies to effectively activate endogenous immune effectors requires consideration of both the available repertoire of immune effectors and the superimposed forces that influence them. Since tumors arise endogenously from normal tissues, they are typically viewed by the immune system as "self." Therefore, the tumor-specific immune response is shaped by the same regulatory mechanisms that prevent immune-mediated attack on autologous tissues.

The T cell repertoire is first established by central tolerance mechanisms that operate by thymic selection and then modulated by peripheral tolerance pathways that control T cell activation (Hogquist *et al.*, 2005; Redmond and Sherman, 2005; Zou, 2006). Lymphocyte progenitors migrate to the thymus during development and throughout postnatal life, differentiate into double positive thymocytes at the thymic cortex, and then undergo positive and negative selection within the cortex and medulla to establish a repertoire of $CD8^+$ cytotoxic T cells and $CD4^+$ helper T cells, respectively (Hogquist *et al.*, 2005). T cells with no or very low affinity for self peptide: major histocompatibility complex (p:MHC) ligands typically die by neglect, whereas those with low affinity for self p:MHC ligands are selected to survive. T cells with the highest affinity for self p:MHC are selected for destruction or neutralization. This is achieved either by active deletion from the T cell repertoire or by apoptotic cell death (Palmer, 2003), attenuation of their affinity by T cell receptor (TCR) editing (Wang *et al.*, 1998; McGargill *et al.*, 2000), or abrogation of their activity by establishing anergy (Hammerling *et al.*, 1991). Alternatively, T cells with high affinity for self p:MHC ligands may be selected for differentiation into a regulatory T cell of various phenotypes ($CD4^+CD25^+$ regulatory T cell (Treg), $CD8\alpha\alpha$ intestinal epithelial lymphocyte, or natural killer T [NKT] cell) (Baldwin *et al.*, 2004, 2005; Hogquist *et al.*, 2005). Effector and regulatory thymocytes selected for survival differentiate to maturity and emigrate from the thymus to take up residence in peripheral tissues. Notably, these selection processes are regulated by the autoimmune regulator (AIRE) gene (Anderson *et al.*, 2002; Kyewski and Derbinski, 2004), which facilitates the ectopic but constitutive expression of most peripheral antigens by thymic epithelial cells (TECs). TECs either present peripheral antigens themselves or transfer them to thymic dendritic cells (DCs) for cross-presentation to migrating thymocytes (Anderson *et al.*, 2002; Gallegos and Bevan, 2004; Gillard and Farr, 2005). Notably, this selection system is imperfect in that not all self antigens are represented in the thymic microenvironment, facilitating the escape of high affinity, self-reactive T cells into peripheral tissues. Accordingly, backup mechanisms of peripheral tolerance must come into play to prevent immune-mediated injury to normal host tissues (Redmond and Sherman, 2005; Zou, 2006). Manipulation of these pathways of peripheral immune tolerance to allow activation of latent high-affinity tumor-specific T cells in the periphery represents one potent opportunity for immune-based cancer therapy. Alternatively, strategies for activating peripheral low-affinity T cells that are specific for tumor antigens can be developed.

B. Peripheral Tolerance

Potentially self-reactive T cells of differing affinities can thus emigrate from the thymus. Outside the thymic environment, at least four factors determine whether these T cells become activated: (i) the level and persistence of antigen (Bevan, 2004; Masopust *et al.*, 2004); (ii) the activation status of relevant antigen-presenting cells (APCs) (Heath and Carbone, 2001; Weninger *et al.*, 2001); (iii) the ability of T cells to "see" their antigen (Pircher *et al.*, 1991; Walker and Abbas, 2001); and (iv) the influence of suppressive immunoregulatory cells (Zou, 2006). Transient antigen exposure activates T cells to eliminate antigen-bearing cells and to differentiate into memory T cells capable of mediating a secondary immune response upon antigen reexposure. Persistent exposure to low or high levels of antigen results in tolerance by deletion, where cells are removed by apoptosis or by activation-induced cell death (AICD). Functionally immature APCs also induce T cell tolerance by anergy or deletion, whereas mature APCs facilitate potent T cell activation by providing a robust network of costimulatory signals. Low-affinity T cells may recognize their antigen but fail to become activated due to failure to exceed the threshold of signaling required to initiate a response. Alternatively, self-reactive T cells may be spatially compartmentalized away from their antigen and physically unable to detect it. Finally, naturally occurring $CD4^+CD25^+$ Tregs are the major force in the peripheral tissues that inhibit the priming of autoimmune T cell responses (and thus antitumor immune responses) and that also inhibit the activity of end effector $CD8^+$ cytotoxic T cells.

Tregs have emerged as a major locus of control for the prevention of autoimmunity and for the controlled downregulation of desired immune responses (Zou, 2006). However, they also potently suppress antitumor immune responses. Tregs represent 5–10% of the peripheral $CD4^+$ T lymphocyte pool in humans and mice. These cells express cytotoxic T lymphocyte-associated antigen 4 (CTLA-4), the glucocorticoid-induced tumor necrosis factor receptor (GITR), and the forkhead/winged helix transcription factor (FoxP3), and they secrete interleukin (IL)-10 and transforming growth factor β (TGF-β). Tregs inhibit $CD8^+$ T cell responses in an IL-2-dependent manner through direct cell-to-cell contact or through the immunosuppressive effects of IL-10 and/or TGF-β. The central role of these cells in suppressing antitumor immune responses identifies them as a major target for therapeutic manipulation in immune-based treatment strategies for cancer.

C. Tumor Microenvironment

Tumor cells themselves also constitute a barrier to the efficacy of immune-based cancer treatments. First, the sheer burden of tumor cells frequently outmatches the magnitude of the tumor-specific T cell response (Ochsenbein et al., 1999; Perez-Diaz et al., 2002). This reflects both the potency of current tumor vaccine formulations and the tendency of clinical investigators to test immunotherapy in patients with advanced cancers and extensive tumor burdens. Notably, current tumor vaccine formulations typically induce a T cell response that is relatively tepid, with tumor antigen-specific T cell precursor frequencies on the order of about 1% or less. In contrast, most vaccines that target foreign antigens induce vigorous immune responses, where the antigen-specific T cell precursor frequency is 10% or more. Second, tumors secrete a number of immunosuppressive factors (Zou, 2005). These include vascular endothelial growth factor (VEGF), IL-10, TGF-β, and prostaglandin E_2 (PGE_2). These cytokines can suppress the activity of DCs, induce the differentiation of Tregs, recruit and activate myeloid suppressor cells (MSCs), or abrogate the activity of tumor-infiltrating lymphocytes (TILs). Third, tumors can express molecules that inhibit the activity of infiltrating T cells by inducing overt apoptosis. These molecules include fasL (CD95L), tumor necrosis factor-related apoptosis-inducing ligand (TRAIL) (Marincola et al., 2000), and the counter-stimulatory molecules B7-H1, B7-H2, B7-H3, and B7-H4 (Khoury and Sayegh, 2004). Furthermore, the genetic instability of the tumor can lead to the dynamic plasticity of its antigen expression profile. Thus, tumors can downregulate the expression of tumor antigens that are the target of an immune-based intervention either spontaneously or in response to therapy as a consequence of immunoediting (Davis et al., 1999; Knutson et al., 2004). Tumor cells may also downregulate the expression of MHC molecules or other components of the antigen processing machinery (i.e., proteasome subunits or the TAP transporter), rendering them less effective with regard to antigen presentation (Marincola et al., 2000). In the aggregate, these latter mechanisms of immunosuppression represent a mechanism for the outgrowth of antigen-loss variant tumors that are intrinsically resistant to antigen-specific immunotherapy, a phenotype that has been correlated with poor clinical outcome (Kageshita et al., 1999; Ogino et al., 2003; Meissner et al., 2005; Anichini et al., 2006).

Within the tumor microenvironment, a variety of immunoregulatory cell types congregate to further shut down the antitumor immune response (Zou, 2005, 2006). Three distinct types of immunoregulatory T cells may suppress TIL. These include (i) naturally occurring Tregs that home to the tumor microenvironment via the interaction of CC-chemokine receptor 4 (CCR4) on the surface of the Treg and CC-chemokine ligand 22 (CCL22) at the tumor site; (ii) Treg-induced from effector $CD4^+CD25^+$ T helper cells by high levels of TGF-β in the tumor microenvironment; and (iii) IL-10-secreting $CCR7^+$ $CD8^+$ Tregs, which also inhibit $CD8^+$ T cell-mediated immunity (Zou, 2006). In addition, the immunosuppressive cytokines VEGF, IL-10, and TGF-β promote the maintenance of immature myeloid DC, which inhibits the activity of $CD8^+$ TIL and promotes the activation and expansion of IL-10-secreting $CD4^+$ Tregs. Immature myeloid DCs accomplish this in part by expressing indoleamine 2,3-dioxygenase (IDO), which catalyzes the oxidative catabolism of tryptophan, a nutrient essential for T cell proliferation and differentiation. In the absence of tryptophan, T cells undergo apoptosis, anergy, or immune deviation. Notably, Tregs may induce IDO in resident DCs through CTLA-4 signaling, converting them to DCs that induce tolerance (Grohmann et al., 2002). Furthermore, IDO-arrested effector T cells can acquire a Treg phenotype, thereby further short-circuiting the immune response (Mellor and Munn, 2003). Plasmacytoid DCs induce $CCR7^+$ $CD8^+$ regulatory T cells (Zou, 2005). Furthermore, immature myeloid DCs can express the counter-regulatory molecule B7-H1, resulting in decreased IL-12 production and reduced immune potency.

MSCs can also accumulate in the tumor microenvironment. Representing a mixed population of immature and mature myeloid cells that express Gr-1, CD31, and CD11b in mice (and possibly CD34 in humans), MSCs develop from bone marrow-derived precursors in response to IL-3, granulocyte-macrophage colony-stimulating factor (GM-CSF), or VEGF. They reversibly inhibit the activation of $CD8^+$ T cells through the activation of inducible nitric acid synthase (iNOS) or arginase 1 (ARG1), which generate nitric oxide (NO) or depletes the environment of arginine respectively (Zou, 2005). NO blocks the activation of IL-2 signaling, resulting in T cell death. L-arginine is an essential nutrient for the effects of IL-2 and for the development of the memory T cell phenotype. Notably, the simultaneous induction of both these enzymes results in T cell apoptosis. Finally, tumor-infiltrating macrophages (TIM) can express high levels of B7-H4 and suppress tumor-associated immunity (Kryczek et al., 2006, 2006).

These multiple mechanisms of immune tolerance are addressed in more detail in other chapters in this book. The important point this chapter will illustrate with specific examples is that it is possible to modulate a number of these immune tolerance mechanisms by incorporating specific chemotherapeutic agents or targeted molecular therapeutics in sequence with a T cell-stimulating vaccine.

III. T CELL ACTIVATION: A RHEOSTAT FOR TUNING IMMUNE RESPONSES

T cell activation occurs in this highly regulated environment and also presents an intricate network of regulatory pathways that converge to determine the overall cell state. At its most basic level, T cell activation requires two major signals. Signal 1 is provided by the interaction of the TCR on the T cell surface with the p:MHC complex on the APC surface and dictates the specificity of the immune response. Signal 2 is provided by the summation of positive and negative signals provided by accessory molecules presented by the T cell and APC. This complex signaling system for T cell activation has two major implications for tumor immunity. First, TCR signaling in the absence of signal 2 turns the T cell off (Jenkins and Schwartz, 1987). Since both normal tissues and transformed cells typically fail to express positive costimulatory molecules, this is an additional mechanism by which tolerance to self (and therefore tumors) is maintained in the periphery. Second, the diversity of costimulatory and counterstimulatory molecules represents a highly integrated network for tuning the magnitude and quality of the T cell response up or down, presenting a potent opportunity for therapeutic manipulation to enhance tumor immunotherapies. Professional APCs (typically DCs) are specially programmed to provide highly coordinated and effective secondary costimulation (Pardoll, 2002).

The second T cell activation signal is provided by a series of molecules in the B7 and tumor necrosis factor receptor (TNFR) families. B7 molecules are a set of receptor-ligand pairs: one pair transmits a positive signal, and its partner pair transmits a negative signal (Greenwald et al., 2005). The prototype is B7–1/B7–2 (CD80/CD86) on the APC surface interacting with CD28 or CTLA-4 on the T cell surface. Signaling through CD28 enhances T cell activation, with negative feedback provided by counter-regulatory signaling through CTLA-4. This negative signaling both raises the threshold for T cell activation and limits T cell expansion (Egen et al., 2002). The balance of signals may be influenced by the preferential recruitment of CD28 by B7–2 and CTLA-4 by B7–1, providing a means by which the B7 phenotype of the presenting APC determines the final strength of T cell activation (Pentcheva-Hoang et al., 2004). Additional B7 family members include B7-DC (PD-L2) and B7-H1 (PD-L1), which transmit positive signals by interacting with undefined positive coreceptors on the T cell surface (Tseng et al., 2001; Shin et al., 2003, 2005; Khoury and Sayegh, 2004), and negative signals by interacting with the PD-1 ligand expressed by the T cell (Freeman et al., 2000; Latchman et al., 2001). B7-H2, the inducible costimulator (ICOS) ligand, is expressed on APC and interacts with ICOS on the T cell surface to promote T helper 2 (Th2) differentiation and humoral immunity (McAdam et al., 2001; Wallin et al., 2001). B7-H3 is also expressed on APC and transmits positive and negative signals in a similar fashion (Chapoval et al., 2001; Suh et al., 2003; Luo et al., 2004; Prasad et al., 2004). B7-H4 interacts with the T cell molecule B and T lymphocyte attenuator 4 (BTLA-4) to inhibit T cell activation (Sica et al., 2003). It is also expressed by breast and renal cancers (among others) (Salceda et al., 2005; Tringler et al., 2005; Krambeck et al., 2006) and by suppressive APCs. Thus, B7-H1, B7-H2, B7-H3, and B7-H4 are all expressed in nonlymphoid tissues (Khoury and Sayegh, 2004). Therefore, signaling mediated by the molecules in this family can impact T cell function at two levels. First, it determines the starting level of effector T cell activation, which is then tuned and matured by further signaling through molecules in the TNFR family. B7 molecules reenter to impact end effector T cell function in the tumor microenvironment itself.

Signaling through proteins in the TNFR family influences the quality of the resulting T cell response once the starting level of T cell activation has been established by B7 molecules. The quality is determined by the serial recruitment of TNFR family members to further control the initiation, expansion, and durability of the resulting T cell response, including signaling through CD40, OX40, 4-1BB, CD27/CD70, CD30, and herpes virus entry mediator (HVEM)/ LIGHT (for homologous to lymphotoxins, shows inducible expression and competes with herpes simplex glycoprotein D for HVEM, a receptor expressed by T lymphocytes) (Watts, 2005). CD40 signaling is essential for T cell-dependent humoral immunity and can substitute for T cell help in priming CD4$^+$ and CD8$^+$ T cell responses (Tong and Stone, 2002). B7/CD28 signaling accelerates the rate and level of recruitment of molecules in the OX40, 4-1BB, and CD30 pathways, thereby fine-tuning the level of T cell activation, the extent of clonal expansion, and the establishment of a memory T cell pool. The HVEM/LIGHT pathway is active in immature DCs and resting T cells and helps to keep the immune response quiescent. The recent identification of HVEM–BTLA-4 interactions as a negative influence on signaling is consistent with this role (Gonzalez et al., 2005; Sedy et al., 2005), as is the downregulation of HVEM/ LIGHT with DC maturation and T cell activation (Watts, 2005). In contrast, signaling through LIGHT to activate and expand NK cells can prime CD8$^+$ T cell-dependent tumor immunity in an NK and interferon (IFN)-γ-dependent fashion, initiating the rejection of established tumors (Yu et al., 2004; Fan et al., 2006).

A. Chemotherapy and Tumor Immunity

Systemic and local pathways of immune tolerance interact with tumor biology to shut tumor immune responses down. It is not surprising, then, that most clinical trials testing the adoptive transfer of antigen-specific T cells or cancer vaccines as a single intervention in patients with advanced, refractory malignancies have failed to show evidence of clinically relevant bioactivity. Since surgery, radiation, and chemotherapy are widely used to treat most established cancers, properly integrating immune-based therapies with these standard modalities to capitalize both on potential synergies and their individually unique modes of action is highly attractive. Carefully choosing the drugs, their dose, and their schedule of administration in terms of timing immune-based therapy with these other modalities is critical for maximizing synergy to achieve optimal therapeutic benefit. Sequencing immune-based therapy after optimal tumor debulking with surgery, radiation, and cytotoxic chemotherapy can mitigate the impact of tumor burden on the antitumor immune response. This achieves a state of minimal residual disease and, therefore, tips the balance between the numbers of cancer cells and tumor-specific T cells in favor of tumor rejection. Here, standard doses of chemotherapy are used, and tumor vaccines are typically given during the time of cytopenia and immune reconstitution or after full marrow recovery. Chemotherapy can also be used to modulate tumor-specific immunity in at least three distinct major ways (Table 12.1). First, chemotherapy-induced cell death by apoptosis can enhance cross-priming, thereby increasing the antitumor T cell response. Second, chemotherapy can further groom the tumor microenvironment by modulating the expression of tumor antigens, costimulatory molecules, and molecules involved in antigen processing and presentation. Third, chemotherapy can be used to manipulate pathways of immune tolerance and regulation. Here it is used strategically at a dose and schedule designed to abrogate specific mechanisms of immune tolerance or to manipulate immunoregulatory processes, like homeostatic proliferation, that normally maintain the peripheral T cell compartment. In the latter case, the peripheral T cell repertoire can be redefined by vaccination or adoptive cell transfer during immune reconstitution to enrich it for tumor-specific T cells.

B. Standard Chemotherapy and Tumor Vaccines: Altering the Tumor Microenvironment

Several studies have shown that vaccination in close proximity to standard dose,

TABLE 12.1 Immune Modulating Activity of Various Chemotherapeutic Drugs

Enhanced cross-priming by apoptosis
Cyclophosphamide
Gemcitabine
Doxorubicin
Paclitaxel
Modulation of CD4+CD25+ regulatory T cells
Cyclophosphamide
Fludarabine
Modulation of myeloid suppressor cells
Gemcitabine
DC migration and maturation
Paclitaxel
Docetaxel
Reversal of immunologic skew
Cyclophosphamide
Paclitaxel
Bleomycin
Melphalan
Reexpression of tumor antigens
5′-Aza-2′Deoxycytidine
5-Fluorouracil
Paclitaxel
Modulation of costimulatory molecules
Melphalan
Mitomycin-C
Cytosine arabinoside
Bleomycin
Antiangiogenic activity
Cyclophosphamide
Paclitaxel
Doxorubicin
Vinblastine

cytotoxic chemotherapy can inhibit vaccine-induced immune responses, allow similar levels of immune priming as vaccination in the absence of chemotherapy, or increase the response to subsequent chemotherapy. Two studies suggest a detrimental effect of standard dose chemotherapy on the activity of tumor vaccines. Patients with advanced carcinoembryonic antigen (CEA)-expressing colorectal cancer were treated with the canary pox vaccine ALVAC-CEA-B7.1, and CEA-specific immune responses were assessed (von Mehren et al., 2001). The number of vaccine-induced CEA-specific T cells was decreased in patients who had received a greater number of prior chemotherapy regimens and in those who had recently received standard dose chemotherapy. In another study, some patients with high-risk stages II and III pancreatic cancer who were managed with a pancreaticoduodenectomy followed by one vaccination (with a GM-CSF-secreting vaccine) developed vaccine-induced mesothelin-specific $CD8^+$ T cells (Jaffee et al., 2001; Thomas et al., 2004). This immune response became undetectable after these same patients completed 6 months of 5-fluorouracil (5-FU)-based chemoradiation and was restored only after three additional vaccinations given after the completion of chemotherapy. These observations suggest that standard chemotherapy can inhibit the activity of tumor vaccines but that this may be overcome with appropriate boosting schedules.

Conversely, standard dose chemotherapy can interact with established vaccine-induced immune responses for clinical benefit. One study tested the activity of vaccination with DCs transduced with an adenoviral vector expressing wild-type p53 in patients with extensive stage small-cell lung cancer (SCLC). Although almost 60% of patients developed p53-specific immune responses with vaccine alone, all of them except one developed progressive disease. Notably, about 62% of patients with progressive disease enjoyed an objective clinical response to subsequent chemotherapy, and this clinical benefit correlated with p53-specific immunity (Antonia et al., 2006). A similar observation has been reported in patients with malignant glioma who were vaccinated with DC-based vaccines and who subsequently received standard chemotherapy (Liu et al., 2006). A third study tested a prime/boost vaccination regimen that primed the immune response with recombinant vaccinia virus (rVV)-expressing prostate-specific antigen (PSA) admixed with rVV-expressing B7.1, followed by boosting the immune response with recombinant fowlpox virus (rF)-expressing PSA with and without concurrent docetaxel (DTX) therapy in patients with hormone-refractory metastatic prostate cancer (Arlen et al., 2006). Vaccination resulted in a 3.33-fold increase in PSA-specific T cell precursors by ELISPOT after 3 months, regardless of concomitant docetaxel treatment. Median progression-free survival for vaccinated patients on docetaxel was 6.1 months compared to 3.7 months for a historical cohort treated with docetaxel alone. Another study tested the combination of three cycles of standard irinotecan/high dose 5-FU/leucovorin and concurrent vaccination with a CEA-derived peptide, followed by weekly vaccination alone in patients with newly diagnosed metastatic colorectal cancer (Weihrauch et al., 2005). Although recall antigen-specific $CD8^+$ T cells decreased by about 14% in these patients, almost half of treated patients demonstrated the induction of CEA-specific immune responses by intracellular cytokine staining assays.

Together, these latter studies show that vaccine-induced T cell responses may not be inhibited by chemotherapy and that previously vaccinated patients may benefit more from subsequent standard dose chemotherapy than those who have never been vaccinated. The differences between the results obtained across the studies are likely related to the extent of existing tumor burden in treated patients, the order of administration of chemotherapy and vaccine, and the chemotherapy drugs used. The augmented clinical benefit with chemotherapy given after vaccination may be due to the enhancement of preexisting immunity by treatment-related apoptosis, resulting in cross-priming by tumor antigens released by chemotherapy-induced cell death. This mechanism has been demonstrated with gemcitabine (GEM) (Nowak et al., 2002, 2003a, 2003b), doxorubicin (DOX) (Casares et al., 2006), cyclophosphamide (CY) (Tong et al., 2001), and paclitaxel (PTX) (Yu et al., 2003) in vivo in murine models. At least one study has suggested the relevance of this mechanism to human cancer treatment. This study used neoadjuvant PTX therapy for locally advanced breast cancer. This treatment induced TILs with the extent of TIL infiltration correlating with clinical response (0% with stable disease, 25% with partial clinical response, and 67% with complete clinical response but residual pathologic disease) (Demaria et al., 2001). Notably, the extent of tumor cell apoptotic response with the first PTX treatment predicted the relationship of TIL with clinical benefit.

In addition to inducing apoptosis, chemotherapy can render the tumor microenvironment more conducive to an effective antitumor immune response. Many cytotoxic drugs have dose- and sequence-dependent antiangiogenic activity. CY, PTX, DOX, or vinblastine given at regular intervals preferentially targets the tumor-associated vasculature over tumor cells themselves (Bocci et al., 2002). This targeting may augment tumor immunity by increasing tumor cell apoptosis and vascular access early in treatment. Chemotherapy can be used to reverse epigenetic changes, reinducing the expression of both tumor antigens themselves and molecules that play a role in antigen processing and presentation. 5′-Aza-2′-deoxycytidine (Coral et al., 2002) and 5-FU (Correale et al., 2003) can, by this mechanism, render some tumor cell lines sensitive to lysis by cytotoxic T lymphocytes (CTLs). Melphalan and mitomycin C have been reported to upregulate the expression of B7 molecules (Sojka et al., 2000; Donepudi et al., 2001), thereby enabling tumor cells themselves to present a second signal for T cell activation. Moreover, in a murine model of acute myelogenous leukemia (AML), cytosine arabinoside (AraC) has been shown to increase the expression of B7–1/B7–2 and decrease the expression of B7-H1 *in vivo*, thereby enhancing CTL-mediated killing (Vereecque et al., 2004). The impact of AraC on the expression of costimulatory molecules was confirmed in a majority of primary cultured human AML cells *in vitro*. As suggested earlier, multiple chemotherapeutic agents can sensitize tumor cells to CTL-mediated apoptosis through Fas- or perforin-granzyme-mediated pathways (Yang and Haluska, 2004).

C. Immune Modulating Chemotherapy and Tumor Vaccines: Altering the Host Milieu

Drugs that are conventionally used for their cytotoxic activity can also be used to potentiate vaccine-induced immune response by mitigating systemic mechanisms of active immune tolerance or by altering the global and/or local environment in which the antitumor immune response develops (Figure 12.1). Many chemotherapy drugs can either increase or decrease the antigen-specific immune response, with the outcome depending on the dose of drug used and the relative timing of drug and antigen exposure (Nigam et al., 1998; Emens et al., 2001). For example, CY given at low doses prior to antigen exposure (usually 1 to 3 days) abrogates immune tolerance, enhancing both humoral and cellular immunity. In contrast, CY given concurrently with, or subsequent to, antigen exposure induces immune tolerance. A number of groups have shown that CY can mitigate the influence of $CD4^+CD25^+$ Tregs, thereby allowing immunity to unfold (Ghiringhelli et al., 2004; Ercolini et al., 2005; Lutsiak et al., 2005; Taieb et al., 2006). CY modulation favors the development of the T helper 1 (Th1) phenotype, thus reversing the immunologic deviation typically associated with preexisting tumor burdens (Machiels et al., 2001). Mitigating the influence of Tregs with CY prior to vaccination enabled the recruitment of high avidity $CD8^+$ T cells to the antitumor immune response in tolerized *neu* transgenic mice, a finding that correlated with tumor rejection (Ercolini et al., 2005). CY can also upregulate the production of type 1 IFNs, promoting the evolution of the $CD44^{hi}$ memory response (Schiavoni et al., 2000). These latter three features of antitumor immunity are characteristic of high quality tumor-specific

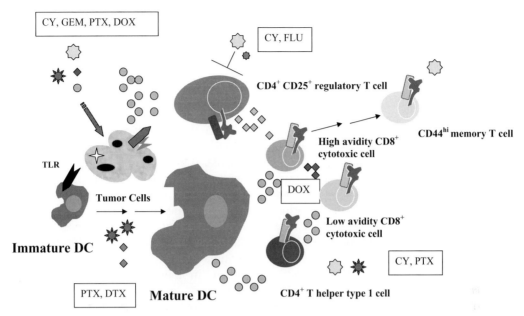

FIGURE 12.1 **Chemotherapy modulating the tumor microenvironment and systemic mechanisms of immune tolerance to enhance tumor immunity.** Both pathways of immune tolerance and suppression and tumor cell biology itself offer multiple loci where chemotherapy can modulate a developing or established immune response. Cell aptoptosis induced by cyclophosphamide (CY), gemcitabine (GEM), paclitaxel (PTX), and doxorubicin (DOX) can enhance the ability of DCs to cross-prime the antitumor immune response. PTX and docetaxel potentiate this process by interacting with TLRs to promote the maturation and activation of DCs. CY and PTX promote the development of the CD4+ Th1 response (rather than the CD4+ Th2 response) optimal for antitumor immunity. Both CY and fludarabine can negate the suppressive influence of CD4+CD25+ regulatory T cells. DOX further promotes the development of the CD8+ T cell response, and CY promotes the establishment of a CD44^{hi} memory cell pool. (See color plate.)

immune responses. Additional studies have shown that CY modulates immunity in part by inhibiting the induction of iNOS (Loeffler *et al.*, 2005). Standard dose fludarabine can also decrease the numbers and function of Tregs in patients with B cell chronic lymphocytic leukemia (CLL) (Beyer *et al.*, 2005). Furthermore, treatment of patients with metastatic colorectal carcinoma with the combination of GEM and FOLFOX 4 (oxaliplatin, 5-fluorouracil, and folinic acid), followed by subcutaneous GM-CSF plus IL-2, resulted in a significant reduction in Tregs in the majority of patients (20 of 29), with a high overall objective response rate of 70% (Correale *et al.*, 2005). Also potentially relevant to this latter study is the observation that standard dose GEM can eliminate MSCs in mice, thereby enhancing the antitumor activity of CD8+ T cells and NK cells (Suzuki *et al.*, 2005). As discussed previously, GEM promotes the development of tumor immunity by inducing tumor cell apoptosis. In contrast to these adjuvant effects, GEM administration inhibits vaccine-induced tumor immunity in tolerant *neu* mice immunized with GM-CSF-secreting tumor vaccines (Jaffee, 2001, unpublished data).

The taxanes, PTX and DTX, also have immunomodulatory activity that extends beyond their ability to induce tumor cell apoptosis. PTX has lipopolysaccharide (LPS)-mimetic activity, binding to Toll-like receptor 4 (TLR-4) expressed by murine DCs and inducing the secretion of proinflammatory cytokines (Kawasaki *et al.*, 2000; Byrd-Leifer *et al.*, 2001). These effects are independent of TLR-4 in humans, but continued dependence on Myd88 suggests the involvement of an alternative TLR

(Wang et al., 2002). Like CY, PTX can reverse the immunologic skew, favoring the differentiation of CD4+ Th1 cells (Machiels et al., 2001). Low dose PTX given 1 day prior to vaccinating tumor-bearing, tolerant *neu* transgenic mice with a GM-CSF-secreting vaccine augments tumor immunity, possibly by modulating TLR-4 signaling in DCs at the time of immune priming (Machiels et al., 2001). These higher immune responses correlate with improved tumor-free survival. PTX also augments vaccine-induced immunity if given prior to immunization with a *neu*-specific virally derived vaccine or a genomically modified fibroblast vaccine in murine models of breast cancer (Eralp et al., 2004; Chopra et al., 2006). The combination of DTX and vaccination with a GM-CSF-secreting vaccine has also been shown to be effective in murine models of B16 melanoma and Lewis lung carcinoma (Chu et al., 2006; Prell et al., 2006).

DOX also has immunomodulatory activity when specifically timed and dosed. In tolerant *neu* transgenic mice, DOX given 1 week after vaccination can augment vaccine-induced immunity; giving DOX prior to vaccination in the same fashion as CY and PTX inhibits vaccine activity in this system (Machiels et al., 2001); this finding is consistent with observations made in a CT26 model of colon cancer (Nigam et al., 1998). In contrast, DOX can enhance vaccine activity if given prior to vaccination in other systems (Eralp et al., 2004). The difference in the ability of these various chemotherapeutics to enhance vaccine activity with regard to timing is striking, and this finding argues for synergy at the time of immune priming for CY (Tregs) and PTX (DC modulation) and at the time of effector T cell activity for DOX (apoptosis and cross-priming). Several clinical trials testing the ability of low doses of chemotherapy to modulate immunity are underway or have just been completed in 2006. Two small clinical trials involving patients with metastatic pancreatic or non-small-cell lung cancer (NSCLC) tested immune-modulating doses of CY given 1 day prior to vaccination with a cell-based GM-CSF-secreting vaccine (Laheru et al., 2005; Schiller et al., 2005). Both studies demonstrated trends toward increased vaccine-activated immunity and clinical benefit with CY-modulated vaccination compared to vaccination alone, with a transient decrement in Treg numbers noted with time after CY treatment in the NSCLC patients. Ongoing clinical trials testing these concepts include a phase I clinical trial testing a GM-CSF-secreting breast tumor vaccine in sequence with CY and DOX in metastatic breast cancer patients (Emens et al., 2004) and a phase III trial testing a GM-CSF-secreting prostate cancer vaccine in men with advanced prostate cancer.

High dose chemotherapy can also synergize with tumor vaccines in multiple ways. It is highly effective in debulking established disease burdens, can eliminate Tregs, and incites lymphopenia-induced homeostatic T cell proliferation (Cho et al., 2000; Goldrath et al., 2000). The adoptive transfer of lymphocytes or vaccination during homeostatic T cell proliferation offers the opportunity to skew the reestablished T cell repertoire toward a desired antigenic specificity (Mackall et al., 1996). Consistent with this, a robust population of melanoma-specific T cells was generated when lymphopenic, melanoma-bearing, RAG-1 deficient mice were vaccinated with a GM-CSF-secreting vaccine. This produced significant tumor regressions (Hu et al., 2002). In more clinically relevant models, vaccine-induced tumor immunity was enhanced when tumor-bearing mice were immunized with GM-CSF-secreting tumor vaccines during early engraftment after syngeneic or allogeneic T cell-depleted bone marrow depletion (Borrello et al., 2000; Teshima et al., 2001) and can be enhanced by donor leukocyte infusion from vaccinated donor mice (Teshima et al., 2002). Similar studies showed that tumor-bearing mice that underwent surgical resection, followed by nonmyeloablative allogeneic stem cell transplantation and donor leukocyte infusions plus vaccination with a GM-CSF-secreting tumor vaccine, developed immune responses capable of lysing metastatic 4T-1 breast tumors (Luznik et al., 2003). This concept has been evaluated in trials employing the adoptive transfer of T lymphocytes (Dudley et al., 2002, 2005) and in clinical studies testing immunization combined with the adoptive transfer of primed lymphocytes during immune reconstitution subsequent to myeloablative chemotherapy. Significant levels of tumor-specific T cells were documented in the studies of adoptive cellular therapy, and these correlated with tumor regressions. Characterizing the kinetics, persistence, and functional quality of tumor antigen-specific T cells after immune reconstitution will be essential to optimize this therapeutic approach.

TABLE 12.2 Immune Modulating Activity of Selected Targeted Cancer Therapies

Modulation of CD4+CD25+ T cells
Denileukin diftitox
Ipilumumab
Bevacizumab
Modulation of T cell costimulation
Ipilimumab
Enhanced cross-priming by apoptosis
Trastuzumab
Rituximab
Augmented antigen processing and presentation
Trastuzumab
Recruitment of innate immune effectors
Trastuzumab
Rituximab
DC migration and maturation
Bevacizumab

IV. IMMUNE MODULATION WITH THERAPEUTIC MONOCLONAL ANTIBODIES

Therapeutic mAbs can be broadly divided into those directed to the tumor cell itself and those that target biologic features that shape the tumor microenvironment (Table 12.2). Tumor-specific mAbs are particularly attractive since they provide a passive means of partially reconstituting the antigen-specific humoral arm of the immune response in the context of active antigen-specific vaccination to induce antitumor T cells. Tumor-specific mAbs already in widespread use for cancer treatment include trastuzumab, rituximab, cetuximab, and bevacizumab. Equally promising is the use of therapeutic mAbs that specifically target the immunologic control checkpoints for T cell activation. Here, mAbs can be used to amplify the positive signals that drive T cell development or to inhibit the negative signals that keep it in check. To date, the only mAb checkpoint modulators that have been tested clinically are those specific for CTLA-4, but there is intense interest in developing others for clinical use.

A. The mAbs Specific for Tumor Cell Biology

1. Trastuzumab

Trastuzumab is a humanized mAb-specific for human epidermal growth factor receptor 2 (HER2)/*neu*, a protooncogene overexpressed by up to 25% of human breast cancers. It is widely used to manage HER2/*neu*-positive disease at every stage except ductal carcinoma *in situ* and has a number of potential immunomodulatory activities. Trastuzumab recruits innate immune effectors to the tumor microenvironment to facilitate antibody-dependent cellular cytotoxicity (ADCC) (Clynes et al., 2000; Gennari et al., 2004; Arnould et al., 2006). It also augments the lytic activity of MHC class I-restricted, HER2/*neu*-specific CTLs against HER2/*neu*-positive targets (zum Buschenfelde et al., 2002; Kono et al., 2004), most likely by enhancing antigen processing and presentation. Trastuzumab has been shown to induce the ubiquitination and degradation of internalized HER2/*neu* molecules, thus increasing proteasome-dependent antigen presentation (Klapper et al., 2000; Castilleja et al., 2001). It interferes with tumor cell growth directly by disrupting pathways that promote growth and metastasis and *in vivo* functions, primarily by inducing apoptosis (Mohsin et al., 2005). Supporting the idea that humoral and cellular immune effectors may exert a more potent antitumor effect than either alone, the passive administration of both HER2/*neu*-specific antibodies and HER2/*neu*-specific CTLs induces a more effective antitumor response than either alone in severe combined immunodeficient mice (Reilly et al., 2001). Furthermore, the passive administration of HER-2/*neu*-specific mAbs to tolerant *neu* transgenic mice in combination with a GM-CSF-secreting cellular vaccine produced a more robust antitumor effect than either alone, curing about 40% of tumor-bearing mice (Wolpoe et al., 2003). Combination therapy augmented the number of vaccine-induced, tumor necrosis factor (TNF)-secreting, HER2/*neu*-specific CTLs as measured by ELISPOT. Interestingly, the mAbs themselves induced *de novo* $CD8^+$ T cell immunity specific for HER2/*neu* (Wolpoe et al., 2003), suggesting that mAb therapy alone may be able to cross-prime the immune response. Others have shown that humoral HER2/*neu*-specific immunity induced after peptide vaccination in patients can inhibit tumor cell growth and signaling (Montgomery et al., 2005) and mediate ADCC in murine models (Jasinska et al., 2003). These observations together provide a compelling argument for testing trastuzumab in combination with HER2/*neu*-targeted vaccination in patients with malignancies dependent on HER2/*neu* signaling.

2. Rituximab

Rituximab is a humanized mAb specific for CD20, a surface molecule expressed by normal B cells and over 95% of B cell leukemias and lymphomas (Olszewski and Grossbanrd, 2004). Like trastuzumab, it mediates ADCC (Clynes et al., 2000). Rituximab modulates signaling through *src* in follicular lymphoma cells, thereby decreasing the production of IL-10 (Vega et al., 2004). This latter activity of rituximab has two major implications. First, IL-10 normally supports the constitutive expression of the signal transducer and activator of transcription 3 (STAT-3), which results in sustained B-cell lymphocyte 2 (bcl-2) expression. Thus, rituximab may abrogate the intrinsic resistance to apoptosis characteristic of follicular lymphomas. Consistent with this concept, rituximab can induce apoptosis in lymphoma cells, resulting in the cross-priming of lymphoma-specific $CD8^+$ T cells (Selenko et al., 2001). Second, rituximab is known to induce profound B cell depletion. B cells have been shown to inhibit tumor-specific $CD8^+$ CTLs in a murine model (Inoue et al., 2006). Thus, rituximab may enhance T cell-mediated tumor immunity by abrogating the immunosuppressive effects of B lymphocytes. Consistent with the idea that B cell depletion does not inhibit and instead may enhance antitumor immune responses, vaccine-induced immunity to an idiotype-specific vaccine was observed in patients with mantle cell lymphoma despite severe B cell depletion due to treatment with rituximab-based chemotherapy (Neelapu et al., 2005).

3. Cetuximab

Cetuximab is a humanized mAb specific for the epidermal growth factor receptor (EGFR). EGFR is a transmembrane tyrosine kinase that is overexpressed by a variety of epithelial tumors, including colorectal, aerodigestive, breast, prostate, and ovarian cancers. It inhibits signaling pathways that promote tumor growth and progression and induces apoptosis. Cetuximab synergizes with PTX to inhibit angiogenesis and induce tumor cell apoptosis (Inoue et al., 2000). Immunomodulatory activity specific to cetuximab has otherwise not been reported.

4. Bevacizumab

Bevacizumab is a humanized mAb specific for VEGF, a cytokine that is both proangiogenic and immunosuppressive. VEGF causes thymic atrophy, inhibits T cell development, and renders DCs less active (Almand et al., 2000; Ohm et al., 2003). The mAbs specific for VEGF can improve the number and function of DCs in tumor-bearing mice, potentiating the effect of DC-based immunotherapy (Gabrilovich et al., 1999). Together, these observations suggest that VEGF blockade may enhance the activity of tumor vaccines or of the function of adoptively transferred tumor-specific lymphocytes.

B. The mAbs Specific for Immunologic Checkpoints

The mAbs that specifically target the nodal checkpoints of T cell activation are under intense investigation as potent immunomodulatory drugs, either as agonists that further push the T cell response forward or as antagonists that release the break on developing T cell activation. The diversity of molecules involved in tuning the T cell response offers multiple opportunities for manipulating the magnitude and quality of the antitumor immune response during immune priming or potentiating immune effector activity in the tumor microenvironment. In hematological malignancies, these antibodies may also modulate tumor cell biology directly.

C. The mAbs Specific for B7-DC

A naturally occurring immunoglobulin M antibody specific for B7-DC can augment DC survival, antigen processing, and IL-12 secretion, thereby facilitating the activation of naive T cells (Radhakrishnan et al., 2003, 2004, 2005). Giving this antibody to mice bearing B16 melanomas activates immunity that mediates tumor rejection and is dependent on both CD4$^+$ and CD8$^+$ T cells. The development of a recombinant human mAb specific for B7-DC has been reported, paving the way for clinical testing (Van Keulen et al., 2006).

D. The mAbs Specific for B7-H1, B7-H2, B7-H3, and B7-H4

The mAbs that abrogate signaling through B7-H1 can modulate immunity both by enhancing immune priming and by facilitating the activity of end immune effectors in the tumor microenvironment. B7-H1 blockade can augment myeloid DC-activated T cell immunity (Curiel et al., 2003). The role of B7-H1 in promoting T cell apoptosis in the tumor microenvironment suggests that B7-H1 blockade could further promote immunity by preventing the destruction of TILs at the tumor site (Dong et al., 2002). Consistent with this idea, treatment with a mAb against B7-H1 augmented the efficacy of adoptive T cell immunotherapy for squamous cell carcinoma in a murine model system. (Strome et al., 2003). The mAbs that modulate signaling through the B7-H2 (ICOS ligand), B7-H3, and B7-H4 pathways have not been tested in vivo although a functional ICOS-specific mAb that augments T cell activation and also induces apoptosis in an ICOS-expressing multiple myeloma cell line in vitro has been reported (Deng et al., 2004). Abrogating the negative influence of these molecules on effector T cell function in the local tumor microenvironment with mAb blockade will be a critical adjunct to immunomodulatory strategies that target immune priming.

E. The mAbs Specific for CD40

CD40 is normally expressed by B cells, and MAbs specific for CD40 are particularly effective against B cell malignancies, promoting tumor cell lysis by direct binding, enhancing antigen presentation by transformed B cells, and increasing tumor-specific CTL activation. These activities may extend to selected tumors of epithelial and mesenchymal derivation, which can also express CD40. Combining αCD40 mAbs with tumor vaccines promotes the maturation of endogenous DC, thereby augmenting the cross-presentation of tumor-associated antigens to break immune tolerance (Diehl et al., 1999; Sotomayor et al., 1999). The mAbs specific for CD40 are entering clinical trials for cancer treatment.

F. The mAbs Specific for OX40

OX40 signaling supports the activation, expansion, and survival of antigen-specific effector CD4$^+$ T cells, thereby promoting humoral immunity, enlarging the CD4$^+$ T cell memory pool, and supporting the development of CD8$^+$ T cell memory (Sugamura et al., 2004). It also directly supports the function of OX40$^+$ CD8$^+$ effector T cells, guiding their extravasation through OX40 ligand-expressing tumor-associated vasculature into the tumor site itself. The mAbs that augment OX40 signaling *in vivo* abrogate CD4$^+$ T cell tolerance (Bansal-Pakala et al., 2001) by preventing peripheral deletion (Maxwell et al., 2000) and relieving the suppressive influence of Tregs (Takeda et al., 2004). Treatment of tumor-bearing mice with mAb specific for OX40 alone results in prolonged tumor-free survival that is dependent on CD4$^+$ and CD8$^+$ T cells in murine models of sarcoma, colon cancer, breast cancer, and glioma (Sugamura et al., 2004). Treatment of tolerant, tumor-bearing *neu* transgenic mice with the combination of OX40 mAbs and a GM-CSF-secreting vaccine surmounted immune tolerance, augmenting HER2/*neu*-specific CD8$^+$ T cell immunity and increasing tumor-free survival (Murata et al., 2006). It is important to note that OX40 costimulation in combination with vaccination recruited long-lived, high-avidity CD8$^+$ T cells specific for the immunodominant HER2/*neu* epitope to the antitumor immune response.

G. The mAbs Specific for 4-1BB

The mAbs specific for 4-1BB as a single intervention can result in the regression of even poorly immunogenic tumors (Melero et al., 1997), an activity dependent on CD4$^+$ and CD8$^+$ T cells as well as NK cells. The α4-1BB treatment supports the activity of preactivated CD8$^+$ T cells by prolonging their survival (May et al., 2002). Interestingly, the combination of mAb specific for CTLA-4 and 4-1BB to treat established MC38 tumors in mice resulted in the CD8$^+$ T cell-mediated rejection of large tumors, with fewer autoimmune manifestations than with either antibody alone (Kocak et al., 2006). This therapy was not effective against B16 tumors, suggesting that the therapy may depend in part on the immunobiology of the tumor itself. In combination with tumor vaccines, mAbs specific for 4-1BB lift immunologic ignorance by promoting CD4$^+$ Th1 activity in an NK cell-dependent fashion and potentiate a preexisting but tepid antitumor immune response to synergistically reject tumors (Wilcox et al., 2002; Ito et al., 2004). A humanized 4-1BB mAb inhibits T cell-dependent antibody responses in non-human primates but has not yet been tested in human clinical trials (Hong et al., 2000).

V. THERAPEUTICS THAT MITIGATE THE INFLUENCE OF CD4$^+$CD25$^+$ TREGS

The central role of Tregs in maintaining peripheral tolerance has led to intense interest in developing ways to abrogate their negative influence on tumor immunity. Therapies targeting surface molecules that are expressed at high levels relative to effector T cells have received the greatest attention and include agents specific for CD25, GITR, CTLA-4, and, as previously discussed, CY. Here, the focus is on drugs that specifically target Tregs; the use of CY has been previously discussed.

A. Drugs Specific for CD25

Multiple murine tumor models have demonstrated that depleting Tregs with a mAb specific for CD25 alone or in combination with IFN-α or IL-12 results in tumor immunity that can retard tumor growth. Human mAbs specific for CD25 antagonize T cell signaling and are used clinically to ameliorate inflammatory and autoimmune disease. Accordingly, for tumor immunotherapy, interest has turned to the use of denileukin diftitox (ONTAK®), approved by the FDA for the treatment of CD25$^+$ cutaneous T cell leukemia and lymphoma. Denileukin diftitox is composed of the full-length IL-2 molecule fused to the enzymatically active and translocating domains of the diphtheria toxin. It is internalized by CD25$^+$ T cells, leading to inhibition of protein synthesis and cell death. In *neu* transgenic mice, the administration of denileukin diftitox effectively depleted Tregs, markedly inhibiting tumor growth (Knutson *et al.*, 2006). This effect was reversed by the adoptive transfer of Tregs and was associated with the induction of HER2/*neu*-specific immunity. The immunomodulatory activity of denileukin diftitox has been tested in two clinical trials. In the first, a single dose given to eight patients with lung, breast, or ovarian cancers reduced peripheral Treg populations and increased effector T cell activation (Zou, 2006). Treatment was associated with CA-125 normalization and the objective regression of visceral disease in one patient with ovarian cancer. In the second, denileukin diftitox was tested in sequence with a tumor-specific vaccination using DCs transfected with total tumor RNA in patients with renal cell carcinoma (RCC) (Dannull *et al.*, 2005). Here, peripheral Tregs were reduced by denileukin diftitox in a dose-dependent manner, and the combination therapy resulted in more tumor-specific effector T cells than either intervention alone.

B. Drugs Specific for GITR

The mAb specific for GITR can also either physically deplete or abrogate the function of Tregs. Treatment with αGITR mAb can protect mice from a B16 tumor challenge and induce the regression of established Meth-A sarcomas or CT26 colon cancers (Zou, 2006). Treatment with two mAbs individually specific for either GITR or CTLA-4 resulted in a synergistic antitumor effect; in contrast, giving a mAb specific for GITR and a mAb specific for CD25 was less effective due to the depletion of CD4$^+$CD25$^+$ effector T cells by αCD25 (Ko *et al.*, 2005). Giving αGITR mAb with a xenogeneic DNA vaccine enhanced CD8$^+$ T cell responses and increased protection from a lethal B16 tumor challenge (Cohen *et al.*, 2006). Notably, CD8$^+$ T cell responses persisted longer and occurred more rapidly with recall stimulation, implying a role for GITR modulation in programming the memory T cell response.

C. Drugs Specific for CTLA-4

Of potential Treg modulators other than CY, targeted drugs specific for CTLA-4 have been the most extensively tested in murine models. As a single intervention, αCTLA-4 mAb can induce CD8$^+$ T cell-dependent regression. Combining CTLA-4 blockade with GM-CSF-secreting vaccination produces a synergistic antitumor effect compared to either alone (Hurwitz *et al.*, 1998; van Elsas *et al.*, 1999). In one model system, treated B16 tumors contained primarily CD4$^+$ T cells that were both FoxP3$^+$ and FoxP3$^-$ but contained few CD8$^+$ T cells (Quezada *et al.*, 2006). Interestingly, CTLA-4 blockade alone recruited both Tregs and effector T cells to the tumor microenvironment, and the addition of vaccination with a GM-CSF-secreting vaccine recruited CD8$^+$ effector T cells, thus inverting the ratio of effector CD8$^+$ T cells and Tregs. The depletion of Tregs with an mAb specific for CD25 with simultaneous CTLA-4 blockade and GM-CSF-secreting vaccination produced synergistic effects in one poorly immunogenic tumor model (Sutmuller *et al.*, 2001).

Multiple clinical trials testing mAbs specific for CTLA-4 have been reported. One study treated nine subjects with either metastatic melanoma or ovarian cancer with single agent ipilimumab (the humanized IgG₁ mAb MDX-010) (Hodi *et al.*, 2003). Five had been previously vaccinated with autologous GM-CSF-secreting tumor cells for ovarian cancer or melanoma, and four had been vaccinated with melanoma peptides specific for gp100 or MAGE1. Three melanoma patients previously treated with GM-CSF-secreting tumor cells developed marked tumor necrosis with associated inflammatory cell infiltrates, and two ovarian cancer patients also vaccinated with GM-CSF-secreting tumor cells responded with stable or declining levels of the tumor marker CA-125. A second study treated 56 patients with metastastic melanoma with weekly infusion of two doses of ipilimumab in the context of vaccination with melanoma gp100 peptide vaccines (Attia *et al.*, 2005). No significant differences in toxicity or bioactivity were noted between the two dose levels. This trial documented an objective response rate of 13%, with two subjects developing a complete response and five a partial response.

Notably, 14 people developed grade III/IV autoimmune toxicity, with 5 of these 14 (36%) demonstrating an associated clinical response. In contrast, only 2 of the 42 patients (5%) who did not develop significant autoimmunity developed a clinical response. In another study, the frequency of autoimmune enterocolitis in 137 patients with metastatic melanoma who received single agent ipilimumab alone or combined with peptide vaccination, and 61 patients with metastastic RCC who received ipilimumab alone was reported (Beck et al., 2006). The overall objective response rate was 14%. Immune breakthrough events included dermatitis, enterocolitis, hypophysitis, uveitis, hepatitis, and nephritis. Enterocolitis was the most frequent major toxicity, occurring in about 21% of subjects. It was typically successfully managed with high dose corticosteroids, with the use of TNFR blockade with infliximab reserved for refractory cases. Objective tumor responses for patients with enterocolitis were 36% for melanoma and 35% for RCC, compared to 11% and 2%, respectively, in patients without autoimmunity. Another study tested ipilimumab in combination with IL-2 in 36 patients with metastatic melanoma (Maker *et al.*, 2005). The objective response rate was 22% and consistent with the additive activity of each individual drug. Toxicities were primarily related to the CTLA-4 blockade. A distinct group of investigators tested escalating doses of ipilimumab in 19 patients with high-risk resected stage III and IV melanomas vaccinated with three melanoma peptides emulsified in Montanide ISA 51 (Sanderson *et al.*, 2005). Grade III gastrointestinal toxicity developed in four patients. There was again an association with toxicity and clinical benefit: 3 of 8 people who developed autoimmunity ultimately experienced a relapse of their disease, whereas 9 of 11 who did not have autoimmunity relapsed. A final study tested a distinct mAb specific for CTLA-4, the fully human IgG₂ mAb CP-675,206 (Ribas *et al.*, 2005). Here, 39 patients with melanoma, RCC, or colon cancer were enrolled to receive one of seven dose levels of the mAb. Autoimmune side effects included colitis, dermatitis, vitiligo, hypophysitis, and thyroiditis, with two complete and two partial responses in patients with melanoma.

VI. ENDOCRINE AND BIOLOGICALLY TARGETED THERAPY

Breast and prostate cancers are unique among the solid tumors since endocrine manipulation is frequently the mainstay of therapy. Endocrine manipulation for breast cancer includes the selective estrogen receptor modulators and destroyers (SERMs and SERDs: tamoxifen or raloxifene and fulvestrant), the aromatase inhibitors (anastrozole, letrozole, and exemestane), and ovarian ablation. Androgen ablation is a common treatment for prostate cancer. Surprisingly, despite a clear role for estrogen in growth and development of lymphocytes (Nalbandian and Kovats, 2005), there is little information available about the potential impact of these therapies on tumor immunity. Tamoxifen and raloxifene have been reported to inhibit the differentiation and LPS-induced maturation of DCs, antagonizing the estrogen receptor to maintain them in an immature state *in vitro* (Nalbandian *et al.*, 2005). Aromatase inhibitors have been reported to sensitize tumor cells to monocyte-mediated ADCC (Braun *et al.*, 2005). While no immunomodulatory effects of ovarian ablation are available, androgen ablation in murine models of prostate cancer have been shown to diminish tolerance (Drake *et al.*, 2005), a finding likely relevant to the human condition (Mercader *et al.*, 2001).

VII. CONCLUSION

The oncology field has made a great deal of progress in understanding the immunobiology of the host–tumor interaction at the molecular level and now stands at an immunotherapeutic crossroads. The ability to maximize the bioactivity and clinical efficacy of immune-based therapies has never been greater. This opportunity rests in strategically combining immunotherapies with both traditional and novel cancer drugs to shape both the global host environment and the local tumor environment and to ameliorate distinct layers of immune tolerance, ultimately supporting a vigorous and sustained antitumor immune response. Within this modified host environment, immunotherapy treatment regimens that (i) combine tumor vaccines or tumor-specific lymphocytes with targeted drugs that amplify the magnitude and quality of end immune effectors and (ii) relieve the normal controls at specific points in the process of T cell activation will be critical for success. Carefully working out the proper dose and sequencing of drugs that go into these complex combination immunotherapy regimens in clinically relevant laboratory models will accelerate clinical development. Future challenges will include both the development of strategies to dissociate the undesired toxicities of successful tumor immunotherapy, such as enterocolitis, from the therapeutic antitumor effect, and the elucidation of novel pathways of therapeutic resistance to immunotherapy.

Acknowledgments

This work was supported by funding from the Department of Defense (W81XWH-04-1-0595), the National Cooperative Drug Discovery Groups (NCDDG) (U19CA72108), the Breast SPORE Program (P50CA88842), the National Institutes of Health (1K23 CA098498 and RO1 CA 93714), the AVON Foundation, and the Cancer Treatment Research Foundation. Dr. Jaffee is the Dana and Albert "Cubby" Broccoli Professor of Oncology.

Conflict of Interest

This chapter describes work using granulocyte-macrophage colony-stimulating factor-secreting tumor vaccines. Under a licensing agreement between Cell Genesys and Johns Hopkins University, the university is entitled to a share of royalty received by the university on sales of products described in this chapter. The terms of this arrangement are being managed by the Johns Hopkins University in accordance with its conflict of interest policies.

References

Almand, B., Resser, J., Lindman, B., Nadaf, S., Clark, J., Kwon, E., Carbone, D., and Gabrilovich, D. (2000). Clinical significance of defective dendritic cell differentiation in cancer. *Clin. Cancer Res.* **6**, 1755–1766.

Anderson, M., Venanzi, E., Klein, L., Chen, Z., Berzins, S., Turley, S., von Boehmer, H., Bronson, R., Dierich, A., Benoist, C., *et al.* (2002). Projection of an immunological self shadow within the thymus by the aire protein. *Science* **298**, 1395–1401.

Anichini, A., Mortarini, R., Nonaka, D., Molla, A., Vegetti, C., Montaldi, E., Wang, X., and Ferrone, S. (2006). Association of antigen-processing machinery and HLA antigen phenotype of melanoma cells with survival in American Joint Committee on Cancer Stage III and IV melanoma patients. *Cancer Res.* **66**, 6405–6411.

Antonia, S., Mirza, N., Fricke, I., Chiappori, A., Thompson, P., Williams, N., Bepler, G., Simon, G., Janssen, W., Lee, J., et al. (2006). Combination of p53 cancer vaccine with chemotherapy in patients with extensive stage small cell lung cancer. *Clin. Cancer Res.* **12**, 878–887.

Arlen, P., Gulley, J. P., C, Skarupa, L., Pazdur, M., Panicali, D., Beetham, P., Tsang, K., Grosenbach, D., Feldman, J., Steinberg, S., et al. (2006). A randomized phase II study of concurrent docetaxel plus vaccine versus vaccine alone in metastatic androgen-independent prostate cancer. *Clin. Cancer Res.* **12**, 1260–1269.

Arnould, L., Gelly, M., Penault-Liorca, F., Benoit, L., Bonnetain, F., Migeon, C., Cabaret, V., Fermeaux, V., Bertheau, P., Garnier, J., et al. (2006). Trastuzumab-based treatment of HER-2-positive breast cancer: An antibody-dependent cellular cytotoxicity mechanism? *Brit. J. Cancer* **94**, 259–267.

Attia, P., Phan, G., Maker, A., Robinson, M., MM, Q., Yang, J., Sherry, R., Topalian, S., Kammula, U., Royal, R., et al. (2005). Autoimmunity correlates with tumor regression in patients with metastatic melanoma treated with anti-cytotoxic T-lymphocyte antigen-4. *J. Clin. Oncol.* **23**, 6043–6053.

Baldwin, T., Hogquist, K., and Jameson, S. (2004). The fourth way? Harnessing aggressive tendencies in the thymus. *J. Immunol.* **173**, 6515–6520.

Bansal-Pakala, P., Jember, A.-H., and Croft, M. (2001). Signaling through OX40 (CD134) breaks peripheral T-cell tolerance. *Nat. Med.* **7**, 907–912.

Beck, K., Blansfield, J., Tran, K., Feldman, A., Hughes, M., Royal, R., Kammula, U., Topalian, S., Sherry, R., Kleiner, D., et al. (2006). Enterocolitis in patients with cancer after antibody blockade of cytotoxic T-lymphocyte-associated antigen 4. *J. Clin. Oncol.* **24**, 2283–2289.

Bevan, M. (2004). Helping the CD8⁺ T-cell response. *Nat. Rev. Immunol.* **4**, 595–602.

Beyer, M., Kochanek, M., Darabi, K., Popov, A., Jensen, M., Endl, E., Knolle, P., Thomas, R., von Bergwelt-Baildon, M., Debey, S., et al. (2005). Reduced frequencies and suppressive function of CD4⁺CD25[hi] regulatory T cells in patients with chronic lymphocytic leukemia after therapy with fludarabine. *Blood* **106**, 2018–2025.

Bocci, G., Nicolaou, K., and Kerbel, R. (2002). Protracted low-dose effects on human endothelial cell proliferation and survival *in vitro* reveal a selective antiangiogenic window for various chemotherapeutic agents. *Cancer Res.* **62**, 6938–6943.

Borrello, I., Sotomayor, E., Rattis, F.-M., Cooke, S., Gu, L., and Levitsky, H. (2000). Sustaining the graft-versus-tumor effect through posttransplant immunization with granulocyte-macrophage colony-stimulating factor (GM-CSF)-producing tumor vaccines. *Blood* **95**, 3011–3019.

Braun, D., Crist, K., Shaheen, F., Staren, E., Andrews, S., and Parker, J. (2005). Aromatase inhibitors increase the sensitivity of human tumor cells to monocyte-mediated, antibody-dependent cellular cytotoxicity. *Amer. J. Surg.* **190**, 570–571.

Byrd-Leifer, C. A., Block, E. F., Takeda, K., Akira, S., and Ding, A. (2001). The role of MyD88 and TLR4 in the LPS-mimetic activity of Taxol. *Eur. J. Immunol.* **31**, 2448–2457.

Casares, N., Pequignot, N., Tesniere, A., Ghiringhelli, F., Roux, S., Chaput, N., Schmitt, E., Hamai, A., Hervas-Stubbs, S., Obeid, M., et al. (2006). Caspase-dependent immunogenicity of doxorubicin-induced tumor cell death. *J. Exper. Med.* **202**, 1691–1701.

Castilleja, A., Ward, N., O'Brian, C., Swearingen, B. N., Swan, E., Gillogly, M., Murray, J., Kudelka, A., Gershenson, D., and Ioannides, C. (2001). Accelerated HER-2 degradation enhances ovarian tumor recognition by CTL. Implications for tumor immunogenicity. *Mol. Cell. Biochem.* **217**, 21–33.

Chapoval, A., Ni, J., Lau, J., Wilcox, R., Flies, D., Liu, D., Dong, H., Sica, G., Zhu, G., Tamada, K., et al. (2001). B7-H3: A costimulatory molecule for T cell activation and IFN-gamma production. *Nat. Immunol.* **2**, 269–274.

Cho, B., Rao, V., Ge, Q., Eisen, H., and Chen, J. (2000). Homeostasis-stimulated proliferation drives naive T cells to differentiate directly into memory T cells. *J. Exper. Med.* **192**, 549–556.

Chopra, A., Kim, T., O-Sullivan, I., Martinez, D., and Cohen, E. (2006). Combined therapy of an established, highly aggressive breast cancer in mice with paclitaxel and a unique DNA-based vaccine. *Inter. J. Cancer* **118**, 2888–2898.

Chu, Y., Wang, L., Yang, G., Ross, H., Urba, W., Prell, R., Jooss, K., Xiong, S., and Hu, H. (2006). Efficacy of GM-CSF-producing tumor vaccine after docetaxel chemotherapy in mice bearing established Lewis lung carcinomas. *J. Immunother.* **29**, 367–380.

Clynes, R., Towers, T., Presta, L., and Ravetch, J. (2000). Inhibitory Fc receptors modulate *in vivo* cytotoxicity against tumor targets. *Nat. Med.* **6**, 443–446.

Cohen, A., Diab, A., Perales, M., Wolchok, J., Rizzuto, G., Merghoub, T., Huggins, D., Liu, C., Turk, M., Restifo, N., et al. (2006). Agonist anti-GITR antibody enhances vaccine-induced CD8⁺ T cell responses and anti-tumor immunity. *Cancer Res.* **66**, 4904–4912.

Coral, S., Sigalotti, L., Altomonte, M., Engelsberg, A., Colizzi, F., Cattarossi, I., Maraskowsky, E., Jager, E., Seliger, B., and Maio, M. (2002). 5-aza-2'-Deoxycytidine-induced expression of functional cancer testis antigens in human renal cell carcinoma: Immunotherapeutic implications. *Clin. Cancer Res.* **8**, 2690–2695.

Correale, P., Aquino, A., Giuliani, A., Pellegrini, M., Micheli, L., Cusi, M., Nencini, C., Petrioli, R., Prete, S., De Vecchis, L., et al. (2003). Treatment of colon and breast carcinoma cells with 5-fluorouracil enhances expression of carcinoembryonic antigen and susceptibility to HLA-A*02/01 restricted, CEA-peptide-specific cytotoxic T cells *in vitro*. *Inter. J. Cancer* **104**, 437–445.

Correale, P., Cusi, M., Tsang, K., Del Vecchio, M., Marsili, S., Placa, M., Intrivici, C., Aquino, A., Micheli, L., Nencini, C., et al. (2005). Chemo-immunotherapy of metastatic colorectal carcinoma with gemcitabine plus FOLFOX 4 followed by subcutaneous granulocyte macrophage colony stimualting factor and interleukin-2 induces strong immunologic and antitumor activity in metastatic colon cancer patients. *J. Clin. Oncol.* **23**, 8950–8958.

Curiel, T., Wei, S., Dong, H., Alvarez, X., Cheng, P., Mottram, P., Krzysiek, R., Knutson, K., Daniel, B., Zimmermann, M., et al. (2003). Blockade of B7-H1 improves myeloid dendritic cell-mediated antitumor immunity. *Nat. Med.* **9**, 562–567.

Dannull, J., Su, Z., Rizzieri, D., Yang, B., Coleman, D., Yancey, D., Zhang, A., Dahm, P., Chao, N., Gilboa, E., et al. (2005). Enhancement of vaccine-mediated antitumor immunity in cancer patients after depletion of regulatory T cells. *J. Clin. Invest.* **115**, 3623–2633.

Davis, T., Czerwinski, D., and Levy, R. (1999). Therapy of B-cell lymphoma with anti-CD20 antibodies can result in the loss of CD20 antigen expression. *Clin. Cancer Res.* **5**, 611–615.

Demaria, S., Volm, M., Shapiro, R., Yee, H., Oratz, R., Formenti, S., Muggia, F., and Symmans, W. (2001). Development of tumor-infiltrating lymphocytes in breast cancer after neoadjuvant paclitaxel chemotherapy. *Clin. Cancer Res.* **7**, 3025–3030.

Deng, Z., Zhu, W., Lu, C., Shi, Q., Ju, S., Ma, H., Xu, Y., and Zhang, X. (2004). An agonist human ICOS monoclonal antibody that induces T cell activation and inhibits the proliferation of a myeloma line. *Hybrid Hybridomics* **23**, 176–182.

Diehl, L., den Boer, A., Schoenberger, S., van der Voort, E., Schumacher, T., Melief, C., Offringa, R., and Toes, R. (1999). CD40 activation *in vivo* overcomes peptide-induced peripheral cytotoxic T-lymphocyte tolerance and augments anti-tumor vaccine efficacy. *Nat. Med.* **5**, 774–779.

Donepudi, M., Raychaudhuri, P., Bluestone, J., and Mokyr, M. (2001). Mechanism of Melphalan-induced B7-1 gene expression in P815 tumor cells. *J. Immunol.* **166**, 6491–6499.

Dong, H., Strome, S., Salomao, D., Tamura, H., Hirano, F., Flies, D., Roche, P., Lu, J., Zhu, G., Tamada, K., et al. (2002). Tumor-associated B7-H1 promotes T-cell apoptosis: A potential mechanism of immune evasion. *Nat. Med.* **8**, 793–800.

Drake, C., Doody, A., Mihalyo, M., Huang, C., Kelleher, E., Ravi, S., Hipkiss, E., Flies, D., Kennedy, E., Long, M., et al. (2005). Androgen ablation mitigates tolerance to a prostate/prostate cancer-restricted antigen. *Cancer Cell* **7**, 239–249.

Dudley, M., Wunderlich, J., Robbins, P., Yang, J., Hwu, P., Schwartwentruber, D., Topalian, S., Sherry, R., Restifo, N., Hibicki, A., et al. (2002). Cancer regression and autoimmunity in patients after clonal repopulation with antitumor lymphocytes. *Science* **298**, 850–854.

Dudley, M., Wunderlich, J., Yang, J., Sherry, R., Topalian, S., Restifo, N., Royal, R., Kammula, U., White, D., Mavroukakis, S., et al. (2005). Adoptive cell transfer therapy following non-myeloablative but lymphodepleting chemotherapy for the treatment of patients with refractory metastastic melanoma. *J. Clin. Oncol.* **23**, 2346–2357.

Egen, J., Kuhns, M., and Allison, J. (2002). CTLA-4: New insights into its biological function and use in tumor immunotherapy. *Nat. Immunol.* **3**, 611–618.

Emens, L., Armstrong, D., Biedrzycki, B., Davidson, N., Davis-Sproul, J., Fetting, J., Jaffee, E., Onners, B., Piantadosi, S., Reilly, R., et al. (2004). A phase I vaccine safety and chemotherapy dose-finding trial of an allogeneic GM-CSF-secreting breast cancer vaccine given in a specifically timed sequence with immunomodulatory doses of cyclophosphamide and doxorubicin. *Human Gene Ther.* **15**, 313–337.

Emens, L. A. (2005). Trastuzumab: Targeted therapy for the management of HER-2/*neu*-overexpressing metastatic breast cancer. *Amer. J. Ther.* **12**, 243–253.

Emens, L. A., Machiels, J. P., Reilly, R. T., and Jaffee, E. M. (2001). Chemotherapy: Friend or foe to cancer vaccines? *Curr. Opin. Mol. Ther.* **3**, 77–84.

Eralp, Y., Wang, X., Wang, J., Maughan, M., Polo, J., and Lachman, L. (2004). Doxorubicin and paclitaxel enhance the antitumor efficacy of vaccines directed against HER-2/*neu* in a murine mammary carcinoma model. *Breast Cancer Res.* **6**, R275–R283.

Ercolini, A., Ladle, B., Manning, E., Pfannenstiel, L., Armstrong, T., Machiels, J., Bieler, J., Emens, L., Reilly, R., and Jaffee, E. (2005). Recruitment of latent pools of high-avidity $CD8^+$ T cells to the antitumor immune response. *J. Exper. Med.* **201**, 1591–1602.

Fan, Z., Yu, P., Wang, Y., Wang, Y., Fu, M., Liu, W., Sun, Y., and Fu, Y. (2006). NK-cell activation by LIGHT triggers tumor-specific $CD8^+$ T cell immunity to reject established tumors. *Blood* **107**, 1342–1351.

Freeman, G., Long, A., Iwai, Y., Bourque, K., Chernova, T., Nishimura, H., Fitz, L., Malenkovich, N., Okazaki, T., Byrne, M., et al. (2000). Engagement of the PD-1 immunoinhibitory receptor by a novel B7 family member leads to negative regulation of lymphocyte activation. *J. Exper. Med.* **192**, 1027–1034.

Gabrilovich, D., Ishida, T., Nadaf, S., Ohm, J., and Carbone, D. (1999). Antibodies to vascular endothelial growth factor enhance the efficacy of cancer immunotherapy by improving endogenous dendritic cell function. *Clin. Cancer Res.* **5**, 2963–2970.

Gallegos, A., and Bevan, M. (2004). Central tolerance to tissue-specific antigens mediated by direct and indirect antigen presentation. *J. Exper. Med.* **200**, 1039–1049.

Gennari, R., Menard, S., Fagnoni, F., Ponchio, L., Seelsi, M., Tagliabue, E., Castiglioni, F., Villani, L., Magalotti, C., Gibelli, N., et al. (2004). Pilot study of the mechanism of action of preoperative trastuzumab in patients with primary operable breast tumors overexpressing HER-2. *Clin. Cancer Res.* **10**, 5650–5655.

Ghiringhelli, F., Larmonier, N., Schmitt, E., Parcellier, A., Cathelin, D., Garrido, C., Chauffert, B., Solary, E., Bonnotte, B., and Martin, F. (2004). $CD4^+CD25^+$ regulatory T cells suppress tumor immunity but are sensitive to cyclophosphamide which allows immunotherapy of established tumors to be curative. *Europ. J. Immunol.* **34**, 336–344.

Giles, F., Cortes, J., and Kantarjian, H. (2005). Targeting the kinase activity of the BCR-ABL fusion protein in patients with chronic myeloid leukemia. *Curr. Mol. Med.* **5**, 615–623.

Gillard, G., and Farr, A. (2005). Contrasting models of promiscuous gene expression by thymic epithelium. *J. Exper. Med.* **202**, 15–19.

Goldrath, A., Bogatzki, L., and Bevan, M. (2000). Naive T cells transiently aquire a memory-like phenotype during homeostasis-driven proliferation. *J. Exper. Med.* **192**, 557–564.

Gonzalez, L., Loyet, K., Calemine-Fenaux, J., Chauhan, V., Wranik, B., Ouyang, W., and Eaton, D. (2005). A coreceptor interaction between the CD28 and TNF receptor family members B and T lymphocyte attenuator and herpesvirus entry mediator. *Proc. Natl. Acad. Sci. USA* **102**, 1116–11121.

Greenwald, R., Freeman, G., and Sharpe, A. (2005). The B7 family revisited. *Ann. Rev. Immunol.* **23**, 515–548.

Grohmann, U., Orabona, C., Fallarino, F., Vacca, C., Calcinaro, F., Falorni, A., Candeloro, P., Belladonna, M., Bianchi, R., Fioretti, M., et al. (2002). CTLA-4 Ig regulates tryptophan catabolism *in vivo*. *Nat. Immunol.* **3**, 1097–1101.

Hammerling, G., Schonrich, G., Momburg, F., Auphan, N., Malissen, M., Malissen, B., Schmitt-Verhulst, A., and Arnold, B. (1991). Non-deletional mechanisms of peripheral and central tolerance: Studies with transgenic mice with tissue-specific expression of a foreign MHC class I antigen. *Immunol. Rev.* **122**, 47–67.

Heath, W., and Carbone, F. (2001). Cross-presentation, dendritic cells, tolerance, and autoimmunity. *Ann. Rev. Immunol.* **19**, 47–64.

Hodi, F., Mihm, M., Soiffer, R., Haluska, F., Butler, M., Seiden, M., Davis, T., Henry-Spires, R., MacRae, S., Willman, A., et al. (2003). Biologic activity of cytotoxic T lymphocyte-associated antigen 4 antibody blockade in previously vaccinated metastatic melanoma and ovarian carcinoma patients. *Proc. Natl. Acad. Sci. USA* **100**, 4712–4717.

Hogquist, K., Baldwin, T., and Jameson, S. (2005). Central tolerance: Learning self-control in the thymus. *Nat. Rev. Immunol.* **5**, 772–782.

Hong, H., Lee, J., Park, S., Kang, Y., Chang, S., Kim, K., Kim, J., Murthy, K., Payne, J., Yoon, S., et al. (2000). A humanized 4-1BB monoclonal antibody suppresses antigen-induced humoral immune responses in non-human primates. *J. Immunother.* **23**, 613–621.

Hu, H., Poehlein, C., Urba, W., and Fox, B. (2002). Development of antitumor immune responses in reconstituted lymphopenic hosts. *Cancer Res.* **62**, 3914–3919.

Hurwitz, A., Yu, T., Leach, D., and Allison, J. (1998). CTLA-4 blockade synergizes with tumor-derived granulocyte-macrophage colony-stimulating factor for treatment of an experimental mammary carcinoma. *Proc. Natl. Acad. Sci. USA* **18**, 10067–10071.

Inoue, S., Leitner, W., Golding, B., and Scott, D. (2006). Inhibitory effects of B cells on antitumor immunity. *Cancer Res.* **66**, 7741–7747.

Inoue, S., Slaton, J., Perrotte, P., Davis, D., Bruns, C., Hicklin, D., McConkey, D., Sweeney, P., Radinsky, R., and Dinney, C. (2000). Paclitaxel enhances the effects of the anti-epidermal growth factor receptor monoclonal antibody ImClone C225 in mice with metastatic human bladder transitional cell carcinoma. *Clin. Cancer Res.* **6**, 4874–4884.

Ito, F., Li, Q., Shreiner, A., Okuyama, R., Jure-Kunkel, M., Teitz-Tennenbaum, S., and Chang, A. (2004). Anti-CD137 monoclonal antibody administration augments the antitumor efficacy of dendritic cell-based vaccines. *Cancer Res.* **64**, 8411–8419.

Jaffee, E., Hruban, R., Biedrzycki, B., Laheru, D., Schepers, K., Sauter, P., Goemann, M., Coleman, J., Grochow, L., Donehower, R., et al. (2001). Novel allogeneic granulocyte-macrophage colony-stimulating factor-secreting tumor vaccine for pancreatic cancer: A phase I trial of safety and immune activation. *J. Clin. Oncol.* **19**, 145–156.

Jasinska, J., Wagner, S., Radauer, C., Sedivy, R., Brodowicz, T., Wiltschke, C., Breiteneder, H., Pehamberger, H., Scheiner, O., Wiedermann, U., et al. (2003). Inhibition of tumor cell growth by antibodies induced after vaccination with peptides derived from the extracellular domain of HER-2/neu. *Inter. J. Cancer* **107**, 976–983.

Jenkins, M., and Schwartz, R. (1987). Antigen presentation by chemically modified splenocytes induces antigen-specific T cell unresponsiveness *in vitro* and *in vivo*. *J. Exper. Med.* **165**, 302–319.

Kageshita, R., Hirai, S., Ono, T., Hicklin, D., and Ferrone, S. (1999). Down-regulation of HLA class I antigen-processing molecules in malignant melanoma: Association with disease progression. *Amer. J. Pathol.* **154**, 745–754.

Kawasaki, K., Akashi, S., Shimazu, R., Yoshida, S., Miyake, K., and Nishijima, M. (2000). Mouse Toll-like receptor 4.MD-2 complex mediates lipopolysaccharide-mimetic signal transduction by Taxol. *J. Biol. Chem.* **275**, 2251–2254.

Khoury, S., and Sayegh, M. (2004). The roles of the new negative T cell costimulatory pathways in regulating autoimmunity. *Immunity* **20**, 529–538.

Klapper, L., Waterman, H., Sela, M., and Yarden, Y. (2000). Tumor-inhibitory antibodies to HER-2/ErbB-2 may act by recruiting c-Cbl and enhancing ubiquitination of HER-2. *Cancer Res.* **60**, 3384–3388.

Knutson, K., Dang, Y., Lu, H., Lukas, J., Amand, B., Gad, E., Azeke, E., and Disis, M. (2006). IL-2 immunotoxin therapy modulates tumor-associated regulatory T cells and leads to lasting immune-mediated rejection of breast cancers in *neu*-transgenic mice. *J. Immunol.* **177**, 84–91.

Knutson, K. L., Almand, B., Dang, Y., and Disis, M. L. (2004). Neu antigen-negative variants can be generated after neu-specific antibody therapy in *neu* transgenic mice. *Cancer Res.* **64**, 1146–1151.

Ko, K., Yamazaki, K., Nakamura, K. N., T, Hirota, K. Y., T, Shimizu, J., Nomura, T., Chiba, T., and Sakaguchi, S. (2005). Treatment of advanced tumors wtih agonistic anti-GITR mAb and its effects on tumor-infiltrating FoxP3$^+$C25$^+$CD4$^+$ regulatory T cells. *J. Exper. Med.* **202**, 885–891.

Kocak, E., Lute, K., Chang, X., May, K. J., Exten, K., Zhang, H., Abdessalam, S., Lehman, A., Jarjoura, D., Zheng, P., et al. (2006). Combination therapy with anti-CTL antigen-4 and anti-4-1BB antibodies enhances cancer immunity and reduces autoimmunity. *Cancer Res.* **66**, 7276–7284.

Kono, K., Sato, E., Naganuma, H., Takahashi, A., Mimura, K., Nukui, H., and Fujii, H. (2004). Trastuzumab (Herceptin) enhances class I-restricted antigen presentation recognized by HER-2/*neu*-specific T cytotoxic lymphocytes. *Clin. Cancer Res.* **10**, 2538–2544.

Krambeck, A., Thompson, R., Dong, H., Lohse, C., Park, E., Kuntz, S., Leibovich, B., Blute, M., Cheville, J., and Kwon, E. (2006). B7-H4 expression in renal cell carcinoma and tumor vasculature: associations with cancer progression and survival. *Proc. Natl. Acad. Sci. USA* **103**, 10391–10396.

Kryczek, I., Wei, S., Zou, L., Zhu, G., Mottram, P., Xu, H., Chen, L., and Zou, W. (2006). Cutting edge: Induction of B7-H4 on APCs through IL-10: Novel suppressive mode for regulatory T cells. *J. Immunol.* **177**, 40–44.

Kryczek, I., Zou, L., Rodriguez, P., Zhu, G., Wei, S., Mottram, P., Brumlik, M., Cheng, P., Curiel, T., Myers, L., et al. (2006). B7-H4 expression identifies a novel suppressive macrophage population in human ovarian carcinoma. *J. Exper. Med.* **203**, 871–881.

Kyewski, B., and Derbinski, J. (2004). Self-representation in the thymus: An extended view. *Nat. Rev. Immunol.* **4**, 688–698.

Laheru, D., Nemunaitis, J., and Biedrzychi, B. (2005). A feasibility study of a GM-CSF-secreting irradiated whole cell allogeneic vaccine (GVAX) alone or in sequence wtih Cytoxan for patients with locally advanced or metastatic pancreatic cancer. Presented at the Proceedings of the AACR: Pancreatic Cancer 2005—Advances and Challenges, Anaheim, CA, 2005.

Latchman, Y., Wood, C., Chernova, T., Chaudhary, D., Borde, M., Chernova, I., Iwai, Y., Long, A., Brown, J., Nunes, R., et al. (2001). PD-L2 is a second ligand for PD-1 and inhibits T cell activation. *Nat. Immunol.* **2**, 261–268.

Liu, G., Black, K., and Yu, J. (2006). Sensitization of malignant glioma to chemotherapy through dendritic cell vaccination. *Expert Rev. Vacc.* **5**, 233–247.

Loeffler, M., Kruger, J., and Reisfeld, R. (2005). Immunostimulatory effects of low-dose cyclophosphamide are controlled by inducible nitric oxide synthase. *Cancer Res.* **65**, 5027–5030.

Luo, L., Chapoval, A., Flies, D., Zhu, G., Hirano, F., Wang, S., Lau, J., Dong, H., Tamada, K., Flies, A., et al. (2004). B7-H3 enhances tumor immunity *in vivo* by costimulating rapid clonal expansion of antigen-specific CD8$^+$ cytolytic T cells. *J. Immunol.* **173**, 5445–5450.

Lutsiak, M., Semnani, R., De Pascalis, R., Kashmiri, S., Schlom, J., and Sabzevari, H. (2005). Inhibition of CD4$^+$CD25$^+$ T regulatory cell function implicated in enhanced immune response by low dose cyclophosphamide. *Blood* **105**, 2862–2868.

Luznik, L., Slansky, J., Jalla, S., Borrello, I., Levitsky, H., Pardoll, D., and Fuchs, E. (2003). Successful therapy of metastatic cancer using tumor vaccines in mixed allogeneic bone marrow chimeras. *Blood* **101**, 1645–1652.

Machiels, J., Reilly, R., Emens, L., Ercolini, A., Lei, R., Weintraub, D., Okoye, F., and Jaffee, E. (2001). Cyclophosphamide, doxorubicin, and paclitaxel enhance the antitumor immune response of granulocyte/macrophage-colony stimulating factor-secreting whole-cell vaccines in HER-2/*neu* tolerized mice. *Cancer Res.* **61**, 3689–3697.

Mackall, C., Bare, C., Granger, L., Sharrow, S., Titus, J., and Gress, R. (1996). Thymic-independent T cell regeneration occurs via antigen-driven expansion of peripheral T cells resulting in a repertoire that is limited in diversity and prone to skewing. *J. Immunol.* **156**, 4609–4616.

Maker, A., Phan, G., Attia, P., Yang, J., Sherry, R., Topalian, S., Kammula, U., Royal, R., Haworth, L., Levy,

C., et al. (2005). Tumor regression and autoimmunity in patients treated with cytotoxic T lymphocyte-associated antigen-4 blockade and interleukin 2: A phase I/II study. *Annals Surg. Oncol.* **12**, 1005–1016.

Marincola, F., Jaffee, E., Hicklin, D., and Ferrone, S. (2000). Escape of human solid tumors from T cell recognition: Molecular mechanisms and functional significance. *Adv. Immunol.* **74**, 181–273.

Masopust, D., Kaech, K., Wherry, E., and Ahmed, R. (2004). The role of programming in memory T-cell development. *Curr. Opin. Immunol.* **16**, 217–225.

Maxwell, J., Weinberg, A., Prell, R., and Vella, A. (2000). Danger and OX40 receptor signaling synergize to enhance memory T cell survival by inhibiting peripheral deletion. *J. Immunol.* **164**, 107–112.

May, K. J., Chen, L., Zheng, P., and Liu, Y. (2002). Anti-4-1BB monoclonal antibody enhances rejection of large tumor burden by promoting survival but not clonal expansion of tumor-specific CD8$^+$ T cells. *Cancer Res.* **62**, 3459–3465.

McAdam, A., Greenwald, R., Levin, M., Chernova, T., Malenkovich, N., Ling, V., Freeman, G., and Sharpe, A. (2001). ICOS is critical for CD40-mediated antibody class-switching. *Nature* **409**, 102–105.

McGargill, M., Derbinski, J., and Hogquist, K. (2000). Receptor editing in developing T cells. *Nat. Immunol.* **1**, 336–341.

Meissner, M., Reichert, T., Kunkel, M., Gooding, W., Whiteside, T., Ferrone, S., and Seliger, B. (2005). Defects in the human leukocyte antigen class I antigen processing machinery in head and neck squamous cellcarcinoma: Association with clinical outcome. *Clin. Cancer Res.* **11**, 2552–2560.

Melero, I., Shuford, W., Newby, S., Aruffo, A., Ledbetter, J., Hellstrom, K., Mittler, R., and Chen, L. (1997). Monoclonal antibodies against the 4-1BB T-cell activation molecule eradicate established tumors. *Nat. Med.* **3**, 682–685.

Mellor, A., and Munn, D. (2003). Tryptophan catabolism and regulation of adaptive immunity. *J. Immunol.* **170**, 5809–5813.

Mercader, M., Bodner, B., Moser, M., Kwon, P., Park, E., Manecke, R., Ellis, T., Wojcik, E., Yang, D., Flanigan, R., et al. (2001). T cell infiltration of the prostate induced by androgen withdrawal in patients with prostate cancer. *Proc. Natl. Acad. Sci. USA* **98**, 14565–14570.

Mohsin, S., Wiess, H., Gutierrez, M., Chamness, G., Schiff, R., Digiovanna, M., Wang, C., Hilsenbeck, S., Osborne, C., Allred, D., et al. (2005). Neoadjuvant trastuzumab induces apoptosis in primary breast cancers. *J. Clin. Oncol.* **23**, 2460–2468.

Montgomery, R., Makary, E., Schiffman, K., Goodell, V., and Disis, M. (2005). Endogenous anti-HER2 antibodies block HER2 phosphorylation and signaling through extracellular signal-regulated kinase. *Cancer Res.* **65**, 650–656.

Murata, S., Ladle, B., Kim, P., Lutz, E., Wolpoe, M., Ivie, S., Smith, H., Armstrong, T., Emens, L., Jaffee, E., et al. (2006). OX40 costimulation synergizes with GM-CSF whole-cell vaccination to overcome established CD8$^+$ T cell tolerance to an endogenous tumor antigen. *J. Immunol.* **176**, 974–983.

Nalbandian, G., and Kovats, S. (2005). Understanding sex biases in immunity: Effects of estrogen on the differentation and function of antigen presenting cells. *Immunol. Res.* **31**, 91–106.

Nalbandian, G., Pahrkova-Vatchkova, V., Mao, A., Nale, S., and Kovats, S. (2005). The selective estrogen receptor modulators, tamoxifen and raloxifene, impair dendritic cell differentiation and activation. *J. Immunol.* 175, 2666–2675.

Neelapu, S., Kwak, L., Kobrin, C., Reynolds, C., Janik, J., Dunleavy, K., White, T., Harvey, L., Pennington, R., Stetler-Stevenson, M., et al. (2005). Vaccine-induced tumor-specific immunity despite severe B cell depletion in mantle cell lymphoma. *Nat. Med.* **11**, 986–991.

Nigam, A., Yacavone, R., Zahurak, M., Johns, C., Pardoll, D., Piantadosi, S., Levitsky, H., and Nelson, W. (1998). Immunomodulatory properties of antineoplastic drugs administered in conjunction with GM-CSF-secreting cancer cell vaccines. *Inter. J. Cancer* **12**, 161–170.

Nowak, A., Lake, R., Marzo, A., Scott, B., Heath, W., Collins, E., Frelinger, J., and Robinson, B. (2003a). Induction of tumor cell apoptosis *in vivo* increases tumor antigen cross-presentation, cross-priming rather than cross-tolerizing host tumor-specific CD8 T cells. *J. Immunol.* **170**, 4905–4913.

Nowak, A., Robinson, B., and Lake, R. (2003b). Synergy between chemotherapy and immunotherapy in the treatment of established murine solid tumors. *Cancer Res.* **63**, 4490–4496.

Nowak, A., Robinson, B., and Lake, R. (2002). Gemcitabine exerts a selective effect on the humoral immune response: Implications for combination chemo-immunotherapy. *Cancer Res.* **62**, 2353–2358.

Ochsenbein, A., Klenerman, P., Karrer, U., Ludewig, B., Pericin, M., Hengartner, H., and Zinkernagel, R. (1999). Immune surveillance against a solid tumor fails because of immunological ignorance. *Proc. Natl. Acad. Sci. USA* **96**, 2233–2238.

Ogino, T., Bandoh, N., Hayashi, T., Miyokawa, N., Harabuchi, Y., and Ferrone, S. (2003). Association of tapasin and HLA class I antigen down-regulation in primary maxillary sinus squamous cell carcinoma lesions with reduced survival of patients. *Clin. Cancer Res.* **8**, 4043–4051.

Ohm, J., Gabrilovich, D., Sempowski, G., Kisseleva, E., Parman, K., Nadaf, S., and Carbone, D. (2003). VEGF inhibits T-cell development and may contribute to tumor-induced immune suppression. *Blood* **101**, 4878–4886.

Olszewski, A., and Grossbanrd, M. (2004). Empowering targeted therapy: Lessons from rituximab. *Science STKE* **241**, pe30.

Palmer, E. (2003). Negative selection—clearing out the bad apples from the T-cell repertoire. *Nat. Rev. Immunol.* **3**, 383–391.

Pardoll, D. (2002). Spinning molecular immunology into successful immunotherapy. *Nat. Rev. Immunol.* **2**, 227–238.

Pentcheva-Hoang, T., Egen, J., Wojnoonski, K., and Allison, J. (2004). B7–1 and B7–2 selectively recruit CTLA-4 and CD28 to the immunological synapse. *Immunity* **21**, 401–413.

Perez-Diaz, A., Spiess, P., Restifo, N., Matzinger, P., and Marincola, F. (2002). Intensity of the vaccine-elicited immune response determines tumor clearance. *J. Immunol.* **168**, 338–347.

Pircher, H., Rohrer, U., Moskophidis, D., Zinkernagel, R., and Hengartner, H. (1991). Lower receptor avidity required for thymic clonal deletion than for effector T-cell function. *Nature* **351**, 482–485.

Prasad, D., Nguyen, T., Li, Z., Yang, Y., Duong, J., Wang, Y., and Dong, C. (2004). Murine B7-H3 is a negative regulator of T cells. *J. Immunol.* **173**, 2500–2506.

Prell, R., Gearin, L., Simmons, A., Vanroey, M., and Jooss, K. (2006). The anti-tumor efficacy of a GM-CSF-secreting tumor cell vaccine is not inhibited by docetaxel administration. *Cancer Immunol. Immunother.* **12**, 1–9.

Quezada, S., Peggs, K., Curran, M., and Allison, J. (2006). CTLA-4 blockade and GM-CSF combination immunotherapy alters the intratumor balance of effector and regulatory T cells. *J. Clin. Invest.* **116**, 1935–1945.

Radhakrishnan, S., Celis, E., and Pease, L. (2005). B7-DC cross-linking restores antigen uptake and augments antigen-presenting cell function by matured dendritic cells. *Proc. Natl. Acad. Sci. USA* **102**, 11438–11443.

Radhakrishnan, S., Nguyen, L., Ciric, B., Flies, D., Van Keulen, V., Tamada, K., Chen, L., Rodriguez, M., and Pease, L. (2004). Immunotherapeutic potential of B7-DC (PD-L2) cross-linking antibody in conferring antitumor immunity. *Cancer Res.* **64**, 4965–4972.

Radhakrishnan, S., Nguyen, L., Ciric, B., Ure, D., Zhou, B., Tamada, K., Dong, H., Tseng, S., Shin, T., Pardoll, D., *et al.* (2003). Naturally occurring human IgM antibody that binds B7-DC and potentiates T cell stimulation by dendritic cells. *J. Immunol.* **170**, 1830–1838.

Redmond, W., and Sherman, L. (2005). Peripheral tolerance of CD8 lymphocytes. *Immunity* **22**, 275–284.

Reilly, R., Machiels, J., Emens, L., Ercolini, A., Okoye, F., Lei, R., Weintraub, D., and Jaffee, E. (2001). The collaboration of both humoral and cellular HER-2/*neu*-targeted immune responses is required for the complete eradication of HER-2/*neu*-expressing tumors. *Cancer Res.* **61**, 880–883.

Ribas, A., Camacho, L., Lopez-Berestein, G., Pavlov, D., Bulanhagul, C., Millham, R., Comin-Anduix, B., Reuben, J., Seja, E., Parker, C., *et al.* (2005). Antitumor activity in melanoma and anti-self responses in a phase I trial with anti-cytotoxic T lymphocyte-associated antigen-4 monoclonal antibody CP-675,206. *J. Clin. Oncol.* **23**, 8968–8977.

Salceda, S., Tang, T., Kmet, M., Munteanu, A., Ghosh, M., Macina, R., Liu, W., Pilkington, G., and Papkoff, J. (2005). The immunomodulatory protein B7-H4 is overexpressed in breast and ovarian cancers and promotes epithelial cell transformation. *Exper. Cell Res.* **306**, 128–141.

Sanderson, K., Scotland, R., Lee, P., Liu, D., Groshen, S., Snively, J., Sian, S., Nichol, G., Davis, T., Keler, T., *et al.* (2005). Autoimmunity in a phase I trial of a fully human anti-cytotoxic T-lymphocyte-antigen-4 monoclonal antibody with multiple melanoma peptides and Montanide ISA 51 for patients with resected stages III and IV melanoma. *J. Clin. Oncol.* **23**, 741–750.

Schiavoni, G., Mattei, F., Di Puchio, T., Santini, S., Bracci, L., Belardelli, F., and Proietti, E. (2000). Cyclophosphamide induces type I interferon and augments the number of $CD44^{high}$ T lymphocytes in mice: Implications for strategies of chemoimmunotherapy of cancer. *Blood* **95**, 2024–2030.

Schiller, J., Nemunaitis, J., and Ross, H. (2005). A phase 2 randomized study of GM-CSF gene-modified autologous tumor vaccine (CG8123) with and without low dose cyclophosphamide in advanced stage non-small cell lung cancer (NSCLC). Presented at the International Association for the Study of Lung Cancer 2005, Barcelona, Spain, July, 2005.

Sedy, J., Gavrieli, M., Potter, K., Hurchla, M., Lindsley, R., Hildner, K., Scheu, S., Pfeffer, K., Ware, C., Murphy, T., *et al.* (2005). B and T lymphocyte attenuator regulates T cell activation through interaction with herpesvirus entry mediator. *Nat. Immunol.* **6**, 90–98.

Selenko, N., Maidic, O., Draxier, S., Berer, A., Jager, U., Knapp, W., and Stockl, J. (2001). CD20 antibody (C2B8)-induced apoptosis of lymphoma cells promotes phagocytosis by dendritic cells and cross-priming of $CD8^+$ cytotoxic T cells. *Leukemia* **15**, 1619–1626.

Shin, T., Kennedy, G., Gorski, K., Tsuchiya, H., Koseki, H., Azuma, M., Yagita, H., Chen, L., Powell, J., Pardoll, D., *et al.* (2003). Cooperative B7-1/2 (CD80/CD86) and B7-DC costimulation of $CD4^+$ T cells independent of the PD-1 receptor. *J. Exper. Med.* **198**, 31–38.

Shin, T., Yoshimura, K., Shin, T., Crafton, E., Tsuchiya, H., Housseau, F., Koseki, H., Schulick, R., Chen, L., and Pardoll, D. (2005). *In vivo* co-stimulatory role of

B7-DC in tuning T helper cell1 and cytotoxic T lymphocyte responses. *J. Exper. Med.* **201**, 1531–1541.

Sica, G., Choi, I., Zhu, G., Tamada, K., Wang, S., Tamura, H., Chapoval, A., Flies, D., Bajorath, J., and Chen, L. (2003). B7-H4, a molecule of the B7 family, negatively regulates T cell immunity. *Immunity* **18**, 849–861.

Sojka, D., Donepudi, M., Bluestone, J., and Mokyr, M. (2000). Melphalan and other anticancer modalities up-regulate B7–1 gene expression in tumor cells. *J. Immunol.* **164**, 6230–6236.

Sotomayor, E., Borrello, I., Tubb, E., Rattis, F., Bien, H., Lu, Z., Fein, S., Schoenberger, S., and Levitsky, H. (1999). Conversion of tumor-specific CD4+ T-cell tolerance to T-cell priming through *in vivo* ligation of CD40. *Nat. Med.* **5**, 780–787.

Strome, S., Dong, H., Tamura, H., Voss, S., Flies, D., Tamada, K., Salomao, D., Cheville, J., Hirano, F., Lin, W., *et al.* (2003). B7-H1 blockade augments adoptive T cell immunotherapy for squamous cell carcinoma. *Cancer Res.* **63**, 6501–6505.

Sugamura, K., Ishii, N., and Weinberg, A. (2004). Therapeutic targeting of the effector T-cell co-stimulatory molecule OX40. *Nat. Rev. Immunol.* **4**, 420–431.

Suh, W., Gajewska, B., Okada, H., Gronski, M., Bertram, E., Dawicki, W., Duncan, G., Bukczynski, J., Plyte, S., Elia, A., *et al.* (2003). The B7 family member B7-H3 preferentially down-regulates T helper type 1-mediated immune responses. *Nat. Immunol.* **4**, 899–906.

Sutmuller, R., van Duivenvoorde, L., van Elsas, A., Schumacher, T., Wildenberg, M., Allison, J., Toes, R., Offringa, R., and Melief, C. (2001). Synergism of cytotoxic T lymphocyte-associated antigen-4 blockade and depletion of CD25+ regulatory T cells in antitumor therapy reveals alternative pathways for suppression of autoreactive cytotoxic T lymphocyte responses. *J. Exper. Med.* **194**, 823–832.

Suzuki, E., Kapoor, V., Jassar, A., Kaiser, L., and Albelda, S. (2005). Gemcitabine selectively eliminates splenic Gr-1/Cd11b+ myeloid suppressor cells in tumor-bearing animals and enhances antitumor immune activity. *Clinical Cancer Res.* **11**, 6713–6721.

Taieb, J., Chaput, N., Schartz, N., Roux, S., Novault, S., Menard, C., Ghiringhelli, F., Terme, M., Carpentier, A., Darrasse-Jeze, G., *et al.* (2006). Chemoimmunotherapy of tumors: Cyclophosphamide synergizes with exosome based vaccines. *J. Immunol.* **176**, 2722–2729.

Takeda, I., Ine, S., Killeen, N., Ndhlovu, L., Murata, K., Satomi, S., Sugamura, K., and Ishii, N. (2004). Distinct roles for the OX40–OX40 ligand interaction in regulatory and nonregulatory T cells. *J. Immunol.* **172**, 3580–3589.

Teshima, T., Liu, C., Lowler, K., Dranoff, G., and Ferrara, J. (2002). Donor leukocyte infusion from immunized donors increases tumor vaccine efficacy after allogeneic bone marrow transplantation. *Cancer Res.* **62**, 796–800.

Teshima, T., Mach, N., Hill, G., Pan, L., Gillessen, S., Dranoff, G., and Ferrara, J. (2001). Tumor cell vaccine elicits potent antitumor immunity after allogeneic T-cell-depleted bone marrow transplantation. *Cancer Res.* **61**, 162–171.

Thomas, A. M., Santarsiero, L., Lutz, E., Armstrong, T., Chen, Y., Huang, L., Laheru, D., Goggins, M., Hruban, R., and Jaffee, E. (2004). Mesothelin-specific CD8+ T cell responses provide evidence of *in vivo* cross-priming by antigen-presenting cells in vaccinated pancreatic cancer patients. *J. Exper. Med.* **200**, 297–306.

Tong, A., and Stone, M. (2002). Prospects for CD-40-directed experimental therapy of human cancer. *Cancer Gene Ther.* **10**, 1–13.

Tong, Y., Song, W., and Crystal, R. (2001). Combined intratumoral injection of bone-marrow-derived dendritic cells and systemic chemotherapy to treat pre-existing murine tumors. *Cancer Res.* **61**, 7530–7535.

Tringler, B., Zhuo, S., Pilkington, G., Torkko, K., Singh, M., Lucia, M., Heinz, D., Papkoff, J., and Shroyer, K. (2005). B7-H4 is highly expressed in ductal and lobular breast cancer. *Clin. Cancer Res.* **11**, 1842–1848.

Tseng, S., Otsuji, M., Gorski, K., Huang, X., Slansky, J., Pai, S., Shalabi, A., Shin, T., Pardoll, D., and Tsuchiya, H. (2001). B7-DC, a new dendritic cell molecule with potent costimulatory properties for T cells. *J. Exp. Med.* **193**, 839–846.

van Elsas, A., Hurwitz, A., and Allison, J. (1999). Combination immunotherapy of B16 melanoma using anti-cytotoxic T lymphocyte-associated antigen-4 (CTLA-4) and granulocyte-macrophage colony-stimulating factor (GM-CSF)-producing vaccines induces rejection of subcutaneous and metastatic tumors accompanied by autoimmune depigmentation. *J. Exp. Med.* **190**, 355–366.

Van Keulen, V., Ciric, B., Radharkrishnan, S., Heckman, K., Mitsunaga, Y., Iijima, K., Kita, H., Rodriguez, M., and Pease, L. (2006). Immunomodulation using the recombinant monoclonal human B7-DC cross-linking antibody rHIgM12. *Clin. Exp. Immunol.* **143**, 314–321.

Vega, M., Herta-Yepaz, S., Garban, H., Jazirehi, C., Emmanouilides, C., and Bonavida, B. (2004). Rituximab inhibits p38 MAPK activity in 2F7 B NHL and decreases IL-10 transcription: Pivotal role of p38 MAPK in drug resistance. *Oncogene* **23**, 3530–3540.

Vereecque, R., Saudemont, A., and Quesnel, B. (2004). Cytosine arabinoside induces costimulatory molecule expression in acute myeloid leukemia cells. *Leukemia* **18**, 1223–1230.

von Mehren, M., Arlen, P., Gulley, J., Rogatko, A., Cooper, H., Meropol, N., Alpaugh, R., Davey, M., McLaughlin, S., Beard, M., *et al.* (2001). The influence of granulocyte macrophage colony-stimulating factor and prior chemotherapy on the immunological response to a vaccine (ALVAC-CEA B7.1) in patients with metastatic carcinoma. *Clin. Cancer Res.* **7**, 1181–1191.

Walker, L., and Abbas, A. (2001). The enemy within: Keeping self-reactive T cells at bay in the periphery. *Nat. Rev. Immunol.* **21**, 11–19.

Wallin, J., Liang, L., Bakardjiev, A., and Sha, W. (2001). Enhancement of CD8$^+$ T cell responses by ICOS/B7h costimulation. *J. Immunol.* **167**, 123–139.

Wang, F., Huang, C., and Kanagawa, O. (1998). Rapid deletion of rearranged T cell antigen receptor (TCR) Va-Ja segment by secondary rearrangement in the thymus: Role of continuous rearrangement of TCRa chain gene and positive selection in the T cell repertoire formation. *Proc. Natl. Acad. Sci. USA* **95**, 11834–11839.

Wang, J., Kobayashi, M., Han, M., Choi, S., Takano, M., Hashino, S., Tanaka, J., Kondoh, T., Kawamura, K., and Hosokawa, M. (2002). MyD88 is involved in the signalling pathway for Taxol-induced apoptosis and TNF-alpha expression in human myelomonocytic cells. *Brit. J. Hematol.* **11**, 638–645.

Watts, T. (2005). TNF-TNFR family members in costimulation of T cell responses. *Ann. Rev. Immunol.* **23**, 23–68.

Weihrauch, M., Ansen, S., Jurkiewicz, E., Geisen, C., Xia, Z., Anderson, K., Gracien, E., Schmidt, M., Wittig, B., Diehl, V., *et al.* (2005). Phase I/II combined chemoimmunotherapy with carcinoembryonic antigen-derived HLA-A2-restricted CAP-1 peptide and irinotecan, 5-fluorouracil, and leucovorin in patients with primary metastatic colorectal cancer. *Clin. Cancer Res.* **15**, 5993–6001.

Weninger, W., Crowley, M., Manjunath, N., and von Andrian, U. (2001). Migratory properties of naive, effector, and memory CD8$^+$ T cells. *J. Exp. Med.* **194**, 953–966.

Wilcox, R., Flies, D., Zhu, G., Johnson, A., Tamada, K., Chapoval, A., Strome, S., Pease, L., and Chen, L. (2002). Provision of antigen and CD137 signaling breaks immunological ignorance, promoting regression of poorly immunogenic tumors. *J. Clin. Invest.* **109**, 651–659.

Wolpoe, M., Lutz, E., Ercolini, A., Murata, S., Ivie, S., Garrett, E., Emens, L., Jaffee, E., and Reilly, R. (2003). HER-2/*neu*-specific monoclonal antibodies collaborate with HER-2/*neu*-targeted granulocyte macrophage colony-stimulating factor secreting whole cell vaccination to augment CD8$^+$ T cell effector function and tumor-free survival in Her-2/*neu*-transgenic mice. *J. Immunol.* **171**, 2161–2169.

Yang, F., and Haluska, F. (2004). Treatment of melanoma with 5-fluorouracil or dacarbazine *in vitro* sensitizes cells to antigen-specific CTL lysis through perforin/granzyme- and Fas-mediated pathways. *J. Immunol.* **172**, 4599–4608.

Yu, B., Kusmartsev, S., Cheng, F., Paolini, M., Nefedova, Y., Sotomayor, E., and Gabrilovich, D. (2003). Effective combination of chemotherapy and dendritic cell administration for the treatment of advanced stage experimental breast cancer *Clin. Cancer Res.* **9**, 285–294.

Yu, P., Lee, Y., Liu, W., Chin, R., Wang, J., Wang, Y., Schietinger, A., Phillip, A., Schreiber, H., and Fu, Y. (2004). Priming of naive T cells inside tumors leads to eradication of established tumors. *Nat. Immunol.* **5**, 141–149.

Zou, W. (2005). Immunosuppressive networks in the tumor environment and their therapeutic relevance. *Nat. Rev. Cancer* **5**, 263–274.

Zou, W. (2006). Regulatory T cells, tumor immunity, and immunotherapy. *Nat. Rev. Immunol.* **6**, 295–307.

zum Buschenfelde, C., Hermann, C., Schmidt, B., Peschel, C., and Bernhard, H. (2002). Anti-human epidermal growth factor receptor-2 (HER-2) monoclonal antibody Trastuzumab enhances cytolytic activity of Class I-restricted HER-2-specific T lymphocytes against HER-2-overexpressing tumor cells. *Cancer Res.* **62**, 2244–2247.

CHAPTER 13

Immune Stimulatory Features of Classical Chemotherapy

ROBBERT G. VAN DER MOST, ANNA K. NOWAK, AND RICHARD A. LAKE

I. Introduction
II. Tumor Cell Death
   A. Different Pathways to Tumor Cell Death
III. Pathways to Immunogenicity
   A. Immunogenic Apoptosis
   B. Uric Acid as an Endogenous Danger Signal
   C. Genotoxic Stress (DNA Damage) Upregulates NK Cell Receptor Ligands
   D. Immunogenic DNA
   E. Cyclophosphamide: DNA Damage, IFNs, and Tregs
IV. Chemotherapy and the Immune System
   A. Chemotherapy and Lymphopenia
   B. Tregs
   C. Lymphopenia and Homeostatic Proliferation
   D. Breaking Tumor-Driven Immunomodulation
V. A Practical Partnership: Chemotherapy and Immunotherapy
   A. How to Choose the Right Immunotherapy
   B. DC Activation Using CD40–CD40L Ligation
   C. Stimulation of Innate Immunity Through TLRs
   D. Countering Tumor-Driven Immunosuppression
   E. Enhancing the Response at the Effector Site
   F. Cytokines
   G. Hybrids
VI. Effects of Chemotherapy on Human Antitumor Immunity and Chemoimmunotherapy Clinical Trials
   References

It is increasingly clear that anticancer chemotherapy is not an immunological "null" event. Chemotherapy does more than just kill tumor cells, and the oncology field's understanding on the interactions between dying tumor cells, antigen-presenting cells, and antitumor effector cells is increasing rapidly. Chemotherapeutic drugs kill tumor cells and, in the process, increase the amount of tumor antigens that are presented to the immune system. Moreover, the process of apoptotic cell death may in itself provide an immunostimulatory signal. Uric acid, damaged DNA, and heat-shock proteins are molecular flags for immunogenicity, and more such markers will undoubtedly be discovered. Chemotherapy also has a direct effect on the immune system by depleting or inactivating regulatory T cells (Tregs) and triggering homeostatic T cell proliferation. Both have the capacity to enhance antitumor immune responses. From this it follows that there are multiple entry points for immune therapy. Strategies that add a "dangerous" context to tumor antigens (e.g., Toll-like receptor [TLR] ligands) are now at the point of reaching the clinic. In addition, CD40 ligation could bypass $CD4^+$ "help" and directly activate tumor antigen-presenting cells (APCs). The increasing understanding of the molecular details of the complex cytokine communication network also provides ample opportunities for immune intervention. Combination chemoimmunotherapy protocols are now finding their way in clinical trials with encouraging first results.

I. INTRODUCTION

Although patients with cancer may be prescribed chemotherapy for palliation, the primary aim of using cytotoxic drugs is to kill tumor cells. In most cases, the drugs induce apoptosis ("programmed cell death") in their targets. Ideally, a chemotherapeutic drug would only kill tumor cells, but in reality, some collateral damage is inevitable. Since most drugs target aspects of cellular division, normal cells are sensitive when they are dividing. This leads to some of the well-known iatrogenic effects of chemotherapy, including myelosuppression, hair loss, and mucositis. Because these side effects are often dose limiting, increasing specificity is an important aim of rational drug design. The myelosuppressive properties of chemotherapy in particular have historically been seen as a major problem since these would seem to preclude the therapeutic use of antitumor immune responses. However, as this chapter argues, lymphodepletion does not necessarily rule out immunotherapy and could even have a positive role in the generation of antitumor immune responses. In fact, chemotherapeutic drugs exert their influence on the immune system at different levels, affecting tumor antigen presentation, dendritic cell (DC) maturation, and T cell activation. For example, different chemotherapeutic drugs kill tumor cells in different ways, which has profound effects on the immunogenicity of these dying cells. Furthermore, chemotherapy-induced lymphopenia may also deplete regulatory $CD4^+CD25^+$ T cells (Tregs), which normally maintain tolerance against tumor-associated antigens (Figure 13.1). These two examples, which will be discussed in more depth in the following sections, serve to illustrate the complex relationship between chemotherapy and the immune system. The details of this interaction and the potential for immunotherapeutic intervention are only now being appreciated. Chemotherapy remains the cornerstone of cancer treatment, and the growing insight that the two treatment modalities of chemotherapy and immunotherapy are not mutually exclusive but are more likely to be synergistic has important clinical implications.

II. TUMOR CELL DEATH

A. Different Pathways to Tumor Cell Death

How do chemotherapeutic drugs kill cells? Usually they do so by interfering with cellular DNA synthesis or by dysregulating specific metabolic pathways that are necessary for growth or survival of malignant cells (Lake and Robinson, 2005). Tumor cell apoptosis is the end result in many, but certainly not all, cases. Table 13.1 summa-

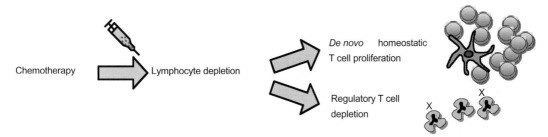

FIGURE 13.1 Cytotoxic chemotherapy affecting the tumor and targeting dividing lymphocytes. These lymphocytes are the very cells that are required to develop an immune response. For this reason, chemotherapy and immunotherapy have been seen as antagonistic. This is, however, only part of the story. Chemotherapy also depletes Tregs and could, in this way, enhance potential immune responses. Furthermore, lymphodepletion triggers homeostatic T cell reconstitution, creating new populations of pre-T cells that need education in the thymic environment. The postchemotherapy phase of immune system reconstitution provides a unique opportunity for therapeutic intervention by shaping the repertoire toward reactivity to tumor antigens.

TABLE 13.1 Overview of Common Cytotoxic Drugs and Their Mechanism of Action

Name	Class	Targeted Pathway	Cell Death
Gemcitabine	DNA chain terminator	DNA synthesis	Apoptosis
Doxorubicin	DNA cross-linker/ Topoisomerase inhibitor	DNA synthesis	Apoptosis
Cisplatin	DNA cross-linker	DNA synthesis	Apoptosis
Pemetrexed	Multitargeted antifolate		Apoptosis
Temozolomide	DNA alkylating	DNA synthesis	Apoptosis
Vincristine	Vinca alkaloid	Microtubule formation prevention	
Cyclophosphamide	DNA alkylating	DNA synthesis	Apoptosis
Taxol	Tubulin interaction	Microtubule depolymerization prevention	Apoptosis
Fludarabine	Nucleoside analog	DNA synthesis	Apoptosis
Coramsine	Plant extract	Membrane integrity	Necrosis
PEP	Plant extract	Membrane integrity	Necrosis

rizes some selected agents and their mechanisms of action.

Coramsine and PEP005 are two chemotherapeutic drugs that kill cells by primary necrosis (Ogbourne et al., 2004; van der Most et al., 2006a, 2006b). This form of cell death seems to result in minimal activation of the immune system.

1. The Immunogenicity of Cell Death

Successful chemotherapy results in massive tumor cell death and, thus, the potential release of large quantities of tumor antigens. How does the immune system respond to this? To address this question, this discussion considers immunogenicity as a function of antigenic *content* and immunogenic *context*. The current understanding is that chemotherapy changes both parameters. Before discussing how tumor immunogenicity is modified by chemotherapy, this section first addresses how the immune system perceives tumor antigens in the absence of therapeutic intervention.

2. Constitutive Cross-Presentation of Tumor Antigens

Tumor cells do not typically present antigens very efficiently to T cells. Instead, as with antigen expressed by virally infected cells, most tumor antigens are perceived by the immune system through a process known as cross-presentation (van der Most et al., 2006a, 2006b). The essence of cross-presentation is that professional antigen-presenting cells (APCs), including DCs, pick up antigen from other cells and process the antigen into small peptides that can be presented to T cells via their own class I presentation route. Thus, antigens from an exogenous source cross into the endogenous MHC class I presentation pathway. A subpopulation of $CD11c^+CD8^+$ DCs plays the major role in this process (Heath et al., 2004). It is not entirely clear how DCs sample antigen and in what form from other cells. The most widely held hypotheses include (i) nibbling from live cells, (ii) phagocytosis from apoptotic cells, (iii) transfer of antigens or peptides via heat-shock proteins, and (iv) transfer of antigens via exosomes. There is evidence that all of these can play some role (reviewed in van der Most et al., 2006a, 2006b). The finding that $CD8^+$ T cell epitopes located in signal sequences, which become cleaved from their parent as peptides soon after their synthesis, are not cross-presented suggests that proteins rather than peptides are the source of transferred antigen (Wolkers et al., 2004). Controversy persists over the role of heat-shock proteins in cross-presentation (Binder and Srivastava, 2005). There are, however, strong indications that phagocytosis of apoptotic bodies plays a major role in cross-presentation, and this clearly links cross-presentation of tumor antigens to chemotherapy. Antigen cross-presentation is involved both in the maintenance of tolerance and in the induction of immune responses. Thus, the same $CD8^+$ DC may present the same antigens in an immunogenic or a tolerogenic fashion, and evidence suggests that the outcome is determined by the context in which the antigen is presented. Typically, an immunogenic or proinflammatory context is provided by members of the group of microbial products referred to as pathogen-associated molecular patterns (PAMPs) (Iwasaki and Medzhitov, 2004). PAMPs include viral structures, such as double-stranded RNA, and bacterial products, such as lipopolysaccharide (LPS), flagellin, and unmethylated CpG-containing DNA motifs. These structures bind to Toll-like receptors (TLRs) and provide strong proinflammatory signals (Table 13.2).

TABLE 13.2 Tregs and Their Ligands

Ligand	Receptor	Source	Pharmacological Analog
Triacylated lipopeptides	TLR-1	Bacteria	
Lipopeptides, peptidoglycan	TLR-2	Bacteria, mycobacteria	Zymosan
Double-stranded RNA	TLR-3	RNA viruses	Poly-I:C
LPS, paclitaxel (Taxol®)	TLR-4	Gram-negative bacteria	
Flagellin	TLR-5	Bacteria	
Diacylated lipopeptides, lipoteichoic acid	TLR-6	Bacteria, mycoplasma	
Single-stranded viral RNA	TLR-7	RNA viruses	Imiquimod, R-837, R-848, loxoribine (guanosine analogs)
Single-stranded viral RNA	TLR-8		Poly-G, R-848
Unmethylated CpG-containing bacterial DNA	TLR-9	DNA viruses, bacteria	CpG-oligonucleotides
Unknown	TLR-10		

Part of the proinflammatory effect of TLR-stimulation is mediated by type I interferons (IFN-α and IFN-β). Evidence indicates that type I IFNs are particularly important for the generation of CD8 T cell responses against cross-presented antigens (Le Bon et al., 2003). Because tumor antigens are predominantly cross-presented, type I IFNs could play a central role in the transformation of antitumor tolerance into antitumor immune responses.

Thus, to generate useful immune responses, tumor antigens must be picked up by DCs for cross-presentation and must then be presented to T cells in an immunogenic context. Chemotherapy is likely to affect both of these processes because different drugs have differential toxicity for cells of the immune system and because they invoke different pathways to cell death.

3. Chemotherapy Augments Antigen Cross-Presentation

How does chemotherapy affect the nature and the amount of antigens for cross-presentation? Results indicate that tumor cell apoptosis induced by chemotherapy increases the amount of cross-presented antigen for a given tumor size (Nowak et al., 2003a). To make this point, Nowak et al. (2003a) performed the following experiment. Mice bearing subcutaneously growing tumors were treated with the antimetabolite named gemcitabine, which reduced the tumor burden by triggering tumor cell death. Adoptive transfer of tumor-specific CD8 T cells was used to quantitate tumor antigen cross-presentation. These cells were labeled with the fluorescent dye 5,6-carboxyfluoroscein succinimidyl (CFSE), and the extent of T cell proliferation was measurable by fluorescence-activated cell sorter (FACS) because the dye was serially diluted with each cell division. The amount of cell division was used as a surrogate marker of the amount of antigen presentation. These experiments revealed that tumor cell apoptosis dramatically increased antigen cross-presentation in the draining lymph nodes. Thus, chemotherapy has the capacity to alter the antigenic content that is cross-presented. It is also important to note that the data also indicate that apoptotic cells are a good source of antigen for cross-presentation. Increasing the total amount of cross-presented antigen could potentially permit other antigens to reach a threshold concentration in the draining lymph nodes that would render them immunogenic. It is possible that any drug that triggers apoptosis will increase levels of antigen cross-presentation, but this remains to be determined.

III. PATHWAYS TO IMMUNOGENICITY

A. Immunogenic Apoptosis

Given that chemotherapy can increase the quantity of cross-presented antigen, two questions follow. First, does the immune system perceive these tumor antigens as immunogenic? Second, are all cytotoxic agents equivalent in their capacity to invoke an immune response? Chemotherapy-induced tumor cell death would not *a priori* be expected to be immunogenic because typical TLR-ligands (e.g., PAMPs) are not likely to be a consistent feature of the tumor environment. Apoptotic cell death is associated with the translocation of phosphatidylserine (PS) from the inner to the outer leaflet of the plasma membrane, where it acts as a molecular flag, signaling the immune system to take no notice. PS promotes uptake by macrophages rather than DCs and stimulates the production of anti-inflammatory factors, such as transforming growth factor β (TGF-β) and interleukin (IL)-10, and blocks IL-12 production. Since billions of cells undergo apoptosis every day as part of normal tissue homeostasis, tolerogenic apoptosis is essential to avoid overwhelming autoimmunity. Necrotic cell death, on the other hand, is associated with infection and tissue destruction and would therefore be expected to promote an immune response.

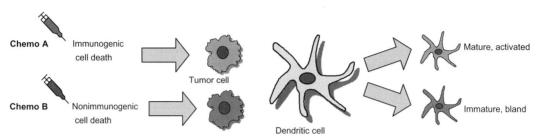

FIGURE 13.2 **Cytotoxic chemotherapy can kill tumor cells but may not evoke an immune response.** It is now clear that the way a chemotherapeutic drug kills a tumor cell determines how that dying cell interacts with the immune system and whether the interaction will lead to an immune response. Immunogenic cell death induces DC maturation, allowing the DC to activate relevant T cells; in contrast, nonimmunogenic cell death is bland and does not activate the DC.

Although there is convincing evidence for this view, there are also strong counterindications, as will be discussed in the following sections. The chapter authors' own view is that the explanation for the apparent discrepancy in the apoptosis/necrosis debate lies in the observation that not all apoptosis is equal. Many different pathways invoke a set of phenomena that have been collectively called apoptosis. The immune system clearly discriminates between bland apoptosis leading to ignorance or tolerance and inflammatory apoptosis leading to immune activation (Figure 13.2).

The finding that apoptotic cell death is an essential prequel for immunogenicity was illustrated in an elegant series of experiments using recombinant alphaviruses (Leitner et al., 2004). Alphaviruses are small RNA viruses that kill the cells that they replicate in by apoptosis. By codelivering a recombinant alphavirus expressing the tumor antigen tyrosine-related protein-1 (TRP-1) with a plasmid encoding the antiapoptotic gene Bcl-X_L, Restifo and coworkers showed that apoptotic cell death was necessary to make the vaccine fully immunogenic. Prevention of apoptosis by Bcl-X_L expression reduced T cell responses against TRP-1. In contrast, the immunogenicity of a conventional DNA plasmid vaccine was increased when apoptosis was prevented by coimmunization with a plasmid expressing Bcl-X_L. An intriguing possibility, put forward by Leitner and Restifo (2003), is that apoptosis increases the immunogenicity of somatic cells as it allows antigens to enter the cross-presentation pathway, but the process curbs the response when transfected DCs also undergo apoptosis. Thus, the impact of adding antiapoptotic genes would depend on the cell type that is targeted. In this view, alphavirus immunogenicity depends on apoptotic somatic cells and activated DCs, and adding Bcl-X_L in this context may reduce the source of cross-presented antigen. In contrast, the immunogenicity of plasmid DNA vaccines may depend more on long-term survival of transfected DCs. One important observation from this study is that recombinant alphavirus vectors provide a tractable model for immunogenic cell death coupled to cross-presented antigen (Figure 13.1).

The obvious caveat to the interpretation of these experiments is that they were based on the generation of immunogenic apoptosis. Apoptotic cell death occurred in the context of viral infection and would therefore likely be accompanied by TLR stimulation. However, chemotherapy does not necessarily result in immunogenic cell death. A series of experiments compared the immunological consequences of doxorubicin and mitomycin-mediated tumor cell death (Casares et al., 2005). These experiments support the view that chemotherapy has the potential to be immunogenic and also demonstrate that not all apoptosis is equal. Vaccination with doxorubicin-killed tumor cells protected mice against tumor challenge, whereas mitomycin-killed cells did not. Moreover, direct intratumoral injection of doxorubicin induced tumor regression. Thus, the immune system perceives doxorubicin-killed cells differently than it perceives mitomycin killed cells. DCs played a key role in this process; whereas antigens from doxorubicin-killed cells were efficiently taken up by DCs, mitomycin killed cells were ignored. How can this difference be explained? As indicated in Table 13.1, both drugs interact with DNA, albeit in different ways. Mitomycin is a DNA alkylating agent, whereas doxorubicin is a DNA cross-linking/intercalating drug. The precise mechanism by which doxorubicin kills cells is not completely understood and may involve topoisomerase II poisoning. In this case, the key difference at the immunological level appears to be the capacity of DC to process and present antigen. Thus, it seems likely that doxorubicin, but not mitomycin, induces the production or release of a molecular flag that induces DCs to phagocytose dying tumor cells. However, the unknown "DC-uptake" signal does not guarantee immunogenicity since cross-presentation is also involved in the maintenance of tolerance. As has been indicated, gemcitabine stimulates DC uptake and cross-presentation without the strong immune responses induced by doxorubicin. Thus, although these studies clearly show that not all apoptosis is equal from the immunological point of view, they also leave several important questions unanswered. One of these questions is the nature of the proinflammatory signals that are associated with chemotherapy-induced cell death. In other words, how does the immune system decode cell death? Known candidates for the endogenous danger signals include uric acid, heat-shock proteins, damaged DNA, and cytokines (Lake and van der Most, 2006).

B. Uric Acid as an Endogenous Danger Signal

Uric acid is an end product of purine metabolism. As such, it is generated in dying cells as they degrade their DNA and RNA. Crystalline uric acid has a strong adjuvant effect with vaccination: its injection results in DC activation and strong CD8 T cell responses (Shi *et al.*, 2003). These effects can be reversed by uric acid depletion, using allopurinol or uricase. The identification of uric acid as a major player in driving the innate immune system (as an endogenous "danger" signal) is supported by the finding that it plays a key role in antitumor responses (Hu *et al.*, 2004). Hu *et al.* (2004) showed that injection of crystalline uric acid close to a tumor accelerated the rejection of the tumor. They also showed that regressing tumors are associated with elevated uric acid levels and that uric acid depletion delayed tumor rejection (Hu *et al.*, 2004). Taken together, these data suggest that the massive tumor cell death occurring during chemotherapy has the potential to release uric acid, which then could provide the basis for immune activation. However, given the proinflammatory nature of uric acid and its link to apoptotic cell death, why are some forms of apoptosis decidedly noninflammatory (bland apoptosis)? The answer is suggested by the observation that the uric acid concentration required to activate DCs coincides with the crystallization point, indicating that crystalline rather than soluble uric acid is the danger signal. This finding helps explain why normal apoptosis in tissue homeostasis is noninflammatory, whereas apoptosis in a chemotherapy-treated tumor can be inflammatory. In addition, this may also provide an explanation for the different modes of cell death seen with different drugs (e.g., doxorubicin and mitomycin). If the cytotoxic effects of different DNA-modifying drugs affect the nucleic acid metabolic pathways in different ways, it may be that this translates in different uric acid levels. The requirement for crystalline uric acid sets a concentration threshold and, therefore, has the potential to amplify differences in the immunogenicity of dying cells.

C. Genotoxic Stress (DNA Damage) Upregulates NK Cell Receptor Ligands

As has been noted, DNA itself can trigger TLR activation and mediate a proinflammatory signal. In addition, damaging DNA invokes a response pathway that is potentially involved in the process by which dying cells are labeled as immunogenic or bland. The DNA damage response, also known as the genotoxic stress pathway, results in initiation of DNA repair functions, cell cycle arrest, or apoptosis, depending on the severity of the damage. Interestingly, this response can also lead to upregulation of the ligands for natural killer, group 2, member D (NKG2D) (Gasser and Raulet, 2006), a stimulatory receptor expressed on natural killer (NK) cells, γδ-T cells, CD8⁺ T cells, and, importantly, on the newly discovered IFN-producing killer DCs (IKDC) (Chan et al., 2006; Taieb et al., 2006b). These "hybrid" cells have been postulated to play a key role in tumor immunosurveillance (Taieb et al., 2006a). Two families of NKG2D ligands can be distinguished: the major histocompatibility complex (MIC) family (MHC class I chain-related A and B, MICA and MICB), only expressed in humans, and the retinoic acid early transcript 1 (RAET-1) family, expressed in mice and man (comprising the *Rae1*, *H60*, and *Mult1* genes in mice). These ligands can trigger cytotoxic responses and production of inflammatory cytokines in NK cells and CD8⁺ T cells, providing a potential mechanism for immune surveillance of cells under genotoxic stress. It is possible that NK cells and CD8 T cells mediate a surveillance function by eliminating precancerous cells via NKG2D ligation. Unsurprisingly, where these pathways are effective, tumors are selected that exhibit evasive strategies, such as expression of soluble (decoy) MICA, selected loss of NKG2D ligands, or constitutive high expression of these ligands. Furthermore, TGF-β, an immunosuppressive cytokine often produced by tumors, downregulates NKG2D expression on NK cells. How chemotherapy impacts on this potentially proinflammatory pathway is still the subject of debate, but the DNA damaging properties of many drugs (including doxorubicin, 5-fluorouracil, and mitomycin) result in increased expression of NKG2D ligands on tumor cells. However, as of this writing, it has never been confirmed whether this pathway can indeed activate NK cells and CD8 T cells after DNA damaging chemotherapy. If this turns out to be so, additional possibilities for immunotherapy aimed at the NKG2D receptor may arise.

The NKG2D signaling pathway is affected in a different way by cisplatin. This drug has suppressed NKG2D expression in lymphocytes. Combination of cisplatin with IL-2 restored NKG2D expression, thereby greatly improving its antitumor efficacy (Li et al., 2002).

D. Immunogenic DNA

DNA can be derivatized by interaction with a chemotherapy agent, and this also may have direct inflammatory effects. Tumor cells killed with the DNA alkylating agents melphalan or chlorambucil were shown to activate DCs, as evidenced by upregulation of costimulatory molecules, production of IL-12, and T cell activation (Rad et al., 2003). These same effects were observed when DCs were incubated with DNA purified from killed tumor cells. These findings now pose several important questions. First, how does a DC sense damaged DNA? The data would imply that DCs have a receptor, perhaps similar to those of TLRs, for such damaged DNA products. Second, it is clear that not all DNA damaging agents generate immunogenic DNA: mitomycin also kills cells by DNA alkylation (albeit with a different specificity then melphalan and chlorambucil), yet mitomycin-triggered cell death is not immunogenic (Casares et al., 2005). The specificity in the process of DNA derivatization is not yet incompletely understood.

E. Cyclophosphamide: DNA Damage, IFNs, and Tregs

When chemotherapeutic drugs are discussed in terms of their immunogenicity, the DNA alkylating agent cyclophosphamide occupies a unique position. The antitumor effects of cyclophosphamide have been attributed to its potential to boost antitumor T cell responses. This has been most clearly demonstrated in animal models. It is noteworthy that the DNA alkylating properties of cyclophosphamide closely resemble those of the related nitrogen mustard compounds melphalan and chlorambucil. Since these drugs have been linked to "immunogenic" DNA (Rad et al., 2003), it is also possible that at least part of cyclophosphamide-mediated immunogenicity resides in damaged DNA.

The immunostimulatory properties of cyclophosphamide have long been recognized but are only now beginning to be understood. Early studies implicated a selective loss of "suppressor T cells" for the immunostimulatory effect. Indeed, suppressor T cells, in their modern reincarnation as regulatory $CD4^+CD25^+$ forkhead family transcriptional regulator $(Foxp3)^+$ Treg cells, do undergo a selective loss of function when exposed to cyclophosphamide (Lutsiak et al., 2005). An aspect of cyclophosphamide that has received less attention is the potential immunogenicity of the apoptotic pathway that it triggers in tumor cells. Interestingly, cyclophosphamide induces a rapid and strong type I IFN response (Schiavoni et al., 2000). This finding has been demonstrated by the systemic upregulation of the IFN response marker Ly6C. Both the type I IFN response and the inactivation of Treg responses are affected when iNOS-deficient mice are treated with cyclophosphamide, suggesting a common mechanism that involves NO activity (Loeffler et al., 2005). The implications of the cyclophosphamide-induced IFN response are twofold. First, it may explain the immunostimulatory properties of cyclophosphamide because type I IFNs strongly stimulate CD8 T cell responses against cross-presented antigens (Le Bon et al., 2003). Second, IFN-α/β production may be the signature response to immunogenic cell death, such as the type of cell death described for doxorubicin, and could be triggered by DC-activating agents, such as uric acid. Finally, cyclophosphamide has been shown to change the phenotype of tumor-infiltrating macrophages from immunosuppressive IL-10 producers into immune activating IFN-γ producers (Ibe et al., 2001), possibly in a NO-dependent fashion (Abe et al., 1998; Loeffler et al., 2005).

In summary, chemotherapy can increase the quantity of cross-presented tumor antigen and can even provide an immunogenic context. However, when discussing the proinflammatory mediators that can be associated with apoptotic cell death, it is important to consider that the default immune response to apoptosis is likely to be tolerogenic and based on the role of apoptosis in normal tissue homeostasis. Cross-presented self-antigens most likely play a key role in this process. Clearly, tumors treated with chemotherapy behave differently, which can be explained by particular features of the specific drug (e.g., production of uric acid, release of damaged DNA products) or by assuming that the massive amount of cell death may overload the normal macrophage-dependent scavenger system leading to a "spillover" of antigen into the DC compartment.

IV. CHEMOTHERAPY AND THE IMMUNE SYSTEM

A. Chemotherapy and Lymphopenia

Chemotherapeutic drugs often cause suppression of normal bone marrow function, with resulting lymphopenia. The obvious disadvantage of this result is that patients on chemotherapy can be more susceptible to opportunistic infections and that the prospects for successful antitumor immunotherapy may appear to be reduced. However, the rapid advances in the understanding of cellular immunology over the last decade have altered this view. It is now clear that lymphodepleting chemotherapy may actually create an immunological environment that is more amenable to the induction of antitumor immune responses even though this may seem counterintuitive. The two key discoveries that have led to this understanding are (i) the reappraisal of Tregs as mediators of peripheral tolerance and (ii) the concept that lymphopenia is followed by a phase of homeostatic T cell proliferation and that these proliferating T cells as a group are more inclined towards self reactivity. Taken together, the implication of these concepts is that lymphodepletion could result in an "immunological reset," thereby creating an opportunity for immunotherapy particularly directed at a tumor that might be expected to resemble normal self tissue.

B. Tregs

T cell-mediated suppression of immune responses has long been a controversial subject. Suppressor T cells were first described in the early 1970s to explain the peripheral regulation of immune responses, for which they were postulated to play a major role in the downregulation of antitumor immune responses. However, the existence and role of these cells was subject to much skepticism until Fehervari and Sakaguchi (2004) described a population of CD4 T cells that was characterized by the expression of the IL-2 receptor α-chain (CD25). It is now clear that these $CD25^+CD4^+$ Tregs, which are further characterized by the expression of glucocorticoid-induced tumor necrosis factor receptor (GITR) family, of cytotoxic T lymphocyte-associated antigen 4 (CTLA-4), and of FoxP3, play a central role in the maintenance of peripheral tolerance. There is now increasing evidence that these Tregs have a regulatory role in antitumor immunity because their depletion results in better antitumor immunity. In patients with invasive breast and pancreatic cancer, $CD4^+CD25^+$ Tregs increased in number in the tumor microenvironment and peripheral blood (Liyanage et al., 2002). Increased frequencies of Tregs are also found in patients with ovarian cancer (Curiel et al., 2004) and in patients with carcinomatous and mesothelioma pleural effusions (DeLong et al., 2005). Tumor-infiltrating Tregs are potent inhibitors of autologous T cell proliferation and may mediate a local immunosuppressive effect (Woo et al., 2002).

Because many chemotherapeutic drugs deplete lymphocytes, it is likely that Tregs will be depleted as well, potentially improving the efficacy of immunotherapy. This has been most convincingly established for cyclophosphamide. Early studies had already linked cyclophosphamide to the depletion of suppressor T cells, and this has been confirmed as a selective loss of Treg function. Intriguingly, cyclophosphamide not only depletes Tregs but, in mouse models, also curtails the suppressive capacity of the surviving Tregs (Lutsiak et al., 2005). Prolonged, lower dose cyclophosphamide (metronomic chemotherapy) also potentiates immunotherapy in murine models despite deleting proliferating tumor-specific cytotoxic T lymphocytes (CTLs) and preserves CD43LO memory $CD8^+$ T cells (Hermans et al., 2003). Cyclophosphamide also directly induces type I IFN production shortly after administration, increasing the number and persistence of cells with the CD44hi memory phenotype (Schiavoni et al., 2000). Clearly, several important questions remain: does cyclophosphamide have the same effect in humans, and what is the molecular basis of the Treg loss-of-function phenotype in mice? The strong proimmunogenic effects of cyclophosphamide, linked to Treg inactivation/depletion (but also to IFN-α/β induction), have led to a renewed interest in this drug and in the mechanisms responsible for its effects. There is some evidence that Treg depletion is not a unique property of cyclophosphamide. A treatment combination including fludarabine led to reduced Treg frequencies, whereas the drug temozolomide specifically depleted CD4 T cells, most likely including Tregs (Su et al., 2004). As of this writing, it is unclear whether these observations translate into enhanced antitumor immunity. It is possible that the negative effects of depletion of other lymphocyte subsets mask the potential positive effects of Treg depletion. This could be solved by cytokine-driven immune reconstitution (e.g., IL-7, IL-15), adoptive transfer of antitumor T cells, or lower dosage of chemotherapeutic drugs. Irrespective of its eventual implementation, it is expected that the interaction between chemotherapy and Tregs will be further investigated and exploited in the near future since Tregs remain an attractive therapeutic target.

C. Lymphopenia and Homeostatic Proliferation

As stated earlier, chemotherapy-induced lymphodepletion is a potential obstacle for the generation of effective antitumor immune responses, even potentially masking beneficial effects of Treg depletion. However, the rapidly expanding understanding of the biology of the immune system suggests that there may even be opportunities for therapy in this setting. The key to this is the phase of homeostatic T cell proliferation that follows a phase of lymphopenia (Klebanoff et al., 2005). The immune system responds to lymphopenia by increased T cell proliferation, driven by a variety of cytokines, including IL-7, IL-15, and IL-21. Proliferation of these newly generated T cells depends on interactions with MHC class I presented self-antigens, which creates a potential bias toward self-responses. The important implication for tumor immunology is that tolerance against the altered self that constitutes that antigenic repertoire of the tumor can be broken during this phase. Proof of principle for the potential therapeutic use of this approach has been provided in animal models and in clinical trials. Severely lymphodepleted animals respond better to adoptive transfer of tumor-specific CD8 T cells than nondepleted animals, and this translates into better antitumor immunity in lymphodepleted and reconstituted animals. The difference was attributed to the "cytokine storm," referring to the high levels of the homeostatic cytokines IL-7 and IL-15, driving proliferation and survival of adoptively transferred cells. Experimental data also suggest that tumor rejection can be mediated by expansion of preexisting T cells recognizing tumor antigens in the regional lymph nodes and that tumor rejection during homeostatic proliferation can lead to long-term tumor-specific memory (Dummer et al., 2004). Dudley and coworkers then put this approach to the test in a melanoma clinical trial, in which patients were treated with a combination of cyclophosphamide and fludarabine and then received autologous melanoma-specific CD8 T cells. These adoptively transferred T cells proliferated vigorously in the patients, leading to extremely high frequencies of melanoma-specific CD8 T cells and remarkable clinical responses (Dudley et al., 2002). A caveat to the interpretation of these data is that, as pointed out previously, cyclophosphamide also triggers a type I IFN response, which could contribute to the observed effects. Exploiting post-chemotherapy homeostatic proliferation has so far only been attempted in the context of adoptively transferred CD8 T cells. Because this may not always be possible or cost effective, the option of using or boosting (using cytokines such as IL-7, IL-15, and IL-21, alone or in combination) the natural phase of homeostatic proliferation is an attractive one and deserves further study. The use of IL-7 for this purpose is being tested in clinical trials. There is some concern that vaccination after having lymphopenia may initiate autoimmunity in mice, arising from observations in mouse models of autoimmune diabetes (King et al.), as well as autoimmunity in humans (Baccala and Theofilopoulos, 2005), but this has not been a significant problem in the tumor models used to date.

D. Breaking Tumor-Driven Immunomodulation

The relationship between a tumor and the immune system is complex but needs to be understood for rational chemoimmunotherapy design. Evidently, tumors have been selected to evade and suppress host antitumor immune responses. Yet, at the same time, tumors are associated with chronic inflammation (e.g., liver cancer and hepatitis B virus, stomach cancer and *Helicobacter pylori*, and, possibly, mesothelioma and asbestos) (de Visser and Coussens, 2006). Intriguingly, tumors secrete immunosuppressive cytokines (e.g., TGF-β) as well as inflammatory cytokines (e.g., IL-1, IL-6). The identification of a novel subset of $CD4^+$ T cells may provide an important clue to help scientists understand this apparent contradiction (Weaver *et al.*, 2006). It is evident that TGF-β by itself is immunosuppressive and stimulates the generation of Tregs. If, however, TGF-β production coincides with secretion of IL-6, $CD4^+$ T cells differentiate into a unique class of IL-17-producing effector T cells rather than into Tregs. IL-17, in turn, recruits neutrophils and granulocytes and has been associated with autoimmune responses but, importantly, also has the capacity to downregulate antitumor $CD8^+$ T cell responses. The IL-12 family member IL-23 (which shares its p40 chain with IL-12) stimulates the expansion of these IL-17-producing $CD4^+$ T cells. Given that many tumors produce TGF-β in combination with IL-6 and IL-23, it can be predicted that the tumor environment may resemble a chronic inflammation in which innate effectors dominate. CD8 T cells are excluded, which may explain why tumors have been selected to express cytokines such as IL-6 and IL-23. Indeed, a landmark study has demonstrated that IL-23 produced in the tumor (but not necessarily by tumor cells themselves) strongly promoted tumor growth, whereas IL-12 had the opposite effect (Langowski *et al.*, 2006). Why is this finding relevant for chemotherapy? Cytotoxic drugs may interfere with this protumor cytokine network in two ways. First, chemical debulking of the tumor mass combined with lymphocyte depletion may remove both the source (i.e., the tumor) and the drivers (IL-17-producing CD4 T cells) of such networks, providing an "immunological reset," which can then be exploited by subsequent immunotherapy. Second, immunogenic cell death, triggered by chemotherapy, may alter the cytokine balance in the tumor such that antitumor CD8 T cell responses may start to dominate. Here, the challenge for immunotherapy is to generate or boost these antitumor T cell responses. Puzzlingly, and illustrating the complexity in this field, IL-23 has also been proposed as a candidate for immunotherapy due to its positive effects on CD8 T cell proliferation and/or survival (Overwijk *et al.*, 2006).

V. A PRACTICAL PARTNERSHIP: CHEMOTHERAPY AND IMMUNOTHERAPY

A. How to Choose the Right Immunotherapy

It has been suggested that the goal of optimal immunotherapy is to generate a *de novo* autoimmune response, in which the tumor antigens are treated as "self." This concept of tumors and autoimmune diseases being at the opposite ends of the immunological spectrum is supported by the role of Tregs in cancer and autoimmunity. Tregs reduce the likelihood of autoimmune responses but may promote tumor growth by blocking antitumor immune responses. However, the concept that cancer can result from chronic inflammation and is supported by the similar roles of IL-17-producing $CD4^+$ T cells in both tumors and autoimmune responses suggests that anti-inflammatory drugs, such as cyclooxygenase 2 (COX-2) inhibitors, could be effective. Interestingly, the recent discovery of the IL-17-secreting $CD4^+$ T cell lineage could provide part of the solution to the paradox of treating cancer by enhancing and reducing immune responses. The IL-17-driven regulatory circuit, involving TGF-β, IL-6, neutrophils, and other proinflammatory mediators, excludes T helper 1 (Th1)-type cytotoxic T cell responses. Thus, IL-17 and associates expose a dichotomy in responses, inflammatory versus cytotoxic. If cancer is seen as the result of chronic, IL-17/IL-23-driven inflammation, the therapeutic answer should be to replace this response by an IL-12 driven cytotoxic response. The effectors to be targeted in such an approach are IFN-γ–producing cytotoxic $CD8^+$ T cells, NK cells, and natural killer T (NKT) cells. However, the lack of correlation between the frequencies of antitumor effector cells in the periphery and clinical responses suggests that successful therapy also involves modifications at the effector site (i.e., in the tumor itself). As will be discussed in the following paragraphs, chemotherapy presents one way to achieve this result.

B. DC Activation Using CD40–CD40L Ligation

The induction and maintenance of $CD8^+$ T cell responses against cross-presented antigens requires $CD4^+$ T cell help. Cognate $CD4^+$ T cells "license" DCs to activate $CD8^+$ T cells by activating the CD40 molecule on DCs. Since antitumor $CD4^+$ T cell responses are often inadequate, it stands to reason that a therapeutic "bypass" in the form of a CD40 agonist could stimulate antitumor $CD8^+$ T cell responses by activating DCs. Indeed, treatment of mice with an agonistic anti-CD40 antibody has modest antitumor effects, and curative responses are difficult to achieve with this approach. However, when anti-CD40 immunotherapy is combined with chemotherapy (gemcitabine), the result is an 80% cure rate (Nowak et al., 2003a, 2003b). Apoptosis of tumor cells is essential for the observed synergy because it is not observed when a gemcitabine-resistant tumor is used, even when the resistant tumor is surgically debulked prior to immunotherapy. The chapter authors hypothesize that the enhanced tumor antigen cross-presentation that results from gemcitabine treatment is responsible for the observed synergy. Thus, chemotherapy provides the antigen, and CD40 ligation provides the immunogenic context, thereby creating a situation of immunogenic cell death. The synergy between gemcitabine and CD40 ligation may be the first example of a "practical partnership" between chemotherapy and immunotherapy. Humanized agonistic anti-CD40 antibodies are currently developed, and it seems clear that human clinical trials based on the synergy between CD40 activation and cytotoxic chemotherapy are warranted.

C. Stimulation of Innate Immunity Through TLRs

As discussed earlier, PAMPs provide very strong proinflammatory signals that are mediated by DCs via TLR ligation (Table 13.2). Thus, TLR activation in a therapeutic setting could provide cross-presented tumor antigens with an immunogenic context and could generate an antitumor T cell response. A pioneering example of this approach is this is Coley's toxin. In the late 1800s, Dr. William Coley showed potent antitumor activity with a mixture of heat-killed *Streptococci* and *Serratia*. It is likely that the antitumor activity of Coley's toxin can be attributed to bacterial DNA, containing unmethylated CpG motifs. The molecular definition of PAMPs is progressing rapidly and pharmacological analogs of several PAMPs (e.g., poly-I:C, CpG-oligodeoxynucleotides, and imiquimod) are now being evaluated for their therapeutic efficacy. Since these compounds mediate their therapeutic efficacy by activating DCs, it seems plausible that, similar to anti-CD40, TLR agonists will be synergistic with chemotherapy. There is increasing evidence to support this view, including the finding that a combination of CpG-oligodeoxynucleotides and gemcitabine showed therapeutic synergism in an orthotopic human pancreatic carcinoma xenograft model (Pratesi et al., 2005). Not all TLR stimulation, however, results in antitumor responses; recent data show that TLR4 stimulation through LPS can promote tumor growth and lead to the acquisition of paclitaxel chemo resistance. These findings may in part be explained by the fact that paclitaxel is an LPS mimetic (Kelly et al., 2006). TLR4 ligation results in the production of proinflammatory cytokines (e.g., IL-6) that may sustain the proinflammatory milieu of the tumor. Note that IL-6 is the key cytokine required for the *de novo* generation of IL-17 secreting $CD4^+$ T cells, which may play a role in the maintenance of the tumor-promoting proinflammatory environment. These results illustrate the delicate balance that exists between antitumor and protumor immunogenic responses. One approach to avoid the pitfall of protumor immunotherapy may be to block the mechanisms that the tumor uses to curb potential antitumor T cell responses.

D. Countering Tumor-Driven Immunosuppression

Tumors that continue to grow are clearly selected to evade host antitumor immune responses, and this presents some formidable obstacles to the development of antitumor immunotherapy. Chemotherapy is one way to approach the problem because chemical debulking should release some of the immunosuppressive pressure. However, chemotherapy rarely induces antitumor immune responses that are sufficiently strong to destroy the remnants of malignant tissue. Thus, additional therapy is needed. As already discussed, $CD25^+CD4^+$ Tregs undoubtedly play a role in immune suppression, and their depletion or inactivation by chemotherapy or immunotherapy should be an important therapeutic goal. Examples of immunotherapeutic Treg depletion include anti-CD25 or anti-GITR antibodies or a fusion protein between IL-2 and diphtheria toxin that targets CD25 (denileukin diftitox, also known as ONTAK®). Combination of these therapies with chemotherapy should provide a therapeutic benefit: increased antigen presentation following chemotherapy, in the absence of Tregs, could powerfully boost antitumor $CD8^+$ T cell responses. As an alternative, neutralization of the immunosuppressive factor TGF-β has been considered. TGF-β is produced by many tumors and plays a key role in the generation of both Tregs and IL-17-producing $CD4^+$ T cells and in the downregulation of $CD8^+$ T cell responses. An approach to neutralizing TGF-β involves the use of neutralizing antibodies, soluble TGF-β receptor decoy molecules (Suzuki et al., 2004). Similar to the approach with Tregs, the combination of TGF-β neutralization with chemical debulking and the resulting increase in the levels of cross-presented antigen lead to good potential for therapeutic synergy. CTLA-4 is another key molecule involved in the maintenance of tolerance, and CTLA-4-blocking antibodies have been used in trials of antitumor therapy. Interestingly, the combination of CTLA-4 blockade with low-dose chemotherapy have potentiated tumor treatment in animal models.

A further mechanism by which tumor cells suppress antitumor immune responses involves a deregulation of tryptophan metabolism, in which the tryptophan catabolizing enzyme indoleamine 2,3-dioxygenase (IDO) plays a key role (Muller and Prendergast, 2005). Activation of IDO is essential to maintain tolerance against the allogeneic fetus during pregnancy, and tumors appear to have hijacked this important mechanism. Many tumors have upregulated IDO, leading to elevated levels of tryptophan metabolites, which in turn deactivates T cells. IDO inhibition by itself turned out to be only marginally efficacious, but combination with chemotherapy was highly synergistic. Since IDO is a well-defined enzyme, it is feasible that small-molecule IDO inhibitors can be developed.

E. Enhancing the Response at the Effector Site

Even when the use of therapy has overcome the hurdles of a lack of $CD4^+$ T cell help, a poor immunogenic context, and tumor immunosuppression, tumors may yet pose a further obstacle to preventing their own destruction. Many tumors upregulate inhibitory members of the B7 family of costimulatory molecules (Greenwald et al., 2005), either constitutively or after exposure to IFN-γ. Newly described family members include B7-H1, B7-H3, and B7-DC. Both B7-H1 and B7-DC (also known as PD-L1 and PD-L2, respectively) have been identified as ligands of the PD-1 molecules expressed on T cells. It is important to note that B7-H1–PD1 ligation has the potential to powerfully inactivate CD8$^+$ T cells, and it is this interaction that has received the most attention so far. Indeed, blockade of B7-H1 and, to a lesser extent, of B7-DC has a profound antitumor effect. Given the multifactorial nature of tumor immune evasion, it can be predicted that combination of B7-H1 blockade with other therapeutic interventions should be synergistic. Chemotherapy, with its dual benefits of tumor debulking and increased antigen cross-presentation, is a prime candidate for such synergistic therapies.

F. Cytokines

Cytokines have always held the promise of being antitumor agents. So far, IL-2, IL-12, and type I IFNs have been trialed as antitumor therapies, but the results have been somewhat disappointing. In hindsight, it can be concluded that an insufficient understanding of the precise biological roles of the cytokines in question may have affected their clinical applications. IL-2 is a case in point. This cytokine was originally defined as a crucial T cell growth factor, produced by CD4$^+$ T cells and consumed by CD8$^+$ T cells. As such, IL-2 was used as an adjuvant to boost CD8$^+$ T cell responses. The reality is more complex, and nowhere is this better illustrated than in a clinical trial in which IL-2 therapy was used after lymphodepleting chemotherapy. Under these conditions, IL-2 stimulated proliferation of Tregs (which are characterized by the expression of CD25, that is, the IL-2R β chain) rather than antitumor CD8$^+$ T cells (Zhang et al., 2005). Clinical use of IL-12 has faced a different problem. Although IL-12 has powerful antitumor effects, its clinical use in humans has been hampered by severe toxicity upon systemic delivery.

The understanding of tumor immunology as well as the list of ILs has grown rapidly since the mid-1990s. As discussed, the functions of cytokines, such as IL-6, IL-17, and IL-23, are only now being understood. Eventually, this will pave the way for a more rational design of cytokine therapies. One important area in which cytokine therapy could be useful is to enhance immune reconstitution after chemotherapy-induced lymphopenia. Accelerated immune system reconstitution remains an important goal, both to facilitate potential antitumor responses and to prevent opportunistic infection. Several cytokines are involved in this homeostatic T cell proliferation. The roles of IL-7 and IL-15 have been best studied. Clinical grade IL-7 is now in clinical trials for this purpose, and IL-15 will undoubtedly follow. IL-21 appears to play a more complex role—it enhances homeostatic T cell proliferation, in synergy with IL-15, and promotes NK cell responses. Thus, combinations of these cytokines may find clinical use in postchemotherapy immune reconstitution. The opportunity for therapy may be to expose the developing repertoire of homeostatically proliferating CD8 T cells to tumor antigens in an immunological context, thereby provoking *de novo* antitumor responses.

It will certainly be interesting to revisit IL-12. Since systemic delivery is not possible due to excessive toxicity, studies will have to focus on local delivery of this cytokine. Since IL-12 also enhances CD8$^+$ T cell homeostatic proliferation, its use in the lymphopenic postchemotherapy environment could be particularly promising.

G. Hybrids

An interesting category in the chemotherapy and immunotherapy partnership regards hybrid therapies (i.e., compounds that have both direct effects on the tumor as well as on the immune system). The archetype for this phenomenon is cyclophosphamide. As already discussed, cyclophosphamide kills tumor cells, inactivates Tregs, and induces a strong type I IFN response. Combined, these features may be responsible for the powerful antitumor effects seen with this drug. The TLR7 agonist imiquimod shares some of these features (Skinner, 2003). Imiquimod as a topical cream is used with success against skin cancers and acts by inducing apoptosis in tumor cells. However, TLR7 ligation also activates DCs and leads to IFN-α/β production and strong immunostimulation. Thus, imiquimod may kill tumor cells, induce enhanced antigen cross-presentation, and provide a strong type I IFN signal. COX-2 inhibitors may also belong to this group. Many tumor cells have upregulated COX-2 expression and are sensitive to the inhibitors. However, COX-2 inhibitors, developed as anti-inflammatory drugs, also have strong immunostimulatory effects. By blocking the production of prostaglandin E_2 (PGE_2), COX-2 inhibitors may be able to break the tumor-promoting inflammatory circuit that is driven by IL-6, IL-17, and TGF-β. Moreover, through their effects on PGE_2, COX-2 inhibitors impede Treg activity, thereby promoting antitumor $CD8^+$ T cell responses. Thus, COX-2 inhibitors are likely to be synergistic with immunotherapies that augment responses against cross-presented antigens (e.g., TLR ligands, IFN-α/β). DeLong et al. provided convincing evidence for this in a mouse model of malignant mesothelioma (DeLong et al., 2003).

VI. EFFECTS OF CHEMOTHERAPY ON HUMAN ANTITUMOR IMMUNITY AND CHEMOIMMUNOTHERAPY CLINICAL TRIALS

The systematic examination of murine and human antitumor immunity and the development of cytotoxic drugs and clinical application of chemotherapy have occurred in parallel, with little cross-talk between the disciplines of tumor immunology and clinical pharmacology or medical oncology. Until very recently, the concept of chemoimmunotherapy was considered unconventional and counterintuitive. Hence, there has been very little work assessing the effects of cytotoxic chemotherapy on human antitumor immunity. Furthermore, the ability to examine antigen-specific antitumor immunity in detail has, until recently, been hampered by the technology available. Measurement of chemotherapy effects on the immune system has been limited to crude techniques, such as assessing change in lymphocyte subset numbers and delayed type hypersensitivity (DTH) reactions. The increasing identification of tumor-specific antigens and the development of tetramer technology and FACS sorting has enabled a more sophisticated examination of this process in humans in the last 5 years. Nevertheless, reports on the use of this technology to examine human immune responses after chemotherapy are scarce.

Much earlier work on the effects of chemotherapy on the immune system was done in children with hematological malignancies to assess the timing of, and response to, vaccination against childhood infectious diseases after treatment for cancer. Because most chemotherapy is given as combinations of two or more agents, the effects of individual drugs have rarely been described. As a single agent, most effects of cyclophosphamide on the human immune response are well-known. The ability of cyclophosphamide to augment immune responses was first demonstrated in 1982 in 22 patients with solid tumors. When full doses of cyclophosphamide were given 3 days before antigen challenge, DTH responses were markedly augmented in comparison with an untreated control group. Similar DTH responses were seen with a lower cyclophosphamide dose; however, only the lower dose potentiated antibody responses (Berd et al., 1982). The same authors then examined whether low-dose cyclophosphamide could potentiate the immune response to autologous tumor cell vaccine in patients with metastatic malignant melanoma. Despite small patient numbers, median DTH responses to autologous tumor cells increased significantly in cyclophosphamide-treated patients, and tumor responses with prolonged disease-free survival were seen in two patients (Berd et al., 1986). Subsequent investigators demonstrated reduction in regulatory T cell activity by cyclophosphamide compared with controls, with inhibition declining with repeated treatments. Numbers of CD8+CD11B+ cells (at that time considered a marker for "suppressor" T cells) were lower in the cyclophosphamide-treated group (Hoon et al., 1990). These results have been confirmed by others (Berd and Mastrangelo, 1988; Livingston et al., 1987).

In the context of these results, and the observation that both antibody and DTH responses correlate with survival, the use of cyclophosphamide as an immunopotentiating agent in advanced cancer has been trialed clinically in patients with different types of cancer, most frequently malignant melanoma (Berd et al., 1988; Berd et al., 1990; Bystryn et al., 1988; Jones et al., 1996; Livingston et al., 1987; Livingston et al., 1994). In patients with metastatic breast cancer, cyclophosphamide has been shown to augment humoral immune responses to tumors, with a suggestion of prolongation of survival (MacLean et al., 1996). A clinical trial focusing on advanced renal cell carcinoma (RCC) used an allogeneic DC vaccine with keyhole limpet haemocyanin (KLH) adjuvant, with or without cyclophosphamide, measuring clinical and laboratory outcomes. Unfortunately, the lack of random assignment to treatment groups limits the conclusions that can be drawn from this study, although all patients with tumor response and one of three with disease stabilization had received cyclophosphamide in addition to vaccine. Clinically, as well as immunologically, these trials prove difficult to interpret due to their small sample sizes, low statistical power, single center nature, and the frequent lack of randomization. An assumption of beneficial effects of cyclophosphamide has been made in some trials (Berd et al., 1990; Miller et al., 1995). The most appropriate surrogate endpoints for trials of chemoimmunotherapy have not been well developed or validated, and it is unclear whether measures, such as DTH and antibody responses, reflect an important antitumor response that will affect survival. Furthermore, the results have often been compared with historical controls, an approach that can only demonstrate feasibility rather than efficacy. Overall, while the results of clinical studies are interesting, they remain inconclusive about the value of cyclophosphamide in combination with tumor vaccines as well as are inconclusive regarding the optimum dosing, timing, and schedule. Well-conducted phase III randomized studies are needed for the oncology field to advance this area.

The immunomodulatory effects of gemcitabine in animal models have been discussed in this chapter. The effects of gemcitabine on the human immune system have been examined in patients with pancreatic cancer. Whereas all subsets of lymphocytes decreased after each dose, these stabilized and returned to normal at the end of treatment. T and B lymphocyte subsets changed in the same direction, and the proportions of T and B lymphocytes were stable. However, no changes in the regulatory T cell subset were observed. The number of IFNα producing T cells increased following gemcitabine infusion, as did the number of cells expressing the activation marker CD69$^+$ (Plate et al., 2005). These effects are likely to be beneficial and may facilitate combined chemoimmunotherapy protocols. Gemcitabine may also stimulate tumor necrosis factor α (TNFα) and IL-2 production (Levitt et al., 2004). There is one reported clinical trial combining gemcitabine and immunotherapy; this is a phase II trial using the multidrug regimen of gemcitabine, oxaliplatin, leucovorin, and 5-fluorouracil (GOLF) in 29 patients with metastatic colorectal cancer, followed by subcutaneous granulocyte-macrophage colony-stimulating factor (GM-CSF) and low-dose IL-2. Excepting gemcitabine, these drugs are in widespread clinical use to treat this disease. Objective response rates were higher than expected for chemotherapy alone (69%), and the average time to tumor progression was 12.5 months, an impressive result for patients with pretreated colorectal cancer (Correale et al., 2005). Translational studies demonstrated reduced Treg numbers in all patients and increased frequency of antitumor CTL precursors in responding patients. Nevertheless, patient selection that is biased toward those who are fit with more indolent disease is likely in a phase II study of a novel yet aggressive approach to treatment. Further clinical trials combining gemcitabine and immunotherapy are under development at our center (Tumour Immunology Group, University of Western Australia) and elsewhere, including a phase III trial of gemcitabine with or without telomerase vaccine in advanced pancreatic cancer (Dr. G. Middleton, personal communication, 2006).

The fluoropyrimidine 5-fluorouracil is a common constituent of active chemotherapy combinations against gastrointestinal malignancies, in both postoperative adjuvant treatment and treatment of metastatic disease. The presence of known tumor antigens, such as carcinoembryonic antigen (CEA), which is expressed on the majority of colorectal cancers, has stimulated much interest in tumor vaccines targeting CEA. Small phase I and II clinical trials have examined the safety and efficacy of chemoimmunotherapy in patients with colon cancer. Weihrauch et al. (2005) combined 5-fluorouracil, irinotecan, and immunotherapy with a CEA-derived peptide and different adjuvants (GM-CSF, DCs, or CpG-containing DNA molecules). A total of 17 patients were treated, and the approach was safe and feasible. An objective tumor response was noted in 6 of 17 patients, not dissimilar to the response rate expected with chemotherapy alone. An increase in CEA-specific $CD8^+$ T cells postvaccination was observed in 8 patients (48%), and an increased proportion of patients showed more T-cell IFN secretion in response to the peptide. There was no effect of treatment on $CD4^+$ and $CD8^+$ frequency. Finally, in a phase II trial, 32 patients with resected and metastatic colon cancer were treated with an anti-idiotypic antibody CEA vaccine, 14 also receiving 5-fluorouracil-based regimens (although this was not randomized) (Foon et al., 1999). Treatment with 5-fluorouracil did not affect immunological or clinical endpoints, although without randomization, the comparison is not entirely valid. All patients generated good humoral and cellular immune responses against the tumor antigen. Combining chemotherapy and anti-CEA vaccination in colorectal cancer will require better characterization of the immunomodulatory effects of 5-fluorouracil and, increasingly, of other drugs, such as oxaliplatin.

The immunomodulatory effects in humans of many other major classes of anticancer drugs, including platinum compounds, anthracyclines, vinca alkaloids, and taxanes, have not been systematically investigated. There is much scope for translational research in this field. Anticancer treatment is increasingly incorporating multiple non-cytotoxic drug classes, including vascular targeting agents such as bevacizumab (anti-VEGF antibody), the epidermal growth factor receptor (EGFR) antagonists, and small-molecule tyrosine kinase inhibitors. Immunotherapy joins an increasingly crowded field of treatment options and will have to establish its place with rigorous preclinical and translational science before moving to well-designed, well-conducted, and rigorously controlled clinical trials.

References

Abe, K., Harada, M., Tamada, K., Ito, O., Li, T., and Nomoto, K. (1998). Early-appearing tumor-infiltrating natural killer cells play an important role in the nitric oxide production of tumor-associated macrophages through their interferon production. *Cancer Immunol. Immunother.* **45**, 225–233.

Baccala, R., and Theofilopoulos, A. N. (2005). The new paradigm of T-cell homeostatic proliferation-induced autoimmunity. *Trends Immunol.* **26**, 5–8.

Berd, D., Maguire, H. C., Jr., and Mastrangelo, M. J. (1986). Induction of cell-mediated immunity to autologous melanoma cells and regression of metastases after treatment with a melanoma cell vaccine preceded by cyclophosphamide. *Cancer Res.* **46**, 2572–2577.

Berd, D., Maguire, H. C., Jr., McCue, P., and Mastrangelo, M. J. (1990). Treatment of metastatic melanoma with an autologous tumor-cell vaccine: Clinical and immunologic results in 64 patients. *J. Clin. Oncol.* **8**, 1858–1867.

Berd, D., and Mastrangelo, M. J. (1988). Effect of low dose cyclophosphamide on the immune system of cancer patients: Depletion of $CD4^+$, $2H4^+$ suppressor-inducer T-cells. *Cancer Res.* **48**, 1671–1675.

Berd, D., Mastrangelo, M. J., Engstrom, P. F., Paul, A., and Maguire, H. (1982). Augmentation of the human immune response by cyclophosphamide. *Cancer Res.* **42**, 4862–4866.

Binder, R. J., and Srivastava, P. K. (2005). Peptides chaperoned by heat-shock proteins are a necessary and sufficient source of antigen in the cross-priming of $CD8^+$ T cells. *Nat. Immunol.* **6**, 593–599.

Bystryn, J. C., Oratz, R., Harris, M. N., Roses, D. F., Golomb, F. M., and Speyer, J. L. (1988). Immunogenicity of a polyvalent melanoma antigen vaccine in humans. *Cancer* **61**, 1065–1070.

Casares, N., Pequignot, M. O., Tesniere, A., Ghirinhelli, F., Roux, S., Chaput, N., Schmitt, E., Hamai, A., Hervas-Stubbs, S., Obeid, M., Coutant, F., Metivier, D., Pichard, E., Aucouturier, P., Pierron, G., Garrido, C., Zitvogel, L., and Kroemer, G. (2005). Caspase-dependent immunogenicity of doxorubicin-induced tumor cell death. *J. Exp. Med.* **202**, 1691–1701.

Chan, C. W., Crafton, E., Fan, H. N., Flook, J., Yoshimura, K., Skarica, M., Brockstedt, D., Dubensky, T. W., Stins, M. F., Lanier, L. L., Pardoll, D. M., Housseau, F. (2006). Interferon-producing killer dendritic cells provide a link between innate and adaptive immunity. *Nat. Med.* **12**, 207–213.

Correale, P., Cusi, M. G., Tsang, K. Y., Del Vecchio, M. T., Marsili, S., La Placa, M., Intrivici, C., Aquino, A., Micheli, L., Nencini, C., Ferrari, F., Giorgi, G., Bonmassar, E., and Francini, G. (2005). Chemo-immunotherapy of metastatic colorectal carcinoma with gemcitabine plus FOLFOX 4 followed by subcutaneous granulocyte macrophage colony-stimulating factor and interleukin-2 induces strong immunologic and antitumor activity in metastatic colon cancer patients. *J. Clin. Oncol.* **23**, 8950–8958.

Curiel, T. J., Coukos, G., Zou, L., Alvarez, X., Cheng, P., Mottram, P., Evdemon-Hogan, M., Conejo-Garcia, J. R., Zhang, L., Burow, M., Zhu, Y., Wei, S., Kryczek, I., Daniel, B., Gordon, A., Myers, L., Lackner, A., Disis, M. L., Knutson, K. L., Chen, L., and Zou, W. (2004). Specific recruitment of regulatory T cells in ovarian carcinoma fosters immune privilege and predicts reduced survival. *Nat. Med.* **10**, 942–949.

de Visser, K. E., and Coussens, L. M. (2006). The inflammatory tumor microenvironment and its impact on cancer development. *Contrib. Microbiol.* **13**, 118–137.

DeLong, P., Carroll, R. G., Henry, A. C., Tanaka, T., Ahmad, S., Leibowitz, M. S., Sterman, D. H., June, C. H., Albelda, S. M., and Vonderheide, R. H. (2005). Regulatory T cells and cytokines in malignant pleural effusions secondary to mesothelioma and carcinoma. *Cancer Biol. Ther.* **4**, 342–346.

DeLong, P., Tanaka, T., Kruklitis, R., Henry, A. C., Kapoor, V., Kaiser, L. R., Sterman, D. H., and Albelda, S. M. (2003). Use of cyclooxygenase-2 inhibition to enhance the efficacy of immunotherapy. *Cancer Res.* **63**, 7845–7852.

Dudley, M. E., Wunderlich, J. R., Robbins, P. F., Yang, J. C., Hwu, P., Schwartzentruber, D. J., Topalian, S. L., Sherry, R., Restifo, N. P., Hubicki, M., Robinson, M. R., Raffeld, M., Duray, P., Seipp, C. A., Rogers-Freezer, L., Morton, K. E., Mavroukakis, S. A., White, D. E., and Rosenberg, S. A. (2002). Cancer regression and autoimmunity in patients after clonal repopulation with antitumor lymphocytes. *Science* **298**, 850–854.

Dummer, W., Niethammer, A. G., Baccala, R., Lawson, B. R., Wagner, N., Reisfeld, R. A., Theofilopoulos, A. N., (2002). T cell homeostatic proliferation elicits effective antitumor autoimmunity. *J. Clin. Invest.* **110**, 157–159.

Fehervari, Z., and Sakaguchi, S. (2004). Development and function of $CD25^+CD4^+$ regulatory T cells. *Curr. Opin. Immunol.* **16**, 203–208.

Foon, K. A., John, W. J., Chakraborty, M., Das, R., Teitelbaum, A., Garrison, J., Kashala, O., Chatterjee, S. K., and Bhattacharya-Chatterjee, M. (1999). Clinical and immune responses in resected colon cancer patients treated with anti-idiotype monoclonal antibody vaccine that mimics the carcinoembryonic antigen. *J. Clin. Oncol.* **17**, 2889–2895.

Gasser, S., and Raulet, D. H. (2006). The DNA damage response arouses the immune system. *Cancer Res.* **66**, 3959–3962.

Greenwald, R. J., Freeman, G. J., and Sharpe, A. H. (2005). The B7 family revisited. *Annu. Rev. Immunol.* **23**, 515–548.

Heath, W. R., Belz, G. T., Behrens, G. M., Smith, C. M., Forehan, S. P., Parish, I. A., Davey, G. M., Wilson, N. S., Carbone, F. R., and Villadangos, J. A. (2004). Cross-presentation, dendritic cell subsets, and the generation of immunity to cellular antigens. *Immunol. Rev.* **199**, 9–26.

Hermans, I. F., Chong, T. W., Palmowski, M. J., Harris, A. L., and Cerundolo, V. (2003). Synergistic effect of metronomic dosing of cyclophosphamide combined with specific antitumor immunotherapy in a murine melanoma model. *Cancer Res.* **63**, 8408–8413.

Hoon, D. S., Foshag, L. J., Nizze, A. S., Bohman, R., and Morton, D. L. (1990). Suppressor cell activity in a randomized trial of patients receiving active specific immunotherapy with melanoma cell vaccine and low dosages of cyclophosphamide. *Cancer Res.* **50**, 5358–5364.

Hu, D. E., Moore, A. M., Thomsen, L. L., and Brindle, K. M. (2004). Uric acid promotes tumor immune rejection. *Cancer Res.* **64**, 5059–5062.

Ibe, S., Qin, Z., Schuler, T., Preiss, S., and Blankenstein, T. (2001). Tumor rejection by disturbing tumor stroma cell interactions. *J. Exp. Med.* **194**, 1549–1559.

Iwasaki, A., and Medzhitov, R. (2004). Toll-like receptor control of the adaptive immune responses. *Nat. Immunol.* **5**, 987–995.

Jones, R. C., Kelley, M., Gupta, R. K., Nizze, J. A., Yee, R., Leopoldo, Z., Qi, K., Stern, S., and Morton, D. L. (1996). Immune response to polyvalent melanoma cell vaccine in AJCC stage III melanoma: An immunologic survival model. *Ann. Surg. Oncol.* **3**, 437–445.

Kelly, M. G., Alvero, A. B., Chen, R., Silasi, D. A., Abrahams, V. M., Chan, S., Visintin, I., Rutherford, T., and Mor, G. (2006). TLR-4 signaling promotes tumor growth and paclitaxel chemoresistance in ovarian cancer. *Cancer Res.* **66**, 3859–3868.

King, C., Ilic, A., Koelsch, K., and Sarvetnick, N. (2004). Homeostatic expansion of T cells during immune insufficiency generates autoimmunity. *Cell* **117**, 265–277.

Klebanoff, C. A., Khong, H. T., Antony, P. A., Palmer, D. C., and Restifo, N. P. (2005). Sinks, suppressors and antigen presenters: How lymphodepletion enhances T cell-mediated tumor immunotherapy. *Trends Immunol.* **26**, 111–117.

Lake, R. A., and Robinson, B. W. S. (2005). Immunotherapy and chemotherapy—a practical partnership. *Nat. Rev. Cancer* **5**, 397–405.

Lake, R. A., and van der Most, R. G. (2006). A better way for a cancer cell to die. *N. Engl. J. Med.* **354**, 2503–2504.

Langowski, J. L., Zhang, X., Wu, L., Mattson, J. D., Chen, T., Smith, K., Basham, B., McClanahan, T., Kastelein, R. A., and Oft, M. (2006). IL-23 promotes tumour incidence and growth. *Nature* **442**, 461–465.

Le Bon, A., Etchart, N., Rossmann, C., Ashton, M., Hou, S., Gewert, D., Borrow, P., and Tough, D. F. (2003). Cross-priming of CD8+ T cells stimulated by virus-induced type I interferon. *Nat. Immunol.* **4**, 1009–1015.

Leitner, W. W., Hwang, L. N., Bergmann-Leitner, E. S., Finkelstein, S. E., Frank, S., and Restifo, N. P. (2004). Apoptosis is essential for the increased efficacy of alphaviral replicase-based DNA vaccines. *Vaccine* **22**, 1537–1544.

Leitner, W. W., and Restifo, N. P. (2003). DNA vaccines and apoptosis: To kill or not to kill? *J. Clin. Invest.* **112**, 22–24.

Levitt, M. L., Kassem, B., Gooding, W. E., Miketic, L. M., Landreneau, R. J., Ferson, P. F., Keenan, R., Yousem, S. A., Lindberg, C. A., Trenn, M. R., Ponas, R. S., Tarasoff, P., Sabatine, J. M., Friberg, D., and Whiteside, T. L. (2004). Phase I study of gemcitabine given weekly as a short infusion for non-small cell lung cancer: Results and possible immune system-related mechanisms. *Lung Cancer* **43**, 335–44.

Li, D., Ronson, B., Guo, M., Liu, S., Bishop, J. S., Van Echo, D. A., and O'Malley, B. W., Jr. (2002). Interleukin 2 gene transfer prevents NKG2D suppression and enhances antitumor efficacy in combination with cisplatin for head and neck squamous cell cancer. *Cancer Res.* **62**, 4023–4028.

Livingston, P. O., Cunningham-Rundles, S., Marfleet, G., Gnecco, C., Wong, G. Y., Schiffman, G., Enker, W. E., and Hoffman, M. K. (1987). Inhibition of suppressor-cell activity by cyclophosphamide in patients with malignant melanoma. *J. Biol. Response Mod.* **6**, 392–403.

Livingston, P. O., Wong, G. Y., Adluri, S., Tao, Y., Padavan, M., Parente, R., Hanlon, C., Calves, M. J., Helling, F., Ritter, G., *et al.* (1994). Improved survival in stage III melanoma patients with GM2 antibodies: A randomized trial of adjuvant vaccination with GM2 ganglioside. *J. Clin. Oncol.* **12**, 1036–1044.

Liyanage, U. K., Moore, T. T., Joo, H. G., Tanaka, Y., Herrmann, V., Doherty, G., Drebin, J. A., Strasberg, S. M., Eberlein, T. J., Goedegebuure, P. S., and Linehan, D. C. (2002). Prevalence of regulatory T cells is increased in peripheral blood and tumor microenvironment of patients with pancreas or breast adenocarcinoma. *J. Immunol.* **169**, 2756–2761.

Loeffler, M., Kruger, J. A., and Reisfeld, R. A. (2005). Immunostimulatory effects of low-dose cyclophosphamide are controlled by inducible nitric oxide synthase. *Cancer Res.* **65**, 5027–5030.

Lutsiak, M. E., Semnani, R. T., De Pascalis, R., Kashmiri, S. V., Schlom, J., and Sabzevari, H. (2005). Inhibition of CD4(+)25+ T regulatory cell function implicated in enhanced immune response by low-dose cyclophosphamide. *Blood* **105**, 2862–2868.

MacLean, G. D., Miles, D. W., Rubens, R. D., Reddish, M. A., and Longenecker, B. M. (1996). Enhancing the effect of THERATOPE STn-KLH cancer vaccine in patients with metastatic breast cancer by pretreatment with low-dose intravenous cyclophosphamide. *J. Immunother. Emphasis Tumor Immunol.* **19**, 309–316.

Miller, K., Abeles, G., Oratz, R., Zeleniuch-Jacquotte, A., Cui, J., Roses, D. F., Harris, M. N., and Bystryn, J. C. (1995). Improved survival of patients with melanoma with an antibody response to immunization to a polyvalent melanoma vaccine. *Cancer* **75**, 495–502.

Muller, A. J., and Prendergast, G. C. (2005). Marrying immunotherapy with chemotherapy: Why say IDO? *Cancer Res* **65**, 8065–8068.

Nowak, A. K., Lake, R. A., Marzo, A. L., Scott, B., Heath, W. R., Collins, E. J., Frelinger, J. A., and Robinson, B. W. (2003a). Induction of tumor cell apoptosis *in vivo* increases tumor antigen cross-presentation, cross-priming rather than cross-tolerizing host tumor-specific CD8 T cells. *J. Immunol.* **170**, 4905–4913.

Nowak, A. K., Robinson, B. W., and Lake, R. A. (2003b). Synergy between chemotherapy and immunotherapy in the treatment of established murine solid tumors. *Cancer Res.* **63**, 4490–4496.

Ogbourne, S. M., Suhrbier, A., Jones, B., Cozzi, S. J., Boyle, G. M., Morris, M., McAlpine, D., Johns, J., Scott, T. M., Sutherland, K. P., Gardner, J. M., Le, T. T., Lenarczyk, A., Aylward, J. H., and Parsons, P. G. (2004). Antitumor activity of 3-ingenyl angelate: Plasma membrane and mitochondrial disruption and necrotic cell death. *Cancer Res.* **64**, 2833–2839.

Overwijk, W. W., de Visser, K. E., Tirion, F. H., de Jong, L. A., Pols, T. W., van der Velden, Y. U., van den Boorn, J. G., Keller, A. M., Buurman, W. A., Theoret, M. R., Blom, B., Restifo, N. P., Kruisbeek, A. M., Kastelein, R. A., and Haanen, J. B. (2006). Immunological and antitumor effects of IL-23 as a cancer vaccine adjuvant. *J. Immunol.* **176**, 5213–5222.

Plate, J. M., Plate, A. E., Shott, S., Bograd, S., and Harris, J. E. (2005). Effect of gemcitabine on immune cells in subjects with adenocarcinoma of the pancreas. *Cancer Immunol. Immunother.* **54**, 915–925.

Pratesi, G., Petrangolini, G., Tortoreto, M., Addis, A., Belluco, S., Rossini, A., Selleri, S., Rumio, C., Menard, S., and Balsari, A. (2005). Therapeutic synergism of gemcitabine and CpG-oligodeoxynucleotides in an orthotopic human pancreatic carcinoma xenograft. *Cancer Res.* **65**, 6388–6393.

Rad, A. N., Pollara, G., Sohaib, S. M., Chiang, C., Chain, B. M., and Katz, D. R. (2003). The differential influence of allogeneic tumor cell death via DNA damage on dendritic cell maturation and antigen presentation. *Cancer Res.* **63**, 5143–5150.

Schiavoni, G., Mattei, F., Di Pucchio, T., Santini, S. M., Bracci, L., Belardelli, F., and Proietti, E. (2000). Cyclophosphamide induces type I interferon and augments the number of CD44(hi) T lymphocytes in mice: Implications for strategies of chemoimmunotherapy of cancer. *Blood* **95**, 2024–2030.

Shi, Y., Evans, J. E., and Rock, K. L. (2003). Molecular identification of a danger signal that alerts the immune system to dying cells. *Nature* **425**, 516–521.

Skinner, R. B., Jr. (2003). Imiquimod. *Dermatol. Clin* **21**, 291–300.

Su, Y. B., Sohn, S., Krown, S. E., Livingston, P. O., Wolchok, J. D., Quinn, C., Williams, L., Foster, T., Sepkowitz, K. A., and Chapman, P. B. (2004). Selective CD4+ lymphopenia in melanoma patients treated with temozolomide: A toxicity with therapeutic implications. *J. Clin. Oncol.* **22**, 610–616.

Suzuki, E., Kapoor, V., Cheung, H. K., Ling, L. E., DeLong, P. A., Kaiser, L. R., and Albelda, S. M. (2004). Soluble type II transforming growth factor-beta receptor inhibits established murine malignant mesothelioma tumor growth by augmenting host antitumor immunity. *Clin. Cancer Res.* **10**, 5907–5918.

Taieb, J., Chaput, N., Schartz, N., Roux, S., Novault, S., Menard, C., Ghiringhelli, F., Terme, M., Carpentier, A. F., Darrasse-Jese, G., Lemonnier, F., and Zitvogel, L. (2006a). Chemoimmunotherapy of tumors: Cyclophosphamide synergizes with exosome based vaccines. *J. Immunol.* **176**, 2722–2729.

Taieb, J., Chaput, N., Menard, C., Apetoh, L., Ullrich, E., Bonmort, M., Pequignot, M., Casares, N., Terme, M., Flament, C., Opolon, P., Lecluse, Y., Metivier, D., Tomasello, E., Vivier, E., Ghiringhelli, F., Martin, F., Klatzmann, D., Poynard, T., Tursz, T., Raposo, G., Yagita, H., Ryffel, B., Kroemer, G., Zitvogel, L. (2006b). A novel dendritic cell subset involved in tumor immunosurveillance. *Nat. Med.* **12**, 214–219.

van der Most, R. G., Currie, A., Robinson, B. W., and Lake, R. A. (2006a). Cranking the immunologic engine with chemotherapy: Using context to drive tumor antigen cross-presentation towards useful antitumor immunity. *Cancer Res* **66**, 601–604.

van der Most, R. G., Himbeck, R., Aarons, S., Carter, S. J., Larma, I., Robinson, C., Currie, A., and Lake, R. A. (2006b). Antitumor efficacy of the novel chemotherapeutic agent coramsine is potentiated by cotreatment with CpG-containing oligodeoxynucleotides. *J. Immunother.* **29**, 134–142.

Weaver, C. T., Harrington, L. E., Mangan, P. R., Gavrieli, M., and Murphy, K. M. (2006). Th17: An effector CD4 T cell lineage with regulatory T cell ties. *Immunity* **24**, 677–688.

Weihrauch, M. R., Ansen, S., Jurkiewicz, E., Geisen, C., Xia, Z., Anderson, K. S., Gracien, E., Schmidt, M., Wittig, B., Diehl, V., Wolf, J., Bohlen, H., and Nadler, L. M. (2005). Phase I/II combined chemoimmunotherapy with carcinoembryonic antigen-derived HLA-A2-restricted CAP-1 peptide and irinotecan, 5-fluorouracil, and leucovorin in patients with primary metastatic colorectal cancer. *Clin. Cancer Res.* **11**, 5993–6001.

Wolkers, M. C., Brouwenstijn, N., Bakker, A. H., Toebes, M., and Schumacher, T. N. (2004). Antigen bias in T cell cross-priming. *Science* **304**, 1314–1317.

Woo, E. Y., Yeh, H., Chu, C. S., Schlienger, K., Carroll, R. G., Riley, J. L., Kaiser, L. R., and June, C. H. (2002). Cutting edge: Regulatory T cells from lung cancer patients directly inhibit autologous T cell proliferation. *J. Immunol.* **168**, 4272–4276.

Zhang, H., Chua, K. S., Guimond, M., Kapoor, V., Brown, M. V., Fleisher, T. A., Long, L. M., Bernstein, D., Hill, B. J., Douek, D. C., Berzofsky, J. A., Carter, C. S., Read, E. J., Helman, L. J., and Mackall, C. L. (2005). Lymphopenia and interleukin-2 therapy alter homeostasis of $CD4^+CD25^+$ regulatory T cells. *Nat. Med.* **11**, 1238–1243.

CHAPTER 14

Dendritic Cells and Coregulatory Signals: Immune Checkpoint Blockade to Stimulate Immunotherapy

DREW PARDOLL

I. Regulation of T Cell Responses to Antigen
   A. Signal 1: Antigen Presentation
   B. Signal 2: Costimulation
   C. Immunologic Checkpoints
II. Regulatory T Cells
III. Immune Checkpoints in the Tumor Microenvironment
IV. Monoclonal Antibodies that Interfere with Coinhibitory Receptors on T Cells
   A. CTLA-4
   B. PD-1
V. What Is the Most Effective Way to Use Checkpoint Inhibitors?
   References

As engineered immunotherapeutics continue to improve, a potency ceiling will be reached owing to the presence of hard-wired inhibitory pathways that negatively regulate lymphocyte responses. Blockade of these so-called immune checkpoints is emerging as one of the most important elements of cancer immunotherapy and promises to reverse the trend of repetitive failures in phase III immunotherapy trials that have attempted to demonstrate anticancer activity of therapeutic vaccines applied as single agents. It is now clear that the quantitative response to antigen is determined by a balance between positive (costimulatory) and negative (checkpoint or regulatory) signaling pathways (Figure 14.1). In the case of T cell responses, a number of these pathways appear to have components that are either exclusively or at least selectively expressed by T cells. Immunologic checkpoints thus represent a major target for pharmacologic intervention. Past efforts in the development of pharmacologic agents that target the immune system have exclusively identified drugs that either inhibit or activate immune responses in a nonspecific, antigen-independent fashion. The discovery

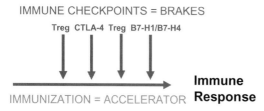

FIGURE 14.1 **The amplitude of the immune response determined by counteracting positive and negative signals.** The immune response is schematically depicted as a balance between "accelerators" and "brakes." The accelerators are antigen (presented to T cells by APCs) and costimulatory signals. The brakes are both cellular (i.e., Tregs) and molecular. A more detailed compendium of costimulatory and inhibitory ligand–receptor pairs are presented in Figure 14.2.

of specific negative regulatory signaling pathways that check immune responses by dampening T cell receptors (TCRs) or costimulatory signaling pathways provides a wonderful opportunity for antigen-specific immunopharmacology by applying drugs or antibodies that block these pathways and using antigen-specific activation stimuli.

I. REGULATION OF T CELL RESPONSES TO ANTIGEN

The revolution in cancer immunology and immunotherapy builds on the understanding of the receptors and signaling pathways that regulate T cell responses. At the fundamental level, three elements, which are described in the following paragraphs, determine T cell responsiveness to an antigen (Figure 14.2).

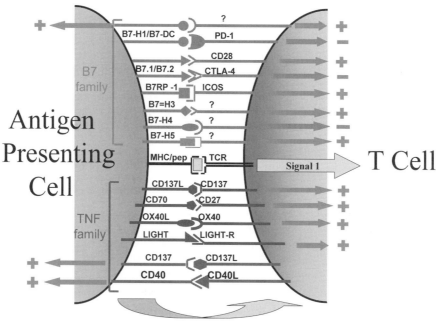

FIGURE 14.2 **The three catagories of costimulatory and coinhibitory signals at the APC–T cell interface.** Shown are the ligand-receptor pairs for the B7 family, TNF family, and cytokine family of signals between the APC and T cell. Not all receptors for some B7 family members have been identified. Some B7 family members interact with both costimulatory receptors and coinhibitory receptors. While many signals are delivered from an APC to a T cell, the T cell also delivers signals to the APC, creating bidirectional cross talk. (See color plate.)

A. Signal 1: Antigen Presentation

The first element, termed *signal 1*, is transmitted by the T cell receptor (TCR) that acts as a signal transducer for external stimuli to initiate T cell activation (Krogsgaard and Davis, 2005). Small peptide fragments derived from proteolysis of antigens are presented to the TCR by major histocompatibility complex (MHC) (human leukocyte antigen [HLA] in humans) molecules expressed by antigen-presenting cells (APCs). The critical APC that activates T cell responses is the dendritic cell (DC) (Banchereau and Steinman, 1998). Therefore, it is appreciated that essentially all vaccines stimulate immune responses through transfer of antigen to DCs, which in turn degrade the antigen into peptides and present those peptides on MHC molecules to the T cell via TCR recognition (Pardoll, 2002).

B. Signal 2: Costimulation

It is clear that the signals generated solely by engagement of the TCR are insufficient to activate T cells to an effector state. In fact, when T cells receive signal 1 through TCR engagement without additional signals, they enter an unresponsive, or anergic state, in which they do not mediate effector function. This represents one mechanism for self tolerance that protects normal tissues from immune destruction and probably also represents a mechanism by which tumor-specific T cells in patients are naturally unresponsive to their tumor, thereby allowing it to grow. The critical second element in T cell activation—collectively referred to as *signal 2*—is delivered by a large number of costimulatory molecules expressed by the APC that interact with costimulatory receptors on the T cell. The prototypical costimulatory molecules are B7–1 and its homolog B7–2. The B7–1 and B7–2 costimulate T cells by interacting with the CD28 receptor on T cells (Acuto and Michel, 2003). Signals delivered by both the TCR (signal 1) and CD28 collaborate to enhance T cell activation. Five additional B7 family members have been identified over the last 5 years (since 2001); these members are B7RP-1 (also called ICOS-L, B7h, B7-H2), B7-H1 (also called PD-L1), B7-DC (also called PD-L2), B7-H3, and B7-H4 (also called B7s, B7x) (Greenwald *et al.*, 2005). Most of these possess additional costimulatory functions and, in some cases, can collaborate with B7–1 and B7–2 to costimulate T cells through receptors distinct from CD28. A second class of costimulatory ligands expressed by APCs are tumor necrosis factor (TNF) family members, including 4-1BB ligand, CD70, and OX40 ligand (Croft 2003a, 2003b; Watts, 2005). Each of these ligands has a cognate TNF receptor family member expressed on activated T cells. The third class of costimulatory signals is provided by cytokines released by activated APCs, such as interleukin (IL)-1, TNF-α, IL-6, and IL-12 (Reis e Sousa, 2006).

C. Immunologic Checkpoints

The final element in T cell regulation is represented by inhibitory pathways, termed *immunologic checkpoints* (Chen, 2004). There are many immunologic checkpoints that serve two purposes. One helps generate and maintain self tolerance among T cells specific for self antigens. The other restrains the amplitude of normal T cell responses so that they do not "overshoot" in their natural response to foreign pathogens. The biologic function of immunologic checkpoints can be readily demonstrated by the dramatically heightened and sometimes pathologic immune responses observed in mice in which these checkpoints have been genetically disrupted. The prototypical immunologic checkpoint is the cytotoxic T lymphocyte-associated antigen 4 (CTLA-4) counter receptor that is expressed when T cells become activated (Alegre *et al.*, 2001; Teft *et al.*, 2006). CTLA-4 binds B7–1 and B7–2 with roughly 20-fold higher affinity than CD28. When naive T cells are presented with antigen on B7–1 and B7–2 expressing APCs, they are costimulated because resting T cells express CD28 but not CTLA-4. Upon activation, CTLA-4 is expressed on T cells. Because of its higher affinity, CTLA-4 outcompetes CD28 for B7–1 and B7–2 binding and acts to downmodulate T cell responses. Two additional B7 family members, B7-H1 (Dong et al., 1999; Freeman et al., 2000) and B7-DC (Tseng et al., 2001; Latchman et al., 2001), also appear to interact with costimulatory and counterregulatory inhibitory receptors. PD-1 (Ishida et al., 1992), which is upregulated on T cells upon activation, appears to be a counter-regulatory immunologic checkpoint when it binds either B7-DC or B7-H1. The activating receptor(s) for B7-DC and B7-H1 have not yet been identified. Because B7-DC is largely expressed on DCs, it likely plays a predominantly costimulatory role in enhancing activation of naive or resting T cells. B7-H1 expressed on DCs can also costimulate T cells. However, in contrast to B7-DC, B7-H1 is expressed on multiple peripheral tissues and importantly on many tumors (Dong et al., 2002). Expression of B7-H1 on tumors represents an important protective mechanism by which they downmodulate the activity of PD-1 expressing tumor-specific T cells. Another B7 family member, B7-H4, appears to mediate a predominantly or exclusively inhibitory function in the immune system (Sica, 2003). B7-H4 is upregulated on many human tumors as well as on tumor-infiltrating myeloid cells (Kryczek et al., 2006; Simon et al., 2006; Tringler et al., 2006). While therapeutic data with B7-H4 blockade in preclinical models are currently limited by a paucity of good blocking antibodies, B7-H4's overexpression in human tumors makes it a strong potential target for antibody blockade.

Checkpoint receptors on the surface of T cells, as well as their ligands, represent the primary clinical targets for development of blocking antibodies as a means of enhancing antitumor immunity and will be discussed in some detail in this chapter. It is important to point out, however, that as signaling pathways for activation of both T cells and APCs are elucidated, additional inhibitory molecules that regulate these pathways have been discovered. A number of these intracellular inhibitory signaling pathways in T cells represent potential targets for pharmacologic intervention. Some of the best candidates include Cbl-b, Cabin, Carubin, certain protein tyrosine phosphatases (PTPs), as well as the tyrosine kinase Csk. Among the phosphatases, SHIP-1, SHP-1, and SHP-2 have all been implicated in downmodulating signaling pathways activated by TCR engagement (Chan et al., 1994; Long, 1999; Gascoigne and Zal, 2004). Csk has been well demonstrated to inhibit or downmodulate TCR signaling through phosphorylation of regulatory tyrosines on the Src family tyrosine kinases, which are critical for T cell activation (Vang et al., 2004). Cbl-b is an adaptor protein that appears to negatively regulate T cell activation by antagonizing CD28-mediated costimulatory pathways. Thus, T cells from Cbl-b knockout mice are hypersensitive to low doses of T cell stimulatory ligands and are furthermore relatively CD28-independent in their activation (Krawczyk et al., 2000; Bachmaier et al., 2000). Cabin is a molecule that appears to have multiple functions, including acting as a scaffold for coordinating transcription factors. Cabin was originally identified as a molecule that binds to and inhibits calcineurin, a critical serine phosphatase that mediates TCR-dependent cytokine activation through dephosphorylation of nuclear factor of activated T cells (NFAT-c), a critical step in nuclear translocation. The calcineurin-inhibiting portion of Cabin has been localized and thus represents an interesting target for pharmacologic intervention (Sun et al., 1998). A regulatory molecule termed Carubin has been shown to coordinately inhibit two arms of TCR signaling—the calcineurin-NFATc pathway and the Ras-MEK-Erk pathway (Fan Pan et al., in press).

On the APC side, Toll-like receptor (TLR) signaling pathways critical to DC activation are regulated at multiple levels by molecules such as interleukin-1 receptor associated kinase-M (IRAK-M) (Naka et al., 2005). In addition, signal transducers and activators of transcription (Stat) signaling pathways, operative in both APCs and T cells, are inhibited by cytokine-inducible SH2 protein (CIS) and suppressors of cytokine signaling (SOCS) proteins that act as negative feedback inhibitors of upstream Stat-activating kinases, such as the janus tyrosine kinases (JAKs) (Ilangumaran et al., 2004; Alexander and Hilton, 2004). Knockouts of these intracellular signaling regulatory molecules demonstrate enhanced immune responses, defining them as promising targets for pharmacologic inhibition to enhance antitumor immunity. However, their intracellular localization precludes antibody inhibition and necessitates the creation of small-molecule inhibitors capable of diffusing across membranes. Most have not yet been seriously addressed, so their "drugability" is not known.

II. REGULATORY T CELLS

In contrast to the specific checkpoint molecules described previously, regulatory T cells (Tregs) represent a cell subset that inhibits immune responses. Natural Tregs generated in the thymus are clearly important in maintaining self-tolerance and intestinal immune homeostasis (Sakaguchi et al., 2001; Shevach, 2002; Maloy and Powrie, 2001). Evidence is accumulating that Tregs can additionally develop in the periphery under certain circumstances (so-called induced Tregs), including cancer (Yamaguchi and Sakaguchi, 2006). Numerous murine studies have demonstrated that Tregs expand in animals with cancer and significantly limit the efficacy of antitumor immune responses induced by vaccines. For example, in a study by van Elsas et al. (1999), a combination of a granulocyte-macrophage colony-stimulating factor (GM-CSF) transduced tumor vaccine plus anti-CTLA-4 was much more effective at eliminating established tumors when animals were treated with anti-IL-2 receptor α antibodies prior to vaccination/anti-CTLA-4 treatment (Sutmuller et al., 2001). The notion of eliminating Tregs came from the surprising finding that while CD4 depletion of animals significantly diminished the ability of GM-CSF vaccine/anti-CTLA-4 to protect animals from subsequent tumor challenge (with B16 melanoma), the opposite effect was observed for therapy of established B16 tumors. These results suggested the idea that CD4 cells predominately played an enhancing helper role when vaccine/anti-CTLA-4 treatment was done prior to tumor inoculation; however, once tumors were established, they induced a dominant population of $CD4^+$ Tregs. Since Tregs typically express IL-2 receptor, it was reasoned that depletion of IL-2 receptor positive cells prior to vaccination/anti-CTLA-4 would eliminate these Tregs. In a second set of studies, Jaffee et al. demonstrated that treatment of mice with low-dose cytoxan prior to vaccination enhanced the ability of HER2/neu/GM-CSF vaccines to protect HER2/neu transgenic mice from challenge with HER2/neu-expressing tumors (Ercolini et al., 2005). As the cytoxan and vaccine treatments were performed prior to the tumor challenge, the enhanced effect of cytoxan could not be explained by a direct antitumor effect. Indeed, low-dose cytoxan treatment has long been touted to inhibit or kill suppressor cells although this effect has also been attributed to creation of lymphoid "space." However, adoptive transfer experiments with $CD4^+IL-2R\alpha^+$ cells from noncytoxan-treated HER2/neu transgenic mice proved that the cytoxan effect was indeed due to inhibition of Tregs. It is likely that the next few years will see many additional demonstrations for an important role of Tregs in blunting or blocking antitumor immunity, likely because they are a natural consequence of tolerance induction.

They represent a very tempting target for inhibition as part of combination immunotherapy strategies.

The gold standard Treg-specific molecule is FoxP3, a forkhead transcription factor (Hori et al., 2003). FoxP3 knockout mice and children with homozygous FoxP3 mutations are depleted of Tregs and develop multisystem autoimmune and hyperimmune syndromes, attesting to the selective immunologic role of FoxP3 in Tregs development. Until 1995, essentially no Treg-specific cell surface molecules had been identified. Thus, Tregs are identified by the coexpression of CD4 and CD25. However, CD25 (the alpha chain of IL-2 receptor) is also expressed on effector T cells undergoing acute activation. The elucidation of Treg-specific cell surface molecules would represent a major handle in modulating their function.

Indeed, two Treg-selective cell surface molecules have been identified through gene expression profiling. One, termed *glucocorticoid-induced tumor necrosis factor receptor* (GITR), is a TNF receptor family member whose expression in mice tracks closely with CD25 and FoxP3 (Shimizu et al., 2002; McHugh et al., 2002). Administration of anti-GITR antibodies enhance antitumor immunity in some murine systems, but since GITR is also expressed on activated effector T cells, it is not clear at which level the anti-GITR antibodies are acting (Ko et al., 2005). It has in fact been suggested that anti-GITR antibodies diminish the susceptibility of effector T cells to suppression by Tregs (Stephens et al., 2004).

LAG-3 is a CD4 homolog that is selectively expressed on the surface of Tregs. Transduction of CD4 T cells with the LAG-3 gene confers suppressor activity upon them, and blockade of LAG-3 on Tregs with a monoclonal antibody inhibits their suppressive activity (Huang et al., 2004). In addition, Tregs from LAG-3 knockout mice fail to regulate the homeostatic expansion of T cells in lymphopenic hosts (Workman and Vignali, 2005). These findings directly implicate LAG-3 in immune suppression mediated by Tregs. While surface expression of LAG-3 is very low on circulating Tregs, it is upregulated on Tregs in tissues and tumors, suggesting that its role may be on activated Tregs at the site of immune suppression. Recently, LAG-3 expression on antigen-specific CD8 cells in tumors has been demonstrated, where it exerts a cell autonomous inhibitory effect independent of its role on CD4 Tregs. These findings suggest that LAG-3 blockade may enhance antitumor immunity at multiple levels (Grosso et al., submitted).

A number of studies have claimed an increase in Tregs in cancer patients. Most of these studies are limited by the absence of specific Treg markers, evidence of antigen specificity, or convincing functional data. To date, the most suggestive study has been performed in patients with ovarian cancer. Curiel et al. (2004) demonstrated an increase in $CD4^+CD25^+$ T cells in ovarian tumors and ascites associated with more advanced stages. Many were $FoxP3^+$ and exhibited suppressive activity *in vitro*. In addition, increased numbers of $CD4^+CD25^+$ T cells in a given stage of disease correlated with shorter overall survival (Curiel et al., 2004). Vieweg et al. performed an interventional study in patients with metastatic renal cancer using a diphtheria toxin conjugated IL-2 (ONTAK®) to eliminate $CD25^+$ Tregs in conjunction with an RNA-transduced DC vaccine (2007). The researchers demonstrated short-term decreases in circulating $CD4^+CD25^+$ cells after therapy associated with an increase in vaccine-induced CTL responses compared with patients receiving the DC vaccine alone (Dannull et al., 2005). These preliminary results are encouraging for future development of more specific inhibitors of Tregs in the future.

III. IMMUNE CHECKPOINTS IN THE TUMOR MICROENVIRONMENT

Mounting evidence supports the idea that tumors actively participate in the modeling of their microenvironment. In most solid tumors, the majority of cells in the tumor mass are not in fact cancer cells but, rather, are nontransformed stromal elements consisting of vasculature, stromal fibroblasts, and hematopoietic cells. The hematopoietic component includes all the elements of innate and adaptive immunity with the capacity to generated potent antitumor immunity. From the tumor's perspective, it is therefore critical that these cells are maintained in a quiescent state such that their potent cytotoxic machinery is not unleashed and directed at the cancer cells in their midst. Tumors appear to employ specific mechanisms to restrain immune activation in their microenvironment by inhibiting the release and sensing of danger signals, thereby converting inflammatory responses to those that could instead potentiate tumor growth (Figure 14.3).

Studies have in fact demonstrated that oncogenic signaling pathways not only promote tumor growth and survival in a cell-intrinsic fashion but also actively modulate their immunologic microenvironment. The best-studied oncogenic signaling pathway to downmodulate destructive immune responses in the tumor microenvironment is the Stat3 pathway. Stat3 is commonly constitutively activated in diverse cancers of both hematopoietic and epithelial origin. Constitutively activated Stat3 enhances tumor cell proliferation and prevents apoptosis (Bowman *et al.*, 2001; Bromberg *et al.*, 1999; Turkson *et al.*, 1998). Indeed, it was found that constitutive Stat3 activity in tumors negatively regulates inflammation, DC activity, and T cell immunity (Wang *et al.*, 2004). The Stat3 signaling pathway in tumor cells appears to accomplish this by inhibiting the production of proinflammatory danger signals and by inducing expression of factors that inhibit functional maturation of DCs and other tumor-infiltrating hematopoietically derived cells responsible for innate immunity. Blockade of constitutive Stat3 signaling in

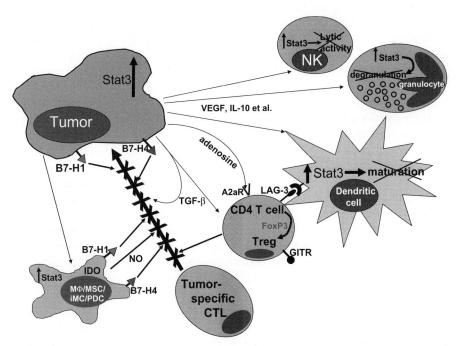

FIGURE 14.3 **Multiple levels of inhibition of antitumor immunity in the tumor microenvironment.** The tumor, via signaling by Stat3 and other pathways, organizes a complex immunologic microenvironment that is highly inhibitory to the generation of both innate and adaptive antitumor responses. These multiple levels of inhibition involve many different cell types and molecular ligands.

tumor cells resulted in the dramatic upregulation of proinflammatory cytokine genes, such as TNF-α and interferon (IFN)-β. Proinflammatory chemokine genes, such as IP-10 and RANTES, were also upregulated. These cytokines began to be produced without any exogenous inductive stimuli when Stat3 was inhibited, indicating that Stat3 signaling restrains a natural propensity of tumors to produce these molecules. It was determined that Stat3 redirects nuclear factor (NF)-κB away from the promoters of proinflammatory genes and onto promoters of antiapoptotic genes. Thus, when Stat3 is blocked in tumor cells, the constitutively active NF-κB activates transcription of proinflammatory genes.

It has also been shown that elements of the tumor microenvironment can promote immune tolerance, at least in part, by bone marrow-derived DCs. Indeed, DCs are immature and functionally impaired in both cancer patients and tumor-bearing animals. Dysfunction of DCs in tumor-bearing hosts may be due to the lack of "danger" signals necessary for DC activation together with factors in the tumor milieu inhibiting functional maturation of DCs (Hawiger et al., 2001; Almand et al., 2000; Vicari et al., 2002). Stat3 activity was also found to promote the production of multiple factors, among them vascular endothelial growth factor (VEGF) and IL-10, that activate Stat3 signaling and inhibit functional DC maturation in culture (Nefedova et al., 2004, 2005; Kortylewski et al., 2005). Inhibition of Stat3 activity in DC progenitors has also been shown to reduce accumulation of immature DCs in the tumor microenvironment. In addition to its role in tumor cells, Stat3 activation in tumor-infiltrating hematopoietic elements appears to be a major checkpoint for innate and adaptive antitumor responses. The role of hematopoietic Stat3 signaling was investigated by staining for phospho-Stat3 in tumor-infiltrating cells as well as inducible hematopoietic knockout of Stat3. Analysis of phospho-Stat3 by flow cytometry reveals that Stat3 is also constitutively activated in tumor-infiltrating NK cells and neutrophils. Induction of Stat3 deletion considerably increased the number of splenic granulocytic lineage cells, and reduced Stat3 signaling in neutrophils enhanced their cytolytic activity against target tumor cells. Furthermore, tumor-infiltrating NK cells from hematopoietically *Stat3* ablated mice demonstrated strongly enhanced cytolytic activity compared to those derived from tumor-free control mice. Many tumor-associated factors, including IL-10, VEGF, and IL-6, are activators of Stat3, and IL-10 is abundantly produced by tumor-associated macrophages in many different tumors. The findings of increased function of purified *Stat3$^{-/-}$* NK cells and neutrophils from tumor-bearing mice together with an increase in Stat3 activity directly in these populations in the tumor indicates that the role of Stat3 signaling in downregulating function of these cell types is at least in part cell intrinsic.

The central role of Stat3 signaling in both tumors and hematopoietic cells as an immunologic checkpoint defines it as an interesting target for antagonism to enhance antitumor immunity. A small-molecule Stat3 inhibitor, CPA-7, was identified. CPA-7 disrupts Stat3 DNA-binding activity, which is followed by a reduction of phospho-Stat3 protein in the treated cells *in vitro* within hours (Turkson et al., 2005). Tumor-infiltrating DCs from the mice receiving CPA-7 displayed considerably reduced phospho-Stat3 compared to vehicle-treated mice. CPA-7 treatment *in vivo* also led to significant growth inhibition of established tumors that was both T cell and NK cell dependent (Kortylewski et al., 2005). While active programs to identify additional direct Stat3 inhibitors are underway, a number of agents that inhibit upstream activators of the Stat3 pathway are already in the clinic. These include antibody and small-molecule inhibitors of epidermal growth factor receptor (EGFR) and HER2. Src also activates Stat3, and new Src tyrosine kinase inhibitors

(TKIs) may block Stat3 signaling (Yu and Jove, 2004). The immunologic consequences of blockade of upstream activators of Stat3 have not yet been evaluated in humans.

In addition to its role in inhibiting the activation and effector function of DCs, granulocytes and NK cells in the tumor microenvironment, Stat3 signaling has also been reported to play a role in guiding immature myeloid cells (iMCs) to differentiate into myeloid suppressor cells (MSCs) rather than DCs with APC activity. The iMCs and MSCs represent a cadre of myeloid cell types, including tumor-associated macrophages (TAM) that share the common feature of inhibiting both the priming and effector function of tumor-reactive T cells (Kusmartsev and Gabrilovich, 2005; Young et al., 2001; Zea et al., 2005; Bronte et al., 2003; Mazzoni et al., 2002). As such, they represent another example of a "cellular checkpoint" operating in the tumor microenvironment. It is still not clear whether these myeloid cell types represent distinct lineages or different states of the same general immune inhibitory cell subset. In mice, iMCs and MSCs are characterized by coexpression of CD11b (considered a macrophage marker) and Gr1 (considered a granulocyte marker) while expressing low or no major histiocompatibility complex (MHC) class II or the CD86 costimulatory molecule. In humans, they are defined as CD33$^+$ but lacking markers of mature macrophages, DCs, or granulocytes and are DR$^-$ (Bronte et al., 2000). A number of molecular species reported to be produced by tumors tend to drive iMC/MSC accumulation. These include IL-6, CSF-1, IL-10, and gangliosides. IL-6 and IL-10 are potent inducers of Stat3 signaling.

A number of mechanisms have been proposed to explain how iMC/MSC can inhibit T cell responses. Most include the production of reaction oxygen species (ROS) and/or reaction nitrogen species (RNS). NO production by iMC/MSC as a result of arginase activity, which is high in these cells, has been well documented, and inhibition of this pathway with a number of drugs can mitigate the inhibitory effects of iMC/MSC (Zea et al., 2005; Bronte et al., 2003; Mazzoni et al., 2002). ROS, including H_2O_2, have been reported to block T cell function associated with the downmodulation of the ζ chain of the TCR signaling complex, a phenomenon well recognized in T cells from cancer patients and associated with generalized T cell unresponsiveness (Schmielau and Finn, 2001). Studies have shown that Viagra®, which inhibits the G protein-coupled receptor that activates the arginase pathway in iMC/MSC, can block the inhibitory activity of iMC/MSC, offering an intriguing opportunity to enhance antitumor immunity with an approved drug that has an extensive safety profile.

Another mediator of T cell unresponsiveness associated with cancer is the production of indoleamine-2,3 dioxygenase (IDO) (Munn et al., 2002; Baban et al., 2005). IDO appears to be produced by DCs either in tumors or in tumor-draining lymph nodes. Interestingly, IDO in DCs has been reported to be induced via backward signaling by B7–1 and B7–2 upon ligation with CTLA-4 (Mellor et al., 2004). The major IDO-producing DC subset is apparently either a plasmacytoid DC (PDC) or a PDC-related cell (Munn et al., 2004). IDO appears to inhibit T cell responses through catabolism of tryptophan. Activated T cells are highly dependent on tryptophan and are therefore sensitive to tryptophan depletion. Thus, Munn et al. (2004) and Mellor et al. (2004) have proposed a bystander mechanism, whereby DCs in the local environment deplete tryptophan via IDO upregulation, thereby inducing metabolic apoptosis in locally activated T cells.

Another inhibitory molecule that has been implicated in blunting antitumor immune responses is transforming growth factor β (TGF-β) (Blobe et al., 2000; Pepper, 1997; Pepper, 1997; Cui et al., 1996; Shariat et al., 2001; Hasegawa et al., 2001; Saito et al., 1999; Shariat et al., 2001; Letterio and Roberts, 1998). TGF-β is produced by a variety of cell types, including cancer cells and, therefore, likely plays an important role in molding the immune microenvironment of the tumor. For most normal epithelial cells, TGF-β is a potent inhibitor of cell proliferation, causing cell cycle arrest in the G1 stage. In many cancer cells, however, mutations in the TGF-β pathway confer resistance to cell cycle inhibition, allowing uncontrolled proliferation. In addition, in cancer cells, the production of TGF-β is increased and may contribute to invasion by promoting the activity of matrix metaloproteinases. *In vivo*, TGF-β directly stimulates angiogenesis; this stimulation can be blocked by anti-TGF-β antibodies. A bimodal role of TGF-β in cancer has been verified in a transgenic animal model using a keratinocyte-targeted overexpression. Initially, these animals are resistant to the development of early-stage or benign skin tumors. However, once tumors form, they progress rapidly to a more aggressive spindle-cell phenotype. Although this clear bimodal pattern of activity is more difficult to identify in a clinical setting, it should be noted that elevated serum TGF-β levels are associated with poor prognosis in a number of malignancies, including prostate cancer, lung cancer, gastric cancer, and bladder cancer.

One additional immune checkpoint operative in the tumor microenvironment is the adenosine A2a receptor. This receptor, which binds adenosine, is expressed on a number of cell types in the innate immune system and limits inflammatory responses in the context of tissue injury (Sitkovsky *et al.*, 2004). This is because injured tissue releases adenosine. The A2a receptor has also been found to be upregulated on T cells under tolerizing conditions and in tumors (Zheng *et al.*, 2007). Tumors release high amounts of adenosine into their microenvironment due to their high rate of cell turnover, hypoxic necrosis, and anaerobic metabolism. Thus, tumor-infiltrating T cells signal actively through their A2aR. This signaling further enforces T cell tolerance and can lead to conversion to a Treg phenotype with induction of Treg genes, such as FoxP3 and LAG-3. While A2aR blockade or knockout alone has little effect on tumor growth in animal models, its inhibition strongly enhances the antitumor effects of vaccination when applied concurrently (Ohta *et al.*, 2006).

IV. MONOCLONAL ANTIBODIES THAT INTERFERE WITH COINHIBITORY RECEPTORS ON T CELLS

As mentioned earlier, checkpoint molecules expressed on the cell membrane are amenable to antibody blockade, and, indeed, the leading edge of clinical checkpoint inhibition involves antibody blockade of CTLA-4 and PD-1. It is therefore worth examining the preclinical and clinical development of blockers of these so-called "coinhibitory" receptors in some detail.

A. CTLA-4

CTLA-4 is highly homologous with the costimulatory receptor CD28, and they are able to bind the same ligands. As already mentioned, the avidity of CTLA-4 for CD80 and CD86 is much stronger than that of CD28 (Alegre *et al.*, 2001). CTLA-4 protein expression is induced when the TCR complex interacts with antigen and, although retained in internal vesicular T cell compartments, emerges into the immune synapse subsequent to antigen encounter. CTLA-4 decreases T cell activation both by outcompeting CD28 for ligand binding and via recruitment of tyrosine and serine/threonine phosphatases (Teft *et al.*, 2006). The most important molecular function of CTLA-4 seems to be the inhibition of CD28 costimulation as evidenced by the uncontrolled lethal CD28-dependent lymphoproliferative/autoimmune syndrome observed in CTLA-4 knockout mice (Tivol *et al.*, 1995; Waterhouse *et al.*, 1995).

In addition, CTLA-4 is constitutively expressed on CD4$^+$CD25$^+$ FoxP3$^+$ Tregs, and it has been proposed to function as a costimulator of their suppressive function based on *in vitro* experiments. It is possible that CTLA-4 expressed by Tregs is involved in inducing tolerance in CD80 expressing DCs, owing to reverse signaling (Fallarino *et al.*, 2003). Indeed, adoptive transfer of CTLA-4$^{-/-}$ T cells into normal mice does not recapitulate the autoimmune syndrome seen in the donor mice, suggesting that the function of CTLA-4 is not only to regulate the activation of effector T cells but also involves activity exerted in a dominant manner by Tregs. However, the role of CTLA-4 in the function of Tregs remains to be definitively proven because patients treated with anti-CTLA-4 monoclonal antibodies (mAbs) do not show significant changes in the number or function of circulating Tregs although affects on Tregs in tumors or tissues in treated patients has not been assessed (Maker *et al.*, 2005).

Monoclonal antibodies that inhibit the function of CTLA-4 are able to increase the CD8 and CD4 therapeutic immune response toward a wide array of transplantable syngeneic murine tumors with a strong therapeutic effect, although antitumor effects of single-agent anti-CTLA-4 mAbs are typically restricted to immunogenic tumors (Leach *et al.*, 1996). Since CTLA-4 is induced on primed cells, potential synergy with tumor vaccination seems plausible; while anti-CTLA-4 treatment alone had little effect on the poorly immunogenic B16 melanoma, it synergized potently with tumor cells expressing GM-CSF against established lesions of this poorly immunogenic melanoma, with concomitant autoimmune vitiligo (van Elsas *et al.*, 1999). Similarly, treatment with anti-CTLA-4 mAbs synergized with vaccination against a prostate-specific antigen (PSA) to induce antitumor effects in a transgenic model of spontaneous prostate cancer (TRAMP mice), with evidence of destructive inflammation in nonmalignant prostate tissue (Hurwitz *et al.*, 2000). Synergy with tumor vaccines has also been documented with synthetic peptide and DC vaccines (Ito *et al.*, 2000; Pedersen *et al.*, 2006; Keler *et al.*, 2003; Hodi and Dranoff, 2006; Ribas *et al.*, 2004).

The antitumor effects of the antibodies are probably not mediated by their effects on Tregs because, as described earlier, anti-CTLA-4 mAb treatment synergizes with Treg depletion (Sutmuller *et al.*, 2001), and the *in vivo* function of Treg does not appear to be downregulated by anti-CTLA-4 treatment (Maker *et al.*, 2005). Therefore, the primary mechanism of action seems to be the prevention of CTLA-4 binding with CD80 or CD86. Murine transplantable tumor cells seem to express low levels of CD80, indicating that blocking this interaction might prevent CD80-mediated activation of CTLA-4 on the cytotoxic T cell that has infiltrated the tumor (Tirapu *et al.*, 2006).

Two different humanized monoclonal antibodies directed to human CTLA-4 are being independently tested in phase III clinical trials for patients with metastatic melanoma. Clinical development of both agents has identified clinical therapeutic responses, some of them of long duration, which are uncommon in the type of cases included in these trials; the trials have also revealed adverse autoimmune effects that have been transient and reversible.

Adverse events in the form of autoimmunity most commonly include pruriginous (itchy) skin eczema with generalized rashes and inflammatory bowel syndrome with moderate to severe diarrhea (Blansfield *et al.*, 2005). Pathological examination of skin and gastrointestinal lesions reveals infiltrates of CD4$^+$ and CD8$^+$ T cells. There have been less frequent cases of hypophysitis causing hypopituitarism, uveitis, and hepatitis (Blansfield *et al.*, 2005; Phan *et al.*, 2003; Robinson *et al.*, 2004; Ribas *et al.*, 2005; Sanderson *et al.*, 2005). It should be noted that while these side effects are attributed to induction of autoimmunity, no responses to defined autoantigens have yet been ascertained. Hence, it is formally possible that some of these, particularly inflammatory bowel syndrome, could represent inappropriate responses to commensal flora.

Treatment of these adverse events has required steroid therapy that conceivably counteracts the antitumor effects of anti-CTLA-4, albeit there are cases of ongoing objective tumor responses during steroid therapy (Attia et al., 2005). The first published clinical trial of CTLA-4 involved a single dose of 3 mg/kg given to patients with melanoma or ovarian cancer. Although these were not objective responses, there were several signs of increased inflammatory responses, one of them in a cerebellar melanoma. However, this lesion became so inflamed that the patient died from intracranial hypertension (Hodi et al., 2003). It was noted that the patients in this trial who had previously been treated with a DC vaccine or autologous tumor cells that express exogenous GM-CSF (GVAX®) had stronger pathological signs of inflammation and necrosis in the tumors. A clinical trial involving 36 patients with malignant melanoma combined a bolus injection of IL-2 with escalating doses of CTLA-4 mAb. Of the treated patients, 22% showed an overall response with three complete responses, and there was no evidence of additive toxicity (Maker et al., 2005). At the present time, there is an extensive series of exploratory phase II clinical trials in which anti-CTLA-4 mAb are being combined with accepted chemotherapy protocols (http://clinicaltrials.gov/).

The data generated in the clinic involves patients with mainly malignant melanoma or renal cell carcinoma although there are patients with a variety of solid tumors who have been included in various clinical trials. The results of more comprehensive trials in patients with malignant melanoma have indicated several findings to date. First, dosing and schedule for maximum efficacy and autoimmune reactions are not established although clinical response is rarely observed at doses less than 3 mg/kg. Improved responses and increased immune toxicity are observed more frequently with multiple doses. Second, clinical responses take weeks if not months to occur, requiring an unusually long time window to evaluate efficacy. Third, autoimmune reactions clearly correlate with clinical efficacy. Fourth, persistent immune infiltrates in the lesions can last months and, therefore, response criteria based only on imaging techniques may fail to fully define efficacy. In a trial combining a previously described gp100 peptide vaccine (Rosenburg et al., 1998), anti-CTLA-4 mAb did not increase the frequency of T cells responding to this antigen even in cases with objective responses, contrary to the expectations from mouse studies. Tregs in peripheral blood are not significantly affected in number or function in treated patients. Thus, there is not yet any useful immune parameter to correlate with efficacy other than tumor infiltration by immune system cells (Reuben et al., 2006). It is still unclear if the anti-CTLA-4 antibodies amplify a latent weak preexisting T lymphocyte response or facilitate activation of naive T cells. Indeed, there are no reports of the antigen specificity of T cells infiltrating the tumors of patients who respond to this treatment.

The clinical responses to anti-CTLA-4 represent an important proof-of-principle regarding the potential efficacy of the checkpoint blockade. The high rate of serious immune toxicities also highlights the potential concerns of inducing immune dysregulation with these agents. As suggested by the preclinical studies previously described, it is possible that combination of anti-CTLA-4 with vaccination might enhance the antitumor effect without increased accompanying toxicity. A recent phase I clinical trial evaluated the effects of an allogeneic prostate cancer GVAX® vaccine (consisting of two irradiated, GM-CSF-transduced human prostate cancer lines) together with increasing doses of anti-CTLA-4 in men with advanced, hormone refractory prostate cancer (Gerritsen et al., 2006). No clinical responses were observed in patients treated with GVAX (every 2 weeks × 12 doses) plus anti-CTLA-4 (every 4 weeks × 6 doses) at the lower dose levels (0.1, 0.3, and 1.0 mg/kg). However, at the higher doses of antibody (3 and 5 mg/kg), five or six patients demonstrated PSA responses of greater than 50% decrease, with two patients' PSA levels dropping to within normal limits. A number of patients had concurrent tumor regressions on bone scan and CT scan. Two of the responses did not occur until more than 5 months after initiation of therapy. Each of the five responding patients demonstrated immune toxicities, with four displaying autoimmune hypophysitis and one experiencing alveolitis. This finding reveals a very different distribution of affected organs than observed in the other anti-CTLA-4 trials and may be due to the fact that all of the patients remained on leutinizing hormone-releasing hormone (LHRH) inhibitors. Nonetheless, since the response rates to GVAX® alone and anti-CTLA-4 alone in this patient population are less than 5% and approximately 10% respectively, the 80% response rate at the higher dose levels suggests a synergistic clinical effect. Larger studies will be necessary to confirm this conclusion.

B. PD-1

PD-1 is another coinhibitory receptor that shows homology with CD28 and whose expression is induced upon activation on $CD4^+$ and $CD8^+$ T cells, B cells, and monocytes. This surface receptor has two known ligands, B7-H1 (PD-L1) and B7-DC (PD-L2) (Dong et al., 1999; Freeman et al., 2000; Tseng et al., 2001; Latchman et al., 2001). PD-1 ligation causes inhibition of T cell activation and proliferation, causing cell cycle arrest without apoptosis. The phenotype of PD-1$^{-/-}$ mice is characterized by cell-mediated organ-specific autoimmunity (Nishimura et al., 2001), consistent with the inhibitory function of PD1 in regulating TCR-CD28-dependent T cell activation. In contrast to CTLA-4$^{-/-}$ mice, which develop lethal multiorgan autoimmunity or hyperimmunity within 3 weeks, PD-1$^{-/-}$ mice generally develop more restricted autoimmune syndromes that are strain dependent and do not occur until beyond 6 months of age. Thus, PD-1 appears to be a more subtle immune checkpoint than CTLA-4.

B7-H1 expression has been described on a variety of human tumor cells (Dong et al., 2002), and expression of this molecule decreases the immunogenicity of mouse tumors in vivo (Dong et al., 2002). In addition, expression of B7-H1 in human malignancies correlates with a poor prognosis (Thompson et al., 2006). B7-DC can deliver costimulatory or coinhibitory signals to primed lymphocytes, depending on the culture conditions. Its expression is highly restricted to activated DCs and is thought to induce T cell costimulation via an unidentified counter-receptor expressed on activated T cells.

Administration of mAbs against PD-1 and B7-H1 has produced CTL-mediated antitumor effects in mice (Hirano et al., 2005). B7-H1 interferes with the effector phase of CTL killing of malignant cells. Interest in this target is fueled by the finding in mice chronically infected by lymphocytic choriomeningitis virus (LCMV) of CD8 clonal exhaustion toward the viral antigens. Exhausted cells expressed high levels of PD-1; blockade of both PD-1 and B7-H1 with mAbs reversed this phenotype in vivo, and the mice regained T cell effective antiviral activity (Barber et al., 2006). No data have been reported on the reversion of clonal exhaustion in T cells in tumor-bearing mice. Phase I clinical trials have been initiated in cancer patients with a fully human anti-PD-1 mAb.

V. WHAT IS THE MOST EFFECTIVE WAY TO USE CHECKPOINT INHIBITORS?

The explosion of information on immune regulatory receptors, ligands, signaling pathways, and cells provides as much challenge as opportunity in the immunotherapy

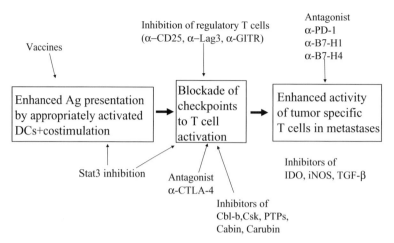

FIGURE 14.4 **The lack of one magic bullet for successful cancer immunotherapy.** It is unlikely that a single intervention will provide successful immunotherapy of established cancers. Three points of intervention are proposed: initiation of the immune response, blockade of checkpoints to T cell activation, and enhanced traffic and function of antitumor T cells to and in the tumor. A number of potential targets discussed in the chapter are listed in this figure.

of cancer. Can inhibitors of these checkpoints be used as single agents, releasing a natural endogenous antitumor immunity? Can they be combined with approved agents or procedures such as radiation therapy, as has been shown preclinically with anti-CTLA-4? Are they best used in conjunction with vaccines? Which pathways are complementary and which are redundant? All of these questions remain to be answered. The oncology field's understanding of the generation and execution of an effective immune response as well as the multiple levels of immune inhibition operative in the setting of cancer suggests that successful immune therapy is a multistep process. It is therefore likely that combinatorial approaches will be most effective (Figure 14.4). Preclinical experiments described in this chapter as well as early clinical experience with anti-CTLA-4 supports this prediction. However, cost and regulatory and feasibility issues dictate that these new agents be initially tested as single agents, hopefully in settings that can reveal activity, even if suboptimal. From there, stepwise development of combinatorial approaches must be contemplated prior to marketing approval of agents as monotherapy. Combinations with vaccines are reasonable because vaccines display little toxicity on their own, and many cancer vaccines with a strong safety record have demonstrated the capacity to induce T cell responses without showing significant clinical benefit as monotherapy. The immune toxicities of anti-CTLA-4 have created concern; however, the rapidly lethal phenotype of knockout mice for CTLA-4 compared with other checkpoint molecules (such as PD-1, B7-H1, and A2aR) suggests that blockade of other checkpoints may carry far lower immune side effects. The next 5 years will truly thrust the field into a new world of immune regulation not previously experienced.

References

Acuto, O., and Michel, F. (2003). CD28-mediated co-stimulation: a quantitative support for TCR signalling. *Nat. Rev. Immunol.* **3**, 939–951.

Alegre, M. L., Frauwirth, K. A., and Thompson, C. B. (2001). T-cell regulation by CD28 and CTLA-4. *Nat. Rev. Immunol.* **1**, 220–228.

Alexander, W. S., and Hilton, D. J. (2004). The role of suppressors of cytokine signaling (SOCS) proteins in regulation of the immune response. *Annu. Rev. Immunol.* **22**, 503–529.

Almand, B., Resser, J. R., Lindman, B., Nasaf, S., Clark, J. I., Kwon, E. D., Carbone, D. P., and Gabrilovich,

D. I. (2000). Clinical significance of defective dendritic cell differentiation in cancer. *Clin. Cancer Res.* **6**, 1755–1766.

Attia, P., Phan, G. Q., Maker, A. V., Robinson, M. R., Quezado, M. M., Yang, J. C., Sherry, R. M., Topalian, S. L., Kammula, U. S., Royal, R. E., Restifo, N. P., Haworth, L. R., Levy, C., Mavroukakis, S. A., Nichol, G., Yellin, M. J., and Rosenberg, S. A. (2005). Autoimmunity correlates with tumor regression in patients with metastatic melanoma treated with anti-cytotoxic T-lymphocyte antigen-4. *J. Clin. Oncol.* **23**, 6043–6053.

Baban, B., Hansen, A. M., Chandler, P. R., Manlapat, A., Bingaman, A., Kahler, D. J., Munn, D. H., and Mellor, A. L. (2005). A minor population of splenic dendritic cells expressing CD19 mediates IDO-dependent T cell suppression via type I IFN signaling following B7 ligation. *Int. Immunol.* **17**, 909–919.

Bachmaier, K., Krawczyk, C., Kozieradzki, I., Kong, Y. Y., Sasaki, T., Oliveira-dos-Santos, A., Mariathasan, S., Bouchard, D., Wakeham, A., Itie, A., Le, J., Ohashi, P. S., Sarosi, I., Nishina, H., Lipkowitz, S., and Penninger, J. M. (2000). Negative regulation of lymphocyte activation and autoimmunity by the molecular adaptor Cbl-b. *Nature* **403**, 211–216.

Banchereau, J., and Steinman, R. M. (1998). Dendritic cells and the control of immunity. *Nature* **392**, 245–252.

Barber, D. L., Wherry, E. J., Masopust, D., Zhu, B., Allison, J. P., Sharpe, A. H., Freeman, G. J., and Ahmed, R. (2006). Restoring function in exhausted CD8 T cells during chronic viral infection. *Nature* **439**, 682–687.

Blansfield, J. A., Beck, K. E., Tran, K., Yang, J. C., Hughes, M. S., Kammula, U. S., Royal, R. E., Topalian, S. L., Haworth, L. R., Levy, C., Rosenberg, S. A., and Sherry, R. M. (2005). Cytotoxic T-lymphocyte-associated antigen-4 blockage can induce autoimmune hypophysitis in patients with metastatic melanoma and renal cancer. *J. Immunother.* **28**, 593–598.

Blobe, G. C., Schiemann, W. P., and Lodish, H. F. (2000). Role of transforming growth factor beta in human disease. *N. Engl. J. Med.* **342**, 1350–1358.

Bowman, T., Broome, M. A., Sinibaldi, D., Wharton, W., Pledger, W. J., Sedivy, J. M., Irby, R., Yeatman, T., Courtneidge, S. A., and Jove, R. (2001). Stat3-mediated Myc expression is required for Src transformation and PDGF-induced mitogenesis. *Proc. Natl. Acad. Sci. USA* **98**, 7319–7324.

Bromberg, J. F., Wrzeszczynska, M. H., Devgan, G., Zhao, Y., Pestell, R. G., Albanese, C., and Darnell, J. E., Jr. (1999). Stat3 as an oncogene. *Cell* **98**, 295–303.

Bronte, V., Serafini, P., De Santo, C., Marigo, I., Tosello, V., Mazzoni, A., Segal, D. M., Staib, C., Lowel, M., Sutter, G., Colombo, M. P., and Zanovello, P. (2003). IL-4-induced arginase 1 suppresses alloreactive T cells in tumor-bearing mice. *J. Immunol.* **170**, 270–278.

Bronte, V., Apolloni, E., Cabrelle, A., Ronca, R., Serafini, P., Zamboni, P., Restifo, N. P., and Zanovello, P. (2000). Identification of a CD11b($^+$)/Gr-1($^+$)/CD31($^+$) myeloid progenitor capable of activating or suppressing CD8($^+$) T cells. *Blood* **96**, 3838–3846.

Chan, A. C., Desai, D. M., and Weiss, A. (1994). The role of protein tyrosine kinases and protein tyrosine phosphatases in T cell antigen receptor signal transduction. *Annu. Rev. Immunol.* **12**, 555–592.

Chen, L. (2004). Co-inhibitory molecules of the B7-CD28 family in the control of T-cell immunity. *Nat. Rev. Immunol.* **4**, 336–347.

Croft, M. (2003a). Costimulation of T cells by OX40, 4-1BB, and CD27. *Cytokine Growth Factor Rev.* **14**, 265–273.

Croft, M. (2003b). Co-stimulatory members of the TNFR family: Keys to effective T-cell immunity? *Nat. Rev. Immunol.* **3**, 609–620.

Cui, W., Fowlis, D. J., Bryson, S., Duffie, E., Ireland, H., Balmain, A., and Akhurst, R. J. (1996). TGFbeta1 inhibits the formation of benign skin tumors, but enhances progression to invasive spindle carcinomas in transgenic mice. *Cell* **86**, 531–542.

Curiel, T. J., Coukos, G., Zou, L., Alvarez, X., Cheng, P., Mottram, P., Evdemon-Hogan, M., Conejo-Garcia, J. R., Zhang, L., Burow, M., Zhu, Y., Wei, S., Kryczek, I., Daniel, B., Gordon, A., Myers, L., Lackner, A., Disis, M. L., Knutson, K. L., Chen, L., and Zou, W. (2004). Specific recruitment of regulatory T cells in ovarian carcinoma fosters immune privilege and predicts reduced survival. *Nat. Med.* **10**, 942–949.

Dannull, J., Su, Z., Rizzieri, D., Yang, B. K., Coleman, D., Yancey, D., Zhang, A., Dahm, P., Chao, N., Gilboa, E., and Vieweg, J. (2005). Enhancement of vaccine-mediated antitumor immunity in cancer patients after depletion of regulatory T cells. *J. Clin. Invest.* **115**, 3623–3633.

Dong, H., Strome, S. E., Salomao, D. R., Tamura, H., Hirano, F., Flies, D.B., Roche, P.C., Lu, J., Zhu, G., Tamada, K., Lennon, V. A., Celis, E., and Chen, L. (2002). Tumor-associated B7-H1 promotes T-cell apoptosis: a potential mechanism of immune evasion. *Nat. Med.* **8**, 793–800.

Dong, H., Zhu, G., Tamada, K., and Chen, L. (1999). B7-H1, a third member of the B7 family, co-stimulates T-cell proliferation and interleukin-10 secretion. *Nat. Med.* **5**, 1365–1369.

Ercolini, A., Ladle, E. H., Manning, E. A., Pfannenstiel, L. W., Armstrong, T., Machiels, J.-P., Bieler, J. G., Emens, L. A., Reilly, T., and Jaffee, E. (2005). Recruitment of latent pools of high-avidity CD8($^+$) T cells to the antitumor immune response. *J. Exp. Med.* **201**, 1591–1602.

Fallarino, F., Grohmann, U., Hwang, K. W., Orabona, C., Vacca, C., Bianchi, R., Belladonna, M. L., Fioretti,

M. C., Alegre, M. L., and Puccetti, P. (2003). Modulation of tryptophan catabolism by regulatory T cells. *Nat. Immunol.* **4**, 1206–1212.

Fan Pan, L. S., Kardian, D. B., Whartenby, K. A., Pardoll, D. M., and Liu, J. O. (2007). Feedback inhibition of calcineurin and Ras by a dual inhibitory protein Carabin. *Nature*, in press.

Fontenot, J. D., Gavin, M. A., and Rudensky, A. Y. (2003). Foxp3 programs the development and function of CD4$^+$CD25$^+$ regulatory T cells. *Nat. Immunol.* **4**, 330–336.

Freeman, G. J., Long, A. J., Iwai, Y., Bourque, K., Chernova, T., Nishimura, H., Fitz, L. J., Malenkovich, N., Okazaki, T., Byrne, M. C., Horton, H. F., Fouser, L., Carter, L., Ling, V., Bowman, M. R., Carreno, B. M., Collins, M., Wood, C. R., and Honjo, T. (2000). Engagement of the PD-1 immunoinhibitory receptor by a novel B7 family member leads to negative regulation of lymphocyte activation. *J. Exp. Med.* **192**, 1027–1034.

Gascoigne, N. R., and Zal, T. (2004). Molecular interactions at the T cell-antigen-presenting cell interface. *Curr. Opin. Immunol.* **16**, 114–119.

Gerritsen, W., van den Eertwegh, A., De Gruijl. T., Giaccone, G., Scheper, R., Lowy, I., *et al.* (2006). A dose-escalation trial of GM-CSF-gene transduced allogeneic prostate cancer cellular immunotherapy in combination with a fully human anti-CTLA antibody (MDX-010, ipilimumab) in patients with metastatic hormone-refractory prostate cancer (mHRPC). *J. Clin. Oncol. ASCO Annu. Meet. Pro. Part I. 2006* **24 Suppl June 20**, 2500.

Greenwald, R. J., Freeman, G. J., and Sharpe, A. H. (2005). The B7 family revisited. *Annu. Rev. Immunol.* **23**, 515–548.

Hasegawa, Y., Takanashi, S., Kanehira, Y., Tsushima, T., Imai, T., and Okumura, K. (2001). Transforming growth factor-beta1 level correlates with angiogenesis, tumor progression, and prognosis in patients with nonsmall-cell lung carcinoma. *Cancer* **91**, 964–971.

Hawiger, D., Inaba, K., Dorsett, Y., Guo, M., Mahnke, K., Rivera, M., Ravetch, J. V., Steinman, R. M., and Nussenzweig, M. C. (2001). Dendritic cells induce peripheral T cell unresponsiveness under steady state conditions *in vivo*. *J. Exp. Med.* **194**, 769–779.

Hirano, F., Kaneko, K., Tamura, H., Dong, H., Wang, S., Ichikawa, M., Rietz, C., Flies, D. B., Lau, J. S., Zhu, G., Tamada, K., and Chen, L. (2005). Blockade of B7-H1 and PD-1 by monoclonal antibodies potentiates cancer therapeutic immunity. *Cancer Res.* **65**, 1089–1096.

Hodi, F. S., and Dranoff, G. (2006). Combinatorial cancer immunotherapy. *Adv. Immunol.* **90**, 341–368.

Hodi, F. S., Mihm, M. C., Soiffer, R. J., Haluska, F. G., Butler, M., Seiden, M. V., Davis, T., Henry-Spires, R., MacRae, S., Willman, A., Padera, R., Jaklitsch, M. T., Shankar, S., Chen, T. C., Korman, A., Allison, J. P., and Dranoff, G. (2003). Biologic activity of cytotoxic T lymphocyte-associated antigen 4 antibody blockade in previously vaccinated metastatic melanoma and ovarian carcinoma patients. *Proc. Natl. Acad. Sci. USA.* **100**, 4712–4717.

Hori, S., Nomura, T., and Sakaguchi, S. (2003). Control of regulatory T cell development by the transcription factor Foxp3. *Science* **299**, 1057–1061.

Huang, C. T., Workman, C. J., Flies, D., Pan, X., Marson, A. L., Zhou, G., Hipkiss, E. L., Ravi, S., Kowalski, J., Levitsky, H. I., Powell, J. D., Pardoll, D. M., Drake, C. G., and Vignali, D. A. (••). Role of LAG-3 in regulatory T cells. *Immunity* **21**, 503–513.

Hurwitz, A. A., Foster, B. A., Kwon, E. D., Truong, T., Choi, E. M., Greenberg, N. M., Burg, M. B., and Allison, J. P. (2000). Combination immunotherapy of primary prostate cancer in a transgenic mouse model using CTLA-4 blockade. *Cancer Res.* **60**, 2444–2448.

Ilangumaran, S., Ramanathan, S., and Rottapel, R. (2004). Regulation of the immune system by SOCS family adaptor proteins. *Semin. Immunol.* **16**, 351–365.

Ishida, Y., Agata, Y., Shibahara, K., and Honjo, T. (1992). Induced expression of PD-1, a novel member of the immunoglobulin gene superfamily, upon programmed cell death. *EMBO J.* **11**, 3887–3895.

Ito, D., Ogasawara, K., Iwabuchi, K., Inuyama, Y., and Onoe, K. (2000). Induction of CTL responses by simultaneous administration of liposomal peptide vaccine with anti-CD40 and anti-CTLA-4 mAb. *J. Immunol.* **164**, 1230–1235.

Keler, T., Halk, E., Vitale, L., O'Neill, T., Blanset, D., Lee, S., Srinivasan, M., Graziano, R. F., Davis, T., Lonberg, N., and Korman, A. (2003). Activity and safety of CTLA-4 blockade combined with vaccines in cynomolgus macaques. *J. Immunol.* **171**, 6251–6259.

Khattri, R., Cox, T., Yasayko, S. A., and Ramsdell, F. (2003). An essential role for Scurfin in CD4$^+$CD25$^+$ T regulatory cells. *Nat. Immunol.* **4**, 337–342.

Ko, K., Yamazaki, S., Nakamura, K., Nishioka, T., Hirota, K., Yamaguchi, T., Shimizu, J., Nomura, T., Chiba, T., and Sakaguchi, S. (2005). Treatment of advanced tumors with agonistic anti-GITR mAb and its effects on tumor-infiltrating Foxp3$^+$CD25$^+$CD4$^+$ regulatory T cells. *J. Exp. Med.* **202**, 885–891.

Kortylewski, M., *et al.* (2005). Inhibiting Stat3 signalling in the hematopoietic system elicits multicomponent antitumor immunity. *Nat. Med.* **11**, 1314–1321.

Krawczyk, C., Bachmaier, K., Sasaki, T., Jones, R. G., Snapper, S. B., Bouchard, D., Kozieradzki, I., Ohashi, P. S., Alt, F. W., and Penninger, J. M. (2000). Cbl-b is a negative regulator of receptor clustering and raft aggregation in T cells. *Immunity* **13**, 463–473.

Krogsgaard, M., and Davis, M. M. (2005). How T cells "see" antigen. *Nat. Immunol.* **6**, 239–245.

Kryczek, I., Zou, L., Rodriguez, P., Zhu, G., Wei, S., Mottram, P., Brumlik, M., Cheng, P., Curiel, T., Myers, L., Lackner, A., Alvarez, X., Ochoa, A., Chen, L., and Zou, W. (2006). B7-H4 expression identifies a novel suppressive macrophage population in human ovarian carcinoma. *J. Exp. Med.* **203**, 871–881.

Kusmartsev, S., and Gabrilovich, D. I. (2005). Role of immature myeloid cells in mechanisms of immune evasion in cancer. *Cancer Immunol. Immunother.* **55**, 1–9.

Latchman, Y., Wood, C. R., Chernova, T., Chaudhary, D., Borde, M., Chernova, I., Iwai, Y., Long, A. J., Brown, J. A., Nunes, R., Greenfield, E. A., Bourque, K., Boussiotis, V. A., Carter, L. L., Carreno, B. M., Malenkovich, N., Nishimura, H., Okazaki, T., Honjo, T., Sharpe, A. H., and Freeman, G. J. (2001). PD-L2 is a second ligand for PD-1 and inhibits T cell activation. *Nat. Immunol.* **2**, 261–268.

Leach, D. R., Krummel, M. F., and Allison, J. P. (1996). Enhancement of antitumor immunity by CTLA-4 blockade. *Science* **271**, 1734–1736.

Letterio, J. J., and Roberts, A. B. (1998). Regulation of immune responses by TGF-beta. *Annu. Rev. Immunol.* **16**, 137–161.

Liew, F. Y., Xu, D., Brint E. K., and O'Neill, L. A. (2005). Negative regulation of Toll-like receptor-mediated immune responses. *Nat. Rev. Immunol.* **5**, 446–458.

Long, E. O. (1999). Regulation of immune responses through inhibitory receptors. *Annu. Rev. Immunol.* **17**, 875–904.

Maker, A. V., Attia, P., and Rosenberg, S. A. (2005). Analysis of the cellular mechanism of antitumor responses and autoimmunity in patients treated with CTLA-4 blockade. *J. Immunol.* **175**, 7746–7754.

Maker, A. V., Phan, G. Q., Attia, P., Yang, J. C., Sherry, R. M., Topalian, S. L., Kammula, U. S., Royal, R. E., Haworth, L. R., Levy, C., Kleiner, D., Mavroukakis, S. A., Yellin, M., and Rosenberg, S. A. (2005). Tumor regression and autoimmunity in patients treated with cytotoxic T lymphocyte-associated antigen 4 blockade and interleukin 2: A phase I/II study. *Ann. Surg. Oncol.* **12**, 1005–1016..

Maloy, K., and Powrie, F. (2001). Regulatory T cells in the control of immune pathology. *Nat. Immunol.* **2**, 816–822.

Mazzoni, A., Bronte, V., Visintin, A., Spitzer, J. H., Apolloni, E., Serafini, P., Zanovello, P., and Segal, D. M. (2002). Myeloid suppressor lines inhibit T cell responses by an NO-dependent mechanism. *J. Immunol.* **168**, 689–695.

McHugh, R. S., Whitters, M. J., Piccirillo, C. A., Young, D. A., Shevach, E. M., Collins, M., and Byrne, M. C. (2002). CD4(+)CD25(+) immunoregulatory T cells: Gene expression analysis reveals a functional role for the glucocorticoid-induced TNF receptor. *Immunity* **16**, 311–323.

Mellor, A. L., Chandler, P., BaLan, B., Hansen, A. M., Marshall, B., Pihkala, J., Waldmann, H., Cobbold, S., Adams, E., and Munn, D. H. (2004). Specific subsets of murine dendritic cells acquire potent T cell regulatory functions following CTLA4-mediated induction of indoleamine 2,3 dioxygenase. *Int. Immunol.* **16**, 1391–1401.

Munn, D. H., Sharma, M. D., Hou, D., Baban, B., Lee, J. R., Antonia, S. J., Messina, J. L., Chandler, P., Koni, P. A., and Mellor, A. L. (2004). Expression of indoleamine 2,3-dioxygenase by plasmacytoid dendritic cells in tumor-draining lymph nodes. *J. Clin. Invest.* **114**, 280–290.

Munn, D. H., Sharma, M. D., Lee, J. R., Jhaver, K. G., Johnson, T. S., Keskin, D. B., Marshall, B., Nomura, T., Toda, M., and Takahashi, T. (2002). Potential regulatory function of human dendritic cells expressing indoleamine 2,3-dioxygenase. *Science* **297**, 1867–1870.

Naka, T., Fujimoto, M., Tsutsui, H., and Yoshimura, A. (2005). Negative regulation of cytokine and TLR signalings by SOCS and others. *Adv. Immunol.* **87**, 61–122.

Nefedova, Y., Nagaraj, S., Rosenbauer, A., Muro-Cacho, C., Sebti, S. M., and Gabrilovich, D. I. (2005). Regulation of dendritic cell differentiation and antitumor immune response in cancer by pharmacologic-selective inhibition of the janus-activated kinase 2/signal transducers and activators of transcription 3 pathway. *Cancer Res.* **65**, 9525–9535.

Nefedova, Y., Huang, M., Kusmartsev, S., Bhattacharya, R., Cheng, P., Salup, R., Jove, R., and Gabrilovich, D. (2004). Hyperactivation of STAT3 is involved in abnormal differentiation of dendritic cells in cancer. *J. Immunol.* **172**, 464–474.

Nishimura, H., Okazaki, T., Tanaka, Y., Nakatani, K., Hara, M., Matsumori, A., Sasayama, S., Mizoguchi, A., Hiai, H., Minato, N., and Honjo, T. (2001). Autoimmune dilated cardiomyopathy in PD-1 receptor-deficient mice. *Science* **291**, 319–322.

Ohta, A., Gorelik, E., Prasad, S. J., Ronchese, F., Lukashev, D., Wong, M. K., Huang, X., Caldwell, S., Liu, K., Smith, P., Chen, J. F., Jackson, E. K., Apasov, S., Abrams, S., and Sitkovsky, M. (2006). A2A adenosine receptor protects tumors from antitumor T cells. *Proc. Natl. Acad. Sci. USA* **103**, 13132–13137.

Pardoll, D. M. (2002). Spinning molecular immunology into successful immunotherapy. *Nat. Rev. Immunol.* **2**, 227–238.

Pedersen, A. E., Buus, S., and Claesson, M. H. (2006). Treatment of transplanted CT26 tumour with dendritic cell vaccine in combination with blockade of vascular endothelial growth factor receptor 2 and CTLA-4. *Cancer Lett.* **235**, 229–238.

Pepper, M. S. (1997). Transforming growth factor-beta: Vasculogenesis, angiogenesis, and vessel wall integrity. *Cytokine Growth Factor Rev.* **8**, 21–43.

Phan, G. Q., Yang, J. C., Sherry, R. M., Hwu, P., Topalian, S. L., Schwartzentruber, D. J., Restifo, N. P., Haworth, L. R., Seipp, C. A., Freezer, L. J., Morton,

K. E., Mavroukakis, S. A., Duray, P. H., Steinberg, S. M., Allison, J. P., Davis, T. A., and Rosenberg, S. A. (2003). Cancer regression and autoimmunity induced by cytotoxic T lymphocyte-associated antigen 4 blockade in patients with metastatic melanoma. *Proc. Natl. Acad. Sci. USA* **100**, 8372–8377.

Reis e Sousa, C. (2006). Dendritic cells in a mature age. *Nat. Rev. Immunol.* **6**, 476–483.

Reuben, J. M., Lee, B. N., Li, C., Gomez-Navarro, J., Bozon, V. A., Parker, C. A., Hernandez, I. M., Gutierrez, C., Lopez-Berestein, G., and Camacho, L. H. (2006). Biologic and immunomodulatory events after CTLA-4 blockade with ticilimumab in patients with advanced malignant melanoma. *Cancer* **106**, 2437–2444.

Ribas, A., Glaspy, J. A., Lee, Y., Dissette, V. B., Seja, E., Vu, H. T., Tchekmedyian, N. S., Oseguera, D., Comin-Anduix, B., Wargo, J. A., Amarnani, S. N., McBride, W. H., Economou, J. S., and Butterfield, L. H. (2004). Role of dendritic cell phenotype, determinant spreading, and negative costimulatory blockade in dendritic cell-based melanoma immunotherapy. *J. Immunother.* **27**, 354–367.

Ribas, A., et al. (2005). Antitumor activity in melanoma and anti-self responses in a phase I trial with the anti-cytotoxic T lymphocyte-associated antigen 4 monoclonal antibody CP-675,206. *J. Clin. Oncol.* **23**, 8968–8977.

Robinson, M. R., Chan, C. C., Yang, J. C., Rubin, B. I., Gracia, G. J., Sen, H. N., Csaky, K. G., and Rosenberg, S. A. (2004). Cytotoxic T lymphocyte-associated antigen 4 blockade in patients with metastatic melanoma: A new cause of uveitis. *J. Immunother.* **27**, 478–479.

Rosenberg, S. A., Yang, J. C., Schwartzentruber, D. J., Hwu, P., Marincola, F. M., Topalian, S. L., Restifo, N. P., Dudley, M. E., Schwarz, S. L., Spiess, P. J., Wunderlich, J. R., Parkhurst, M. R., Kawakami, Y., Seipp, C. A., Einhorn, J. H., and White, D. E. (1998). Immunologic and therapeutic evaluation of a synthetic peptide vaccine for the treatment of patients with metastatic melanoma. *Nat. Med.* **4**, 321–327.

Saito, H., Tsujitani, S., Oka, S., Kondo, A., Ikeguchi, M., Maeta, M., and Kaibara, N. (1999). The expression of transforming growth factor-beta1 is significantly correlated with the expression of vascular endothelial growth factor and poor prognosis of patients with advanced gastric carcinoma. *Cancer* **86**, 1455–1462.

Sakaguchi, S., Sakaguchi, N., Shimizu, J., Yamazaki, S., Sakihama, T., Itoh, M., Kuniyasu, Y., Nomura, T., Toda, M., and Takahashi, T. (2001). Immunologic tolerance maintained by CD35$^+$CD4$^+$ regulatory cells: Their common role in controlling autoimmunity, tumor immunity, and transplantation tolerance. *Immunol. Rev.* **182**, 18–32.

Sanderson, K., Scotland, R., Lee, P., Liu, D., Groshen, S., Snively, J., Sian, S., Nichol, G., Davis, T., Keler, T., Yellin, M., and Weber, J. (2005). Autoimmunity in a phase I trial of a fully human anti-cytotoxic T-lymphocyte antigen-4 monoclonal antibody with multiple melanoma peptides and Montanide ISA 51 for patients with resected stages III and IV melanoma. *J. Clin. Oncol.* **23**, 741–750.

Schmielau, J., and Finn, O. J. (2001). Activated granulocytes and granulocyte-derived hydrogen peroxide are the underlying mechanism of suppression of T-cell function in advanced cancer patients. *Cancer Res.* **61**, 4756–4760.

Serafini, P., et al. Enhanced antitumor immunity after pharmacologic inhibition of myeloid suppressor cell arginase with Viagra. *J. Exp. Med.*, in press.

Shariat, S. F., et al. (2001). Preoperative plasma levels of transforming growth factor beta(1) (TGF-beta(1)) strongly predict progression in patients undergoing radical prostatectomy. *J. Clin. Oncol.* **19**, 2856–2864.

Shevach, E. (2002). CD4$^+$ CD25$^+$ supppressor T cells: More questions than answers. *Nat. Rev. Immunol.* **2**, 389–400.

Shimizu, J., Yamazaki, S., Takahashi, T., Ishida, Y., and Sakaguchi, S. (2002). Stimulation of CD25($^+$)CD4($^+$) regulatory T cells through GITR breaks immunological self-tolerance. *Nat. Immunol.* **3**, 135–142.

Sica, G. L., Choi, I. H., Zhu, G., Tamada, K., Wang, S. D., Tamura, H., Chapoval, A. I., Flies, D. B., Bajorath, J., and Chen, L. (2003). B7-H4, a molecule of the B7 family, negatively regulates T cell immunity. *Immunity* **18**, 849–861.

Simon, I., Zhuo, S., Corral, L., Diamandis, E. P., Sarno, M. J., Wolfert, R. L., and Kim, N. W. (2006). B7-h4 is a novel membrane-bound protein and a candidate serum and tissue biomarker for ovarian cancer. *Cancer Res.* **66**, 1570–1575.

Sitkovsky, M. V., Lukashev, D., Apasov, S., Kojima, H., Koshiba, M., Caldwell, C., Ohta, A., and Thiel, M. (2004). Physiological control of immune response and inflammatory tissue damage by hypoxia-inducible factors and adenosine A2A receptors. *Annu. Rev. Immunol.* **22**, 657–682.

Stephens, G. L., McHugh, R. S., Whitters, M. J., Young, D. A., Luxenberg, D., Carreno, B. M., Collins, M., and Shevach, E. M. (2004). Engagement of glucocorticoid-induced TNFR family-related receptor on effector T cells by its ligand mediates resistance to suppression by CD4$^+$CD25$^+$ T cells. *J. Immunol.* **173**, 5008–5020.

Sun, L., Youn, H. D., Loh, C., Stolow, M., He, W., and Liu, J. O. (1998). Cabin 1, a negative regulator for calcineurin signaling in T lymphocytes. *Immunity* **8**, 703–711.

Sutmuller, R. P., van Duivenvoorde, L. M., van Elsas, A., Schumacher, T. N., Wildenberg, M. E., Allison, J. P., Toes, R. E., Offringa, R., and Melief, C. J. (2001).

Synergism of cytotoxic T lymphocyte-associated antigen 4 blockade and depletion of CD25(+) regulatory T cells in antitumor therapy reveals alternative pathways for suppression of autoreactive cytotoxic T lymphocyte responses. *J. Exp. Med.* **194**, 823–832.

Teft, W. A., Kirchhof, M. G., and Madrenas, J. A. (2006). Molecular perspective of CTLA-4 function. *Annu. Rev. Immunol.* **24**, 65–97.

Thompson, R. H., Kuntz, S. M., Leibovich, B. C., Dong, H., Lohse, C. M., Webster, W. S., Sengupta, S., Frank, I., Parker, A. S., Zincke, H., Blute, M. L., Sebo, T. J., Cheville, J. C., and Kwon, E. D. (2006). Tumor B7-H1 is associated with poor prognosis in renal cell carcinoma patients with long-term follow-up. *Cancer Res.* **66**, 3381–3385.

Tirapu, I., Huarte, E., Guiducci, C., Arina, A., Zaratiegui, M., Murillo, O., Gonzalez, A., Berasain, C., Berraondo, P., Fortes, P., Prieto, J., Colombo, M. P., Chen, L., and Melero, I. (2006). Low surface expression of B7–1 (CD80) is an immunoescape mechanism of colon carcinoma. *Cancer Res.* **66**, 2442–2450.

Tivol, E. A., et al. (1995). Loss of CTLA-4 leads to massive lymphoproliferation and fatal multiorgan tissue destruction, revealing a critical negative regulatory role of CTLA-4. *Immunity* **3**, 541–547.

Tringler, B., Liu, W., Corral, L., Torkko, K. C., Enomoto, T., Davidson, S., Lucia, M. S., Heinz, D. E., Papkoff, J., and Shroyer, K. R. (2006). B7-H4 overexpression in ovarian tumors. *Gynecol. Oncol.* **100**, 44–52.

Tseng, S. Y., Otsuji, M., Gorski, K., Huang, X., Slansky, J. E., Pai, S. I., Shalabi, A., Shin, T., Pardoll, D. M., and Tsuchiya, H. (2001). B7-DC, a new dendritic cell molecule with potent costimulatory properties for T cells. *J. Exp. Med.* **193**, 839–846.

Turkson, J., Zhang, S., Mora, L. B., Burns, A., Sebti, S., Jove, R. (2005). A novel platinum compound inhibits constitutive Stat3 signaling and induces cell cycle arrest and apoptosis of malignant cells. *J. Biol. Chem.* **280**, 32979–32988.

Turkson, J., Bowman, T., Garcia, R., Caldenhoven, E., De Groot, R. P., and Jove, R. (1998). Stat3 activation by Src induces specific gene regulation and is required for cell transformation. *Mol. Cell Biol.* **18**, 2545–2552.

van Elsas, A., Hurwitz, A. A., and Allison, J. P. (1999). Combination immunotherapy of B16 melanoma using anti-cytotoxic T lymphocyte-associated antigen 4 (CTLA-4) and granulocyte/macrophage colony-stimulating factor (GM-CSF)-producing vaccines induces rejection of subcutaneous and metastatic tumors accompanied by autoimmune depigmentation. *J. Exp. Med.* **190**, 355–366.

Vang, T., Abrahamsen, H., Myklebust, S., Enserink, J., Prydz, H., Mustelin, T., Amarzguioui, M., and Tasken, K. (2004). Knockdown of C-terminal Src kinase by siRNA-mediated RNA interference augments T cell receptor signaling in mature T cells. *Eur. J. Immunol.* **34**, 2191–2199.

Vicari, A. P., Caux, C., and Trinchieri, G. (2002). Tumour escape from immune surveillance through dendritic cell inactivation. *Semin. Cancer Biol.* **12**, 33–42.

Viewag, J., Su, Z., Dahm, P., and Kusmartsev, R. (2007). Reversal of tumor-mediated immunosuppression. *Clin. Cancer Res.* **13**, 727–735.

Wang, T., Niu, G., Kortylewski, M., Burdelya, L., Shain, K., Zhang, S., Bhattacharya, R., Gabrilovich, D., Heller, R., Coppola, D., Dalton, W., Jore, R., Pardoll, D., and Yu, H. (2004). Regulation of the innate and adaptive immune responses by Stat-3 signaling in tumor cells. *Nat. Med.* **10**, 48–54.

Waterhouse, P., Penninger, J. M., Timms, E., Wakeham, A., Shahinian, A., Lee, K. P., Thompson, C. B., Griesser, H., and Mak, T. W. (1995). Lymphoproliferative disorders with early lethality in mice deficient in Ctla-4. *Science* **270**, 985–988.

Watts, T. H. (2005). TNF/TNFR family members in costimulation of T cell responses. *Annu. Rev. Immunol.* **23**, 23–68.

Workman, C. J., and Vignali, D. A. (2005). Negative regulation of T cell homeostasis by lymphocyte activation gene-3 (CD223). *J. Immunol.* **174**, 688–695.

Yamaguchi, T., and Sakaguchi, S. (2006). Regulatory T cells in immune surveillance and treatment of cancer. *Semin. Cancer Biol.* **16**, 115–123.

Young, M. R., et al. (2001). Human squamous cell carcinomas of the head and neck chemoattract immune suppressive CD34(+) progenitor cells. *Hum. Immunol.* **62**, 332–341.

Yu, H., and Jove, R. (2004). The STATs of cancer—new molecular targets come of age. *Nat. Rev. Cancer.* **4**, 97–105.

Zarek, P. E., Huang, C.-T., Lutz, E. R., Kowalski, J., Horton, M. R., Linden, J., Drake, C. G., and Powell, J. D. (2007). Tissue-derived adenosine promotes peripheral T cell tolerance by stimulating the A2A receptor.

Zheng, Y., Collins, S. L., Lutz, M. A., Allen, A. N., Kole, T. P., Zarek, P. E., and Powell, J. D. (2007). A role for mammalian target of rapamycin in regulating T cell activation versus anergy. *J. Immunol.* **178**, 2163–2170.

Zea, A. H., Rodriguez, P. C., Atkins, M. B., Hernandez, C., Signoretti, S., Zabaleta, J., McDermott, D., Quiceno, D., Youmans, A., O'Neill, A., Mier, J., and Ochoa, A. C. (2005). Arginase-producing myeloid suppressor cells in renal cell carcinoma patients: A mechanism of tumor evasion. *Cancer Res.*, **65**, 3044–3048.

CHAPTER

15

Regulatory T cells in Tumor Immunity: Role of Toll-Like Receptors

RONG-FU WANG

I. Introduction
II. Immune Cells in Immunosurveillance and Tumor Destruction
III. TLRs and Their Signaling Pathways
IV. TLRs in Innate Immunity, Inflammation, and Cancer Development
V. Tumor-Infiltrating Immune Cells in the Tumor Microenvironment
    A. Tumor-Infiltrating Macrophages and DCs
    B. $CD4^+$ and $CD8^+$ Tregs in Cancer
VI. Molecular Marker for $CD4^+$ Tregs
VII. Antigen Specificity of $CD4^+$ Tregs
VIII. Suppressive Mechanisms of Tregs
IX. Functional Regulation of Tregs and Effector Cells by TLR Signaling
X. Implications for Enhancing Antitumor Immunity
    A. Depletion of Tregs and Reversal of Cell Function
    B. Combination of Cancer Peptide Vaccination with Poly-G Treatment
    C. The Stimulation of Effector T Cells with GITR-Specific Antibodies
XI. Conclusion
    References

Innate and adaptive immunity play important roles in immunosurveillance and tumor destruction. However, increasing evidence suggests that tumor-infiltrating immune cells may have dual functions: inhibiting and promoting tumor growth and progression. Although innate immune responses are an important first-line of defense against cancer and infectious pathogens, a failure in their ability to clear infectious agents, injured tissues, or malignant cells may lead to chronic inflammation. In addition, tumor-infiltrating immune cells may convert themselves into immune suppressive cells. $CD4^+$ regulatory T cells (Tregs), which normally prevent autoimmune diseases by suppressing host immune responses, are recruited to tumor sites to inhibit antitumor immunity and promote tumor growth. Notably, elevated proportions of Tregs are present in various types of cancer where they mediate immune suppression. Besides naturally occurring $CD4^+CD25^+$ Tregs, tumor-specific Tregs have also been identified in tumor-infiltrating lymphocytes isolated from cancer patients. These tumor-specific Tregs inhibit immune responses after activation by ligands expressed on tumor cells. Therefore, Tregs at tumor sites have detrimental effects on active immunotherapies directed to cancer. Toll-like receptor (TLR)-mediated recognition of invading pathogens can launch powerful innate and adaptive immune responses, prompting the investigation of whether TLR8 signaling can reverse the suppressive function of Tregs. This chapter considers how TLR signaling is linked to the functional control of Tregs, discussing new tactics to improve cancer therapy by relieving Treg-mediated suppression of antigen-specific antitumor immunity.

I. INTRODUCTION

Adoptive transfer of tumor-reactive tumor-infiltrating lymphocytes (TILs) plus interleukin (IL)-2 can effectively eradicate tumor masses in a minority of treated patients (Wang and Rosenberg, 1999). The clinical response rate was further improved by treating cancer patients with lymphocyte-depleting chemotherapeutic agents prior to T cell transfer (Dudley et al., 2002), suggesting the existence of a population of lymphocytes that limit TIL activity and supporting the concept that tumor-reactive T cells can mediate tumor regression *in vivo*. Identification of tumor antigens recognized by T cells derived from melanoma and other cancers is believed to have great potential to develop effective vaccines (Wang and Rosenberg, 1999), but this knowledge has yet to yield clinical benefit (Rosenberg et al., 2004). One plausible explanation is that solid tumors are embedded in a stromal environment that limits immune controls, consisting of both immune cells, such as macrophages and lymphocytes, as well as nonimmune cells, such as endothelial cells and fibroblasts. Indeed, studies suggest that inflammatory and innate immune cells present at tumor sites promote rather than inhibit cancer development and progression (Greten et al., 2004). Moreover, Tregs at tumor sites potently suppress $CD4^+$ and $CD8^+$ T cell responses, thereby promoting tumor growth as well. Clearly, a better understanding of immune cells and their function in the tumor microenvironment is needed to develop more effective approaches to cancer therapy, in chemotherapeutic, immunotherapeutic, and vaccine settings. This chapter provides an overview of the current understanding of Tregs and their regulation in cancer by innate immune cells and tumor-infiltrating dendritic cells (DCs), also touching on roles for macrophages and γδ-T cells.

II. IMMUNE CELLS IN IMMUNOSURVEILLANCE AND TUMOR DESTRUCTION

A large body of evidence supports the concept of cancer immunosurveillance, originally proposed by Burnet and Thomas about 50 years ago (Dunn et al., 2004). Innate immunity is the first line of host defense against pathogens and transformed tumor cells. Innate immune cells, including natural killer (NK), natural killer T (NKT), and γδ-T cells, have been shown to play a critical role in protecting the host against malignant cells (Dunn et al., 2004; Smyth et al., 2001). Both macrophages and DCs function as major sensors for invading pathogens and transformed tumor cells via a limited number of germ line-encoded pattern recognition receptors (PRRs)—the so-called Toll-like receptors (TLRs)—that play critical roles in modulating inflammation and immune responses (Akira and Takeda, 2004). Adaptive immunity is involved in the elimination of pathogens and transformed tumor cells in the late phase of host defense, generating more specific immunity as well as immunological memory.

Besides immune cells, soluble factors, such as interferon (IFN)-γ, tumor necrosis factor-related apoptosis-inducing ligand (TRAIL), and tumor necrosis factor α (TNF-α), also provide host immunity against cancer cells. Several studies showed a correlation between the level of tumor-infiltrating immune cells, in particular $CD8^+$ T cells, and patient survival (Sato et al., 2005; Zhang et al., 2003), while the presence of other immune cells, such as Tregs, confers a poor prognosis (Curiel et al., 2004). Together, these studies outline the importance of innate and adaptive immunity in cancer immunosurveillance but also as critical supportive components of the tumor microenvironment.

From early stages of disease, there is a continuous dynamic battle between immune cells and tumor cells, which, in some circumstances, can clearly favor the growth of the latter. The failure of initial immune responses to control infections/tissue damage leads to chronic inflammation. Although the mechanisms by which chronic inflammation directly contributes to cancer are poorly understood, activation of the innate immune response (particularly through mechanisms involving the nuclear factor [NF]-κB pathway) by TLR-mediated recognition of invading pathogens or damaged tissues serves as a link between chronic inflammation and cancer (Clevers, 2004; Condeelis and Pollard, 2006; Karin and Greten, 2005). Infections and tissue damage by bacteria and viruses in the host can contribute significantly to long-term inflammation, leading to the recruitment of more immune cells. If acute immune responses completely clear infections, inflammation is resolved and disappears. However, if initial immune responses fail to control infections/tissue injury or transformed cells, chronic inflammation results. In this setting, many cytokines and chemokines released by immune cells promote angiogenesis and tumor growth (Condeelis and Pollard, 2006; Coussens and Werb, 2002), and chronic stimulation of T cells by tumor cells and tumor-infiltrating macrophages/DCs may lead to the generation or accumulation of Tregs at tumor sites. Thus, immune cells can either inhibit or promote tumor growth, depending in part on inflammatory conditions in the tumor microenvironment.

III. TLRs AND THEIR SIGNALING PATHWAYS

TLRs have emerged as a critical component of innate immune system to detect microbial infection and to activate DC maturation programs for the induction of adaptive immune responses (Iwasaki and Medzhitov, 2004). TLRs are type 1 integral membrane glycoproteins characterized by the extracellular domains containing varying numbers of leucine-rich-repeat (LRR) motifs and a cytoplasmic Toll/IL-1R homology (TIR) domain. At least 13 TLRs have been identified in humans and mice. These TLRs recognize a limited but highly conserved set of molecular structures characteristic of infectious agents that are called pathogen-associated molecular patterns (PAMPs). For example, TLR4 recognizes bacterial lipopolysaccharide (LPS), which is unique to gram-negative bacteria, whereas TLR2 recognizes bacterial peptidoglycan, which is found in gram-positive bacteria. TLR3, TLR7, TLR8, and TLR9 recognize nucleic acids: TLR3 recognizes double-stranded RNA produced during viral infection, TLR9 recognizes unmethylated CpG DNA motifs that are highly represented in prokaryotic genomes and DNA viruses (Takeda and Akira, 2005), and TLR7 and TLR8 recognize single-stranded viral RNAs or guanosine-related analogs (loxoribine and imidazoquinoline) (Heil et al., 2004). TLRs are expressed in many types of cells, including different subsets of DCs, macrophages, T cells, neutrophils, eosinophils, mast cells, monocytes, epithelial cells, and tumor cells (Iwasaki and Medzhitov, 2004). Interestingly, TLRs are expressed on different cellular membranes, with TLR2, 4, 5, and 6, residing on the cell surface and TLR3, 7, 8, and 9 residing in internal endosomal compartments (Akira and Takeda, 2004).

Most TLRs use MyD88 as the sole receptor-binding adaptor protein for signal transduction. Thus, MyD88 is essential for signaling from most TLRs to downstream signaling molecules in the cell. Recent intensive work in the field has unraveled the effector pathways acting downstream of TLRs, highlighting a convergence on the transcription factors NF-κB, IRF-5, and IRF-7, which are activated by TLR signaling-trafficking processes. Upon TLR ligand binding, MyD88 forms a complex with the IRAK kinase that leads to its activation. Tumor necrosis factor receptor (TNFR)-associated factor 6 (TRAF6) is activated by MyD88–IRAK complexes. TRAF6 recruits a ubiquitination E_2 enzyme consisting of ubiguitin-conjugating enzyme E2 and E2 variant 1 (UBC13 and UEV1A), catalyzing the formation of K63-linked polyubiquitin on TRAF6 itself and on NF-κB essential modulator (NEMO), a subunit of the NF-κB-activating inhibitor of NF-κB kinase (IKK) complex. Meanwhile, the recruitment of transforming growth factor β (TGF-β)-activated kinase 1 (TAK1) and its binding proteins (TAB1, TAB2, and TAB3) leads to the phosphorylation of IKK-β and MAP kinase 6 (MKK6) and then to activation of NF-κB and mitogen-activated protein (MAP) kinase pathways. These pathways are critically important for the production of inflammatory cytokines. Findings suggest that MyD88 may also interact with the IFN-regulated transcription factors IRF-5 and IRF-7 for induction of proinflammatory cytokines or type I IFN responses (IFN-α/β) (Honda et al., 2005a, 2005b). In contrast, TLR3 relies on a MyD88-independent pathway that involves the adapter protein TRIF, which mediates production of IFN-β in response to pathogen recognition. Activation of TLR3 leads to recruitment of receptor-interacting protein 1 (RIP1), TRAF3, and TRAF6, which activates the TRAF family-member-associated NF-κB activator (TANK) binding kinase 1 (TBK1) and/or inducible IkB kinase (IKK-i), which directly phosphorylate IRF3 and IRF7 for the production of type I IFN cytokines (Akira et al., 2006).

IV. TLRs IN INNATE IMMUNITY, INFLAMMATION, AND CANCER DEVELOPMENT

Although TLRs are critically involved in pathogen recognition and antibacterial or viral immunity, how TLRs link inflammation and cancer still remains uncertain. Studies show a significant association between prostate cancer and TLR4 sequence variants, suggesting that TLR sequence polymorphism is an important risk factor of prostate cancer (Zheng et al., 2004). One possible explanation is that TLRs are sensors of immune cells such as macrophages, DCs, and T cells. The activation of TLRs by bacteria, viruses, or damaged tissues will trigger the NF-κB pathway to secrete proinflammatory cytokines or chemokines, and chronic inflammation involving NF-κB activation has been linked to cancer development and growth. Indeed, accumulating evidence supports the concept that innate immune responses link chronic inflammation to cancer (Karin et al., 2006). For example, chronic inflammation in the stomach induced by *Helicobacter pylori* infection is a leading cause of gastric cancer, while inflammatory bowel diseases, including ulcerative colitis and Crohn's disease, are closely associated with increased risk of colon cancer (Karin et al., 2006). Similarly, hepatitis B and C viruses (HBV and HCV) infection in the liver are leading risk factors in liver cancer (Karin et al., 2006). Thus, innate immunity generated through the TLR and NF-κB pathways may create inflammatory environments to recruit immune cells, sustaining chronic inflammatory states that lead to the production of immune suppressive cells, which

V. TUMOR-INFILTRATING IMMUNE CELLS IN THE TUMOR MICROENVIRONMENT

Immune/inflammatory cells in a developing neoplasm include a diverse array of leukocytes: macrophages, DCs, neutrophils, eosinophils, mast cells, and αβ and γδ effector and regulatory T cells. These cells are capable of producing a spectrum of cytokines, chemokines, and cytotoxic mediators, including reactive oxygen and nitrogen species, proteases and matrix metalloproteinase, and soluble context-dependent mediators of cell killing, such as TNF-α, TRAIL, and IFNs. The role of these various factors in tumor immunity or tolerance is discussed in the following paragraphs.

A. Tumor-Infiltrating Macrophages and DCs

Once bone-marrow-derived hematopoietic progenitor cells or monocytes are recruited to inflammatory sites or tissues, they can differentiate into macrophages, DCs, and Langerhans DCs, depending upon the cytokine and tissue environment. As discussed in detail in Chapter 16, tumor-associated macrophages (TAMs) are a significant component of inflammatory cells at tumor sites and display a dual function (Condeelis and Pollard, 2006; Coussens and Werb, 2002). Although they can kill tumor cells following activation by IL-2 and IL-12, TAMs can also promote tumor growth (Coussens and Werb, 2002). Disruption of the NF-κB pathway in myeloid cells or epithelium by cell-specific deletion of a floxed IKKβ allele (IKK-β$^{F/F}$) affects tumor initiation and subsequent tumor growth (Karin and Greten, 2005) because this pathway has at least two roles in tumorigenesis: preventing the death of cells with malignant potential and stimulating the production of proinflammatory cytokines in immune cells. Tumor-infiltrating DCs, which possess an immature phenotype compatible with tumor growth promotion, can migrate to tumor vessels and contribute independently to neovascularization by facilitating the growth of tumors expressing high levels of the angiogenesis vascular endothelial growth factor (VEGF)-A (Conejo-Garcia et al., 2004). Immature DCs tend to induce immune tolerance rather than immunity either by deleting reactive T cells or by inducing Tregs (Steinman et al., 2003). Finally, myeloid suppressor cells at tumor sites may stimulate Tregs and thus inhibit antitumor immunity, as discussed further in Chapter 17.

B. CD4$^+$ and CD8$^+$ Tregs in Cancer

CD4$^+$ Tregs normally comprise a small subset (5–6%) of the overall CD4$^+$ T cell population. Besides naturally occurring CD4$^+$CD25$^+$ Tregs, other CD4$^+$ Tregs include Tr-1 cells, which secrete IFN-γ and IL-10, and T helper 3 (Th3) cells, which secrete high levels of TGF-β, IL-4, and IL-10. The existence of CD4$^+$ T cell-mediated immune suppression has long been observed in animal tumor models and human cancer (Berendt and North, 1980; Mukherji et al., 1989) although such cells are not well characterized due to the lack of reliable molecular markers. Studies demonstrate that an elevated proportion of CD4$^+$CD25$^+$ Tregs in the total CD4$^+$ T cell populations is a common occurrence in a variety of human cancers, including lung and breast cancers as well as ovarian tumors and human melanoma (Wang et al., 2006). In addition, antigen-specific CD4$^+$ Tregs have been studied at tumor sites (Wang et al., 2004., 2005), showing that these Tregs can suppress the proliferation of naive CD4$^+$ T cells and inhibit the secretion of IL-2 by CD4$^+$ effector cells upon activation by tumor-specific ligands. These findings reinforce the concept that Tregs limit tumor-associated antigen (TAA)-dependent activation of T cell immunity in cancer.

In addition to CD4$^+$ Tregs, a class of CD8$^+$ Tregs also has been identified as mediators of antigen-dependent immune suppression (Jiang and Chess, 2004). In contrast to naturally occurring CD4$^+$CD25$^+$ Tregs, the common feature of CD8$^+$ Tregs is that they are generated or induced only after antigen priming (Cantor, 2004; Jiang and Chess, 2004; LVAD et al., 2005). Other than this feature, which distinguishes them from CD4+ Tregs, very little is known about how CD8$^+$ Tregs can be generated by tumor cells or tumor-infiltrating DCs or whether they may suppress immune cells, such as CD4$^+$, CD8$^+$ effector cells, and stimulatory DCs at tumor sites.

VI. MOLECULAR MARKER FOR CD4$^+$ TREGS

Expression in T cells of the cell surface molecule CD25—encoding IL-2 receptor protein—has been used as a useful marker for Tregs. However, CD25 expression is also expressed by certain activated nonregulatory effector lymphocytes, and it is not essentially linked to Treg cell function. Other molecules, including the TNF-family molecule glucocorticoid-induced tumor necrosis factor receptor (GITR) and cytotoxic T-lymphocyte antigen 4 (CTLA-4), may serve as markers for Tregs (Sakaguchi, 2004). Notably, the forkhead box protein P3 FoxP3 has emerged as a highly specific marker of CD4$^+$ Tregs in both mice (Font not et al., 2003; Hori et al., 2003; Khat Tri et al., 2003) and humans (Walker et al., 2003; Wang et al., 2004). FoxP3 expression correlates better with suppressive activity of Tregs than does CD25 expression. Moreover, mutation of FoxP3 causes a loss of Tregs and the production of an X-linked recessive inflammatory disease in Scurfy mutant mice. Last, ectopic expression of FoxP3 in vitro and in vivo is capable of converting naive murine CD4$^+$ T cells to Tregs. Using a knock-in approach, two groups showed that the expression of FoxP3 is highly restricted to the Treg subset of T cells and correlates with suppressor activity, irrespective of CD25 expression (Fontenot et al., 2005; Wan and Flaval, 2005). These studies suggest that FoxP3 is a specific marker for murine Tregs. One caveat in the human system is that while FoxP3 is highly expressed in human CD4$^+$ Tregs, its expression is not solely restricted to these cells. While ectopic expression of FoxP3 in human CD4$^+$CD25$^-$ T cells is not sufficient to generate Tregs, stimulation of CD4$^+$CD25$^-$ T cells with anti-CD3 and CD28 antibodies or antigen-specific ligands leads to upregulation of FoxP3 and generation of T cells that clearly have suppressive function. Although CD4$^+$CD25$^+$ T cells are found normal in patients with various FoxP3 mutations, the suppressive activity of CD4$^+$CD25$^+$ Tregs depends on the types of mutations in FoxP3. CD127 has been found as a marker that inversely correlates with FoxP3 expression and the suppressive activity of Tregs.

VII. ANTIGEN SPECIFICITY OF CD4$^+$ TREGS

Despite the important roles of Tregs in cancer and many other diseases, little is known about the physiological target antigens recognized by these cells. Increasing evidence suggests the requirement of tumor-specific and pathogen-specific ligands for activating Tregs (Shish et al., 2004; Wang et al., 2004). However, the identity of these ligands remains largely unknown. Many such tumor/antigen-specific CD4$^+$ Treg cell lines have been generated, allowing the identification of their ligands. A cancer-testis antigen (LAGE1) and an antigen recognized by Treg cells (ARTC1) are examples of tumor-specific ligands recognized by melanoma-derived CD4$^+$ Tregs, indicating that antigen-specific CD4$^+$ Tregs are present at tumor sites and mediate antigen-specific and local immune suppression of antitumor immunity (Wang et al., 2004, 2005). DNA-like 2, an autoantigen identified by recombinant expression cloning (SEREX), may also serve as a ligand that is capable of activating CD4+CD25+ Tregs for immune suppression in mice (Nishioka *et al.*, 2003). Together, these studies suggest the existence of antigen-specific Tregs in cancer and other diseases. Further studies are needed for identification of these ligands that are capable of stimulating Tregs *in vivo*.

VIII. SUPPRESSIVE MECHANISMS OF TREGS

CD4+ Tregs require antigen-specific activation or polyclonal TCR stimulation to exert their suppressive function. Once they are activated, they can suppress CD4+ and CD8+ T cells in an antigen-nonspecific manner. Several mechanisms have been proposed to explain how CD4+ Tregs inhibit CD4+ effector T cells (Sakaguchi, 2004; Shevach, 2002). Most naturally occurring CD4+CD25+ Tregs and antigen-specific Tregs inhibit target cell proliferation and function through a cell-to-cell contact-dependent mechanism, while some antigen-induced Tregs suppress immune responses through soluble factors, including IL-10 and/or a TGF-β-dependent mechanism (Sakaguchi, 2004; Shevach, 2002). Therefore, more than one mechanism of CD4+ Treg cell-mediated suppression appears to operate *in vitro* and *in vivo*.

Most tumor-specific CD4+ Tregs suppress immune responses (proliferation and IL-2 secretion of naive or effector T cells) through a cell-contact mechanism (Wang *et al.*, 2004, 2005). Although membrane-bound TGF-β expressed on the surface of Tregs has been implicated as a molecule required for immune suppression, the precise mechanism responsible for Treg cell-mediated suppression remains poorly understood. While initial experiments using an anti-GITR antibody suggested that GITR may mediate the suppressive activity of Tregs (Shimizu *et al.*, 2002), further experiments by several groups showed that both anti-GITR and GITR ligand act mainly on effector T cells, stimulating them to resist suppression mediated by Tregs (Shevach and Stephens, 2006). Thus, the molecules that mediate immune suppression through the cell-to-cell contact mechanism remain to be determined.

IX. FUNCTIONAL REGULATION OF TREGS AND EFFECTOR CELLS BY TLR SIGNALING

To overcome immune suppression mediated by Tregs, some investigators have depleted CD4+CD25+ T cells using an anti-CD25+ antibody to illustrate enhanced antitumor responses in mouse models of cancer (Shimizu *et al.*, 1999; Sutmuller *et al.*, 2001). However, in the setting of human vaccine trials, this strategy would likely be flawed because CD25 is expressed on both CD4+CD25+ Tregs and newly activated effector T cells. Indeed, clinical trials using ONTAK® (an IL-2-toxin fusion protein) show either inefficient elimination of Tregs (Attia *et al.*, 2005) or depletion of both Tregs and activated effector cells (Dannull *et al.*, 2005). Therefore, new strategies for overcoming the suppressive function of Tregs should be developed for human cancer immunotherapy.

Since TLRs play an important role in sensing microbial infection and initiating both innate and adaptive immunity, it can be argued that TLR signaling might reverse the suppressive activity of human Tregs, perhaps identifying a tactic to trigger T cell immunity by derepressing Treg activity. With this in mind, it has been demonstrated that activating TLR signaling in DCs can render naive T cells refractory to suppression by Tregs in mice (Pasare and Medzhitov, 2003). To establish that TLR signaling can directly affect the suppressive function of Tregs, it has been shown that TLR ligand provided by a poly-G10 oligonucleotide (G*GGGG*G*G*G*G*G) was sufficient to reverse the suppressive function of Tregs in the absence of DCs (Peng *et al.*, 2005). Interestingly, the reversal depended on a short stretch of guanosines but not on CpG motifs, suggesting a TLR9-independent mechanism of action.

Using sRNA-mediated knockdown, it was shown that TLR8 and the MyD88 signaling pathway are essential for poly-G oligonucleotide to reverse human Tregs function (Peng et al., 2005). Consistent with this result, natural ligands for human TLR8, such as ssRNA40 and ssRNA33 derived from HIV viral sequences (Heil et al., 2004), completely reversed the suppressive function of Tregs. Moreover, ligands for other human TLRs failed to reverse the suppressive function of Tregs, suggesting that TLR8 activation is specifically linked to regulation of Treg cell function. Other observations indicate that activation of TLR5 on DCs and effector T cells by its ligand flagellin can enhance proliferation and IL-2 secretion of Tregs (Crellin et al., 2005). Thus, stimulation of human Tregs with flagellin increases rather than reverses the suppressive function (Crellin et al., 2005). Similarly, a study has shown that TLR2 activation by the ligand Pam3Cys directly increases the proliferation of murine Tregs, temporally reversing their suppressive function (Sutmuller et al., 2006). Interestingly, poly-G oligonucleotides did not reverse the suppressive activity of murine Tregs, consistent with other evidence that TLR8 does not function in mice. In summary, these studies indicate that the mechanisms for TLR-mediated regulation of the suppression of Tregs differs to some extent in humans and mice. Further studies are needed to understand conserved versus nonconserved roles of other TLRs on murine and human Tregs and effector cells.

X. IMPLICATIONS FOR ENHANCING ANTITUMOR IMMUNITY

Following the discussion in the previous paragraphs, several strategies may be applicable to boosting immune responses in cancer. In particular, manipulating the dynamic balance between Tregs and effector cells may be an effective way to achieve this goal, as listed here.

A. Depletion of Tregs and Reversal of Cell Function

Tregs can be depleted with specific antibodies, or TLR ligands can reverse Tregs function. Strategies to limit the number or function of Tregs may have therapeutic potential in cancer treatment, but efforts to use ONTAK® fusion protein for this purpose may not work effectively in patients; new and more specific antibodies are needed for such immunological interventions. Treatment of cancer patients with TLR8 ligands, such as poly-G oligonucleotides, offers an intriguing alternative to reverse the suppressive function of Tregs, perhaps acting by converting suppressive Tregs into nonsuppressive T helper cells.

B. Combination of Cancer Peptide Vaccination with Poly-G Treatment

Cancer peptide vaccination can be combined with poly-G treatment. Strong antigen-specific T cell responses can be induced by vaccinating patients with antigenic peptides derived from tumor antigens such as NY-ESO-1, but Tregs may limit the extent of such responses. CO treatment with poly-G oligonucleotides offers one tactic to control the suppressive function of Tregs in this setting.

C. The Stimulation of Effector T Cells with GITR-Specific Antibodies

Effector T cells can be stimulated with GITR-specific antibodies. An alternative approach to overcome Tregs-mediated suppression is to overstimulate effector cells with anti-GITR or anti-CD28, rendering the effector cells resistant to suppression by Tregs. In animal models of cancer, anti-GITR antibody has been shown to enhance antitumor immunity (Shevach and Stephens, 2006). Human studies to confirm anticancer effects have yet to be performed. Anti-CTLA-4 and anti-PD1 antibodies have also been used to block negative signaling in effector cells, as another tactic to relieve suppression and shift the balance toward effector T cell responses. Ideally, the combined use of multiple agents or strategies in clinical settings to defeat Tregs suppression may by inducing powerful antitumor immune responses increase the efficacy of a variety of cancer immunotherapeutic strategies.

XI. CONCLUSION

Innate immunity and $CD4^+$ Tregs are key regulators of immune responses in cancer. New strategies that simultaneously stimulate $CD4^+$ effector T cells while inhibiting or depleting $CD4^+$ Tregs may improve the outcome of cancer immunotherapy. Although the suppressive function of Tregs can be reversed (e.g., in human cells through manipulation of TLR8 signaling), the downstream signaling pathways involving TLR8 or other players to the functional regulation of Tregs are not yet known. Indeed, the critical molecules responsible for immune suppression mediated by different subsets of Tregs remain largely unknown. Identifying and characterizing key pathways including those controlled by TLR ligands hold much promise for developing new molecular-targeted therapies for cancer treatment. In particular, the ultimate success of vaccine strategies may lie in relieving Treg cell suppression functions to increase therapeutic efficacy.

Acknowledgments

The author apologizes to those researchers whose work has not been cited due to space limitations. The work of the authors is supported in part by grants from the NIH, American Cancer Society, and Cancer Research Institute. The author declares no competing financial interests.

References

Akira, S., and Takeda, K. (2004). Toll-like receptor signalling. *Nat. Rev. Immunol.* **4**, 499–511.

Akira, S., Uematsu, S., and Takeuchi, O. (2006). Pathogen recognition and innate immunity. *Cell* **124**, 783–801.

Attia, P., Maker, A. V., Haworth, L. R., Rogers-Freezer, L., and Rosenberg, S. A. (2005). Inability of a fusion protein of IL-2 and diphtheria toxin (denileukin diftitox, DAB389IL-2, ONTAK) to eliminate regulatory T lymphocytes in patients with melanoma. *J. Immunother.* **28**, 582–592.

Berendt, M. J., and North, R. J. (1980). T-cell-mediated suppression of anti-tumor immunity. An explanation for progressive growth of an immunogenic tumor. *J. Exp. Med.* **151**, 69–80.

Cantor, H. (2004). Reviving suppression? *Nat. Immunol.* **5**, 347–349.

Clevers, H. (2004). At the crossroads of inflammation and cancer. *Cell* **118**, 671–674.

Condeelis, J., and Pollard, J. W. (2006). Macrophages: Obligate partners for tumor cell migration, invasion, and metastasis. *Cell* **124**, 263–266.

Conejo-Garcia, J. R., Benencia, F., Courreges, M. C., Kang, E., Mohamed-Hadley, A., Buckanovich, R. J., Holtz, D. O., Jenkins, A., Na, H., Zhang, L., *et al.* (2004). Tumor-infiltrating dendritic cell precursors recruited by a beta-defensin contribute to vasculogenesis under the influence of VEGF-A. *Nat. Med.* **10**, 950–958.

Coussens, L. M., and Werb, Z. (2002). Inflammation and cancer. *Nature* **420**, 860–867.

Crellin, N. K., Garcia, R. V., Hadisfar, O., Allan, S. E., Steiner, T. S., and Levings, M. K. (2005). Human $CD4^+$ T cells express TLR5 and its ligand flagellin enhances the suppressive capacity and expression of FOXP3 in $CD4^+CD25^+$ T regulatory cells. *J. Immunol.* **175**, 8051–8059.

Curiel, T. J., Coukos, G., Zou, L., Alvarez, X., Cheng, P., Mottram, P., Evdemon-Hogan, M., Conejo-Garcia, J. R., Zhang, L., Burow, M., *et al.* (2004). Specific recruitment of regulatory T cells in ovarian carcinoma fosters immune privilege and predicts reduced survival. *Nat. Med.* **10**, 942–949.

Dannull, J., Su, Z., Rizzieri, D., Yang, B. K., Coleman, D., Yancey, D., Zhang, A., Dahm, P., Chao, N., Gilboa, E., and Vieweg, J. (2005). Enhancement of vaccine-mediated antitumor immunity in cancer patients after depletion of regulatory T cells. *J. Clin. Invest.* **115**, 3623–3633.

Dudley, M. E., Wunderlich, J. R., Robbins, P. F., Yang, J. C., Hwu, P., Schwartzentruber, D. J., Topalian, S. L., Sherry, R., Restifo, N. P., Hubicki, A. M., *et al.* (2002). Cancer regression and autoimmunity in patients after clonal repopulation with antitumor lymphocytes. *Science* **298**, 850–854.

Dunn, G. P., Old, L. J., and Schreiber, R. D. (2004). The immunobiology of cancer immunosurveillance and immunoediting. *Immunity* **21**, 137–148.

Fontenot, J. D., Gavin, M. A., and Rudensky, A. Y. (2003). Foxp3 programs the development and function of CD4$^+$CD25$^+$ regulatory T cells. *Nat. Immunol.* **4**, 330–336.

Fontenot, J. D., Rasmussen, J. P., Williams, L. M., Dooley, J. L., Farr, A. G., and Rudensky, A. Y. (2005). Regulatory T cell lineage specification by the forkhead transcription factor foxp3. *Immunity* **22**, 329–341.

Greten, F. R., Eckmann, L., Greten, T. F., Park, J. M., Li, Z. W., Egan, L. J., Kagnoff, M. F., and Karin, M. (2004). IKKbeta links inflammation and tumorigenesis in a mouse model of colitis-associated cancer. *Cell* **118**, 285–296.

Heil, F., Hemmi, H., Hochrein, H., Ampenberger, F., Kirschning, C., Akira, S., Lipford, G., Wagner, H., and Bauer, S. (2004). Species-specific recognition of single-stranded RNA via Toll-like receptor 7 and 8. *Science* **303**, 1526–1529.

Honda, K., Ohba, Y., Yanai, H., Negishi, H., Mizutani, T., Takaoka, A., Taya, C., and Taniguchi, T. (2005a). Spatiotemporal regulation of MyD88-IRF-7 signalling for robust type-I interferon induction. *Nature* **434**, 1035–1040.

Honda, K., Yanai, H., Negishi, H., Asagiri, M., Sato, M., Mizutani, T., Shimada, N., Ohba, Y., Takaoka, A., Yoshida, N., and Taniguchi, T. (2005b). IRF-7 is the master regulator of type-I interferon-dependent immune responses. *Nature* **434**, 772–777.

Hori, S., Nomura, T., and Sakaguchi, S. (2003). Control of regulatory T cell development by the transcription factor foxp3. *Science* **299**, 1057–1061.

Hsieh, C. S., Liang, Y., Tyznik, A. J., Self, S. G., Liggitt, D., and Rudensky, A. Y. (2004). Recognition of the peripheral self by naturally arising CD25$^+$ CD4$^+$ T cell receptors. *Immunity* **21**, 267–277.

Iwasaki, A., and Medzhitov, R. (2004). Toll-like receptor control of the adaptive immune responses. *Nat. Immunol.* **5**, 987–995.

Jiang, H., and Chess, L. (2004). An integrated view of suppressor T cell subsets in immunoregulation. *J. Clin. Invest.* **114**, 1198–1208.

Karin, M., and Greten, F. R. (2005). NF-kappaB: Linking inflammation and immunity to cancer development and progression. *Nat. Rev. Immunol.* **5**, 749–759.

Karin, M., Lawrence, T., and Nizet, V. (2006). Innate immunity gone awry: Linking microbial infections to chronic inflammation and cancer. *Cell* **124**, 823–835.

Khattri, R., Cox, T., Yasayko, S. A., and Ramsdell, F. (2003). An essential role for Scurfin in CD4($^+$)CD25($^+$) T regulatory cells. *Nat. Immunol.* **4**, 337–342.

Mukherji, B., Guha, A., Chakraborty, N. G., Sivanandham, M., Nashed, A. L., Sporn, J. R., and Ergin, M. T. (1989). Clonal analysis of cytotoxic and regulatory T cell responses against human melanoma. *J. Exp. Med.* **169**, 1961–1976.

Nishikawa, H., Kato, T., Tanida, K., Hiasa, A., Tawara, I., Ikeda, H., Ikarashi, Y., Wakasugi, H., Kronenberg, M., Nakayama, T., et al. (2003). CD4$^+$ CD25$^+$ T cells responding to serologically defined autoantigens suppress antitumor immune responses. *Proc. Natl. Acad. Sci. USA* **100**, 10902–10906.

Pasare, C., and Medzhitov, R. (2003). Toll pathway-dependent blockade of CD4$^+$CD25$^+$ T cell-mediated suppression by dendritic cells. *Science* **299**, 1033–1036.

Peng, G., Guo, Z., Kiniwa, Y., Voo, K. S., Peng, W., Fu, T., Wang, D. Y., Li, Y., Wang, H. Y., and Wang, R.-F. (2005). Toll-like receptor 8 mediated-reversal of CD4$^+$ regulatory T cell function. *Science* **309**, 1380–1384.

Rosenberg, S. A., Yang, J. C., and Restifo, N. P. (2004). Cancer immunotherapy: Moving beyond current vaccines. *Nat. Med.* **10**, 909–915.

Sakaguchi, S. (2004). Naturally arising CD4$^+$ regulatory T cells for immunologic self-tolerance and negative control of immune responses. *Annu. Rev. Immunol.* **22**, 531–562.

Sato, E., Olson, S. H., Ahn, J., Bundy, B., Nishikawa, H., Qian, F., Jungbluth, A. A., Frosina, D., Gnjatic, S., Ambrosone, C., et al. (2005). Intraepithelial CD8$^+$ tumor-infiltrating lymphocytes and a high CD8$^+$/regulatory T cell ratio are associated with favorable prognosis in ovarian cancer. *Proc. Natl. Acad. Sci. USA* **102**, 18538–18543.

Shevach, E. M. (2002). CD4$^+$ CD25$^+$ suppressor T cells: More questions than answers. *Nat. Rev. Immunol.* **2**, 389–400.

Shevach, E. M., and Stephens, G. L. (2006). The GITR-GITRL interaction: Co-stimulation or contrasuppression of regulatory activity? *Nat. Rev. Immunol.* **6**, 613–618.

Shimizu, J., Yamazaki, S., and Sakaguchi, S. (1999). Induction of tumor immunity by removing CD25$^+$CD4$^+$ T cells: A common basis between tumor immunity and autoimmunity. *J. Immunol.* **163**, 5211–5218.

Shimizu, J., Yamazaki, S., Takahashi, T., Ishida, Y., and Sakaguchi, S. (2002). Stimulation of CD25$^+$ CD4$^+$ regulatory T cells through GITR breaks immunological self-tolerance. *Nat. Immunol.* **3**, 135–142.

Smyth, M. J., Crowe, N. Y., and Godfrey, D. I. (2001). NK cells and NKT cells collaborate in host protection from methylcholanthrene-induced fibrosarcoma. *Int. Immunol.* **13**, 459–463.

Steinman, R. M., Hawiger, D., and Nussenzweig, M. C. (2003). Tolerogenic dendritic cells. *Annu. Rev. Immunol.* **21**, 685–711.

Sutmuller, R. P., den Brok, M. H., Kramer, M., Bennink, E. J., Toonen, L. W., Kullberg, B. J., Joosten, L. A., Akira, S., Netea, M. G., and Adema, G. J. (2006). Toll-like receptor 2 controls expansion and function of regulatory T cells. *J. Clin. Invest.* **116**, 485–494.

Sutmuller, R. P., van Duivenvoorde, L. M., van Elsas, A., Schumacher, T. N., Wildenberg, M. E., Allison, J. P., Toes, R. E., Offringa, R., and Melief, C. J. (2001). Synergism of cytotoxic T lymphocyte-associated antigen 4 blockade and depletion of CD25$^+$ regulatory T cells in antitumor therapy reveals alternative pathways for suppression of autoreactive cytotoxic T lymphocyte responses. *J. Exp. Med.* **194**, 823–832.

Takeda, K., and Akira, S. (2005). Toll-like receptors in innate immunity. *Int. Immunol.* **17**, 1–14.

Vlad, G., Cortesini, R., and Suciu-Foca, N. (2005). License to heal: Bidirectional interaction of antigen-specific regulatory T cells and tolerogenic APC. *J. Immunol.* **174**, 5907–5914.

Walker, M. R., Kasprowicz, D. J., Gersuk, V. H., Benard, A., Van Landeghen, M., Buckner, J. H., and Ziegler, S. F. (2003). Induction of FoxP3 and acquisition of T regulatory activity by stimulated human CD4$^+$CD25- T cells. *J. Clin. Invest.* **112**, 1437–1443.

Wan, Y. Y., and Flavell, R. A. (2005). Identifying Foxp3-expressing suppressor T cells with a bicistronic reporter. *Proc. Natl. Acad. Sci. USA* **102**, 5126–5131.

Wang, H. Y., Deen A. Lee, Guangyong Peng, Zhong Guo, Yanchun Li, Yukiko Kiniwa, Shevach, E. M., and Wang, R.-F. (2004). Tumor-specific human CD4$^+$ regulatory T cells and their ligands: Implication for immunotherapy. *Immunity* **20**, 107–118.

Wang, H. Y., Peng, G., Guo, Z., Shevach, E. M., and Wang, R.-F. (2005). Recognition of a new ARTC1 peptide ligand uniquely expressed in tumor cells by antigen-specific CD4$^+$ gegulatory T cells. *J. Immunol.* **174**, 2661–2670.

Wang, R. F., Peng, G., and Wang, H. Y. (2006). Regulatory T cells and Toll-like receptors in tumor immunity. *Semin. Immunol.* **18**, 136–142.

Wang, R. F., and Rosenberg, S. A. (1999). Human tumor antigens for cancer vaccine development. *Immunol. Rev.* **170**, 85–100.

Zhang, L., Conejo-Garcia, J. R., Katsaros, D., Gimotty, P. A., Massobrio, M., Regnani, G., Makrigiannakis, A., Gray, H., Schlienger, K., Liebman, M. N., *et al.* (2003). Intratumoral T cells, recurrence, and survival in epithelial ovarian cancer. *N. Engl. J. Med.* **348**, 203–213.

Zheng, S. L., Augustsson-Balter, K., Chang, B., Hedelin, M., Li, L., Adami, H. O., Bensen, J., Li, G., Johnasson, J. E., Turner, A. R., *et al.* (2004). Sequence variants of Toll-like receptor 4 are associated with prostate cancer risk: Results from the cancer prostate in Sweden Study. *Cancer Res.* **64**, 2918–2922.

CHAPTER 16

Tumor-Associated Macrophages in Cancer Growth and Progression

ALBERTO MANTOVANI, PAOLA ALLAVENA, AND ANTONIO SICA

I. Introduction
II. Macrophage Polarization
III. Macrophage Recruitment at the Tumor Site
IV. Tam Expression of Selected M2 Protumoral Functions
V. Modulation of Adaptive Immunity by Tams
VI. Targeting Tams
   A. Activation
   B. Recruitment
   C. Angiogenesis
   D. Survival
   E. Matrix Remodeling
   F. Effector Molecules
VII. Concluding Remarks
   References

Macrophage infiltration is a major component of chronic inflammation that is strongly linked to tumor development and progression. Mononuclear phagocytes can generally undergo diverse forms of polarized activation. Tumor-associated macrophages (TAMs) that have immunoregulatory and immunosuppressive activity produce growth factors that stimulate angiogenesis, remodel tissues, and facilitate invasion and metastasis. TAMs and related immature myeloid suppressor cells generally have properties of M2 macrophage populations that are supportive to tumors. Thus, TAMs and related cell types have immunoregulatory functions that mediate various aspects of key cancer-promoting inflammatory reactions.

I. INTRODUCTION

As early as the nineteenth century, it was perceived that cancer is linked to inflammation (Balkwill et al., 2001). Although this perception waned for a long time, the twenty-first century has seen a renaissance of the inflammation and cancer connection stemming from different lines of work and leading to a generally accepted paradigm (Balkwill et al., 2001, 2005; Coussens et al., 2002). Epidemiological studies have revealed that chronic inflammation predisposes an individual to different forms of cancer, such as colon, prostate, and liver cancers, and that usage of nonsteroidal anti-inflammatory agents can protect against the emergence of various tumors. An inflammatory component is present in the microenvironment of most neoplastic tissues, including those not causally related to an obvious inflammatory process. Hallmarks of cancer-associated inflammation include the infiltration of white blood cells, the presence of polypeptide messengers of inflammation (cytokines and chemokines), and the occurrence of tissue remodeling and angiogenesis.

Strong evidence suggests that cancer-associated inflammation promotes tumor growth and progression (Balkwill et al., 2001, 2005; Coussens et al., 2002). By the late 1970s, it was found that tumor growth is promoted by tumor-associated macrophages (TAMs), a major leukocyte population present in tumors (Balkwill et al., 2001, 2005; Coussens et al., 2002 Mantovani et al., 2002). Accordingly, in many but not all human tumors, a high frequency of infiltrating TAMs is associated with poor prognosis. Interestingly, this pathological finding has reemerged in the postgenomic era: genes associated with leukocyte or macrophage infiltration (e.g., CD68) are part of the molecular signatures that herald poor prognosis in lymphomas and breast carcinomas (Paik et al., 2004). Gene-modified mice and cell transfer have provided direct evidence for the protumor function of myeloid cells and their effector molecules. These results raise the interesting possibility of targeting myelomonocytic cells associated with cancer as an innovative therapeutic strategy. This chapter will review key properties of TAMs, emphasizing genetic evidence and emerging targets for therapeutic intervention.

II. MACROPHAGE POLARIZATION

Heterogeneity and plasticity are hallmarks of cells belonging to the monocyte–macrophage lineage (Gordon, 2003; Mantovani et al., 2002., 2004). Lineage-defined populations of mononuclear phagocytes have not been identified; however, what has been described is the short-lived stage of circulating precursor monocyte subsets characterized by differential expression of the Fcγ-RIII receptor (CD16) or of chemokine receptors (CCR2, CX3CR1, and CCR8) and by different functional properties. Once in tissues, macrophages acquire distinct morphological and functional properties directed by the tissue (e.g., the lung alveolar macrophage) and immunological microenvironment.

In response to cytokines and microbial products, mononuclear phagocytes express specialized and polarized functional properties (Gordon, 2003; Mantovani et al., 2004, Ghassabech et al., 2006). Mirroring the T helper 1 and T helper 2 (Th1/Th2) nomenclature, many refer to polarized macrophages as M1 and M2 cells. Classically activated M1 macrophages have long been known to be induced by interferon (IFN)-γ alone or in concert with microbial stimuli (e.g., lipopolysaccharides [LPS]) or cytokines (e.g., tumor necrosis factor [TNF] and granulocyte-macrophage colony-stimulating factor [GM-CSF]). Interleukin (IL)-4 and IL-13 were subsequently found to be more than simple inhibitors of macrophage activation and to induce an alternative M2 form of macrophage activation (Gordon, 2003). M2 is a generic name for various forms of macrophage activation other than the classic M1, including cells exposed to IL-4 or IL-13, immune complexes, IL-10, glucocorticoid, or secosteroid hormones (Mantovani et al., 2004).

In general, M1 cells have an IL-12^{high}, IL-23^{high}, IL-10^{low} phenotype; are efficient producers of effector molecules (reactive oxygen and nitrogen intermediates) and inflammatory cytokines (IL-1β, TNF, IL-6); participate as inducer and effector cells in polarized Th1 responses; and mediate resistance against intracellular parasites and tumors. In contrast, the various forms of M2 macrophages share an IL-12^{low}, IL-23^{low}, IL-10^{high} phenotype with variable capacity to produce inflammatory cytokines depending on the signal utilized. M2 cells generally have high levels of scavenger, mannose, and galactose-type receptors, and the arginine metabolism is shifted to ornithine and polyamines. Differential regulation of components of the IL-1 system (Dinarello, 2005) occurs in polarized macrophages, with low IL-1β and low caspase 1, high IL-1Rα, and high decoy type II receptor in M2 cells. M1 and the various forms of M2 cells have distinct chemokine and chemokine receptor repertoires (Mantovani et al., 2004). In general, M2 cells participate in polarized Th2 reactions; promote killing and encapsulation of parasites (Noel et al., 2004); are present in established tumors; promote progression, tissue repair, and remodeling (Wynn, 2004); and have immunoregulatory functions (Mantovani et al., 2004). Immature myeloid suppressor cells have functional properties and a transcriptional profile related to M2 cells (Biswas et al., 2006; Ghassabeh et al., 2006). Moreover, polarization of neutrophil functions has also been reported (Tsuda et al., 2004).

Profiling techniques and genetic approaches have been put to the test and shed new light on the M1/M2 paradigm. Transcriptional profiling has offered a comprehensive picture of the genetic programs activated in polarized macrophages, led to the discovery of new polarization-associated genes (e.g., Fizz and YM-1), tested the validity of the paradigm *in vivo* in selected diseases (Takahashi et al., 2004; Desnues et al., 2005; Biswas et al., 2006), and questioned the generality of some assumptions. For instance, arginase is unexpectedly not expressed prominently in human IL-4-induced M2 cells (Scotton et al., 2005). M2 cells express high levels of the chitinase-like YM-1. Chitinases represent an antiparasite strategy conserved in evolution, and there is now evidence that acidic mammalian chitinase induced by IL-13 in macrophages is an important mediator of type II inflammation (Zhu et al., 2004).

III. MACROPHAGE RECRUITMENT AT THE TUMOR SITE

Since the first observation by Rudolf Virchow, who noticed the infiltration of leukocytes into malignant tissues and suggested that cancers arise at regions of chronic inflammation, the origin of TAMs has been studied in terms of recruitment, survival, and proliferation. TAMs derive from circulating monocytes and are recruited at the tumor site by a tumor-derived chemotactic factor for monocytes, originally described by this group (Bottazzi et al., 1983) and later identified as the chemokine ligand 2/monocyte chemotactic protein-1 (CCL2/MCP-1) (Yoshimura et al., 1989; Matsushima et al., 1999) (Figure 16.1). Evidence supporting a pivotal role of chemokines in the recruitment of monocytes in neoplastic tissues includes correlation between production and infiltration in murine and human tumors, passive immunization, and gene modification (Rollins et al., 1999). In addition, the central role of chemokines in shaping the tumor microenvironment is supported by the observation that tumors are generally characterized by the constitutive expression of chemokines belonging to the inducible realm (Mantovani et al., 1999). The molecular mechanisms accounting for the constitutive expression of chemokines by cancer cells have been defined only for CXCL1 and involve nuclear factor (NF)-κB activation by NF-κB-inducing kinase (Yang et al., 2001).

CCL2 is probably the most frequently found CC chemokine in tumors. Most human carcinomas produce CCL2 (Table

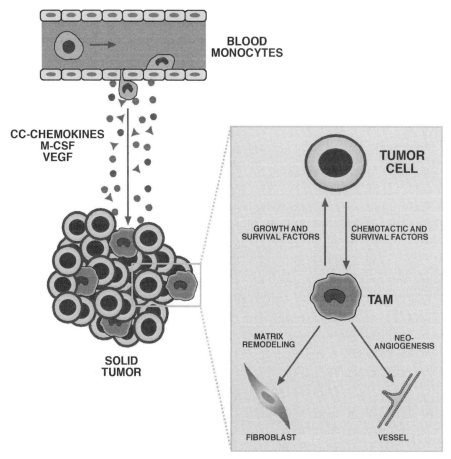

FIGURE 16.1 **TAM recruitment and symbiosis in tumors.** Circulating blood monocytes are actively recruited to the tumor site by tumor-derived chemotactic factors, such as chemokines (e.g., CCL2), MCSF, and VEGF. In the tumor microenvironment, monocytes differentiate into TAMs that establish a symbiotic relationship with tumor cells: tumor-derived factors positively modulate TAM survival, and TAM-derived factors promote tumor cell proliferation, survival, matrix deposition, tissue remodeling, and neoangiogenesis. Figure modified from Sica et al. (2006).

16.1), and its levels of expression correlate with the increased infiltration of macrophages (Mantovani et al., 2002; Balkwill et al., 2004; Conti et al., 2004). Interestingly, CCL2 production has also been detected in TAMs, indicating the existence of an amplification loop for their recruitment (Mantovani et al., 2002; Ueno et al., 2000). Other CC chemokines related to CCL2, such as CCL7 and CCL8, are also produced by tumors and shown to recruit monocytes (van Damme et al., 1992).

Along with the supposed protumoral role of TAMs, the local production of chemokines and the extent of TAMs infiltration have been studied as prognostic factors. For example, in human breast and esophageal cancers, CCL2 levels correlated with the extent of macrophage infiltration, lymph node metastasis, and clinical aggressiveness (Azenshtein et al., 2002; Saji et al., 2001). In an experimental model of nontumorigenic melanoma, a low level of CCL2 secretion with "physiological" accumulation of TAM promoted tumor formation, while a high level of CCL2 secretion resulted in massive macrophage infiltration into the tumor mass and in its destruction (Nesbit

TABLE 16.1 Tumor- and/or Stroma-Derived Chemokines

Ligand	Producing Tumor
CXC family	
CXCL1/Groa	Colon carcinoma (Li et al., 2004)
CXCL8/IL-8	Melanoma (Haghnegahdar et al., 2000), breast (Azenshtein et al., 2005)
CXCL9/Mig	Hodgkin's disease (Teruya et al., 2000)
CXCL10/IP-10	Hodgkin's lymphoma and nasopharyngeal carcinoma (Teichmann et al., 2005)
CXCL12/SDF-1	Melanoma (Scala et al., 2005); prostate, breast, ovary (Rollins et al., 1999); pancreas (Marchesi et al., 2004)
CXCL13/BCA1	Non-Hodgkin's B cell lymphoma (Smith et al., 2003)
CC family	
CCL1/I-309	Adult T cell leucemia (Ruckes et al., 2001)
CCL2/MCP-1	Pancreas (Saji et al., 2001); sarcomas, gliomas, lung, breast, cervix, ovary, melanoma (Mantovani et al., 1999)
CCL3/MIP-1a	Schwann cell tumors (Mori et al., 2004)
CCL3LI/LD78b	Glioblastoma (Kouno et al. 2004)
CCL5/RANTES	Breast (Ueno et al., 2000); melanoma (Payne et al., 2002)
CCL6	Non-small-cell lung (Yi et al., 2003)
CCL7/MCP-3	Osteosarcoma (Conti et al., 2004)
CCL8/MCP-2	Osteosarcoma (Conti et al., 2004)
CCL11/eotaxin	T-cell lymphoma (Kleinhans et al., 2003)
CCL17/TARC	Lymphoma (Vermeer et al., 2002)
CCL18/PARC	Ovary (Schutyser et al., 2002)
CCL22/MDC	Ovary (Sakaguchi et al., 2005)
CCL28/MEC	Hodgkin's disease (Hanamoto et al., 2004)

et al., 2001). In pancreatic cancer patients, high serum levels of CCL2 were associated with a more favorable prognosis and with a lower proliferative index of tumor cells (Monti et al., 2003). These biphasic effects of CCL2 are consistent with the "macrophage balance" hypothesis (Mantovani et al., 1992) and emphasize the concept that levels of macrophage infiltration similar to those observed in human malignant lesions express protumor activity (Bingle et al., 2000).

A variety of other chemokines have been detected in neoplastic tissues as products of either tumor cells or stromal elements (Table 16.1). These molecules play an important role in tumor progression by direct stimulation of neoplastic growth, promotion of inflammation, and induction of angiogenesis. In spite of constitutive production of neutrophil chemotactic proteins by tumor cells, by CXCL8, and by related chemokines, neutrophils are not a major and obvious constituent of the leukocyte infiltrate. However, these cells, though present in minute numbers, may play a key role in triggering and sustaining the inflammatory cascade. Macrophages are also recruited by molecules other than chemokines. In particular, tumor-derived cytokines interacting with tyrosine kinase receptors, such as vascular endothelial growth factor (VEGF) and macrophage colony-stimulating factor (MCSF) (Lin et al., 2001; Duyndam et al., 2002), promote macrophage recruitment as well as macrophage survival and proliferation, the latter generally limited to murine TAMs (Mantovani et al., 2002; Lin et al., 2001; Duyndam et al., 2002) (Figure 16.1). Using genetic approaches, it has been demonstrated that depletion of MCSF markedly decreases the infiltration of macrophages at the tumor site, and this correlates with a significant delay in tumor progression. By contrast, overexpression of MCSF by tumor cells dramatically increased macrophage recruitment, and this was correlated with accelerated tumor growth (Lin *et al.*, 2001; Nowicki *et al.*, 1996; Aharinejad *et al.*, 2002). MCSF overexpression is common among tumors of the reproductive system, including ovarian, uterine, breast, and prostate cancers, and correlates with poor prognosis (Pollard *et al.*, 2004). Placenta-derived growth factor (PlGF), a molecule related to VEGF in terms of structure and receptor usage, has been reported to promote the survival of TAMs (Adini *et al.*, 2002).

IV. TAM EXPRESSION OF SELECTED M2 PROTUMORAL FUNCTIONS

The cytokine network expressed at the tumor site plays a central role in the orientation and differentiation of recruited mononuclear phagocytes, thus contributing to direct the local immune system away from antitumor functions (Mantovani *et al.*, 2002). This idea is supported by both preclinical and clinical observations (Bingle *et al.*, 2000; Goerdt *et al.*, 1999) that clearly demonstrate an association between macrophage number/density and prognosis in a variety of murine and human malignancies.

The immunosuppressive cytokines IL-10 and transforming growth factor β (TGF-β) are produced by both ovarian cancer cells and TAMs (Mantovani *et al.*, 2002). IL-10 promotes the differentiation of monocytes to mature macrophages and blocks their differentiation to DCs (Allavena *et al.*, 2000) (Figure 16.2). Thus, a gradient of tumor-derived IL-10 may account for differentiation along the DC pathway versus the macrophage pathway in different microanatomical localizations in a tumor. Such

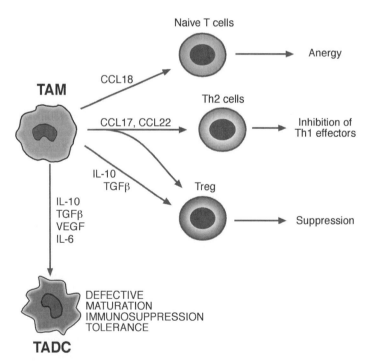

FIGURE 16.2 **TAMs suppressing adaptive immunity.** TAMs produce cytokines that negatively modify the outcome of a potential antitumor response. IL-10, IL-6, VEGF, and TGF-β inhibit the maturation and activation of tumor-associated dendritic cells (TADCs). IL-10, TGF-β, and selected chemokines act on T helper 2 (Th2)-polarized lymphocytes and T regulatory cells (Tregs), which are ineffective in antitumor immunity and suppress antitumor responses. Figure modified from Sica *et al.* (2006). (See color plate.)

situation was observed in papillary carcinoma of the thyroid, where TAMs are evenly distributed throughout the tissue, in contrast to DCs that are present in the periphery (Scarpino et al., 2000). In breast carcinoma, DCs with a mature phenotype (DC-LAMP⁺) were localized in peritumoral areas, while immature DCs were inside the tumor (Bell et al., 1999). Interestingly, it was shown that Stat3 is constitutively activated in tumor cells (Wang et al., 2004) and in diverse tumor-infiltrating immune cells (Kortylewski et al., 2005), leading to inhibition of the production of several proinflammatory cytokines and chemokines and to the release of factors that suppress DC maturation. It is noteworthy that ablating Stat3 in hematopoietic cells triggers an intrinsic immune-surveillance system that inhibits tumor growth and metastasis (Kortylewski et al., 2005).

As previously discussed, IL-10 promotes the M2c alternative pathway of macrophage activation and induces TAMs to express M2-related functions. Indeed, under many aspects, TAMs summarize a number of functions expressed by M2 macrophages; they are involved in tuning inflammatory responses and adaptive immunity, scavenge debris, promote angiogenesis, as well as tissue remodeling and repair. The production of IL-10, TGF-β, and prostaglandin E_2 (PGE2) by cancer cells and TAMs (Mantovani et al., 2002) contributes to a general suppression of antitumor activities (Figure 16.2).

In situ in ovarian cancers, TAMs are poor producers of NO (DiNapoli et al., 1996), and only a minority of macrophages localized at the periphery in a minority of tumors scored positive for iNOS (Klimp et al., 2001). Moreover, in contrast to M1 polarized macrophages, TAMs have been shown to be poor producers of reactive oxygen species (ROS), consistent with the hypothesis that these cells represent a skewed M2 population (Klimp et al., 2001).

TAMs also express low levels of inflammatory cytokines (e.g., IL-12, IL-1β, TNF-α, IL-6) (Mantovani et al., 2002). Activation of NF-κB is a necessary event promoting transcription of several proinflammatory genes. Previous studies (Sica et al., 2000) indicated that TAMs display defective NF-κB activation in response to the M1-polarizing signal LPS, and similar results were observed in response to the proinflammatory cytokines TNF-α and IL-1β (Biswas et al., 2006; unpublished observation). Thus, in terms of cytotoxicity and expression of inflammatory cytokines, TAMs resemble the M2 macrophages. Unexpectedly, TAMs display a high level of interferon regulatory factor (IRF)-3/Stat1 activation, which may be part of the molecular events promoting TAM-mediated T cell deletion (Kusmartsev et al., 2005).

In agreement with the M2 signature, TAMs also express high levels of both the scavenger receptor-A (SR-A) (Biswas et al., 2006) and the mannose receptor (MR) (Allavena et al., 2005). Further, TAMs are poor antigen-presenting cells (APCs) (Mantovani et al., 2002).

Arginase expression in TAMs has not been studied. However, it has been proposed that the carbohydrate-binding protein galectin-1, which is abundantly expressed by ovarian cancer (van den Brule et al., 2003) and shows specific anti-inflammatory effects, tunes the classic pathway of L-arginine, resulting in a strong inhibition of the nitric oxide production by lipopolysaccharide-activated macrophages.

Angiogenesis is an M2-associated function that represents a key event in tumor growth and progression. In several studies in human cancer, TAM accumulation has been associated with angiogenesis and with the production of angiogenic factors, such as VEGF and platelet-derived endothelial cell growth factor (PDEGF) (Mantovani et al., 2002). Moreover, in human cervical cancer, VEGF-C production by TAMs was proposed to play a role in peritumoral lymphoangiogenesis and subsequent dissemination of cancer cells with formation of lymphatic metastasis (Schoppmann et al.,

2002). In addition, TAMs participate in the proangiogenic process by producing the angiogenic factor thymidine phosphorylase (TP), which promotes endothelial cell migration *in vitro* and whose levels of expression are associated with tumor neovascularization (Hotchkiss *et al.*, 2003). TAMs contribute to tumor progression also by producing proangiogenic and tumor-inducing chemokines, such as CCL2 (Vicari *et al.*, 2002). Moreover, TAMs accumulate in hypoxic regions of tumors, and hypoxia triggers a proangiogenic program in these cells (see following paragraphs). Therefore, macrophages recruited *in situ* represent an indirect pathway of amplification of angiogenesis, in concert with angiogenic molecules directly produced by tumor cells. On the antiangiogenic side, in a murine model, GM-CSF released from a primary tumor upregulated TAM-derived metalloelastase and angiostatin production, thus suppressing tumor growth of metastases (Dong *et al.*, 1998).

TAMs also express additional molecules that affect tumor cell proliferation, angiogenesis, and dissolution of connective tissues, including epidermal growth factor (EGF), members of the fibroblast growth factor (FGF) family, TGF-β, VEGF, and various chemokines. In lung cancer, TAMs may favor tumor progression by contributing to stroma formation and angiogenesis through their release of PDGF, in conjunction with TGF-β1 production by cancer cells (Mantovani *et al.*, 2002). Macrophages can produce enzymes and inhibitors that regulate the digestion of the extracellular matrix, such as matrix metalloproteinase (MMP), plasmin, urokinase-type plasminogen activator (uPA), and the uPA receptor. Direct evidence has been presented that MMP-9 derived from hematopoietic cells of host origin contributes to skin carcinogenesis (Coussens *et al.*, 2000). Chemokines have been shown to induce gene expression of various MMPs and, in particular, MMP-9 production, along with the uPA receptor (Locati *et al.*, 2002). Evidence suggests that MMP-9 has complex effects beyond matrix degradation, including promotion of the angiogenesis switch and release of growth factors (Coussen *et al.*, 2000).

The mechanisms responsible for the M2 polarization of TAMs have not been completely defined. Data derived from ovarian and pancreatic tumors point to tumor-derived signals that promote M2 differentiation of mononuclear phagocytes (Hagemann *et al.*, 2006; Marchesi, 2005).

V. MODULATION OF ADAPTIVE IMMUNITY BY TAMS

It has long been known that TAMs have poor antigen-presenting capacity and can actually suppress T cell activation and proliferation (Mantovani *et al.*, 2002). The suppressive mediators produced by TAMs include prostaglandins, IL-10 and TGF-β, and indoleamine 2,3-dioxygenase (IDO) (Mantovani *et al.*, 2004). Moreover, TAMs are unable to produce IL-12, even upon stimulation by IFN-γ and LPS (Sica *et al.*, 2000). With this cytokine profile, which is characteristic of M2 macrophages, TAMs are unable to trigger Th1-polarized immune responses but rather induce T regulatory cells (Tregs) (Figure 16.2). Tregs possess a characteristic anergic phenotype and strongly suppress the activity of effector T cells and other inflammatory cells, such as monocytes. Infiltrating Tregs strongly affect the tumor microenvironment by producing a high level of immunosuppressive cytokines (IL-10, TGF-β) (Jarnicki *et al.*, 2006). Suppression of T-cell mediated antitumor activity by Tregs is associated with increased tumor growth and, hence, decreased survival (Sakaguchi *et al.*, 2005). For instance, in patients with advanced ovarian cancer, an increase in the number of functionally active Tregs present in the ascites was predictive of reduced survival (Curiel *et al.*, 2004). Immature myeloid suppressor cells present in the neoplastic tissue of some tumors have been shown to potently inhibit

T cell responses (Bronte et al., 2003). The relationship, if any, of immature myeloid suppressor cells with TAMs remains to be defined.

The complex network of chemokines present at the tumor site can play a role also in the induction of the adaptive immunity. Chemokines also regulate the amplification of polarized T cell responses (Figure 16.2). Some chemokines may enhance specific host immunity against tumors; however, other chemokines may contribute to escape from the immune system, by recruiting Th2 effectors and Tregs (Mantovani et al., 2004). As mentioned earlier, in addition to being a target for chemokines, TAMs are a source of a selected set of these mediators: CCL2, CCL17, CCL18, and CCL22. CCL18 was recently identified as the most abundant chemokine in human ovarian ascites fluid. When the source of CCL18 was investigated, it was tracked to TAMs, with no production by ovarian carcinoma cells (Schutyser et al., 2002). CCL18 is a CC chemokine produced constitutively by immature DCs and inducible in macrophages by IL-4, IL-13, and IL-10. Since IL-4 and IL-13 are not expressed in substantial amounts in ovarian cancer, it is likely that IL-10, produced by tumor cells and macrophages themselves, accounts for CCL18 production by TAMs. CCL18 is an attractant for naive T cells by interacting with an unidentified receptor (Adema et al., 1997). Attraction of naive T cells in a peripheral microenvironment dominated by M2 macrophages and immature DCs is likely to induce T cell anergy.

Work in gene-modified mice has shown that CCL2 can orient specific immunity in a Th2 direction. Although the exact mechanism for this action has not been defined, it may include stimulation of IL-10 production in macrophages (Gu et al., 2000). Overall, TAM-derived chemokines most frequently recruit effector T cell, but the recruitment is inefficient to mount a protective antitumor immunity. TAMs also produce chemokine, specifically attracting T cells with immunosuppressive functions.

VI. TARGETING TAMS

A. Activation

Defective NF-κB activation in TAMs correlates with impaired expression of NF-κB-dependent inflammatory functions (e.g., expression of cytotoxic mediators, NO) and cytokines, including TNF-α, IL-1, and IL-12 (Mantovani et al., 2002; Sica et al., 2000; Torroella et al., 2005). Restoration of NF-κB activity in TAMs is therefore a potential strategy to restore M1 inflammation and intratumoral cytotoxicity. In agreement, recent evidence indicates that restoration of an M1 phenotype in TAMs may provide therapeutic benefit in tumor-bearing mice. In particular, combination of CpG plus an anti-IL-10 receptor antibody switched infiltrating macrophages from M2 to M1 and triggered innate response, debulking large tumors within 16 hours (Guiducci et al., 2005). It is likely that this treatment may restore NF-κB activation and inflammatory functions by TAMs. Moreover, TAMs from STAT6$^{-/-}$ tumor-bearing mice display an M1 phenotype, with a low level of arginase and a high level of NO. As a result, these mice immunologically rejected spontaneous mammary carcinoma (Sihna et al., 2005). These data suggest that switching the TAM phenotype from M2 to M1 during tumor progression may promote antitumor activities. In this regard, the SHP-1 phosphatase was shown to play a critical role in programming macrophage M1 versus M2 functions. Mice deficient for SHP-1 display a skewed development away from M1 macrophages, which have high iNOS levels and produce NO, toward M2 macrophages, which have high arginase levels and produce ornithine (Rauh et al., 2004).

Reports have identified a myeloid M2-biased cell population in lymphoid organs and peripheral tissues of tumor-bearing hosts, referred to as the myeloid suppressor cells (MSCs), which are suggested to contribute to the immunosuppressive phenotype (Bronte et al., 2005a). These cells are phenotypically distinct from TAM and are characterized by the expression of the Gr-1 and CD11b markers. MSCs use two enzymes involved in the arginine metabolism to control T cell response: inducible nitric oxide synthase and arginase 1, which deplete the milieu of arginine and cause peroxinitrite generation, as well as lack of CD3-ζ chain expression and T cell apoptosis. In prostate cancer, selective antagonists of these two enzymes were proved beneficial in restoring T cell-mediated cytotoxicity (Bronte et al., 2005b). Based on the observation that constitutively activated Stat3 play a pivotal role in human tumor malignancy, the discovery of selective inhibitors of this pathway appears a promising strategy with antitumor activity against human and murine cancer cells in mice (Blaskovich et al., 2003).

B. Recruitment

Chemokines and chemokine receptors are a prime target for the development of innovative therapeutic strategies in the control of inflammatory disorders. Study results suggest that chemokine inhibitors could affect tumor growth by reducing macrophage infiltration. Preliminary results in CCL2/MCP-1 gene-targeted mice suggest that this chemokine can indeed promote progression in a HER2/neu-driven spontaneous mammary carcinoma model (Conti et al., 2004). Thus, available information suggests that chemokines represent a valuable therapeutic target in neoplasia. Colony stimulating factor (CSF)-1 was identified as an important regulator of mammary tumor progression to metastasis, by regulating infiltration, survival, and proliferation of TAMs. Transgenic expression of CSF-1 in mammary epithelium led to the acceleration of the late stages of carcinoma and increased lung metastasis, suggesting that agents directed at CSF-1/CSF-1R activity could have important therapeutic effects (Pollard, 2004; Aharinejad et al., 2002).

Goswami et al. (2005) described a new role of TAMs in promoting invasion of breast carcinoma cells via a CSF-1/EGF paracrine loop. Thus, disruption of this circuit by blockade of either EGF receptor or CSF-1 receptor signaling may represent a new therapeutic strategy to inhibit both macrophage and tumor cell migration and invasion.

One study found that anti-MCSF antibodies, which interfere with TAM recruitment in different mammary carcinoma models, restored susceptibility *in vivo* to combination chemotherapy, implying a role of TAMs in chemoresistance of tumors (Paulus et al., 2006).

Genes of the Wnt family play a crucial role in cellular proliferation, migration, and tissue patterning during embryonic development. Activation of Wnt5a member in macrophages cocultured with breast cancer cells was shown to induce cancer cell invasion through a TNF-α-mediated induction of MMP-7. This novel circuit links the migration-regulating Wnt pathway with the proteolytic cascade, both mechanisms being indispensable for successful invasion. In addition, the Wnt antagonist dickkopf-1 has been reported to inhibit cancer cell invasiveness (Pukrop et al., 2006).

Study findings have also shed new light on the links between certain TAM chemokines and genetic events that cause cancer. The CXCR4 receptor lies downstream of the vonHippel/Lindau/hypoxia-inducible factor (HIF) axis. Transfer of activated *ras* into the cervical carcinoma line HeLa induces IL-8/CXCL8 production that is sufficient to promote angiogenesis and progression. Moreover, a frequent early and sufficient gene rearrangement that causes papillary thyroid carcinoma (Ret-PTC) activates an inflammatory genetic program that includes CXCR4 and inflammatory chemokines in primary human thyrocytes (Borello et al., 2005). The emerging direct connections between oncogenes, inflammatory mediators, and the chemokine system provide a strong impetus for exploration of the anticancer potential of anti-inflammatory strategies. It was further demonstrated in non-small-cell lung cancer (NSCLC) that mutation of the tumor suppressor gene PTEN results in upregulation of HIF-1 activity and ultimately in HIF-1-dependent transcription of the CXCR4 gene, which provides a mechanistic basis for the upregulation of CXCR4 expression and promotion of metastasis formation (Phillips et al., 2005). It appears, therefore, that targeting HIF-1 activity may disrupt the HIF-1/CXCR4 pathway and affect TAM accumulation as well as cancer cell spreading and survival. The HIF-1-inducible VEGF is commonly produced by tumors and elicits monocyte migration. There is evidence that VEGF can significantly contribute to macrophage recruitment in tumors. Along with CSF-1, this molecule also promotes macrophage survival and proliferation. Due to the localization of TAMs into the hypoxic regions of tumors, viral vectors were used to transduce macrophages with therapeutic genes, such as IFN-γ, that were activated only in low oxygen conditions (Carta et al., 2001). These works present promising approaches that use macrophages as vehicles to deliver gene therapy in regions of tumor hypoxia.

C. Angiogenesis

VEGF is a potent angiogenic factor as well as a monocyte attractant that contributes to TAM recruitment. TAMs promote angiogenesis, and there is evidence that inhibition of TAM recruitment plays an important role in antiangiogenic strategies. In addition, the angiogenic program established by hypoxia may also rely on the increased expression of CXCR4 by TAMs and endothelial cells (Schioppa et al., 2003). Intratumoral injection of CXCR4 antagonists, such as the bicyclam AMD3100, may potentially work as in vivo inhibitors of tumor angiogenesis. Linomide, an antiangiogenic agent, caused significant reduction of the tumor volume in a murine prostate cancer model by inhibiting the stimulatory effects of TAM on tumor angiogenesis (Joseph et al., 1998). Based on this observation, the effects of linomide or other antiangiogenic drugs on the expression of proangiogenic and antiangiogenic molecules by TAMs may define useful targets for anticancer therapy.

D. Survival

Antitumor agents with selective cytotoxic activity on monocyte/macrophages would be ideal therapeutic tools for their combined action on tumor cells and TAMs. It has been reported that trabectedin (Yondelis®), a natural product derived from the marine organism Ecteinascidia turbinata that has potent antitumor activity (Sessa et al., 2005), is specifically cytotoxic to macrophages and TAMs while sparing lymphocytes. This compound inhibits NF-Y, a transcription factor of major importance for differentiation of mononuclear phagocytes. In addition, trabectedin inhibits the production of CCL2 and IL-6 both by TAMs and tumor cells (Allavena et al., 2005). These anti-inflammatory properties of trabectedin may be an extended mechanism of its antitumor activity. Finally, proinflammatory cytokines (e.g., IL-1 and TNF-α), expressed by infiltrating leukocytes, can activate NF-κB in cancer cells and contribute to their proliferation, survival, and metastasis (Balkwill et al., 2001, 2005), thus representing potential anticancer targets.

E. Matrix Remodeling

Macrophages are major source of proteases, and, interestingly, MMP-9 has been found to be preferentially expressed in M2 versus M1 macrophages (Martinez et al., 2006). TAMs produce several matrix MMPs (e.g., MMP-2, MMP-9) that degrade proteins of the extracellular matrix and also produce activators of MMPs, such as chemokines (de Visser et al., 2006). Inhibition of this molecular pattern may prevent degradation of extracellular matrix as well as tumor cell invasion and migration. TAMs or neutrophil-derived proteases (e.g., MMP-9 or cathepsin B) stimulate cancer invasion and metastasis (Hanahan et al., 2000; Condeelis et al., 2006; Vasiljeva et al., 2006).

The biphosphonate zoledronic acid is a prototypical MMP inhibitor. In cervical cancer, this compound has suppressed MMP-9 expression by infiltrating macrophages and has inhibited metalloprotease activity, reducing angiogenesis and cervical carcinogenesis (Giraudo et al., 2004). The halogenated bisphosphonate derivative clodronate is a macrophage toxin that depletes selected macrophage populations. Given the clinical usage of this agent and similar agents, it is important to assess whether they may have potential as TAM toxins. In support of this hypothesis, clodronate encapsulated in liposomes efficiently depleted TAMs in murine teratocarcinoma and human rhabdomyosarcoma mouse tumor models, resulting in significant inhibition of tumor growth (Zeisberger et al., 2006).

The secreted protein is acidic and rich in cysteine (SPARC) and has gained much interest in cancer research, being either upregulated or downregulated in progressing tumors. SPARC produced by macrophages present in tumor stroma can modulate collagen density, leukocytes, and blood vessel infiltration (Sangaletti et al., 2003).

F. Effector Molecules

Cyclooxygenase (COX) is a key enzyme in the prostanoid biosynthetic pathway. COX-2 is upregulated by activated oncogenes (i.e., **-catenin, MET) but is also produced by TAMs in response to tumor-derived factors (e.g., mucin in colon cancer). The usage of COX-2 inhibitors in the form of nonsteroidal anti-inflammatory drugs is associated with reduced risk of diverse tumors (colorectal, esophageal, lung, stomach, and ovary). Indeed, it is believed that selective COX-2 inhibitors will be useful additions to combination cancer therapy (Colombo et al., 2005).

Suppressor of cytokine signaling 1 (SOCS1) deficiency is associated to IFN-γ-dependent spontaneous development of colorectal carcinomas. Under these conditions, accumulation of aberrantly activated TAMs is observed *in situ*, and these cells account for expression of carcinogenesis-related enzymes COX-2 and iNOS (Hanada et al., 2006).

The IFN-γ-inducible enzyme IDO has become a well-known suppressor of T cell activation. It catalyzes the initial rate-limiting step in tryptophan catabolism, which leads to the biosynthesis of nicotinamide adenine dinucleotide. By depleting tryptophan from the local microenvironment, IDO blocks activation of T lymphocytes. Ectopic expression of IDO in tumor cells has been shown to inhibit T cell responses (Mellor et al., 2002). It has been shown that inhibition of IDO may cooperate with cytotoxic agents to elicit regression of established tumors and may increase the efficacy of cancer immunotherapy (Muller et al., 2005).

Prostate cancer is strongly linked to inflammation based on epidemiological and molecular analysis (Balkwill et al., 2005). TAMs mediate hormone resistance in prostate cancer by a nuclear receptor derepression pathway. TAMs in prostate cancer convert selective androgen receptor antagonists/modulators into agonists through an IL-1 pathway (Zhu et al., 2006).

TNF promotes asbestos carcinogenesis by blocking death of mesothelial cells via the NF-κB pathway (Yang et al., 2006). Figure 16.3 summarizes therapeutic approaches to prevent TAMs protumoral functions.

VII. CONCLUDING REMARKS

Although the presence of TAMs has been long considered evidence of a host response against the growing tumor, it has become increasingly clear that TAMs are actually active players in the process of tumor progression and invasion. Molecular and

VII. CONCLUDING REMARKS

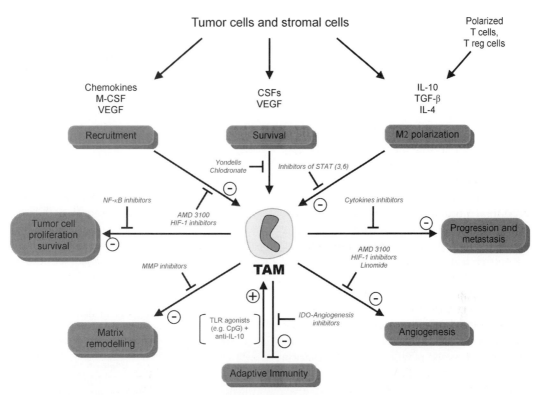

FIGURE 16.3 Multifaceted therapeutic approaches to prevent protumoral functions of TAMs. TAMs promote tumor progression by promoting angiogenesis and suppressing adaptive immunity, matrix remodeling, tumor progression, and metastasis. The figure summarizes strategies to block protumoral functions of TAMs (−) or rescue antitumor activities by converting TAMs to M1 macrophages (+). Cytotoxic drugs (e.g., Yondelis®) may decrease TAM number and prevent protumoral functions. A similar result may be obtained by limiting TAM recruitment (Linomide®, HIF-1 inhibitors, AMD3100). Restoration of M1 immunity (Stat3 and Stat6 inhibitors; anti-IL-10 plus CpG; IDO inhibitors) would provide cytotoxic activity and reactivation of Th1-specific antitumor immunity. Inhibition of both proinflammatory cytokines and growth factors expression (NF-κB inhibitors) may disrupt inflammatory circuits supporting tumor growth and progression. MMP inhibitors would prevent cancer cells spread and metastasis. Finally, inhibitors of TAM-mediated angiogenesis (Linomide®, HIF-1 inhibitors, AMD3100) would restrain blood supply and inhibit tumor growth. MCSF, macrophage colony-stimulating factor; VEGF, vascular endothelial growth factor; CSFs, colony stimulating factors; IL, interleukin; TGF-β, transforming growth factor β; IDO, indoleamine 2,3-dioxygenase; MMP inhibitors, matrix metalloproteinase inhibitors; TLR agonists, Toll-like receptor agonists; Stat, signal transducers and activators of transcription; NF-κB, nuclear factor kappa B.

biological studies have been supported by a large number of clinical studies that found a significant correlation between the high macrophage content of tumors and poor patient prognosis. TAMs share many similarities with prototypical polarized populations of M2 mononuclear phagocytes in terms of their expression and functions. In parallel to the known properties of M2 macrophage populations, several lines of evidence suggest that TAMs promote tumor progression and metastasis by activating circuits that regulate tumor growth, adaptive immunity, stroma formation, and angiogenesis. Analysis of the mechanisms mediating this phenotype involve detective NF-κB activation (Sica et al., 2000), an event likely responsible for the inability of TAMs to mount effective M1 inflammatory responses (Mantovani et al., 2002; Biswas et al., 2006). Clarification of the mechanisms that promote differentiation of macrophages toward the M2 phenotype is expected to disclose valuable new therapeutic tactics to attack tumors.

Several studies have defined key molecules and pathways that drive recruitment and activation of TAMs, and the TAM transcriptome in a murine fibrosarcoma has been defined (Biswas et al., 2006). However, it remains the case that TAM functions have only been characterized significantly in animal models; at the time of this writing, there is at best only a partial phenotypic characterization of TAMs in human cancers. Moreover, it remains unknown whether the different tumor microenvironments that are likely established by different tumor types drive functionally distinct phenotypes in TAMs, thereby generating specific protumor or antitumor activities. For example, as new players in inflammation have expressed on TAMs and other cells, findings have suggested a generally protumoral function for Toll-like receptors (TLR) in the cancer microenvironment (Mantovani et al., 2006).

In closing, mechanisms of TAM recruitment, activation, and polarization offer important new areas to attack tumor immune suppression, and the functional heterogenicity of TAMs in tumors continues to represent an important area of investigation for innovative anticancer strategies.

Acknowledgments

This work was supported by Associazione Italiana Ricerca sul Cancro (AIRC), Italy; by the European Community and by Ministero Istruzione Università Ricerca (MIUR), Italy; and by the Istituto Superiore Sanitá (ISS).

References

Adema, G. J., Hartgers, F., Verstraten, R., de Vries. E., Marland, G., Menon, S., Foster, J., Xu, Y., Nooyen, P., McClanahan, T., Bacon, K. B., and Figdor, C. G. (1997). A dendritic-cell-derived C-C chemokine that preferentially attracts naive T cells. *Nature* **387**, 713–717.

Adini, T., Kornaga, F., Firoozbakht, F., and Benjamin, L. E. (2002). Placental growth factor is a survival factor for tumor endothelial cells and macrophages. *Cancer Res.* **62**, 2749–2752.

Aharinejad, S., Abraham, D., Paulus, P., Abri, H., Hofmann, M., Grossschmidt, K., Schafer, R., Stanley, E. R., and Hofbauer, R. (2002). Colony-stimulating factor-1 antisense treatment suppresses growth of human tumor xenografts in mice. *Cancer Res.* **62**, 5317–5324.

Allavena, P., Sica, A., Vecchi, A., Locati, M., Sozzani, S., and Mantovani, A. (2000). The chemokine receptor switch paradigm and dendritic cell migration: Its significance in tumor tissues. *Immunol. Rev.* **177**, 141–149.

Allavena, P., Signorelli, M., Chieppa, M., Erba, E., Bianchi, G., Marchesi, F., Olimpio, C. O., Bonardi, C., Garbi, A., Lissoni, A., de Braud, F., Jimeno, J., and D'Incalci, M. (2005) Anti-inflammatory properties of the novel antitumor agent Yondelis (trabectedin) inhibition of macrophage differentiation and cytokine production. *Cancer Res.* **65**, 2964–2971.

Azenshtein, E., Luboshits, G., Shina, S., Neumark, E., Shahbazian, D., Weil, M., Wigler, N., Keydar, I., and Ben-Baruch, A. (2002). The CC chemokine RANTES in breast carcinoma progression: Regulation of expression and potential mechanisms of promalignant activity, *Cancer Res.* **62**, 1093–1102.

Azenshtein, E., Meshel, T., Shina, S., Barak, N., Keydar, I., and Ben-Baruch, A. (2005). The angiogenic factors CXCL8 and VEGF in breast cancer: Regulation by an array of pro-malignancy factors. *Cancer Lett.* **217**, 73–86.

Balkwill, F. (2004) Cancer and the chemokine network. *Nat. Rev. Cancer* **4**, 540–550.

Balkwill, F., and Mantovani, A. (2001). Inflammation and cancer: Back to Virchow? *Lancet* **357**, 539–545.

Balkwill, F., Charles, K. A., and Mantovani, A. (2005). Smoldering and polarized inflammation in the initiation and promotion of malignant disease. *Cancer Cell* **7**, 211–217.

Bell, D., Chomarat, P., Broyles, D., Netto, G., Harb, G. M., Lebecque, S., Valladeau, J., Davoust, J., Palucka, K. A., and Banchereau, J. (1999). In breast carcinoma tissue, immature dendritic cells reside within the tumor, whereas mature dendritic cells are located in peritumoral areas. *J. Exp. Med.* **190**, 1417–1426.

Bingle, L., Brown, N. J., and Lewis, C. E. (2000). The role of tumor-associated macrophages in tumor progression: Implications for new anticancer therapies. *J. Pathol.* **196**, 254–265.

Biswas, S. K., Gangi, L., Paul, S., Schioppa, T., Saccani, A., Sironi, M., Bottazzi, B., Doni, A., Vincenzo, B., Pasqualini, F., Vago, L., Nebuloni, M., Mantovani, A., and Sica, A. (2006). A distinct and unique transcriptional program expressed by tumor-associated macrophages (defective NF-kappaB and enhanced IRF-3/STAT1 activation). *Blood* **107**, 2112–2122.

Blaskovich, M. A., Sun, J., Cantor, A., Turkson, J., Jove, R., and Sebti, S. M. (2003). Discovery of JSI-124 (cucurbitacin I), a selective Janus kinase/signal transducer and activator of transcription 3 signaling pathway inhibitor with potent antitumor activity against human and murine cancer cells in mice. *Cancer Res.* **63**, 1270–1279.

Borrello, M. G., Alberti, L., Fischer, A., Degl'innocenti, D., Ferrario, C., Gariboldi, M., Marchesi, F., Allavena, P., Greco, A., Collini, P., Pilotti, S., Cassinelli, G., Bressan, P., Fugazzola, L., Mantovani, A., and Pierotti, M. A. (2005). Induction of a proinflammatory program in normal human thyrocytes by the RET/PTC1 oncogene. *Proc. Natl. Acad. Sci. USA* **102**, 14825–14830.

Bottazzi, B., Polentarutti, N., Acero, R., Balsari, A., Boraschi, D., Ghezzi, P., Salmona, M., and Mantovani, A. (1983). Regulation of the macrophage content of neoplasms by chemoattractants. *Science* **220**, 210–212.

Bronte, V., Serafini, P., Mazzoni, A., Segal, D. M., and Zanovello, P. (2003). L-arginine metabolism in myeloid cells controls T-lymphocyte functions. *Trends Immunol.* **24**, 302–306.

Bronte, V., and Zanovello, P. (2005a). Regulation of immune responses by L-arginine metabolism. *Nat. Rev. Immunol.* **5**, 641–654.

Bronte, V., Kasic, T., Gri, G., Gallana, K., Borsellino, G., Marigo, I., Battistini, L., Iafrate, M., Prayer-Galetti, T., Pagano, F., and Viola, A. (2005b). Boosting antitumor responses of T lymphocytes infiltrating human prostate cancers. *J. Exp. Med.* **201**, 1257–1268.

Carta, L., Pastorino, S., Melillo, G., Bosco, M. C., Massazza, S., and Varesio, L. (2001) Engineering of macrophages to produce IFN-gamma in response to hypoxia. *J. Immunol.* **166**, 5374–5380.

Colombo, M. P., and Mantovani, A. (2005). Targeting myelomonocytic cells to revert inflammation-dependent cancer promotion. *Cancer Res.* **65**, 9113–9116.

Condeelis, J., and Pollard, J. W. (2006). Macrophages: Obligate partners for tumor cell migration, invasion, and metastasis. *Cell* **124**, 263–266

Conti, I., and Rollins, B. J. (2004) CCL2 (monocyte chemoattractant protein-1) and cancer. *Semin. Cancer Biol.* **14**, 149–154.

Coussens, L. M., and Werb, Z. (2002). Inflammation and cancer. *Nature* **420**, 860–867.

Coussens, L. M., Tinkle, C. L., Hanahan, D., and Werb, Z. (2000). MMP-9 supplied by bone marrow-derived cells contributes to skin carcinogenesis. *Cell* **103**, 481–490.

Curiel, T. J., Coukos, G., Zou, L., Alvarez, X., Cheng, P., Mottram, P., Evdemon-Hogan, M., Conejo-Garcia, J. R., Zhang, L., Burow, M., Zhu, Y., Wei, S., Kryczek, I., Daniel, B., Gordon, A., Myers, L., Lackner, A., Disis, M. L., Knutson, K. L., Chen, L., and Zou, W. (2004). Specific recruitment of regulatory T cells in ovarian carcinoma fosters immune privilege and predicts reduced survival. *Nat. Med.* **10**, 942–949.

de Visser, K. E., Eichten, A., and Coussens, L. M. (2006). Paradoxical roles of the immune system during cancer development. *Nat. Rev. Cancer* **6**, 24–37.

Desnues, B., Lepidi, H., Raoult, D., and Mege, J. L. (2005). Whipple disease: Intestinal infiltrating cells exhibit a transcriptional pattern of M2/alternatively activated macrophages. *J. Infect. Dis.* **192**, 1642–1646.

Dinapoli, M. R., Calderon, C. L., and Lopez, D. M. (1996). The altered tumoricidal capacity of macrophages isolated from tumor-bearing mice is related to reduced expression of the inducible nitric oxide synthase gene. *J. Exp. Med.* **183**, 1323–1329.

Dinarello, C. A. (2005). Blocking IL-1 in systemic inflammation. *J. Exp. Med.* **201**, 1355–1359.

Dong, Z., J. Yoneda, Kumar, R., and Fidler, I. J. (1998). Angiostatin-mediated suppression of cancer metastases by primary neoplasms engineered to produce granulocyte/macrophage colony-stimulating factor. *J. Exp. Med.* **188**, 755–763.

Ghassabeh, G. H., De Baetselier, P., Brys, L., Noel, W., Van Ginderachter, J. A., Meerschaut, S., Beschin, A., Brombacher, F., and Raes, G. (2006). Identification of a common gene signature for type II cytokine-associated myeloid cells elicited *in vivo* in different pathologic conditions. *Blood* **108**, 575–583.

Giraudo, E., Inoue, M., and Hanahan, D. (2004). An amino-bisphosphonate targets MMP-9-expressing macrophages and angiogenesis to impair cervical carcinogenesis. *J. Clin. Invest.* **114**, 623–633.

Goerdt, S., and Orfanos, C. E. (1999). Other functions, other genes: Alternative activation of antigen-presenting cells. *Immunity* **10**, 137–142.

Gordon, S. (2003). Alternative activation of macrophages. *Nat. Rev. Immunol.* **3**, 23–35.

Goswami, S., Sahai, E., Wyckoff, J. B., Cammer, M., Cox, D., Pixley, F. J., Stanley, E. R., Segall, J. E., and Condeelis, J. S. (2005). Macrophages promote the invasion of breast carcinoma cells via a colony-stimulating factor-1/epidermal growth factor paracrine loop. *Cancer Res.* **65**, 5278–5283

Gu, L., Tseng, S., Horner, R. M., Tam, C., Loda, M., and Rollins, B. J. (2000), Control of TH2 polarization by the chemokine monocyte chemoattractant protein-1, *Nature* **404**, 407–411.

Guiducci, C., Vicari, A. P., Sangaletti, S., Trinchieri, G., and Colombo, M. P. (2005). Redirecting *in vivo* elicited tumor infiltrating macrophages and dendritic cells towards tumor rejection. *Cancer Res.* **65**, 3437–3446.

Hagemann, T., Wilson, J., Burke, F., Kulbe, H., Li, N. F., Pluddemann, A., Charles, K., Gordon, S., and Balkwill, F. R. (2006). Ovarian cancer cells polarize macrophages toward a tumor-associated phenotype. *J. Immunol.* **176**, 5023–5032.

Haghnegahdar, H., Du, J., Wang, D., Strieter, R. M., Burdick, M. D., Nanney, L. B., Cardwell, N., Luan, J., Shattuck-Brandt, R., and Richmond, A. (2000). The tumorigenic and angiogenic effects of MGSA/GRO proteins in melanoma. *J. Leukoc. Biol.* **67**, 53–62.

Hanada, T., Kobayashi, T., Chinen, T., Saeki, K., Takaki, H., Koga, K., Minoda, Y., Sanada, T., Yoshioka, T., Mimata, H., Kato, S., and Yoshimura, A. (2006). IFN-gamma-dependent, spontaneous development of colorectal carcinomas in SOCS1-deficient mice. *J. Exp. Med.* **203**, 1391–1397.

Hanahan, D., and Weinberg, R. A. (2000). The hallmarks of cancer. *Cell* **100**, 57–70

Hanamoto, H., Nakayama, T., Miyazato, H., Takegawa, S., Hieshima, K., Tatsumi, Y., Kanamaru, A., and Yoshie, O. (2004). Expression of CCL28 by Reed-Sternberg cells defines a major subtype of classical Hodgkin's disease with frequent infiltration of eosinophils and/or plasma cells. *Am. J. Pathol.* **164**, 997–1006.

Hotchkiss, A., Ashton, A. W., Klein, R. S., Lenzi, M. L., Zhu, G. H., and Schwartz, E. L. (2003). Mechanisms by which tumor cells and monocytes expressing the angiogenic factor thymidine phosphorylase mediate human endothelial cell migration. *Cancer Res.* **63**, 527–533.

Jarnicki, A. G., Lysaght, J., Todryk, S., and Mills, K. H. (2006). Suppression of antitumor immunity by IL-10 and TGF-beta-producing T cells infiltrating the growing tumor: Influence of tumor environment on the induction of CD4+ and CD8+ regulatory T cells. *J. Immunol.* **177**, 896–904.

Joseph, I. B., and Isaacs, J. T. (1998). Macrophage role in the anti-prostate cancer response to one class of antiangiogenic agents. *J. Natl. Cancer. Inst.* **90**, 1648–1653

Kleinhans, M., Tun-Kyi, A., Gilliet, M., Kadin, M. E., Dummer, R., Burg, G., and Nestle, F. O. (2003). Functional expression of the eotaxin receptor CCR3 in CD30+ cutaneous T-cell lymphoma. *Blood* **101**, 1487–1493.

Klimp, A. H., Hollema, H., Kempinga, C., van der Zee, A. G., de Vries, E. G., and Daemen, T. (2001). Expression of cyclooxygenase-2 and inducible nitric oxide synthase in human ovarian tumors and tumor-associated macrophages. *Cancer Res.* **61**, 7305–7309.

Kortylewski, M., Kujawski, M., Wang, T., Wei, S., Zhang, S., Pilon-Thomas, S., Niu, G., Kay, H., Mule, J., Kerr, W. G., Jove, R., Pardoll, D., and Yu, H. (2005). Inhibiting Stat3 signaling in the hematopoietic system elicits multicomponent antitumor immunity. *Nat. Med.* **11**, 1314–1321.

Kouno, J., Nagai, H., Nagahata, T., Onda, M., Yamaguchi, H., Adachi, K., Takahashi, H., Teramoto, A., and Emi, M. (2004). Up-regulation of CC chemokine, CCL3L1, and receptors, CCR3, CCR5 in human glioblastoma that promotes cell growth. *J. Neurooncol.* **70**, 301–307.

Kusmartsev, S., and Gabrilovich, D. I. (2005). STAT1 signaling regulates tumor-associated macrophage-mediated T cell deletion. *J. Immunol.* **174**, 4880–4891.

Li, A., Varney, M. L., and Singh, R. K. (2004). Constitutive expression of growth regulated oncogene (gro) in human colon carcinoma cells with different metastatic potential and its role in regulating their metastatic phenotype. *Clin. Exp. Metastasis* **21**, 571–579.

Locati, M., Deuschle, U., Massardi, M. L., Martinez, F. O., Sironi, M., Sozzani, S., Bartfai, T., and Mantovani, A. (2002). Analysis of the gene expression profile activated by the CC chemokine ligand 5/RANTES and by lipopolysaccharide in human monocytes. *J. Immunol.* **168**, 3557–3562

Mantovani, A., Sozzani, S., Locati, M., Allavena, P., and Sica, A. (2002). Macrophage polarization: Tumor-associated macrophages as a paradigm for polarized M2 mononuclear phagocytes. *Trends Immunol.* **23**, 549–555.

Mantovani, A., Sica, A., Sozzani, S., Allavena, P., Vecchi, A., and Locati, M. (2004). The chemokine system in diverse forms of macrophage activation and polarization. *Trends Immunol.* **25**, 677–686.

Mantovani, A. (1999). The chemokine system: Redundancy for robust outputs. *Immunol. Today* **20**, 254–257.

Mantovani, A., Bottazzi, B., Colotta, F., Sozzani, S., and Ruco, L. (1992). The origin and function of tumor-associated macrophages. *Immunol. Today* **13**, 265–270.

Mantovani, A., and Garlanda, C. (2006). Inflammation and multiple myeloma: The Toll connection. *Leukemia* **20**, 937–938.

Marchesi, F., Monti, P., Leone, B. E., Zerbi, A., Vecchi, A. (2006). Piemonti, L., Mantovani, A., and Allavena, P. (2004). Increased survival, proliferation, and migration in metastatic human pancreatic tumor cells expressing functional CXCR4. *Cancer Res.* **64**, 8420–8427.

Martinez, F. O., Gordon, S., Locati, M., and Mantovani, A. (2006). Transcriptional profiling of the human monocyte-macrophage differentiation and polarization: New molecules and patterns of gene expression. *J. Immunol.*, **177**, 7303–7311.

Matsushima, K., Larsen, C. G., DuBois, G. C., and Oppenheim, J. J. (1999). Purification and characterization of a novel monocyte chemotactic and activating factor produced by a human myelomonocytic cell line. *J. Exp. Med.* **169**, 1485–1490.

Mellor, A. L., Keskin, D. B., Johnson, T., Chandler, P., and Munn, D. H. (2002). Cells expressing indoleamine 2,3-dioxygenase inhibit T cell responses. *J. Immunol.* **168**, 3771–3776.

Monti, P., Leone, B. E., Marchesi, F., Balzano, G., Zerbi, A., Scaltrini, F., Pasquali, C., Calori, G., Pessi, F., Sperti, C., Di Carlo, V., Allavena, P., and Piemonti, L. (2003). The CC chemokine MCP-1/CCL2 in pancreatic cancer progression: Regulation of expression and potential mechanisms of antimalignant activity. *Cancer Res.* **63**, 7451–7461.

Mori, K., Chano, T., Yamamoto, K., Matsusue, Y., and Okabe, H. (2004). Expression of macrophage inflammatory protein-1alpha in Schwann cell tumors. *Neuropathology* **24**, 131–135.

Muller, A. J., DuHadaway, J. B., Donover, P. S., Sutanto-Ward, E., and Prendergast, G. C. (2005). Inhibition of indoleamine 2,3-dioxygenase, an immunoregulatory target of the cancer suppression gene Bin1, potentiates cancer chemotherapy. *Nat. Med.* **11**, 312–319.

Nesbit, M., Schaider, H., Miller, T. H., and Herlyn, M. (2001). Low-level monocyte chemoattractant protein-1 stimulation of monocytes leads to tumor formation in nontumorigenic melanoma cells. *J. Immunol.* **166**, 6483–6490.

Noel, W., Raes, G., Hassanzadeh-Ghassabeh, G., De Baetselier, P., and Beschin, A. (2004). Alternatively activated macrophages during parasite infections. *Trends Parasitol.* **20**, 126–133.

Paik, S., Shak, S., Tang, G., Kim, C., Baker, J., Cronin, M., Baehner, F. L., Walker, M. G., Watson, D., Park, T., Hiller, W., Fisher, E. R., Wickerham, D. L., Bryant, J., and Wolmark, N. (2004). A multigene assay to predict recurrence of tamoxifen-treated, node-negative breast cancer. *N. Engl. J. Med.* **351**, 2817–2826.

Paulus, P., Stanley, E. R., Schafer, R., Abraham, D., and Aharinejad, S. (2006). Colony-stimulating factor-1 antibody reverses chemoresistance in human MCF-7 breast cancer xenografts. *Cancer Res.* **66**, 4349–4356

Payne, A. S., and Cornelius, L. A. (2002). The role of chemokines in melanoma tumor growth and metastasis. *J. Invest. Dermatol.* **118**, 915–922.

Phillips, R. J., Mestas, J., Gharaee-Kermani, M., Burdick, M. D., Sica, A., Belperio, J. A., Keane, M. P., and Strieter, R. M. (2005). Epidermal growth factor and hypoxia-induced expression of CXC chemokine receptor 4 on non-small cell lung cancer cells is regulated by the phosphatidylinositol 3-kinase/PTEN/AKT/mammalian target of rapamycin signaling pathway and activation of hypoxia inducible factor-1alpha. *J. Biol. Chem.* **280**, 22473–22481.

Pollard, J. W. (2004). Tumour-educated macrophages promote tumor progression and metastasis. *Nat. Rev. Cancer* **4**, 71–78.

Pukrop, T., Klemm, F., Hagemann, T., Gradl, D., Schulz, M., Siemes, S., Trumper, L., and Binder, C. (2006). Wnt 5a signaling is critical for macrophage-induced invasion of breast cancer cell lines. *Proc. Natl. Acad. Sci. USA* **103**, 5454–5459.

Rauh, M. J., Sly, L. M., Kalesnikoff, J., Hughes, M. R., Cao, L. P., Lam, V., and Krystal, G. (2004). The role of SHIP1 in macrophage programming and activation. *Biochem. Soc. Trans.* **32**, 785–788.

Rollins, B. (1999). *Chemokines and Cancer*. Totowa, NJ: Humana Press.

Ruckes, T., Saul, D., Van Snick, J., Hermine, O., and Grassmann, R. (2001). Autocrine antiapoptotic stimulation of cultured adult T-cell leukemia cells by overexpression of the chemokine I-309. *Blood* **98**, 1150–1159.

Saji, H., Koike, M., Yamori, T., Saji, S., Seiki, M., Matsushima, K., and Toi, M. (2001). Significant correlation of monocyte chemoattractant protein-1 expression with neovascularization and progression of breast carcinoma. *Cancer* **92**, 1085–1091.

Sakaguchi, S. (2005). Naturally arising Foxp3-expressing $CD25^+CD4^+$ regulatory T cells in immunological tolerance to self and non-self. *Nat. Immunol.* **6**, 345–352.

Sangaletti, S., Stoppacciaro, A., Guiducci, C., Torrisi, M. R., and Colombo, M. P. (2003). Leukocyte, rather than tumor-produced SPARC, determines stroma and collagen type IV deposition in mammary carcinoma. *J. Exp. Med.* **198**, 1475–1485

Scala, S., Ottaiano, A., Ascierto, P. A., Cavalli, M., Simeone, E., Giuliano, P., Napolitano, M., Franco, R., Botti, G., and Castello, G. (2005). Expression of CXCR4 predicts poor prognosis in patients with malignant melanoma. *Clin. Cancer Res.* **11**, 1835–1841.

Scarpino, S., Stoppacciaro, A., Ballerini, F., Marchesi, M., Prat, M., Stella, M. C., Sozzani, S., Allavena, P., Mantovani, A., and Ruco, L. P. (2000). Papillary carcinoma of the thyroid: Hepatocyte growth factor (HGF) stimulates tumor cells to release chemokines active in recruiting dendritic cells. *Am. J. Pathol.* **156**, 831–837.

Schioppa, T., Uranchimeg, B., Saccani, A., Biswas, S. K., Doni, A., Rapisarda, A., Bernasconi, S., Saccani, S., Nebuloni, M., Vago, L., Mantovani, A., Melillo, G., and Sica, A. (2003) Regulation of the chemokine receptor CXCR4 by hypoxia. *J. Exp. Med.* **198**, 1391–1402

Schoppmann, S. F., Birner, P., Stockl, J., Kalt, R., Ullrich, R., Caucig, C., Kriehuber, E., Nagy, K., Alitalo, K., and Kerjaschki, D. (2002). Tumor-associated macrophages express lymphatic endothelial growth factors and are related to peritumoral lymphoangiogenesis. *Am. J. Pathol.* **161**, 947–956.

Schutyser, E., Struyf, S., Proost, P., Opdenakker, G., Laureys, G., Verhasselt, B., Peperstraete, L., van de Putte, I., Saccani, A., Allavena, P., Mantovani, A., and van Damme, J. (2002). Identification of biologically active chemokine isoforms from ascitic fluid and elevated levels of CCL18/pulmonary and activation-regulated chemokine in ovarian carcinoma. *J. Biol. Chem.* **277**, 24584–24593.

Scotton, C. J., Martinez, F. O., Smelt, M. J., Sironi, M., Locati, M., Mantovani, A., and Sozzani, S. (2005). Transcriptional profiling reveals complex regulation of the monocyte IL-1 beta system by IL-13. *J. Immunol.* **174**, 834–845.

Sessa, C., De Braud, F., Perotti, A., Bauer, J., Curigliano, G., Noberasco, C., Zanaboni, F., Gianni, L., Marsoni, S., Jimeno, J., D'Incalci, M., Dall'O, E., and Colombo, N. (2005). Trabectedin for women with ovarian carcinoma after treatment with platinum and taxanes fails. *J. Clin. Oncol.* **23**, 1867–1874.

Sica, A., Saccani, A., Bottazzi, B., Polentarutti, N., Vecchi, A., van Damme, J., and Mantovani, A. (2000). Autocrine production of IL-10 mediates defective IL-12 production and NF-kappa B activation in tumor-associated macrophages. *J. Immunol.* **164**, 762–767.

Sica, A., Schioppa, T., Mantovani, A., Allavena, P. (2006). Tumour-associated macrophages are a distinct M2 polarised population promoting tumor progression: Potential targets of anti-cancer therapy. *Eur. J. Cancer* **42**, 717–727.

Sinha, P., Clements, V. K., and Ostrand-Rosenberg, S. (2005). Reduction of myeloid-derived suppressor cells and induction of M1 macrophages facilitate the rejection of established metastatic disease. *J. Immunol.* **174**, 636–645.

Smith, J. R., Braziel, R. M., Paoletti, S., Lipp, M., Uguccioni, M., and Rosenbaum, J. T. (2003). Expression of B-cell-attracting chemokine 1 (CXCL13) by malignant lymphocytes and vascular endothelium in primary central nervous system lymphoma. *Blood* **101**, 815–821.

Takahashi, H., Tsuda, Y., Takeuchi, D., Kobayashi, M., Herndon, D. N., and Suzuki, F. (2004). Influence of systemic inflammatory response syndrome on host resistance against bacterial infections. *Crit. Care Med.* **32**, 1879–1885.

Teichmann, M., Meyer, B., Beck, A., and Niedobitek, G. (2005). Expression of the interferon-inducible chemokine IP-10 (CXCL10), a chemokine with proposed anti-neoplastic functions, in Hodgkin lymphoma and nasopharyngeal carcinoma. *J. Pathol.* **20**, 68–75.

Teruya-Feldstein, J., Tosato, G., and Jaffe, E. S. (2000). The role of chemokines in Hodgkin's disease. *Leuk. Lymphoma* **38**, 363–371.

Torroella-Kouri, M., Ma, X., Perry, G., Ivanova, M., Cejas, P. J., Owen, J. L., Iragavarapu-Charyulu, V., and Lopez, D. M. (2005). Diminished expression of transcription factors nuclear factor kappaB and CCAAT/enhancer binding protein underlies a novel tumor evasion mechanism affecting macrophages of mammary tumor-bearing mice. *Cancer Res.* **65**, 10578–10584.

Tsuda, Y., Takahashi, H., Kobayashi, M., Hanafusa, T., Herndon, D. N., and Suzuki, F. (2004). Three different neutrophil subsets exhibited in mice with different susceptibilities to infection by methicillin-resistant *Staphylococcus aureus*. *Immunity* **21**, 215–226.

Ueno, T., Toi, M., Saji, H., Muta, M., Bando, H., Kuroi, K., Koike, M., Inadera, H., and Matsushima, K. (2000). Significance of macrophage chemoattractant protein-1 in macrophage recruitment, angiogenesis, and survival in human breast cancer. *Clin. Cancer Res.* **6**, 3282–3289.

van Damme, J., Proost, P., Lenaerts, J. P., and Opdenakker, G. (1992). Structural and functional identification of two human, tumor-derived monocyte chemotactic proteins (MCP-2 and MCP-3) belonging to the chemokine family. *J. Exp. Med.* **176**, 59–65.

van den Brule F., Califice, S., Garnier, F., Fernandez, P. L., Berchuck, A., and Castronovo, V. (2003). Galectin-1 accumulation in the ovary carcinoma peritumoral stroma is induced by ovary carcinoma cells and affects both cancer cell proliferation and adhesion to laminin-1 and fibronectin. *Lab. Invest.* **83**, 377–386.

Vasiljeva, O., Papazoglou, A., Kruger, A., Brodoefel, H., Korovin, M., Deussing, J., Augustin, N., Nielsen, B. S., Almholt, K., Bogyo, M., Peters, C., and Reinheckel, T. (2006). Tumor cell-derived and macrophage-derived cathepsin B promotes progression and lung metastasis of mammary cancer. *Cancer Res.* **66**, 5242–5250.

Vermeer, M. H., Dukers, D. F., ten Berge, R. L., Bloemena, E., Wu, L., Vos, W., de Vries, E., Tensen, C. P., Meijer, C. J., and Willemze, R. (2002). Differential expression of thymus and activation regulated chemokine and its receptor CCR4 in nodal and cutaneous anaplastic large-cell lymphomas and Hodgkin's disease. *Mod. Pathol.* **15**, 838–844.

Vicari, A. P., and Caux, C. (2002). Chemokines in cancer. *Cytokine Growth Factor Rev.* **13**, 143–154.

Wang, T., Niu, G., Kortylewski, M., Burdelya, L., Shain, K., Zhang, S., Bhattacharya, R., Gabrilovich, D., Heller, R., Coppola, D., Dalton, W., Jove, R., Pardoll, D., and Yu, H. (2004). Regulation of the innate and adaptive immune responses by Stat-3 signaling in tumor cells. *Nat. Med.* **10**, 48–54.

Wynn, T. A. (2004). Fibrotic disease and the T(H)1/T(H)2 paradigm. *Nat. Rev. Immunol.* **4**, 583–594.

Yang, J., and Richmond, A. (2001). Constitutive IkappaB kinase activity correlates with nuclear factor-kappaB activation in human melanoma cells. *Cancer Res.* **61**, 4901–4909.

Yang, H., Bocchetta, M., Kroczynska, B., Elmishad, A. G., Chen, Y., Liu, Z., Bubici, C., Mossman, B. T., Pass, H. I., Testa, J. R., Franzoso, G., and Carbone, M. (2006). TNF-alpha inhibits asbestos-induced cytotoxicity via a NF-kappaB-dependent pathway, a possible mechanism for asbestos-induced oncogenesis. *Proc. Natl. Acad. Sci. USA* **103**, 10397–10402.

Yi, F., Jaffe, R., and Prochownik, E. V. (2003). The CCL6 chemokine is differentially regulated by c-Myc and L-Myc, and promotes tumorigenesis and metastasis. *Cancer Res.* **63**, 2923–2932.

Yoshimura, T., Robinson, E. A., Tanaka, S., Appella, E., Kuratsu, J., and Leonard, E. J. (1989). Purification and aminoacid analysis of two human glioma-derived monocyte chemoattractants. *J. Exp. Med.* **169**, 1449–1459.

Zeisberger, S. M., Odermatt, B., Marty, C., Zehnder-Fjallman, A. H., Ballmer-Hofer, K., and Schwendener, R. A. (2006). Clodronate-liposome-mediated depletion of tumor-associated macrophages: A new and highly effective antiangiogenic therapy approach. *Br. J. Cancer* **95**, 272–281.

Zhu, Z., Zheng, T., Homer, R. J., Kim, Y. K., Chen, N. Y., Cohn, L., Hamid, Q., and Elias, J. A. (2004). Acidic mammalian chitinase in asthmatic Th2 inflammation and IL-13 pathway activation. *Science* **304**, 1678–1682.

CHAPTER

17

Tumor-Associated Myeloid-Derived Suppressor Cells

STEPHANIE K. BUNT, ERICA M. HANSON, PRATIMA SINHA,
MINU K. SRIVASTAVA, VIRGINIA K. CLEMENTS, AND
SUZANNE OSTRAND-ROSENBERG

I. Introduction
II. Multiple Suppressive Mechanisms that Contribute to Immunosuppression in Individuals with Tumors
III. MDSCs as a Key Cell Population that Mediates Tumor-Induced Immunosuppression
    A. MDSCs Accumulation in Patients and Tumor-Bearing Animals
    B. MDSCs' Possible Role in Healthy Individuals
    C. MDSCs' Role in Facilitating Tumor Progression
    D. Phenotypic Diversity and Multiple Subpopulations for MDSCs
IV. MDSCs' Use of Mechanisms to Mediate Effects on Multiple Target Cells
    A. MDSC Alteration of DC, Macrophage, NK Cell, and B Cell Functions
    B. MDSC Inhibition of $CD4^+$ and $CD8^+$ T Cell Activation
    C. MDSC Mediation of Suppression by Producing Inducible NO Synthase and Arginase
    D. MDSC Facilitation of Tregs Development
    E. MDSC Enhancement of Tumor Progression by Promoting Tumor Angiogenesis
V. MDSC Induction by Tumor-Derived Cytokines and Growth Factors
    A. GM-CSF Induction of MDSCs
    B. PGE_2 Induction of MDSCs
    C. Proinflammatory Cytokine Induction of MDSCs
VI. MDSC Linking of Inflammation and Tumor Progression
VII. Agents Responsible for Reducing MDSC Levels
    A. Differentiation Agent Reduction of MDSCs
    B. Reduction of MDSC Levels by Chemotherapeutic Drugs
VIII. Conclusions: Implications for Immunotherapy
    References
    Further Reading

Tumor progression in many patients and experimental animals with cancer is frequently associated with the expansion of a cell population of myeloid origin. These myeloid-derived suppressor cells (MDSCs) have potent immunosuppressive activity and inhibit both adaptive and innate immunity by preventing the activation of $CD4^+$ and $CD8^+$ T cells, reducing the number of mature dendritic cells (DCs), suppressing natural killer (NK) cell cytotoxicity, and by skewing immunity toward a type 2 phenotype that is compatible with cancer. MDSCs mediate their suppressive activity principally through multiple mechanisms, including the production of arginase and/or inducible nitric oxide synthase, the promotion of tumor angiogenesis, and the induction of regulatory T cells (Tregs). Tumor-secreted factors, such as vascular endothelial growth factor (VEGF) and granulocyte-macrophage colony-stimulating factor (GM-CSF) as well as proinflammatory mediators (e.g., interleukin (IL)-1β, IL-6, and prostaglandin E_2), induce the accumulation and retention of MDSCs. This latter association with inflammation has led to the novel hypothesis that MDSCs may provide a linkage between inflammation and cancer, acting to promote tumor progression by blocking immunosurveillance and antitumor immunity. Because of their potent suppressive activity and their presence in many patients with cancer, MDSCs promote tumor progression and impede cancer immunotherapy. This chapter describes the induction, characteristics, and effector mechanisms of MDSCs, reviews the evidence demonstrating how their suppressive activity promotes tumor progression, and explores some of the strategies developed to eliminate this heterogeneous population of detrimental cells. Because of MDSCs' pivotal role in tumor progression, a better understanding of these cells is essential for improving the efficacy of cancer immunotherapies.

I. INTRODUCTION

The concept of cancer immunosurveillance suggests that the immune system may protect against the development and growth of malignancies (Dunn *et al.*, 2002). Despite evidence supporting this concept, transformed cells arise, and cancers grow progressively. These contradictory observations may be due to immunosuppressive agents induced by transformed cells as well as by an immunosuppressive microenvironment that precedes and favors malignant transformation. Accordingly, tumor-induced immunosuppression may play a pivotal role in the initiation and growth of cancers. After a brief overview of the mechanisms implicated in tumor-induced immunosuppression, this chapter will focus on a relatively newly identified cell population, called myeloid-derived suppressor cells (MDSCs). MDSCs are commonly found in many patients and experimental animals with cancer and are gaining notoriety as potent inhibitors of antitumor immunity and as facilitators of tumor progression.

II. MULTIPLE SUPPRESSIVE MECHANISMS THAT CONTRIBUTE TO IMMUNOSUPPRESSION IN INDIVIDUALS WITH TUMORS

Tumor progression is associated with the development of a profound systemic immune dysfunction. Deficiencies in professional antigen-presenting cells (APCs), defects in T cell and B cell responses, and the production of various suppressive soluble factors have been observed in tumor-bearing mice and patients with various malignancies (Finn, 2003). Cancer-associated immune suppression can be modulated by factors secreted by the tumor or the tumor microenvironment or by suppressive immune effector cells.

Tumor-associated immune dysfunction is mediated by various cytokines and immunomodulatory enzymes that suppress the antitumor immune response. Certain cytokines, such as transforming growth factor β (TGF-β), inhibit interleukin (IL)-2-dependent T cell proliferation as well as B cell and macrophage proliferation (Letterio *et al.*, 1998) and induce suppressive regulatory T cells (Tregs) (Taylor *et al.*, 2006). Production of TGF-β by macrophages also plays a central role in the inflammatory response (Letterio et al., 1998), which can be immunosuppressive in a tumor microenvironment. Immunosuppressive enzymes, such as indoleamine 2,3-dioxygenase (IDO), have been implicated in the suppression of the immune system (Muller et al., 2005). IDO plays an important role in immune tolerance to the developing fetus; however, it is also overexpressed in many malignancies leading to immune suppression (Muller et al., 2005).

In addition to soluble factors, various cell populations mediate tumor-associated immune suppression, including (i) Tregs, (ii) plasmacytoid dendritic cells (PDCs), (iii) macrophages, and (iv) MDSCs. Similar to factors such as IDO, these cells can have both a protective role and an inhibitory role, depending on the disease situation. For example, T cell-mediated suppression by Tregs, while essential for protection against autoimmunity, facilitates tumor progression (Shimizu et al., 1999). Interestingly, the transcription factor forkhead box 3 (Foxp3), which is a marker of Tregs and a modulator of the suppressive function of these cells (Fontenot et al., 2003), is at least partially regulated by the immunosuppressive factor TGF-β (Marie et al., 2005). Tregs are also thought to be induced by PDCs (Moseman et al., 2004). This specialized cell population typically regulates inflammation, thereby linking innate and adaptive immunity. PDCs in their immature state also promote tolerance by inducing T cell anergy, and their presence in the tumor microenvironment prevents the induction of an antitumor response (reviewed in Colonna et al., 2004). Macrophages, while important for many immune responses, can both facilitate and inhibit antitumor immunity. Although M1 macrophages, which are tumoricidal because they produce inducible nitric oxide synthase (iNOS), favor tumor rejection, M2 macrophages are typically present within solid tumors (reviewed in Mantovani et al., 2002), and they suppress T cell function by the production of arginase (Rodriguez et al., 2004) or IL-10 plus TGFβ (Giuducci et al., 2005). Therefore, multiple effector molecules and cell populations that regulate autoimmunity may be detrimental in individuals with cancer because they counteract potentially beneficial antitumor immune responses.

III. MDSCs AS A KEY CELL POPULATION THAT MEDIATES TUMOR-INDUCED IMMUNOSUPPRESSION

MDSCs are a population of immature cells of myeloid origin that induce potent tumor-associated immune suppression in both cancer patients and in experimental animals with malignant tumors. They mediate suppression via multiple mechanisms that limit the number of mature APCs, block T cell activation, skew macrophage activity, and inhibit NK cytotoxicity.

Active immunotherapies require a functional immune system. Therefore, the presence of MDSCs limits the efficacy of potential therapies based on activating the host's immune system. The compelling evidence supporting a pivotal role for MDSCs in systemic immune dysfunction highlights the need for a better understanding of their role in suppression and the conditions that support their accumulation.

A. MDSCs Accumulation in Patients and Tumor-Bearing Animals

MDSCs originate in the bone marrow as $Gr1^+CD11b^+CD31^+$ hematopoietic precursor cells. Normally produced colony-stimulating factors, such as granulocyte-macrophage colony-stimulating factor (GM-CSF), stimulate the production of precursor cells and induce the migration of cells that have lost CD31 but retained Gr1 and CD11b from the bone marrow to the secondary lymphoid organs. These $Gr1^+CD11b^+$ double-positive cells are immature myeloid cells, and under normal differentiation condi-

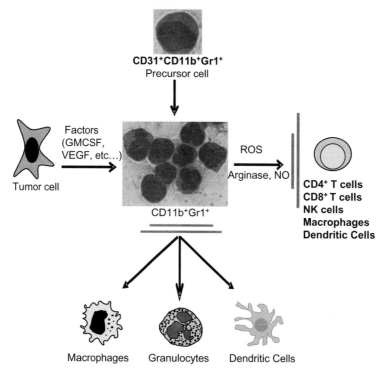

FIGURE 17.1 Tumor-derived factors that block cell differentiation in the myeloid lineage and that induce the accumulation of Gr1+CD11b+ MDSCs. Under normal conditions, myeloid precursor cells differentiate into DCs, macrophages, and granulocytes. The presence of tumor and tumor-derived factors blocks this differentiation pathway and leads to the accumulation of immature CD11b+Gr1+ myeloid cells, called MDSCs. MDSCs suppress T cells and other immune cells through a variety of mechanisms, including the production of ROS and arginase. The center panel shows purified MDSCs from the blood of mice with the 4T1 mammary carcinoma. This mixture of cells with single and multilobed nuclei is characteristic of MDSCs in the blood of tumor-bearing mice. Splenic MDSCs from tumor-bearing mice typically have large, single-lobed nuclei. (See color plate.)

tions will differentiate into mature dendritic cells (DCs), macrophages, and/or granulocytes that express CD11c, CD11b, or Gr1 markers, respectively. In tumor-bearing individuals, the presence of tumor-derived factors blocks the differentiation of the immature Gr1+CD11b+ cells, thereby resulting in their accumulation (reviewed in Kusmartsev et al., 2002, 2006; Serafini et al., 2004). Figure 17.1 shows a schematic representation of this differentiation pathway and its perturbation by tumor-secreted factors.

Cells of myeloid origin with suppressive activity were first identified in the mid-1980s in tumor-free mice as "natural suppressor cells," which were hypothesized to be important for neonatal tolerance. These cells were shown to inhibit T cell proliferation and the generation of cytotoxic T lymphocytes (CTLs) in a major histocompatibility complex (MHC)- and antigen-independent manner (Strober, 1984). The same natural suppressor cells were also reported to inhibit the growth of several tumor cell lines (Sugiura et al., 1990). There was little investigation of these cells until the late 1990s, when suppressor cells with myeloid characteristics were identified in patients with head and neck cancer (Young et al., 1997) and in mice with experimental tumors (Bronte et al., 1998). These suppressor cells had diverse phenotypes and lacked markers for mature T cells, B cells, natural killer (NK) cells, and macrophages but expressed monocyte-myeloid markers. These suppressor cell populations were called "inhibitory macrophages," (Bronte et al., 2000) "immature myeloid cells," (ImCs) (Almand et al., 2001; Kusmartsev et al., 2004), or myeloid suppressor cells (MSCs) (Huang et al., 2006; Sinha et al., 2005a,b; Zea et al., 2005). The term *myeloid-derived suppressor cell* or *MDSC* has been adopted by researchers in this field to refer to these cells (Gabrilovich et al., 2007). MDSCs were ultimately defined as Gr1$^+$Mac-1$^+$ (Mac-1 = CD11b) cells that, when depleted *in vitro* or *in vivo*, restored CD8$^+$ T cell activation and CTL activity in response to specific vaccination (Bronte et al., 1998).

The immature myeloid cell (ImC) status of the MDSCs was fur-ther demonstrated by the ability of MDSCs from tumor-bearing mice to differentiate in culture into adherent macrophage-like suppressor cells that expressed some markers of mature macrophages, such as F4/80, but did not express other characteristic markers, such as MHC II or costimulatory molecules (Bronte et al., 2000). Coculture with T helper 1 (Th1) cytokines further matured MDSCs into potent APCs, while the addition of T helper 2 (Th2) cytokines enhanced the cells' suppressive activity. Coculture with GM-CSF and IL-4 induced the expression of MHC II, costimulatory molecules, and CD11c and conferred the ability to activate T cells, consistent with their differentiation to DCs (Bronte et al., 2000; Li et al., 2004).

In parallel with the mouse studies, cells expressing the myeloid lineage marker CD34$^+$ were identified in patients. These cells were reported to be increased fivefold in patients with early stages of head and neck squamous cell carcinoma (HNSCC), breast carcinoma, and non-small-cell lung cancer (NSCLC) (Almand et al., 2001), and 65% of patients with HNSCC were reported to have elevated levels of CD34$^+$ cells (Pak et al., 1995). The CD34$^+$ cell population was identified as suppressor cells because their depletion in freshly dissociated tumor specimens resulted in increased IL-2 production, presumably by the intratumoral T cells, while the addition of CD34$^+$ cells decreased IL-2 production (Pak et al., 1995). HNSCC patients with elevated levels of CD34$^+$ cells had correspondingly low levels of CD8$^+$ T cells, and the presence of CD34$^+$ cells was higher in patients whose tumors recurred or metastasized after surgical removal of the primary tumor (Young et al., 1997). The CD34$^+$ suppressor cells were induced by tumor-secreted GM-CSF, formed colonies, and differentiated into granulocytes and monocytes in soft agar, further confirming their myeloid lineage (Pak et al., 1995). Therefore, although human and mouse MDSCs express different markers, they are both immature myeloid-derived populations with suppressive activity, and, hence, they are considered counterparts.

B. MDSCs' Possible Role in Healthy Individuals

Low levels of Gr1$^+$CD11b$^+$ cells are present in healthy individuals, suggesting that MDSCs may be important for autoimmunity homeostasis and in transplant rejection. For example, Gr1$^+$CD11b$^+$ cells from proinsulin non-obese diabetic (NOD) mice, in which proinsulin is an autoantigen, suppress the development of diabetes (Steptoe et al., 2005) when transferred into wild-type mice. The Gr1$^+$CD11b$^+$ cells are thought to express proinsulin peptides and therefore induce T cell tolerance and limit inappropriate self responses (Steptoe et al., 2005). Likewise, Gr1$^+$CD11b$^+$ cells that block allogeneic T cell responses may prevent graft versus host disease (GvHD) in transplant recipients. Support for the later hypothesis comes from studies with Src homology 2 domain-containing inositol-5-phosphatase (SHIP)$^{-/-}$ mice that do not develop GvHD and that have elevated levels of MDSCs, which may prevent T cell priming (Ghansah et al., 2004). The survival of cardiac grafts in rats has also been reported to be enhanced by the presence of immature bone marrow-derived myeloid cells that suppress T cell activation following their coculture with

GM-CSF and lipopolysaccharide (LPS) (Valente et al., 1997). Therefore, MDSCs are present at low levels in healthy individuals, where their role may be to attenuate autoimmune responses as well as prevent GvHD.

C. MDSCs' Role in Facilitating Tumor Progression

In contrast to what occurs in a healthy individual, MDSCs can accumulate to very high levels in the spleen and blood in a tumor-bearing individual. Their quantity is usually directly proportional to tumor burden, with different types of tumors inducing different absolute levels. MDSCs are thought to directly enhance tumor progression because tumors grow more quickly in mice inoculated with a mixture of tumor cells and Gr1$^+$CD11b$^+$ cells than in mice inoculated with tumor cells alone (Yang et al., 2004). Likewise, antibody depletion of Gr1$^+$ or CD11b$^+$ cells results in tumor rejection or a reduced growth rate (Seung et al., 1995) and protects against tumor recurrence (Terabe et al., 2003).

There is a single report indicating that MDSCs favor tumor rejection (Pelaez et al., 2001). In this study, antitumor immunity was observed when tumor-bearing mice were treated with cyclophosphamide plus Percoll® gradient-enriched nitric oxide (NO)-secreting CD31$^+$Ly6C$^+$ cells from tumor-bearing mice. However, the putative MDSCs were not purified in these experiments, and the authors did not demonstrate that the CD31$^+$Ly6C$^+$ cells had suppressive activity, so the protective effect in this study might be attributed to a contaminating cell population/factor.

D. Phenotypic Diversity and Multiple Subpopulations for MDSCs

Most investigators have studied MDSCs that are localized to the spleen or blood of tumor-bearing individuals, characterizing them by their ability to inhibit T cell activation and by their expression of myeloid-lineage markers as detected by immunofluorescence and flow cytometry. The consensus is that MDSCs include a broad repertoire of myeloid-derived cells at various stages of maturity. Mouse MDSCs have most frequently been identified by their cell surface coexpression of CD11b and Gr1 (also known as Ly6-G). The latter is a differentiation antigen and glycosylphosphatidylinositol- (GPI) anchored protein that is predominantly expressed by granulocytes (Fleming et al., 1993). CD11b, which is also the α–M subunit of the cell surface protein Mac-1, is expressed on macrophages, granulocytes, and some NK cells.

MDSCs in humans are now phenotypically identified by their cell surface expression of CD11b, CD15 (the human equivalent of murine Gr1), and their lack of CD14 expression (Almand et al., 2001; Zea et al., 2005) (Table 17.1). In earlier studies, they were identified in patients with head and neck cancer based on their suppressive activity and their expression of CD34, a marker of human hematopoietic progenitor cells (Garrity et al., 1997). Immature myeloid cells are identified as negative for the lineage markers CD3, CD19, CD56, and CD14 and for MHC II; they are positive for CD15 and CD33, common myeloid markers (Almand et al., 2001; Kusmartsev et al.; 2002, Mirza et al., 2006).

In addition to their expression of granulocyte markers and CD11b, MDSCs express a variety of cell surface markers that differ depending on the type of tumor that induced them. Most investigators concur that MDSCs express MHC class I but not MHC class II molecules (Almand et al., 2001; Bunt et al., 2006; Zea et al., 2005) although low levels of MHC class II have been described on MDSCs obtained from mice with some types of tumors (Melani et al., 2003; Rodriguez et al., 2005). Likewise, costimulatory molecules, including CD80, CD86, and CD40, are present on MDSCs obtained from mice with some tumors (Sinha et al., 2005a; Terabe et al., 2003), while MDSCs from mice with different types of

TABLE 17.1 Phenotype, Target Cells, and Effector Molecules of MDSCs Induced by Mouse and Human Tumors

	MDSC Surface Markers	Target Cells	Effector Molecules	References
Mouse tumors				
4T1 (mammary carcinoma)	Gr1$^+$CD11b$^+$F4/80$^-$ CD11c$^-$	CD4$^+$ T cells, CD8$^+$ T cells	Arginase, ROS	Bunt et al. (2006); Sinha et al. (2005)
SaI (sarcoma)	Gr1$^+$CD11b$^+$	–	–	
TS/A (mammary carcinoma)	Gr1$^+$CD11b$^+$F4/80$^+$	CD8$^+$ T cells	NO	Bronte et al. (1999); Pericle et al. (1997)
CMS4 (fibrosarcoma)	Mac-1$^+$Gr1$^+$	T cells		Salvadori et al. (2000)
MCA-26 (colon carcinoma)	Gr1$^+$CD11b$^+$F4/80$^+$	CD4$^+$ T cells	NO, MMP9	Li et al. (2004); Yang et al. (2000)
	Gr1$^+$CD11b$^+$	Naive and activated T cells	NO	Kusmartsev et al. (2000)
	Gr1$^+$CD115$^+$ F4/80$^+$	Tregs	IL-10	Huang et al. (2006)
15–12RM (fibrosarcoma)	Gr1$^+$CD115$^+$ F4/80$^+$	CD8$^+$ T cells	TGF-β	Terabe et al. (2003)
CT26 (colon carcinoma)				Bronte et al. (2000); Bronte et al. (1999)
A20 (B cell lymphoma)	Gr1$^+$CD11b$^+$	Activated CD4$^+$ T cells	NO	Serafini et al. (2004); Ostrand-Rosenberg, unpublished observations (2004)
B78H1 (melanoma)				Serafini et al. (2004)
TC-1 (lung carcinoma)	Gr1$^+$CD11b$^+$	CD8$^+$ T cells, NK cells	–	Suzuki et al. (2005)
3LL (lung carcinoma)	Gr1$^+$CD11b$^+$CD34$^+$	–	MMP9, arginase	Yang et al. (2004) Rodriguez et al. (2004) Young et al., 1992
BALB-neuT (mammary carcinoma)	Gr1$^+$CD11b$^+$CD31$^+$ MHCII$^+$CD11c$^-$	Allogeneic T cells	VEGF	Melani et al. (2003)
C3 (cervical carcinoma)	Gr1$^+$CD11b$^+$	CD8$^+$ T cells	ROS, H_2O_2, arginase	Kusmartsev et al. (2004)
TRAMP (prostate carcinoma)	Gr1$^+$CD11b$^+$	–	–	Ostrand-Rosenberg, unpublished observations (2004)
MethA (fibrosarcoma)	Gr1$^+$CD11b$^+$MHCII$^-$	CD8$^+$ T cells	NO	Gabrilovich et al. (2001)
Human tumors				
RCC	CD11b$^+$CD14$^-$, CD15$^+$MHCII$^-$	T cells	Arginase	Zea et al. (2005)
RCC	Lin$^-$HLA-DR$^-$CD33$^+$		–	Mirza et al. (2006)
HNSCC	CD34$^+$	T cells	–	Young et al. (2001)
HNSCC Breast carcinoma NSCLC	CD33$^+$Lin$^-$HLA-DR$^-$	CD4$^+$ T cells, CD8$^+$ T cells, dendritic cells	–	Almand et al. (2001)
Pancreatic carcinoma, breast carcinoma, colon carcinoma	CD15$^+$	T cells	H_2O_2	Schmielau et al. (2001)

tumors do not express these molecules (Melani et al., 2003). A variety of molecules associated with maturation have also been observed on some populations of MDSCs. These molecules include F4/80, a marker of mature macrophages (Bronte et al., 1999; Huang et al., 2006; Li et al., 2004); the DC markers CD11c and DEC205; the B cell marker B220; and the T cell markers CD4 and CD8 (Bunt et al., 2006; Sinha et al., 2005a). Although the progenitor stem cell marker CD34 is usually absent on murine MDSCs (Bunt et al., 2006; Gabrilovich et al., 2001), CD31, another marker for myeloid progenitor cells, has been observed in some (Bronte et al., 2000) but not other populations of MDSCs (Bunt et al., 2006; Sinha et al., 2005a). Table 17.1 lists the MDSCs induced by different types of tumors and the markers they express.

In addition to primary, explanted MDSCs, cloned myeloid suppressor cell lines have been prepared from the spleens of BALB/c mice inoculated with TS/A mammary carcinoma cells and from mice immunized with recombinant vaccinia virus-encoding mouse IL-2 (Bronte et al., 2000). These clonal lines suppress T cell activation and express CD11b, CD11c, F4/80, and CD86 costimulatory molecules (Mazzoni et al., 2002).

The discrepancy in expression of cell surface markers suggests that MDSCs include multiple subpopulations of myeloid-derived cells with suppressive activity that are at various stages of maturity and/or are in diverging lineages. Differential expression of maturation markers, such as MHC II and costimulatory molecules, and of lineage markers, such as F4/80, supports this view (Almand et al., 2001, Bronte et al., 1999). Since MDSCs are at least partially induced by tumor-secreted factors, it is likely that different tumors, which secrete different cytokines and chemokines, induce phenotypically distinct MDSCs.

The term *myeloid-derived suppressor cells* has been adopted for these cells, as stated earlier, because it is an inclusive term for describing this heterogeneous cell population and because the characteristics of MDSCs do not precisely match any particular type of myeloid cell (Gabrilovich et al., 2007). For example, although MDSCs express Gr1 and CD11b, both of which are markers of classical neutrophils, MDSCs have lower levels of Gr1 than neutrophils (Rossner et al., 2005; Terabe et al., 2003). Likewise, their nuclear morphology does not correspond to a particular myeloid cell population. Splenic MDSCs typically have a large, single nucleus characteristic of immature myeloid cells (Sinha et al., 2005; Terabe et al., 2003), while MDSCs in the blood have either a large single nucleus or a multilobed nucleus (Schmielau et al., 2001) (e.g., see center panel of Figure 17.1). Additional studies are needed to clarify whether the heterogeneity of MDSCs is the result of an expansion of individual immature myeloid cells or a dysregulation of myelopoiesis.

Because MDSCs express markers that are also expressed by other cell types, they may be confused with other cells of myeloid origin that also have suppressive activity. For example, M2 macrophages are characterized by their expression of CD11b and their production of arginase. They promote tumor progression and are localized to solid tumors (Mantovani et al., 2002; Rodriguez et al., 2005). If potential suppressor cells are identified exclusively by their expression of a single phenotypic marker, such as CD11b, they may be mature M2 macrophages rather than immature MDSCs.

Given the large repertoire of markers that are used to phenotypically characterize MDSCs in mice and patients, additional experiments with larger numbers of individuals with the same types of tumors are needed to identify the optimal markers and to determine if different types of tumors or tumor growth conditions induce different subpopulations of MDSCs. Regardless of whether future studies reveal subpopulations of MDSCs that are identifiable by discrete cell surface or internal markers, it is likely that MDSCs are a family of cells that includes multiple, and possibly distinct, subpopulations.

IV. MDSCs' USE OF MECHANISMS TO MEDIATE EFFECTS ON MULTIPLE TARGET CELLS

In addition to their phenotypic heterogeneity, MDSCs are also heterogeneous with respect to the target cells they inhibit and their suppressive mechanisms. These differences are associated with the source of the MDSCs and possibly reflect differential activity by MDSC subpopulations. Despite these differences, the suppressive function of MDSCs on T cell activation requires direct contact with the target cells because the addition of MDSC supernatants to T cell cultures does not inhibit T cell proliferation (Almand et al., 2001). Likewise, if T cells are physically separated from MDSCs by a semipermeable membrane, they are not suppressed (Bronte et al., 1999; Gabrilovich et al., 2001; Sinha et al., 2005a,b). The following sections describe the suppressive mechanisms attributed to MDSCs and the target cells that are suppressed. This information is summarized schematically in Figure 17.2.

A. MDSC Alteration of DC, Macrophage, NK Cell, and B Cell Functions

A significant reduction in functional DCs has been reported in cancer patients and tumor-bearing mice, suggesting that the block in myeloid cell development, which results in the accumulation of MDSCs, may reduce the number of mature DCs. DCs in these patients and animal models express low levels of MHC II and costimulatory molecules and were defective in clustering with allogeneic T cells (Tas et al., 1993). In cancer patients, DC

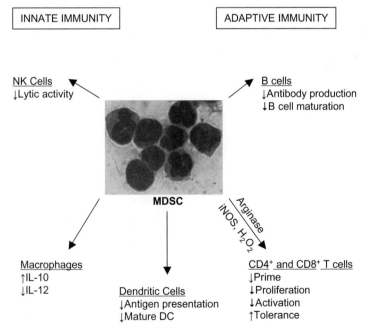

FIGURE 17.2 **MDSC suppression of innate and adaptive immune systems.** MDSCs inhibit innate immunity by suppressing NK cell-mediated lysis and by polarizing macrophages toward a type 2 phenotype. MDSCs inhibit adaptive immunity by suppressing the activation and proliferation of T cells and antibody production by B cells. In addition, MDSCs limit the availability of mature and functional DCs, which bridge the gap between innate and adaptive immunity. (See color plate.)

activity and quantity decrease as a function of progressive disease (Almand et al., 2000), and the quantity of mature DCs inversely correlates with the number of immature myeloid cells, suggesting that DC maturation is impaired (Almand et al., 2000). In tumor-bearing mice, the generation of functional mature DCs is suppressed, and DCs that are present may be deficient in lymphoid homing and retention (Gabrilovich et al., 1996). In addition, DCs from tumor-bearing mice are poor APCs and do not efficiently activate T cells (Gabrilovich et al., 1996). However, DC abnormalities do not always occur when MDSCs are present, since DCs in mice with very elevated levels of MDSCs, due to either transplantable (Danna et al., 2004) or spontaneous (Melani et al., 2003) breast tumors, are neither dysfunctional nor reduced in quantity, respectively. MDSCs may mediate their effects on DCs either directly through cell-to-cell contact or indirectly by influencing DC maturation. The findings that immature myeloid lineage cells accumulate and that present mature DCs are deficient in MHC II and costimulatory molecules are most consistent with an indirect mechanism. Additional experiments clarifying the physical relationship between MDSCs and DCs are needed to definitively answer this question.

MDSCs also impair NK cell activity and may diminish antibody production by B lymphocytes. Inhibition of NK lytic activity was noted when splenic Gr1$^+$CD11b$^+$ cells from mice with TC-1 and 4T1 tumors were cocultured in vitro with NK cells (Suzuki et al., 2005; S. Bunt and S. Ostrand-Rosenberg, unpublished observations, 2005). Although direct evidence for MDSC inhibition of antibody production is not available, antigen-specific antibody responses are impaired in tumor-bearing animals with high levels of MDSCs (Danna et al., 2004; Pericle et al., 1997). Moreover, antibody production returns to normal when MDSC levels are reduced following surgical removal of the primary tumor (Danna et al., 2004; Sinha et al., 2005). Although these data are consistent with the concept that MDSCs modulate antibody responses, additional studies are clearly needed to corroborate this hypothesis.

Evidence suggests that MDSCs may also impact macrophage polarization in tumor-bearing animals. Cocultures of Gr1$^+$CD11b$^+$ MDSCs from mice bearing 4T1 mammary tumors with autologous peritoneal macrophages produced elevated levels of IL-10 and reduced levels of IL-12, relative to cultures of macrophages alone, implying that MDSCs polarize immunity toward a type 2 response (Sinha et al., 2007). Although it is unclear if the MDSCs or macrophages are producing the increased quantities of IL-10, these results demonstrate that MDSCs may also suppress tumor immunity by skewing it toward a type 2 response, which favors tumor progression.

B. MDSC Inhibition of CD4$^+$ and CD8$^+$ T Cell Activation

Studies in the early 1990s demonstrated the ability of "natural suppressor cells" from the bone marrow of tumor-bearing mice to impair T cell proliferation in mixed lymphocyte reactions (Schmidt-Wolf et al., 1992). Later studies with better characterized MDSCs confirmed the earlier results and further demonstrated that MDSCs can prevent a GM-CSF vaccine from priming and boosting CD4$^+$ T cells (Serafini et al., 2004). Although low doses of GM-CSF induced an expansion of antigen-specific CD4$^+$ T cells, higher doses of GM-CSF resulted in the expansion of MDSCs and inhibited T cell proliferation.

Two lines of evidence indicate that MDSCs act on naive rather than on activated T cells. T cells that have been preactivated with anti-CD3 antibodies proliferate in the presence of MDSCs, indicating that activated T cells are not responsive to MDSC suppression (Kusmartsev et al., 2000). Likewise, cultured transgenic T cells that are activated by specific antigen are not suppressed if MDSCs are added to the cultures more than 18 hours after addition of antigen (E. Hanson and S. Ostrand-Rosenberg, unpublished observations, 2006).

Although suppression of CD4$^+$ and CD8$^+$ T lymphocytes has been observed, not all populations of MDSC inhibit both cytotoxic and helper T cells. For example, MDSCs from mice with the 4T1 mammary carcinoma are equally suppressive toward both CD4$^+$ and CD8$^+$ T cells (Bunt et al., 2006; Sinha et al., 2005a,b), whereas MDSCs from mice bearing sarcomas induced by the chemical carcinogen MethA suppress CD8$^+$ but not CD4$^+$ T cells (Gabrilovich et al., 2001). Similar dichotomies have been found for MDSCs in patients with HNSCC, NSCLC, and breast carcinoma (Almand et al., 2001). Table 17.1 summarizes the activity of MDSCs induced by different tumors on CD4$^+$ and CD8$^+$ T cells. The variations in suppressive activity in these various studies may be due to phenotypic variation in the MDSCs, distinct MDSC subpopulations, and/or multiple suppressive mechanisms.

Various studies indicated that MDSCs suppress CD8$^+$ T cells via an antigen-specific and MHC I-restricted mechanism. For example, MDSCs in the presence of a specific peptide suppressed transgenic T cells reactive with that peptide but not transgenic T cells reactive with another peptide (Gabrilovich et al., 2001). Antibody blockade of MHC I molecules eliminated the suppression, confirming the role of MHC I in MDSC suppression and supporting the concept that MDSCs suppress via an antigen-specific mechanism (Gabrilovich et al., 2001). Despite these convincing findings, there is disagreement concerning whether MDSCs are antigen-specific suppressors because they also inhibit T cells activated by CD3/CD28 (Kusmartsev et al., 2000) and ConA (Angulo et al., 1995). MDSC suppression of CD4$^+$ T cells is less likely to be antigen-specific and MHC II restricted because most MDSCs do not express MHC II (Bunt et al., 2006; Gabrilovich et al., 2001; Sinha et al., 2005a). The apparent discrepancy in antigen specificity and the requirements for MHC between suppression of CD8$^+$ versus CD4$^+$ T cells strongly suggests that MDSCs mediate their effects through multiple mechanisms. This hypothesis is also supported by the observations that MDSCs inhibit NK and possibly macrophage activity since neither of these cells are antigen specific.

C. MDSC Mediation of Suppression by Producing Inducible NO Synthase and Arginase

MDSCs mediate suppression through two pathways, both of which involve catabolism of the essential amino acid L-arginine. In the first mechanism, MDSCs release arginase when they contact their target cells. The released arginase degrades L-arginine to urea and L-ornithine, thereby blocking T cell proliferation and growth in a localized manner. In the second mechanism, MDSCs produce iNOS, which converts L-arginine to NO and citrulline (Bronte et al., 2003). The net effect of NO release is to block the phosphorylation of Janus kinases 1 and 3 (Jak1/3), signal transducers and activators of transcription 5 (Stat5) (Pericle et al., 1997), and Erk and Akt (Bronte et al., 2003), molecules whose phosphorylation is essential for activation of the IL-2 receptor signaling pathway (Mazzoni et al., 2002). Although NO probably does not induce T cell apoptosis, insofar as its effect is reversible within the first 24–48 hours (Mazzoni et al., 2002), it clearly has the potential to do so since NO can convert superoxide to peroxynitrite, which is sufficient to trigger T cell apoptosis (Bronte et al., 2003).

Studies of the addition of NO inhibitors to mixtures of T cells and explanted MDSCs or MDSC lines can partially (Kusmartsev et al., 2000) or completely (Serafini et al., 2004) reverse T cell suppression (Mazzoni et al., 2002). Genetic support for a role of NO in MDSC function is provided by the observation that Gr1$^+$CD11b$^+$ cells from iNOS-deficient mice are not suppressive (Mazzoni et al., 2002). However, NO inhibitors do not restore T cell activation for all mouse MDSCs (Sinha et al., 2005a, 2005b). Furthermore, NO inhibitors are ineffective at reversing MDSC suppression, a finding indicated by the limited number of human cancer patients who have been tested (Almand et al., 2001). Therefore, iNOS-mediated catabolism of L-arginine and the resulting production of NO mediate suppression by some but not all MDSCs, offering further evidence that MDSCs comprise a heterogeneous population of cells.

In addition to NO, other reactive oxygen species (ROS), such as hydrogen peroxide (H_2O_2), hydroxyl radicals (OH^-), and peroxynitrite ($ONOO^-$), may also play a role in MDSC-mediated suppression. ROS production is elevated in $Gr1^+CD11b^+$ cells from tumor-bearing versus tumor-free mice (Kusmartsev et al., 2004, Sinha et al., 2005a,b). Specifically, H_2O_2 and peroxynitrite production are increased in MDSCs from tumor-bearing mice, and inhibitors to these reactive species inhibited the antigen-specific inhibitory effect of MDSCs (Kusmartsev et al., 2004). However, no difference in superoxide production was observed among MDSCs from tumor-bearing and tumor-free mice in studies that used a dye to selectively measure superoxide production (Kusmartsev et al., 2004; Sinha et al., 2005a,b).

Studies with arginase inhibitors have demonstrated that some MDSCs suppress T cell activation by the production of arginase instead of NO (Bunt et al., 2006; Kusmartsev et al., 2004; Rodriguez et al., 2005; Sinha et al., 2005a,b). Increased arginase activity is present in the serum and tumor milieu of breast and colorectal cancer patients (Porembska et al., 2003, 2002) and also in immunosuppressive $CD11b^+CD15^+CD14^-$ peripheral blood cells of patients with renal cell carcinoma (RCC) (Zea et al., 2005). Arginase also appears to constitute the mechanism of suppression for MDSC lines that have enhanced suppressive properties due to their exposure to IL-4 because their suppressive activity is reduced by arginase inhibitors (Bronte et al., 2003). As these MDSCs simultaneously produce arginase and iNOS, these effector molecules are thought to act synergistically to mediate suppression (Bronte et al., 2003).

A major effect of the depletion of L-arginine through the production of arginase is to reduce CD3-ζ expression in the targeted T cells (Rodriguez et al., 2005; Sinha et al., 2005b). CD3-ζ is an essential component of the T cell receptor (TCR) that is required for the intracellular signaling events needed to initiate T cell activation (Bronte et al., 2003). T cells from tumor-bearing mice and from cancer patients frequently have reduced CD3-ζ chain expression that correlates with tumor progression (Aoe et al., 1995). Notably, depletion of MDSCs or addition of arginase inhibitors to cultures of MDSCs and T cells restores CD3-ζ chain expression and T cell reactivity (Zea et al., 2005). In addition to its effects on T cells, arginase overexpression may contribute to immune suppression by blocking B cell maturation (de Jonge et al., 2002), thereby decreasing antibody production.

D. MDSC Facilitation of Tregs Development

In addition to their direct suppression of T cell activation, MDSCs induce $Foxp3^+$ Tregs that are also immunosuppressive (Huang et al., 2006). Although NO is not involved in MDSC induction of Tregs, MDSC-secreted TGF-β may be involved because immature $CD11c^+CD11b^+MHC\ II^+$ cells isolated from tumor-bearing mice secrete TGF-β, which stimulates the expansion of naturally occurring Tregs (Ghiringhelli et al., 2005). $CD11b^+Gr1^+$ cells that facilitate transplant tolerance have also been shown to promote the generation of IL-10-secreting, antigen-specific Tregs (MacDonald et al., 2005). TGF-β has only been identified in some MDSC populations, consistent with the concept that MDSC subpopulations exist and have varying capacities to induce Tregs (Terabe et al., 2003; Young et al., 1992). Still, there have been few reports showing that MDSCs can induce

Tregs, such that additional studies are needed to verify if there are factors other than TGF-β that regulate Treg accumulation and if the ability to induce Tregs correlates with particular MDSC phenotypes.

E. MDSC Enhancement of Tumor Progression by Promoting Tumor Angiogenesis

In addition to their profound effects on cells of the immune system, MDSCs also regulate tumor progression by enhancing tumor angiogenesis. In BALB/neuT mice that develop spontaneous multifocal mammary carcinomas, the level of VEGF in the serum is directly proportional to the percentage of MDSCs in the peripheral blood (Melani et al., 2003). Studies in mice indicate that this correlation of MDSC and VEGF levels reflects the MDSC role in angiogenesis. Coinjection of MDSCs derived from mice with MC26 tumors along with 3LL tumor cells promotes increased 3LL growth as well as increased tumor angiogenesis and vasculature, as compared to 3LL injection alone (Yang et al., 2004). These MDSCs had increased production of matrix metalloproteinase 9 (MMP9), which regulates VEGF. If the coinjected MDSCs were derived from MMP9 null mice, then 3LL tumor growth was suppressed, consistent with the concept that MDSC production of MMP9 upregulates VEGF, which in turn promotes tumor angiogenesis. Although similar studies have not been performed with MDSCs from cancer patients, VEGF concentrations in the plasma of human cancer patients are proportional to the patients' levels of MDSCs (Almand et al., 2000), suggesting that MDSCs may also drive VEGF levels in the human setting.

V. MDSC INDUCTION BY TUMOR-DERIVED CYTOKINES AND GROWTH FACTORS

Multiple diverse molecules have been implicated in the induction and accumulation of MDSCs in tumor-bearing individuals, including GM-CSF, VEGF, prostaglandin E_2 (PGE_2), IL-1β, IL-6, and Stat3. These factors may be secreted by the primary tumor or by the tumor microenvironment in response to other tumor-derived molecules. Evidence supports the concept that MDSC accumulation results from tumor-derived factors that increase the rates of progenitor cell release from the bone marrow (Song et al., 2005); however, alternative mechanisms, such as proliferation of MDSCs in the periphery, have not yet been excluded.

A. GM-CSF Induction of MDSCs

Colony stimulating factors, such as GM-CSF, promote normal myelopoiesis as well as the induction and accumulation of MDSCs. Many tumors, such as breast, ovarian, lung, and head and neck cancers, secrete high levels of colony-stimulating factors, and an excess of these factors is thought to contribute to the elevated levels of MDSC and tumor progression in cancer patients (Bronte et al., 1999; Young et al., 1997). However, GM-CSF also induces the *ex vivo* differentiation of MDSCs into mature APCs and has the potential to reduce immunosuppression (Bronte et al., 1999). Therefore, as for most cytokines, the amount of GM-CSF, the context of production, and the surrounding microenvironment are critical variables that determine if the cytokine is helpful or deleterious to the host.

VEGF mediates MDSC induction and accumulation by impairing hematopoiesis and preventing DC maturation (Gabrilovich et al., 1996). It is produced by almost all tumors, and there is a correlation between high levels of VEGF and low levels of mature DC (Saito et al., 1998). VEGF mediates its effects by binding to $CD34^+$ precursor cells, leading to the inhibition of the transcription factor NF-κB, which results in defective DC differentiation (Oyama et al., 1998). Elevated VEGF production in patients with HNSCC also recruits additional $CD34^+$

progenitor cells (i.e., MDSCs) to the tumor site (Young et al., 2001).

B. PGE$_2$ Induction of MDSCs

PGE$_2$, a product of the cyclooxygenase 2 (COX-2) pathway, is an important mediator of angiogenesis, cell survival, and proliferation that may prevent DC maturation (Haas et al., 2006). A role for PGE$_2$ in the suppressive effects of "natural suppressor cells" was first hypothesized more than 20 years ago (Strober, 1984). Studies using a murine bone marrow *in vitro* culture system support this hypothesis and demonstrate that PGE$_2$ induces the differentiation and accumulation of immunosuppressive MDSCs via interaction with the PGE$_2$ receptors EP2 and EP4 that are present on precursors of Gr1$^+$CD11b$^+$ MDSC (Sinha et al., 2007). Additional support for the role of PGE$_2$ comes from studies with mouse lung carcinoma cells. These and many other tumor cells secrete PGE$_2$ as a result of their constitutive expression of COX-2. Coculture of CD11b$^+$ cells with tumor cell supernatants containing PGE$_2$ induces arginase expression and confers immune suppression activity (Rodriguez et al., 2005). Although PGE$_2$ induces the accumulation of MDSCs, Gr1$^+$CD11b$^+$ cells with suppressive activity are also induced from bone marrow precursors in the absence of PGE$_2$, indicating that PGE$_2$ is only one of multiple mechanisms that upregulates MDSCs (Sinha et al., 2007).

C. Proinflammatory Cytokine Induction of MDSCs

The proinflammatory cytokines IL-1β and IL-6 also promote the induction and accumulation of systemic MDSCs. Tumor-secreted IL-1β increases the rate of release of bone marrow precursor cells into the blood and heightens MDSC accumulation (Song et al., 2005). The incremental MDSCs induced by IL-1β are phenotypically and functionally distinct from MDSCs induced in the absence of the cytokine, and the presence of additional MDSCs are not due to increased tumor burden (Bunt et al., 2006). IL-6, when secreted by transfected mouse tumor cells, also significantly enhances the accumulation of MDSCs (S. Bunt and S. Ostrand-Rosenberg, unpublished observations, 2007). It is likely that proinflammatory cytokines mediate their effects by accentuating the accumulation of immature myeloid cells because tumor cell-derived IL-6 inhibits differentiation of CD34$^+$ stem cells into functional DCs (Menetrier-Caux et al., 1998). Further support for a direct role of IL-6 in MDSC induction comes from studies showing that IL-6 signals through the activation of the Jak2/Stat3 pathway and that hyperactivation of Stat3 and its increased expression results in the abnormal differentiation of myeloid cells and an increase in MDSCs in tumor-bearing animals (Nefedova et al., 2004). Inhibition of Jak2/Stat3 signaling enhances DC differentiation and significantly reduces the accumulation of MDSCs, further confirming a causative role for IL-6 in MDSC activation (Nefedova et al., 2005). This causative association between proinflammatory cytokines and MDSCs suggests that immune suppression may be heightened in individuals whose tumors are accompanied by inflammation.

VI. MDSC LINKING OF INFLAMMATION AND TUMOR PROGRESSION

Pathologists and oncologists have long observed that chronic inflammation is frequently associated with the onset and progression of cancer. For example, chronic ulcerative colitis persisting for 30–40 years is associated with a 30% increased risk of colon carcinoma (Ekbom et al., 1990). Cancer has also been attributed to inflammation resulting from infection, and it is estimated that 15% of the global cancer burden can be attributed to inflammation due to infection (reviewed in Balkwill et al., 2001).

The relationship between inflammation and cancer is under intense study, and multiple mechanisms for the correlation have been proposed, including DNA damage, stimulation of angiogenesis, recruitment of leukocytes (i.e., tumor-associated macrophages [TAMs]), and enhanced growth and survival (reviewed in Balkwill et al., 2001). The observations that proinflammatory cytokines promote tumor progression and induce MDSCs suggest that this suppressive cell population may also be responsible for the correlation between inflammation and cancer. According to this hypothesis, chronic inflammation causes an increase in MDSCs that inhibits immune surveillance and antitumor immunity, thereby facilitating malignant cell transformation and proliferation (Bunt et al., 2006). Figure 17.3 illustrates this novel hypothesis.

Multiple lines of evidence support a causal relationship between inflammation and MDSC accumulation. For example, during the early stages of helminth infection in mice, MDSCs accumulate in the peritoneal cavity and suppress T cells through NO production (Brys et al., 2005). Similarly, other infectious agents or vaccination with recombinant viruses also promote accumulation of MDSCs (Serafini et al., 2004). The effects of the proinflammatory agents IL-1β, IL-6, and PGE_2 on MDSC accumulation discussed in Section V directly show that inflammation increases MDSC levels while promoting tumor growth (Bunt et al., 2006; Sinha et al., 2007; Song et al., 2005). Similarly, the finding that treatment of mice with IL-1 β-secreting tumors with an IL-1 receptor antagonist significantly reduces MDSC levels (Song et al., 2005) also demonstrates that inflammation contributes to MDSC production. If MDSC accumulation is responsible for the association between inflammation and cancer as indicated by these studies, then methods for reducing MDSCs may be therapeutically beneficial.

VII. AGENTS RESPONSIBLE FOR REDUCING MDSC LEVELS

As described in preceding sections, tumor-induced MDSCs inhibit both innate and adaptive immunity, thereby promoting early tumor growth and impeding cancer immunotherapy. Reduction or elimination of MDSCs has already been shown to reduce the rate of tumor progression (Seung et al., 1995) and to improve immune-based cancer therapies (De Santo et al., 2005; Kusmartsev et al., 2003; Suzuki et al., 2005). Therefore, the development of methodologies to reverse MDSC accumulation has become a high priority for therapeutic development.

Because some MDSCs are induced by tumor-secreted factors that are directly proportional to tumor burden, debulking of the primary tumor should conceivably reduce MDSCs. Although this approach has not been tested in cancer patients, surgical removal of a primary tumor in mice significantly reduces MDSC levels even if residual metastatic disease is present (Sinha et al., 2005a,b). Tumor debulking also restores T cell function (Danna et al., 2004) and the expression of T cell signaling molecules; in addition, it may confer protection against tumor rechallenge (Salvadori et al., 2000). In contrast to MDSCs induced by tumor-secreted factors that partially regress after primary tumor removal, inflammation-induced MDSCs remain elevated (Bunt et al., 2006). Therefore, tumor debulking may facilitate a reduction in tumor-induced MDSCs, but if nontumor-derived induction factors are present, suppression may persist.

A. Differentiation Agent Reduction of MDSCs

Several agents have been used to induce differentiation of MDSCs into mature APCs. These agents include GM-CSF and IL-4. When added to cultures of $Gr1^+CD11b^+$ MDSCs, GM-CSF and IL-4 lead to retention of CD11b but loss of Gr1 expression,

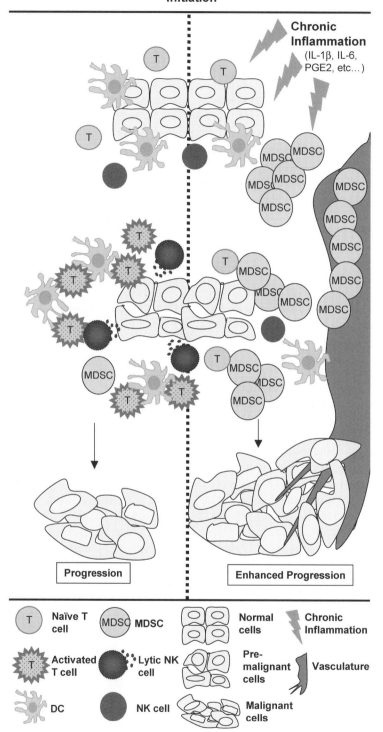

FIGURE 17.3 Inflammation promotion of tumor growth by inducing the accumulation of MDSCs that suppress antitumor immunity. A recent hypothesis proposes that immune suppression caused by MDSCs may be responsible for the correlation between inflammation and malignancy (Bunt et al., 2006). In the absence of inflammation, cells that are transformed following initiation by carcinogens are destroyed by immune surveillance. However, if initiation is accompanied by inflammation, then MDSCs accumulate and prevent the immune system from destroying the transformed cells by blocking T cell activation, suppressing NK cell-mediated lysis, reducing DC antigen presentation, and skewing immunity toward a type 2 response. As premalignant cells become malignant, the inflammatory microenvironment sustains the accumulation of MDSCs, which further inhibits antitumor immunity and facilitates tumor progression. (See color plate.)

implying that MDSCs matured into CD11b⁺ macrophages (Gabrilovich et al., 2001). Treatment of MDSCs in vitro with GM-CSF and TNF-α caused MDSCs to mature into DCs, and treatment of tumor-bearing mice with a GM-CSF vaccine significantly increased the number of CD11c⁺ DCs present (Li et al., 2004). However, as already discussed in Section V of this chapter, if GM-CSF is given at high levels, it instead induces the accumulation of MDSCs (Bronte et al., 1999; Young et al., 1997).

Differentiation agents have also been used in cancer patients to alleviate the accumulation of MDSCs. Treatment of lineage-negative and (human leukocyte antigen) HLA-DR-negative immature myeloid cells in vitro with macrophage colony-stimulating factor (MCSF) led to an increase in CD14⁺ cells, consistent with MDSCs differentiating into monocytes and/or macrophages (Almand et al., 2001). All-trans retinoic acid (ATRA), a derivative of vitamin A that regulates cell differentiation, has also been shown to mature MDSCs and reduce suppressive activity. Addition of ATRA to blood cell fractions from cancer patients enriched for DCs significantly reduced the number of MDSCs and increased mature DCs capable of stimulating T cell responses (Almand et al., 2001). Renal cell carcinoma patients treated with ATRA experienced reductions in MDSCs and improved antigen-specific T cell responses (Mirza et al., 2006). Similarly, ATRA treatment of tumor-bearing mice restored T cell responses and reduced MDSCs in the spleen, lymph nodes, and bone marrow by inducing differentiation to mature DCs, macrophages, and granulocytes (Kusmartsev et al., 2003).

The vitamin derivative 1-alpha 25-dihydroxyvitamin $D_{(3)}$ [$(1,25)(OH)_{(2)}$], a biologically active metabolite of vitamin D3, also stimulates myeloid cell differentiation. In clinical studies and experimental animal models, vitamin D_3 has been reported to diminish the levels of immunosuppressive myeloid-derived cells and enhance the effectiveness of tumor immunotherapy. When mice with Lewis lung carcinoma were treated with vitamin D_3, CD34⁺ MDSC levels were reduced, and tumor-specific T cell activity was enhanced (Wiers et al., 2000). In a phase IB clinical trial of patients with HNSCC, the oral administration of vitamin D_3 promoted the differentiation of myeloid cells, reduced the number of immunosuppressive CD34⁺ cells in peripheral blood, increased plasma levels of IL-12 and IFN-γ, and improved T cell blastogenesis (Lathers et al., 2004, 2001).

B. Reduction of MDSC Levels by Chemotherapeutic Drugs

MDSC levels have also been reduced in tumor-bearing mice by chemotherapeutic drugs. Administration of gemcitabine, an inhibitor of DNA synthesis that induces apoptosis, is commonly used in combination chemotherapy protocols in patients with a variety of cancers (Edelman et al., 2001). In mouse models of primary tumors, gemcitabine selectively eliminates MDSCs without reducing potentially beneficial immune cells and increases tumor-specific T cell-mediated immunity and NK cell activation in response to vaccines (Suzuki et al., 2005). Similar decreases in MDSCs were observed following gemcitabine treatment of mice with metastasis although MDSC levels eventually returned as the metastatic disease progressed (P. Sinha, V. Clements, S. Bunt, S. Albelda, and S. Ostrand-Rosenberg, unpublished observations, 2007).

The observation that PGE_2 induces MDSC accumulation has led to the hypothesis that inhibitors of inflammation, which block COX-2 and COX-2-dependent production of PGE_2, may limit MDSC induction. This hypothesis is supported by studies demonstrating that COX-2 inhibitors and nonsteroidal anti-inflammatory drugs reduce the risk of various cancers, delay tumor progression (Harris et al., 2003; Howe et al., 2002; Roh et al., 2004), and increase T cell responses along with the efficacy of cancer vaccines (Haas *et al.*, 2006). Further support for the utility of COX-2 inhibitors as therapeutic anti-MDSC agents comes from studies in which PGE_2 receptor antagonists reduced MDSC induction in cultures of bone marrow stem cells induced for MDSC differentiation by PGE_2 (Sinha *et al.*, 2007).

VIII. CONCLUSIONS: IMPLICATIONS FOR IMMUNOTHERAPY

MDSCs are present in many cancer patients where they facilitate tumor progression and pose a significant impediment to immunotherapy. Although they were first identified in the mid-1980s, their potent suppressive effects and widespread presence in many cancer patients have only been appreciated within the past five years. Statistics indicate that the objective response rate of cancer patients to cancer vaccine treatments is 3.3% (reviewed in Gattinoni *et al.*, 2006). This response rate is not only dismal but also perplexing because these immune therapies had shown significant efficacy in experimental animals. Although the presence of MDSCs (and other suppressor cell populations and factors) in cancer patients may not be uniquely responsible for the failures of cancer vaccines, they may be a significant contributing factor. Indeed, the demonstration of "efficacy" in experimental animals is typically based on the successful therapy of small, short-term established tumors. Since MDSCs typically accumulate and increase as a function of tumor burden, they may not be a complicating factor in most animal studies, and therefore, the animal studies do not adequately model established disease in patients. In contrast, MDSCs are likely to be present at high levels in most patients receiving immunotherapy because these experimental strategies are tested largely on advanced-stage cancers in patients with extensive tumor burden. Therefore, it is likely that to be efficacious, immunotherapy must be combined with the reversal of systemic immune suppression, such as that mediated by MDSCs.

Although several approaches for eliminating or reducing MDSCs have been developed and tested, the heterogeneity of MDSCs makes it likely that there are multiple subpopulations of MDSCs. Thus, it seems unlikely that a single drug will be effective against all subpopulations of MDSCs. It is also not clear if recurring populations of MDSCs will remain responsive to any effective agent. Given these issues, it is likely that multiple agents and/or approaches will be necessary to control MDSC accumulation.

Tumor-induced inflammation in the local environment and chronic inflammation resulting from nontumor secreted factors are also likely to significantly impact immunotherapy due to induction of MDSCs. Efforts have focused on eliminating inflammation in cancer patients because inflammation is associated with enhanced tumor progression. However, it has not been appreciated until recently that inflammation-induced MDSCs may have enhanced suppressive activity and are more persistent than MDSCs induced by other mechanisms (Bunt *et al.*, 2006). Although this longevity and enhanced suppressive activity will likely complicate immunotherapy, elimination of the inflammatory milieu should still reduce MDSC accumulation and thereby reduce tumor progression. Although there remain many unanswered questions concerning the diversity, origin, persistence, activity, and control of MDSCs, the identification of this potently immunosuppressive cell population and the recognition of its critical role in facilitating tumor progression offer an important step toward the development of successful immunotherapies for cancer treatment.

Acknowledgments

The authors thank their colleagues who communicated unpublished data for inclusion in this review. Unpublished work from the authors' laboratory was supported by NIH grants R01115880, R01CA52527, and R0184232 and by the Susan G. Komen Foundation Grant BCTR053885. S. K. B. was supported by a DOD Breast Cancer Program predoctoral fellowship (#W81XWH-05-1-0276). E. M. H. was supported by an NIH postdoctoral fellowship (#1F32CA119768-01A1).

References

Almand, B., Clark, J. I., Nikitina, E., van Beynen, J., English, N. R., Knight, S. C., Carbone, D. P., and Gabrilovich, D. I. (2001). Increased production of immature myeloid cells in cancer patients: A mechanism of immunosuppression in cancer. *J. Immunol.* **166**, 678–689.

Almand, B., Resser, J. R., Lindman, B., Nadaf, S., Clark, J. I., Kwon, E. D., Carbone, D. P., and Gabrilovich, D. I. (2000). Clinical significance of defective dendritic cell differentiation in cancer. *Clin. Cancer Res.* **6**, 1755–1766.

Angulo, I., Rodriguez, R., Garcia, B., Medina, M., Navarro, J., and Subiza, J. L. (1995). Involvement of nitric oxide in bone marrow-derived natural suppressor activity. Its dependence on IFN-gamma. *J. Immunol.* **155**, 15–26.

Aoe, T., Okamoto, Y., and Saito, T. (1995). Activated macrophages induce structural abnormalities of the T cell receptor-CD3 complex. *J. Exp. Med.* **181**, 1881–1886.

Balkwill, F., and Mantovani, A. (2001). Inflammation and cancer: Back to Virchow? *Lancet* **357**, 539–545.

Bronte, V., Apolloni, E., Cabrelle, A., Ronca, R., Serafini, P., Zamboni, P., Restifo, N. P., and Zanovello, P. (2000). Identification of a CD11b(+)/Gr-1(+)/CD31(+) myeloid progenitor capable of activating or suppressing CD8(+) T cells. *Blood* **96**, 3838–3846.

Bronte, V., Chappell, D. B., Apolloni, E., Cabrelle, A., Wang, M., Hwu, P., and Restifo, N. P. (1999). Unopposed production of granulocyte-macrophage colony-stimulating factor by tumors inhibits CD8+ T cell responses by dysregulating antigen-presenting cell maturation. *J. Immunol.* **162**, 5728–5737.

Bronte, V., Serafini, P., Mazzoni, A., Segal, D. M., and Zanovello, P. (2003). L-arginine metabolism in myeloid cells controls T-lymphocyte functions. *Trends Immunol.* **24**, 302–306.

Bronte, V., Wang, M., Overwijk, W. W., Surman, D. R., Pericle, F., Rosenberg, S. A., and Restifo, N. P. (1998). Apoptotic death of CD8+ T lymphocytes after immunization: Induction of a suppressive population of Mac-1+/Gr-1+ cells. *J. Immunol.* **161**, 5313–5320.

Brys, L., Beschin, A., Raes, G., Ghassabeh, G. H., Noel, W., Brandt, J., Brombacher, F., and De Baetselier, P. (2005). Reactive oxygen species and 12/15-lipoxygenase contribute to the antiproliferative capacity of alternatively activated myeloid cells elicited during helminth infection. *J. Immunol.* **174**, 6095–6104.

Bunt, S. K., Sinha, P., Clements, V. K., Leips, J., and Ostrand-Rosenberg, S. (2006). Inflammation induces myeloid-derived suppressor cells that facilitate tumor progression. *J. Immunol.* **176**, 284–290.

Colonna, M., Trinchieri, G., and Liu, Y. J. (2004). Plasmacytoid dendritic cells in immunity. *Nat. Immunol.* **5**, 1219–1226.

Danna, E. A., Sinha, P., Gilbert, M., Clements, V. K., Pulaski, B. A., and Ostrand-Rosenberg, S. (2004). Surgical removal of primary tumor reverses tumor-induced immunosuppression despite the presence of metastatic disease. *Cancer Res.* **64**, 2205–2211.

de Jonge, W. J., Kwikkers, K. L., te Velde, A. A., van Deventer, S. J., Nolte, M. A., Mebius, R. E., Ruijter, J. M., Lamers, M. C., and Lamers, W. H. (2002). Arginine deficiency affects early B cell maturation and lymphoid organ development in transgenic mice. *J. Clin. Invest.* **110**, 1539–1548.

De Santo, C., Serafini, P., Marigo, I., Dolcetti, L., Bolla, M., Del Soldato, P., Melani, C., Guiducci, C., Colombo, M. P., Iezzi, M., Musiani, P., Zanovello, P., and Bronte, V. (2005). Nitroaspirin corrects immune dysfunction in tumor-bearing hosts and promotes tumor eradication by cancer vaccination. *Proc. Natl. Acad. Sci. USA* **102**, 4185–4190.

Dunn, G. P., Bruce, A. T., Ikeda, H., Old, L. J., and Schreiber, R. D. (2002). Cancer immunoediting: From immunosurveillance to tumor escape. *Nat. Immunol.* **3**, 991–998.

Edelman, M. J., Gandara, D. R., Lau, D. H., Lara, P., Lauder, I. J., and Tracy, D. (2001). Sequential combination chemotherapy in patients with advanced non-small-cell lung carcinoma: Carboplatin and gemcitabine followed by paclitaxel. *Cancer* **92**, 146–152.

Ekbom, A., Helmick, C., Zack, M., and Adami, H. O. (1990). Ulcerative colitis and colorectal cancer. A population-based study. *N. Engl. J. Med.* **323**, 1228–1233.

Finn, O. J. (2003). Cancer vaccines: Between the idea and the reality. *Nat. Rev. Immunol.* **3**, 630–641.

Fleming, T. J., Fleming, M. L., and Malek, T. R. (1993). Selective expression of Ly-6G on myeloid lineage cells in mouse bone marrow. RB6-8C5 mAb to granulocyte-differentiation antigen (Gr-1) detects members of the Ly-6 family. *J. Immunol.* **151**, 2399–2408.

Fontenot, J. D., Gavin, M. A., and Rudensky, A. Y. (2003). Foxp3 programs the development and function of $CD4^+CD25^+$ regulatory T cells. *Nat. Immunol.* **4**, 330–336.

Gabrilovich, D., Bronte, V., Chen, S. H., Colombo, M. P., Ochoa, A. C., Ostrand-Rosenberg, S., and Schreiber, H. (2006). The terminology issue for myeloid-suppressor cells. *Cancer Res.* **67**, 245.

Gabrilovich, D. I., Chen, H. L., Girgis, K. R., Cunningham, H. T., Meny, G. M., Nadaf, S., Kavanaugh, D., and Carbone, D. P. (1996). Production of vascular endothelial growth factor by human tumors inhibits the functional maturation of dendritic cells. *Nat. Med.* **2**, 1096–1103.

Gabrilovich, D. I., Ciernik, I. F., and Carbone, D. P. (1996). Dendritic cells in antitumor immune responses. I. Defective antigen presentation in tumor-bearing hosts. *Cell Immunol.* **170**, 101–110.

Gabrilovich, D. I., Nadaf, S., Corak, J., Berzofsky, J. A., and Carbone, D. P. (1996). Dendritic cells in antitumor immune responses. II. Dendritic cells grown from bone marrow precursors, but not mature DC from tumor-bearing mice, are effective antigen carriers in the therapy of established tumors. *Cell Immunol.* **170**, 111–119.

Gabrilovich, D. I., Velders, M. P., Sotomayor, E. M., and Kast, W. M. (2001). Mechanism of immune dysfunction in cancer mediated by immature Gr-1+ myeloid cells. *J. Immunol.* **166**, 5398–5406.

Garrity, T., Pandit, R., Wright, M. A., Benefield, J., Keni, S., and Young, M. R. (1997). Increased presence of CD34$^+$ cells in the peripheral blood of head and neck cancer patients and their differentiation into dendritic cells. *Int. J. Cancer* **73**, 663–669.

Gattinoni, L., Powell, D. J., Jr., Rosenberg, S. A., and Restifo, N. P. (2006). Adoptive immunotherapy for cancer: building on success. *Nat. Rev. Immunol.* **6**, 383–393.

Ghansah, T., Paraiso, K. H., Highfill, S., Desponts, C., May, S., McIntosh, J. K., Wang, J. W., Ninos, J., Brayer, J., Cheng, F., Sotomayor, E., and Kerr, W. G. (2004). Expansion of myeloid suppressor cells in SHIP-deficient mice represses allogeneic T cell responses. *J. Immunol.* **173**, 7324–7330.

Ghiringhelli, F., Puig, P. E., Roux, S., Parcellier, A., Schmitt, E., Solary, E., Kroemer, G., Martin, F., Chauffert, B., and Zitvogel, L. (2005). Tumor cells convert immature myeloid dendritic cells into TGF-beta-secreting cells inducing CD4$^+$CD25$^+$ regulatory T cell proliferation. *J. Exp. Med.* **202**, 919–929.

Guiducci, C., Vicari, A. P., Sangaletti, S., Trinchieri, G., and Colombo, M. P. (2005). Redirecting *in vivo* elicited tumor infiltrating macrophages and dendritic cells towards tumor rejection. *Cancer Res.* **65**, 3437–3446.

Haas, A. R., Sun, J., Vachani, A., Wallace, A. F., Silverberg, M., Kapoor, V., and Albelda, S. M. (2006). Cycloxygenase-2 inhibition augments the efficacy of a cancer vaccine. *Clin. Cancer Res.* **12**, 214–222.

Harris, R. E., Chlebowski, R. T., Jackson, R. D., Frid, D. J., Ascenseo, J. L., Anderson, G., Loar, A., Rodabough, R. J., White, E., and McTiernan, A. (2003). Breast cancer and nonsteroidal anti-inflammatory drugs: Prospective results from the Women's Health Initiative. *Cancer Res.* **63**, 6096–6101.

Howe, L. R., Subbaramaiah, K., Patel, J., Masferrer, J. L., Deora, A., Hudis, C., Thaler, H. T., Muller, W. J., Du, B., Brown, A. M., and Dannenberg, A. J. (2002). Celecoxib, a selective cyclooxygenase 2 inhibitor, protects against human epidermal growth factor receptor 2 (HER-2)/neu-induced breast cancer. *Cancer Res.* **62**, 5405–5407.

Huang, B., Pan, P. Y., Li, Q., Sato, A. I., Levy, D. E., Bromberg, J., Divino, C. M., and Chen, S. H. (2006). Gr-1$^+$CD115$^+$ immature myeloid suppressor cells mediate the development of tumor-induced T regulatory cells and T-cell anergy in tumor-bearing host. *Cancer Res.* **66**, 1123–1131.

Kusmartsev, S., Cheng, F., Yu, B., Nefedova, Y., Sotomayor, E., Lush, R., and Gabrilovich, D. (2003). All-trans-retinoic acid eliminates immature myeloid cells from tumor-bearing mice and improves the effect of vaccination. *Cancer Res.* **63**, 4441–4449.

Kusmartsev, S., and Gabrilovich, D. I. (2002). Immature myeloid cells and cancer-associated immune suppression. *Cancer Immunol. Immunother.* **51**, 293–298.

Kusmartsev, S., and Gabrilovich, D. I. (2006). Role of immature myeloid cells in mechanisms of immune evasion in cancer. *Cancer Immunol. Immunother.* **55**, 237–245.

Kusmartsev, S., Nefedova, Y., Yoder, D., and Gabrilovich, D. I. (2004). Antigen-specific inhibition of CD8$^+$ T cell response by immature myeloid cells in cancer is mediated by reactive oxygen species. *J. Immunol.* **172**, 989–999.

Kusmartsev, S. A., Li, Y., and Chen, S. H. (2000). Gr-1$^+$ myeloid cells derived from tumor-bearing mice inhibit primary T cell activation induced through CD3/CD28 costimulation. *J. Immunol.* **165**, 779–785.

Lathers, D. M., Clark, J. I., Achille, N. J., and Young, M. R. (2004). Phase 1B study to improve immune responses in head and neck cancer patients using escalating doses of 25-hydroxyvitamin D_3. *Cancer Immunol. Immunother.* **53**, 422–430.

Lathers, D. M., Clark, J. I., Achille, N. J., and Young, M. R. (2001). Phase IB study of 25-hydroxyvitamin $D_{(3)}$ treatment to diminish suppressor cells in head and neck cancer patients. *Hum. Immunol.* **62**, 1282–1293.

Letterio, J. J., and Roberts, A. B. (1998). Regulation of immune responses by TGF-beta. *Annu. Rev. Immunol.* **16**, 137–161.

Li, Q., Pan, P. Y., Gu, P., Xu, D., and Chen, S. H. (2004). Role of immature myeloid Gr-1$^+$ cells in the development of antitumor immunity. *Cancer Res.* **64**, 1130–1139.

MacDonald, K. P., Rowe, V., Clouston, A. D., Welply, J. K., Kuns, R. D., Ferrara, J. L., Thomas, R., and Hill, G. R. (2005). Cytokine expanded myeloid precursors function as regulatory antigen-presenting cells and promote tolerance through IL-10-producing regulatory T cells. *J. Immunol.* **174**, 1841–1850.

Mantovani, A., Sozzani, S., Locati, M., Allavena, P., and Sica, A. (2002). Macrophage polarization: Tumor-associated macrophages as a paradigm for polarized M2 mononuclear phagocytes. *Trends Immunol.* **23**, 549–555.

Marie, J. C., Letterio, J. J., Gavin, M., and Rudensky, A. Y. (2005). TGF-beta1 maintains suppressor function and Foxp3 expression in CD4$^+$CD25$^+$ regulatory T cells. *J. Exp. Med.* **201**, 1061–1067.

Mazzoni, A., Bronte, V., Visintin, A., Spitzer, J. H., Apolloni, E., Serafini, P., Zanovello, P., and Segal, D. M. (2002). Myeloid suppressor lines inhibit T cell responses by an NO-dependent mechanism. *J. Immunol.* **168**, 689–695.

Melani, C., Chiodoni, C., Forni, G., and Colombo, M. P. (2003). Myeloid cell expansion elicited by the progression of spontaneous mammary carcinomas in c-erbB-2 transgenic BALB/c mice suppresses immune reactivity. *Blood* **102**, 2138–2145.

Menetrier-Caux, C., Montmain, G., Dieu, M. C., Bain, C., Favrot, M. C., Caux, C., and Blay, J. Y. (1998). Inhibition of the differentiation of dendritic cells from CD34($^+$) progenitors by tumor cells: Role of interleukin-6 and macrophage colony-stimulating factor. *Blood* **92**, 4778–4791.

Mirza, N., Fishman, M., Fricke, I., Dunn, M., Neuger, A., Frost, T., Lush, R., Antonia, S., and Gabrilovich, D. (2006). All-trans-retinoic acid improves differentiation of myeloid cells and immune response in cancer patients. *Cancer Res.* **66**, 9299–9307.

Moseman, E. A., Liang, X., Dawson, A. J., Panoskaltsis-Mortari, A., Krieg, A. M., Liu, Y. J., Blazar, B. R., and Chen, W. (2004). Human plasmacytoid dendritic cells activated by CpG oligodeoxynucleotides induce the generation of CD4$^+$CD25$^+$ regulatory T cells. *J. Immunol.* **173**, 4433–4442.

Muller, A. J., DuHadaway, J. B., Donover, P. S., Sutanto-Ward, E., and Prendergast, G. C. (2005). Inhibition of indoleamine 2,3-dioxygenase, an immunoregulatory target of the cancer suppression gene *Bin1*, potentiates cancer chemotherapy. *Nat. Med.* **11**, 312–319.

Muller, A. J., and Prendergast, G. C. (2005). Marrying immunotherapy with chemotherapy: Why say IDO? *Cancer Res.* **65**, 8065–8068.

Nefedova, Y., Huang, M., Kusmartsev, S., Bhattacharya, R., Cheng, P., Salup, R., Jove, R., and Gabrilovich, D. (2004). Hyperactivation of STAT3 is involved in abnormal differentiation of dendritic cells in cancer. *J. Immunol.* **172**, 464–474.

Nefedova, Y., Nagaraj, S., Rosenbauer, A., Muro-Cacho, C., Sebti, S. M., and Gabrilovich, D. I. (2005). Regulation of dendritic cell differentiation and antitumor immune response in cancer by pharmacologic-selective inhibition of the janus-activated kinase 2/signal transducers and activators of transcription 3 pathway. *Cancer Res.* **65**, 9525–9535.

Oyama, T., Ran, S., Ishida, T., Nadaf, S., Kerr, L., Carbone, D. P., and Gabrilovich, D. I. (1998). Vascular endothelial growth factor affects dendritic cell maturation through the inhibition of nuclear factor-kappa B activation in hemopoietic progenitor cells. *J. Immunol.* **160**, 1224–1232.

Pak, A. S., Wright, M. A., Matthews, J. P., Collins, S. L., Petruzzelli, G. J., and Young, M. R.. (1995). Mechanisms of immune suppression in patients with head and neck cancer: Presence of CD34($^+$) cells which suppress immune functions within cancers that secrete granulocyte-macrophage colony-stimulating factor. *Clin. Cancer Res.* **1**, 95–103.

Pelaez, B., Campillo, J. A., Lopez-Asenjo, J. A., and Subiza, J. L. (2001). Cyclophosphamide induces the development of early myeloid cells suppressing tumor cell growth by a nitric oxide-dependent mechanism. *J. Immunol.* **166**, 6608–6615.

Pericle, F., Kirken, R. A., Bronte, V., Sconocchia, G., DaSilva, L., and. Segal, D. M. (1997). Immunocompromised tumor-bearing mice show a selective loss of STAT5a/b expression in T and B lymphocytes. *J. Immunol.* **159**, 2580–2585.

Porembska, Z., Luboinski, G., Chrzanowska, A., Mielczarek, M., Magnuska, J., and Baranczyk-Kuzma, A. (2003). Arginase in patients with breast cancer. *Clin. Chim. Acta* **328**, 105–111.

Porembska, Z., Skwarek, A., Mielczarek, M., and Baranczyk-Kuzma, A. (2002). Serum arginase activity in postsurgical monitoring of patients with colorectal carcinoma. *Cancer* **94**, 2930–2934.

Rodriguez, P. C., Hernandez, C. P., Quiceno, D., Dubinett, S. M., Zabaleta, J., Ochoa, J. B., Gilbert, J., and Ochoa, A. C. (2005). Arginase I in myeloid suppressor cells is induced by COX-2 in lung carcinoma. *J. Exp. Med.* **202**, 931–939.

Rodriguez, P. C., Quiceno, D. G., Zabaleta, J., Ortiz, B., Zea, A. H., Piazuelo, M. B., Delgado, A., Correa, P., Brayer, J., Sotomayor, E. M., Antonia, S., Ochoa, J. B., and Ochoa, A. C. (2004). Arginase I production in the tumor microenvironment by mature myeloid cells inhibits T-cell receptor expression and antigen-specific T-cell responses. *Cancer Res.* **64**, 5839–5849.

Roh, J. L., Sung, M. W., Park, S. W., Heo, D. S., Lee, D. W., and Kim, K. H. (2004). Celecoxib can prevent tumor growth and distant metastasis in postoperative setting. *Cancer Res.* **64**, 3230–3235.

Rossner, S., Voigtlander, C., Wiethe, C., Hanig, J., Seifarth, C., and Lutz, M. B. (2005). Myeloid dendritic cell precursors generated from bone marrow suppress T cell responses via cell contact and nitric oxide production in vitro. *Eur. J. Immunol.* **35**, 3533–3544.

Saito, H., Tsujitani, S., Ikeguchi, M., Maeta, M., and Kaibara, N. (1998). Relationship between the expression of vascular endothelial growth factor and the density of dendritic cells in gastric adenocarcinoma tissue. *Br. J. Cancer* **78**, 1573–1577.

Salvadori, S., Martinelli, G., and Zier, K. (2000). Resection of solid tumors reverses T cell defects and restores protective immunity. *J. Immunol.* **164**, 2214–2220.

Schmidt-Wolf, I. G., Dejbakhsh-Jones, S., Ginzton, N., Greenberg, P., and Strober, S. (1992). T-cell subsets and suppressor cells in human bone marrow. *Blood* **80**, 3242–3250.

Schmielau, J., and Finn, O. J. (2001). Activated granulocytes and granulocyte-derived hydrogen peroxide are the underlying mechanism of suppression of T-cell function in advanced cancer patients. *Cancer Res.* **61**, 4756–4760.

Serafini, P., Carbley, R., Noonan, K. A., Tan, G., Bronte, V., and Borrello, I. (2004). High-dose granulocyte-macrophage colony-stimulating factor-producing vaccines impair the immune response through the recruitment of myeloid suppressor cells. *Cancer Res.* **64**, 6337–6343.

Serafini, P., De Santo, C., Marigo, I., Cingarlini, S., Dolcetti, L., Gallina, G., Zanovello, P., and Bronte, V. (2004). Derangement of immune responses by myeloid suppressor cells. *Cancer Immunol. Immunother.* **53**, 64–72.

Seung, L. P., Rowley, D. A., Dubey, P., and Schreiber, H. (1995). Synergy between T-cell immunity and inhibition of paracrine stimulation causes tumor rejection. *Proc. Natl. Acad. Sci. USA* **92**, 6254–6258.

Shimizu, J., Yamazaki, S., and Sakaguchi, S. (1999). Induction of tumor immunity by removing $CD25^+CD4^+$ T cells: A common basis between tumor immunity and autoimmunity. *J. Immunol.* **163**, 5211–5218.

Sinha, P., Clements, V. K., and Ostrand-Rosenberg, S. (2005a). Reduction of myeloid-derived suppressor cells and induction of M1 macrophages facilitate the rejection of established metastatic disease. *J. Immunol.* **174**, 636–645.

Sinha, P., Clements, V. K., and Ostrand-Rosenberg, S. (2005b). Interleukin-13-regulated M2 macrophages in combination with myeloid suppressor cells block immune surveillance against metastasis. *Cancer Res.* **65**, 11743–11751.

Sinha, P., Clements, V. K., Fulton, A. M., and Ostrand-Rosenberg, S. (2007). Prostaglandin E2 promotes tumor progression by inducing myeloid-derired suppressor cells. *Cancer Res.* In press.

Song, X., Krelin, Y., Dvorkin, T., Bjorkdahl, O., Segal, S., Dinarello, C. A., Voronov, E., and Apte, R. N. (2005). $CD11b^+/Gr-1^+$ immature myeloid cells mediate suppression of T cells in mice bearing tumors of IL-1beta-secreting cells. *J. Immunol.* **175**, 8200–8208.

Steptoe, R. J., Ritchie, J. M., Jones, L. K., and Harrison, L. C. (2005). Autoimmune diabetes is suppressed by transfer of proinsulin-encoding $Gr-1^+$ myeloid progenitor cells that differentiate *in vivo* into resting dendritic cells. *Diabetes* **54**, 434–442.

Strober, S. (1984). Natural suppressor (NS) cells, neonatal tolerance, and total lymphoid irradiation: Exploring obscure relationships. *Annu. Rev. Immunol.* **2**, 219–237.

Sugiura, K., Inaba, M., Ogata, H., Yasumuzu, R., Sardina, E. E., Inaba, K., Kuma, S., Good, R. A., and Ikehara, S. (1990). Inhibition of tumor cell proliferation by natural suppressor cells present in murine bone marrow. *Cancer Res.* **50**, 2582–2586.

Suzuki, E., Kapoor, V., Jassar, A. S., Kaiser, L. R., and Albelda, S. M. (2005). Gemcitabine selectively eliminates splenic $Gr-1^+/CD11b^+$ myeloid suppressor cells in tumor-bearing animals and enhances antitumor immune activity. *Clin. Cancer Res.* **11**, 6713–6721.

Tas, M. P., Simons, P. J., Balm, F. J., and Drexhage, H. A. (1993). Depressed monocyte polarization and clustering of dendritic cells in patients with head and neck cancer: *In vitro* restoration of this immunosuppression by thymic hormones. *Cancer Immunol. Immunother.* **36**, 108–114.

Taylor, A., Verhagen, J., Blaser, K., Akdis, M., and Akdis, C. A. (2006). Mechanisms of immune suppression by interleukin-10 and transforming growth factor-beta: The role of T regulatory cells. *Immunology* **117**, 433–442.

Terabe, M., Matsui, S., Park, J. M., Mamura, M., Noben-Trauth, N., Donaldson, D. D., Chen, W., Wahl, S. M., Ledbetter, S., Pratt, B., Letterio, J. J., Paul, W. E., and Berzofsky, J. A. (2003). Transforming growth factor-beta production and myeloid cells are an effector mechanism through which CD1d-restricted T cells block cytotoxic T lymphocyte-mediated tumor immunosurveillance: Abrogation prevents tumor recurrence. *J. Exp. Med.* **198**, 1741–1752.

Valente, J. F., Ogle, C. K., Alexander, J. W., Li, B. G., Custer, D. A., Noel, J. G., and Ogle, J. D. (1997). Bone marrow and splenocyte coculture-generated cells enhance allograft survival. *Transplantation* **64**, 114–123.

Wiers, K. M., Lathers, D. M., Wright, M. A., and Young, M. R. (2000). Vitamin D_3 treatment to diminish the levels of immune suppressive $CD34^+$ cells increases the effectiveness of adoptive immunotherapy. *J. Immunother.* **23**, 115–124.

Yang, L., DeBusk, L. M., Fukuda, K., Fingleton, B. Green-Jarvis, B., Shyr, Y., Matrisian, L. M., Carbone, D. P., and Lin, P. C. (2004). Expansion of myeloid immune suppressor Gr^+CD11b^+ cells in tumor-bearing host directly promotes tumor angiogenesis. *Cancer Cell* **6**, 409–421.

Young, M. R., Petruzzelli, G. J., Kolesiak, K., Achille, N., Lathers, D. M., and Gabrilovich, D. I. (2001). Human squamous cell carcinomas of the head and neck chemoattract immune suppressive

CD34(⁺) progenitor cells. *Hum. Immunol.* **62**, 332–341.

Young, M. R., Wright, M. A., Coogan, M., Young, M. E., and Bagash, J. (1992). Tumor-derived cytokines induce bone marrow suppressor cells that mediate immunosuppression through transforming growth factor beta. *Cancer Immunol. Immunother.* **35**, 14–18.

Young, M. R., Wright, M. A., Lozano, Y., Prechel, M. M., Benefield, J., Leonetti, J. P., Collins, S. L., and Petruzzelli, G. J. (1997). Increased recurrence and metastasis in patients whose primary head and neck squamous cell carcinomas secreted granulocyte-macrophage colony-stimulating factor and contained CD34⁺ natural suppressor cells. *Int. J. Cancer* **74**, 69–74.

Zea, A. H., Rodriguez, P. C., Atkins, M. B., Hernandez, C., Signoretti, S., Zabaleta, J., McDermott, D., Quiceno, D., Youmans, A., O'Neill, A., Mier, J., and Ochoa, A. C. (2005). Arginase-producing myeloid suppressor cells in renal cell carcinoma patients: A mechanism of tumor evasion. *Cancer Res.* **65**, 3044–3048.

Further Reading

Coussens, L. M., and Werb, Z. (2002). Inflammation and cancer. *Nature* **420**, 860–867.

Dunn, G. P., Old, L. J., and Schreiber, R. D. (2004). The immunobiology of cancer immunosurveillance and immunoediting. *Immunity* **21**, 137–148.

Rodriguez, P. C., and Ochoa, A. C. (2006). T cell dysfunction in cancer: Role of myeloid cells and tumor cells regulating amino acid availability and oxidative stress. *Semin. Cancer Biol.* **16**, 66–72.

Shacter, E., and Weitzman, S. A. (2002). Chronic inflammation and cancer. *Oncology (Williston Park)* **16**, 217–226, 229; discussion, 230–212.

Whiteside, T. L. (2006). Immune suppression in cancer: Effects on immune cells, mechanisms and future therapeutic intervention. *Semin. Cancer Biol.* **16**, 3–15.

CHAPTER 18

Programmed Death Ligand-1 and Galectin-1: Pieces in the Puzzle of Tumor-Immune Escape

GABRIEL A. RABINOVICH AND THOMAS F. GAJEWSKI

I. Programmed Death Ligand 1 and Programmed Death 1 Interactions
   A. Programmed Death 1 as an Inhibitory Receptor
   B. Expression and Regulation of Programmed Death Ligand 1 on Tumor Cells
   C. PD-1/PD-L1 Blockade and Antitumor Immunity
   D. Evidence for an Alternative Receptor for PD-L1/PD-L2
   E. Future Directions and Unresolved Issues
II. Galectin 1
   A. Galectins: Multifunctional Glycan-Binding Proteins with Immunoregulatory Properties
   B. Gal-1: Biochemistry and Cellular Biology
   C. Expression of Gal-1 on Tumors
   D. Role of Gal-1 in Negative Regulation of T Cell Responses
   E. Gal-1 in Tumor-Immune Escape
   F. Gal-1 as a Target for Anticancer Therapies: Status, Future Directions, and Unresolved Issues
References
Further Reading

Despite major advances in the understanding of tumor rejection mechanisms, the successful translation of basic and preclinical information into an effective tumor immunotherapy in the clinic has continued to struggle. It has become increasingly apparent that one important obstacle is the activation of immunosuppressive mechanisms and inhibitory pathways that act in concert to thwart immune effector responses. This chapter discusses biochemical and functional aspects of two inhibitory signals, stimulated by the cellular ligands programmed death ligand 1 (PDL-1) and galectin 1, that help create an immunosuppressive network in the tumor microenvironment. Evidence is mounting that these inhibitory signals can influence tumor progression by modulating the activation, differentiation, and survival of effector T cells. Further investigation into their roles in the induction and maintenance of immune tolerance to tumor antigens may reveal appropriate frameworks to develop strategies for overcoming these barriers for cancer immunotherapy.

I. PROGRAMMED DEATH LIGAND 1 AND PROGRAMMED DEATH 1 INTERACTIONS

A. Programmed Death 1 as an Inhibitory Receptor

Programmed death 1 (PD-1) is a 55-kD member of the Ig superfamily that was cloned originally from a T cell hybridoma undergoing apoptosis (Ishida *et al.*, 1992). The extracellular domain is 23% identical to cytotoxic T lymphocyte-associated antigen 4 (CTLA-4), a known inhibitory receptor expressed on activated T cells (Thompson and Allison, 1997). In normal tissues, PD-1 is expressed on approximately one-third of $CD4^-CD8^-$ thymocytes, arguing for a potential role in thymic selection. Indeed, absence of PD-1 on $CD8^+$ T cell receptor (TCR) Tg populations appears to allow emergence of $CD8^-$ TCR Tg T cells in the periphery, arguing that PD-1 signaling in the thymus alters the threshold for positive selection (Blank *et al.*, 2003). Direct ligation of PD-1 in thymocytes also has been shown to inhibit positive selection (Keir *et al.*, 2005). However, it is believed that the activity of PD-1 on peripheral lymphocytes is the most critical for regulating immune responses. PD-1 surface expression is minimally detected on naive T cells and B cells. However, it is inducibly expressed on both lymphocyte populations following activation, suggesting that activated T and B cells are perhaps the main cellular targets for PD-1 signaling.

The clearest evidence regarding the functional role of PD-1 has come from studies of PD-1-deficient mice. $PD-1^{-/-}$ mice on the C57BL/6 genetic background develop splenomegaly with age as well as increased total serum immunoglobulin levels (Nishimura *et al.*, 1998). Over time, arthritis and glomerulonephritis with Ig deposition are also seen. On the BALB/c background, a lethal autoimmune cardiomyopathy has been reported to occur that is T cell and B cell dependent (Nishimura *et al.*, 2001). Together, these observations argue that the dominant functional role of PD-1 is inhibitory for immune responses.

Additional supportive data regarding the inhibitory function for PD-1 have come from *in vitro* studies. The engagement of PD-1 by its natural ligands has been demonstrated to inhibit T cell proliferation and both T helper 1 (Th1) and T helper 2 (Th2) cytokine production. Intercrossing $2C/RAG2^{-/-}$ mice with PD-1-deficient mice has allowed examination of the role of PD-1 in a monoclonal population of $CD8^+$ T cells. Primed PD-1-deficient $CD8^+$ T cells show augmented cytokine production and cytolysis compare to wild-type T cells *in vitro* (Blank *et al.*, 2004), suggesting that effector functions are particularly regulated by PD-1 engagement.

The mechanism by which PD-1 exerts its inhibitory effect on activated T cells is only partially elucidated. The PD-1 cytoplasmic tail contains an immunoreceptor, a tyrosine inhibition motif (ITIM), and an immunoreceptor tyrosine-based switch motif (ITSM), which in other receptors can mediate recruitment of phosphatases. Indeed, in B cell lymphoma cell lines, the protein tyrosine phosphatase Src homology 2 domain-containing phosphatase 2 (SHP-2) has been shown to be recruited to tyrosine phosphorylated PD-1 (Okazaki *et al.*, 2001). In Jurkat T cell leukemia cells, coengagement of PD-1 along with the TCR resulted in SHP-2 phosphorylation, suggesting a similar mechanism is operating in T cells. Recruitment and activation of SHP-2 would be expected to favor dephosphorylation of signaling intermediates involved in early TCR/CD28 signal transduction.

However, extensive biochemical analysis of the signaling alterations mediated by PD-1 in normal lymphocytes has not been completed.

Outside of its expression and function on activated T cells, PD-1 also has been found to be expressed by certain subtypes of T cell lymphoma (Dorfman *et al.*, 2006). Thus, targeting PD-1 could be relevant as a direct therapeutic approach for these lymphoma subsets.

B. Expression and Regulation of Programmed Death Ligand 1 on Tumor Cells

Two ligands for PD-1 have been identified, PD-L1/B7-H1 and PD-L2/B7-DC. Programmed death ligand 1 (PD-L1) was cloned from an expressed sequence tags (EST) database based on homology to the B7 family of CD28 ligands (Dong *et al.*, 1999). Programmed death ligand 2 (PD-L2) was cloned from a library of genes preferentially expressed in dendritic cells (DCs) over macrophages (Tseng *et al.*, 2001). In contrast to the CD28 ligands B7–1 and B7–2, which are expressed largely by antigen-presenting cell (APC) populations, transcripts for PD-1 ligands, particularly PD-L1, have been found in many tissue types. PD-L1 mRNA has been detected in thymus, heart, liver, brain, kidney, lung, muscle, spleen, stomach, and testis. This diverse distribution of expression suggests that PD-L1 may serve to dampen immune responses in peripheral tissues. However, detection of mRNA does not always predict expression of PD-L1 protein, and evidence is emerging that PD-L1 protein expression is in part regulated at the posttranscriptional level, perhaps through regulated translation. PD-L1 protein is inducibly expressed on most antigen-stimulated T cells and B cells. In addition, various parenchymal cell types have been shown to upregulate PD-L1 in response to interferon (IFN)-γ. This latter result suggests that the function of PD-L1 in normal physiology may be to limit the duration of the effector phase of a cellular immune response upon IFN-γ release in peripheral tissues. Consistent with this notion, IFN-γ response elements have been identified in the PD-L1 promoter. PD-L1 protein expression has been found in muscle biopsies from patients with inflammatory myopathies and in other situations of chronic inflammation. PD-L1 also has been suggested to contribute to immune tolerance of the semiallogeneic fetus during pregnancy.

The wide potential for PD-L1 expression in many normal cell types suggests that the malignant counterpart of those cells may retain the ability to resist immune responses through expression of PD-L1. Indeed, expression of PD-L1 protein, either *de novo* or following exposure to IFN-γ, has been reported in human cancers of the lung, head and neck, stomach, colon, liver, and bladder, among others (Dong and Chen, 2003). In the chapter authors' own laboratory, all mouse tumor cell lines tested and all human melanoma cell lines examined show surface expression of PD-L1 in response to IFN-γ. In addition, in the same laboratory, variable mRNA and protein expression of PD-L1 have been observed in the majority of fresh biopsies of metastatic melanoma lesions from patients. These observations have suggested a potential role for PD-L1 in immune escape of many tumor types in both murine and human systems.

C. PD-1/PD-L1 Blockade and Antitumor Immunity

The expression of PD-L1 on tumor cells and the apparent negative regulatory function of PD-1 have suggested that interference with PD-L1/PD-1 interactions might potentiate the effector phase of an antitumor immune response and thus improve immune-mediated tumor destruction (Figure 18.1). Transfection of P815 cells to

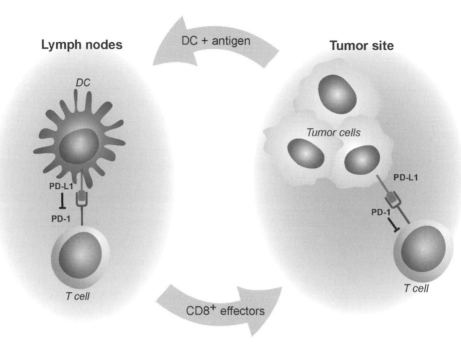

FIGURE 18.1   **Model for PD-L1/PD-1 interactions in limiting antitumor T cell responses.** While some effects of PD-L1 could be mediated through expression by APCs in lymphoid compartments, the dominant effect is thought to be at the effector phase within the tumor microenvironment.

constitutively express PD-L1 rendered them relatively resistant to lysis and promoted more aggressive growth *in vivo*, which was reversed by an anti-PD-L1 Ab (Dong *et al.*, 2002). In the chapter authors' laboratory, PD-1-deficient TCR Tg T cells were found to be superior at cytokine production in response to tumor cells *in vitro* and also at tumor rejection *in vivo*, compared to wild-type TCR Tg T cells (Blank *et al.*, 2004). This was efficacious even when CTLA-4-deficient T cells failed to reject the tumor, arguing that PD-1 blockade may be more critical than CTLA-4 blockade in certain tumor types. Interestingly, interference with PD-1 engagement also has been reported to decrease the development of distant metastasis in the B16 melanoma and CT26 colon cancer models in mice (Iwai *et al.*, 2005). Research has indicated that blockade of either PD-1 or PD-L1 also could potently improve antitumor immunity in combination with anti-4-1BB monoclonal antibody (mAb) *in vivo*.

The successful clinical development of neutralizing anti-CTLA-4 mAb as a novel cancer therapeutic targeting one T cell inhibitory receptor (Phan *et al.*, 2003) has motivated development of analogous anti-PD-1 mAbs for clinical translation. *In vitro*, human T cell responses against renal cell carcinoma cell lines have been shown to be augmented when PD-L1/PD-1 interactions were neutralized (Blank *et al.*, 2006), supporting the negative regulatory role for PD-1 on human T cells as had been observed in murine systems. A fully human antihuman PD-1 mAb is being developed by Ono Pharmaceutical and Medarex, and that is planned to enter early-phase clinical trials in 2007.

An alternative approach for interfering with PD-1 engagement by PD-L1 that has been explored in murine studies utilizes a gene transfer vector encoding a soluble form of the extracellular domain of PD-1 itself (He *et al.*, 2005). Delivery of this vector has been shown to promote improved antitumor immunity alone and in combination with an HSP70 vaccine in mice. This and other similar approaches for blocking PD-1 engagement using a soluble receptor fusion protein thus represent a viable alternative strategy for interfering with PD-L1/PD-1 interactions clinically.

Interestingly, data from infectious disease models has generated additional evidence for an important inhibitory role for PD-1 in another setting of persistence of antigen, that of chronic viral infections. In a chronic form of lymphocytic choriomeningitis virus (LCMV) infection in mice, $CD8^+$ T cells normally become "exhausted," resulting in incomplete viral clearance. The exhausted T cells in this model were shown to express PD-1, and their function was restored upon PD-1 blockade (Barber et al., 2006). In humans, PD-1 has been shown to be upregulated and functional on HIV-specific T cells (Day et al., 2006). These observations point to a more generalized role for PD-1 in limiting T cell function in situations of antigen persistence and suggest that therapeutic opportunities for PD-1 blockade may extend beyond just antitumor immunity.

D. Evidence for an Alternative Receptor for PD-L1/PD-L2

Indirect evidence has suggested the possible existence of a second receptor that can be ligated by PD-L2 (and perhaps PD-L1) and that may deliver a positive costimulatory signal to T cells rather than an inhibitory one. Some data have suggested that PD-L2/B7-DC can be costimulatory rather than inhibitory in certain model systems *in vivo*. Transfection of tumor cells to express PD-L2 led to improved recognition *in vitro* and superior tumor rejection *in vivo* (Liu et al., 2003). Interestingly, this effect did not depend on PD-1 expression by T cells, and a PD-L2 fusion protein was found to bind to the surface of PD-1-deficient T cells. Some studies of PD-L1/B7-H1 also have revealed an apparent positive effect on T cell function (Dong et al., 1999). These and other similar observations suggest an additional receptor exists that is distinct from PD-1 and that may provide a positive signal to activated T cells, much like the CD28/CTLA-4 counterbalance. Because of this concern, it seems prudent to target the PD-1 receptor itself for the first attempts of clinical translation, as it is clearly providing an inhibitory signal in the vast majority of experimental settings.

E. Future Directions and Unresolved Issues

Based on the evidence just described, there is growing interest in interfering with PD-L1/PD-1 interactions as a strategy to facilitate immune-mediated tumor rejection in the clinic. It will be of interest to determine the effects of neutralizing human antihuman PD-1 mAbs as they enter clinical trials in patients with advanced cancer. Ultimately, it may be necessary to combine PD-1 blockade with strategies to increase the frequency of tumor-specific T cells, such as through vaccination or adoptive T cell transfer. In addition, observations have indicated that multiple immune escape mechanisms may coexist in metastatic tumors. The chapter authors have found that most metastatic melanomas expressing PD-L1 also frequently contain $FoxP3^+$ regulatory T cells as well as transcripts for indoleamine 2,3-dioxygenase (IDO). Thus, it may be necessary ultimately to counter multiple inhibitory mechanisms to achieve optimal therapeutic efficacy in patients.

There are several aspects of PD-L1/PD-1 interactions relevant for antitumor immunity that have been incompletely characterized. The biochemical and/or molecular alterations in cancer cells that enable constitutive expression of PD-L1 in many tumors have not been described. One could imagine that targeting such signaling pathways could diminish PD-L1 expression on tumor cells, thus improving antitumor immunity. Similarly, the signals leading to

PD-1 expression and/or inhibitory function on T cells are not fully understood. In theory, blocking PD-1 expression and/or function on T cells also could have therapeutic potential. Finally, although much work has focused on PD-1 as it inhibits T cell responses, the role of PD-1 on other lymphocyte subsets that could contribute to antitumor immunity, including B cells, NK cells, and NKT cells, has not been thoroughly explored.

II. GALECTIN 1

A. Galectins: Multifunctional Glycan-Binding Proteins with Immunoregulatory Properties

Protein–glycan interactions can control essential physiological processes, including cell–cell and cell–matrix interactions, lymphocyte migration, and cytokine secretion (Ohtsubo and Marth, 2006). Galectins, a family of evolutionarily conserved glycan-binding proteins, are defined by a conserved carbohydrate-recognition domain (CRD) with a canonical amino acid sequence and affinity for β-galactosides (Camby et al., 2006). These β-galactoside-binding proteins have emerged as novel regulators of immune cell homeostasis and inflammation (Rabinovich et al., 2002a). To date, 15 mammalian galectins have been identified, 11 of which have human orthologs. These can be subdivided into three groups: one-CRD galectins (including galectins 1, 2, 5-, 7, 10, 11, 13, 14, and 15); two-CRD galectins (galectins 4, 6, 8-, 9, and 12); and the unique "chimera-type" galectin 3 that contains a single CRD fused to unusual tandem repeats of short amino acid stretches (Camby et al., 2006). Many galectins are either bivalent or multivalent with regard to their carbohydrate-binding activities: some one-CRD galectins exist as dimers, two-CRD galectins have two carbohydrate-binding sites in tandem, and galectin 3 forms oligomers when it binds to multivalent carbohydrates (Figure 18.2). Cross-linkage of cell-surface receptors by galectins can trigger transmembrane signaling events through which diverse processes, such as apoptosis, cytokine secretion, and cell migration, are modulated (Liu and Rabinovich, 2005). Remarkably, the responsiveness of cells to individual members of the galectin family can fluctuate depending on the repertoire of potentially glycosylated molecules expressed on the cell surface and the activities of specific glycosyltransferases that are responsible for generating galectin ligands. These variables can dramatically change according to the differentiation and activation stage of the cells (Rabinovich et al., 2002a). This chapter will focus on the role of galectin 1 (Gal-1), the first identified

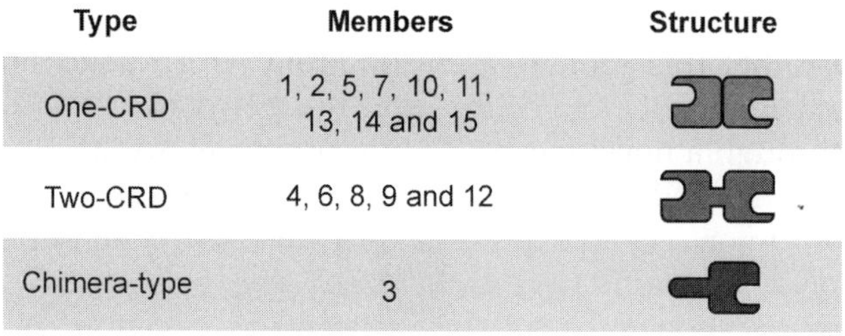

Type	Members	Structure
One-CRD	1, 2, 5, 7, 10, 11, 13, 14 and 15	
Two-CRD	4, 6, 8, 9 and 12	
Chimera-type	3	

FIGURE 18.2 The galectin family. Galectins are a family of carbohydrate-binding proteins characterized by affinity for β-galactoside-containing saccharides and conserved CRDs of about 130 amino acids that are responsible for carbohydrate binding. They can be subdivided into three groups: those containing one CRD (galectins 1, 2, 5, 7, 10, 13, 14, and 15); those containing two distinct CRDs in tandem, connected by a linker of up to 70 amino acids (galectins 4, 6, 8, 9, and 12); and galectin 3, which consists of unusual tandem repeats of proline- and glycine-rich short stretches (a total of about 120 amino acids) fused onto the CRD. Galectins can cross-link cell surface glycoconjugates and trigger a cascade of transmembrane signaling events.

member of this protein family, in immunoregulation and tumor-immune escape.

B. Gal-1: Biochemistry and Cellular Biology

Gal-1, a β-galactoside-binding protein widely expressed in human tissues, may occur as a monomer as well as a noncovalent homodimer composed of 14.5-kDa subunits (Camby et al., 2006). One of the main properties of the homodimeric Gal-1 protein is that it spontaneously dissociates at low concentrations (Kda~7 μM) into a monomeric form that is still able to bind carbohydrates but with lower affinity. Interestingly, evidence indicates that the monomeric and dimeric forms of Gal-1 are associated with different and, in some cases, contrasting biological functions. Although this protein binds preferentially to glycoconjugates containing the ubiquitous disaccharide N-acetyllactosamine (Gal β1–3/4 GlcNAc), binding to individual lactosamine units is of relatively low affinity, and it is the arrangement of lactosamine disaccharides in repeating chains (polylactosamine) that increases the binding activity (Rabinovich et al., 2002a).

Gal-1 lacks recognizable secretion signal sequences and does not pass through the standard endoplasmic reticulum/Golgi pathway (Camby et al., 2006). In addition, it shows characteristics of typical cytoplasmic proteins, including an acetylated N terminus and the lack of glycosylation. Nevertheless, it is well-known that this protein, as well as other members of the galectin family, is secreted through a novel mechanism distinct from classical vesicle-mediated exocytosis, which has not been thoroughly explored.

C. Expression of Gal-1 on Tumors

Expression of Gal-1 has been well documented in cancer cells and cancer-associated stromal cells of different tumor types, including astrocytoma, melanoma, prostate, thyroid, colon, breast, and ovarian carcinomas, and also in hematological malignancies (van den Brüle et al., 2004; Camby et al., 2006). Interestingly, in most cases, this expression correlates with the aggressiveness of these tumors and the acquisition of metastatic phenotype (Liu and Rabinovich, 2005).

How does Gal-1 contribute to tumor progression? Emerging evidence indicates that Gal-1 has diverse functions in several aspects of cancer biology, including the regulation of tumor transformation, cell cycle regulation, and apoptosis. Furthermore, this glycan-binding protein may also contribute to tumor metastasis through modulation of cell adhesion, migration, and invasiveness (Liu and Rabinovich, 2005). In addition, accumulating evidence indicates that Gal-1 acts as a negative regulator of T cell immunity and contributes to tumor cell evasion of T cell responses (Rubinstein et al., 2004; Le et al., 2005). This discussion will focus on the role of Gal-1 as an immunosuppressive mediator employed by tumors to evade immune responses.

D. Role of Gal-1 in Negative Regulation of T Cell Responses

1. Control of T Cell Survival

Research over the past decade has demonstrated that Gal-1 plays a pivotal role in the regulation of T cell homeostasis by regulating T cell survival, activation, and cytokine synthesis (Rabinovich et al., 2002a) (Figure 17.3). Through interaction with specific carbohydrate ligands on T cell surface glycoconjugates, this glycan-binding protein can induce cell cycle arrest and promote apoptosis of developing thymocytes and peripheral T lymphocytes (Perillo et al., 1997; Blaser et al., 1998; Rabinovich et al., 2002b). Different glycoproteins on the surface of activated T cells, including CD45, CD43, CD7, CD3, and CD2, appear to be primary receptors for Gal-1 (Pace et al., 2000).

Susceptibility to Gal-1-induced cell death is tightly controlled by the expression of

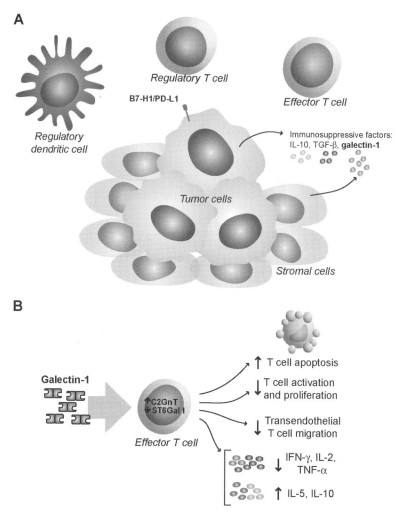

FIGURE 18.3 **Role of Gal-1 in tumor-induced immunosuppression.** Tumors employ a plethora of immunosuppressive mechanisms that may act in concert to counteract effective immune responses. These include activation of negative costimulatory signals in the tumor microenvironment (CTLA-4/B7, PD-1/PD-L1), recruitment of regulatory T cells, and elaboration of immunosuppressive factors (IL-10, TGF-β, Gal-1). Gal-1 can suppress antitumor responses through different mechanisms, including induction of effector T cell apoptosis, partial blockade of T cell activation, inhibition of transendothelial T cell migration, and suppression of proinflammatory cytokine secretion. Susceptibility to Gal-1-induced cell death is tightly controlled by the expression of specific glycosyltransferases, including the core-2-β-1,6-N-acetylglucosaminyltransferase (C2GnT) and the ST6Gal-1 sialyltransferase, that are regulated during T cell development, activation, and differentiation.

specific glycosyltransferases that are regulated during T cell development, activation, and differentiation (Galvan et al., 2000; Amano et al., 2003). It has been shown that T cells lacking the core-2-β-1,6-N-acetylglucosaminyltransferase (C2GnT) are resistant to Gal-1-induced cell death. This enzyme is responsible for creating branched structures on O-glycans of T cell surface glycoproteins, such as CD45 (Galvan et al., 2000). Consistent with these findings, studies have shown that haploinsufficiency of C2GnT-I results in altered cellular glycosylation and resistance to Gal-1-induced cell death in T lymphoma cells (Cabrera et al., 2006). In addition, other glycosyltransferases can also act to reduce Gal-1 binding by directly or indirectly masking Gal-1 saccharide ligands. In this context, addition of α2,6-linked sialic acids to lactosamine units by the ST6Gal-1 sialyltransferase has been shown to block Gal-1 binding and cell death (Amano et al., 2003).

The signal transduction events triggered by Gal-1 are still controversial. It has been demonstrated that Gal-1-induced T cell death requires the activation of specific transcription factors (i.e., AP-1) (Rabinovich et al., 2000), cytochrome c release, and modulation of caspases-8 and -3 (Matarrese et al., 2005). However, another report showed that apoptosis induced by Gal-1 in a T cell line is not dependent on the activation of caspase-3 or on cytochrome c release (Hahn et al., 2004). Interestingly, another study proposed a model in which Gal-1 induces the release of ceramide, which in turn promotes downstream events, including decreased Bcl-2 protein expression, depolarization of mitochondria, and activation of caspase-9 and -3 (Ion et al., 2006). All downstream events required the presence of two tyrosine kinases, p56^{lck} and ZAP-70. On the other hand, other studies have shown that Gal-1 promotes phosphatidylserine exposure in different cell types (including activated neutrophils and T cells) with no apparent signs of apoptosis (Dias Baruffi et al., 2003). Therefore, it seems evident that Gal-1 may trigger distinct death pathways or different apoptosis end points in different cell types.

In contrast to the proapoptotic role of Gal-1 on activated T cells, Endharti et al. (2005) demonstrated that secretion of this protein by stromal cells is capable of supporting the survival of naive T cells without promoting proliferation. In this regard, it has been found that murine DCs transduced with an adenoviral vector expressing Gal-1 can elicit contrasting results when exposed to resting or activated T cells. While this protein stimulated activation of naive T cells, it rapidly triggered apoptosis of activated T cells (Perone et al., 2006). Thus, Gal-1 might trigger different signals (i.e., apoptosis, proliferation, or survival) depending on a number of factors, including the activation state of the cells, the spatiotemporal expression of specific glycosyltransferases, the nature of the target cells (primary cell cultures or immortalized T cell lines), and different biochemical or biophysical parameters (i.e., equilibrium between the monomeric or dimeric protein forms).

2. Control of Cytokine Synthesis

Compelling evidence indicates an essential role of Gal-1 on the control of cytokine synthesis and secretion, suggesting that this lectin may influence a variety of physiological processes, including T cell activation, differentiation, and homing. In addition, the ability to control cytokine production endows this sugar-binding protein with the capacity to positively or negatively interfere with pathological processes such as chronic inflammation and cancer (Figure 17.3).

Gal-1 has been shown to selectively block secretion of Th1 and proinflammatory cytokines in vitro, including IL-2, IFN-γ, and tumor necrosis factor (TNF)-α (Rabinovich et al., 1999a; Vespa et al., 1999). In addition, studies in experimental models of chronic inflammation and autoimmunity showed the ability of Gal-1 to skew the balance toward a Th2-type cytokine profile in vivo, with decreased levels of IFN-γ and increased secretion of IL-5 or IL-10 by pathogenic T cells (Rabinovich et al., 1999b; Santucci et al., 2003; Toscano et al., 2006). Accordingly, van der Leij et al. (2007) reported a marked increase in IL-10 secretion from nonactivated and activated CD4$^+$ and CD8$^+$ T cells following exposure to Gal-1. Interestingly, Gal-1 treatment at early phases of a pathogenic autoimmune response has been found to promote the expansion of IL-10-producing regulatory T cells in lymph nodes from treated mice (Toscano et al.,

2006). In addition, increased levels of apoptosis were detected in lymph nodes from mice treated with recombinant Gal-1 during the efferent phase of autoimmune diseases, confirming the ability of Gal-1 to induce T cell apoptosis *in vivo* (Rabinovich et al., 1999b; Toscano et al., 2006). These results underscore the ability of this endogenous lectin to counteract Th1-mediated responses through different but potentially overlapping anti-inflammatory mechanisms. Furthermore, these findings might also have implications in the regulation of tumor-immune escape, given the ability of regulatory T cells to restrain antitumor immunity.

Taken together, these data suggest that under distinct physiological or pathological conditions, Gal-1 may provide inhibitory signals to control immune cell homeostasis and regulate inflammation following an antigenic challenge. The mechanisms underlying Gal-1-mediated regulation of cytokine production and the association of this effect with the regulation of T cell apoptosis still remain to be elucidated.

3. Regulation of T Cell Adhesion and Migration

In addition to its role in modulating T cell survival, proliferation, and cytokine synthesis, Gal-1 (at low concentrations) has the ability to inhibit T cell adhesion to extracellular matrix glycoproteins, such as laminin and fibronectin, without any evidence of T cell apoptosis, as studies have demonstrated (Rabinovich et al., 1999a). Similarly, it was found that Gal-1 specifically inhibits T cell migration across endothelial cells (He and Baum, 2006). Interestingly, this effect was also independent of the ability of Gal-1 to promote T cell death because T cells lacking core 2 O-glycans (which are critical for apoptosis) still showed reduced T cell migration across endothelial cells when exposed to this carbohydrate-binding protein. Since endothelial cells express very high levels of Gal-1, inhibition of T cell adhesion and migration may represent a novel anti-inflammatory effect of this protein that may have profound implications in tumor immunity.

4. Modulation of Immune Cell Activation and Immunological Synapse

Gal-1 has been also shown to influence T cell activation and differentiation and modulate the immunological synapse. Vespa *et al.* (1999) found that Gal-1 inhibits TCR signals and antagonizes TCR-induced IL-2 production in a murine T cell hybridoma clone. Interestingly, the same group further demonstrated that Gal-1 induces partial TCR-ζ chain phosphorylation and is able to antagonize full TCR responses, including the production of IL-2 (Chung et al., 2000).

Regarding the B cell compartment, Schiff *et al.* (2002) demonstrated that Gal-1 can act as a stromal ligand of the pre-B cell receptor and contribute to synapse formation between pre-BCR and stromal cells. Thus, by acting at early events of TCR or B cell receptor (BCR) signaling, Gal-1 can also control lymphocyte development and activation.

E. Gal-1 in Tumor-Immune Escape

The association between Gal-1 expression in different tumor types and the aggressiveness of these tumors (Liu and Rabinovich, 2005) prompted the chapter authors' investigation of Gal-1's role in tumor cell evasion of T cell responses. The authors hypothesized that tumor cells may impair T cell effector functions through secretion of Gal-1 and that this mechanism may contribute in tilting the balance toward an immunosuppressive environment at the tumor site. By a combination of *in vitro* and *in vivo* experiments, the authors established a link between Gal-1-mediated immunoregulation and its contribution to tumor-immune escape (Figure 18.3). Human and mouse melanoma cells secreted substantial levels of Gal-1, which contributed to the immunosuppressive activity of these tumor cells. In addition, blockade of the inhibitory effects of Gal-1 in the tumor tissue resulted in reduced tumor size and augmented stimulation of a tumor-specific T cell response *in vivo*. This effect required intact CD4$^+$ and CD8$^+$ T cell compartments since simultaneous depletion of these T cell subpopulations *in vivo* completely abrogated the antitumor response induced by knockdown transfectants (Rubinstein *et al.*, 2004). Furthermore, tumor-secreted Gal-1 induced *in situ* apoptosis of tumor-infiltrating lymphocytes, and interruption of this proapoptotic pathway in the tumor microenvironment resulted in the generation of an otherwise repressed tumor-specific Th1 response in tumor-draining lymph nodes (Rubinstein *et al.*, 2004).

Supporting these findings, Le *et al.* (2005) have shown that Gal-1 can act as a molecular link between tumor hypoxia and tumor-immune privilege. Using proteomic analysis, Le *et al.* (2005) demonstrated that Gal-1 is overexpressed following exposure of tumor cells to low oxygen levels. In addition, they found a strong inverse correlation between Gal-1 expression and T cell recruitment in tumor sections corresponding to head and neck squamous cell carcinoma (HNSCC) patients (Le *et al.*, 2005). Taken together, these results support the concept that Gal-1 contributes to immune privilege of tumors by modulating survival and differentiation of effector T cells. Nevertheless, the overall effects of Gal-1 in tumor progression could be multifactorial, including its ability to promote tumor escape, homotypic and heterotypic cell adhesion, tumor cell migration, and cell cycle regulation (Liu and Rabinovich, 2005).

F. Gal-1 as a Target for Anticancer Therapies: Status, Future Directions, and Unresolved Issues

Given its contribution to tumor progression, Gal-1 has emerged as a promising molecular target for cancer therapy. Therefore, a current challenge is the development of potent and selective small molecule inhibitors of Gal-1 to delay tumor progression and stimulate antitumor T cell responses. In fact, molecules with such properties have already been designed for Gal-1 and other galectins; these include synthetic glycoamines, glycodendrimers, or natural products such as pectins (Glinsky *et al.*, 1996; Andre *et al.*, 2001; Nangia Makker *et al.*, 2002; Rabinovich *et al.*, 2006; Tejler *et al.*, 2006). The emergence of novel high-throughput screening platforms, including glycan arrays (Blixt *et al.*, 2004), will facilitate the characterization and profiling of the specificity of a diverse range of galectins and the identification of more specific and potent inhibitors of galectin–carbohydrate interactions. In addition, studies are being conducted to elucidate the antitumor effects of neutralizing anti-Gal-1 mAbs in different preclinical models.

Similarly to the unresolved issues mentioned for PD-L1, several aspects of Gal-1 remain to be addressed. First, there is still scarce information regarding the molecular mechanisms and exogenous stimuli that regulate Gal-1 expression in tumor cells. In this regard, evidence indicates that Gal-1 expression in tumor cells can be modulated by differentiating, chemotherapeutic, and antimetastatic agents, including retinoic acid, TGF-β, and cyclophosphamide (Lu *et al.*, 2000; Zacarias Fluck *et al.*, 2006; Daroqui *et al.*, 2006). Second, although there is compelling evidence about the activity of Gal-1 as a T cell inhibitor, the impact of this carbohydrate-binding protein on other lymphocyte subsets that could contribute to antitumor immunity, including DCs, natural killer (NK) cells, and natural killer T (NKT) cells, has not been thoroughly explored. Finally, future studies should be conducted to examine the association between Gal-1 and other mechanisms of tumor-immune escape, including PD-1/PD-L1, FoxP3$^+$ regulatory T cells (Tregs), and IDO, with the final goal of targeting different mechanisms of tumor-immune escape to "release the breaks" of tumor immunity.

In conclusion, although several inhibitory checkpoints in antitumor immune responses, including PD-L1 and Gal-1, remain incompletely characterized in terms of the oncology field's complete understanding of their multifunctional modes of action, sufficient evidence has been accumulated demonstrating their importance so that new therapeutic approaches directed toward modulating their activities are being developed.

Acknowledgments

We apologize for limiting citation of many excellent studies due to space limitations. Work in the G.A.R. laboratory is supported by grants from the Cancer Research Institute (USA), Mizutani Foundation for Glycoscience (Japan), Agencia de Promoción Científica y Tecnológica (PICT 2003-05-13787) (Argentina), Fundación Sales (Argentina), Fundación Bunge & Born (Argentina), Fundación Florencio Fiorini (Argentina), and the University of Buenos Aires (UBACYT-M091) (Argentina). G.A.R. is a member of the National Research Council (CONICET, Argentina). Work in the T. F. G. laboratory is supported by P01 CA97296, R01 AI47919, and R01 CA118153 from the National Institutes of Health (NIH), a Burroughs Wellcome Clinical Scientist Award in Translational Research, and a grant from the Ludwig Trust. We thank M. A. Toscano for her generous support and thank all the members of our laboratories for productive discussions.

References

Amano, M., Galvan, M., He, J., and Baum, L. G. (2003). The ST6Gal I sialyltransferase selectively modifies N-glycans on CD45 to negatively regulate galectin-1-induced CD45 clustering, phosphatase modulation and T cell death. *J. Biol. Chem.* **278**, 7469–7475.

Andre, S., Pieters, R. J., Vrasidas, I., Kaltner, H., Kuwabara, I., Liu, F-T., Liskamp, R. M., and Gabius, H.-J. (2001). Wedgelike glycodendrimers as inhibitors of binding of mammalian galectins to glycoproteins, lactose, maxiclusters, and cell surface glycoconjugates. *Chembiochem.* **2**, 822–830.

Barber, D. L., Wherry, E. J., Masopust, D., Zhu, B., Allison, J. P., Sharpe, A. H., Freeman, G. J., and Ahmed, R. (2006). Restoring function in exhausted CD8 T cells during chronic viral infection. *Nature* **439**, 682–687.

Blank, C., Brown, I., Marks, R., Nishimura, H., Honjo, T., and Gajewski, T. F. (2003). Absence of programmed death receptor 1 alters thymic development and enhances generation of CD4/CD8 double-negative TCR-transgenic T cells. *J. Immunol.* **171**, 4574–4581.

Blank, C., Brown, I., Peterson, A. C., Spiotto, M., Iwai, Y., Honjo, T., and Gajewski, T. F. (2004). PD-L1/B7H-1 inhibits the effector phase of tumor rejection by T cell receptor (TCR) transgenic CD8+ T cells. *Cancer Res.* **64**, 1140–1145.

Blank, C., Kuball, J., Voelkl, S., Wiendl, H., Becker, B., Walter, B., Majdic, O., Gajewski, T. F., Theobald, M., Andreesen, R., and Mackensen, A. (2006). Blockade of PD-L1 (B7-H1) augments human tumor-specific T cell responses *in vitro*. *Int. J. Cancer* **119**, 317–327.

Blaser, C., Kaufmann, M., Muller, C., Zimmermann, C., Wells, V., Malluci, L., and Pircher, H. (1998). β-galactoside-binding protein secreted by activated T cells inhibits antigen-induced proliferation of T cells. *Eur. J. Immunol.* **28**, 2311–2319.

Blixt, O., Head, S., Mondala, T., Scanlan, C., Huflejt, M.E., Alvarez, R., Bryan, M. C., Fazio, F., Calarese, D., Stevens, J., et al. (2004). Printed covalent glycan array for ligand profiling of diverse glycan-binding proteins. *Proc. Natl. Acad. Sci. USA* **101**, 17933–17938.

Cabrera, P. V., Amano, M., Mitoma, J., Chan, J., Said, J., Fukuda, M., and Baum, L. G. (2006). Haploinsufficiency of C2GnT-1 glycosyltransferase renders T lymphoma cells resistant to cell death. *Blood* **108**, 2399–2406.

Camby, I., Mercier, M.L., Lefranc, F., and Kiss. R. (2006). Galectin-1: A small protein with major functions. *Glycobiology*, Epub in press.

Chung, C. D., Patel, V. P., Moran, M., Lewis, L. A., and Miceli, M. C. (2000). Galectin-1 induces partial TCR ζ-chain phosphorylation and antagonizes processive TCR signal transduction. *J. Immunol.* **165**, 3722–3729.

Daroqui, M. C., Ilarregui, J. M., Rubinstein, N., Salatino, M., Toscano, M. A., Vazquez, P., Bakin, A., Puricelli, L., Bal de Kier Joffe, E., and Rabinovich, G. A. (2006). Regulation of galectin-1 expression by transforming growth factor β1: Implications for tumor-immune escape. *Cancer Immunol. Immunother*, Epub in press.

Day, C. L., Kaufmann, D. E., Kiepiela, P., Brown, J. A., Moodley, E. S., Reddy, S., Mackey, E. W., Miller, J. D., Leslie, A. J., DePierres, C., et al. (2006). PD-1 expression on HIV-specific T cells is associated with T-cell exhaustion and disease progression. *Nature* **443**, 350–354.

Dong, H., and Chen, L. (2003). B7-H1 pathway and its role in the evasion of tumor immunity. *J. Mol. Med.* **81**, 281–287.

Dong, H., Strome, S. E., Salomao, D. R., Tamura, H., Hirano, F., Flies, D. B., Roche, P. C., Lu, J., Zhu, G., Tamada, K., et al. (2002). Tumor-associated B7-H1 promotes T-cell apoptosis: A potential mechanism of immune evasion. Nat. Med. 8, 793–800.

Dong, H., Zhu, G., Tamada, K., and Chen, L. (1999). B7-H1, a third member of the B7 family, co-stimulates T-cell proliferation and interleukin-10 secretion. Nat. Med. 5, 1365–1369.

Dorfman, D. M., Brown, J. A., Shahsafaei, A., and Freeman, G. J. (2006). Programmed death-1 (PD-1) is a marker of germinal center-associated T cells and angioimmunoblastic T-cell lymphoma. Am. J. Surg. Pathol. 30, 802–810.

Galvan, M., Tsuboi, S., Fukuda, M., and Baum, L. G. (2000). Expression of a specific glycosyltransferase enzyme regulates T cell death mediated by galectin-1. J. Biol. Chem. 275, 16730–16737.

Gauthier, L, Rossi, B., Roux F., Termine, E., and Schiff, C. (2002). Galectin-1 is a stromal cell ligand of the pre-B cell receptor (BCR) implicated in synapse formation between pre-B and stromal cells and in pre-BCR triggering. Proc. Natl. Acad. Sci. USA 99, 13014–13019.

Glinsky, G. V., Price, J. E., Glinsky, V. V., Mossine, V. V., Kiriakova, G., and Metcalf, J. B. (1996). Inhibition of human breast cancer metastasis in nude mice by synthetic glycoamines. Cancer Res. 56, 5319–5324.

Hahn, H. P., Pang, M., He, J., Hernandez, J. D., Yang, R. Y., Li, L. Y., Wang, X., Liu, F. T., and Baum, L. G. (2004). Galectin-1 induces nuclear translocation of endonuclease G in caspase- and cytochrome c-independent T cell death. Cell Death Diff. 11, 1277–1286.

He, J., and Baum, L.G. (2006). Endothelial cell expression of galectin-1 induced by prostate cancer cells inhibits T-cell transendothelial migration. Lab. Invest. 86, 578–590.

He, L., Zhang, G., He, Y., Zhu, H., Zhang, H., and Feng, Z. (2005). Blockade of B7-H1 with sPD-1 improves immunity against murine hepatocarcinoma. Anticancer Res. 25, 3309–3313.

Ion, G., Fajka-Boja, R., Kovacs, F., Szebeni, G., Gombos, I., Czibula, A., Matko, J., and Monostori, E. (2006). Acid sphingomyelinase-mediated release of ceramide is essential to trigger the mitochondrial pathway of apoptosis by galectin-1. Cell Signal, Epub in press.

Ishida, Y., Agata, Y., Shibahara, K., and Honjo, T. (1992). Induced expression of PD-1, a novel member of the immunoglobulin gene superfamily, upon programmed cell death. EMBO J. 11, 3887–3895.

Iwai, Y., Terawaki, S., and Honjo, T. (2005). PD-1 blockade inhibits hematogenous spread of poorly immunogenic tumor cells by enhanced recruitment of effector T cells. Int. Immunol. 17, 133–144.

Keir, M. E., Latchman, Y. E., Freeman, G. J., and Sharpe, A. H. (2005). Programmed death-1 (PD-1): PD-ligand 1 interactions inhibit TCR-mediated positive selection of thymocytes. J. Immunol. 175, 7372–7379.

Lahm, H., Andre, S., Hoeflich, A., Kaltner, H., Siebert, H.C., Sordat, B., von der Lieth, C. W., Wolf, E., and Gabius, H.-J. (2004). Tumor galectinology: Insights into the complex network of a family of endogenous lectins. Glycoconj. J. 20, 227–238.

Le, Q.T., Shi, G., Cao, H., Nelson, D. W., Wang, Y., Chen, E. Y., Zhao, S., Kong, C., Richardson, D., O'Byrne, K. J., et al. (2005). Galectin-1: A link between tumor hypoxia and tumor immune privilege. J. Clin. Oncol. 23, 8932–8941.

Liu, F. T., and Rabinovich, G. A. (2005). Galectins as modulators of tumour progression. Nat. Rev. Cancer 5, 29–41.

Liu, X., Gao, J. X., Wen, J., Yin, L., Li, O., Zuo, T., Gajewski, T. F., Fu, Y. X., Zheng, P., and Liu, Y. (2003). B7DC/PDL2 promotes tumor immunity by a PD-1-independent mechanism. J. Exp. Med. 197, 1721–1730.

Lu, Y., Lotan, D., and Lotan, R. (2000). Differential regulation of constitutive and retinoic acid-induced galectin-1 gene transcription in murine embryonal carcinoma and myoblastic cells. Biochim. Biophys. Acta. 1491, 13–19.

Matarrese, P., Tinari, A., Mormone, E., Bianco, G. A., Toscano, M. A., Ascione, B., Rabinovich, G. A., and Malorni, W. (2005). Galectin-1 sensitizes resting human T lymphocytes to Fas (CD95)-mediated cell death via mitochondrial hyperpolarization, budding and fission. J. Biol. Chem. 280, 6969–6985.

Nangia-Makker, P., Hogan, V., Honjo, Y., Baccarini, S., Tait, L., Breaslier, R., and Raz, A. (2002). Inhibition of human cancer cell growth and metastasis in nude mice by oral intake of modified citrus pectin. J. Natl. Cancer Inst. 94, 854–1862.

Nishimura, H., Minato, N., Nakano, T., and Honjo, T. (1998). Immunological studies on PD-1 deficient mice: Implication of PD-1 as a negative regulator for B cell responses. Int. Immunol. 10, 1563–1572.

Nishimura, H., Okazaki, T., Tanaka, Y., Nakatani, K., Hara, M., Matsumori, A., Sasayama, S., Mizoguchi, A., Hiai, H., Minato, N., and Honjo, T. (2001). Autoimmune dilated cardiomyopathy in PD-1 receptor-deficient mice. Science 291, 319–322.

Ohtsubo, K., and Marth, J.D. (2006). Glycosylation in cellular mechanisms of health and disease. Cell 126, 855–867.

Okazaki, T., Maeda, A., Nishimura, H., Kurosaki, T., and Honjo, T. (2001). PD-1 immunoreceptor inhibits B cell receptor-mediated signaling by recruiting src homology 2-domain-containing tyrosine phosphatase 2 to phosphotyrosine. Proc. Natl. Acad. Sci. USA 98, 13866–13871.

Pace, K. E., Lee, C., Stewart, P. L., and Baum, L. G. (1999). Restricted receptor segregation into membrane microdomains occurs on human T cells during apoptosis induced by galectin-1. J. Immunol. 163, 3801–3811.

Perillo, N. L., Uittenbogaart, C. H., Nguyen, J. T., and Baum, L. G. (1997). Galectin-1, an endogenous lectin produced by thymic epithelial cells, induces apoptosis of human thymocytes. *J. Exp. Med.* **97**, 1851–1858.

Perone, M. J., Larregina, A. T., Shufesky, W. J., Papworth, G. D., Sullivan, M. L., Zahorchak, A. F., Stolz, D. B., Baum, L. G., Watkins, S. C., Thomson, A. W., *et al.* (2006). Transgenic galectin-1 induces maturation of dendritic cells that elicit contrasting responses in naive and activated T cells. *J. Immunol.* **176**, 7207–7220.

Phan, G. Q., Yang, J. C., Sherry, R. M., Hwu, P., Topalian, S. L., Schwartzentruber, D. J., Restifo, N. P., Haworth, L. R., Seipp, C. A., Freezer, L. J., *et al.* (2003). Cancer regression and autoimmunity induced by cytotoxic T lymphocyte-associated antigen 4 blockade in patients with metastatic melanoma. *Proc. Natl. Acad. Sci. USA* **100**, 8372–8377.

Rabinovich, G. A., Alonso, C. R., Sotomayor, C. E., Durand, S., Bocco, J. L., and Riera, C. M. (2000). Molecular mechanisms implicated in galectin-1-induced apoptosis: Activation of the AP-1 transcription factor and downregulation of Bcl-2. *Cell Death Diff.* **7**, 747–753.

Rabinovich, G. A., Ariel, A., Hershkoviz, R., Hirabayashi, J., Kasai, K. I., and Lider, O. (YEAR). Specific inhibition of T-cell adhesion to extracellular matrix and pro-inflammatory cytokine secretion by human recombinant galectin-1. *Immunology* **97**,100–106.

Rabinovich, G. A., Baum, L.G., Tinari, N., Paganelli, R., Natoli, C., Liu, F. T., and Iacobelli, S. (2002a). Galectins and their ligands: Amplifiers, silencers or tuners of the inflammatory response *Trends Immunol.* **23**, 313–320.

Rabinovich, G. A., Cumashi, A., Bianco, G. A., Ciavardelli, D., Iurisci, I., D Egidio, M., Piccolo, E., Tinari, N., Nifantiev, N., and Iacobelli, S. (2006). Synthetic lactulose amines: Novel class of anticancer agents that induce tumor-cell apoptosis and inhibit galectin-mediated homotypic cell aggregation and endothelial cell morphogenesis. *Glycobiology* **16**, 210–220.

Rabinovich, G. A., Daly, G., Dreja, H., Tailor, H., Riera, C. M, Hirabayashi, J., and Chernajovsky, Y. (1999). Recombinant galectin-1 and its genetic delivery suppress collagen-induced arthritis via T cell apoptosis. *J. Exp. Med.* **190**, 385–397.

Rabinovich, G. A., Ramhorst, R. E., Rubinstein, N., Corigliano, A., Daroqui, M. C., Kier-Joffe, E. B., and Fainboim, L. (2002b). Induction of allogeneic T-cell hyporesponsiveness by galectin-1-mediated apoptotic and non-apoptotic mechanisms. *Cell Death Diff.* **9**, 661–670.

Rubinstein, N., Alvarez, M., Zwirner, N. W., Toscano, M. A., Ilarregui, J. M., Bravo, I., Mordoh, J., Fainboim, L., Podhajcer, O. L., and Rabinovich, G. A. (2004). Targeted inhibition of galectin-1 gene expression in tumor cells results in heightened T cell-mediated rejection: A potential mechanism of tumor-immune privilege. *Cancer Cell* **5**, 241–251.

Santucci, L., Fiorucci, S., Rubinstein, N., Mencarelli, A., Palazzetti, B., Federici, B., Rabinovich, G.A., and Morelli, A. (2003). Galectin-1 suppresses experimental colitis in mice. *Gastroenterology* **124**, 1381–1394.

Tejler, J., Tullberg, E., Frejd, T., Leffler, H., and Nilsson, U. J. (2006). Synthesis of multivalent lactose derivatives by 1,2-dipllar cycloadditions: Selective galectin-1 inhibition. *Carbohydr. Res.* **341**, 1353–1362.

Thompson, C. B., and Allison, J. P. (1997). The emerging role of CTLA-4 as an immune attenuator. *Immunity* **7**, 445–450.

Toscano, M. A., Commodaro, A. G., Bianco, G. A., Ilarregui, J. M., Liberman, A., Serra, H. M., Hirabayashi, J., Rizzo, L. V., and Rabinovich, G. A. (2006). Galectin-1 suppresses autoimmune retinal disease by promoting concomitant T helper (Th)2- and T regulatory mediated anti-inflammatory responses. *J. Immunol.* **176**, 6323–6332.

Tseng, S. Y., Otsuji, M., Gorski, K., Huang, X., Slansky, J. E., Pai, S. I., Shalabi, A., Shin, T., Pardoll, D. M., and Tsuchiya, H. (2001). B7-DC, a new dendritic cell molecule with potent costimulatory properties for T cells. *J. Exp. Med.* **193**, 839–846.

Van den Brule, F., Califice, S., and Castronovo, V. (2004). Expression of galectins in cancer: A critical review. *Glycoconj. J.* **19**, 537–542.

Van der Leij, J., van den Berg, A., Harms, G., Eschbach, H., Vos, H., Zwiers, P., van Weeghel, R., Groen, H., Poppema, S., and Visser, L. (2007). Strongly enhanced IL-10 production using stable galectin-1 homodimers. *Mol. Immunol.* **44**, 506–513.

Vespa, G. N., Lewis, L. A., Kozak, K. R., Moran, M., Nguyen, J. T., Baum, L. G., and Miceli, M. C. (1999). Galectin-1 specifically modulates TCR signals to enhance TCR apoptosis but inhibits IL-2 production and proliferation. *J. Immunol.* **162**, 799–806.

Zacarias Fluck, M. F., Rico, M. J., Gervasoni, S. I., Ilarregui, J. M., Toscano, M. A., Rabinovich, G. A., and Scharovsky, O. G. (2006). Low-dose cyclophosphamide modulates galectin-1 expression and function in an experimental rat lymphoma model. *Cancer Immunol. Immunother.*, Epub in press.

Further Reading

CFG. Web site for The Consortium for Functional Glycomics. Retrieved February 2007: http://www.functionalglycomics.org/fg/

FFCA. Web site for Forum: Carbohydrates Coming of Age. Retrieved February 2007: http://www.gak.co.jp/FCCA/index.html

CHAPTER

19

Indoleamine 2,3-Dioxygenase in Immune Escape: Regulation and Therapeutic Inhibition

ALEXANDER J. MULLER AND GEORGE C. PRENDERGAST

I. Introduction
II. IDO Function in T Cell Regulation
III. Complex Control of IDO by Immune Regulatory Factors
IV. Immune Tolerance Via IDO in Dendritic Cells
V. IDO Dysregulation in Cancer Cells
VI. IDO as a Target for Therapeutic Intervention
VII. Discovery and Development of IDO Inhibitors
VIII. Conclusion
References
Further Reading

Indoleamine 2,3-dioxygenase (IDO) is one of two enzymes that degrade the essential amino acid tryptophan in mammals by catalyzing the initial, rate-limiting step in the pathway that produces nicotinamide adenine dinucleotide (NAD). Interest in IDO and the tryptophan catabolic pathway it feeds into has grown rapidly with the discovery that IDO activity is critical for generating tolerance to foreign antigens in a variety of tissue microenvironments. In cancer, IDO is overexpressed in both tumor cells and stromal immune cells where it promotes the establishment of peripheral tolerance to tumor antigens. By helping tumor cells evade immune attack, IDO contributes to pathological inflammatory states that permit tumor survival and outgrowth.

In preclinical studies, small-molecule inhibitors of IDO can reverse this mechanism of immunosuppression that primarily targets T cells. Notably, in combination treatment regimens, IDO inhibition strongly leverages the efficacy of classical cancer chemotherapeutic agents, causing the regression of tumors that are otherwise largely resistant to treatment. These findings strongly support the clinical development of IDO inhibitors for cancer. After presenting a historical background to the discovery and early studies of this enzyme, this chapter focuses on work that defines IDO as an important mediator of peripheral tolerance, presents findings about IDO dysregulation in cancer cells, and explains the development of anticancer modalities that employ

IDO inhibitors, the first of which is expected to enter clinical trials in late 2007.

I. INTRODUCTION

The treatment of metastatic cancers is a major clinical challenge. Regimens that involve chemotherapy and other systemic modalities in many cases provide only a limited benefit to approximately 50% of cancer patients in developed countries who present with advanced disease at diagnosis. Similarly, these regimens ultimately fail patients who relapse with disseminated disease following the initial treatment of primary tumors. It has long been recognized that tumors display immunogenic tumor antigens yet escape immune destruction, somehow evading, subverting, or perhaps even reprogramming the immune system for their own benefit. The phenomenon of immune escape is central to tumor cell survival (Dunn *et al.*, 2002), but its basis has remained poorly understood, in part because its contribution as a critical trait of cancer was not fully appreciated by cancer geneticists until recently (Prendergast and Jaffee, in press). While an appropriately activated immune system can eradicate cancer, even when it is aggressive and disseminated, spontaneous occurrences of such events are rare. This has prompted development of numerous peptide- and cell-based anticancer therapies aimed at stimulating an antitumor immune response, for example, through administration of cytokines, tumor-associated antigen peptide vaccines, dendritic cell (DC) vaccines, and adoptive transfer of tumor antigen-specific effector T cells expanded *ex vivo* from cancer patients (Dudley *et al.*, 2002; Figdor *et al.*, 2004; Finn, 2003; Gilboa, 2004; Melief *et al.*, 2000; O'Neill *et al.*, 2004; Rosenberg *et al.*, 1988). These therapies, which are conceptually based on stimulating components of the immune system to elicit an effective response, may not, however, be sufficient to overcome tumoral immune escape if pathological immune tolerance is dominant in cancer patients as has been proposed (Zou *et al.*, 2005). In cases where tolerance is achieved by active principles of immune suppression, then relieving these suppression mechanisms may be an essential component of therapeutic efficacy. In short, to "get on the gas" of immune activation, it may be necessary to "get off the brakes" of immune suppression to achieve effective cancer treatment.

As described by Kim in Chapter 2 of this book, immune escape is the last stage of a three-stage process that has been termed *immunoediting* (Dunn *et al.*, 2002). This process is engaged when precancerous lesions are detected by the immune system, following the genetic damage that is responsible for initiating cancer development and for arousing the immune system (Gasser and Raulet, 2006). Briefly, the first stage of immunoediting involves initial control of a tumor by the immune system, termed *immunosurveillance*. At this point, no clinical evidence of cancer would be apparent. The second stage involves the evolution of a state in tumor cells that allows them to colonize in a hospitable tissue microenvironment, effectively reaching a draw in the battle with the immune system that continues to hold the tumor at bay; this process is referred to as *immune equilibrium*. At this point, there would still be no apparent clinical evidence of cancer although such dormant lesions might be revealed by certain patient imaging techniques. Evidence derived from organ transplant patients supports the existence of such lesions, which can clearly persist for many years before emerging as cancer (MacKie *et al.*, 2003). The third and final stage is called *immune escape* and involves the evolution of a state in the tumor in which tumor cells are able to effectively evade, suppress, or overcome control by the immune system. At this point, clinical disease becomes apparent with the victory of the tumor over the immune system. Viewed from this perspective, understanding the mechanisms of immune escape is critical to the study of cancer pathophysiology since they distinguish preclinical dormant lesions from clinically relevant disease; the development of effective methods to attack these tumoral immune escape mechanisms holds the prospect of managing cancer by converting it from a clinical challenge to a subclinical dis-order that may persist but not in a life-threatening state.

In the twenty-first century, there have been significant advances in understanding how tumors escape the immune system (Kim et al., 2006; Zou, 2005). Intriguingly, many immune escape mechanisms are configured as active immune suppression by the tumor or by stromal cells under the influence of the tumor, implying that continuous activity is required. This finding is provocative because it implies that disrupting these mechanisms of immune suppression could derepress (activate) the immune system, enabling it to attack the tumor. Such mechanisms may offer particularly attractive targets for therapeutic intervention with small-molecule drugs, which have distinct advantages over the biological agents that are the early twenty-first century norm for immunotherapeutic strategies (Muller and Scherle, 2006). Of the mechanisms that have been described to date, one with considerable practical appeal involves the tryptophan catabolizing enzyme indoleamine 2,3-dioxygenase (IDO) (Muller and Prendergast, 2005).

Although IDO is clearly important outside of the pathophysiological subtext of cancer, the discovery of this enzyme is actually rooted in initial observations made in cancer patients. Elevated tryptophan catabolism was first reported as common in patients with bladder cancer in the 1950s (Boyland and Williams, 1956). By the 1960s, elevated levels of tryptophan catabolites had been documented in the urine of patients with a variety of malignancies including leukemia, Hodgkin's disease, prostate cancer, and breast cancer (Ambanelli and Rubino, 1962; Chabner et al., 1970; Ivanova, 1959, 1964; Rose, 1967; Wolf et al., 1968). At that time, the hepatic enzyme tryptophan dioxygenase (TDO2; EC 1.13.11.11) was known to catalyze the catabolism of dietary tryptophan, having been the first inducible mammalian enzyme to be isolated in the 1930s (Kotake and Masayama, 1937; Taylor and Feng, 1991). However, no increase in TDO2 activity was detected in cancer patients who presented elevated tryptophan catabolites (Gailani et al., 1973), implying the existence of a second enzyme.

The isolation of IDO (EC 1.13.11.42; originally D-tryptophan pyrrolase), the non-hepatic tryptophan catabolizing enzyme, was first reported in 1963 (Higuchi and Hayaishi, 1967; Higuchi et al., 1963). Notably, while IDO catalyzes the same reaction as its hepatic relative TDO2—the conversion of tryptophan to N-formyl-kynurenine—these two enzymes are otherwise remarkably dissimilar (Shimizu et al., 1978). Whereas active TDO2 is a homotetramer of 320 kD, IDO is a monomeric enzyme of 41 kD that is antigenically distinct (Watanabe et al., 1981) and lacking in amino acid sequence similarity. In addition, IDO has less stringent substrate specificity, cleaving a number of indole-containing compounds not recognized by the hepatic enzyme. Last, IDO is a heme-containing enzyme that utilizes superoxide anion for activity, whereas TDO2 does not use superoxide to donate oxygen in the tryptophan catabolic reaction.

Structural and enzymological studies have revealed several interesting features of IDO. Enzymological studies indicate that an electron donor, such as methylene blue, is critical to achieve full activity *in vitro*, a role that *in vivo* is thought to be assumed by tetrahydrobiopterin or flavin cofactors. The binding site on the enzyme for the putative cofactor is distinct from the substrate binding site (Sono, 1989), implying the potential for allosteric regulation and possibly opportunities for developing noncompetitive enzymatic inhibitors (in addition to the more classical substrate-competitive inhibitors). Crystallographic studies of human IDO reveal a two-domain structure of alpha-helical domains with the heme group located in between (Sugimoto et al., 2006). Notably, these findings suggest that strict shape requirements in the catalytic site are required, not for substrate binding but instead for abstraction of a proton from the substrate by iron-bound dioxygen in the first step of the reaction (Sugimoto et al., 2006). This detail of the reaction mechanism is important because it is distinct from that used by other monooxygenases (e.g., cytochrome P450), filling a gap in the understanding of heme chemistry. In terms of small-molecule inhibitor development, the biochemical differences of IDO relative to TDO2 and other monooxygenase inhibitors are useful because they increase the likelihood of identifying IDO-specific inhibitors.

Mammalian genomes include the IDO-encoding gene *INDO* but also a newly identified relative termed *IDO2* or *INDOL1* (G.C.P., 2006, LIMR, unpublished observations). Human *INDO* comprises 10 exons spanning ~15 kb at chromosome 8p12–11 that encodes a 403 amino acid polypeptide of ~41 kD (Kadoya et al., 1992; Najfeld et al., 1993). Mouse *Indo* is syntenic and similar in its genomic organization; however, the gene product diverges somewhat at the primary amino acid sequence level from human *INDO*, sharing only 63% identity. The *INDO*-related gene *IDO2* was identified by inspection of sequences immediately downstream of *INDO* in the human genome (G.C.P., 2006, LIMR, unpublished observations). In this region, the database has been erroneously annotated, referring to a set of partial *INDO*-related sequences identified in a large-scale human cDNA sequencing effort (Ota, 2004) and assigned the anonymous designation *LOC169355* which in 2006 was changed to *INDOL1* (*INDO-like-1*). This NCBI Reference Sequence (NM_194294.1) has been permanently suppressed in Genebank because there is insufficient support for the transcript and the protein is not supported by current protein homology data. Correction of the erroneous annotation by trial-and-error exon searches revealed the presence of a 420-amino acid open reading frame that is 44% identical to IDO at the primary sequence level. While it is not yet known whether this open reading frame is enzymatically active, it conserves all the residues in IDO that have been defined as critical for tryptophan binding and catabolism (Sugimoto et al., 2006). The predicted IDO2 proteins in mouse and human are more closely conserved than the mouse and human IDO proteins, displaying 73% identity at the primary sequence level. As in the human genome, the mouse *Ido2* gene is located immediately downstream of *Indo*. Interestingly, *in silico* SAGE and promoter analyses suggest that *IDO2* expression may be very narrowly limited, mainly to certain classes of immune DCs (G.C.P., 2006, LIMR, unpublished observations).

In contrast to the biochemical and genetic knowledge about IDO that has accumulated over the years since its discovery, a precise understanding of its physiological function has remained obscure because of the fact that mammals mostly salvage rather than synthesize NAD to meet their metabolic needs. Why has IDO been evolutionary conserved in mammals? Initial clues as to its function were suggested in the late 1970s by findings from Hayaishi and his colleagues that IDO expression was stimulated strongly by viral infection or interferon treatment (Yoshida et al., 1981, 1979). In the late 1990s, a conceptual breakthrough emerged from the discovery by Munn, Mellor, and their colleagues that IDO activity was important to prevent allogeneic fetal rejection by maternal T cell immunity (Munn et al., 1998). Following the implications of this report, the Munn and Mellor groups and others have subsequently conducted a series of seminal studies, as discussed in the following paragraphs, that have definitively established the important physiological role of IDO in immune control.

II. IDO FUNCTION IN T CELL REGULATION

The discovery that *INDO* is strongly induced at the transcriptional level by the T cell stimulatory cytokine interferon (IFN)-γ prompted a reinterpretation of the longstanding clinical observations that tryptophan catabolites are elevated significantly in the urine of cancer patients (Rose, 1967). Briefly, the idea was that IDO levels driven by IFN-γ activity in cancer patients were responsible for their elevated levels of tryptophan catabolites. Consistent with this idea, IDO activity was later reported to be elevated in lung tumors (Yasui *et al.*, 1986). Given the antitumor properties of IFN-γ, a consensus emerged around the interpretation that IDO acted functionally in the manner of a tumor suppressor, contributing to the antitumor effects of IFN-γ activity by starving growing tumor cells of tryptophan (Ozaki *et al.*, 1988).

Subsequently, groundbreaking work by Mellor and Munn (1999) established the possibility that IDO might mediate a specific regulatory function elicited through the preferential sensitivity of T cells to tryptophan deprivation (Munn *et al.*, 1999), leading to impairment of antigen-dependent T cell activation in microenvironments where IDO activation results in reduced tryptophan levels. The ability of IDO to promote immune tolerance to "foreign" antigens was most dramatically illustrated by the ability of the specific bioactive IDO inhibitor 1-methyl-tryptophan (1MT) (Cady and Sono, 1991) to elicit MHC-restricted T cell-mediated rejection of allogeneic mouse concepti (Mellor *et al.*, 2001; Munn *et al.*, 1998), a result that has been confirmed in the chapter authors' laboratories (Muller *et al.*, 2005a). In cancer, this interpretation implied that IDO can be pro-oncogenic by limiting the eradication of tumor cells that occurs through immune-based recognition of "foreign" tumor antigens.

In the last few years, the concept that tryptophan catabolism regulates T cell immunity has been corroborated widely in many laboratories. Upregulation of IDO in antigen-presenting cells (APCs) that occurs in response to IFN-γ, which is produced by activated T cells, suggests that IDO participates in a negative feedback loop that regulates T cell activation. In particular, the antigen-presenting function of DCs is dramatically affected by IDO activation, such that events that would normally activate T cells are rendered benign or nonactivating, perhaps contributing to T cell anergy (Grohmann *et al.*, 2003b; Mellor and Munn, 2004). With regard to the role of IDO in cancer, these findings have produced a radical rethinking of what elevated tryptophan catabolism means to the developing tumor, by introducing the concept that IDO activity may be a way for tumors to promote pathological tolerization in order to defeat antitumor immunity and facilitating immune escape. Research has broadened these concepts by illustrating how deficits in IDO activity can contribute to autoimmune disorders in contradistinction to how surfeits in IDO activity can contribute to immune suppression in cancer or chronic viral disease (Munn, 2006).

III. COMPLEX CONTROL OF IDO BY IMMUNE REGULATORY FACTORS

A comprehensive review of how IDO is regulated in DCs, likely a major site of action, has appeared in work by Mellor and Munn (2004). The cytokine IFN-γ is a major inducer of IDO, especially in APCs, including macrophages and DCs (Carlin *et al.*, 1987, 1989; Hwu *et al.*, 2000; Takikawa *et al.*, 1999). Transcriptional induction of the *INDO* gene is mediated through the Jak/Stat pathway, in particular Jak1 and Stat1α (Du *et al.*, 2000). Stat1α appears to act to induce *INDO* gene expression both directly through binding of GAS sites within the *INDO* promoter as well as indirectly through induction of interferon regulatory factor 1 (IRF-1), which binds the *INDO* promoter at two interferon-stimulated response element (ISRE) sites (Chon et al., 1996, 1995; Du et al., 2000; Konan and Taylor, 1996; Robinson et al., 2005). Nuclear factor (NF)-κB also contributes to *INDO* induction (Du et al., 2000). In particular, IFN-γ and tumor necrosis factor (TNF)-1 (which signals through NF-κB) appear to act synergistically to induce expression of IRF-1 through a novel composite binding element for both Stat1α and NF-κB in the IRF-1 promoter (termed a *GAS/κB element*), which combines a GAS element overlapped by a nonconsensus site for NF-κB (Pine, 1997).

Studies indicate that IDO is integrated in a complex milieu of factors that promote immune tolerance. One important regulator of expression may be the proinflammatory prostaglandin E_2 (PGE_2), which is elevated frequently during cancer progression as a result of activation of cyclooxygenase 2 (COX-2). IDO is induced by PGE_2, consistent with the role of COX-2 in promoting immune suppression. Interestingly, while PGE_2 is employed widely as an *in vitro* maturation factor for DCs, treatment of these cells with PGE_2 has been reported to elevate IDO expression approximately 100-fold (Braun et al., 2005). Whether such preparations may compromise the desired immune stimulatory activity of DCs used in the setting of cancer vaccines is unclear: PGE_2 treatment is sufficient to induce an IDO message, but the induction of IDO enzymatic activity appears to require an additional signal(s) that can be triggered by exposure to TNF or agonists of Toll-like receptors (TLRs). In support of the concept that IDO mediates some of the immune inhibitory effects of COX-2 and PGE_2, induction of IDO activity can be blocked *in vitro* by COX-2 inhibitors, such as aspirin, indomethacin, and phenylbutazone, but not by anti-inflammatory agents that do not affect prostaglandin (Sayama et al., 1981). The relationship between PGE_2 and IDO is clearly complex insofar as IDO activity can affect the ratio of prostaglandin synthesis (Marshall et al., 2001).

Other important immune regulatory agents that can influence IDO activity are nitric oxide (NO) and transforming growth factor-β (TGF-β). IDO and inducible nitric oxide synthase (iNOS) appear to be mutually antagonistic in DC-based studies (Chiarugi et al., 2003; Fujigaki et al., 2006, 2002). The production of NO by iNOS prevents the IFN-γ-induced expression of IDO (Alberati-Giani et al., 1997), interferes directly with its enzymatic activity (Alberati-Giani et al., 1997; Daubener et al., 1999; Thomas et al., 1994), and promotes its proteolytic degradation (Hucke et al., 2004). NO can directly inactivate IDO by binding directly to the heme iron, which under lowered pH conditions induces iron–His bond rupture and the formation of a 5C NO-bound derivative that is associated with protein conformational changes that may be sufficient to target the protein for ubiquitination and proteosomal degradation (Samelson-Jones and Yeh, 2006). In the non-obese diabetic (NOD) mouse model, *in vivo* evidence suggests that IFN-γ signaling is impaired as the result of nitration of the downstream Stat1 transcription factor by peroxynitrite, which is derived from NO and superoxide. This impairment can be overcome by cytotoxic T lymphocyte-associated antigen 4 (CTLA-4)-Ig treatment, which, by promoting phosphatase and tensin homolog (PTEN) activity, relieves the negative regulation that phosphorylated Akt imposes on FoxO3a-mediated transcription of superoxide dismutase (SOD2), which degrades peroxynitrite (Fallarino et al., 2004). Through this complex route, the blockade to activation of IDO gene expression, to which iNOS contributes through peroxynitrite-mediated nitration of Stat1, is relieved. Two implications of the configuration of this mechanism are the following. First, NO agonists will tend to reverse immunosuppression at the level of DCs in cancer, which should benefit treatment. Second, small-molecule inhibitors of Akt that are being developed as anticancer therapeutics will tend to heighten immunosuppression by phenocopying this effect of

CTLA-4-Ig on IDO expression. Findings suggest that Akt inhibition may also heighten the invasive capability of cancer cells (Yoeli-Lerner and Toker, 2006). Thus, for cancer treatment, the desirable proapoptotic quality of Akt inhibitors may be balanced by their undesirable proinvasive and immunosuppressive properties. TGF-β has also been reported to antagonize IFN-γ-mediated induction of IDO expression, in this case in fibroblasts (Yuan et al., 1998). This finding appears to run counter to immunosuppressive activity ascribed to TGF-β but is consistent with its ability to antagonize positively regulated targets of IFN-γ. The balance between the effects of TGF-β and IFN-γ signaling on IDO expression in different cells of tumor and peripheral microenvironments provides a complex mechanism for local control of IDO activity.

IV. IMMUNE TOLERANCE VIA IDO IN DENDRITIC CELLS

Although the default mode for immature DCs appears to be tolerogenic, even more effective suppression may be achieved with appropriate maturation (Mellor, 2005; Mellor et al., 2004). A number of DC subgroups in mice have been ascribed tolerogenic properties (Baban et al., 2005; Belz et al., 2002; Homann et al., 2002; Kronin et al., 1996; Wakkach et al., 2003). Whether any of these represent distinct lineages or alternative differentiation outcomes for a common precursor has yet to be determined; however, the tumor itself plays a key role in determining the type of response that tumor-associated DCs will elicit (Ghiringhelli et al., 2005; Liu et al., 2005).

Local catabolism of tryptophan through induction of the enzyme IDO in DCs has been proposed as a mechanism for inducing tolerance (Mellor and Munn, 2004). In addition to being directly tolerogenic, mature DCs have the capacity to expand regulatory T cells (Tregs) (Moser, 2003). B7 costimulatory signals from DCs appear to play a critical role in Treg development as both CD28-null and B7–1/B7–2-double null mice were found to exhibit a markedly reduced population of Tregs (Salomon et al., 2000) and constitutively expressed B7 costimulatory molecules maintain self tolerance through suppression of T cell activation by sustaining a Treg population (Lohr et al., 2003). $CD4^+CD25^-$ T cells transferred into congenic animals can be converted into Treg-like cells in vivo, but this does not occur if the recipient mice lack expression of B7 costimulatory molecules (Liang et al., 2005). Reciprocally, CTLA-4 binding of B7 molecules on DCs induces IDO (Fallarino et al., 2003; Grohmann et al., 2002; Munn et al., 2002).

Accumulating evidence indicates that $CD4^+CD25^+$ Tregs play an indispensable role in the maintenance of negative control over pathological as well as physiological immune responses (Bluestone and Abbas, 2003; Piccirillo and Shevach, 2004; Sakaguchi, 2004) and that removal of Tregs not only elicits autoimmune diseases but also enhances responses to non-self antigens, including xenogenic proteins and allografts (Wood and Sakaguchi, 2003). In mice, the absence of Tregs has been reported to lead gestational failure due to immunological rejection (Aluvihare et al., 2004), and adoptive transfer of pregnancy-induced Tregs can protect against fetal rejection in abortion-prone mice (Zenclussen et al., 2005).

CTLA-4 is a major signaling molecule for Tregs, and IDO has been implicated to be an important downstream effector for CTLA-4-mediated immune tolerance. This mechanism is summarized in Figure 19.1. The first in vivo evidence for this was the observation that, in a diabetic mouse model, the ability of administered CTLA-4-Ig to effectively suppress immune rejection of pancreatic islet allografts was defeated by concurrent treatment with the IDO inhibitor 1MT (Grohmann et al., 2003b). This study further suggested that CTLA-4-Ig-mediated tolerance occurs through a heterodox mechanism of "reverse" signaling through B7 molecules on APCs, which

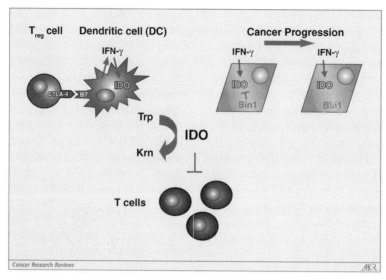

FIGURE 19.1 Mechanisms of IDO-induced tumoral immune escape. IDO expression in local immune stroma or tumor cells has been implicated in promoting immune tolerance. IDO is upregulated in antigen-presenting DCs by autocrine IFN-γ released as a result of Treg-induced CTLA-4/B7-dependent cell–cell signaling. Local tryptophan catabolism limits the proliferation and survival of T cells that would otherwise be activated by tumor antigens presented by the DCs. This mechanism may operate in tumor-draining lymph nodes. In tumor cells, attenuation of the suppressor gene *Bin1* leads to superactivation of IDO expression by IFN-γ, directly suppressing activation of T cells in the local tumor environment. Blocking IDO activity systemically with small-molecule inhibitors (e.g., 1MT) reverses T cell suppression that occurs as a result of tryptophan catabolism in both settings. From Muller and Prendergast (2005).

promotes IFN-γ to induce IDO. CD4+CD25+ Tregs that constitutively express CTLA-4 on their surface have likewise been shown to promote IDO activity in DCs through a CTLA-4 dependent mechanism (Fallarino et al., 2003), and defects in CTLA-4 signaling that promote IDO activity have been implicated in the failure of IFN-γ to activate the tolerogenic properties of DCs derived from female NOD mice early in prediabetes (Fallarino et al., 2004). As mentioned earlier, *in vivo* evidence from this model indicates that IFN-γ signaling in DCs is impaired as the result of NO-dependent nitration of Stat1, which can be overcome by CTLA-4-Ig treatment via Akt/FoxO3a-mediated activation of SOD2 (Fallarino et al., 2004).

Other work has confirmed that systemic administration of CTLA-4-Ig results in IDO upregulation in DC subsets, including CD8α+ DCs. Systemic CTLA-4-Ig can also block clonal expansion and CTL activity of H-2Kb-specific T cells from TCR transgenic mice adoptively transferred into H-2Kb hosts. This effect of CTLA-4-Ig was abrogated by treatment with 1MT or by using IDO-null knockout mice as hosts (Mellor et al., 2003). This latter control is especially informative because it argues against suggestions that the effects of 1MT may be mediated through targets other than IDO (Terness et al., 2006), thereby offering direct evidence that the functionally relevant target of 1MT is IDO and not other tryptophan- or indoleamine-binding proteins. Data indicating that IDO mediates biological effects of CTLA-4-Ig run counter to the idea that CTLA-4 is directly antagonistic to CD28 on T cells through outcompeting CD28 for access to B7 ligand, inducing immunosuppressive cytokines, or directly interfering with CD28-mediated and/or TCR-mediated signaling (Sharpe and Freeman, 2002). The different mechanisms need not be mutually exclusive, however, especially considering that loss of IDO activity, either pharmacologically or genetically, does not completely phenocopy genetic loss of CTLA-4. Thus, IDO does not account for all of the immunological activity of CTLA-4. Notably, IDO-null knockout mice have not been observed to exhibit spontaneous autoimmune disease. This observation suggests that IDO is not required for homeostatic maintenance of central or peripheral tolerance to self antigens. In contrast, acquired tolerance is defective in IDO-null mice, suggesting that induction of IDO activity might play a role in the acquisition of tolerance to neoantigens (Mellor et al., 2003). Becoming tolerized to neoantigens is critical to maintaining pregnancy and may be similarly important to permit progression of early neoplastic lesions that present mutated oncoproteins to the immune system.

Much of the published work on the role of IDO in DCs has come from investigating the tolerogenic mechanism elicited by $CD11c^+CD8\alpha^+$ DCs in the context of challenge with self/tumor antigens. One model that has been used for immune tolerance experiments is a delayed type hypersensitivity (DTH) skin test model, in which mice receiving peptide-pulsed DCs were assayed for MHC class I-restricted reactivity by challenging with peptide in the footpad. Fractionated $CD8\alpha^-$ myeloid DCs presented with tumor/self peptide were actively immunogenic but could be effectively inhibited by reintroduction of a minority population of $CD8\alpha^+$ lymphoid DCs. The Th1-asssociated cytokines IL-12 and IFN-γ were found to work at cross purposes in this system, with tolerogenic suppression relieved by exposure of the $CD8\alpha^-$ population to IL-12 but reestablished by exposure of the $CD8\alpha^+$ population to IFN-γ (Grohmann et al., 2000; Grohmann et al., 1999). Addition of the IDO inhibitory compound 1MT suppressed the ability of IFN-γ to overcome the adjuvant effect of IL-12. Induction of IDO activity by IFN-γ in $CD8\alpha^+$ cells, which exhibit a significant basal level of IDO expression, has been shown to be regulated both at the level of expression as well as in the posttranslational period (Fallarino et al., 2002b). Interestingly, in the $CD8\alpha^-$ population, both the basal and induced levels of IDO expression appear to be comparable with $CD8\alpha^+$ cell IDO levels, but IDO activity remained suppressed in response to IFN-γ. Responses to engagement of cell surface molecules on DCs were found to play a key role in determining whether DC activity would be immunogenic or tolerogenic. Engagement of CD40 with agonistic antibody blocked IFN-γ-mediated induction of IDO activity in otherwise protolerogenic $CD8\alpha^+$ DCs and induced these cells to actively prime rather than suppress CTL responses to self/tumor peptide in the DTH model (Grohmann et al., 2001). The agonistic CTLA-4-Ig receptor/antibody fusion protein had an opposite effect as otherwise proimmunogenic $CD8\alpha^-$ DCs treated with CTLA-4-Ig-induced IDO activity and were rendered tolerogenic (Grohmann et al., 2003a). Together, these findings highlight the functional plasticity of these different DC subsets.

Like CTLA-4-Ig, CD28-Ig binds B7 molecules except with the opposite effect of promoting immunogenicity (Orabona et al., 2004). DCs treated with CD28-Ig exhibited early and sustained production of IL-6 that was not induced by CTLA-4-Ig. IL-6 upregulates SOCS3, which inhibits the Stat-dependent IFN-γ signaling required for IDO induction; consistent with these signaling connections, silencing of SOCS3 expression in DCs caused CD28-Ig to elicit a CTLA-4-Ig-like tolerogenic response (Orabona et al., 2005a, 2004). Gene expression profiling in $CD8\alpha^+$ DCs has revealed that the *Tyrobp*-encoded DAP12 protein, which is controlled by IFN-γ via negative transcriptional regulation imposed by IRF-8, is important for posttranslational suppression of IDO activity (Orabona et al., 2005b). DAP12 belongs to the family of immunoreceptor tyrosine activation motif (ITAM)-bearing membrane adaptor molecules that associate with the transmembrane regions of activating receptors in natural killer (NK) and myeloid cells. CD8α+ DC from transgenic mice overexpressing the DAP12 protein exhibited impaired tolerogenic function, while CD8α+ DCs lacking DAP12 function exhibited increased IDO-dependent tolerogenic activity (Orabona et al., 2005b). These observations argue that DAP12 acts as a modifier of IDO activity.

In addition to DCs, other immune cells that may utilize IDO for immunosuppression include macrophages, granulocytes, and neutrophils. The first indications of IDO involvement in the suppression of T cell activation were obtained from *in vitro* studies of macrophages exposed to colony-stimulating factor 1 (CSF-1), which induced IDO activity in the context of a mixed lymphocyte reaction (MLR). Expression of CSF-1 in tumors has been implicated in polarizing macrophages toward an M2 suppressor phenotype (Mantovani et al., 2002). Studies in the Stat6 knockout mouse support a role for both myeloid-derived suppressor cells (MDSCs) and tumor-associated macrophages (TAMs) in tumor metastasis. Stat6 knockout mice are resistant to metastases produced by isogenic tumor grafts of the highly metastatic breast cancer cell line 4T1. The resistance observed in these animals has been linked to diminished MDSC induction coupled with a reduction in M2 TAMs. In CD-1 knockout mice, which lack IL-13-producing NKT cells, polarization toward an M2 phenotype is no longer supported after implantation of 4T1 cells, and the resulting production of cytotoxic M1 macrophages is associated with rejection of 4T1 tumors (Sinha et al., 2005a). Thus, a permissive environment for activated T cells combined with a reduction in MDSCs is sufficient to result in effective immune surveillance and rejection of metastatic 4T1 tumors. Local catabolism of the amino acid arginine through induction of the ARG1 enzyme is one mechanism by which M2 macrophages can promote tumor tolerance through suppression of effector T cell responses (Sinha et al., 2005b). In addition, when cultured in the presence of CSF-1, macrophages display expression of IDO (Munn et al., 1999), which also suppresses effector T cell responses. The production of TGF-β and IL-10 by TAMs would also perpetuate polarization toward a protolerogenic state (Mantovani et al., 2002).

Precisely how IDO-mediated tryptophan catabolism elicits immune tolerance remains somewhat uncertain given evidence for different but not necessarily mutually exclusive mechanisms of action. Some studies in DCs have indicated that induction of IDO activity, triggered by exposure to either IFNγ or CTLA-4-Ig, may be necessary for acquiring rather than directly eliciting the suppressor phenotype (Grohmann et al., 2000; Orabona et al., 2005b). The former possibility is consistent with the finding that DC maturation in response to TNF or lipopolysaccharide (LPS) treatment is suppressed by 1MT along with expression of the chemokine receptors CCR5 and CXCR4, which mediate tumor migration and infiltration (Hwang et al., 2005). Most attention, however, has focused on the direct suppression of T cells by IDO-mediated tryptophan catabolism. The first studies examining how IDO suppresses T cell immunity proposed the core concept that starving T cells of tryptophan, an essential amino acid, limits their ability to be activated by appropriately presented antigen due to the imposition of a block to cell division, which is required for activation (Munn et al., 1999). However, other studies have proposed that the catabolites produced by the IDO pathway can trigger preferential T cell apoptosis (Fallarino et al., 2002a), and *in vitro* experiments meant to demonstrate which of these two mechanisms is more relevant to immunological suppression have produced contradictory results (Muller et al., 2005b).

Recent *in vivo* data are likewise incongruent. There is genetic evidence that signaling through the GCN2 kinase pathway, which responds to environmental depletion of tryptophan, elicits the arrest response in T cells: T cells from GCN2-null mice are no longer responsive to IDO-expressing DCs, arguing that IDO-mediated induction of growth arrest and anergy in responding

T cells is signaled by stress signals mediated by GCN2 kinase (Munn et al., 2005). On the other hand, systemic treatment of mice with the IDO catabolite mimetic N-(3,4-dimethoxycinnamoyl) anthranilic acid (3,4-DAA) has been reported to ameliorate symptoms in a mouse model of multiple sclerosis in conjunction with a reduction in inflammatory foci in the brain and spinal cord (Platten et al., 2005). In this study, 3,4-DAA interfered with IFN-γ-induced Stat1 signaling in a microglial cell line and suppressed the activation of APCs in vivo, but no data on its effects on T cells independent of APCs were presented. One possible explanation consistent with these two in vivo reports is that depletion of tryptophan might have a direct effect in suppressing T cell activation, while the accumulation of tryptophan catabolites might act indirectly to further impair T cell activation by suppressing immunogenic APCs. A 2006 study has reported that IDO-expressing DCs in mice can suppress $CD8^+$ T cell activity through suppression of ζ-chain expression and can induce naive $CD4^+$ T cells to develop into $Foxp3^+$ Tregs (Fallarino et al., 2006). These effects require combined tryptophan depletion and exposure to tryptophan catabolites and are mediated through GCN2 kinase. Not only may this finding address some of the confusion surrounding the physiological mechanism of IDO-mediated immune suppression, but it is also intriguing given the relationship that has previously been established between CTLA-4-expressing Tregs and their ability to induce IDO-mediated immune suppression through responsive DCs; this finding suggests there might be a feed-forward mechanism in which the tolerogenic DCs go on to recruit the development of additional Tregs though the action of IDO.

V. IDO DYSREGULATION IN CANCER CELLS

Malignant development is accompanied by a breakdown of normal cellular physiology. This process involves the acquisition of cell-intrinsic traits, which include immortalization, growth sufficiency, insensitivity to growth inhibitory signals, and apoptosis resistance, as well as the acquisition of cell-extrinsic traits, which include angiogenesis, invasive capability, metastatic capacity, and immune escape. Immune escape can be viewed as a terminal feature of immunoediting, which consists of surveillance, equilibrium, and, ultimately, escape of tumor cells from effective control by the immune system (Dunn et al., 2004). Immune escape has recently gained wide recognition among cancer geneticists as a crucial feature of malignant development: according to Shankaran et al. (2001), there is compelling evidence in knockout mice (IFN-γ, Stat1, and Rag2 knockouts) that genetic ablation of T cell immunity is sufficient to increase tumor incidence. It has become clear that there is dynamic interplay in the interaction of genetically plastic tumor cells with the innate and adaptive arms of the immune system. On the one hand, inflammatory properties provided by immune cells can provide a supportive tumor microenvironment (Balkwill et al., 2005). On the other hand, immune surveillance limits malignant growth but also provides a selective pressure for the evolution of mechanisms capable of subverting or evading the immune response that tumor antigens elicit (Zou, 2005).

IDO activation represents one mechanism evolved by tumor cells to escape the immune system. The survival benefit of this mechanism to a tumor cell that has evolved is balanced by the cost of depriving itself of an essential amino acid. Thus, like other pro-oncogenic alterations that accumulate in cancer cells, IDO activation represents a stochastic event whose value is determined by the particular setting of transformation and immunoediting pathways that are relevant to a particular malignancy. As mentioned previously, tryptophan catabolites are significantly elevated in the urine of cancer patients, which is reversed upon surgical tumor reductive therapy (Rose, 1967).

While it has not been proven that IDO elevation is responsible for this phenomenon, it seems likely given the common elevation of IDO in human tumors (Uyttenhove et al., 2003).

How does IDO become deregulated in cancer cells? One mechanism that has been reported involves *Bin1*, a gene that is frequently attenuated in breast cancer, prostate cancer, melanoma, neuroblastoma, and other cancers (Ge et al., 1999, 2000a, 2000b; Tajiri et al., 2003). *Bin1* was initially identified in a two-hybrid screen for Myc-interacting proteins (Sakamuro et al., 1996). Numerous investigations of *Bin1*, also known as *Amphiphysin II*, have suggested a function for this gene in cancer suppression (DuHadaway et al., 2001; Elliott et al., 2000, 1999; Galderisi et al., 1999; Ge et al., 1999, 2000a, 2000b; Sakamuro et al., 1996; Tajiri et al., 2003). Genetic studies in mice targeted for homologous deletion of *Bin1* indicate that it facilitates apoptosis and limits proliferation and immune escape in oncogenically transformed cells (Muller et al., 2004, 2005a). Along with the *Bin3* gene, *Bin1* is one of two evolutionarily conserved members of an adapter protein family termed *Bin/amphiphysin/Rvs (BAR) adapters*. This family of adapters is named for the presence of a signature fold called the BAR domain that can mediate interaction with curved vesicular membranes (Peter et al., 2003). *Bin1* messages are alternately spliced to generate at least 10 isoforms that differentially localize to diverse nuclear, cytosolic, and membranous sites in cells (Butler et al., 1997; DuHadaway et al., 2003; Kadlec and Pendergast, 1997; Ramjaun et al., 1997; Sparks et al., 1996; Tsutsui et al., 1997; Wechsler-Reya et al., 1997). Only two of these splice isoforms are ubiquitously expressed, whereas the remainder are restricted to specific terminally differentiated tissues, including neurons and skeletal muscle cells. In addition, only the ubiquitous and muscle-specific isoforms of *Bin1*, which can access the nucleus, display anticancer properties. BAR proteins make diverse interactions in cells, and a simple and readily classifiable function for *Bin1* has yet to emerge. However, existing information suggests that the ubiquitous *Bin1* isoforms and certain other nucleocytosolic BAR family adapter proteins may act not only in trafficking processes at vesicular membranes but also in transcriptional processes at the nucleus (Elliott et al., 1999; Miaczynska et al., 2004). For example, studies offer some genetic support for the notion that *Bin1* may modify the efficiency of nuclear trafficking or function of the NF-κB and Stat transcription factors (Muller et al., 2004, 2005a). These connections are interesting in the present context given the important roles these factors play in immunity as well as cancer.

Investigation of how *Bin1* loss facilitates the outgrowth of oncogenically transformed cells has identified immune tolerance via IDO activation in the transformed cells as a mechanistic explanation (Muller et al., 2005a). This mechanism is summarized in Figure 19.1. Targeted deletion of *Bin1* in mouse cells resulted in superinduction of IDO gene expression by IFN-γ. Transformation of *Bin1*-null and *Bin1*-expressing mouse embryo keratinocytes with *c-Myc+Ras* oncogenes produced cell lines with similar *in vitro* growth properties. However, when these cells were introduced into syngeneic animals, the *Bin1*-null cells formed large tumors, whereas the *Bin1*-expressing cells formed only indolent nodules. This dichotomy reflected a difference in immune response to the cells, as *Bin1*-expressing cells were capable of producing rapidly growing tumors when introduced into either athymic nude mice or syngeneic mice depleted of CD4$^+$/CD8$^+$ T cells. Treatment with the IDO inhibitor 1MT suppressed the outgrowth of *Bin1*-null tumors in syngeneic mice, but this effect was absent in nude mice and immunodepleted syngeneic animals. Taken together, the findings showed how the deregulation of IDO by *Bin1* loss promoted tumorigenicity by enabling immune escape. Given the frequent attenuation of *Bin1* expression and the frequent overexpression of IDO in human cancers, it will be important to further evaluate the relationship between these two events.

Several studies have suggested that IDO overexpression is associated with poor prognosis in cancer. In a small study of ovarian cancers, immunohistochemical overexpression of IDO in tumor sections correlated inversely with patient survival, such that a higher extent of IDO immunohistochemical staining in tumors was associated with poor survival outcomes (Okamoto et al., 2005). For sporadic, focal, and diffusely staining tumors, the 50% survival of patients was 41, 17, and 11 months, respectively. In contrast, all patients with tumors classified as negative for IDO staining in this study survived at least 5 years after surgery (Okamoto et al., 2005). A similar trend was found in a larger study of IDO in colorectal cancer, where levels and activity of the enzyme were examined by reverse transcription polymerase chain reaction (RT-PCR), immunohistochemistry, and high performance liquid chromatography (HPLC) analyses (Brandacher et al., 2006). IDO expression and enzyme activity in colon cancer cells were strictly dependent on IFN-γ stimulation. High IDO expression was associated with increased incidence of liver metastases and with reduced infiltration of tumors by $CD3^+$ T cells, as compared to tissues expressing low IDO. In addition, IDO was associated significantly with reduced survival, and comparative analyses assigned this finding as an independent prognostic variable (Brandacher et al., 2006). Interestingly, the same group has observed elevated IDO in inflammatory bowel disease, a condition that is associated with a significant increase in the risk of colon cancer (Wolf et al., 2004). Other studies capture the possibility that IDO expression in stromal cells may have prognostic impact. In one case, a small study of small-cell lung cancer (SCLC) revealed IDO overexpression in eosinophil granulocytes that infiltrated the tumor, rather than in DCs or tumor cells, and that the level of IDO-positive eosinophil infiltrate was associated with poor survival (Astigiano et al., 2005). In another study, IDO positive-staining in DCs of the tumor-draining lymph nodes of melanoma patients was correlated with poor prognosis (Lee et al., 2003; Munn et al., 2004). While it will be important to extend these early observations, they all exhibit a similar trend that is consistent with the expectation that, by facilitating tumoral immune escape, IDO activation may pro-vide a powerful driver of malignant progression.

VI. IDO AS A TARGET FOR THERAPEUTIC INTERVENTION

A small number of studies have offered evidence that IDO inhibition with 1MT or other small-molecule inhibitors can exert antitumor effects. Initial evidence was offered in 2002 that the IDO inhibitor 1MT could retard the growth of mouse lung carcinoma cells engrafted onto a syngeneic host (Friberg et al., 2002). Similar results were obtained as part of an investigation to assess the ramifications of IDO overexpression that was detected in a wide range of human tumors (Uyttenhove et al., 2003). In this study, ectopic overexpression of IDO in an established tumor cell line was shown to be sufficient to promote tumor formation in animals preimmunized against a specific tumor antigen, and 1MT partially suppressed tumor outgrowth in this context. In established autochthonous (spontaneously arising) mammary tumors in the MMTV-*neu*/HER2 transgenic mouse model of breast cancer, it has likewise shown that 1MT treatment can retard tumor growth (Muller et al., 2005a), although, by itself, 1MT was unable to elicit tumor regression, as shown previously in the tumor cell graft models, suggesting that IDO inhibition may produce limited antitumor efficacy when applied as a monotherapy.

In contrast, the delivery of 1MT in combination with a variety of classical cytotoxic chemotherapeutic agents elicited regression of established MMTV-*neu*/HER2 tumors, which responded poorly to any single-agent therapy (Muller et al., 2005a). In each case, the observed regressions were unlikely to result from a drug–drug interaction, that is, by 1MT acting to raise the effective dose of the cytotoxic agent because efficacy was increased in the absence of increased side effects (e.g., neuropathy produced by paclitaxel, which is displayed by hind leg dragging in affected mice). Immunodepletion of $CD4^+$ or $CD8^+$ T cells from the mice before treatment abolished the combinatorial efficacy observed in this model, confirming the expectation that 1MT acted indirectly through activation of T cell-mediated antitumor immunity. The chapter authors have extended these observations using novel small-molecule inhibitors that they have identified, including several thiohydantoin derivatives of tryptophan. For example, continuous administration of methylthiohydantoin-tryptophan was found to display a same pattern of antitumor properties as 1MT, retarding the growth of MMTV-neu/HER2 tumors and eliciting tumor regressions in combination with paclitaxel in the absence of increased side effects (Muller et al., 2005a). The chapter authors have also observed that efficacy in achieving regressions can be replicated by oral dosing of 1MT at 400 mg/kg two times per day (bid), again in the absence of any detectable side effects. Strikingly, as little as 4–5 days of 1MT administration on a 2-week trial is sufficient to produce substantial regressions when combined with chemotherapy in the model (Hou et al., 2007). In future work, to validate IDO as the target of putative inhibitors such as 1MT, it will be important to show that additional bioactive inhibitors have similar antitumor properties, that antitumor and pharmacodynamic responses can be correlated appropriately, and that genetic manipulations of IDO can alter the response to putative inhibitors. However, taken together, the existing results offer an initial step in validating IDO for drug development in the context of a cytotoxic combination treatment modality.

VII. DISCOVERY AND DEVELOPMENT OF IDO INHIBITORS

IDO has a number of appealing features as a target for drug development. First, IDO is a single-chain catalytic enzyme with a well-defined biochemistry. Unlike many proposed therapeutic targets in cancer, this means that IDO is very tractable for discovery and development of small-molecule inhibitors. Second, the other known tryptophan catabolizing enzyme on the kynurenine pathway, TDO2, is structurally distinct from IDO and has a much more restricted pattern of expression and substrate specificity, which mitigates "off-target'" issues usually posed by novel agents. Third, bioactive and orally bioavailable "lead" inhibitors exist that can serve as useful tools for preclinical validation studies. Fourth, an *Indo* gene "knockout" mouse that has been constructed is reported to be viable and healthy (Baban et al., 2004). While further analysis is necessary, this observation encourages the notion that IDO inhibitors will not produce unmanageable, mechanism-based toxicities. Fifth, pharmacodynamic evaluation of IDO inhibitors can be performed by examining blood serum levels of tryptophan and kynurenine, the chief substrate and downstream product of the IDO reaction, respectively. Last, small-molecule inhibitors of IDO offer logistical and cost advantages compared to biological or cell-based therapeutic alternatives to modulating T cell immunity.

The rational design and development of new inhibitory compounds requires understanding the IDO active site and catalytic mechanism. Proposed models for the processes at work in the active site have been developed based on mechanistic studies

(Malachowski et al., 2005). The publication of an X-ray crystal structure for IDO complexed with a simple inhibitor will greatly facilitate this work (Sugimoto et al., 2006). Alternately, screening for novel inhibitors is likely to identify novel structural series to evaluate. Through this route, the chapter authors' groups together have identified the natural product brassinin as an IDO inhibitor and evaluated brassinin derivatives for *in vitro* potency and cell-based activity (Gaspari et al., 2006). Brassinin is a phytoalexin-type compound found in cruciferous vegetables that has reported chemopreventative activity in preclinical models for breast and colon cancers in rodent models (Mehta et al., 1995; Park and Pezzuto, 2002).

A series of derivatives from the core brassinin structure has been used to probe the relationship between inhibitors and the IDO site, that is, perform a structure-activity relationship (SAR) analysis (Gaspari et al., 2006). Among the conclusions drawn, the indole core was determined not to be essential for enzyme inhibitory activity. This is consistent with the known promiscuity of the active site in IDO (Cady and Sono, 1991) and broadens the spectrum of potential inhibitory compounds. On the other hand, the dithiocarbamate segment of brassinin was identified as an optimized moiety for inhibition that is proposed to chelate the heme iron at the active site of IDO. Of the large number of derivatives evaluated, the most potent were only ~1 µM, suggesting that it may be difficult to achieve significant improvements in potency in this particular structural class. High-throughput screening of comprehensive compound libraries remains the most effective way to identify new structural series. In addition, IDO inhibitory compounds representing diverse structural classes have been reported in a unique yeast screen (Vottero et al., 2006). Consistent with SAR-based evidence that the indole core is not critical for enzyme inhibition, at least one of the non-indole compounds identified in the yeast screen exhibited submicromolar potency. Given the intriguing features of IDO as an immune modulator in cancer and other diseases, future efforts to identify IDO inhibitors may broaden quickly.

VIII. CONCLUSION

In a relatively short period, IDO has emerged to become recognized as a major regulator of the immune system. IDO has been strongly implicated pathophysiologically in tumoral immune tolerance and immune escape and appears to be widely overexpressed in cancer at the level of tumor cells and/or tumor-associated immune cells. IDO has a variety of characteristics that make it appealing as a target for cancer drug development. To date, preclinical validation of IDO inhibitors suggests they may offer the greatest promise in combination with classical cytotoxic drugs, but their potential to heighten the response to active immunotherapeutic agents, such as TLR ligands or tumor vaccines, is also important to consider. Given the provocative preclinical findings that have emerged from studies of agents targeting IDO and the IDO pathway, one would expect therapeutic interest in this pathway to continue to grow.

Acknowledgments

Work in the authors' laboratories is supported by the DoD Breast Cancer Research program; the State of Pennsylvania Department of Health; the Lance Armstrong and Concern Foundations (A. J. M.); National Institutes of Health (NIH) grants CA82222, CA100123, and CA109542 (G.C.P.); and the Lankenau Hospital Foundation (A. J. M. and G. C. P.). The authors declare competing financial interests. G. C. P. and A. J. M. are significant stockholders and G. C. P. is a member of the scientific advisory board at New Link Genetics Corporation, a biotechnology company that has licensed IDO intellectual property created by the authors, described in patents WO 2004 093871 "Novel methods for the treatment of cancer" (pending) and WO 2004 094409 "Novel IDO inhibitors and methods of use" (pending).

References

Alberati-Giani, D., Malherbe, P., Ricciardi-Castagnoli, P., Kohler, C., Denis-Donini, S., and Cesura, A. M. (1997). Differential regulation of indoleamine 2,3-dioxygenase expression by nitric oxide and inflammatory mediators in IFN-gamma-activated murine macrophages and microglial cells. *J. Immunol.* **159**, 419–426.

Aluvihare, V. R., Kallikourdis, M., and Betz, A. G. (2004). Regulatory T cells mediate maternal tolerance to the fetus. *Nat. Immunol.* **5**, 266–271.

Ambanelli, U., and Rubino, A. (1962). Some aspects of tryptophan–nicotinic acid chain in Hodgkin's disease. Relative roles of tryptophan loading and vitamin supplementation on urinary excretion of metabolites. *Haematol. Lat.* **5**, 49–73.

Astigiano, S., Morandi, B., Costa, R., Mastracci, L., D'Agostino, A., Ratto, G. B., Melioli, G., and Frumento, G. (2005). Eosinophil granulocytes account for indoleamine 2,3-dioxygenase-mediated immune escape in human non-small cell lung cancer. *Neoplasia* **7**, 390–396.

Baban, B., Chandler, P., McCool, D., Marshall, B., Munn, D. H., and Mellor, A. L. (2004). Indoleamine 2,3-dioxygenase expression is restricted to fetal trophoblast giant cells during murine gestation and is maternal genome specific. *J. Reprod. Immunol.* **61**, 67–77.

Baban, B., Hansen, A. M., Chandler, P. R., Manlapat, A., Bingaman, A., Kahler, D. J., Munn, D. H., and Mellor, A. L. (2005). A minor population of splenic dendritic cells expressing CD19 mediates IDO-dependent T cell suppression via type I IFN signaling following B7 ligation. *Int. Immunol.* **17**, 909–919.

Balkwill, F., Charles, K. A., and Mantovani, A. (2005). Smoldering and polarized inflammation in the initiation and promotion of malignant disease. *Cancer Cell* **7**, 211–217.

Belz, G. T., Behrens, G. M., Smith, C. M., Miller, J. F., Jones, C., Lejon, K., Fathman, C. G., Mueller, S. N., Shortman, K., Carbone, F. R., *et al.* (2002). The CD8alpha(+) dendritic cell is responsible for inducing peripheral self-tolerance to tissue-associated antigens. *J. Exp. Med.* **196**, 1099–1104.

Bluestone, J. A., and Abbas, A. K. (2003). Natural versus adaptive regulatory T cells. *Nat. Rev. Immunol.* **3**, 253–257.

Boyland, E., and Williams, D. C. (1956). The metabolism of tryptophan 2. The metabolism of tryptophan in patients suffering from cancer of the bladder. *Biochem. J.* **64**, 578–582.

Brandacher, G., Perathoner, A., Ladurner, R., Schneeberger, S., Obrist, P., Winkler, C., Werner, E. R., Werner-Felmayer, G., Weiss, H. G., Gobel, G., *et al.* (2006). Prognostic value of indoleamine 2,3-dioxygenase expression in colorectal cancer: Effect on tumor-infiltrating T cells. *Clin. Cancer Res.* **12**, 1144–1151.

Braun, D., Longman, R. S., and Albert, M. L. (2005). A two-step induction of indoleamine 2,3-dioxygenase (IDO) activity during dendritic-cell maturation. *Blood* **106**, 2375–2381.

Butler, M. H., David, C., Ochoa, G.-C., Freyberg, Z., Daniell, L., Grabs, D., Cremona, O., and De Camilli, P. (1997). Amphiphysin II (SH3P9; BIN1), a member of the amphiphysin/RVS family, is concentrated in the cortical cytomatrix of axon initial segments and nodes of Ravier in brain and around T tubules in skeletal muscle. *J. Cell Biol.* **137**, 1355–1367.

Cady, S. G., and Sono, M. (1991). 1-methyl-DL-tryptophan, beta-(3-benzofuranyl)-DL-alanine (the oxygen analog of tryptophan), and beta-[3-benzo(b)thienyl]-DL-alanine (the sulfur analog of tryptophan) are competitive inhibitors for indoleamine 2,3-dioxygenase. *Arch. Biochem. Biophys.* **291**, 326–333.

Carlin, J. M., Borden, E. C., Sondel, P. M., and Byrne, G. I. (1987). Biologic-response-modifier-induced indoleamine 2,3-dioxygenase activity in human peripheral blood mononuclear cell cultures. *J. Immunol.* **139**, 2414–2418.

Carlin, J. M., Borden, E. C., Sondel, P. M., and Byrne, G. I. (1989). Interferon-induced indoleamine 2,3-dioxygenase activity in human mononuclear phagocytes. *J. Leukoc. Biol.* **45**, 29–34.

Chabner, B. A., DeVita, V. T., Livingston, D. M., and Oliverio, V. T. (1970). Abnormalities of tryptophan metabolism and plasma pyridoxal phosphate in Hodgkin's disease. *N. Engl. J. Med.* **282**, 838–843.

Chiarugi, A., Rovida, E., Dello Sbarba, P., and Moroni, F. (2003). Tryptophan availability selectively limits NO-synthase induction in macrophages. *J. Leukoc. Biol.* **73**, 172–177.

Chon, S. Y., Hassanain, H. H., and Gupta, S. L. (1996). Cooperative role of interferon regulatory factor 1 and p91 (STAT1) response elements in interferon-gamma-inducible expression of human indoleamine 2,3-dioxygenase gene. *J. Biol. Chem.* **271**, 17247–17252.

Chon, S. Y., Hassanain, H. H., Pine, R., and Gupta, S. L. (1995). Involvement of two regulatory elements in interferon-gamma-regulated expression of human indoleamine 2,3-dioxygenase gene. *J. Interferon Cytokine Res.* **15**, 517–526.

Daubener, W., Posdziech, V., Hadding, U., and MacKenzie, C. R. (1999). Inducible anti-parasitic effector mechanisms in human uroepithelial cells: Tryptophan degradation vs. NO production. *Med. Microbiol. Immunol. (Berl)* **187**, 143–147.

Du, M. X., Sotero-Esteva, W. D., and Taylor, M. W. (2000). Analysis of transcription factors regulating induction of indoleamine 2,3-dioxygenase by IFN-gamma. *J. Interferon Cytokine Res.* **20**, 133–142.

Dudley, M. E., Wunderlich, J. R., Robbins, P. F., Yang, J. C., Hwu, P., Schwartzentruber, D. J., Topalian, S. L., Sherry, R., Restifo, N. P., Hubicki, A. M., et al. (2002). Cancer regression and autoimmunity in patients after clonal repopulation with antitumor lymphocytes. *Science* **298**, 850–854.

DuHadaway, J. B., Lynch, F. J., Brisbay, S., Bueso-Ramos, C., Troncoso, P., McDonnell, T., and Prendergast, G. C. (2003). Immunohistochemical analysis of *Bin1*/Amphiphysin II in human tissues: Diverse sites of nuclear expression and losses in prostate cancer. *J. Cell. Biochem.* **88**, 635–642.

DuHadaway, J. B., Sakamuro, D., Ewert, D. L., and Prendergast, G. C. (2001). *Bin1* mediates apoptosis by *c-Myc* in transformed primary cells. *Cancer Res.* **16**, 3151–3156.

Dunn, G. P., Bruce, A. T., Ikeda, H., Old, L. J., and Schreiber, R. D. (2002). Cancer immunoediting: From immunosurveillance to tumor escape. *Nat. Immunol.* **3**, 991–998.

Dunn, G. P., Old, L. J., and Schreiber, R. D. (2004). The immunobiology of cancer immunosurveillance and immunoediting. *Immunity* **21**, 137–148.

Elliott, K., Ge, K., Du, W., and Prendergast, G. C. (2000). The *c-Myc*-interacting adapter protein *Bin1* activates a caspase-independent cell death program. *Oncogene* **19**, 4669–4684.

Elliott, K., Sakamuro, D., Basu, A., Du, W., Wunner, W., Staller, P., Gaubatz, S., Zhang, H., Prochownik, E., Eilers, M., et al. (1999). *Bin1* functionally interacts with *Myc* in cells and inhibits cell proliferation by multiple mechanisms. *Oncogene* **18**, 3564–3573.

Fallarino, F., Bianchi, R., Orabona, C., Vacca, C., Belladonna, M. L., Fioretti, M. C., Serreze, D. V., Grohmann, U., and Puccetti, P. (2004). CTLA-4-Ig activates forkhead transcription factors and protects dendritic cells from oxidative stress in non-obese diabetic mice. *J. Exp. Med.* **200**, 1051–1062.

Fallarino, F., Grohmann, U., Hwang, K. W., Orabona, C., Vacca, C., Bianchi, R., Belladonna, M. L., Fioretti, M. C., Alegre, M. L., and Puccetti, P. (2003). Modulation of tryptophan catabolism by regulatory T cells. *Nat. Immunol.* **4**, 1206–1212.

Fallarino, F., Grohmann, U., Vacca, C., Bianchi, R., Orabona, C., Spreca, A., Fioretti, M. C., and Puccetti, P. (2002a). T cell apoptosis by tryptophan catabolism. *Cell Death Differ.* **9**, 1069–1077.

Fallarino, F., Grohmann, U., You, S., McGrath, B. C., Cavener, D. R., Vacca, C., Orabona, C., Bianchi, R., Belladonna, M. L., Volpi, C., et al. (2006). The combined effects of tryptophan starvation and tryptophan catabolites down-regulate T cell receptor zeta-chain and induce a regulatory phenotype in naive T cells. *J. Immunol.* **176**, 6752–6761.

Fallarino, F., Vacca, C., Orabona, C., Belladonna, M. L., Bianchi, R., Marshall, B., Keskin, D. B., Mellor, A. L., Fioretti, M. C., Grohmann, U., et al. (2002b). Functional expression of indoleamine 2,3-dioxygenase by murine CD8 alpha($^+$) dendritic cells. *Int. Immunol.* **14**, 65–68.

Figdor, C. G., de Vries, I. J., Lesterhuis, W. J., and Melief, C. J. (2004). Dendritic cell immunotherapy: Mapping the way. *Nat. Med.* **10**, 475–480.

Finn, O. J. (2003). Cancer vaccines: Between the idea and the reality. *Nat. Rev. Immunol.* **3**, 630–641.

Friberg, M., Jennings, R., Alsarraj, M., Dessureault, S., Cantor, A., Extermann, M., Mellor, A. L., Munn, D. H., and Antonia, S. J. (2002). Indoleamine 2,3-dioxygenase contributes to tumor cell evasion of T cell-mediated rejection. *Int. J. Cancer* **101**, 151–155.

Fujigaki, H., Saito, K., Lin, F., Fujigaki, S., Takahashi, K., Martin, B. M., Chen, C. Y., Masuda, J., Kowalak, J., Takikawa, O., et al. (2006). Nitration and inactivation of IDO by peroxynitrite. *J. Immunol.* **176**, 372–379.

Fujigaki, S., Saito, K., Takemura, M., Maekawa, N., Yamada, Y., Wada, H., and Seishima, M. (2002). L-tryptophan-L-kynurenine pathway metabolism accelerated by *Toxoplasma gondii* infection is abolished in gamma interferon-gene-deficient mice: Cross-regulation between inducible nitric oxide synthase and indoleamine-2,3-dioxygenase. *Infect. Immun.* **70**, 779–786.

Gailani, S., Murphy, G., Kenny, G., Nussbaum, A., and Silvernail, P. (1973). Studies on tryptophen metabolism in patients with bladder cancer. *Cancer Res.* **33**, 1071–1077.

Galderisi, U., Di Bernardo, G., Cipollaro, M., Jori, F. P., Piegari, E., Cascino, A., Peluso, G., and Melone, M. A. B. (1999). Induction of apoptosis and differentiation in neuroblastoma and astrocytoma cells by the overexpression of *Bin1*, a novel *Myc* interacting protein. *J. Cell. Biochem.* **74**, 313–322.

Gaspari, P., Banerjee, T., Malachowski, W. P., Muller, A. J., Prendergast, G. C., Duhadaway, J., Bennett, S., and Donovan, A. M. (2006). Structure-activity study of brassinin derivatives as indoleamine 2,3-dioxygenase inhibitors. *J. Med. Chem.* **49**, 684–692.

Gasser, S., and Raulet, D. H. (2006). The DNA damage response arouses the immune system. *Cancer Res.* **66**, 3959–3962.

Ge, K., DuHadaway, J., Du, W., Herlyn, M., Rodeck, U., and Prendergast, G. C. (1999). Mechanism for elimination of a tumor suppressor: Aberrant splicing of a brain-specific exon causes loss of function of *Bin1* in melanoma. *Proc. Natl. Acad. Sci. USA* **96**, 9689–9694.

Ge, K., DuHadaway, J., Sakamuro, D., Wechsler-Reya, R., Reynolds, C., and Prendergast, G. C. (2000a). Losses of the tumor suppressor *Bin1* in breast carcinoma are frequent and reflect deficits in a programmed cell death capacity. *Int. J. Cancer* **85**, 376–383.

Ge, K., Minhas, F., DuHadaway, J., Mao, N.-C., Wilson, D., Sakamuro, D., Buccafusca, R., Nelson, P., Malkowicz, S. B., Tomaszewski, J. T., et al. (2000b). Loss of heterozygosity and tumor suppressor activity of Bin1 in prostate carcinoma. *Int. J. Cancer* **86**, 155–161.

Ghiringhelli, F., Puig, P. E., Roux, S., Parcellier, A., Schmitt, E., Solary, E., Kroemer, G., Martin, F., Chauffert, B., and Zitvogel, L. (2005). Tumor cells convert immature myeloid dendritic cells into TGF-beta-secreting cells inducing CD4+CD25+ regulatory T cell proliferation. *J. Exp. Med.* **202**, 919–929.

Gilboa, E. (2004). The promise of cancer vaccines. *Nat. Rev. Cancer* **4**, 401–411.

Grohmann, U., Bianchi, R., Belladonna, M. L., Silla, S., Fallarino, F., Fioretti, M. C., and Puccetti, P. (2000). IFN-gamma inhibits presentation of a tumor/self peptide by CD8 alpha- dendritic cells via potentiation of the CD8 alpha+ subset. *J. Immunol.* **165**, 1357–1363.

Grohmann, U., Bianchi, R., Belladonna, M. L., Vacca, C., Silla, S., Ayroldi, E., Fioretti, M. C., and Puccetti, P. (1999). IL-12 acts selectively on CD8 alpha-dendritic cells to enhance presentation of a tumor peptide *in vivo*. *J. Immunol.* **163**, 3100–3105.

Grohmann, U., Bianchi, R., Orabona, C., Fallarino, F., Vacca, C., Micheletti, A., Fioretti, M. C., and Puccetti, P. (2003a). Functional plasticity of dendritic cell subsets as mediated by CD40 versus B7 activation. *J. Immunol.* **171**, 2581–2587.

Grohmann, U., Fallarino, F., Bianchi, R., Belladonna, M. L., Vacca, C., Orabona, C., Uyttenhove, C., Fioretti, M. C., and Puccetti, P. (2001). IL-6 inhibits the tolerogenic function of CD8 alpha+ dendritic cells expressing indoleamine 2,3-dioxygenase. *J. Immunol.* **167**, 708–714.

Grohmann, U., Fallarino, F., and Puccetti, P. (2003b). Tolerance, DCs and tryptophan: Much ado about IDO. *Trends Immunol.* **24**, 242–248.

Grohmann, U., Orabona, C., Fallarino, F., Vacca, C., Calcinaro, F., Falorni, A., Candeloro, P., Belladonna, M. L., Bianchi, R., Fioretti, M. C., et al. (2002). CTLA-4-Ig regulates tryptophan catabolism *in vivo*. *Nat. Immunol.* **3**, 1097–1101.

Higuchi, K., and Hayaishi, O. (1967). Enzymic formation of D-kynurenine from D-tryptophan. *Arch. Biochem. Biophys.* **120**, 397–403.

Higuchi, K., Kuno, S., and Hayaishi, O. (1963). Enzymatic formation of D-kynurenine. Abstract in *Federation Proc.* **22**, 243.

Homann, D., Jahreis, A., Wolfe, T., Hughes, A., Coon, B., van Stipdonk, M. J., Prilliman, K. R., Schoenberger, S. P., and von Herrath, M. G. (2002). CD40L blockade prevents autoimmune diabetes by induction of bitypic NK/DC regulatory cells. *Immunity* **16**, 403–415.

Hou, D. Y., Muller, A. J., Sharma, M. D., DuHadaway, J., Banerjee, T., Johnson, M., Mellor, A. L., Prendergast, G. C., and Munn, D. H. (2007) Inhibition of indoleamine 2,3-dioxygenase in dendritic cells by stereoisomers of 1-methyl-tryptophan correlates with antitumor responses. *Cancer Res.* **67**, 792–801.

Hucke, C., MacKenzie, C. R., Adjogble, K. D., Takikawa, O., and Daubener, W. (2004). Nitric oxide-mediated regulation of gamma interferon-induced bacteriostasis: Inhibition and degradation of human indoleamine 2,3-dioxygenase. *Infect. Immun.* **72**, 2723–2730.

Hwang, S. L., Chung, N. P., Chan, J. K., and Lin, C. L. (2005). Indoleamine 2, 3-dioxygenase (IDO) is essential for dendritic cell activation and chemotactic responsiveness to chemokines. *Cell Res.* **15**, 167–175.

Hwu, P., Du, M. X., Lapointe, R., Do, M., Taylor, M. W., and Young, H. A. (2000). Indoleamine 2,3-dioxygenase production by human dendritic cells results in the inhibition of T cell proliferation. *J. Immunol.* **164**, 3596–3599.

Ivanova, V. D. (1959). [Studies on tryptophan metabolites in the blood and urine of patients with leukemia.]. *Probl. Gematol. Pereliv. Krovi.* **4**, 18–21.

Ivanova, V. D. (1964). Disorders of tryptophan metabolism in leukaemia. *Acta Unio. Int. Contra Cancrum* **20**, 1085–1086.

Kadlec, L., and Pendergast, A.-M. (1997). The amphiphysin-like protein 1 (ALP1) interacts functionally with the cABL tyrosine kinase and may play a role in cytoskeletal regulation. *Proc. Natl. Acad. Sci. USA* **94**, 12390–12395.

Kadoya, A., Tone, S., Maeda, H., Minatogawa, Y., and Kido, R. (1992). Gene structure of human indoleamine 2,3-dioxygenase. *Biochem. Biophys. Res. Commun.* **189**, 530–536.

Kim, R., Emi, M., Tanabe, K., and Arihiro, K. (2006). Tumor-driven evolution of immunosuppressive networks during malignant progression. *Cancer Res.* **66**, 5527–5536.

Konan, K. V., and Taylor, M. W. (1996). Importance of the two interferon-stimulated response element (ISRE) sequences in the regulation of the human indoleamine 2,3-dioxygenase gene. *J. Biol. Chem.* **271**, 19140–19145.

Kotake, Y., and Masayama, T. (1937). Uber den mechanismus der kynurenine-bildung aus tryptophan. *Hoppe-Seyler's Z. Physiol. Chem.* **243**, 237–244.

Kronin, V., Winkel, K., Suss, G., Kelso, A., Heath, W., Kirberg, J., von Boehmer, H., and Shortman, K. (1996). A subclass of dendritic cells regulates the response of naive CD8 T cells by limiting their IL-2 production. *J. Immunol.* **157**, 3819–3827.

Lee, J. R., Dalton, R. R., Messina, J. L., Sharma, M. D., Smith, D. M., Burgess, R. E., Mazzella, F., Antonia, S. J., Mellor, A. L., and Munn, D. H. (2003). Pattern of recruitment of immunoregulatory antigen-presenting cells in malignant melanoma. *Lab. Invest.* **83**, 1457–1466.

Liang, S., Alard, P., Zhao, Y., Parnell, S., Clark, S. L., and Kosiewicz, M. M. (2005). Conversion of CD4+CD25- cells into CD4+CD25+ regulatory T cells *in vivo* requires B7 costimulation, but not the thymus. *J. Exp. Med.* **201**, 127–137.

Liu, Y., Bi, X., Xu, S., and Xiang, J. (2005). Tumor-infiltrating dendritic cell subsets of progressive or regressive tumors induce suppressive or protective immune responses. *Cancer Res.* **65**, 4955–4962.

Lohr, J., Knoechel, B., Jiang, S., Sharpe, A. H., and Abbas, A. K. (2003). The inhibitory function of B7 costimulators in T cell responses to foreign and self-antigens. *Nat. Immunol.* **4**, 664–669.

MacKie, R. M., Reid, R., and Junor, B. (2003). Fatal melanoma transferred in a donated kidney 16 years after melanoma surgery. *N. Engl. J. Med.* **348**, 567–568.

Malachowski, W. P., Metz, R., Prendergast, G. C., and Muller, A. J. (2005). A new cancer immunosuppression target: Indoleamine 2,3-dioxygenase (IDO). A review of the IDO mechanism, inhibition, and therapeutic applications. *Drugs Fut.* **30**, 897–813.

Mantovani, A., Sozzani, S., Locati, M., Allavena, P., and Sica, A. (2002). Macrophage polarization: Tumor-associated macrophages as a paradigm for polarized M2 mononuclear phagocytes. *Trends Immunol.* **23**, 549–555.

Marshall, B., Keskin, D. B., and Mellor, A. L. (2001). Regulation of prostaglandin synthesis and cell adhesion by a tryptophan catabolizing enzyme. *BMC Biochem.* **2**, 5.

Mehta, R. G., Liu, J., Constantinou, A., Thomas, C. F., Hawthorne, M., You, M., Gerhuser, C., Pezzuto, J. M., Moon, R. C., and Moriarty, R. M. (1995). Cancer chemopreventive activity of brassinin, a phytoalexin from cabbage. *Carcinogenesis* **16**, 399–404.

Melief, C. J., Toes, R. E., Medema, J. P., van der Burg, S. H., Ossendorp, F., and Offringa, R. (2000). Strategies for immunotherapy of cancer. *Adv. Immunol.* **75**, 235–282.

Mellor, A. (2005). Indoleamine 2,3 dioxygenase and regulation of T cell immunity. *Biochem. Biophys. Res. Commun.* **338**, 20–24.

Mellor, A. L., Baban, B., Chandler, P., Marshall, B., Jhaver, K., Hansen, A., Koni, P. A., Iwashima, M., and Munn, D. H. (2003). Cutting edge: Induced indoleamine 2,3 dioxygenase expression in dendritic cell subsets suppresses T cell clonal expansion. *J. Immunol.* **171**, 1652–1655.

Mellor, A. L., Chandler, P., Baban, B., Hansen, A. M., Marshall, B., Pihkala, J., Waldmann, H., Cobbold, S., Adams, E., and Munn, D. H. (2004). Specific subsets of murine dendritic cells acquire potent T cell regulatory functions following CTLA4-mediated induction of indoleamine 2,3 dioxygenase. *Int. Immunol.* **16**, 1391–1401.

Mellor, A. L., and Munn, D. H. (1999). Tryptophan catabolism and T-cell tolerance: Immunosuppression by starvation? *Immunol. Today* **20**, 469–473.

Mellor, A. L., and Munn, D. H. (2004), IDO expression by dendritic cells: Tolerance and tryptophan catabolism. *Nat. Rev. Immunol.* **4**, 762–774.

Mellor, A. L., Sivakumar, J., Chandler, P., K., S., Molina, H., Mao, D., and Munn, D. H. (2001). Prevention of T cell-driven complement activation and inflammation by tryptophan catabolism during pregnancy. *Nat. Immunol.* **2**, 64–68.

Miaczynska, M., Christoforidis, S., Giner, A., Shevchenko, A., Uttenweiler-Joseph, S., Habermann, B., Wilm, M., Parton, R. G., and Zerial, M. (2004). APPL proteins link Rab5 to nuclear signal transduction via an endosomal compartment. *Cell* **116**, 445–456.

Moser, M. (2003). Dendritic cells in immunity and tolerance-do they display opposite functions? *Immunity* **19**, 5–8.

Muller, A. J., DuHadaway, J. B., Donover, P. S., Sutanto-Ward, E., and Prendergast, G. C. (2004). Targeted deletion of the suppressor gene Bin1/Amphiphysin2 enhances the malignant character of transformed cells. *Cancer Biol. Ther.* **3**, 1236–1242.

Muller, A. J., DuHadaway, J. B., Sutanto-Ward, E., Donover, P. S., and Prendergast, G. C. (2005a). Inhibition of indoleamine 2,3-dioxygenase, an immunomodulatory target of the tumor suppressor gene Bin1, potentiates cancer chemotherapy. *Nature Med.* **11**, 312–319.

Muller, A. J., Malachowski, W. P., and Prendergast, G. C. (2005b). Indoleamine 2,3-dioxygenase in cancer: Targeting pathological immune tolerance with small-molecule inhibitors. *Expert Opin. Ther. Targets* **9**, 831–849.

Muller, A. J., and Prendergast, G. C. (2005). Marrying immunotherapy with chemotherapy: Why say IDO? *Cancer Res.* **65**, 8065–8068.

Muller, A. J., and Scherle, P. A. (2006). Targeting the mechanisms of tumoral immune tolerance with small-molecule inhibitors. *Nat. Rev. Cancer* **6**, 613–625.

Munn, D. H. (2006). Indoleamine 2,3-dioxygenase, tumor-induced tolerance and counter-regulation. *Curr. Opin. Immunol.* **18**, 220–225.

Munn, D. H., Shafizadeh, E., Attwood, J. T., Bondarev, I., Pashine, A., and Mellor, A. L. (1999). Inhibition of T cell proliferation by macrophage tryptophan catabolism. *J. Exp. Med.* **189**, 1363–1372.

Munn, D. H., Sharma, M. D., Baban, B., Harding, H. P., Zhang, Y., Ron, D., and Mellor, A. L. (2005). GCN2 kinase in T cells mediates proliferative arrest and anergy induction in response to indoleamine 2,3-dioxygenase. *Immunity* **22**, 633–642.

Munn, D. H., Sharma, M. D., Hou, D., Baban, B., Lee, J. R., Antonia, S. J., Messina, J. L., Chandler, P., Koni, P. A., and Mellor, A. L. (2004). Expression of indoleamine 2,3-dioxygenase by plasmacytoid dendritic cells in tumor-draining lymph nodes. *J. Clin. Invest.* **114**, 280–290.

Munn, D. H., Sharma, M. D., Lee, J. R., Jhaver, K. G., Johnson, T. S., Keskin, D. B., Marshall, B., Chandler, P., Antonia, S. J., Burgess, R., et al. (2002). Potential regulatory function of human dendritic cells expressing indoleamine 2,3-dioxygenase. *Science* **297**, 1867–1870.

Munn, D. H., Zhou, M., Attwood, J. T., Bondarev, I., Conway, S. J., Marshall, B., Brown, C., and Mellor, A. L. (1998). Prevention of allogeneic fetal rejection by tryptophan catabolism. *Science* **281**, 1191–1193.

Najfeld, V., Menninger, J., Muhleman, D., Comings, D. E., and Gupta, S. L. (1993). Localization of indoleamine 2,3-dioxygenase gene (INDO) to chromosome 8p12->p11 by fluorescent in situ hybridization. *Cytogenet. Cell. Genet.* **64**, 231–232.

O'Neill, D. W., Adams, S., and Bhardwaj, N. (2004). Manipulating dendritic cell biology for the active immunotherapy of cancer. *Blood* **104**, 2235–2246.

Okamoto, A., Nikaido, T., Ochiai, K., Takakura, S., Saito, M., Aoki, Y., Ishii, N., Yanaihara, N., Yamada, K., Takikawa, O., et al. (2005). Indoleamine 2,3-dioxygenase serves as a marker of poor prognosis in gene expression profiles of serous ovarian cancer cells. *Clin. Cancer Res.* **11**, 6030–6039.

Orabona, C., Belladonna, M. L., Vacca, C., Bianchi, R., Fallarino, F., Volpi, C., Gizzi, S., Fioretti, M. C., Grohmann, U., and Puccetti, P. (2005a). Cutting edge: Silencing suppressor of cytokine signaling 3 expression in dendritic cells turns CD28-Ig from immune adjuvant to suppressant. *J. Immunol.* **174**, 6582–6586.

Orabona, C., Grohmann, U., Belladonna, M. L., Fallarino, F., Vacca, C., Bianchi, R., Bozza, S., Volpi, C., Salomon, B. L., Fioretti, M. C., et al. (2004). CD28 induces immunostimulatory signals in dendritic cells via CD80 and CD86. *Nat. Immunol.* **5**, 1134–1142.

Orabona, C., Tomasello, E., Fallarino, F., Bianchi, R., Volpi, C., Bellocchio, S., Romani, L., Fioretti, M. C., Vivier, E., Puccetti, P., et al. (2005b). Enhanced tryptophan catabolism in the absence of the molecular adapter DAP12. *Eur. J. Immunol.* **35**, 3111–3118.

Ota, T., Suzuki, Y., Nishikawa, T., Otsuki, T., Sugiyama, T., Irie, R., Wakamatsu, A., Hayashi, K., Sato, H., Nagai, K., et al. (2004) Complete sequencing and characterization of 21,243 full-length human cDNAs. *Nat. Genet.* **36**, 40–45.

Ozaki, Y., Edelstein, M. P., and Duch, D. S. (1988). Induction of indoleamine 2,3-dioxygenase: A mechanism of the antitumor activity of interferon gamma. *Proc. Natl. Acad. Sci. USA* **85**, 1242–1246.

Park, E. J., and Pezzuto, J. M. (2002). Botanicals in cancer chemoprevention. *Cancer Metastasis Rev.* **21**, 231–255.

Peter, B. J., Kent, H. M., Mills, I. G., Vallis, Y., Butler, P. J., Evans, P. R., and McMahon, H. T. (2003). BAR domains as sensors of membrane curvature: The amphiphysin BAR structure. *Science* **303**, 495–499.

Piccirillo, C. A., and Shevach, E. M. (2004). Naturally-occurring $CD4^+CD25^+$ immunoregulatory T cells: Central players in the arena of peripheral tolerance. *Semin. Immunol.* **16**, 81–88.

Pine, R. (1997). Convergence of TNFalpha and IFN-gamma signalling pathways through synergistic induction of IRF-1/ISGF-2 is mediated by a composite GAS/kappaB promoter element. *Nucleic Acids Res.* **25**, 4346–4354.

Platten, M., Ho, P. P., Youssef, S., Fontoura, P., Garren, H., Hur, E. M., Gupta, R., Lee, L. Y., Kidd, B. A., Robinson, W. H., et al. (2005). Treatment of autoimmune neuroinflammation with a synthetic tryptophan metabolite. *Science* **310**, 850–855.

Prendergast, G. C., and Jaffee, E. M. Cancer immunologists and molecular cell biologists: Why we didn't talk then but need to now. *Cancer Res.* In press.

Ramjaun, A. R., Micheva, K. D., Bouchelet, I., and McPherson, P. S. (1997). Identification and characterization of a nerve terminal-enriched amphiphysin isoform. *J. Biol. Chem.* **272**, 16700–16706.

Robinson, C. M., Hale, P. T., and Carlin, J. M. (2005). The role of IFN-gamma and TNF-alpha-responsive regulatory elements in the synergistic induction of indoleamine dioxygenase. *J. Interferon Cytokine Res.* **25**, 20–30.

Rose, D. P. (1967). Tryptophan metabolism in carcinoma of the breast. *Lancet* **1**, 239–241.

Rosenberg, S. A., Packard, B. S., Aebersold, P. M., Solomon, D., Topalian, S. L., Toy, S. T., Simon, P., Lotze, M. T., Yang, J. C., Seipp, C. A., et al. (1988). Use of tumor-infiltrating lymphocytes and interleukin-2 in the immunotherapy of patients with metastatic melanoma. A preliminary report. *N. Engl. J. Med.* **319**, 1676–1680.

Sakaguchi, S. (2004). Naturally arising $CD4^+$ regulatory T cells for immunologic self-tolerance and negative control of immune responses. *Annu. Rev. Immunol.* **22**, 531–562.

Sakamuro, D., Elliott, K., Wechsler-Reya, R., and Prendergast, G. C. (1996). BIN1 is a novel MYC-interacting protein with features of a tumor suppressor. *Nature Genet.* **14**, 69–77.

Salomon, B., Lenschow, D. J., Rhee, L., Ashourian, N., Singh, B., Sharpe, A., and Bluestone, J. A. (2000). B7/CD28 costimulation is essential for the homeostasis of the $CD4^+CD25^+$ immunoregulatory T cells that control autoimmune diabetes. *Immunity* **12**, 431–440.

Samelson-Jones, B. J., and Yeh, S. R. (2006). Interactions between nitric oxide and indoleamine 2,3-dioxygenase. *Biochemistry* **45**, 8527–8538.

Sayama, S., Yoshida, R., Oku, T., Imanishi, J., Kishida, T., and Hayaishi, O. (1981). Inhibition of interferon-mediated induction of indoleamine 2,3-dioxygenase in mouse lung by inhibitors of prostaglandin biosynthesis. *Proc. Natl. Acad. Sci. USA* **78**, 7327–7330.

Shankaran, V., Ikeda, H., Bruce, A. T., White, J. M., Swanson, P. E., Old, L. J., and Schreiber, R. D. (2001). IFNgamma and lymphocytes prevent primary tumour development and shape tumour immunogenicity. *Nature* **410**, 1107–1111.

Sharpe, A. H., and Freeman, G. J. (2002). The B7-CD28 superfamily. *Nat. Rev. Immunol.* **2**, 116–126.

Shimizu, T., Nomiyama, S., Hirata, F., and Hayaishi, O. (1978). Indoleamine 2,3-dioxygenase. Purification and some properties. *J. Biol. Chem.* **253**, 4700–4706.

Sinha, P., Clements, V. K., and Ostrand-Rosenberg, S. (2005a). Interleukin-13-regulated M2 macrophages in combination with myeloid suppressor cells block immune surveillance against metastasis. *Cancer Res.* **65**, 11743–11751.

Sinha, P., Clements, V. K., and Ostrand-Rosenberg, S. (2005b). Reduction of myeloid-derived suppressor cells and induction of M1 macrophages facilitate the rejection of established metastatic disease. *J. Immunol.* **174**, 636–645.

Sono, M. (1989). Enzyme kinetic and spectroscopic studies of inhibitor and effector interactions with indoleamine 2,3-dioxygenase. 2. Evidence for the existence of another binding site in the enzyme for indole derivative effectors. *Biochemistry* **28**, 5400–5407.

Sparks, A. B., Hoffman, N. G., McConnell, S. J., Fowlkes, D. M., and Kay, B. K. (1996). Cloning of ligand targets: Systematic isolation of SH3 domain-containing proteins. *Nat. Biotech.* **14**, 741–744.

Sugimoto, H., Oda, S. I., Otsuki, T., Hino, T., Yoshida, T., and Shiro, Y. (2006). Crystal structure of human indoleamine 2,3-dioxygenase: Catalytic mechanism of O2 incorporation by a heme-containing dioxygenase. *Proc. Natl. Acad. Sci. USA* **103**, 2311–2316.

Tajiri, T., Liu, X., Thompson, P. M., Tanaka, S., Suita, S., Zhao, H., Maris, J. M., Prendergast, G. C., and Hogarty, M. D. (2003). Expression of a MYCN-interacting isoform of the tumor suppressor BIN1 is reduced in neuroblastomas with unfavorable biological features. *Clin. Cancer Res.* **9**, 3345–3355.

Takikawa, O., Tagawa, Y., Iwakura, Y., Yoshida, R., and Truscott, R. J. (1999). Interferon-gamma-dependent/independent expression of indoleamine 2,3-dioxygenase. Studies with interferon-gamma-knockout mice. *Adv. Exp. Med. Biol.* **467**, 553–557.

Taylor, M. W., and Feng, G. S. (1991). Relationship between interferon-gamma, indoleamine 2,3-dioxygenase, and tryptophan catabolism. *FASEB J.* **5**, 2516–2522.

Terness, P., Chuang, J. J., and Opelz, G. (2006). The immunoregulatory role of IDO-producing human dendritic cells revisited. *Trends Immunol.* **27**, 68–73.

Thomas, S. R., Mohr, D., and Stocker, R. (1994). Nitric oxide inhibits indoleamine 2,3-dioxygenase activity in interferon-gamma primed mononuclear phagocytes. *J. Biol. Chem.* **269**, 14457–14464.

Tsutsui, K., Maeda, Y., Tsutsui, K., Seki, S., and Tokunaga, A. (1997). cDNA cloning of a novel amphiphysin isoform and tissue-specific expression of its multiple splice variants. *Biochem. Biophys. Res. Comm.* **236**, 178–183.

Uyttenhove, C., Pilotte, L., Theate, I., Stroobant, V., Colau, D., Parmentier, N., Boon, T., and Van Den Eynde, B. J. (2003). Evidence for a tumoral immune resistance mechanism based on tryptophan degradation by indoleamine 2,3-dioxygenase. *Nat. Med.* **9**, 1269–1274.

Vottero, E., Balgi, A., Woods, K., Tugendreich, S., Melese, T., Andersen, R. J., Mauk, A. G., and Roberge, M. (2006). Inhibitors of human indoleamine 2,3-dioxygenase identified with a target-based screen in yeast. *Biotech. J.* **1**, 282–288.

Wakkach, A., Fournier, N., Brun, V., Breittmayer, J. P., Cottrez, F., and Groux, H. (2003). Characterization of dendritic cells that induce tolerance and T regulatory 1 cell differentiation in vivo. *Immunity* **18**, 605–617.

Watanabe, Y., Yoshida, R., Sono, M., and Hayaishi, O. (1981). Immunohistochemical localization of indoleamine 2,3-dioxygenase in the argyrophilic cells of rabbit duodenum and thyroid gland. *J. Histochem. Cytochem.* **29**, 623–632.

Wechsler-Reya, R., Sakamuro, D., Zhang, J., Duhadaway, J., and Prendergast, G. C. (1997). Structural analysis of the human BIN1 gene: Evidence for tissue-specific transcriptional regulation and alternate RNA splicing. *J. Biol. Chem.* **272**, 31453–31458.

Wolf, A. M., Wolf, D., Rumpold, H., Moschen, A. R., Kaser, A., Obrist, P., Fuchs, D., Brandacher, G., Winkler, C., Geboes, K., *et al.* (2004). Overexpression of indoleamine 2,3-dioxygenase in human inflammatory bowel disease. *Clin. Immunol.* **113**, 47–55.

Wolf, H., Madsen, P. O., and Price, J. M. (1968). Studies on the metabolism of tryptophan in patients with benign prostatic hypertrophy or cancer of the prostate. *J. Urol.* **100**, 537–543.

Wood, K. J., and Sakaguchi, S. (2003). Regulatory T cells in transplantation tolerance. *Nat. Rev. Immunol.* **3**, 199–210.

Yasui, H., Takai, K., Yoshida, R., and Hayaishi, O. (1986). Interferon enhances tryptophan metabolism by inducing pulmonary indoleamine 2,3-dioxygenase: Its possible occurrence in cancer patients. *Proc. Natl. Acad. Sci. USA* **83**, 6622–6626.

Yoeli-Lerner, M., and Toker, A. (2006). Akt/PKB signaling in cancer: A function in cell motility and invasion. *Cell Cycle* **5**, 603–605.

Yoshida, R., Imanishi, J., Oku, T., Kishida, T., and Hayaishi, O. (1981). Induction of pulmonary indoleamine 2,3-dioxygenase by interferon. *Proc. Natl. Acad. Sci. USA* **78**, 129–132.

Yoshida, R., Urade, Y., Tokuda, M., and Hayaishi, O. (1979). Induction of indoleamine 2,3-dioxygenase in mouse lung during virus infection. *Proc. Natl. Acad. Sci. USA* **76**, 4084–4086.

Yuan, W., Collado-Hidalgo, A., Yufit, T., Taylor, M., and Varga, J. (1998). Modulation of cellular tryptophan metabolism in human fibroblasts by transforming growth factor-beta: Selective inhibition of indoleamine 2,3-dioxygenase and tryptophanyl-tRNA synthetase gene expression. *J. Cell. Physiol.* **177**, 174–186.

Zenclussen, A. C., Gerlof, K., Zenclussen, M. L., Sollwedel, A., Bertoja, A. Z., Ritter, T., Kotsch, K., Leber, J., and Volk, H. D. (2005). Abnormal T-cell reactivity against paternal antigens in spontaneous abortion: adoptive transfer of pregnancy-induced CD4$^+$CD25$^+$ T regulatory cells prevents fetal rejection in a murine abortion model. *Am. J. Pathol.* **166**, 811–822.

Zou, W. (2005). Immunosuppressive networks in the tumour environment and their therapeutic relevance. *Nat. Rev. Cancer* **5**, 263–274.

Further Reading

Fallarino, F., and Puccetti, P. (2006). Toll-like receptor 9-mediated induction of the immunosuppressive pathway of tryptophan catabolism. *Eur. J. Immunol.* **36**, 8–11.

*Discusses how activation of Toll-like receptor 9 (TLR9) signaling by CpG oligonucleotides can lead to IDO activation and immunosuppression. This review highlights how IDO induction by CpG oligonucleotides may limit their potential as adjuvants for cancer therapeutics and vaccines.

Gejewski, T. F. (2006). Identifying and overcoming immune resistance mechanisms in the melanoma tumor microenvironment. *Clin. Cancer Res.* **12**, 2326s–2330s.

*Highlights melanoma as an illustration of how immunosuppression can thwart the activity of tumor antigen-specific T cells, through the influence of T cell anergy, CD25$^+$ regulatory T cells, and overexpression of immune suppressive genes, such as PD-L1 and IDO.

Johns Hopkins University and the Institute of Bioinformatics. Human Protein Reference Database. Indoleamine 2,3-dioxygenase. Retrieved February 2007: http://www.hprd.org/protein/00935

Munn, D. H., Sharma, M. D., Lee, J. R., Jhaver, K. G., Johnson, T. S., Keskin, D. B., Marshall, B., Chandler, P., Antonia, S. J., Burgess, R., Slingluff, C. L., Jr., and Mellor, A. L. Potential regulatory function of human dendritic cell expressing indoleamine 2,3-dioxygenase. *Sci. Signal Trans. Virt. J. (STKE)*. Retrieved: http://stke.sciencemag.org/cgi/content/abstract/sci;297/5588/1867.

Schrocksnadel, K., Wirleitner, B., Winkler, C., and Fuchs, D. (2006). Monitoring tryptophan metabolism in chronic immune activation. *Clin. Chim. Acta* **364**, 82–90.

*Discusses how IDO activation in cancer and other chronic immune suppressed states may be decreasing serum tryptophan and also affecting serotonin biosynthesis, thereby contributing to impaired quality of life and depressed moods.

Swiss Institute of Bioinformatics. SwissProt: In silico analysis of proteins. Indoleamine 2,3-dioxygenase. Retrieved February 2007: http://ca.expasy.org/uniprot/P14902

CHAPTER 20

Arginase, Nitric Oxide Synthase, and Novel Inhibitors of L-Arginine Metabolism in Immune Modulation

SUSANNA MANDRUZZATO, SIMONE MOCELLIN, AND VINCENZO BRONTE

I. Introduction
II. NOS: Genes, Regulation, and Activity
III. ARG: Genes, Regulation, and Activity
IV. Immunoregulatory Activities of ARG and NOS
    A. Mechanisms of NOS-Dependent Immunoregulation
    B. Mechanisms of ARG-Dependent Immunoregulation
    C. ARG and NOS Cooperation in Immunoregulation: An Emerging Concept
V. Possible Physiological Role for L-ARG Metabolism in Immunity Control
VI. NOS in Cancer
VII. ARG in Cancer
VIII. ARG and NOS Inhibitors: A Novel Class of Immune Adjuvants?
IX. Conclusion and Perspectives
    References
    Further Reading

In higher organisms, the control of amino acid metabolism is emerging as an evolutionarily preserved strategy for limiting the expansion of actively proliferating cells, including antigen-activated T lymphocytes, and tumor cells have surreptitiously adopted it to avoid or restrain attack by the immune system. Tumor growth is often associated with an altered metabolism of the amino acid L-arginine (L-Arg) by the enzymes nitric oxide synthase (NOS) and arginase (ARG). Once activated, either in tumor-recruited myeloid cells or the cancerous cells, both enzymes can alter the function of antitumor T lymphocytes. The prevalence of one enzyme pathway or cooperation between both pathways appears to depend on the tumor type and the genetic background of the host. Experimental findings indicate that when either one of the two enzymes is active, the net effect on T lymphocytes can be attributed to cell cycle arrest, whereas the concomitant activation of both enzymes in the same environment can lead to T cell death by apoptosis. Moreover, immune regulation by L-Arg metabolism is not antigen specific but requires activation of T cells through their clonotypic T cell receptor in order to be susceptible to these inhibitory circuits. This chapter will discuss exceptions to these simple rules and provide a description of novel compounds that can deactivate these metabolic pathways in tumor-bearing hosts and thus help to restore immune reactivity against cancer.

I. INTRODUCTION

A successful immune response against offending insults to the body relies on the orderly interaction between cells of the innate and adaptive immune systems. Although T lymphocytes bearing a clonotypic receptor for the antigen are needed for the complete elimination of the offending insult, other cells assist them in antigen clearance, destruction of intracellular pathogens, triggering of vascular and inflammatory responses, production of collagen, and tissue remodeling. These "accessory cells" are also important in guiding lymphocyte activation, and in this regard, one can schematically picture two main classes of accessory cells: (i) those deciding whether an immune response needs to be generated and whether naive T lymphocytes need to be activated by the antigen (gatekeepers) and (ii) those ensuring that activated T lymphocytes are correctly turned off at the end of their job to avoid damage and accumulation of useless cells (caretakers). Dendritic cells (DCs) are the perfect gatekeepers since they are strategically placed at the ports of antigen entry and express an array of molecules that can aid the priming of naive T lymphocytes. On the other hand, they can employ L-tryptophan (L-Trp) metabolism via indoleamine 2,3-deoxygenase (IDO) to limit T lymphocyte reactivity if necessary, to tolerize rather than activate.

The myelomonocytic system, on the other hand, possesses the qualities to accomplish the caretaker role. Cells belonging to this differentiation lineage are released from pools circulating in the blood and from hematopoietic organs in a number proportional to the offending insult and are equipped with a number of tools that are extremely efficient in killing invading pathogens but are also useful to silence activated T lymphocytes. A pathological expansion of myelomonocytic cells in the bone marrow, secondary lymphoid organs, and blood can be seen under different pathological conditions: during acute and chronic immune responses to pathogens; during periods of immune stress leading to extensive T lymphocyte activation, such as exposure to superantigens or certain parasite products; during altered hematopoiesis by radiation, graft-versus-host reaction, or chemotherapy; during the body's reaction to autoimmune diseases; and, most important to the aim of this chapter, during tumor growth/development. These cells have been termed *myeloid-derived suppressor cells* (MDSCs), and they have an uncanny ability to impair T lymphocyte functions (Serafini *et al.*, 2003, 2006), as discussed in Chapter 17.

Myelomonocytic cells exploit the metabolism of another amino acid, L-arginine (L-Arg), to restrain lymphocyte activation. The enzymes metabolizing L-Arg that are critical for this function, nitric oxide synthase (NOS) and arginase (ARG), are under the control of T helper 1 (Th1)- and T helper 2 (Th2)-type cytokines that direct cellular and humoral adaptive immune responses, respectively. The other two enzymes catabolizing L-Arg (i.e., arginine:glycine amidinotransferase and arginine decarboxylase) have not been shown to possess immunoregulatory properties to date. L-Arg is not an essential amino acid for adult mammals, but its demand can exceed the endogenous biosynthesis in situations of intense stress, in cancer patients, following liver transplantation, or in severe trauma (Barbul, 1990). Findings suggest that enhanced L-Arg metabolism in myeloid cells can result in the impairment of lymphocyte responses to antigen. Moreover, tumor cells themselves can alter L-Arg metabolism in the microenvironment to their own advantage.

II. NOS: GENES, REGULATION, AND ACTIVITY

NO production by NOS has been extensively reviewed (Wu and Morris, 1998; Alderton et al., 2001; Bogdan, 2001; Bronte and Zanovello, 2005), so this section will discuss only basic aspects that are relevant to the understanding of the NOS immunoregulatory activity in cancer. There are three isoforms of the heme-containing enzymes catalyzing the synthesis of NO from L-Arg, namely, the neuronal (nNOS/NOS1), inducible (iNOS/NOS2), and endothelial (eNOS/NOS3) NO synthases (Michel and Feron, 1997; Alderton et al., 2001). NOS1 and NOS2 are constitutively expressed in some cell types, mainly neurons and endothelial cells, respectively. Following a rise in intracellular calcium, these constitutive NOS isoforms can be activated as a result of calmodulin binding and can only produce nanomolar concentrations of NO for seconds or minutes. They may also be activated and/or inhibited by phosphorylation through various protein kinases. For example, the catalytic activity of NOS3 is augmented by phosphorylation of a C-terminal serine residue through the Akt pathway (Dimmeler et al., 1999), which is often activated in malignant cells.

Two main domains can be identified in all NOS isoforms: an N-terminal oxygenase domain and a C-terminal reductase domain. Electrons donated by conversion of NADPH to NADP are transferred to the oxygenase domain through the redox chain involving the electron carriers flavin adenine dinucleotide (FAD) and flavin mononucleotide (FMN). The oxygenase domain uses the iron heam and BH4 to catalyze the reaction between O_2 and L-Arg to generate citrulline and NO. The presence of bound calmodulin is necessary to ensure electron flow (Alderton et al., 2001).

Although several cell types (e.g., neutrophils, hepatocytes, cardiac myocytes) can express NOS2, mouse macrophages represent the typical source of this enzyme (Bogdan, 2001). NOS2 activation is considered a hallmark of classically activated macrophages (i.e., macrophages that are mediators of the delayed type hypersensitivity [DTH] response and are endowed with antitumor properties, as opposed to alternatively activated macrophages) (Gordon, 2003). The expression of NOS2 can be transcriptionally upregulated by proinflammatory cytokines (e.g., interferons [IFNs], interleukin [IL]-1, IL-2, tumor necrosis factor [TNF]-α), bacterial lipopolysaccharide (LPS), and hypoxia, an important inducer of tumor angiogenesis (Kleinert et al., 2003). NOS2 gene expression depends on the activation of transcription factors, such as nuclear factor (NF)-κB, Janus kinase 3/signal transducers and activators of transcription 1 (Jak3/Stat1), and c-Jun NH_2-terminal kinase (JNK) (Ganster et al., 2001). NOS2 transcription is instead downregulated by steroids (Shinoda et al., 2003), anti-inflammatory cytokines (e.g., transforming growth factor-beta [TGF-β], IL-10) (Wink et al., 1998; Chen et al., 2003), p53 (Ambs et al., 1998d), and NO itself (Hinz et al., 2000). To a lesser degree, NOS1 and NOS3 can also be regulated at the transcriptional level (Forstermann et al., 1998; Mocellin et al., 2004b).

Differences have been described between human and mouse cells in regard to NOS2 activity. Human macrophages respond to stimuli different than LPS + IFN-γ (very active on rodent macrophages) because they are rather susceptible to IFN-α, chemokines, or the combination of CD23 + IL-4 (Vouldoukis et al., 1995; Sharara et al., 1997; Villalta et al., 1998). These discrepancies have incited an ongoing debate about the possibility that NOS2 is not active in human macrophages (Schneemann and Schoedon, 2002); however, expression of mRNA and/ or protein has been detected in human blood mononuclear cells and neutrophils under different conditions (Bogdan, 2001).

Unlike NOS1 and NOS3, NOS2 displays a high affinity for calmodulin, which is tightly bound at physiological concentrations of calcium. Moreover, NOS2 protein is permanently active throughout its life (hours or days) and generates micromolar concentrations of NO in the absence of changes in calcium levels (Michel and Feron, 1997). NOS2 knockout mice are viable, fertile, and without evident histopathological abnormalities but are susceptible to infection with the parasite *Leishmania major*. However, knockout mice are capable of developing a stronger Th1-type immune response than wild-type mice. Moreover, mutant mice were resistant to LPS-induced mortality (Wei et al., 1995).

Importantly, polymorphisms of the promoter regions of both NOS2 and NOS3 have been linked to a higher risk of cancer development, greater tumor aggressiveness, and poorer survival in humans (Medeiros et al., 2002; Tatemichi et al., 2005).

III. ARG: GENES, REGULATION, AND ACTIVITY

ARG is a manganese metalloenzyme that catalyzes the hydrolysis of L-Arg to form L-ornithine (L-Orn) and urea. L-Orn is further processed to give rise to L-proline (L-Pro) via ornithine aminotransferase (OAT) and polyamines via the ornithine decarboxylase (ODC) pathway (Wu and Morris, 1998; Bogdan, 2001). In mammals, two genetically distinct isoenzymes have been identified that differ in tissue distribution and subcellular location. Arginase 1 (ARG1, also known as liver-type) is found predominantly in hepatocytes, where it catalyzes the final cytosolic step of the urea cycle; arginase 2 (ARG2, also known as kidney-type) is more widely distributed in numerous tissue (e.g., kidneys, small intestine, brain, skeletal muscle, liver), is localized to the mitochondrial matrix in the cell, and does not appear to function in the urea cycle (Ash, 2004). The human type 1 and type 2 ARGs are related by 58% sequence identity. The human *Arg1* gene gives rise, by alternative splicing, to 12 different transcripts, putatively encoding 11 different protein isoforms; the *Arg2* gene encodes 4 different transcripts by alternative splicing, putatively encoding 4 different protein isoforms. The function and relevance of these variants have not been investigated in detail.

The family of ARGs is highly conserved across species, and ARG activity from pathogens can interfere with host L-Arg metabolism and vice versa. For example, the *Helicobacter pylori* gene *rocF* encodes a constitutively active ARG that consumes L-Arg from the medium and completely prevents NO production by NOS2 in macrophages, conferring an advantage for bacterial growth (Gobert et al., 2001). In fact, whereas wild-type *H. pylori* is resistant to macrophage-derived NO, the *rocF*-deficient strain is efficiently killed and eliminated by activated macrophages. On the other hand, *Leishmania major*-infected BALB/c macrophages exposed to Th2-type cytokines express ARG (likely ARG1) and support the intracellular growth of the parasites, a permissive status that was reversed by the addition of the physiologic ARG inhibitor N^G-hydroxy-L-Arg (NOHA) (Iniesta et al., 2001).

ARG is expressed in mammalian cells of the innate immune system, where it might have assumed important biological functions during evolution. Whereas ARG1 expression in mouse hepatocytes is constitutive, in resting rodent macrophages *Arg1* mRNA, protein, and enzyme activity are undetectable and are upregulated several orders of magnitude by inducers of the transcription factor Stat6, such as the cytokines IL-4 and IL-13 (Munder *et al.*, 1998, 1999; Mills *et al.*, 2000). ARG1 activation has been proposed as one of the specific markers of alternatively activated mouse macrophages, which are important mediators of allergic responses, control of parasitic infections, and wound repair and fibrosis, and have been found in tumor infiltrates of various human tumors where they have been suspected to promote tumorigenesis (Mantovani *et al.*, 2002; Gordon, 2003). The properties of tumor-infiltrating macrophages are discussed in detail in Chapter 16.

Other cytokines including TGF-β (Boutard *et al.*, 1995), the macrophage-stimulating protein (MSP) acting on the receptor RON (Morrison and Correll, 2002), and GM-CSF (Jost *et al.*, 2003) can induce the enzyme activity in different cell types of the innate immune system. ARG1 is also regulated by phosphodiesterase (PDE) inhibitors with rather divergent effects. Inhibitors of PDE4, the predominant PDE in macrophages, elevate cyclin adenosine monophosphate (cAMP) levels, which results in ARG1 (and to a lesser extent ARG2) induction in RAW 264.7 cells and human alveolar macrophages stimulated with IL-4 or TGF-β (Erdely *et al.*, 2006). Conversely, PDE5 inhibitors, such as sildenafil, vardenafil, and tadalafil, increase intracellular concentrations of cGMP and down-regulate ARG1 activity and expression in tumor-infiltrating CD11b$^+$ MDSCs (Serafini *et al.*, 2006). LPS, dexamethasone, and hypoxia induce ARG expression with a certain variability in terms of induced isoforms (i.e., ARG1 versus ARG2) in various cell types (Morris, 2002).

ARG1 regulation by IL-4 and IL-13 in human macrophages has been disputed on the account that IL-4 plus IL-13 does not induce an increase in *Arg1* mRNA in monocytes isolated from the buffy coat of normal volunteers (Raes *et al.*, 2005). However, human alveolar macrophages can respond to IL-4 in the presence of increased cAMP levels (Erdely *et al.*, 2006), suggesting that rather than a general difference between mice and humans, a difference in anatomic localization and/or stimuli responsiveness accounts for the diversity between the species, analogous to the observations for NO production (see earlier discussion in this chapter). A divergence in ARG localization during evolution might indeed exist, however, since ARG1 has been shown to be constitutively present in human granulocytes and has localized selectively to the azurophil granules (Munder *et al.*, 2005).

The 5′ flanking region of the mouse *Arg1* gene is responsible for the transcriptional response to IL-4, cAMP, TGF-β, dexamethasone, and LPS (Morris, 2002). An enhancer element about 3 KB from the basal *Arg1* promoter assembles a multimeric complex comprising Stat6, PU.1, CEBP-β transcription factors and an as of yet unidentified protein synthesized *ex novo* following Stat6 activation (Pauleau *et al.*, 2004). In particular, Stat6 binding to the promoter requires an adjacent CEBP-β to confer responsiveness to IL-4 in mouse macrophages (Gray *et al.*, 2005). CEBP-β is also the transcription factor regulating ARG1 induction by glucocorticoids and glucan in rat hepatocytes (Gotoh *et al.*, 1997), and it is also essential for mediating the activity of cAMP, LPS, and hypoxia on ARG1 expression (Albina *et al.*, 2005).

ARG2, indicated originally as kidney-type, has been considered for a long time to be a mitochondrial enzyme constitutively expressed in different cell types and dedicated to L-Pro synthesis because of its close proximity to ornithine amino transferase (OAT), also located in the mitochondria.

Investigations have challenged this division, and it is clear that both ARG1 and ARG2 regulate polyamine synthesis and can be constitutive and inducible in cells of the innate immune system. ARG2 is activated in macrophages by infection with *H. pylori* (Gobert *et al.*, 2002) and is also induced in Jurkat cells and macrophages infected with Sendai virus through a mechanism independent of type I IFN production but requiring the control of the interferon regulatory factor 3 (IRF3) (Grandvaux *et al.*, 2005). Increased spermine production after ARG2 activation inhibited viral replication and induced apoptosis of infected cells, a mechanism that requires further investigation in view of the known ability of polyamines to promote rather than inhibit cell proliferation.

ARG1 might be part of a functional unit including membrane transporters of L-Arg. In various responses both proteins are, in fact, coregulated by the same signals and stimuli. Mouse cationic amino acid transporters (mCATs) are integral membrane proteins with the function of transporting L-Arg, L-lysine (L-Lys), and L-Orn from the extracellular to intracellular environment. This transport is pH independent, sensitive to stimulation, and saturable at circulating plasma concentrations of cationic amino acids. Among the four related proteins, CAT-2A, CAT-2B, CAT-3, and CAT-4, identified in different mammalian species, CAT-2A is predominantly expressed in liver, whereas CAT-2B is usually induced under inflammatory conditions in a variety of cells and tissues, including T cells, macrophages, lung tissue, and testis tissue (Mann *et al.*, 2003).

ARG1 knockout mice carrying a nonfunctional *Arg1* gene lacked liver ARG activity and died between postnatal days 10 and 14 with severe symptoms of hyperammonemia (Iyer *et al.*, 2002). On the other hand, targeted disruption of the *Arg2* gene showed that homozygous *Arg2*-deficient mice were viable and apparently normal but had elevated plasma L-Arg levels, which suggests that this enzyme also plays a role in L-Arg homeostasis (Shi *et al.*, 2001).

IV. IMMUNOREGULATORY ACTIVITIES OF ARG AND NOS

Depending on the local tissue involved in an immune response, the genetic background of the mouse (skewing of Th1/Th2 balance), and the disease state, the products of L-Arg metabolism not only exert protective or damaging effects on tissues but also regulate activities of innate and specific immune compartments. It is clear that this regulatory role can strongly influence the outcome of infectious and inflammatory pathologies, autoimmune disorders, and cancer. In this sense, NOS and ARG can enter immunoregulatory circuits independently but they can also collaborate, especially under circumstances related to cancer growth. The prevalence of one enzyme pathway over the other appears to be related to the type of tumor and the genetic background of the mice, which dictates the Th1 versus Th2 orientation of the immune response. As a rule of thumb, when one of the two enzymes is active, the net effect on T lymphocytes can be attributed to cell cycle arrest, whereas the concomitant activation of both enzymes in the same environment can lead to T cell death by apoptosis. Moreover, immune regulation by L-Arg metabolism is not antigen specific but requires that T cells be activated through their clonotypic T cell receptor (TCR) to be susceptible to the inhibitory activity of ARG- and NOS-dependent pathways. Exceptions to these simple rules do exist and will be discussed in the following paragraphs.

An intriguing aspect of L-Arg metabolism is the number of molecules and products that can be released extracellularly: NO, NOHA, L-Orn, polyamines, N^G-monomethyl-L-Arg monoacetate (L-NMMA), and even the ARG enzyme itself can be found in the extracellular space, where they act as messengers or biological response modifiers that modulate the functions of nearby cells (Wu and Morris, 1998). NOHA can be released by macrophages during intense NOS2 activity (Hecker et al., 1995), inhibiting ARG and stimulating further release of NO by NOS (Buga et al., 1998), superoxide anions, or hemoproteins (peroxidases and cytochrome p450) (Wu and Morris, 1998) in neighboring cells during inflammatory reactions. NO causes S-nitrosylation of the cysteine present in the ODC active site (Bauer et al., 1999), resulting in its enzymatic inhibition, arrest of polyamine production, and polyamine-driven cell proliferation (Buga et al., 1998). Conversely, polyamines produced by ODC are known NOS inhibitors (Blachier et al., 1997). L-Orn and eosinophil-derived cationic proteins can impede CAT-2B-dependent transport of L-Arg, leading to decreased NO production in patients with asthma (Meurs et al., 2003).

A. Mechanisms of NOS-Dependent Immunoregulation

NO, produced at high levels in the immune system by NOS2, has protective effects during infection but also damaging properties during autoimmune responses. To date, there is no doubt that NO is a critical player in different immunological responses other than pathogens and tissue assault since it regulates cytokine production, leukocyte chemotactic responses, cell survival, and thymic education (MacMicking et al., 1997; Bogdan et al., 2000a,b, 2001). This multifunctional molecule is also involved in the pathogenesis of infectious diseases, the degenerative effects of chronic inflammatory diseases, and the growth of tumors (Moncada et al., 1991).

NOS2-derived NO (the term NO is used here collectively for all its reactive intermediates) plays a dual role in an immune response by exerting protective and toxic effects in parallel. The first immunoregulatory activity assigned to NO was the ability to impair the proliferation of T lymphocytes. For many years, in fact, NO production was considered the key mechanism by which macrophages inhibited T lymphocyte proliferative responses to antigens and mitogens (MacMicking et al., 1997; Weinberg, 1998). This inhibition might account, at least partially, for the immunosuppressed state seen in certain infectious diseases, malignancies, and graft-versus-host reactions, but it is also important for the control of inflammatory processes or to delete autoreactive T lymphocytes (Koblish et al., 1998; Kolb and Kolb-Bachofen, 1998; Bobe et al., 1999). The immunoregulatory role of NO has been unveiled by the use of NOS inhibitors, such as L-NMMA, or by experiments showing that inhibitory cells from NOS2-deficient mice lack immunosuppressive properties.

NO interferes with T lymphocyte activation by altering the signaling cascade downstream of the binding of IL-2 to the IL-2 receptor (IL-2R). This activity might depend on both S-nitrosylation of critical cysteine residues or activation of soluble guanylate cyclase (sGC) and cGMP-dependent protein kinase (cGK) (Bingisser et al., 1998; Duhe et al., 1998; Fischer et al., 2001). The final results of this process are that the phosphorylation and activation of signaling intermediates in the IL-2R pathway (JAK1, JAK3, Stat5, Erk, and Akt) are inhibited in T lymphocytes by NO (Bingisser et al., 1998; Mazzoni et al., 2002). NO can also affect the stability of IL-2 mRNA and its release by activated human T lymphocytes (Fischer et al., 2001; Macphail et al., 2003), again suggesting that NO action on T lymphocytes translates into a reversible proliferative arrest. However, a direct proapoptotic effect on T lymphocytes has been observed with high concentrations of NO, and it has been considered important in the regulation of T cell maturation in the thymus as well as T cell growth in the periphery (Kolb and Kolb-Bachofen, 1998). T cell death has been associated with different mechanisms, such as p53 accumulation, Fas- or TNF-receptor activated signaling, as well as caspase-independent pathways (Hildeman *et al.*, 1999; Mannick *et al.*, 1999; Macphail *et al.*, 2003).

The ability to influence the production of proinflammatory cytokines (including IFN-γ) by various cell types represents a second immunoregulatory property of NO (Bogdan *et al.*, 2000a,b). For several cytokines, however, conflicting results have been reported, and the pathway subject to NO regulation remains to be defined. Controversy also exists with respect to the net effect of NO on Th1/Th2 balance, even if different studies have indicated a possible selective activity of exogenous NO on the cytokine released by established Th1 lines. By interfering prevalently with the IL-2-dependent signaling pathways, low levels of NO control mainly Th1 lymphocyte proliferation/function (Bauer *et al.*, 1997; Niedbala *et al.*, 1999). Under these conditions, NO could also favor Th2-cell functions by upregulating IL-4 production. However, high levels of NO can affect both Th1 and Th2 lymphocytes and induce their apoptosis, possibly in conjunction with other NO$^-$ and O$_2^-$ derived metabolites as discussed in the next sections.

In acute bacterial or parasitic infections, NO is described as a resistance factor elaborated by the host to fight pathogens, whereas persistent infection with parasites, such as *Trypanosoma, Chlamydia, Schistosoma,* or *Leishmania*, are commonly characterized by an increase in ARG expression concomitantly with the prevalent production of IL-4, IL-10, and TGF-β (Vincendeau *et al.*, 2003); however, results obtained in models of helminth *Candida albicans* and *Trypanosoma cruzi* infections have demonstrated the existence of an immunosuppressive mechanism depending on IFN-γ-induced secretion of NO by myeloid cells that colonize the spleen during the acute phase of infection (Terrazas *et al.*, 2001; Goni *et al.*, 2002; Mencacci *et al.*, 2002).

An important but still controversial point concerns the expression and activity of NOS in T lymphocytes. The expression of the NOS2 isoform has been described in mouse T lymphoblasts, Jurkat cells, and human T lymphocytes infiltrating grafts in immunodeficient mice (Mannick *et al.*, 1999; Koh *et al.*, 2004; Vig *et al.*, 2004). In many human studies, the cells analyzed were of neoplastic derivation, raising the question whether NO production could depend on the acquisition of a malignant phenotype, a rather frequent chance discussed in following paragraphs. Indeed, human T cells and peripheral lymphocytes were shown to express NOS2 mRNA only upon infection with human T cell leukemia virus (Mori *et al.*, 1999). A 2006 study, however, has reopened this controversial debate showing that human T cells produce NO via NOS3 following TCR-mediated Ca^{2+} release and phosphotidylinositol 3-kinase (PI3K) activation. Interestingly, NOS3 overexpression increased extracellular signal-regulated kinase (ERK) phosphorylation, altered CD3 distribution in the immunological synapse, and stimulated IFN-γ release while depressing IL-2 production (Ibiza *et al.*, 2006). Endogenous NO production during T lymphocyte activation might thus be involved in limiting the inappropriate expansion of T lymphocytes while maintaining some effector functions, such as the ability to release IFN-γ, thus functioning as an inner rheostat (see following paragraphs).

B. Mechanisms of ARG-Dependent Immunoregulation

An unexpected role of ARG has also been demonstrated in the suppression of the immune response. Although research in the late twentieth century has produced a large amount of experimental evidence, the first observation that ARG can suppress an immune response under certain conditions dates back to 1977 (Kung *et al.*, 1977). Since then, the mechanisms exhibited by ARG to induce suppression of T lymphocytes have been extensively studied (Bronte and Zanovello, 2005).

The effect of ARG on the immune system depends mostly on the depletion of L-Arg from the intracellular and extracellular environments rather than on production of metabolites through the downstream metabolic pathways involving ODC and OAT. Mouse macrophages stimulated *in vitro* with IL-4 + IL-13 upregulate ARG1 and CAT-2B transporter, leading to an enhanced consumption of extracellular L-Arg levels (Rodriguez et al., 2003, 2004). Decreased levels of L-Arg have been detected in wounds and tumors as well as in individuals with liver transplants, in individuals who have just undergone surgery, and in individuals with acute bacterial peritonitis (Rodriguez et al., 2002). ARG1 activation in tumor-infiltrating $CD11b^+$, $Gr-1^-$, $CD16/32^+$, and $F4/80^-$ myeloid cells, also expressing high levels of CAT-2B transporter, consumed L-Arg and inhibited the re-expression of the CD3-ζ chain in T cells stimulated by antigen, thus impairing their proliferation (Rodriguez et al., 2003, 2004). The CD3-ζ chain is the main signal transduction element of the T cell receptor and is required for the correct assembly of the receptor complex (Baniyash, 2004). Loss of CD3-ζ chain in circulating peripheral blood lymphocytes, in fact, is a hallmark of patients who experience a reduction in blood L-Arg (possibly through ARG activation), in pathologies as varied as cancer, chronic infections, liver transplantation, and trauma (Rodriguez et al., 2002; Baniyash, 2004). Alteration of CD3-ζ chain levels in T cells could thus represent an important mechanism for tumor escape *in vivo*.

ARG1 is expressed in immunosuppressive granulocytes expanded in patients with renal cell carcinomas (Zea et al., 2005), and in a mouse lung carcinoma model, the ARG inhibitor N-hydroxy-nor-L-Arg (Nor-NOHA) was shown to slow the growth of an experimental lung carcinoma in a dose-dependent fashion (Rodriguez et al., 2004). Genetic and pharmacological inhibition of cyclooxygenase 2 (COX-2), but not COX-1, also blocked ARG1 activity in tumor-infiltrating cells. COX-2 was expressed in lung cancer cells and generated PGE_2, which ultimately induced ARG1 in myeloid cells through the PGE_2 receptor E-prostanoid 4. Analogously to Nor-NOHA, COX-2 inhibitors enhanced a lymphocyte-mediated antitumor response (Rodriguez et al., 2005).

As previously mentioned, ARG1 is constitutively expressed in human granulocytes (Munder et al., 2005), where it has been demonstrated that ARG1 release can induce a profound suppression of T cell proliferation and cytokine synthesis (Munder et al., 2006). It thus appears that, despite the different prevalent localization in mouse and human cells of the innate immune system, the immunoregulatory properties of ARG1 were conserved during evolution.

Analogously to tumors, parasites might also have developed strategies to overcome T lymphocyte recognition by altering L-Arg metabolism. Patients infected with *H. pylori* can develop chronic gastritis in the absence of protective immunity. Human T lymphocytes stimulated in the presence of a crude extract of *H. pylori* had a reduced proliferation that correlated with a decreased CD3-ζ chain expression (Zabaleta et al., 2004). Interestingly, when the extract was derived from the mutant strain of *H. pylori* $rocF(^-)$, which lacks the ARG gene, alterations of T lymphocytes were absent, indicating a close relationship between *H. pylori*-induced lymphocyte dysfunction and *H. pylori* ARG (Zabaleta et al., 2004).

The biochemical bases for linking T lymphocyte proliferative arrest, loss of CD3-ζ chain, and L-Arg starvation are only partially known. The signaling elements GCN2 kinase and mTOR function as amino acid sensors in mammalian cells, and there is a hypothesis that suggests they play a role in regulating the mRNA stability of crucial molecules in T cells, including the CD3-ζ chain (Bronte and Zanovello, 2005). GCN2 kinase phosphorylation via L-Arg deprivation was shown to occur in astrocytes, similar to how activation of GCN2 kinase mediates proliferative arrest and T cell anergy induction in response to L-Trp deprivation by IDO (Lee et al., 2003; Munn et al., 2005). T cells stimulated and cultured in the absence of L-Arg undergo a proliferative arrest at the G_0–G_1 phase of the cell cycle. Signaling through GCN2 kinase is triggered during L-Arg starvation and causes the inability of T cells to upregulate cyclin D3 and cyclin-dependent kinase 4 (Rodriguez et al., 2007; A. Ochoa, personal communication). These findings might explain how GCN2 can control T lymphocyte proliferation; to date, however, no direct link between L-Arg deprivation and CD3-ζ chain downregulation by the GCN2 and/or mTOR pathways has been established.

B lymphocytes are also severely affected by chronic L-Arg deficiency. Transgenic mice that overexpress ARG1 in their enterocytes suffer from a 30–35% reduction in circulating L-Arg concentration, retardation of hair and muscle growth, and irregular development of the lymphoid tissues. Whereas T cell cellularity is not affected, B cells undergo a maturation block at the transition from the pro- to pre-B cell stage in the bone marrow (de Jonge et al., 2002). As a result, the number of B lymphocytes is reduced in peripheral lymphoid organs and small intestine, serum IgM levels are decreased, and the architecture of lymphoid organs, especially Peyer's patches, is profoundly compromised. The reason for this selective effect on B lymphocyte maturation is not known, and models in which ARG1 expression is targeted to other immunological sites are required to understand the effects of more localized L-Arg consumption.

C. ARG and NOS Cooperation in Immunoregulation: An Emerging Concept

One interesting discovery made in the early twenty-first century is that ARG and NOS can cooperate to restrain T lymphocyte functions in tumor-bearing hosts by altering the production of reactive nitrogen species (RNS) and reactive oxygen species (ROS). The concept has been proposed that the dual coexpression of ARG and NOS might be a singular property of MDSCs, or at least a part of this heterogeneous population of myeloid suppressors is induced by tumors (Bronte and Zanovello, 2005). The molecular bases for the synergism between these enzymes are not entirely known. Actually, whether ARG and NOS are active in the same environment, let alone in the same cells, is still debated. A number of cross-regulatory circuits, in fact, have been described that might limit the activity of one enzyme when the other is functional (reviewed in Bronte and Zanovello, 2005). For example, NOHA, an intermediate during the biosynthesis of NO by NOS, is a physiologic inhibitor of ARG. Despite the plethora of regulatory pathways, there is no definitive conclusion on whether induction of ARG activity limits the availability of L-Arg as a substrate for NOS to the point of impairing NO production. IL-13 pretreatment of mouse peritoneal macrophages stimulated by IFN-γ and LPS caused downregulation of NOS2 protein (but not mRNA), a process dependent on L-Arg depletion by IL-13-induced ARG1 (El-Gayar et al., 2003). Low extracellular L-Arg concentration, overexpression of ARG, or a reduction of the capacity for L-Arg uptake can decrease intracellular L-Arg and halt the translation of NOS2 mRNA, a phenomenon known as the *arginine paradox* (Lee et al., 2003). However, in other experiments, ARG induction and depletion of cytosolic L-Arg content in macrophages did not affect their NO production (Fligger et al., 1999).

1. Peroxynitrite Generation

ARG coexpression could actually have an important supplementary effect (i.e., the possible production of superoxide [O_2^-] from the NOS2 reductase domain). It has been advanced that at low L-Arg environmental concentration, NOS2 function shifts from a prevalent NO production to O_2^- production (Xia and Zweier, 1997; Xia et al., 1998; Bronte et al., 2003). ARG1 might thus function as controller of this NOS2 conversion by decreasing L-Arg content in the microenvironment where T cells are activated, a hypothesis substantiated by a number of findings in mouse and human tumors (Bronte et al., 2003; Rodriguez et al., 2003, 2004; Bronte et al., 2005; De Santo et al., 2005; Zea et al., 2005). Once generated, O_2^- reacts immediately with residual NO, leading to formation of peroxynitrites (ONOO$^-$), highly reactive oxidizing agents that damage different biological targets (Schopfer et al., 2003; Radi, 2004). Peroxynitrites are no longer considered simply toxic byproducts but are currently viewed as potential intracellular and intercellular messengers because they can diffuse through cell membranes and induce post-translational protein modifications by nitrating protein-associated tyrosines (Schopfer et al., 2003; Radi, 2004). Nitrotyrosines are detected in apoptotic thymocytes, suggesting that ONOO$^-$ can be involved in thymic-dependent T cell education *in vivo* (Moulian et al., 2001). Nitrotyrosines are also found in tissue sections of hyperplastic human lymph nodes obtained from surgical resections of patients with lung and colon cancer (Brito et al., 1999) and in human prostate adenocarcinomas (Bronte et al., 2005).

Activated T lymphocytes are more prone to ONOO$^-$-induced death than resting T cells. Peroxynitrite-induced apoptosis of T cells activated by phytohemagglutinin (PHA) or anti-CD3 (but not by phorbol esters) likely requires inhibi-tion of protein tyrosine phosphorylation via nitration of tyrosine residues (Brito et al., 1999). At high doses, ONOO$^-$ can directly control cell death by nitration of the protein voltage-dependent anion channel, a component of the mitochondrial permeability transition pore (Brookes et al., 2000; Aulak et al., 2001). However, ONOO$^-$-induced inhibition of T lymphocytes is not invariably associated with cell apoptosis. In human prostate cancer, the neoplastic but not the normal prostate epithelium overexpressed ARG2 and NOS1, and addition of a combination of enzyme-specific inhibitors to prostate cultures was sufficient to decrease the nitrotyrosine content in tumor-infiltrating lymphocytes (TILs) and to restore their local responsiveness to various stimuli (Bronte et al., 2005). This is an important finding for the develop-ment of therapeutic interventions aimed at restoring lymphocyte responsiveness in tumor-bearing hosts.

2. Peroxynitrites and Autoimmunity

Paradoxically, findings in non-obese diabetic (NOD) mice also suggest that excessive nitrotyrosine formation might be linked to autoimmune disease. NOD mice develop progressive autoimmune diabetes that shares some similarity with the human disease. From 4 weeks of age, mononuclear cells start to infiltrate the islet of the pancreas (insulitis), causing a progressive destruction of insulin-secreting β cells, which ultimately leads to clinical signs of diabetes at 12 weeks of age. Macrophages are among the first cells to seed the pancreata and, together with autoreactive T lymphocytes, are considered the first offensive line for tissue destruction. Among various disorders described in NOD mice, some appear to be related to L-Arg and L-Trp metabolism, including the propensity of NOD mouse bone marrow to produce CD11b$^+$/Gr-1$^+$ (putative MDSCs) rather than DCs in response to GM-CSF (Morin et al., 2003), the overexpression of ARG during insulitis (Rothe et al., 2002), and the increased production of ONOO$^-$ by myeloid cells, including the cells infiltrating the pancreas (Suarez-Pinzon et al., 1997; Grohmann et al., 2003). These abnormalities have a direct link with the pathology. Administration of the ONOO$^-$ scavenger guanidinoethyldisulphide (GED) at age 5 weeks decreased the incidence of diabetes from 80 to 17% (Suarez-Pinzon et al., 2001). Peroxynitrites are also produced in response to

IFN-γ by CD8$^+$ DCs isolated from NOD mice at 4 but not at 8 weeks of age (Grohmann et al., 2003). CD8$^+$ DCs constitute a peculiar regulatory population that exploits L-Trp metabolism to induce tolerance. In normal mice, IFN-γ activates the enzyme IDO in CD8$^+$ DCs through a Stat1-dependent pathway (Grohmann et al., 2000, 2002). In NOD mice, Stat1 is nitrated by ONOO$^-$ and is not able to properly drive IDO expression in response to IFN-γ. The ONOO$^-$ inhibitor GED but not the NOS inhibitor L-NMMA restored the normal pathway, raising the question about the route of ONOO$^-$ production in these cells. Since L-NMMA does not reduce O$_2^-$ generation by the NOS2 reductase domain, an effect that can be obtained only by increasing L-Arg concentrations (Xia and Zweier, 1997; Xia et al., 1998), it would be interesting to examine the role of ARG-driven depletion of L-Arg in this model. However, it is clear that at an age (4 weeks) critical for shaping the repertoire of autoreactive T cells, DCs are impaired in their ability to activate IDO-dependent tolerogenic pathways.

These findings suggest that L-Arg and L-Trp metabolism cannot operate in the same cell due to the existence of cross-inhibitory circuits. IDO is inhibited by NO and ONOO$^-$, terminal effector molecules generated by L-Arg metabolism, whereas L-Trp cytosolic depletion by IDO blocks IFN-γ-dependent NOS2 expression in macrophages (Chiarugi et al., 2003; Grohmann et al., 2003; Hucke et al., 2004). These findings might help to explain the association between cancer and autoimmune paraneoplastic syndromes possibly characterized by an extensive production of peroxynitrites by myeloid cells (or tumors themselves) inhibiting the tolerogenic pathways and unleashing autoreactive lymphocytes.

3. Hydrogen Peroxide Generation

ARG activation was associated with hydrogen peroxide (H$_2$O$_2$) production by MDSCs isolated from the spleen of mice bearing a transplantable fibrosarcoma. In this model, MDSCs presented class I-restricted epitopes directly to CD8$^+$ T lymphocytes and caused the inhibition of their IFN-γ release through contact-dependent production of H$_2$O$_2$ (Kusmartsev et al., 2004). Although the molecular mechanisms are entirely unknown, the link between ARG and H$_2$O$_2$ has been confirmed in other tumor models (Kusmartsev et al., 2000; Sinha et al., 2005). ROS released from CD11b$^+$/Gr-1$^+$ cells were shown to alter CD3-ζ chain expression and function in mice (Otsuji et al., 1996) and in patients with advanced-stage cancer where H$_2$O$_2$ was found to be produced by an expanded pool of circulating, low-density granulocytes (Schmielau and Finn, 2001). These granulocytes are also positive for ARG1 (S. Mandruzzato, unpublished results, 2007). H$_2$O$_2$ originates from the conversion of superoxide in contact with protons in water according to the simple formula: $2\,O_2^- + 2H^+ \leftrightarrow H_2O_2 + O_2$ (Reth, 2002). This spontaneous reaction occurs when superoxide is produced by both oxygen-dependent and oxygen-independent pathways. H$_2$O$_2$ is electrically neutral and stable, and it diffuses across membranes, thus representing another potential intracellular and intercellular messenger (Reth, 2002). Analogously to ONOO$^-$, H$_2$O$_2$ can induce apoptosis of antigen-activated T lymphocytes although through different mechanisms (i.e., by downregulating intracellular Bcl-2 and increasing plasma membrane levels of Fas-L via NF-κB) (Hildeman et al., 2003). Human central memory and effector memory T cells (CD45RA$^-$CCR7$^+$ and CD45RA$^-$CCR7$^-$, respectively) are more sensitive to H$_2$O$_2$ than naive T cells and respond to lower concentrations of this oxidative molecule by undergoing mitochondrial and caspase-dependent apoptosis in the absence of stimulation (Pauleau et al., 2004). The association between H$_2$O$_2$ and ARG1, however, requires further studies since ARG1 was shown to be upregulated after treating porcine coronary arterioles with H$_2$O$_2$ through a process requiring the formation of hydroxyl radicals from H_2O_2 (Thengchaisri et al., 2006). This finding suggests that ARG1 activation might follow and not precede H_2O_2 generation.

V. POSSIBLE PHYSIOLOGICAL ROLE FOR L-ARG METABOLISM IN IMMUNITY CONTROL

Many of the data discussed in previous paragraphs were obtained in pathological settings. However, the impact of ROS and RNS on T lymphocyte functions can be important in controlling normal responses to antigen. Indeed, evidence points to a specific role in modulating the contraction of $CD8^+$ T cells during normal immune responses. The contraction phase follows the massive clonal expansion of naive $CD8^+$ T cells that encountered pathogen that was captured, processed, and presented by antigen presenting cells. During antigen-driven activation, quiescent cells are turned into effector T cells that secrete cytokines and eliminate infected host cells within a few weeks. In the contraction phase, the majority of antigen-specific effector cells die by apoptosis during the next 2–3 weeks, and only a minority of cells (~5–10%) survives as the long-lived memory cell population.

Memory T cells have an increased ability to survive *in vivo* compared with naive T cells; this ability arises from memory T cells being able to bind and respond to the homeostatic cytokine IL-7 (Schluns and Lefrancois, 2003). Selective expression of IL-7Rα on effector $CD8^+$ T cells identifies memory cell precursors. Whereas the molecular bases for the long-term survival of memory cells have been partially unveiled, the factor(s) regulating contraction of $CD8^+$ effectors still remains obscure. The contraction phase is not dependent on antigen clearance, as one can intuitively guess, but it appears to be an event programmed during the initial phases of infection (Badovinac et al., 2002, 2004). The contraction phase is impaired in mice lacking either IFN-γ or its receptor and requires the acquisition of suppressive functions by a population of $CD11b^+$ cells that counter-regulate $CD8^+$ T cells (Badovinac et al., 2002, 2004; Sercan et al., 2006). How $CD11b^+$ cells conditioned by IFN-γ contribute to the contraction phase is still not completely clear, but several lines of evidence point to RNS and ROS released by these cells as final mediators. T cell death during the contraction phase is also inhibited in mice lacking NOS2, resulting in a higher frequency of memory T cells following immunization (Vig et al., 2004). Further, $CD11b^+/Gr-1^+$ MDSCs activated by IFN-γ produce reactive species that then affect T cells during apoptosis induced by superantigens (Cauley et al., 2000). In agreement with a role for ROS/RNS in controlling the execution of the program leading to immunological memory, the $ONOO^-$ scavenger Mn(III)tetrakis(4-benzoic acid) porphyrin chloride (MnTBAP) enhanced memory responses by blocking postactivation death and contraction of $CD8^+$ T cells in mice immunized either with the antigen ovalbumin or infected with lymphocytic choriomeningitis virus (Laniewski and Grayson, 2004; Vig et al., 2004).

It must be pointed out, however, that activation also increases the amount of reactive species (H_2O_2, NO, and O_2^-) in the very same T lymphocytes (Hildeman et al., 1999; Devadas et al., 2002) and that ROS produced by T lymphocytes can regulate their apoptosis during the contraction phase, possibly via modulation of Bcl-2 and Fas-L (Hildeman et al., 2003). Interestingly enough, in one of these studies, MnTABP protected T lymphocytes from apoptotic death (Hildeman et al., 1999). Production of ROS/RNS might thus be the major rheostat regulating adaptive immune responses. It is not entirely clear which population is the most important contributor to ROS/RNS production: T lymphocytes (inner rheostat) or myeloid cells (external rheostat). For this reason, the use of inhibitors that can act both on T cells and myeloid cells, such as MnTABP, certainly cannot help to address this issue.

VI. NOS IN CANCER

Several investigators have reported the expression of NOS2 by malignant cells or in the tumor microenvironment, both at the mRNA and protein levels. In breast carcinoma, an initial study suggested that NOS2 activity is higher in less differentiated tumors (Thomsen et al., 1995) and was detected predominantly in tumor-infiltrating macrophages. Subsequently, other studies have demonstrated that NOS2 is expressed also by breast carcinoma cells, and it correlates with tumor stage and microvessel density (Reveneau et al., 1999; Vakkala et al., 2000a; Loibl et al., 2002, 2005).

In addition to breast cancer, NOS2 is markedly expressed in approximately 60% of human colon adenomas and in 20–25% of colon carcinomas, while expression is either low or absent in the surrounding normal tissues (Ambs et al., 1998b; Nosho et al., 2005). Similarly, in human ovarian cancer and melanoma, NOS2 activity was localized to tumor cells but not to normal ovarian tissue and melanocytes, respectively (Thomsen et al., 1994; Massi et al., 2001). Other conditions that express NOS2 are cancers of the head and neck (Park et al., 2003b), esophagus (Wilson et al., 1998), lung (Ambs et al., 1998a), prostate (Klotz et al., 1998), and bladder (Swana et al., 1999); pancreatic (Hajri et al., 1998) carcinomas and Kaposi's sarcoma (Weninger et al., 1998); brain tumors (Cobbs et al., 1995); mesothelioma (Marrogi et al., 2000); and hematological malignancies (Roman et al., 2000; Mendes et al., 2001). Regarding the other two NOS isoenzymes, in a series of 54 cases of breast carcinoma, NOS3 was expressed in 33 samples (61%), and its degree of expression strongly correlated with that of NOS2 (Loibl et al., 2002). In another series ($n = 80$), both NOS2 and NOS3 were found not only in the surrounding stroma but also in carcinoma cells (Vakkala et al., 2000b). Finally, NOS1 has been detected in some oligodendroglioma and neuroblastoma cell lines (Cobbs et al., 1995).

These findings, together with the observation that NOS (particularly NOS2) expression levels correlate with poor clinical outcome of patients affected with different types of cancer, such as melanoma (Ekmekcioglu et al., 2000) and breast (Loibl et al., 2005), ovarian (Raspollini et al., 2004), head and neck (Gallo et al., 1998), and colorectal (Nozoe et al., 2002) carcinomas, have led several investigators to consider NO as a potential mediator of tumor development and progression. Indeed, various RNS can damage DNA through a variety of different mechanisms (Felley-Bosco, 1998; Wink et al., 1998; Marnett et al., 2003). N_2O_3 is a strong nitrosating agent (which mediates the addition of NO^+ to nucleophiles) that can deaminate DNA bases (e.g., cytosine to uracil, guanine to xanthine), causing a variety of point mutations. N_2O_3 can also react with secondary amines to form carcinogenic N-nitrosamines, which can damage DNA by alkylation. Peroxynitrite, a strong nitrating species, can also react directly with DNA and lead to mutations and single-strand breaks (Routledge, 2000; Marnett et al., 2003). RNS-mediated DNA base damage can be measured as conversion of guanine into 8-nitroguanine and 8-oxoguanine. In humans, H. Pylori infection, which is recognized as an etiologic factor for human gastric lymphoma and carcinoma, has been associated with high levels of 8-oxoguanine in urine (Witherell et al., 1998) and gastric mucosa (Ma et al., 2004). Moreover, 8-oxoguanine levels are higher in gastric cancer rather than in normal surrounding tissues (Chang et al., 2004).

Nitrotyrosine is the footprint of $ONOO^-$-mediated nitration, and the presence of nitrotyrosine has been identified by immunohistochemistry in *H. pylori* gastritis (Kai et al., 2004) and several human cancers (e.g., cholangiocarcinoma) (Jaiswal et al., 2000), including pancreatic (Vickers et al., 1999; Bronte et al., 2005) and esophageal carcinomas (Kato et al., 2000). Similarly, patients with lung cancer not only exhale more NO but also have higher plasma levels of nitrated proteins compared to controls (Pignatelli et al., 2001; Kisley et al., 2002).

The production of NO in the tumor microenvironment is believed to promote tumor growth by stimulating angiogenesis (Bing et al., 2001; Morbidelli et al., 2004), a multistage process regulated by the release of angiogenic factors (e.g., vascular endothelial growth factor [VEGF], integrins, basic-fibroblast growth factor [b-FGF], and hypoxia-inducible factor 1 [HIF-1]) that are produced by malignant cells, stromal fibroblasts, and tumor-infiltrating macrophages. A functional NO/cGMP pathway is required for VEGF to promote angiogenesis (Ziche et al., 1997; Gallo et al., 1998). Accordingly, *in vitro* NOS inhibition decreases endothelial cell proliferation (Mocellin et al., 2004b), and L-Arg depletion inhibits endothelial cell tubularization and capillary formation (Park et al., 2003a). NO-dependent upregulation of VEGF, likely by mRNA stabilization (Chin et al., 1997), corresponds to an increased vascularization in xenograft tumors (Ambs et al., 1998). In addition, in human colon and gastric carcinoma specimens, NOS2 expression levels correlate with both VEGF levels and microvessel density (Cianchi et al., 2003; Ichinoe et al., 2004). Moreover, human carcinoma cells transfected with a murine NOS2 cDNA cassette show increased tumor growth and neovascularization in a nude mouse xenograft model (Thomsen and Miles, 1998).

Another mechanism by which NO may promote tumor growth is by modulating the production of prostaglandins (PGs). NO can activate COX-2 (Marrogi et al., 2000; Park et al., 2003b; Cianchi et al., 2004; Mollace et al., 2005) that, by generating PGs, promotes angiogenesis and inhibits apoptosis (Thun et al., 2002). HIF-1, an oxygen-regulated transcriptional activator that plays a central role in tumor angiogenesis, is known to increase the expression of several angiogenesis-related genes, including NOS2 (Semenza, 2003); in a positive feedback circuit, NO can activate HIF-1 (Kimura et al., 2000). Furthermore, since NO induces HIF-1 expression under nonhypoxic conditions (Kasuno et al., 2004), it has been postulated that high levels of NO in the tumor microenvironment might have the same effect as hypoxia in inducing angiogenesis (Semenza, 2001). It must be pointed out that hypoxia is also an important stimulus for ARG induction, and it might be the major factor responsible for the frequent coexpression of NOS2 and ARG2 in the tumor microenvironment (S. Mandruzzato, unpublished results, 2007).

NO may favor tumor progression also by enhancing tissue invasion (Lala and Chakraborty, 2001). Studies examining the signaling mechanisms underlying this phenomenon show that both tumor migration and endothelial cell migration are reduced by NOS inhibition/knockdown (Siegert et al., 2002; Jadeski et al., 2003; Lopez-Rivera et al., 2005). RNS appear to be involved in NO-mediated induction of matrix metalloproteinase (MMP) activation/overexpression (Okamoto et al., 2001; Ishii et al., 2003), a key feature of cancer invasiveness. Although some knockout models support these findings, others demonstrate that macrophage-derived NO can contribute to tumor growth control and that NO might be fundamental in tumor development but not in tumor progression (Table 20.1). These sometimes conflicting results indicate that the role of NOS2 during tumor development is highly complex and still not completely understood. Both promoting and deterring actions can be observed, presumably depending on the local concentration of NOS2 in the tumor microenvironment.

VII. ARG IN CANCER

Although ARG analysis in mouse and human cancers was initiated more recently in comparison to NOS, it appears that ARG1 and ARG2 are both found in the tumor microenvironment but are distributed in different compartments. From an overall scrutiny of the available literature, the emerging picture is that ARG1 is mainly expressed in tumor-infiltrating myeloid cells, whereas ARG2 is detected primarily in cancerous cells. Although many studies lack immunohistochemical evidence (a real limit in this field), the frequent finding of ARG in cancer suggests an important role for this enzyme in tumor biology and development. ARG-dependent tumor-promoting actions can range from proangiogenetic activity, lymphocyte suppression, assistance in tumor cell proliferation, and stroma remodeling, all properties that have been assigned to tumor-associated macrophages with an alternative activation profile (Mantovani et al., 2002; Balkwill et al., 2005). Several examples, both in preclinical tumor models and in clinical studies, emphasize this multifaceted ARG activity.

Intratumoral ARG induction can help growing tumors basically in two ways: by suppressing TIL response and by providing cancerous cells with polyamines. L-Orn produced by ARG is used by ODC to form polyamines, essential nutrients for mammalian cell proliferation and differentiation. In a mouse model, macrophages transfected with the rat *Arg1* gene enhanced *in vitro* and *in vivo* proliferation of cocultured tumor cells (Chang et al., 2001), and ARG-driven polyamine synthesis in macrophages was shown to enhance tumor vascularization (Davel et al., 2002). The *Arg1* gene was found, by a genome-wide approach, to be overexpressed in two murine cell lines that are highly metastatic and consistently elevated in pulmonary metastases compared to the primary tumor in two different mouse metastatic models, suggesting that ARG may also be important for tumor colonization of the lungs (Margalit et al., 2003). Additional proofs of the involvement of ARG1 in mechanisms of immune system suppression in cancer patients come from the work of Zea et al. (2005), showing that peripheral blood mononuclear cells from renal cell carcinoma patients have a significant increase in ARG activity. Moreover, they showed that this increase in activity was restricted to a specific subset of cells with polymorphonuclear granulocyte morphology and surface phenotype (Zea et al., 2005).

In the late 1990s and the early twenty-first century, a significant association between ARG and cancer has been documented in human tumors. Since polyamines are needed by rapidly dividing cells and tissues, the reaction catalyzed by ARG might meet the requirement for increased metabolism; in fact, increased ARG2 activity has been found in human cell lines (Buga et al., 1998; Singh et al., 2000) as well as in human breast (Porembska et al., 2003) and gastric cancers (Wu et al., 1996). High ARG activity has been described also in patients with various malignancies, including gastric, colon, breast, prostate, and lung cancers, either in the serum or in the tumor tissue. In these studies, ARG activity has been determined by enzymatic assays that do not discriminate between the different isoforms of the enzyme; therefore, it is not known whether ARG activity can be ascribed to ARG1 or ARG2. The increase in serum ARG activity has been associated with disease progression in colorectal and breast cancer, and, moreover, ARG activity has been proposed as a diagnostic tool to monitor patients with colorectal carcinoma (Suer Gokmen et al., 1999; Keskinege et al., 2001; del Ara et al., 2002; Porembska et al., 2002, 2003; Polat et al., 2003; Cederbaum et al., 2004; Mielczarek et al., 2006).

As already anticipated, the role of ARG2 is even more obscure. Using a genomic approach, *Arg2* was found to be differentially expressed between follicular thyroid carcinoma (FTC) and benign follicular

TABLE 20.1 Nitric Oxide and Cancer: Knockout Models[1]

Gene Experimental Model	Findings Comments/Interpretation Reference
NOS2 Animal: mice C57BL Tumor: B16 melanoma, M5076 ovarian sarcoma (syngeneic) NOS2$^{-/-}$: tumor regression (melanoma) or progression (sarcoma), ↓ VEGF, ↓ angiogenesis (melanoma) NOS2$^{+/+}$: macrophages killing sarcoma but not melanoma cells, sarcoma (but not melanoma) cells producing NO NOS2 is crucial in macrophage-mediated killing of sensitive tumors (sarcoma) and in cancer progression of resistant tumors (melanoma). Shi et al. (2000) Wang et al. (2001)	NOS2 Animal: mice C57BL Tumor: H7 pancreatic carcinoma (syngeneic) transfected with IFN-β NOS2-/-: ↑ tumor progression Antitumor effects of IFN-β are NO-dependent. Wang et al. (2001)
NOS2 Animal: mice B6/129P Tumor: urethane-induced lung cancer NOS2$^{-/-}$: ↓ tumor development, ↓ VEGF, ↓ angiogenesis; COX expression: unaffected by NOS2 NOS2 is important in lung cancerogenesis and favors tumor angiogenesis through VEGF (but not COX) overexpression. Kisley et al. (2002)	NOS2, p53 Animal: mice (C57BL) Tumor: spontaneously developing lymphoma (p53$^{-/-}$) and sarcoma (p53$^{+/-}$) NOS2$^{-/-}$: ↑ tumor development, ↑ p21/waf1, (↓ proliferation), ↑ TRAIL and FAS-L (↑ apoptosis) p53$^{+/+}$: ↓ NOS2 (transcriptional repression by p53) Functional NOS2 protects from tumorigenesis, and p53-knockout mice are an animal model of the cancer-prone human Li-Fraumeni syndrome. Hussain et al. (2004)
NOS2 Animal: APC$^{MIN/+}$ (polyposis-prone) mice Tumor: colon adenomas (colon carcinoma precursors) NOS2$^{-/-}$: ↓ tumor development NOS2$^{+/+}$ treated with NOS2-inhibitor (aminoguanidine): ↓ tumor development NOS2 favors colon tumorigenesis. Ahn and Ohshima (2001)	NOS2, p53 Animal: mice (C57BL) Tumor: spontaneously developing thymic and nonthymic lymphomas NOS2$^{-/-}$: ↓ thymic lymphoma development, ↑ nonthymic lymphoma development The microenvironment profoundly affects the relationship between NO and lymphomagenesis. Tatemichi et al., 2004.
NOS2 Animal: APC$^{MIN/+}$ (polyposis-prone) mice Tumor: colon adenomas (colon carcinoma precursors) NOS2$^{-/-}$: ↑ tumor development COX expression: unaffected by NOS2 NOS2 plays a protective role with respect to colon tumorigenesis. Scott et al. (2001)	eNOS Animal: mice C57BL Tumor: B16F1 melanoma (syngeneic) eNOS$^{-/-}$ (and L-NAME treatment): ↓ lung metastases Endothelial cells have a role in controlling tumor metastasis. Qiu et al. (2003)
NOS2 Animal: mice (C57BL) Tumor: NOS2$^{-/-}$ MDA-induced fibrosarcoma cell lines injected subcutaneously and intravenously NOS2$^{-/-}$: ↑ tumor progression NOS2$^{+/+}$: NOS2 expression in tumor nodules by stromal cells Physiological expression of NOS2 in host cells (mainly macrophages) inhibits tumor growth and metastasis. Wei et al. (2003)	eNOS Animal: C57BL Tumor: Lewis lung carcinoma eNOS$^{-/-}$: ↓ antitumor activity of cavtratin (antiangiogenic agent) Cavtratin antitumor effects depend upon eNOS inhibition. Gratton et al. (2003)
NOS2 Animal: mice (C57BL) Tumor: polyomavirus middle T antigen-mediated breast cancerogenesis NOS2$^{-/-}$: ↓ tumor development, no effect on tumor progression (angiogenesis, metastatization rate) NOS2 plays an important role only in the early events of breast tumorigenesis. Ellies et al. (2003)	p53 Animal: athymic mice Tumor: p53$^{-/-}$ human colon carcinoma transfected with NOS2 p53$^{-/-}$: ↓ antitumor activity of radiotherapy NO and ionizing radiation act synergistically to induce p53-dependent apoptosis. Cook et al. (2004)
	IFN-γ Animal: mice C57BL Tumor: H7 pancreatic carcinoma (syngeneic) IFN-γ$^{-/-}$: ↑ tumor progression, ↓ NOS2 expression IFN-γ secretion by host cells mediates tumor killing via NOS2 expression. Wang et al. (2001)

[1]Adapted from Mocellin et al., 2006.

thyroid adenoma (FTA), with an average increase in expression of at least fivefold in nearly all the FTCs tested (Cerutti et al., 2006). Moreover, in this study, Arg2 was one of the four genes that was found to be statistically the most consistent marker for FTC.

Deregulated ARG expression and activity have been associated with a number of pathological conditions other than cancer. One of the most interesting developments concerning L-Arg homeostasis is the discovery of an increase in the expression and activity of ARG in many tissues and cell types in response to a variety of cytokines, in particular those that are Th2 derived. Increased ARG activity has been associated with several inflammatory conditions, such as asthma, arthritis, hyperoxic lung injury, psoriasis, and autoimmune disease (Meurs et al., 2003).

VIII. ARG AND NOS INHIBITORS: A NOVEL CLASS OF IMMUNE ADJUVANTS?

Although many cancers arise *de novo* without an identifiable predisposing disease, chronic inflammation is strongly suspected to favor tumor development (Balkwill and Mantovani, 2001; Coussens and Werb, 2002; Hussain et al., 2003; Clevers, 2004). Several *in vitro* models have shown mutagenic and carcinogenic effects of ROS and RNS produced by activated immune cells (e.g., macrophages, neutrophils, eosinophils) expressing enzymes such as myeloperoxidase, eosinophil peroxidase, and NOS2 (Ohshima et al., 2003). The epidemiological association between inflammatory diseases (e.g., chronic gastritis, bowel inflammatory disease, interstitial lung disease) and gastrointestinal or lung cancers further supports this pathogenetic hypothesis (Macarthur et al., 2004). These findings—together with the promising results obtained with nonsteroidal anti-inflammatory drugs (NSAIDs) in colon cancer chemoprevention (Sandler et al., 2003)—strengthen the notion that inflammation and cancerogenesis share some metabolic pathways, such as those generating high levels of ROS and RNS (Hussain et al., 2000; Ricchi et al., 2003). Because NOS2 overexpression and NO production are cardinal features of inflamed tissues (Lundberg et al., 1997; Coleman, 2001; Laroux et al., 2001; Guzik et al., 2003), these observations suggest that NO may play a fundamental role in the initiation of cancer arising within a background of chronic inflammation, thereby supporting the rationale for the use of anti-NO agents as antineoplastic drugs (Rao, 2004). A similar role for ARG is suspected but not proven due to the absence of proper knockout mice.

Selective NOS and ARG inhibitors or molecules interfering with the generation of RNS/ROS from the combined activities of NOS/ARG have proven beneficial in controlling myeloid-dependent suppression *in vitro* (Bronte et al., 2003; Rodriguez et al., 2004; De Santo et al., 2005; Kusmartsev and Gabrilovich, 2005; Sinha et al., 2005). However, several considerations have motivated a recent interest in dual or multiple inhibitors. Many of the single inhibitors are not selective, are plagued by important side effects when administered *in vivo*, and are difficult to bring to the clinic. Moreover, it is important to note that inhibiting only one arm of L-Arg metabolic pathways might not be sufficient to exert a therapeutic effect in all the different tumors. In addition, the contemporaneous biosynthesis of NO and PGs in various tissues and cells has received much attention since the turn of the twenty-first century because of the possible involvement of the metabolites of these pathways in pathophysiological mechanisms of various chronic inflammatory disorders, including cancer. In fact, increasing evidence suggests that there is a substantial cross talk between NO and PG biosynthetic circuits that involve both negative and positive feedback loops mediated by reaction end products, including the very same NO,

PGs, and other cyclic nucleotides. This strict link was further affirmed by the demonstration that NOS2 binds specifically to COX-2 and activates it by S-nitrosylation of critical cysteine residues (Kim *et al.*, 2005). Together, these results suggest that drugs blocking NOS2–COX-2 activities and their interaction might have a therapeutic benefit greater than that of single inhibitors.

The idea of coupling aspirin and a NO-donating moiety in the same structure was originally inspired by the need for reducing the side effects of the former (Wallace *et al.*, 2002). NO-donating aspirin was shown to possess both antiproliferative and cytocidal effects on colon cancer cell lines depending on the concentration tested (Tesei *et al.*, 2003). *In vivo*, NO-donating aspirin was also able to reduce gastrointestinal tumorigenesis in transgenic *Min* mice, chemically induced colon carcinogenesis in rats, and pancreatic cancer development in hamsters (Williams *et al.*, 2004; Ouyang *et al.*, 2006; Rao *et al.*, 2006). Interestingly, NO-donating aspirin was shown to inhibit NOS and ARG activity (both *in vitro* and *in vivo*), suppress generation of RNS/ROS at the tumor site, and abrogate myeloid cell-dependent suppression *in vivo*. In some mouse tumor models, NO-aspirin did not show direct antitumor activity but synergized with a recombinant cancer vaccine in inducing prevention and even treatment of established tumors (De Santo *et al.*, 2005). One of the chapter authors recently found that NO-aspirin might also increase the efficacy of tumor antigen-specific CD8$^+$ T cells adoptively transferred to tumor-bearing mice (V. B., unpublished, 2006). NO-releasing drugs might thus represent a potent and completely novel immune adjuvant, acting by relieving the suppressive mechanisms negatively affecting T lymphocytes in tumor-bearing hosts.

Prostate cancer represents an interesting target for NO-donating drug therapy for several reasons. Prostate inflammation and prostate cancer development have been related to each other (Nelson *et al.*, 2004). Moreover, prostate-specific antigens (PSAs) have been identified and represent suitable targets for immunotherapy, but, unfortunately, T lymphocytes infiltrating human prostate cancers are rendered unresponsive in the tumor site by the combined expression of ARG and NOS (Bronte *et al.*, 2005). In fact, prostate cancer is associated with one of the many tumors that overexpresses both ARG and NOS within the cancerous cells. This inhibitory mechanism, however, can be corrected *in vitro* by drugs interfering with ARG and NOS, thus allowing the full restoration of cytotoxic activity of T lymphocytes against the prostate cancer cells (Bronte *et al.*, 2005). Initial studies indicated that NO-donating aspirin and derivates could exert the same effect in human prostate cancer cultures (V. B., unpublished, 2006).

The mechanism of NO-donating aspirin action as a modulator of ARG and NOS activity has not yet been completely clarified. Negative feedback inhibition of both NOS enzyme activity and expression by NO-aspirin that has been demonstrated previously (Mariotto *et al.*, 1995) likely depends on the NO donation because this negative feedback was not observed after treatment with a dinitrated derivative that lacks just the nitroester group on the linker and, therefore, is unable to deliver NO. ARG inhibition, on the other hand, could be attributed to the aspirin and aromatic spacer portion of the compound. NO-donating aspirin inhibited ARG enzymatic activity *in vitro*, but the IC$_{50}$ was greater than tenfold the dose administered to tumor-bearing mice; a dose too high to hypothesize an *in vivo* direct activity of this salicylic portion of the NO-donating aspirin on ARG enzyme. Perhaps more likely is that salicylates operate indirectly by inhibiting the intracellular Stat6-mediated signals triggered by some ARG inducers, such as IL-4 and IL-13 (Perez *et al.*, 2002). In recent work, we have demonstrated that expression of IL-4Rα and the binding of IL-13 are critically required for ARG induction and suppressive activity of splenic and tumor-infiltrating MDSCs (Gallina *et al.*, in press).

As discussed earlier, regulation of ARG at the tumor site is likely to be more composite because other factors, such as hypoxia and cyclooxygenase activity, might sustain the enzyme activity. In fact, evidence suggests that ARG expression and function in tumor-infiltrating myeloid cells might be under the control of COX-2, an enzyme that is overexpressed in different human and mouse tumors. In a mouse model of lung cancer, signaling through the PGE_2 receptor in tumor-infiltrating myeloid cells was necessary for ARG induction, and pharmacological interference by COX-2 inhibitors resulted in ARG downregulation and stimulation of an otherwise silent lymphocyte-mediated antitumor response (Rodriguez *et al.*, 2005). NO-aspirin might thus act on COX-2 *in vivo*, but this hypothesis needs further testing since NO-donating COX-2 inhibitors were less effective than NO-donating aspirin or not effective at all in the chapter authors' *in vitro* screening (V. B., unpublished). It must be pointed out that one property common to different compounds altering the myeloid-induced lymphocyte dysfunction is the capability to block the production of RNS/ROS (De Santo *et al.*, 2005).

Given the ability of PDE5 inhibitors (sildenafil, tadalafil, and vardenafil) to inhibit ARG1 and NOS activity in MDSCs, these compounds were tested as potential adjuvants for immunotherapy. PDE5 inhibition reverted tumor-induced immunosuppressive mechanisms, restored immune surveillance in mice, and allowed the spontaneous generation of a measurable antitumor immune response that significantly delayed tumor progression in the absence of any immunotherapeutic approach. Moreover, by removing MDSC-dependent suppression, PDE5 inhibitors enhanced intratumoral T cell infiltration and activation and improved the antitumor efficacy of adoptive T cell therapy (P. Serafini *et al.*, submitted). When tested *in vitro*, PDE5 inhibitors restored T cell proliferation of PBMCs from multiple myeloma and head and neck cancer patients. These results have a potential clinical translation since PDE5 inhibitors are safe and effective agents already in the clinic.

IX. CONCLUSION AND PERSPECTIVES

As discussed in different sections of this chapter, several issues need to be investigated to clarify further the role of the L-Arg metabolizing enzyme in the immune dysfunctions found in tumor-bearing hosts. The major experimental limitation, at this stage, is the lack of suitable models of ARG deficiency in cells of the immune system that would conclusively confirm the data obtained with small inhibitors, which of course might have effects on other intracellular pathways. However, the initial observations discussed in this chapter contributed to defining the potential benefit of modulating L-Arg metabolism in cancer patients. The use of novel adjuvants is perceived as a major improvement in the field of cancer immunotherapy. Active and passive immunotherapies have been exploited as a rational approach based on the exquisite selectivity of the immune system against tumor-associated antigens. An unprecedented number of clinical trials have definitely shown that T and B lymphocytes that recognize tumors can be activated, both *in vitro* and *in vivo*, in patients bearing solid and hematological tumors. Unfortunately, the number of objective clinical responses remains unsatisfactory (Mocellin *et al.*, 2004a,c; Rosenberg *et al.*, 2004).

Immunotherapy adjuvants can be defined not only as those molecules affecting the priming of the immune response (i.e., the classic meaning of the word *adjuvant*) but, in a broader sense, all the substances and treatments that enhance the efficacy of immunotherapy. Many such adjuvants are being developed to relieve the restraints on the antitumor immune response. Molecules affecting the metabolism of either L-Arg or L-Trp amino acid certainly belong to this category, and they could be tested in clinical trials. The chapter authors anticipate that the impact of these combined approaches (i.e., the combination of novel adjuvants with immunotherapy) will be quite significant. One dramatic example is offered by the demonstration that lymphodepletion followed by adoptive cell transfer of tumor-specific T lymphocytes can induce about 50% clinical response in melanoma patients, likely dependent on the elimination of T regulatory lymphocytes and/or interference with negative regulatory signals to T cells (Dudley et al., 2005). These findings offer a harbinger of the impact that reversing immunosuppression in cancer may ultimately exert.

Acknowledgments

This work has been supported by grants from the Italian Ministry of Health (Ricerca Finalizzata); FIRB-MIUR (project # RBAU01935A); MIUR-CNR (progetto strategico oncologia); Fondazione Cariverona, Bando 2004 "Integrazione tra tecnologia e sviluppo di settore—Bando per progetti di ricerca a indirizzo biomedico"; and Italian Association for Cancer Research (AIRC).

Disclaimer

The opinions expressed by the authors in this article do not necessarily reflect the opinions of the companies with which the authors have collaborated in the past. The use of trade names is for identification only and does not constitute endorsement by the authors or their institutions. The authors declare no conflicting interests.

References

Ahn, B., and Ohshima, H. (2001). Suppression of intestinal polyposis in Apc(Min/+) mice by inhibiting nitric oxide production. *Cancer Res.* **61**, 8357–8360.

Albina, J. E., Mahoney, E. J., Daley, J. M., Wesche, D. E., Morris, S. M., Jr., and Reichner, J. S. (2005). Macrophage arginase regulation by CCAAT/enhancer-binding protein beta. *Shock* **23**, 168–172.

Alderton, W. K., Cooper C. E., and Knowles R. G. (2001). Nitric oxide synthases: Structure, function and inhibition. *Biochem. J.* **357**, 593–615.

Ambs, S., Bennett, W. P., Merriam, W. G., Ogunfusika, M. O., Oser, S. M., Khan, M. A., Jones R. T., and Harris C. C. (1998a). Vascular endothelial growth factor and nitric oxide synthase expression in human lung cancer and the relation to p53. *Br. J. Cancer* **78**, 233–239.

Ambs, S., Merriam, W. G., Bennett, W. P., Felley-Bosco, E., Ogunfusika, M. O., Oser, S. M., Klein, S., Shields, P. G., Billiar, T. R., and Harris, C. C. (1998b). Frequent nitric oxide synthase-2 expression in human colon adenomas: Implication for tumor angiogenesis and colon cancer progression. *Cancer Res.* **58**, 334–341.

Ambs, S., Merriam, W. G., Ogunfusika, M. O., Bennett, W. P., Ishibe, N., Hussain, S. P., Tzeng, E. E., Geller, D. A., Billiar, T. R., and Harris, C. C. (1998c). p53 and vascular endothelial growth factor regulate tumor growth of NOS2-expressing human carcinoma cells. *Nat. Med.* **4**, 1371–1376.

Ambs, S., Ogunfusika, M. O., Merriam, W. G., Bennett, W. P., Billiar, T. R., and Harris, C. C. (1998d). Upregulation of inducible nitric oxide synthase expression in cancer-prone p53 knockout mice. *Proc. Natl. Acad. Sci. USA* **95**, 8823–8828.

Ash, D. E. (2004). Structure and function of arginases. *J. Nutr.* **134**, 2760S–2764S; discussion 2765S–2767S.

Aulak, K. S., Miyagi, M., Yan, L., West, K. A., Massillon, D., Crabb, J. W., and Stuehr, D. J. (2001). Proteomic method identifies proteins nitrated *in vivo* during inflammatory challenge. *Proc. Natl. Acad. Sci. USA* **98**, 12056–12061.

Badovinac, V. P., Porter, B. B., and Harty, J. T. (2002). Programmed contraction of CD8(+) T cells after infection. *Nat. Immunol.* **3**, 619–626.

Badovinac, V. P., Porter, B. B., and Harty, J. T. (2004). CD8+ T cell contraction is controlled by early inflammation. *Nat. Immunol.* **5**, 809–817.

Balkwill, F., K. Charles, A., and Mantovani, A. (2005). Smoldering and polarized inflammation in the initiation and promotion of malignant disease. *Cancer Cell* **7**, 211–217.

Balkwill, F., and Mantovani, A. (2001). Inflammation and cancer: Back to Virchow? *Lancet* **357**, 539–545.

Baniyash, M. (2004). TCR zeta-chain downregulation: Curtailing an excessive inflammatory immune response. *Nat. Rev. Immunol.* **4**, 675–687.

Barbul, A. (1990). Arginine and immune function. *Nutrition* **6**, 53–58; discussion 59–62.

Bauer, H., Jung, T., Tsikas, D., Stichtenoth, D. O., Frolich, J. C., and Neumann, C. (1997). Nitric oxide inhibits the secretion of T-helper 1- and T-helper

2-associated cytokines in activated human T cells. *Immunology* **90**, 205–211.

Bauer, P. M., Fukuto, J. M., Buga, G. M., Pegg, A. E., and Ignarro, L. J. (1999). Nitric oxide inhibits ornithine decarboxylase by S-nitrosylation. *Biochem. Biophys. Res. Commun.* **262**, 355–358.

Bing, R. J., Miyataka, M., Rich, K. A., Hanson, N., Wang, X., Slosser H. D., and Shi, S. R. (2001). Nitric oxide, prostanoids, cyclooxygenase, and angiogenesis in colon and breast cancer. *Clin. Cancer Res.* **7**, 3385–3392.

Bingisser, R. M., Tilbrook, P. A., Holt P. G., and Kees, U. R. (1998). Macrophage-derived nitric oxide regulates T cell activation via reversible disruption of the Jak3/STAT5 signaling pathway. *J. Immunol.* **160**, 5729–5734.

Blachier, F., Mignon, A., and Soubrane, O. (1997). Polyamines inhibit lipopolysaccharide-induced nitric oxide synthase activity in rat liver cytosol. *Nitric Oxide* **1**, 268–272.

Bobe, P., Benihoud, K., Grandjon, D., Opolon, P., Pritchard, L. L., and Huchet, R. (1999). Nitric oxide mediation of active immunosuppression associated with graft-versus-host reaction. *Blood* **94**, 1028–1037.

Bogdan, C. (2001). Nitric oxide and the immune response. *Nat. Immunol.* **2**, 907–916.

Bogdan, C., Rollinghoff, M., and Diefenbach, A. (2000a). Reactive oxygen and reactive nitrogen intermediates in innate and specific immunity. *Curr. Opin. Immunol.* **12**, 64–76.

Bogdan, C., Rollinghoff, M., and Diefenbach, A. (2000b). The role of nitric oxide in innate immunity. *Immunol. Rev.* **173**, 17–26.

Boutard, V., Havouis, R., Fouqueray, B., Philippe, C., Moulinoux, J. P., and, Baud, L. (1995). Transforming growth factor-beta stimulates arginase activity in macrophages. Implications for the regulation of macrophage cytotoxicity. *J. Immunol.* **155**, 2077–2084.

Brito, C., Naviliat, M., Tiscornia, A. C., Vuillier, F., Gualco, G., Dighiero, G., Radi, R., and Cayota, A. M. (1999). Peroxynitrite inhibits T lymphocyte activation and proliferation by promoting impairment of tyrosine phosphorylation and peroxynitrite- driven apoptotic death. *J. Immunol.* **162**, 3356–3366.

Bronte, V., Kasic, T., Gri, G., Gallana, K., Borsellino, G., Marigo, I., Battistini, L., Iafrate, M., Prayer-Galetti, T., Pagano, F., *et al.* (2005). Boosting anti-tumor responses of T lymphocytes infiltrating human prostate cancers. *J. Exp. Med.* **201**, 1257–1268.

Bronte, V., Serafini, P., De Santo, C., Marigo, I., Tosello, V., Mazzoni, A., Segal, D. M., Staib, C., Lowel, M., Sutter, G., *et al.* (2003). IL-4-induced arginase 1 suppresses alloreactive T cells in tumor-bearing mice. *J. Immunol.* **170**, 270–278.

Bronte, V., and Zanovello, P. (2005). Regulation of immune responses by L-arginine metabolism. *Nature Rev. Immunol.* **5**, 641–654.

Brookes, P. S., Salinas, E. P., Darley-Usmar, K., Eiserich, J. P., Freeman, B. A., Darley-Usmar, V. M., and Anderson, P. G. (2000). Concentration-dependent effects of nitric oxide on mitochondrial permeability transition and cytochrome c release. *J. Biol. Chem.* **275**, 20474–20479.

Buga, G. M., Wei, L. H., Bauer, P. M., Fukuto, J. M., and Ignarro, L. J. (1998). NG-hydroxy-L-arginine and nitric oxide inhibit Caco-2 tumor cell proliferation by distinct mechanisms. *Am. J. Physiol.* **275**, R1256–1264.

Cauley, L. S., Miller, E. E., Yen, M., and Swain, S. L. (2000). Superantigen-induced CD4 T cell tolerance mediated by myeloid cells and IFN-gamma. *J. Immunol.* **165**, 6056–6066.

Cederbaum, S. D., Yu, H., Grody, W. W., Kern, R. M., Yoo, P., and Iyer, R. K. (2004). Arginases I and II: Do their functions overlap? *Mol Genet Metab* **81 Suppl 1**, S38–44.

Cerutti, J. M., Latini, F. R., Nakabashi, C., Delcelo, R., Andrade, V. P., Amadei, M. J., Maciel, R. M., Hojaij, F. C., Hollis, D., Shoemaker, J., *et al.* (2006). Diagnosis of suspicious thyroid nodules using four protein biomarkers. *Clin. Cancer Res.* **12**, 3311–3318.

Chang, C. I., Liao, J. C., and Kuo, L. (2001). Macrophage arginase promotes tumor cell growth and suppresses nitric oxide-mediated tumor cytotoxicity. *Cancer Res.* **61**, 1100–1106.

Chang, C. S., Chen, W. N., Lin, H. H., Wu, C. C., and Wang, C. J. (2004). Increased oxidative DNA damage, inducible nitric oxide synthase, nuclear factor kappaB expression and enhanced antiapoptosis-related proteins in *Helicobacter pylori*-infected non-cardiac gastric adenocarcinoma. *World J. Gastroenterol.* **10**, 2232–2240.

Chen, Y. H., Layne, M. D., Chung, S. W., Ejima, K., Baron, R. M., Yet, S. F., and Perrella, M. A. (2003). Elk-3 is a transcriptional repressor of nitric-oxide synthase 2. *J. Biol. Chem.* **278**, 39572–39577.

Chiarugi, A., Rovida, E., Dello Sbarba, P., and Moroni, F. (2003). Tryptophan availability selectively limits NO-synthase induction in macrophages. *J. Leukoc. Biol.* **73**, 172–177.

Chin, K., Kurashima, Y., Ogura, T., Tajiri, H., Yoshida, S., and Esumi, H. (1997). Induction of vascular endothelial growth factor by nitric oxide in human glioblastoma and hepatocellular carcinoma cells. *Oncogene* **15**, 437–442.

Cianchi, F., Cortesini, C., Fantappie, O., Messerini, L., Sardi, I., Lasagna, N., Perna, F., Fabbroni, V., Di Felice, A., Perigli, G., *et al.* (2004). Cyclooxygenase-2 activation mediates the proangiogenic effect of nitric oxide in colorectal cancer. *Clin. Cancer Res.* **10**, 2694–2704.

Cianchi, F., Cortesini, C., Fantappie, O., Messerini, L., Schiavone, N., Vannacci, A., Nistri, S., Sardi, I., Baroni, G., Marzocca, C., *et al.* (2003). Inducible nitric oxide synthase expression in human colorectal cancer: Correlation with tumor angiogenesis. *Am. J. Pathol.* **162**, 793–801.

Clevers, H. (2004). At the crossroads of inflammation and cancer. *Cell* **118**, 671–674.

Cobbs, C. S., Brenman, J. E., Aldape, K. D., Bredt, D. S., and Israel, M. A. (1995). Expression of nitric oxide synthase in human central nervous system tumors. *Cancer Res.* **55**, 727–730.

Coleman, J. W. (2001). Nitric oxide in immunity and inflammation. *Int. Immunopharmacol.* **1**, 1397–1406.

Cook, T., Wang, Z., Alber, S., Liu, K., Watkins, S. C., Vodovotz, Y., Billiar, T. R., and Blumberg, D. (2004). Nitric oxide and ionizing radiation synergistically promote apoptosis and growth inhibition of cancer by activating p53. *Cancer Res.* **64**, 8015–8021.

Coussens, L. M., and Werb, Z. (2002). Inflammation and cancer. *Nature* **420**, 860–867.

Davel, L. E., Jasnis, M. A, de la Torre, E., Gotoh, T., Diament, M., Magenta, G., Sacerdote de Lustig, E., and Sales, M. E. (2002). Arginine metabolic pathways involved in the modulation of tumor-induced angiogenesis by macrophages. *FEBS Lett.* **532**, 216–220.

de Jonge, W. J., Kwikkers, K. L., te Velde, A. A., van Deventer, S. J., Nolte, M. A., Mebius, R. E., Ruijter, J. M., Lamers, M. C., and Lamers, W. H. (2002). Arginine deficiency affects early B cell maturation and lymphoid organ development in transgenic mice. *J. Clin. Invest.* **110**, 1539–1548.

De Santo, C., Serafini, P., Marigo, I., Dolcetti, L., Bolla, M., Del Soldato, P., Melani, C., Guiducci, C., Colombo, M. P., Iezzi, M., et al. (2005). Nitroaspirin corrects immune dysfunction in tumor-bearing hosts and promotes tumor eradication by cancer vaccination. *Proc. Natl. Acad. Sci. USA* **102**, 4185–4190.

del Ara, R. M., Gonzalez-Polo, R. A., A. Caro, del Amo, E., Palomo, L., Hernandez, E., Soler, G., and Fuentes, J. M. (2002). Diagnostic performance of arginase activity in colorectal cancer. *Clin. Exp. Med.* **2**, 53–57.

Devadas, S., Zaritskaya, L., Rhee, S. G., Oberley, L., and Williams, M. S. (2002). Discrete generation of superoxide and hydrogen peroxide by T cell receptor stimulation: Selective regulation of mitogen-activated protein kinase activation and fas ligand expression. *J. Exp. Med.* **195**, 59–70.

Dimmeler, S., Fleming, I., Fisslthaler, B., Hermann, C., Busse, R., and Zeiher, A. M. (1999). Activation of nitric oxide synthase in endothelial cells by Akt-dependent phosphorylation. *Nature* **399**, 601–605.

Dudley, M. E., Wunderlich, J. R., Yang, J. C., Sherry, R. M., Topalian, S. L., Restifo, N. P., Royal, R. E., Kammula, U., White, D. E., Mavroukakis, S. A., et al. (2005). Adoptive cell transfer therapy following non-myeloablative but lymphodepleting chemotherapy for the treatment of patients with refractory metastatic melanoma. *J. Clin. Oncol.* **23**, 2346–2357.

Duhe, R. J., Evans, G. A., Erwin, R. A., Kirken, R. A., Cox, G. W., and Farrar, W. L. (1998). Nitric oxide and thiol redox regulation of Janus kinase activity. *Proc. Natl. Acad. Sci. USA* **95**, 126–131.

Ekmekcioglu, S., Ellerhorst, J., Smid, C. M., Prieto, V. G., Munsell, M., Buzaid, A. C., and Grimm, E. A. (2000). Inducible nitric oxide synthase and nitrotyrosine in human metastatic melanoma tumors correlate with poor survival. *Clin. Cancer Res.* **6**, 4768–4775.

El-Gayar, S., Thuring-Nahler, H., Pfeilschifter, J., Rollinghoff, M., and Bogdan, C. (2003). Translational control of inducible nitric oxide synthase by IL-13 and arginine availability in inflammatory macrophages. *J. Immunol.* **171**, 4561–4568.

Ellies, L. G., Fishman, M., Hardison, J., Kleeman, J., Maglione, J. E., Manner, C. K., Cardiff, R. D., and MacLeod, C. L. (2003). Mammary tumor latency is increased in mice lacking the inducible nitric oxide synthase. *Int. J. Cancer* **106**, 1–7.

Erdely, A., Kepka-Lenhart, D., Clark, M., Zeidler-Erdely, P., Poljakovic, M., Calhoun, W. J., and Morris, S. M., Jr. (2006). Inhibition of phosphodiesterase 4 amplifies cytokine-dependent induction of arginase in macrophages. *Am. J. Physiol. Lung Cell Mol. Physiol.* **290**, L534–539.

Felley-Bosco, E. (1998). Role of nitric oxide in genotoxicity: Implication for carcinogenesis. *Cancer Metastasis Rev.* **17**, 25–37.

Fischer, T. A., Palmetshofer, A., Gambaryan, S., Butt, E., Jassoy, C., Walter, U., Sopper, S., and Lohmann, S. M. (2001). Activation of cGMP-dependent protein kinase Ibeta inhibits interleukin 2 release and proliferation of T cell receptor-stimulated human peripheral T cells. *J. Biol. Chem.* **276**, 5967–5974.

Fligger, J., Blum, J., and Jungi, T. W. (1999). Induction of intracellular arginase activity does not diminish the capacity of macrophages to produce nitric oxide in vitro. *Immunobiology* **200**, 169–186.

Forstermann, U., Boissel, J. P., and Kleinert, H. (1998). Expressional control of the "constitutive" isoforms of nitric oxide synthase (NOS I and NOS III). *FASEB J.* **12**, 773–790.

Gallina, G., Dolcetti, L., Serafini, P., Desanto, C., Marigo, I., Colombo, M. P., Basio, G., Brombacher, F., Borrello, I., Zanocello, P., Bicciato, S., and Bronte, V. (2006). Tumors induce a subset of inflammatory monocytes with immunosuppressive activity on CD8 T cells. *J. Clin. Invest.* **10**, 2777–2790.

Gallo, O., Masini, E., Morbidelli, L., Franchi, A., Fini-Storchi, I., Vergari, W. A., and Ziche, M. (1998). Role of nitric oxide in angiogenesis and tumor progression in head and neck cancer. *J. Natl. Cancer Inst.* **90**, 587–596.

Ganster, R. W., Taylor, B. S., Shao, L., and Geller, D. A. (2001). Complex regulation of human inducible nitric oxide synthase gene transcription by Stat 1 and NF-kappa B. *Proc. Natl. Acad. Sci. USA* **98**, 8638–8643.

Gobert, A. P., Cheng, Y., Wang, J. Y., Boucher, J. L., Iyer, R. K., Cederbaum, S. D., Casero, R. A., Jr., Newton, J. C., and Wilson, K. T. (2002). Helicobacter pylori induces macrophage apoptosis by activation of arginase II. *J. Immunol.* **168**, 4692–4700.

Gobert, A. P., McGee, D. J., Akhtar, M., Mendz, G. L., Newton, J. C., Cheng, Y., Mobley, H. L., and Wilson, K. T. (2001). Helicobacter pylori arginase inhibits nitric oxide production by eukaryotic cells: A strategy for bacterial survival. *Proc. Natl. Acad. Sci. USA* **98**, 13844–13849.

Goni, O., Alcaide, P., and Fresno, M. (2002). Immunosuppression during acute Trypanosoma cruzi infection: Involvement of Ly6G (Gr1(+))CD11b(+) immature myeloid suppressor cells. *Int. Immunol.* **14**, 1125–1134.

Gordon, S. (2003). Alternative activation of macrophages. *Nat. Rev. Immunol.* **3**, 23–35.

Gotoh, T., Chowdhury, S., Takiguchi, M., and Mori, M. (1997). The glucocorticoid-responsive gene cascade. Activation of the rat arginase gene through induction of C/EBPbeta. *J. Biol. Chem.* **272**, 3694–3698.

Grandvaux, N., Gaboriau, F., Harris, J., tenOever, B. R., Lin, R., and Hiscott, J. (2005). Regulation of arginase II by interferon regulatory factor 3 and the involvement of polyamines in the antiviral response. *FEBS J.* **272**, 3120–3131.

Gratton, J. P., Lin, M. I., Yu, J., Weiss, E. D., Jiang, Z. L., Fairchild, T. A., Iwakiri, Y., Groszmann, R., Claffey, K. P., Cheng, Y. C., et al. (2003). Selective inhibition of tumor microvascular permeability by cavtratin blocks tumor progression in mice. *Cancer Cell* **4**, 31–39.

Gray, M. J., Poljakovic, M., Kepka-Lenhart, D., and Morris, S. M., Jr. (2005). Induction of arginase I transcription by IL-4 requires a composite DNA response element for STAT6 and C/EBPbeta. *Gene* **353**, 98–106.

Grohmann, U., Bianchi, R., Belladonna, M. L., Silla, S., Fallarino, F., Fioretti, M. C., and Puccetti, P. (2000). IFN-gamma inhibits presentation of a tumor/self peptide by CD8 alpha-dendritic cells via potentiation of the CD8 alpha+ subset. *J. Immunol.* **165**, 1357–1363.

Grohmann, U., Fallarino, F., Bianchi, R., Orabona, C., Vacca, C., Fioretti, M. C., and Puccetti, P. (2003). A defect in tryptophan catabolism impairs tolerance in nonobese diabetic mice. *J. Exp. Med.* **198**, 153–160.

Grohmann, U., Orabona, C., Fallarino, F., Vacca, C., Calcinaro, F., Falorni, A., Candeloro, P., Belladonna, M. L., Bianchi, R., Fioretti, M. C., et al. (2002). CTLA-4-Ig regulates tryptophan catabolism in vivo. *Nat. Immunol.* **3**, 1097–1101.

Guzik, T. J., Korbut, R., and Adamek-Guzik, T. (2003). Nitric oxide and superoxide in inflammation and immune regulation. *J. Physiol. Pharmacol.* **54**, 469–487.

Hajri, A., Metzger, E., Vallat, F., Coffy, S., Flatter, E., Evrard, S., Marescaux, J., and Aprahamian, M. (1998). Role of nitric oxide in pancreatic tumour growth: In vivo and in vitro studies. *Br. J. Cancer* **78**, 841–849.

Hecker, M., Nematollahi, H., Hey, C., Busse, R., and Racke, K. (1995). Inhibition of arginase by NG-hydroxy-L-arginine in alveolar macrophages: Implications for the utilization of L-arginine for nitric oxide synthesis. *FEBS Lett.* **359**, 251–254.

Hildeman, D. A., Mitchell, T., Aronow, B., Wojciechowski, S., Kappler, J., and Marrack, P. (2003). Control of Bcl-2 expression by reactive oxygen species. *Proc. Natl. Acad. Sci. USA* **100**, 15035–15040.

Hildeman, D. A., Mitchell, T., Kappler, J., and Marrack, P. (2003). T cell apoptosis and reactive oxygen species. *J. Clin. Invest.* **111**, 575–581.

Hildeman, D. A., Mitchell, T., Teague, T. K., Henson, P., Day, B. J., Kappler, J., and Marrack, P. C. (1999). Reactive oxygen species regulate activation-induced T cell apoptosis. *Immunity* **10**, 735–744.

Hinz, B., Brune, K., and Pahl, A. (2000). Nitric oxide inhibits inducible nitric oxide synthase mRNA expression in RAW 264.7 macrophages. *Biochem. Biophys. Res. Commun.* **271**, 353–357.

Hucke, C., MacKenzie, C. R., Adjogble, K. D., Takikawa, O., and Daubener, W. (2004). Nitric oxide-mediated regulation of gamma interferon-induced bacteriostasis: Inhibition and degradation of human indoleamine 2,3-dioxygenase. *Infect. Immun.* **72**, 2723–2730.

Hussain, S. P., Amstad, P., Raja, K., Ambs, S., Nagashima, M., Bennett, W. P., Shields, P. G., Ham, A. J., Swenberg, J. A., Marrogi, A. J., et al. (2000). Increased p53 mutation load in noncancerous colon tissue from ulcerative colitis: A cancer-prone chronic inflammatory disease. *Cancer Res.* **60**, 3333–3337.

Hussain, S. P., Hofseth, L. J., and Harris, C. C. (2003). Radical causes of cancer. *Nat. Rev. Cancer* **3**, 276–285.

Hussain, S. P., Trivers, G. E., Hofseth, L. J., He, P., Shaikh, I., Mechanic, L. E., Doja, S., Jiang, W., Subleski, J., Shorts, L., et al. (2004). Nitric oxide, a mediator of inflammation, suppresses tumorigenesis. *Cancer Res.* **64**, 6849–6853.

Ibiza, S., Victor, V. M., Bosca, I., Ortega, A., Urzainqui, A., O'Connor, E., Sanchez-Madrid, J. F., Esplugues, J. V., and Serrador, J. M. (2006). Endothelial nitric oxide synthase regulates T cell receptor signaling at the immunological synapse. *Immunity* **24**, 753–765.

Ichinoe, M., Mikami, T., Shiraishi, H., and Okayasu, I. (2004). High microvascular density is correlated with high VEGF, iNOS and COX-2 expression in penetrating growth-type early gastric carcinomas. *Histopathology* **45**, 612–618.

Iniesta, V., Gomez-Nieto, L. C., and Corraliza, I. (2001). The inhibition of arginase by N(omega)-hydroxy-l-arginine controls the growth of Leishmania inside macrophages. *J. Exp. Med.* **193**, 777–784.

Ishii, Y., Ogura, T., Tatemichi, M., Fujisawa, H., Otsuka, F., and Esumi, H. (2003). Induction of matrix metalloproteinase gene transcription by nitric oxide and mechanisms of MMP-1 gene induction in human melanoma cell lines. *Int. J. Cancer* **103**, 161–168.

Iyer, R. K., Yoo, P. K., Kern, R. M., Rozengurt, N., Tsoa, R., O'Brien, W. E., Yu, H., Grody, W. W., and Cederbaum, S. D. (2002). Mouse model for human arginase deficiency. *Mol. Cell. Biol.* **22**, 4491–4498.

Jadeski, L. C., Chakraborty, C., and Lala, P. K. (2003). Nitric oxide-mediated promotion of mammary tumour cell migration requires sequential activation of nitric oxide synthase, guanylate cyclase and mitogen-activated protein kinase. *Int. J. Cancer* **106**, 496–504.

Jaiswal, M., LaRusso, N. F., Burgart, L. J., and Gores, G. J. (2000). Inflammatory cytokines induce DNA damage and inhibit DNA repair in cholangiocarcinoma cells by a nitric oxide-dependent mechanism. *Cancer Res.* **60**, 184–190.

Jost, M. M., Ninci, E., Meder, B., Kempf, C., Van Royen, N., Hua, J., Berger, B., Hoefer, I., Modolell, M., and Buschmann, I. (2003). Divergent effects of GM-CSF and TGFbeta1 on bone marrow-derived macrophage arginase-1 activity, MCP-1 expression, and matrix metalloproteinase-12: A potential role during arteriogenesis. *FASEB J.* **17**, 2281–2283.

Kai, H., Ito, M., Kitadai, Y., Tanaka, S., Haruma, K., and Chayama, K. (2004). Chronic gastritis with expression of inducible nitric oxide synthase is associated with high expression of interleukin-6 and hypergastrinaemia. *Aliment. Pharmacol. Ther.* **19**, 1309–1314.

Kasuno, K., Takabuchi, S., Fukuda, K., Kizaka-Kondoh, S., Yodoi, J., Adachi, T., Semenza, G. L., and Hirota, K. (2004). Nitric oxide induces hypoxia-inducible factor 1 activation that is dependent on MAPK and phosphatidylinositol 3-kinase signaling. *J. Biol. Chem.* **279**, 2550–2558.

Kato, H., Miyazaki, T., Yoshikawa, M., Nakajima, M., Fukai, Y., Tajima, K., Masuda, N., Tsutsumi, S., Tsukada, K., Nakajima, T., et al. (2000). Nitrotyrosine in esophageal squamous cell carcinoma and relevance to p53 expression. *Cancer Lett.* **153**, 121–127.

Keskinege, A., Elgun, S., and Yilmaz, E. (2001). Possible implications of arginase and diamine oxidase in prostatic carcinoma. *Cancer Detect. Prev.* **25**, 76–79.

Kim, S. F., Huri, D. A., and Snyder, S. H. (2005). Inducible nitric oxide synthase binds, S-nitrosylates, and activates cyclooxygenase-2. *Science* **310**, 1966–1970.

Kimura, H., Weisz, A., Kurashima, Y., Hashimoto, K., Ogura, T., D'Acquisto, F., Addeo, R., Makuuchi, M., and Esumi, H. (2000). Hypoxia response element of the human vascular endothelial growth factor gene mediates transcriptional regulation by nitric oxide: Control of hypoxia-inducible factor-1 activity by nitric oxide. *Blood* **95**, 189–197.

Kisley, L. R., Barrett, B. S., Bauer, A. K., Dwyer-Nield, L. D., Barthel, B., Meyer, A. M., Thompson, D. C., and Malkinson, A. M. (2002). Genetic ablation of inducible nitric oxide synthase decreases mouse lung tumorigenesis. *Cancer Res.* **62**, 6850–6856.

Kleinert, H., Schwarz P. M., and Forstermann, U. (2003). Regulation of the expression of inducible nitric oxide synthase. *Biol. Chem.* **384**, 1343–1364.

Klotz, T., Bloch, W., Volberg, C., Engelmann, U., and Addicks, K. (1998). Selective expression of inducible nitric oxide synthase in human prostate carcinoma. *Cancer* **82**, 1897–1903.

Koblish, H. K., Hunter, C. A., Wysocka, M., Trinchieri, G., and Lee, W. M. (1998). Immune suppression by recombinant interleukin (rIL)-12 involves interferon gamma induction of nitric oxide synthase 2 (iNOS) activity: Inhibitors of NO generation reveal the extent of rIL-12 vaccine adjuvant effect. *J. Exp. Med.* **188**, 1603–1610.

Koh, K. P., Wang, Y., Yi, T., Shiao, S. L., Lorber, M. I., Sessa, W. C., Tellides, G., and Pober, J. S. (2004). T cell-mediated vascular dysfunction of human allografts results from IFN-gamma dysregulation of NO synthase. *J. Clin. Invest.* **114**, 846–856.

Kolb, H., and Kolb-Bachofen, V. (1998). Nitric oxide in autoimmune disease: Cytotoxic or regulatory mediator? *Immunol. Today* **19**, 556–561.

Kung, J. T., Brooks, S. B., Jakway, J. P., Leonard, L. L., and Talmage, D. W. (1977). Suppression of *in vitro* cytotoxic response by macrophages due to induced arginase. *J. Exp. Med.* **146**, 665–672.

Kusmartsev, S., and Gabrilovich, D. I. (2005). STAT1 Signaling Regulates Tumor-Associated Macrophage-Mediated T Cell Deletion. *J. Immunol.* **174**, 4880–4891.

Kusmartsev, S., Nefedova, Y., Yoder, D., and Gabrilovich, D. I. (2004). Antigen-specific inhibition of $CD8^+$ T cell response by immature myeloid cells in cancer is mediated by reactive oxygen species. *J. Immunol.* **172**, 989–999.

Kusmartsev, S. A., Li, Y., and Chen, S. H. (2000). $Gr-1^+$ myeloid cells derived from tumor-bearing mice inhibit primary T cell activation induced through CD3/CD28 costimulation. *J. Immunol.* **165**, 779–785.

Lala, P. K., and Chakraborty, C. (2001). Role of nitric oxide in carcinogenesis and tumour progression. *Lancet Oncol.* **2**, 149–156.

Laniewski, N. G., and Grayson, J. M. (2004). Antioxidant treatment reduces expansion and contraction of antigen-specific $CD8^+$ T cells during primary but not secondary viral infection. *J. Virol.* **78**, 11246–11257.

Laroux, F. S., Pavlick, K. P., Hines, I. N., Kawachi, S., Harada, H., Bharwani, S., Hoffman, J. M., and Grisham, M. B. (2001). Role of nitric oxide in inflammation. *Acta Physiol. Scand.* **173**, 113–118.

Lee, J., Ryu, H., Ferrante, R. J., Morris, S. M., Jr., and Ratan, R. R. (2003). Translational control of inducible nitric oxide synthase expression by arginine can explain the arginine paradox. *Proc. Natl. Acad. Sci. USA* **100**, 4843–4848.

Loibl, S., Buck, A., Strank, C., von Minckwitz, G., Roller, M., Sinn, H. P., Schini-Kerth, V., Solbach, C., Strebhardt, K., and Kaufmann, M. (2005). The role of early expression of inducible nitric oxide synthase in human breast cancer. *Eur. J. Cancer* **41**, 265–271.

Loibl, S., von Minckwitz, G., Weber, S., Sinn, H. P., Schini-Kerth, V. B., Lobysheva, I., Nepveu, F., Wolf, G., Strebhardt, K., and Kaufmann, M. (2002). Expression of endothelial and inducible nitric oxide synthase in benign and malignant lesions of the breast and measurement of nitric oxide using electron paramagnetic resonance spectroscopy. *Cancer* **95**, 1191–1198.

Lopez-Rivera, E., Lizarbe, T. R., Martinez-Moreno, M., Lopez-Novoa, J. M., Rodriguez-Barbero, A., Rodrigo, J., Fernandez, A. P., Alvarez-Barrientos, A., Lamas, S., and Zaragoza, C. (2005). Matrix metalloproteinase 13 mediates nitric oxide activation of endothelial cell migration. *Proc. Natl. Acad. Sci. USA* **102**, 3685–3690.

Lundberg, J. O., Lundberg, J. M., Alving, K., and Weitzberg, E. (1997). Nitric oxide and inflammation: The answer is blowing in the wind. *Nat. Med.* **3**, 30–31.

Ma, N., Adachi, Y., Hiraku, Y., Horiki, N., Horiike, S., Imoto, I., Pinlaor, S., Murata, M., Semba, R., and Kawanishi, S. (2004). Accumulation of 8-nitrogunine in human gastric epithelium induced by *Helicobacter pylori* infection. *Biochem. Biophys. Res. Commun.* **319**, 506–510.

Macarthur, M., Hold, G. L., and El-Omar, E. M. (2004). Inflammation and Cancer II. Role of chronic inflammation and cytokine gene polymorphisms in the pathogenesis of gastrointestinal malignancy. *Am. J. Physiol. Gastrointest. Liver Physiol.* **286**, G515–520.

MacMicking, J., Xie, Q. W., and Nathan, C. (1997). Nitric oxide and macrophage function. *Annu. Rev. Immunol.* **15**, 323–350.

Macphail, S. E., Gibney, C. A., Brooks, B. M., Booth, C. G., Flanagan, B. F., and Coleman, J. W. (2003). Nitric oxide regulation of human peripheral blood mononuclear cells: Critical time dependence and selectivity for cytokine versus chemokine expression. *J. Immunol.* **171**, 4809–4815.

Mann, G. E., Yudilevich, D. L., and Sobrevia, L. (2003). Regulation of amino acid and glucose transporters in endothelial and smooth muscle cells. *Physiol. Rev.* **83**, 183–252.

Mannick, J. B., Hausladen, A., Liu, L., Hess, D. T., Zeng, M., Miao, Q. X., Kane, L. S., Gow, A. J., and Stamler, J. S. (1999). Fas-induced caspase denitrosylation. *Science* **284**, 651–654.

Mantovani, A., Sozzani, S., Locati, M., Allavena, P., and Sica, A. (2002). Macrophage polarization: Tumor-associated macrophages as a paradigm for polarized M2 mononuclear phagocytes. *Trends Immunol.* **23**, 549–555.

Margalit, O., Eisenbach, L., Amariglio, N., Kaminski, N., Harmelin, A., Pfeffer, R., Shohat, M., Rechavi, G., and Berger, R. (2003). Overexpression of a set of genes, including WISP-1, common to pulmonary metastases of both mouse D122 Lewis lung carcinoma and B16-F10.9 melanoma cell lines. *Br. J. Cancer* **89**, 314–319.

Mariotto, S., Cuzzolin, L., Adami, A., Del Soldato, P., Suzuki, H., and Benoni, G. (1995). Effect of a new non-steroidal anti-inflammatory drug, nitroflurbiprofen, on the expression of inducible nitric oxide synthase in rat neutrophils. *Br. J. Pharmacol.* **115**, 225–226.

Marnett, L. J., Riggins, J. N., and West, J. D. (2003). Endogenous generation of reactive oxidants and electrophiles and their reactions with DNA and protein. *J. Clin. Invest.* **111**, 583–593.

Marrogi, A., Pass, H. I., Khan, M., Metheny-Barlow, L. J., Harris, C. C., and Gerwin, B. I. (2000). Human mesothelioma samples overexpress both cyclooxygenase-2 (COX-2) and inducible nitric oxide synthase (NOS2): *In vitro* antiproliferative effects of a COX-2 inhibitor. *Cancer Res.* **60**, 3696–3700.

Massi, D., Franchi, A., Sardi, I., Magnelli, L., Paglierani, M., Borgognoni, L., Maria Reali, U., and Santucci, M. (2001). Inducible nitric oxide synthase expression in benign and malignant cutaneous melanocytic lesions. *J. Pathol.* **194**, 194–200.

Mazzoni, A., Bronte, V., Visintin, A., Spitzer, J. H., Apolloni, E., Serafini, P., Zanovello, P., and Segal, D. M. (2002). Myeloid suppressor lines inhibit T cell responses by an NO-dependent mechanism. *J. Immunol.* **168**, 689–695.

Medeiros, R. M., Morais, A., Vasconcelos, A., Costa, S., Pinto, D., Oliveira, J., Ferreira, P., and Lopes, C. (2002). Outcome in prostate cancer: Association with endothelial nitric oxide synthase Glu-Asp298 polymorphism at exon 7. *Clin. Cancer Res.* **8**, 3433–3437.

Mencacci, A., Montagnoli, C., Bacci, A., Cenci, E., Pitzurra, L., Spreca, A., Kopf, M., Sharpe, A. H., and Romani, L. (2002). CD80$^+$Gr-1$^+$ myeloid cells inhibit development of antifungal Th1 immunity in mice with candidiasis. *J. Immunol.* **169**, 3180–3190.

Mendes, R. V., Martins, A. R., de Nucci, G., Murad, F., and Soares, F. A. (2001). Expression of nitric oxide synthase isoforms and nitrotyrosine immunoreactivity by B-cell non-Hodgkin's lymphomas and multiple myeloma. *Histopathology* **39**, 172–178.

Meurs, H., Maarsingh, H., and Zaagsma, J. (2003). Arginase and asthma: Novel insights into nitric oxide homeostasis and airway hyperresponsiveness. *Trends Pharmacol. Sci.* **24**, 450–455.

Michel, T., and Feron, O. (1997). Nitric oxide synthases: Which, where, how, and why? *J. Clin. Invest.* **100**, 2146–2152.

Mielczarek, M., Chrzanowska, A., Scibior, D., Skwarek, A., Ashamiss, F., Lewandowska, F., and Baranczyk-Kuzma, A. (2006). Arginase as a useful factor for the diagnosis of colorectal cancer liver metastases. *Int. J. Biol. Markers* **21**, 40–44.

Mills, C. D., Kincaid, K., Alt, J. M., Heilman, M. J., and Hill, A. M. (2000). M-1/M-2 macrophages and the Th1/Th2 paradigm. *J. Immunol.* **164**, 6166–6173.

Mocellin, S., Mandruzzato, S., Bronte, V., Lise, M., and Nitti, D. (2004a). Part I: Vaccines for solid tumours. *Lancet Oncol.* **5**, 681–689.

Mocellin, S., Provenzano, M., Rossi, C. R., Pilati, P., Scalerta, R., Lise, M., and Nitti, D. (2004b). Induction of endothelial nitric oxide synthase expression by melanoma sensitizes endothelial cells to tumor necrosis factor-driven cytotoxicity. *Clin. Cancer Res.* **10**, 6879–6886.

Mocellin, S., Semenzato, G., Mandruzzato, S., and Riccardo Rossi, C. (2004c). Part II: Vaccines for haematological malignant disorders. *Lancet Oncol.* **5**, 727–737.

Mocellin, S., Bronte, V., and Nitti, D. (2006). Nitric oxide, a double-edged sword in cancer biology: Searching for therapeutic opportunities. *Med. Res. Rev.*, In press.

Mollace, V., Muscoli, C., Masini, E., Cuzzocrea, S., and Salvemini, D. (2005). Modulation of prostaglandin biosynthesis by nitric oxide and nitric oxide donors. *Pharmacol. Rev.* **57**, 217–252.

Moncada, S., Palmer, R. M., and Higgs, E. A. (1991). Nitric oxide: Physiology, pathophysiology, and pharmacology. *Pharmacol. Rev.* **43**, 109–142.

Morbidelli, L., Donnini, S., and Ziche, M. (2004). Role of nitric oxide in tumor angiogenesis. *Cancer Treat. Res.* **117**, 155–167.

Mori, N., Nunokawa, Y., Yamada, Y., Ikeda, S., Tomonaga, M., and Yamamoto, N. (1999). Expression of human inducible nitric oxide synthase gene in T-cell lines infected with human T-cell leukemia virus type-I and primary adult T-cell leukemia cells. *Blood* **94**, 2862–2870.

Morin, J., Chimenes, A., Boitard, C., Berthier, R., and Boudaly, S. (2003). Granulocyte-dendritic cell unbalance in the non-obese diabetic mice. *Cell. Immunol.* **223**, 13–25.

Morris, S. M., Jr. (2002). Regulation of enzymes of the urea cycle and arginine metabolism. *Annu. Rev. Nutr.* **22**, 87–105.

Morrison, A. C., and Correll, P. H. (2002). Activation of the stem cell-derived tyrosine kinase/RON receptor tyrosine kinase by macrophage-stimulating protein results in the induction of arginase activity in murine peritoneal macrophages. *J. Immunol.* **168**, 853–860.

Moulian, N., Truffault, F., Gaudry-Talarmain, Y. M., Serraf, A., and Berrih-Aknin, S. (2001). In vivo and in vitro apoptosis of human thymocytes are associated with nitrotyrosine formation. *Blood* **97**, 3521–3530.

Munder, M., Eichmann, K., and Modolell, M. (1998). Alternative metabolic states in murine macrophages reflected by the nitric oxide synthase/arginase balance: Competitive regulation by $CD4^+$ T cells correlates with Th1/Th2 phenotype. *J. Immunol.* **160**, 5347–5354.

Munder, M., Eichmann, K., Moran, J. M., Centeno, F., Soler, G., and Modolell, M. (1999). Th1/Th2-regulated expression of arginase isoforms in murine macrophages and dendritic cells. *J. Immunol.* **163**, 3771–3777.

Munder, M., Mollinedo, F., Calafat, J., Canchado, J., Gil-Lamaignere, C., Fuentes, J. M., Luckner, C., Doschko, G., Soler, G., Eichmann, K., et al. (2005). Arginase I is constitutively expressed in human granulocytes and participates in fungicidal activity. *Blood* **105**, 2549–2556.

Munder, M., Schneider, H., Luckner, C., Giese, T., Langhans, C. D., Fuentes, J., Kropf, P., Mueller, I., Kolb, A., Modolell, M., et al. (2006). Suppression of T cell functions by human granulocyte arginase. *Blood* **108**, 1627–1636.

Munn, D. H., Sharma, M. D., Baban, B., Harding, H. P., Zhang, Y., Ron, D., and Mellor, A. L. (2005). GCN2 kinase in T cells mediates proliferative arrest and anergy induction in response to indoleamine 2,3-dioxygenase. *Immunity* **22**, 633–642.

Nelson, W. G., De Marzo, A. M., DeWeese, T. L., and Isaacs, W. B. (2004). The role of inflammation in the pathogenesis of prostate cancer. *J. Urol.* **172**, S6–11; discussion S11–12.

Niedbala, W., Wei, X. Q., Piedrafita, D., Xu, D., and Liew, F. Y. (1999). Effects of nitric oxide on the induction and differentiation of Th1 cells. *Eur. J. Immunol.* **29**, 2498–2505.

Nosho, K., Yamamoto, H., Adachi, Y., Endo, T., Hinoda, Y., and Imai, K. (2005). Gene expression profiling of colorectal adenomas and early invasive carcinomas by cDNA array analysis. *Br. J. Cancer* **92**, 1193–1200.

Nozoe, T., Yasuda, M., Honda, M., Inutsuka, S., and Korenaga, D. (2002). Immunohistochemical expression of cytokine induced nitric oxide synthase in colorectal carcinoma. *Oncol. Rep.* **9**, 521–524.

Ohshima, H., Tatemichi, M., and Sawa, T. (2003). Chemical basis of inflammation-induced carcinogenesis. *Arch. Biochem. Biophys.* **417**, 3–11.

Okamoto, T., Akaike, T., Sawa, T., Miyamoto, Y., van der Vliet, A., and Maeda, H. (2001). Activation of matrix metalloproteinases by peroxynitrite-induced protein S-glutathiolation via disulfide S-oxide formation. *J. Biol. Chem.* **276**, 29596–29602.

Otsuji, M., Kimura, Y., Aoe, T., Okamoto, Y., and Saito, T. (1996). Oxidative stress by tumor-derived macrophages suppresses the expression of CD3 zeta chain of T-cell receptor complex and antigen-specific T-cell responses. *Proc. Natl. Acad. Sci. USA* **93**, 13119–13124.

Ouyang, N., Williams, J. L., Tsioulias, G. J., Gao, J., Iatropoulos, M. J., Kopelovich, L., Kashfi, K., and Rigas, B. (2006). Nitric oxide-donating aspirin prevents pancreatic cancer in a hamster tumor model. *Cancer Res.* **66**, 4503–4511.

Park, I. S., Kang, S. W., Shin, Y. J., Chae, K. Y., Park, M. O., Kim, M. Y., Wheatley, D. N., and Min, B. H. (2003a). Arginine deiminase: A potential inhibitor of angiogenesis and tumour growth. *Br. J. Cancer* **89**, 907–914.

Park, S. W., Lee, S. G., Song, S. H., Heo, D. S., Park, B. J., Lee, D. W., Kim, K. H., and Sung, M. W. (2003b). The effect of nitric oxide on cyclooxygenase-2 (COX-2) overexpression in head and neck cancer cell lines. *Int. J. Cancer* **107**, 729–738.

Pauleau, A. L., Rutschman, R., Lang, R., Pernis, A., Watowich, S. S., and Murray, P. J. (2004). Enhancer-mediated control of macrophage-specific arginase I expression. *J. Immunol.* **172**, 7565–7573.

Perez, G. M., Melo, M., Keegan, A. D., and Zamorano, J. (2002). Aspirin and salicylates inhibit the IL-4- and IL-13-induced activation of STAT6. *J. Immunol.* **168**, 1428–1434.

Pignatelli, B., Li, C. Q., Boffetta, P., Chen, Q., Ahrens, W., Nyberg, F., Mukeria, A., Bruske-Hohlfeld, I., Fortes, C., Constantinescu, V., et al. (2001). Nitrated and oxidized plasma proteins in smokers and lung cancer patients. *Cancer Res.* **61**, 778–784.

Polat, M. F., Taysi, S., Polat, S., Boyuk, A., and Bakan, E. (2003). Elevated serum arginase activity levels in patients with breast cancer. *Surg. Today* **33**, 655–661.

Porembska, Z., Luboinski, G., Chrzanowska, A., Mielczarek, M., Magnuska, J., and Baranczyk-Kuzma, A. (2003). Arginase in patients with breast cancer. *Clin. Chim. Acta* **328**, 105–111.

Porembska, Z., Skwarek, A., Mielczarek, A., and Baranczyk-Kuzma, A. (2002). Serum arginase activity in postsurgical monitoring of patients with colorectal carcinoma. *Cancer* **94**, 2930–2934.

Qiu, H., Orr, F. W., Jensen, D., Wang, H. H., McIntosh, A. R., Hasinoff, B. B., Nance, D. M., Pylypas, S., Qi, K., Song, C., et al. (2003). Arrest of B16 melanoma cells in the mouse pulmonary microcirculation induces endothelial nitric oxide synthase-dependent nitric oxide release that is cytotoxic to the tumor cells. *Am. J. Pathol.* **162**, 403–412.

Radi, R. (2004). Nitric oxide, oxidants, and protein tyrosine nitration. *Proc. Natl. Acad. Sci. USA* **101**, 4003–4008.

Raes, G., Van den Bergh, R., De Baetselier, P., Ghassabeh, G. H., Scotton, C., Locati, M., Mantovani, A., and Sozzani, S. (2005). Arginase-1 and Ym1 are markers for murine, but not human, alternatively activated myeloid cells. *J. Immunol.* **174**, 6561; author reply 6561–6562.

Rao, C. V. (2004). Nitric oxide signaling in colon cancer chemoprevention. *Mutat. Res.* **555**, 107–119.

Rao, C. V., Reddy, B. S., Steele, V. E., Wang, C. X., Liu, X., Ouyang, N., Patlolla, J. M., Simi, B., Kopelovich, L., and Rigas, B. (2006). Nitric oxide-releasing aspirin and indomethacin are potent inhibitors against colon cancer in azoxymethane-treated rats: Effects on molecular targets. *Mol. Cancer Ther.* **5**, 1530–1538.

Raspollini, M. R., Amunni, G., Villanucci, A., Boddi, V., Baroni, G., Taddei, A., and Taddei, G. L. (2004). Expression of inducible nitric oxide synthase and cyclooxygenase-2 in ovarian cancer: Correlation with clinical outcome. *Gynecol. Oncol.* **92**, 806–812.

Reth, M. (2002). Hydrogen peroxide as second messenger in lymphocyte activation. *Nat. Immunol.* **3**, 1129–1134.

Reveneau, S., Arnould, L., Jolimoy, G., Hilpert, S., Lejeune, P., Saint-Giorgio, V., Belichard, C., and Jeannin, J. F. (1999). Nitric oxide synthase in human breast cancer is associated with tumor grade, proliferation rate, and expression of progesterone receptors. *Lab Invest.* **79**, 1215–1225.

Ricchi, P., Zarrilli, R., Di Palma, A., and Acquaviva, A. M. (2003). Nonsteroidal anti-inflammatory drugs in colorectal cancer: From prevention to therapy. *Br. J. Cancer* **88**, 803–807.

Rodriguez, P. C., Hernandez, C. P., Quiceno, D., Dubinett, S. M., Zabaleta, J., Ochoa, J. B., Gilbert, J., and Ochoa, A. C. (2005). Arginase I in myeloid suppressor cells is induced by COX-2 in lung carcinoma. *J. Exp. Med.* **202**, 931–939.

Rodriguez, P. C., Quiceno, D. G., Zabaleta, J., Ortiz, B., Zea, A. H., Piazuelo, M. B., Delgado, A., Correa, P., Brayer, J., Sotomayor, E. M., et al. (2004). Arginase I production in the tumor microenvironment by mature myeloid cells inhibits T-cell receptor expression and antigen-specific T-cell responses. *Cancer Res.* **64**, 5839–5849.

Rodriguez, P. C., Zea, A. H., Culotta, K. S., Zabaleta, J., Ochoa, J. B., and Ochoa, A. C. (2002). Regulation of T cell receptor CD3zeta chain expression by L-arginine. *J. Biol. Chem.* **277**, 21123–21129.

Rodriguez, P. C., Zea, A. H., DeSalvo, J., Culotta, K. S., Zabaleta, J., Quiceno, D. G., Ochoa, J. B., and Ochoa, A. C. (2003). L-arginine consumption by macrophages modulates the expression of CD3 zeta chain in T lymphocytes. *J. Immunol.* **171**, 1232–1239.

Rodriguez, P. C., Quiceno, D. G., and Ochoa, A. C. (2007). L-Arginine availability regulates T-lymphocyte cell-cycle progression. *Blood* **109**, 1568–1573.

Roman, V., Zhao, H., Fourneau, J. M., Marconi, A., Dugas, N., Dugas, B., Sigaux, F., and Kolb, J. P. (2000). Expression of a functional inducible nitric oxide synthase in hairy cell leukaemia and ESKOL cell line. *Leukemia* **14**, 696–705.

Rosenberg, S. A., Yang, J. C., and Restifo, N. P. (2004). Cancer immunotherapy: Moving beyond current vaccines. *Nat. Med.* **10**, 909–915.

Rothe, H., Hausmann, A., and Kolb, H. (2002). Immunoregulation during disease progression in prediabetic NOD mice: Inverse expression of arginase and prostaglandin H synthase 2 vs. interleukin-15. *Horm. Metab. Res.* **34**, 7–12.

Routledge, M. N. (2000). Mutations induced by reactive nitrogen oxide species in the supF forward mutation assay. *Mutat. Res.* **450**, 95–105.

Sandler, R. S., Halabi, S., Baron, J. A., Budinger, S., Paskett, E., Keresztes, R., Petrelli, N., Pipas, J. M., Karp, D. D., Loprinzi, C. L., *et al.* (2003). A randomized trial of aspirin to prevent colorectal adenomas in patients with previous colorectal cancer. *N. Engl. J. Med.* **348**, 883–890.

Schluns, K. S., and Lefrancois, L. (2003). Cytokine control of memory T-cell development and survival. *Nat. Rev. Immunol.* **3**, 269–279.

Schmielau, J., and Finn, O. J. (2001). Activated granulocytes and granulocyte-derived hydrogen peroxide are the underlying mechanism of suppression of T-cell function in advanced cancer patients. *Cancer Res.* **61**, 4756–4760.

Schneemann, M., and Schoedon, G. (2002). Species differences in macrophage NO production are important. *Nat. Immunol.* **3**, 102.

Schopfer, F. J., Baker, P. R., and Freeman, B. A. (2003). NO-dependent protein nitration: A cell signaling event or an oxidative inflammatory response? *Trends Biochem. Sci.* **28**, 646–654.

Scott, D. J., Hull, M. A., Cartwright, E. J., Lam, W. K., Tisbury, A., Poulsom, R., Markham, A. F., Bonifer, C., and Coletta, P. L. (2001). Lack of inducible nitric oxide synthase promotes intestinal tumorigenesis in the Apc(Min/+) mouse. *Gastroenterology* **121**, 889–899.

Semenza, G. L. (2001). HIF-1 and mechanisms of hypoxia sensing. *Curr. Opin. Cell Biol.* **13**, 167–171.

Semenza, G. L. (2003). Targeting HIF-1 for cancer therapy. *Nat. Rev. Cancer* **3**, 721–732.

Serafini, P., Borrello, I., and Bronte, V. (2006). Myeloid suppressor cells in cancer: Recruitment, phenotype, properties, and mechanisms of immune suppression. *Semin. Cancer Biol.* **16**, 53–65.

Serafini, P., De Santo, C., Marigo, I., Cingarlini, S., Dolcetti, L., Gallina, G., Zanovello, P., and Bronte, V. (2003). Derangement of immune responses by myeloid suppressor cells. *Cancer Immunol. Immunother.* **53**, 64–72.

Serafini, P., Meckel, K., Kelso, M., Noonan, K., Califano, J., Koch, W., Dolcetti, L., Bronte, V., and Borrello, I. (2006). Phosphoresterase-5 inhibition angments endogenous antitumor immunity by reducing myeloid-derived suppressor cell function. *J. Exp. Red.* **27**, 2691–2702.

Sercan, O., Hammerling, G. J., Arnold, B., and Schuler, T. (2006). Innate immune cells contribute to the IFN-gamma-dependent regulation of antigen-specific CD8+ T cell homeostasis. *J. Immunol.* **176**, 735–739.

Sharara, A. I., Perkins, D. J., Misukonis, M. A., Chan, S. U., Dominitz, J. A., and Weinberg, J. B. (1997). Interferon (IFN)-alpha activation of human blood mononuclear cells *in vitro* and *in vivo* for nitric oxide synthase (NOS) type 2 mRNA and protein expression: Possible relationship of induced NOS2 to the anti-hepatitis C effects of IFN-alpha in vivo. *J. Exp. Med.* **186**, 1495–1502.

Shi, O., Morris, S. M., Jr., Zoghbi, H., Porter, C. W., and O'Brien, W. E. (2001). Generation of a mouse model for arginase II deficiency by targeted disruption of the arginase II gene. *Mol. Cell. Biol.* **21**, 811–813.

Shi, Q., Xiong, Q., Wang, B., Le, X., Khan, N. A., and Xie, K. (2000). Influence of nitric oxide synthase II gene disruption on tumor growth and metastasis. *Cancer Res.* **60**, 2579–2583.

Shinoda, J., McLaughlin, K. E., Bell, H. S., Swaroop, G. R., Yamaguchi, S., Holmes, M. C., and Whittle, I. R. (2003). Molecular mechanisms underlying dexamethasone inhibition of iNOS expression and activity in C6 glioma cells. *Glia* **42**, 68–76.

Siegert, A., Rosenberg, C., Schmitt, W. D., Denkert, C., and Hauptmann, S. (2002). Nitric oxide of human colorectal adenocarcinoma cell lines promotes tumour cell invasion. *Br. J. Cancer* **86**, 1310–1315.

Singh, R., Pervin, S., Karimi, A., Cederbaum, S., and Chaudhuri, G. (2000). Arginase activity in human breast cancer cell lines: N(omega)-hydroxy-L-arginine selectively inhibits cell proliferation and induces apoptosis in MDA-MB-468 cells. *Cancer Res.* **60**, 3305–3312.

Sinha, P., Clements, V. K., and Ostrand-Rosenberg, S. (2005). Reduction of myeloid-derived suppressor cells and induction of M1 macrophages facilitate the rejection of established metastatic disease. *J. Immunol.* **174**, 636–645.

Suarez-Pinzon, W. L., Mabley, J. G., Strynadka, K., Power, R. F., Szabo, C., and Rabinovitch, A. (2001). An inhibitor of inducible nitric oxide synthase and scavenger of peroxynitrite prevents diabetes development in NOD mice. *J. Autoimmun.* **16**, 449–455.

Suarez-Pinzon, W. L., Szabo, C., and Rabinovitch, A. (1997). Development of autoimmune diabetes in NOD mice is associated with the formation of peroxynitrite in pancreatic islet beta-cells. *Diabetes* **46**, 907–911.

Suer Gokmen, S., Yoruk, Y., Cakir, E., Yorulmaz, F., and Gulen, S. (1999). Arginase and ornithine, as markers in human non-small cell lung carcinoma. *Cancer Biochem. Biophys.* **17**, 125–131.

Swana, H. S., Smith, S. D., Perrotta, P. L., Saito, N., Wheeler, M. A., and Weiss, R. M. (1999). Inducible nitric oxide synthase with transitional cell carcinoma of the bladder. *J. Urol.* **161**, 630–634.

Tatemichi, M., Sawa, T., Gilibert, I., Tazawa, H., Katoh, T., and Ohshima, H. (2005). Increased risk of intestinal type of gastric adenocarcinoma in Japanese women associated with long forms of CCTTT pentanucleotide repeat in the inducible nitric oxide synthase promoter. *Cancer Lett.* **217**, 197–202.

Tatemichi, M., Tazawa, H., Masuda, M., Saleem, M., Wada, S., Donehower, L. A., Ohgaki, H., and Ohshima, H. (2004). Suppression of thymic lymphomas and increased nonthymic lymphomagenesis in Trp53-deficient mice lacking inducible nitric oxide synthase gene. *Int. J. Cancer* **111**, 819–828.

Terrazas, L. I., Walsh, K. L., Piskorska, D., McGuire, E., and Harn, D. A., Jr. (2001). The schistosome oligosaccharide lacto-N-neotetraose expands Gr1($^+$) cells that secrete anti-inflammatory cytokines and inhibit proliferation of naive CD4($^+$) cells: A potential mechanism for immune polarization in helminth infections. *J. Immunol.* **167**, 5294–5303.

Tesei, A., Ricotti, L., Ulivi, P., Medri, L., Amadori, D., and Zoli, W. (2003). NCX 4016, a nitric oxide-releasing aspirin derivative, exhibits a significant antiproliferative effect and alters cell cycle progression in human colon adenocarcinoma cell lines. *Int. J. Oncol.* **22**, 1297–1302.

Thengchaisri, N., Hein, T. W., Wang, W., Xu, X., Li, Z., Fossum, T. W., and Kuo, L. (2006). Upregulation of arginase by H2O2 impairs endothelium-dependent nitric oxide-mediated dilation of coronary arterioles. *Arterioscler. Thromb. Vasc. Biol.*

Thomsen, L. L., Lawton, F. G., Knowles, R. G., Beesley, J. E., Riveros-Moreno, V., and Moncada, S. (1994). Nitric oxide synthase activity in human gynecological cancer. *Cancer Res.* **54**, 1352–1354.

Thomsen, L. L., and Miles, D. W. (1998). Role of nitric oxide in tumour progression: Lessons from human tumours. *Cancer Metastasis Rev.* **17**, 107–118.

Thomsen, L. L., Miles, D. W., Happerfield, L., Bobrow, L. G., Knowles, R. G., and Moncada, S. (1995). Nitric oxide synthase activity in human breast cancer. *Br. J. Cancer* **72**, 41–44.

Thun, M. J., Henley, S. J., and Patrono, C. (2002). Nonsteroidal anti-inflammatory drugs as anticancer agents: Mechanistic, pharmacologic, and clinical issues. *J. Natl. Cancer Inst.* **94**, 252–266.

Vakkala, M., Kahlos, K., Lakari, E., Paakko, P., Kinnula, V., and Soini, Y. (2000a). Inducible nitric oxide synthase expression, apoptosis, and angiogenesis in *in situ* and invasive breast carcinomas. *Clin. Cancer Res.* **6**, 2408–2416.

Vakkala, M., Paakko, P., and Soini, Y. (2000b). eNOS expression is associated with the estrogen and progesterone receptor status in invasive breast carcinoma. *Int. J. Oncol.* **17**, 667–671.

Vickers, S. M., MacMillan-Crow, L. A., Green, M., Ellis, C., and Thompson, J. A. (1999). Association of increased immunostaining for inducible nitric oxide synthase and nitrotyrosine with fibroblast growth factor transformation in pancreatic cancer. *Arch. Surg.* **134**, 245–251.

Vig, M., Srivastava, S., Kandpal, U., Sade, H., Lewis, V., Sarin, A., George, A., Bal, V., Durdik, J. M., and Rath, S. (2004). Inducible nitric oxide synthase in T cells regulates T cell death and immune memory. *J. Clin. Invest.* **113**, 1734–1742.

Villalta, F., Zhang, Y., Bibb, K. E., Kappes, J. C., and Lima, M. F. (1998). The cysteine-cysteine family of chemokines RANTES, MIP-1alpha, and MIP-1beta induce trypanocidal activity in human macrophages via nitric oxide. *Infect. Immun.* **66**, 4690–4695.

Vincendeau, P., Gobert, A. P., Daulouede, S., Moynet, D., and Djavad Mossalayi, M. (2003). Arginases in parasitic diseases. *Trends Parasitol.* **19**, 9–12.

Vouldoukis, I., Riveros-Moreno, V., Dugas, B., Ouaaz, F., Becherel, P., Debre, P., Moncada, S., and Mossalayi, M. D. (1995). The killing of *Leishmania* major by human macrophages is mediated by nitric oxide induced after ligation of the Fc epsilon RII/CD23 surface antigen. *Proc. Natl. Acad. Sci. USA* **92**, 7804–7808.

Wallace, J. L., Ignarro, L. J., and Fiorucci, S. (2002). Potential cardioprotective actions of no-releasing aspirin. *Nat. Rev. Drug Discov.* **1**, 375–382.

Wei, D., Richardson, E. L., Zhu, K., Wang, L., Le, X., He, Y., Huang, S., and Xie, K. (2003). Direct demonstration of negative regulation of tumor growth and metastasis by host-inducible nitric oxide synthase. *Cancer Res.* **63**, 3855–3859.

Wei, X. Q., Charles, I. G., Smith, A., Ure, J., Feng, G. J., Huang, F. P., Xu, D., Muller, W., Moncada, S., and Liew, F. W. (1995). Altered immune responses in mice lacking inducible nitric oxide synthase. *Nature* **375**, 408–411.

Weinberg, J. B. (1998). Nitric oxide production and nitric oxide synthase type 2 expression by human mononuclear phagocytes: A review. *Mol. Med.* **4**, 557–591.

Weninger, W., Rendl, M., Pammer, J., Mildner, M., Tschugguel, W., Schneeberger, C., Sturzl, M., and Tschachler, E. (1998). Nitric oxide synthases in Kaposi's sarcoma are expressed predominantly by vessels and tissue macrophages. *Lab. Invest.* **78**, 949–955.

Williams, J. L., Kashfi, K., Ouyang, N., del Soldato, P., Kopelovich, L., and Rigas, B. (2004). NO-donating aspirin inhibits intestinal carcinogenesis in *Min* (APC(*Min/+*)) mice. *Biochem. Biophys. Res. Commun.* **313**, 784–788.

Wilson, K. T., Fu, S., Ramanujam, K. S., and Meltzer, S. J. (1998). Increased expression of inducible nitric oxide synthase and cyclooxygenase-2 in Barrett's esophagus and associated adenocarcinomas. *Cancer Res.* **58**, 2929–2934.

Wink, D. A., Vodovotz, Y., Laval, J., Laval, F., Dewhirst, M. W., and Mitchell, J. B. (1998). The multifaceted roles of nitric oxide in cancer. *Carcinogenesis* **19**, 711–721.

Witherell, H. L., Hiatt, R. A., Replogle, M., and Parsonnet, J. (1998). *Helicobacter pylori* infection and urinary excretion of 8-hydroxy-2-deoxyguanosine, an oxidative DNA adduct. *Cancer Epidemiol. Biomarkers Prev.* **7**, 91–96.

Wu, C. W., Chung, W. W., Chi, C. W., Kao, H. L., Lui, W. Y., P'Eng, F. K., and Wang, S. R. (1996). Immunohistochemical study of arginase in cancer of the stomach. *Virchows Arch.* **428**, 325–331.

Wu, G., and Morris, S. M., Jr. (1998). Arginine metabolism: Nitric oxide and beyond. *Biochem. J.* **336**, 1–17.

Xia, Y., Roman, L. J., Masters, B. S., and Zweier, J. L. (1998). Inducible nitric-oxide synthase generates superoxide from the reductase domain. *J. Biol. Chem.* **273**, 22635–22639.

Xia, Y., and Zweier, J. L. (1997). Superoxide and peroxynitrite generation from inducible nitric oxide synthase in macrophages. *Proc. Natl. Acad. Sci. USA* **94**, 6954–6958.

Zabaleta, J., McGee, D. J., Zea, A. H., Hernandez, C. P., Rodriguez, P. C., Sierra, R. A., Correa, P., and Ochoa, A. C. (2004). *Helicobacter pylori* arginase inhibits T cell proliferation and reduces the expression of the TCR zeta-chain (CD3zeta). *J. Immunol.* **173**, 586–593.

Zea, A. H., Rodriguez, P. C., Atkins, M. B., Hernandez, C., Signoretti, S., Zabaleta, J., McDermott, D., Quiceno, D., Youmans, A., O'Neill, A., *et al.* (2005). Arginase-producing myeloid suppressor cells in renal cell carcinoma patients: A mechanism of tumor evasion. *Cancer Res.* **65**, 3044–3048.

Ziche, M., Morbidelli, L., Choudhuri, R., Zhang, H. T., Donnini, S., Granger, H. J., and Bicknell, R. (1997). Nitric oxide synthase lies downstream from vascular endothelial growth factor-induced but not basic fibroblast growth factor-induced angiogenesis. *J. Clin. Invest.* **99**, 2625–2634.

Further Reading

Wang, B., Xiong, Q., Shi, Q., Le, X., Abbruzzese, J. L., and Xie, K. (2001). Intact nitric oxide synthase II gene is required for interferon-beta-mediated suppression of growth and metastasis of pancreatic adenocarcinoma. *Cancer Res.* **61**, 71–75.

Wang, B., Xiong, Q. Shi, Q., Le, X., and Xie, K. (2001). Genetic disruption of host interferon-gamma drastically enhances the metastasis of pancreatic adenocarcinoma through impaired expression of inducible nitric oxide synthase. *Oncogene* **20**, 6930–6937.

Wang, B., Xiong, Q., Shi, Q., Tan, D., Le, X., and Xie, K. (2001). Genetic disruption of host nitric oxide synthase II gene impairs melanoma-induced angiogenesis and suppresses pleural effusion. *Int. J. Cancer* **91**, 607–611.

ARG in the web:
http://www.hprd.org/protein/01947
http://www.godatabase.org/cgi-bin/amigo/go.cgi?action=query&view=query&query=arginase&search_constraint=gp
http://www.brenda.uni-koeln.de/php/result_flat.php4?ecno=3.5.3.1
http://www.ncbi.nlm.nih.gov/IEB/Research/Acembly/av.cgi?db=35g&c=Gene&l=ARG1
http://www.ncbi.nlm.nih.gov/IEB/Research/Acembly/av.cgi?exdb=AceView&db=35g&term=ARG2&submit=Go

ARG inhibitors:
http://cgmp.blauplanet.com/tool/arginase.html

Arginine metabolism:
KEGG pathway: http://www.ergo-light.com/ERGO/CGI/show_kegg_map.cgi?request=PAINT_MAP_WITH_ECS&user=&map=map00330&ecgroup=2.6.1.21
KEGG pathway website: http://www.genome.ad.jp/kegg/metabolism.html

NOS on the web:
http://metallo.scripps.edu/PROMISE/NOS.html
http://www.wxumac.demon.co.uk/
http://www.ihop-net.org/UniPub/iHOP/
http://www.godatabase.org/cgi-bin/amigo/go.cgi?action=query&view=query&query=nitric+oxide+synthase&search_constraint=gp
http://www.ncbi.nlm.nih.gov/Structure/mmdb/mmdbsrv.cgi?form=6&db=t&Dopt=s&uid=12498

NOS knockout mice:
http://www.bioscience.org/knockout/inos.htm
http://www.jax.org/
http://sageke.sciencemag.org/resources/experimental/transgenic/

Genes and gene expression profiles:
http://www.nslij-genetics.org/search_omim.html
http://www.ncbi.nlm.nih.gov/entrez/query.fcgi?db=geo
http://www.ihop-net.org/UniPub/iHOP/

Nitroaspirin: http://ctd.mdibl.org/voc.go;jsessionid=7DC23382D6A5FF4CA8C0A802E55D897E?voc=chem&acc=C102148

http://www.nicox.com/

Index

Absorption, optimization, 153
Adapter protein, definition, 63
Adaptive antitumor response
    cytokines, 36–37
    cytotoxic mechanisms
        LIGHT, 38
        perforin, 37–38
        TRAIL, 38
    innate immunity interactions, 38–39
    overview, 33–34
    regulatory T cells, 35
    T cells, 34–35
    targets, 35–36
ADCC, see Antibody-dependent cell-mediated cytotoxicity
Adenosine A2a receptor, immune checkpoint in tumor microenvironment, 266
Allograft transplantation, immunotherapy prospects, 76–77
Alloreactive, definition, 63
All-trans retinoic acid, myeloid suppressor cell targeting, 325
αβ T cell
    adaptive antitumor response, 34–35
    immune surveillance role, 14–15
Amplimexon, features, 130

Angiogenesis
    myeloid suppressor cell promotion, 321
    vascular endothelial growth factor, see V
Antibody-dependent cell-mediated cytotoxicity, immunotherapy, 76
Antigen-presenting cell, see also specific cells
    cancer-induced immunosuppression, 174
    T cell interface signals, 258–261
Apoptosis
    chemotherapy induction, 237
    immunogenic apoptosis, 239–241
Arginase
    expression regulation, 373–374
    function, 372
    gene, 372–373
    immunoregulatory activity, 374–375, 377–378
    isoforms, 372, 374
    knockout mice, 374
    myeloid suppressor cell expression and effects, 319–320
    nitric oxide synthase interactions in immunoregulation
        autoimmune disease role, 379–380
        hydrogen peroxide generation, 380–381

    overview, 378
    peroxynitrite generation, 379
    physiological role for arginine metabolism in immunity control, 381–382
    therapeutic targeting, 386–389
    tumor expression, 384, 386
Aromatase inhibitors, immunity effects, 224–225
ARRY-142886, features, 133
Aspirin, nitric oxide donors, 387–388
ATRA, see All-trans retinoic acid
Aurora kinases, therapeutic targeting, 123–124
AZM475271, features, 136

B7 molecules
    costimulation, 259
    immune-modulating activity of targeting antibodies, 221
    immunologic checkpoint, 260
    T cell activation signals, 212–213
B7-H1
    T cell activation studies, 88
    tumor expression, 87–88
B7-H3, inhibitory function, 73
B7-H4
    cytokine regulation of expression, 89–90
    T cell inhibitory actions, 88–89

BAR adapters, indoleamine
2,3-dioxygenase
expression regulation,
358
B cell
myeloid suppressor cell
effects on function, 318
tolerogenic pressure on
immunity to cancer,
188–189
Bevacizumab, immune-
modulating activity,
220–221
Bin1, indoleamine 2,3-
dioxygenase expression
regulation, 358
Bioavailability, calculation, 154
BMS-354825, features, 136
Bryostatin-1
clinical trials, 123
mechanism of action, 122
Busulfan, features, 104

Cage, John, 3
Camptothecin, features, 106
Cancer stem cell,
immunotherapy
targeting, 176
Cancer therapy, goal, 5
Cancer vaccine
chemotherapy combination
host milieu alterations,
216–219
tumor microenvironment
effects, 214–216
clinical trials and
challenges, 171, 199–200
cytotoxic T lymphocyte
response, 186–187
DNA vaccines
clinical barriers in
humans, 197
overview, 195–196
performance optimization,
197–198
prime/boost strategies,
198–199
T cell activation, 196–197
goals, 171–172
Id protein vaccines, 186
immune response to
conventional vaccines,
189–193

poly-G oligonucleotide
combination therapy,
284
rationale, 184–185
spontaneous immunity to
cancer, 187
strategies
dendritic cell-based
vaccines, 194–195
overview, 185, 193–194
peptide vaccines, 194
tolerogenic pressure on
immunity to cancer
B cells, 188–189
T cells, 188
toxicology, 163–164
tumor antigen identification,
168–169
types, 172
Capecitabine, features, 110–111
Carboplatin, features, 108
Carmustine, features, 105–106
CD3-ζ
cancer-induced
immunosuppression
effects on
immunotherapy, 173
loss in immunoediting,
22–23
CD4$^+$ T cell
interleukin-10 regulation, 51
interleukin-23 and Th17 cell
regulation, 51
myeloid suppressor cell
effects on function,
318–319
repertoire, 209
transforming growth factor-
β regulation, 48–49, 55
CD4$^+$CD25$^+$ T cell, *see*
Regulatory T cell
CD8$^+$ T cell
cancer vaccine response,
186–187
interleukin-10 regulation, 51
myeloid suppressor cell
effects on function,
318–319
repertoire, 209
transforming growth factor-
β regulation, 47–48
CD20, therapeutic targeting,
170

CD25, *see* Regulatory T cell
CD30, therapeutic targeting,
170
CD40
chemotherapy combination
with monoclonal
antibodies, 247
immune-modulating activity
of targeting antibodies,
221–222
CD52, therapeutic targeting,
170
CD226, *see* DNAX-accessory
molecule-1
CD244, *see* 2B4 receptor
CDKs, *see* Cyclin-dependent
kinases
Cell cycle
phases and chemotherapy
disruption, 102–103
regulation, 119, 121–122
small molecule inhibitors,
122–124
Cetuximab, immune-
modulating activity, 220
Chemokines
macrophage recruitment to
tumor site, 291–293
regulatory T cell receptors,
90–93
Chemotherapy, *see* Cytotoxic
chemotherapy
Chlorambucil, features, 104
CI-1033, features, 128
CI-1040, features, 132–134
Cisplatin, features, 108
Cladribine, features, 110
Clearance, pharmacokinetics,
156
Clinical trial design
cancer vaccines, 199–200
small molecule signal
transduction inhibitors
combination therapy,
139–140
endpoints, 138–139
overview, 136–137
patent selection, 137–138
COX-2 inhibitors, *see*
Cyclooxygenase-2
inhibitors
CpG motif
DNA vaccines, 196

CTLA-4, *see* Cytotoxic T lymphocyte-associated antigen 4
Cyclin-dependent kinases
  cell cycle regulation, 121
  inhibitors, 121–122
  therapeutic targeting, 122–123
Cyclooxygenase-2 inhibitors
  immunostimulation, 250
  myeloid suppressor cell targeting, 325
  tumor-associated macrophage targeting, 300
Cyclophosphamide
  features, 105
  immunostimulation, 242–243
  tumor immune response effects, 216, 218
Cytabarine, features, 111
Cytotoxic chemotherapy, *see also specific drugs*
  alkylating agents
    historical perspective, 103
    mechanism of action, 103–104
    types, 104–106
  cancer vaccine combination
    host milieu alterations, 216–219
    tumor microenvironment effects, 214–216
  cell cycle phases and disruption, 102–103
  classification, 101
  DNA topoisomerase inhibitors, 106–108
  folate antagonists, 109
  immune stimulation
    cyclophosphamide, 242–243
    genotoxic stress and natural killer cell receptor ligand upregulation, 241–242
    immunogenic apoptosis, 239–241
    immunogenic DNA, 242
    overview, 103, 235–236
    regulatory T cell depletion, 244–245
    tumor antigen cross-presentation augmentation, 239

uric acid as endogenous danger signal, 241
  immunotherapy combination
    B7 molecule targeting, 248–249
    clinical trials, 250–252
    cytokines, 249
    dendritic cell activation, 247
    hybrid therapies, 249–250
    immunotherapy selection factors, 246–247
    Toll-like receptor stimulation, 247–248
    tumor-driven immunosuppression countering, 248
    lymphopenia induction, 243–244
  mechanisms of tumor cell death, 236–237
  platinum compounds, 109
  purine antagonists, 109–112
  regimens, 113–115
  resistance, 102
  taxanes, 112–113
  tumor-driven immunomodulation interference, 245–246
  tumor immunity effects, 213–214
  vinca compounds, 112
Cytotoxic T lymphocyte, *see* CD8$^+$ T cell
Cytotoxic T lymphocyte-associated antigen 4
  cancer-induced immunosuppression effects on immunotherapy, 175
  functions, 266–267
  immunologic checkpoint, 259–260
  indoleamine 2,3-dioxygenase a downstream effector, 353–355
  monoclonal antibodies
    adverse events, 267–268
    clinical trials, 267–269
    combination therapy, 270

regulatory T cell targeting, 223–224, 261, 266–269
regulatory T cell function, 90

Dacarbazine, features, 106
Dactinomycin, features, 108
Daunorubicin, features, 107–108
DC, *see* Dendritic cell
Dendritic cell
  cancer vaccines, 172, 194–195
  cancer-induced immunosuppression effects on immunotherapy, 174
  dysfunction in tumor microenvironment, 264
  immunoediting elimination phase role, 19–20
  immunogenic apoptosis, 239–241
  indoleamine 2,3-dioxygenase expression and immune tolerance mechanisms, 353–357
  interferon-producing killer dendritic cell, 38
  mature versus immature cells in tumor microenvironment, 83–85
  myeloid dendritic cell, 84–85
  myeloid suppressor cell effects on function, 317–318
  plasmacytoid dendritic cell, 85
  T cell interface signals, 258–261
Distribution, pharmacokinetics, 154–155
DNAM-1, *see* DNAX-accessory molecule-1
DNA vaccine
  clinical barriers in humans, 197
  overview, 195–196
  performance optimization, 197–198
  prime/boost strategies, 198–199
  T cell activation, 196–197

DNAX-accessory molecule-1, activation, 71–72
Docetaxel
  features, 113
  tumor immune response effects, 217
Doxorubicin
  features, 107
  tumor immune response effects, 218

EGFR, see Epidermal growth factor receptor
Ehrlich, P., 10–11
EKB-569, features, 128
Electrocardiogram, toxicology, 161–162
Epidermal growth factor receptor
  classification, 125
  signaling, 125–126
  therapeutic targeting, 119–120, 126–128, 170
Epigenetic surveillance, definition, 19
Erlotinib, features, 127–128
Etoposide, features, 106
Everolimus, features, 130
Excretion, pharmacokinetics, 155–156
Exposure, pharmacokinetics, 152, 157

Fas/FasL system
  cancer immunosuppression role, 23
  immune surveillance role, 13
Flavopiridol
  clinical trials, 122–123
  mechanism of action, 122
Fludarabine, features, 110
4-1BB, immune-modulating activity targeting antibodies, 222
5-Fluorouracil, features, 110
FoxP3
  immune suppression in cancer, 311
  regulatory T cell expression, 262, 282

Galectin 1
  carbohydrate recognition domain, 338
  functions, 339
  structure, 339
  T cell negative regulation
    activation and development, 342
    adhesion and migration regulation, 342
    cytokine synthesis control, 341–342
    survival control, 339–341
    therapeutic targeting, 343–344
  tumor expression, 339
  tumor immune escape role, 342–343
γδ T cell
  immune surveillance role, 12, 14–15
  innate antitumor response, 32–33
GCN2, T cell signaling, 378
Gefitinib, features, 127
Gemcitabine
  features, 111
  immunomodulatory effects, 251
GITR, see Glucocorticoid-inducible tumor necrosis factor receptor
Glucocorticoid-inducible tumor necrosis factor receptor
  monoclonal antibodies
    effector T cell stimulation, 284–285
    regulatory T cell targeting, 223, 262
    regulatory T cell expression, 262, 282
GM-CSF, see Granulocyte-macrophage colony-stimulating factor
Graft-versus-host disease
  immunotherapy utilization, 76–77
Graft-versus-leukemia, immunotherapy utilization, 76–77
Granulocyte, innate antitumor response, 33
Granulocyte-macrophage colony-stimulating factor
  adaptive antitumor response, 37
  myeloid suppressor cell induction, 321–322
GVHD, see Graft-versus-host disease

Her, see Epidermal growth factor receptor
HKI-272, features, 127
Human leukocyte antigen antigen presentation, 259
Hydrogen peroxide
  arginase activation and generation, 380–381
Hydroxyurea, features, 111

Id, cancer vaccines, 186
Idarubicin, features, 108
IDO, see Indoleamine 2,3-dioxygenase
IFN-α, see Interferon-α
IFN-β, see Interferon-β
IFN-γ, see Interferon-γ
Ifosfamide, features, 105
ILs, see specific interleukins
Imiquimod, mechanism of action, 250
Immune surveillance
  adaptive immunity, see Adaptive antitumor response
  clinical evidence
    organ transplant-related cancer, 17–18
    tumor-infiltrating lymphocytes, 17
  definition, 348
  history of study, 11
  innate immunity, see Innate antitumor response
  knockout mouse studies, 11–17
  nonimmunological surveillance, 18–19
Immunoediting
  definition, 348
  mediators, 10
  stages
    elimination, 19–20
    equilibrium, 20–22, 348
    escape
      immunological ignorance and tolerance in tumors, 24

signal transduction
alteration in effector
cells, 22–23
tumor-derived soluble
factors, 23–24
overview, 10
Immunologic checkpoints,
overview, 259–261
Immunotherapy
autoimmunity induction, 176
cancer stem cell targeting,
176
cancer-induced
immunosuppression
interference
antigen-presenting cell
dysfunction, 174
regulatory T cells, 174–175
T cell suppression, 173–174
chemotherapy combination
B7 molecule targeting,
248–249
clinical trials, 250–252
cytokines, 249
dendritic cell activation,
247
hybrid therapies, 249–250
immunotherapy selection
factors, 246–247
Toll-like receptor
stimulation, 247–248
tumor-driven
immunosuppression
countering, 248
mouse models and clinical
relevance, 175–176
passive immunotherapy
monoclonal antibodies,
169–170
T cell therapy, 170–171
prospects, 177–178
rationale, 208–209
tumor antigen identification,
168–169
vaccines, see Cancer vaccines
Indoleamine 2,3-dioxygenase
cancer dysregulation,
357–359
expression regulation in
immune tolerance,
351–353
function, 347, 349
gene, 350

immune suppression in
cancer
dendritic cell
mechanisms, 353–357
overview, 311
myeloid dendritic cell, 85
structure, 349–350
T cell regulation, 350–351
therapeutic targeting
inhibitor development,
360–361
1MT, 359–360
rationale, 359–360
tryptophan metabolites in
cancer, 349
tumor-associated
macrophage targeting,
300
tumor-driven
immunosuppression,
248, 265
Inflammation
myeloid suppressor cell
linkage with tumor
progression, 322–323
Toll-like receptors and
cancer development
role, 280
Innate antitumor response
adaptive immunity
interactions, 38–39
γδ T cell, 32–33
granulocyte, 33
macrophage, 33
natural killer cell, 31–32
natural killer T cell,
31–32
overview, 30–31
Interferon-α
adaptive antitumor
response, 36
immune surveillance role,
15–17
Interferon-β
adaptive antitumor
response, 36
immune surveillance role,
15–17
Interferon-γ
adaptive antitumor
response, 36
immune surveillance role,
12–13

immunoediting role
elimination phase, 19–20
equilibrium phase, 20–22
Interleukin-1, myeloid
suppressor cell
induction, 322
Interleukin-2
adaptive antitumor
response, 36
natural killer cell receptor
induction for
immunotherapy, 76
Interleukin-4, arginase
expression regulation,
373–374, 388
Interleukin-6, myeloid
suppressor cell
induction, 322
Interleukin-10
antitumor activity, 52
$CD4^+$ T cell regulation, 51
$CD8^+$ T cell regulation, 51
functional overview, 50
macrophage M2 protumoral
functions, 295
natural killer cell regulation,
51
Interleukin-12
adaptive antitumor
response, 36
natural killer cell receptor
induction for
immunotherapy, 76
Interleukin-13, arginase
expression regulation,
373–374, 388
Interleukin-15, adaptive
antitumor response, 37
Interleukin-17
adaptive antitumor
response, 37
natural killer cell receptor
induction for
immunotherapy, 76
Interleukin-18, adaptive
antitumor response, 36
Interleukin-21, adaptive
antitumor response, 37
Interleukin-23
adaptive antitumor
response, 37
antitumor immune response
regulation, 53–54

functional overview, 52
Th17 cell regulation, 51
Irinotecan, features, 106–107

Killer immunoglobulin-like receptors
  activation, 70–71
  inhibitory killer immunoglobulin-like receptors, 72–74

LAG-3, regulatory T cell expression, 262
Lapatinib, features, 127
LIGHT, adaptive antitumor response, 38
Ly49 receptor
  features, 74
  immunotherapy approaches
Lymphopenia, chemotherapy induction, 243–245

Macrophage
  immunoediting elimination phase role, 19
  indoleamine 2,3-dioxygenase expression and immune suppression, 356
  innate antitumor response, 33
  polarization of types, 290–291
  therapeutic targeting
    activation, 297–298
    angiogenesis, 299
    effector molecules, 300
    matrix remodeling, 299–300
    prospects, 300–302
    recruitment, 298–299
    survival, 299
  tumor-associated macrophage
    adaptive immunity modulation, 296–297
    M2 protumoral functions, 294–296
    overview, 88–89, 281, 289–290
    recruitment, 291–294
Mammalian target of rapamycin
  function, 128
  signaling, 128–129
  therapeutic targeting, 129–131
MAPK, see Mitogen-activated protein kinase
Matrix metalloproteinases
  macrophage M2 protumoral functions, 296
  therapeutic targeting, 299–300
Mechlorethamine, features, 105
Melphalan, features, 104
Memory T cell
  immune response to conventional vaccines, 190–192
  subsets in persistent infection, 192–193
6-Mercaptopurine, features, 109–110
Methotrexate, features, 109
Microenvironment, see Tumor microenvironment
Mitogen-activated protein kinase
  signaling cascade, 131–132
  therapeutic targeting, 132–135
Mitoxantrone, features, 108
MMPs, see Matrix metalloproteinases
Monoclonal antibodies, see also specific targets
  immune-modulating activity
    B7 molecule-targeting antibodies, 221
    bevacizumab, 220–221
    CD40-targeting antibodies, 221–222
    cetuximab, 220
    4-1BB-targeting antibodies, 222
    OX40-targeting antibodies, 222
    rituximab, 220
    T cell activation-targeting antibodies, 21
    trastuzumab, 219–220
  passive immunotherapy, 169–170
  regulatory T cell targets
    CD25, 222–223
    cytotoxic T lymphocyte-associated antigen 4, 223–224, 261, 266–269
    glucocorticoid-inducible tumor necrosis factor receptor, 223
MSC, see Myeloid suppressor cell
mTOR, see Mammalian target of rapamycin
Myeloid dendritic cell, see Dendritic cell
Myeloid suppressor cell
  abundance in cancer, 310–313
  immune suppression in cancer
    immune cell interactions, 317–321
    overview, 311
  immunotherapy implications, 326
  induction
    granulocyte-macrophage colony-stimulating factor, 321–322
    interleukin-1, 322
    interleukin-6, 322
    prostaglandin E_2, 322
  inflammation linkage with tumor progression, 322–323
  normal functions, 313–314
  phenotypic diversity and subpopulations, 314–317
  therapeutic targeting
    chemotherapy, 325–326
    differentiation agents, 323–324
    rationale, 323
  tumor microenvironment, 212, 265
  tumor progression role, 314

Natural killer cell
  definition, 84
  functional overview, 63–64
  genotoxic stress and receptor ligand upregulation, 241–242
  immune surveillance role, 11–12
  immunoediting elimination phase role, 19
  immunotherapy approaches
    allograft transplantation, 76–77

antibody-dependent cell-
    mediated cytotoxicity,
    76
  cytokine induction of
    receptors, 76
  NKG2D ligand retention
    on tumor cells, 75
  inhibitory receptors
    B7-H3, 73–74
    inhibitory killer
      immunoglobulin-like
      receptors, 72–74
    NKG2A, 73
  innate antitumor response,
    31–32
  interleukin-10 regulation,
    51
  Ly49 receptor, 74
  myeloid suppressor cell
    effects on function, 318
  receptors
    activation, 65–67
    DNAX-accessory
      molecule-1, 71–72
    killer immunoglobulin-
      like receptors, 70–71
    NKG2C, 69
    NKG2D, 65–69
    NKp30, 70
    NKp44, 70
    NKp46, 70
    prospects for study, 78–79
    2B4 receptor, 72
  transforming growth factor-
    β regulation, 49
Natural killer T cell
  immune surveillance role,
    11–12
  immunoediting elimination
    phase role, 19
  innate antitumor response,
    31–32
Necrosis, chemotherapy
  induction, 237
NF-κB, see Nuclear factor-κB
Nitric oxide
  donors, 387–388
  immunoregulation, 375–376
  indoleamine 2,3-
    dioxygenase expression
    regulation, 352
  tumor progression role,
    383–384

Nitric oxide synthase
  arginase interactions in
    immunoregulation
    autoimmune disease role,
      379–380
    hydrogen peroxide
      generation, 380–381
    overview, 378
    peroxynitrite generation,
      379
  immunoregulatory activity,
    374–377
  inducible enzyme
    expression regulation,
      372
    myeloid suppressor cell
      expression and effects,
      319–320
    sources, 371–372
  isoforms, 371
  knockout mice, 384–385
  physiological role for arginine
    metabolism in immunity
    control, 381–382
  polymorphisms, 372
  structure, 371
  T cell expression, 376
  therapeutic targeting,
    386–389
  tumor expression, 382–383
NK cell, see Natural killer
  cell
NKG2A, inhibitory function,
  73
NKG2C, activation, 69
NKG2D
  activation, 65–69
  ligand retention on tumor
    cells as immunotherapy,
    75
  ligands and innate
    antitumor response,
    30–31
NKp30, activation, 70
NKp44, activation, 70
NKp46, activation, 70
NOS, see Nitric oxide synthase
Nuclear factor-κB, indoleamine
  2,3-dioxygenase
  induction, 351–352
1MT, indoleamine 2,3-
  dioxygenase inhibition,
  359–360

Organ transplant-related
  cancer, immune
  surveillance clinical
  evidence, 17–18
OX40, immune-modulating
  activity targeting
  antibodies, 222
Oxaliplatin, features, 109
p53, nonimmunological
  surveillance, 18–19
Paclitaxel
  features, 112–113
  tumor immune response
    effects, 217–218
PD-1, see Programmed death 1
PD0325901, features, 132, 134
PDGFR, see Platelet-derived
  growth factor receptor
Pemetrexed, features, 111
Perforin
  adaptive antitumor
    response, 37–38
  immune surveillance role,
    13–14
Peroxynitrite
  autoimmune disease role,
    379–380
  generation, 379
  nitrotyrosine formation,
    379–380, 383
Pharmacokinetics
  absorption optimization, 153
  bioavailability calculation,
    154
  clearance, 156
  clinical concerns, 165
  definition, 151
  distribution, 154–155
  drug discovery, 150–151
  excretion, 155–156
  exposure, 152, 157
  safety margin, 153
  testing, 152–154, 156–157
  therapeutic index, 153
Phosphatidylserine, cancer
  immunosuppression
  role, 23
Plasmacytoid dendritic cell, see
  Dendritic cell
Platelet-derived growth factor
  receptor, therapeutic
  targeting, 119–120

PLK-1, *see* Polo-like kinase 1
Polo-like kinase 1, therapeutic targeting, 123–124
Poly-G oligonucleotides, cancer vaccine combination therapy, 284
Procarbazine, features, 106
Programmed death 1
  alternative receptor evidence, 337
  blockade and antitumor immunity, 335–337
  functions, 269, 334
  ligand expression and regulation on tumor cells, 335
  mechanism of T cell inhibition, 334–335
  monoclonal antibody targeting, 269
  prospects for study, 337–338
  structure, 334
Prostaglandin E_2
  indoleamine 2,3-dioxygenase expression regulation, 352
  myeloid suppressor cell induction, 322
  receptor antagonists in therapeutic targeting, 325–326

RAG-2, immune surveillance role, 14
Raloxifene, immunity effects, 224–225
Rapamycin, features, 129–130
Reactive nitrogen species
  arginase-nitric oxide synthase interactions in immunoregulation
    autoimmune disease role, 379–380
    overview, 378
    peroxynitrite generation, 379
Reactive oxygen species
  arginase-nitric oxide synthase interactions in immunoregulation
    hydrogen peroxide generation, 380–381
    overview, 378

Receptor tyrosine kinases, families and features, 124–125
Regulatory T cell
  abundance in cancer, 262
  adaptive antitumor response, 35
  antigen specificity of $CD4^+CD25^+$ cells, 282–283
  cancer-induced immunosuppression effects on immunotherapy, 174–175
  $CD8^+$ versus $CD4^+$ cells, 281–282
  chemotherapy depletion, 244–245
  conventional T cells
    imbalance in tumors, 91–92
    interactions in bone marrow, 90–91
  cyclooxygenase-2 inhibitor effects, 250
  depletion rationale in therapy, 284
  immunosuppressive mechanisms, 283
  markers, 282
  myeloid suppressor cell effects on development, 320–321
  stimulation by self-antigens, 35
  therapeutic targeting
    monoclonal antibodies
      CD25, 222–223
      cytotoxic T lymphocyte-associated antigen 4, 223–224, 261, 266–269
      glucocorticoid-inducible tumor necrosis factor receptor, 223, 262
    overview, 210, 222
  Toll-like receptors and functional regulation, 283–284
Rituximab, immune-modulating activity, 220
ROS, *see* Reactive oxygen species

Safety margin, pharmacokinetics, 153
SHIP, *see* Src homology 2-containing inositol-5-phosphatase
SKI-606, features, 136
sMICA, *see* Soluble major histocompatibility complex class I chain-related A proteins
Soluble major histocompatibility complex class I chain-related A proteins, cancer immunosuppression role, 23, 45
Src kinase
  functions, 135–136
  therapeutic targeting, 136
Stat3, signaling defects in tumor microenvironment cells, 264–265

Tamoxifen, immunity effects, 224–225
T cell, *see also* $\alpha\beta$ T cell; $CD4^+$ T cell; $CD8^+$ T cell; $\gamma\delta$ T cell; Memory T cell; Natural killer T cell; Regulatory T cell
  activation signals, 212–213
  antigen-presenting cell interface signals, 258–261
  cancer-induced immunosuppression effects on immunotherapy, 173–174
  DNA vaccine activation, 196–197
  nitric oxide effects, 375–376
  tolerogenic pressure on immunity to cancer, 188
  tolerance
    central tolerance, 209–210
    peripheral tolerance, 210
T cell receptor, signaling loss in immunoediting, 22–23
T cell therapy, passive immunotherapy, 170–171
T cell vaccine
  immune response, 193
  peptide vaccines, 194

TCR, see T cell receptor
Temozolomide, features, 105
Temsirolimus, features, 130–131
TGF-β, see Transforming growth factor-β
T helper cell, see CD4+ T cell
Therapeutic index
 pharmacokinetics, 153
 toxicology, 164–165
TIL, see Tumor-infiltrating lymphocyte
TLRs, see Toll-like receptors
Tolerance
 central tolerance, 209–210
 peripheral tolerance, 210
 tolerogenic pressure on immunity to cancer
  B cells, 188–189
  T cells, 188
Toll-like receptors
 immunologic checkpoint, 261
 immunosurveillance role, 278
 inflammation and cancer development role, 280
 regulatory T cell functional regulation, 283–284
 signaling, 279–280
 stimulation combination with chemotherapy, 247–248
 structure, 279
 types and ligands, 238–239, 279
Topotecan, features, 106–107
Toxicology
 biopharmaceutical agents, 163
 cancer vaccines, 163–164
 cardiovascular safety, 161, 162
 clinical concerns, 165
 drug discovery, 150–151, 159–160
 genotoxicity, 162–163
 off-target effects, 161
 preclinical drug development, 158–159
 risk assessment, 164
 target validation, 160–161
 therapeutic index, 164–165

TRAIL, see Tumor necrosis factor-related apoptosis-inducing ligand
Transforming growth factor-β
 functional overview, 46
 immune suppression in cancer, 310
 indoleamine 2,3-dioxygenase expression regulation, 352–353
 therapeutic targeting, 49–50
 tumor defects, 266
 tumor immune tolerance regulation
  natural killer cell response to tumors, 49
  T cell regulation
   CD4+ T cell response to tumors, 48–49, 55
   CD8+ T cell response to tumors, 47–48
  overview, 46–47
Trastuzumab
 Her2/neu targeting, 118
 immune-modulating activity, 219–220
Treg, see Regulatory T cell
Tumor antigens
 antibody-inducing cancer vaccine, 186
 classification, 185
 cross-presentation
  chemotherapy augmentation, 239
  constitutive, 238–239
 identification, 168–169
Tumor-associated macrophage, see Macrophage
Tumor-infiltrating lymphocyte
 immune surveillance clinical evidence, 17
 suppression by tumor microenvironment, 211
Tumor microenvironment
 cancer vaccine and chemotherapy combination effects, 214–216

 immune checkpoints, 262–266
 immunosuppressive factors, 211
 mature versus immature dendritic cells, 83–85, 211
 myeloid suppressor cells, 212
 tumor-infiltrating cells
  dendritic cells, 281
  macrophages, 281
  regulatory T cells, 281–282
 tumor-infiltrating lymphocyte suppression, 211
Tumor necrosis factor-related apoptosis-inducing ligand, adaptive antitumor response, 38
2B4 receptor, activation, 72

Uric acid, endogenous danger signal, 241

Vascular endothelial growth factor
 antitumor immune response regulation, 54–55
 cancer immunosuppression role, 23
 functional overview, 54
 macrophage M2 protumoral functions, 295
 myeloid suppressor cell expression, 321
 therapeutic targeting, 119–120, 170, 299
VEGF, see Vascular endothelial growth factor
Vinblastine, features, 112
Vincristine, features, 112
Vinorelbine, features, 112
Virchow, Rudolf, 4
Vitamin D_3, myeloid suppressor cell targeting, 325

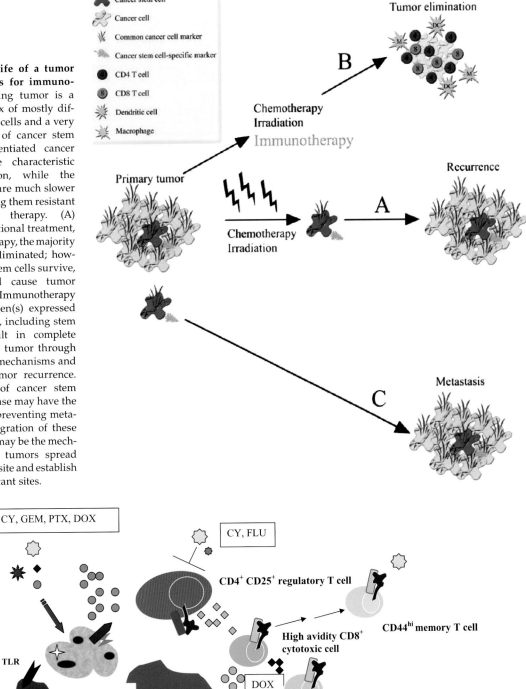

FIGURE 10.1 Life of a tumor and opportunities for immunotherapy. A growing tumor is a heterogeneous mix of mostly differentiated cancer cells and a very small population of cancer stem cells. The differentiated cancer cells exhibit the characteristic rapid proliferation, while the cancer stem cells are much slower at dividing, making them resistant to conventional therapy. (A) Following conventional treatment, such as chemotherapy, the majority of the tumor is eliminated; however, the cancer stem cells survive, differentiate, and cause tumor recurrence. (B) Immunotherapy that targets antigen(s) expressed by all cancer cells, including stem cells, could result in complete eradication of the tumor through immune effector mechanisms and prevention of tumor recurrence. (C) Elimination of cancer stem cells early in disease may have the added benefit of preventing metastasis because migration of these cancer stem cells may be the mechanism by which tumors spread from the primary site and establish themselves at distant sites.

FIGURE 12.1 Chemotherapy modulating the tumor microenvironment and systemic mechanisms of immune tolerance to enhance tumor immunity. Both pathways of immune tolerance and suppression and tumor cell biology itself offer multiple loci where chemotherapy can modulate a developing or established immune response. Cell aptoptosis induced by cyclophosphamide (CY), gemcitabine (GEM), paclitaxel (PTX), and doxorubicin (DOX) can enhance the ability of DCs to cross-prime the antitumor immune response. PTX and docetaxel potentiate this process by interacting with TLRs to promote the maturation and activation of DCs. CY and PTX promote the development of the CD4+ Th1 response (rather than the CD4+ Th2 response) optimal for antitumor immunity. Both CY and fludarabine can negate the suppressive influence of CD4+CD25+ regulatory T cells. DOX further promotes the development of the CD8+ T cell response, and CY promotes the establishment of a CD44^{hi} memory cell pool.

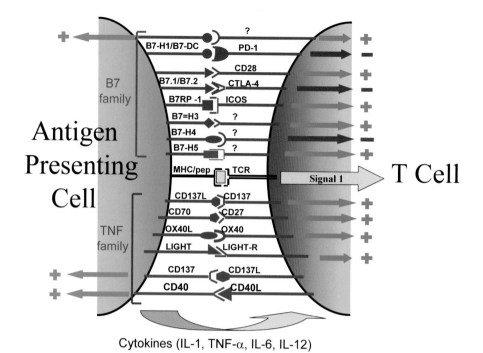

FIGURE 14.2 **The three catagories of costimulatory and coinhibitory signals at the APC–T cell interface.** Shown are the ligand-receptor pairs for the B7 family, TNF family, and cytokine family of signals between the APC and T cell. Not all receptors for some B7 family members have been identified. Some B7 family members interact with both costimulatory receptors and coinhibitory receptors. While many signals are delivered from an APC to a T cell, the T cell also delivers signals to the APC, creating bidirectional cross talk.

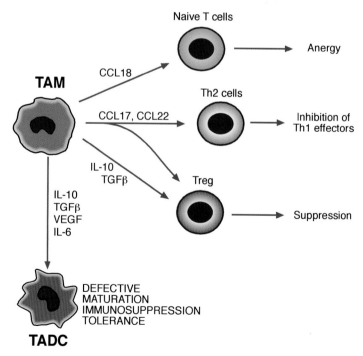

FIGURE 16.2 **TAMs suppressing adaptive immunity.** TAMs produce cytokines that negatively modify the outcome of a potential antitumor response. IL-10, IL-6, VEGF, and TGF-β inhibit the maturation and activation of tumor-associated dendritic cells (TADCs). IL-10, TGF-β, and selected chemokines act on T helper 2 (Th2)-polarized lymphocytes and T regulatory cells (Tregs), which are ineffective in antitumor immunity and suppress antitumor responses. Figure modified from Sica et al. (2006).

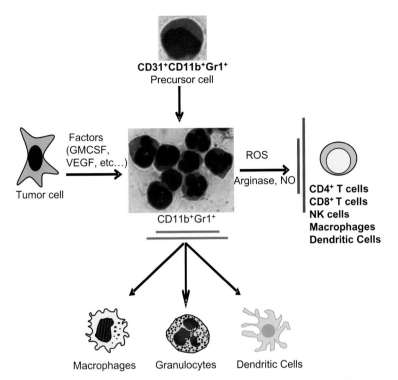

FIGURE 17.1 **Tumor-derived factors that block cell differentiation in the myeloid lineage and that induce the accumulation of Gr1⁺CD11b⁺ MDSCs.** Under normal conditions, myeloid precursor cells differentiate into DCs, macrophages, and granulocytes. The presence of tumor and tumor-derived factors blocks this differentiation pathway and leads to the accumulation of immature CD11b⁺Gr1⁺ myeloid cells, called MDSCs. MDSCs suppress T cells and other immune cells through a variety of mechanisms, including the production of ROS and arginase. The center panel shows purified MDSCs from the blood of mice with the 4T1 mammary carcinoma. This mixture of cells with single and multilobed nuclei is characteristic of MDSCs in the blood of tumor-bearing mice. Splenic MDSCs from tumor-bearing mice typically have large, single-lobed nuclei.

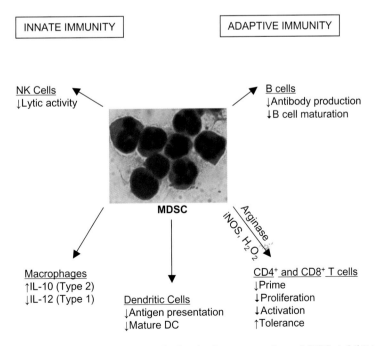

FIGURE 17.2 **MDSC suppression of innate and adaptive immune systems.** MDSCs inhibit innate immunity by suppressing NK cell-mediated lysis and by polarizing macrophages toward a type 2 phenotype. MDSCs inhibit adaptive immunity by suppressing the activation and proliferation of T cells and antibody production by B cells. In addition, MDSCs limit the availability of mature and functional DCs, which bridge the gap between innate and adaptive immunity.

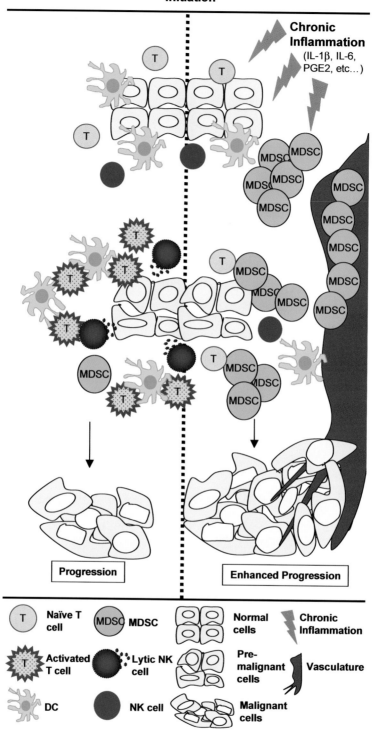

FIGURE 17.3 **Inflammation promotion of tumor growth by inducing the accumulation of MDSCs that suppress antitumor immunity.** A recent hypothesis proposes that immune suppression caused by MDSCs may be responsible for the correlation between inflammation and malignancy (Bunt et al., 2006). In the absence of inflammation, cells that are transformed following initiation by carcinogens are destroyed by immune surveillance. However, if initiation is accompanied by inflammation, then MDSCs accumulate and prevent the immune system from destroying the transformed cells by blocking T cell activation, suppressing NK cell-mediated lysis, reducing DC antigen presentation, and skewing immunity toward a type 2 response. As premalignant cells become malignant, the inflammatory microenvironment sustains the accumulation of MDSCs, which further inhibits antitumor immunity and facilitates tumor progression.